The Practical Handbook
of
COMPOST ENGINEERING

Roger T. Haug

The Practical Handbook
of
COMPOST ENGINEERING

Roger T. Haug

Library of Congress Cataloging-in-Publication Data

Haug, Roger Tim.
    The practical handbook of compost engineering / by Roger T. Haug.
        p. cm.
    Includes bibliographical references and index.
    ISBN 0-87371-373-7
    1. Compost. 2. Refuse and refuse disposal—Biodegradation.
I. Title.
TD796.5.H38   1993
631.8′75—dc20

92-33533
CIP

© 1993 by CRC Press LLC
Lewis Publishers is an imprint of CRC Press LLC

No claim to original U.S. Government works
International Standard Book Number 0-87371-373-7
Library of Congress Card Number 92-33533
Printed in the United States of America   7 8 9 10 11 12 13 14 15
Printed on acid-free paper

THE LIBRARY OF COMPOST ENGINEERING SOFTWARE™

The computer simulation models presented in Chapters 11 to 15, Chapter 17, and in Appendices C to F are available from the author. Write or FAX to the address below for additional information on the library of models and purchase prices.

Roger T. Haug Engineers, Inc.
Attn: Compost Software
5119 Newton Street
Torrance, CA 90505

Phone and FAX (310) 378-3155

Dedication

To Peggy my wife and lifelong love
You are the world to me

To my sons David and Jimmy
Studying to become engineers
Thanks for being such good sons

To my Mother and Father
Thanks for my wonderful upbringing
And your forever love and support

Author

Roger T. Haug is a consulting engineer with 20 years experience in the composting of biosolids and other waste materials. He is currently Division Engineer for the Solids Technology and Resource Recovery (STRR) Division for the City of Los Angeles. The STRR Division is responsible for planning, design, construction management, and startup of all biosolids handling facilities for the City of Los Angeles, a 10-year, $500 million program.

Dr. Haug earned M.S. and Ph.D. degrees in Environmental Engineering from Stanford University and a B.S. in Civil Engineering, magna cum laude, from Loyola University. He is a registered Professional Engineer. He was formerly Assistant Professor of Civil Engineering and Environmental Sciences at Loyola Marymount University, where he still maintains a position as Adjunct Professor.

A member of the American Society of Civil Engineers, Water Environment Federation, and Air & Waste Management Association, Dr. Haug has published extensively in professional journals and serves on the Editorial Board for *BioCycle, Journal of Waste Recycling*. He was a member of the ASCE Subcommittee on Composting and has worked extensively with consulting engineering firms and municipal governments on design and operation of composting and integrated waste management systems.

Foreword

Paris throws five millions a year into the sea. And this without metaphor. How, and in what manner? day and night. With what object? without any object. With what thought? without thinking of it. For what return? for nothing. By means of what organ? by means of its intestine? its sewer....

Science, after long experiment, now knows that the most fertilising and the most effective of manures is that of man. The Chinese, we must say to our shame, knew it before us. No Chinese peasant...goes to the city without carrying back, at the two ends of his bamboo, two buckets full of what we call filth. Thanks to human fertilisation, the earth in China is still as young as in the days of Abraham. Chinese wheat yields a hundred and twenty fold. There is no guano comparable in fertility to the detritus of capital. A great city is the most powerful of stercoraries. To employ the city to enrich the plain would be a sure success. If our gold is filth, on the other hand, our filth is gold.

What is done with this filth, gold? It is swept into the abyss.

We fit out convoys of ships, at great expense, to gather up at the south pole the droppings of petrels and penguins, and the incalculable element of wealth which we have under our own hand, we send to the sea. All the human and animal manure which the world loses, restored to the land instead of being thrown into the water, would suffice to nourish the world.

These heaps of garbage at the corners of the stone blocks, these tumbrils of mire jolting through the streets at night, these horrid scavengers' carts, these fetid streams of subterranean slime which the pavement hides from you, do you know what all this is? It is the flowering meadow, it is the green grass, it is marjoram and thyme and sage, it is game, it is cattle, it is the satisfied low of huge oxen at evening, it is perfumed hay, it is golden corn, it is bread on your table, it is warm blood in your veins, it is health, it is joy, it is life...

Put that into the great crucible; your abundance shall spring from it. The nutrition of the plains makes the nourishment of men.

from *Les Miserables*

by Victor Hugo, 1862

Preface

I find myself sitting here in front of the computer with mixed emotions. I am relieved that this second edition of my original work is nearly finished and satisfied that the new work is a major improvement. At the same time there is a sense of sadness that the effort is drawing to a close. I suppose it's a bit like running a race. There is a thrill at finishing but some regret that the effort has ended. There is also a nagging frustration that I cannot include all of the work I would like. The bright side is that there's plenty to work on for the next edition.

My first book on composting was titled *Compost Engineering — Principles and Practice* and was published in 1980. My first thought for this new work was to keep the original title, but change the subtitle to Theory, Practice, and Art. My publisher pointed out that the more one uses words like "Theory" the smaller the readership becomes. There is a great diversity of people and backgrounds in the composting industry, and I believe this book has something for everyone. The publisher's suggestion was *The Practical Handbook of Compost Engineering*. I like it and I think we'll stick with it for what will hopefully be many editions to come.

My introduction to composting came in 1972 when I was an engineering faculty member at Loyola University. My father then worked for the Los Angeles County Sanitation Districts and was instrumental in implementing a windrow composting facility on anaerobically digested sludge. A key factor which made the composting operation successful was my father's idea of recycling dry product and blending with dewatered cake to condition the starting mixture. The experiments began in February 1972. Within the first few days of composting, previously odorous material was converted to an aerobic condition. Results were so impressive that the districts were commended by the county for solving a major community odor problem. Somewhat earlier my father had directed pilot studies of MSW composting using the Eweson drum digester.

Needless to say, composting was a favorite topic at family meetings, much to the dismay of wives and children. I owe a debt to my father, Mr. Lester Haug, for introducing me to this subject, providing fresh ideas and insights when needed, and acting as the sounding board for numerous theories.

Over the years I have had the fortune of working in positions that kept me actively involved in the composting industry. I directed research activities on composting for the LA/OMA Project in the late 1970s and worked on many sludge composting projects in the 1980s. Like many others, I observed the evolution of sludge composting until it became a proven alternative for sludge management. Based on the sound principles developed for sludge, composting is now enjoying an explosion of activity with other substrates such as yard wastes, industrial wastes, and solid waste fractions. In the 1990s my position with the Solids Technology and Resource Recovery Division, City of Los Angeles, keeps me active on the composting frontier. The future looks bright — composting will continue to contribute to integrated waste management plans.

The goal of my first book was to produce a more fundamental engineering approach to the analysis of composting. I think that goal was at least partially reached. Perhaps the crowning achievement of that effort was the integration of thermodynamics and process kinetics into a first generation, simulation model of sludge composting. This new work greatly improves on the original and extends it to all composting substrates. The book has been reorganized and nearly every chapter completely rewritten. This attests to the tremendous advances made in our understanding of composting in the last decade. The simulation models have been greatly advanced over the original work, to the point that I consider them second generation versions. The entire composting process, including high rate and curing phases, can now be modeled. The new simulation models take full advantage of that wonderful new invention, the personal computer, and are available directly from the author.

I want to express my appreciation to the many colleagues and friends I have come to know in the composting business. They have made this a very enjoyable adventure. Science and industries advance on the hard work of human beings. Composting is no exception. It is filled with many wonderful people who share a desire to solve problems and to contribute in some small way to a better tomorrow for our children. I want to acknowledge my special "composting" friends Dick Kuchenrither, Chuck Murray, Larry Hentz, George Miller, Libby Tortorici, John Donovan, Patty Psaris, Tim Shea, Dan Clark, Paul Gilbert, Jerry Goldstein, Nora Goldstein, John Walker, Jon Hay, and Bob Bastian. I also want to thank the late Ed Lewis, the father of my current publisher Brian Lewis, for taking a chance on me and publishing my first book. It's been a great honor to work with you all.

Roger T. Haug
Torrance, CA

Contents

CHAPTER 1

Introduction

COMPOSTING — DEFINITIONS AND OBJECTIVES

There is no universally accepted definition of composting. This text uses a practical definition of the process. Composting is *the biological decomposition and stabilization of organic substrates, under conditions that allow development of thermophilic temperatures as a result of biologically produced heat, to produce a final product that is stable, free of pathogens and plant seeds, and can be beneficially applied to land.* Thus, composting is a form of waste stabilization, but one that requires special conditions of moisture and aeration to produce thermophilic temperatures. The latter are generally considered to be above about 113°F (45°C). Maintenance of thermophilic temperatures is the primary mechanism for pathogen inactivation and seed destruction.

Most biological stabilization and conversion processes deal with dilute aqueous solutions, and only limited temperature elevations are possible. Thermophilic temperatures in aqueous solutions can be achieved if substrate concentrations are high and special provisions for aeration are employed. Aside from such special cases, composting is usually applied to solid or semisolid materials, making composting somewhat unique among the biological stabilization processes used in sanitary and biochemical engineering.

Aerobic composting is the decomposition of organic substrates in the presence of oxygen (air). The main products of biological metabolism are carbon dioxide, water, and heat. Anaerobic composting is the biological decomposition of organic substrates in the absence of oxygen. Metabolic end products of anaerobic decomposition are methane, carbon dioxide, and numerous low molecular weight intermediates such as organic acids and alcohols. Anaerobic composting releases significantly less energy per weight of organic decomposed compared to aerobic composting. Also, anaerobic composting has a higher odor potential because of the nature of many intermediate metabolites. For these reasons almost all engineered compost systems are aerobic. Mass transfer limitations, however, may cause anaerobic zones in otherwise aerobic systems. Such subtlety aside, this book will deal primarily with aerobic systems because of their commercial importance to man.

The objectives of composting have traditionally been to biologically convert putrescible organics into a stabilized form and to destroy organisms pathogenic to humans. Composting is also capable of destroying plant diseases, weed seeds, insects, and insect eggs. Odor potential from compost is greatly reduced because organics that remain after proper composting are relatively stable with low rates of decomposition. Composting can also effect considerable drying, which has particular value with wet substrates such as municipal and industrial sludges. Decomposition of substrate organics together with drying during composting can reduce the cost of subsequent handling and increase the attractiveness of compost for reuse or disposal.

Organic composts can accomplish a number of beneficial purposes when applied to the land. First, compost can serve as a source of organic matter for maintaining or building supplies of soil humus, necessary for proper soil structure and moisture holding capacity. Second, compost can improve the growth and vigor of crops in commercial agriculture and home related uses. Stable compost can reduce plant pathogens and improve plant resistance to disease. Colonization by beneficial microorganisms during the latter stages of composting appears to be responsible for inducing disease suppression. Third, compost contains valuable nutrients including nitrogen, phosphorus, and a variety of essential trace elements. The nutrient content of compost is related to the quality of the original organic substrate. However, most composts are too low in nutrients to be classified as fertilizers. Their main use is as a soil conditioner, mulch, top dressing, or organic base with fertilizer amendments. On the other hand, nutrients such as nitrogen are organically bound and slowly released throughout the growing season, making them less susceptible to loss by leaching than soluble fertilizers.

Like the process of composting, there is no universal agreement of what makes something a compost. The following is a workable definition as used in this text. Compost is *an organic soil conditioner that has been stabilized to a humus-like product, that is free of viable human and plant pathogens and plant seeds, that does not attract insects or vectors, that can be handled and stored without nuisance, and that is beneficial to the growth of plants.*

ENGINEERING WITH MICROBES

The most common engineering application of microbes is to treat or convert substrates in aqueous solution. Suspended growth reactors, such as the activated sludge process, or fixed film reactors, such as the trickling filter, anaerobic filter, and rotating biological contactor, are widely used for treatment of municipal and industrial liquid wastes. Suspended cultures of microbes are used for fermentations to produce ethanol, antibiotics, and medicines. Biochemical engineering is well developed and it is possible to design and operate such systems using a reasoned, engineered approach.

There are a number of biological processes used on solid or semisolid materials including fermentation and ripening of cheese, production of silage, and, of course, composting. At least in the case of composting, a reasoned, fundamental approach to analysis and design of new facilities and operation of existing ones has not been fully developed. The first edition of this book, published in 1980, presented a "first generation" approach to biochemical engineering analysis of the process. This second edition continues the evolution and considerably advances the tools available for analysis compared to the first edition.

Almost every book on the subject of composting begins with the statement that composting is an ancient art, practiced by man since before the dawn of recorded history. Although the evidence does suggest that man has had a long affair with composting, fundamental scientific studies of the process have generally occurred in the past four decades. Our abilities to engineer the process and to work with the numerous competing forces within a composting matrix are only now developing. In other words, the theory of composting may be understood and most of the forces involved may be known, yet engineering is still often conducted using a "handbook" approach with little knowledge of how to control these forces to achieve the final end product. It is a goal of this book to develop a more fundamental approach to analysis and design, one that relies as much as possible on first principles of physics, chemistry, biology, thermodynamics, and kinetics.

COMPOSTING SUBSTRATES

The quantity of substrates potentially suitable for composting is indeed large. Klass[1] estimated the solid and semisolid organic wastes generated and collected in the U.S. shown in Table 1.1. Major categories include refuse, manure, agricultural crop residues, food wastes, logging and wood residues, and municipal and industrial sludges. The collected fraction totals over 125 million metric tons/year. By way of comparison, Riley[2] estimated that 135 billion metric tons of biomass, mostly wild and not controlled by man, are produced on the earth each year. Mother nature takes care of most of this biomass herself, but considerable quantities remain that can use the help of man and in turn can be of help to man.

Municipal Biosolids

The U.S. Environmental Protection Agency (EPA) estimated in 1989 that approximately 15,300 publicly owned treatment works (POTWs) generate 7.7 million dry metric tons of biosolids annually, equivalent to about 64 lb dry solids (ds) annually for each individual in the U.S.[3] The volume of sludge is expected to double by the year 2000 due to population growth, stricter wastewater treatment requirements, and improvements in POTW operation. The sewage sludge generated each year was estimated to fill 186,000 railroad cars, enough to span half the country.

In 1989–90 the U.S. EPA undertook a massive survey to determine disposal practices for sewage sludges.[4] This became known as the National Sewage Sludge Survey (NSSS). Participants in the NSSS were selected from the estimated 11,407 POTWs in the U.S. identified as using at least secondary wastewater treatment. The frequency of use for the surveyed management practices were as follows:

land application	31.1%
surface disposal	27.6
other	17.1
colandfill	16.2
distribution and marketing	2.7
incineration	2.6
monofill	1.8
ocean disposal	1.0

Composting falls under the category of "distribution and marketing" and accounted for <3% of all POTWs. By contrast, the combined practices of monofilling, co-landfilling with refuse, ocean disposal, and surface disposal account for almost 50% of all POTWs. Pressures exist to reduce the latter practice. Therefore, new supplies of sludge should continue to be directed toward composting.

Table 1.1.   Estimates of Organic Wastes Generated and Collected in the U.S. in 1980

Waste Type	Generated[a]	Collected[a]
Agricultural crops and food wastes	355	21
Manure	180	24
Urban refuse	115	65
Logging and wood-manufacturing residues	50	5
Miscellaneous organic wastes	45	5
Industrial wastes	40	5
Municipal sewage solids	11	2
Total	796	127

Source: Klass.[1]

[a] Units are 10^6 metric tons/year dry wt (values rounded from original reference).

Estimated biosolids production for the countries of the European Community (EC), North America, Japan, and Australia are presented in Table 1.2. Bruce et al.[5] determined that provision of primary sewage treatment raises the national sludge production to about 1 metric ton ds/year for every 50 people, equivalent to about 20 kg/year per capita. Upgrading primary treatment to include secondary treatment raises sludge production by at least 50% to about 30 kg/year per capita. This is essentially equivalent to the 64 lb/year per person (29 kg/year per capita) estimated by EPA for the U.S. The EC countries included in Table 1.2 have an estimated 1990 population of 375 million and produce about 6.2 million dry metric tons/year of raw sludges. The U.S. has an estimated 1990 population of 250 million and produces about 7.7 million dry metric mtpyr. This is a vast resource of material, much of which is suitable for composting.

Powlesland et al.[6] reported the routes used for biosolids management by the EC countries shown in Table 1.3. Significant quantities of biosolids are landfilled, while ocean disposal is still practiced by the UK, Spain, and some others. Many of these practices will be effected by recent EC directives. The most important is the EC Urban Wastewater Directive[7] which requires member states to provide secondary treatment of all sewage from populations of more than 15,000 by the year 2000 and bans sludge disposal to sea by 1998. The EC Environmental Protection Act of 1990[8] requires that management practices use "Best Available Techniques Not Entailing Excessive Cost" (BATNEEC). These directives will increase the quantities of sludge and pressure agencies toward reuse alternatives including composting.

Industrial Sludges

Suler[9] prepared the partial list of organic industrial sludges produced in the U.S. shown in Table 1.4. Composting has been applied to most industrial sludges at one time or another. Currently, there is considerable interest in using wastes from a number of these industries, particularly the food processing industry, as composting substrates.

Manures

Manures represent a major class of wet organics. Production per head and representative moisture contents are presented in Table 1.5. Manures are usually wet and produced in impressive amounts. Using data from Table 1.5, a single dairy cow can produce about 2200

Table 1.2.   Biosolids Production in Europe, North America, Japan, and Australia

Country	Population (millions)	Raw Sludge Production[a]	
		Estimate (year)	Estimate Maximum[b]
Europe			
Austria	7.5	140 (1980)	225
Belgium	9.8	70 (1979)	295
U.K.	57.2	1500 (1977)	1720
Denmark	5.1	130 (1977)	155
Finland	4.8	130 (1980)	145
France	56.4	840 (1980)	1690
Germany	79.5	2200 (1977)	2390
Greece	10	3	300
Ireland	3.6	20 (1980)	110
Italy	57.6	500 (1979)	1730
Luxembourg	0.4	11 (1979)	12
Netherlands	14.9	230 (1979)	450
Norway	4.1	55 (1978)	125
Portugal	10.5		315
Spain	39.6	45 (1982)	1190
Sweden	8.3	210 (1982)	250
Switzerland	6.4	150 (1980)	190
North America			
U.S.	248.7	7700 (1988)	
Canada	24	500 (1982)	720
Japan		1000 (1987)	
Australia	17	263 (1990)	

Source: Riley,[2] USA Today,[25] Spinosa et al.,[26] and Rawlinson et al.[27]

[a] Units are 10^3 metric tons ds/year.
[b] Calculated assuming an average 30 kg/year per capita raw sludge production for the whole population.

Table 1.3.   Disposal of Sewage Biosolids to Different Outlets in the European Community in 1984

Country	Total[a]	Land Application[a]	Landfill[a]	Combust[a]	Sea[a]	Other[a]
Belgium	29	6	15	6	0	1
Denmark	150	68	68	15	0	0
France	850	230	451	170	0	0
Germany	2180	632	1286	196	0	65
Greece	15	0	15	0	0	0
Ireland	24	7	4	0	11	2
Italy	800	272	440	88	0	0
Luxembourg	15	11	3	0	0	2
Netherlands	202	127	55	6	12	2
Spain	281	174	28	0	79	0
U.K.	1118	458	101	34	335	235

Source: Powlesland et al.[6]

[a] Units are 10^3 metric tons ds/year.

Table 1.4.   Organic Industrial Sludges Produced in the U.S.

Industry	Sludge Production[a]	Sludge Type
Food	650,000	Organic sludge mostly readily degradable
Textile	300,000	Mostly organic and composed of cotton, wool, synthetic fibers, dyes, and sizing
Pulp and paper	2,000,000	Mostly carbonaceous containing cellulose fibers and other biomass
Pharmaceutical	200,000	Varies and includes fungal mycelia from antibiotic production
Petroleum	850,000	Petroleum related sludges

Source: Suler.[9]

[a] Units are dry metric ton/year.

Table 1.5.   Estimated Daily Manure Production by Man and Various Farm Animals

Source	Daily Raw Manure Production (wet weight)[a]		Moisture[a] %
	lb	kg	
Man	0.33	0.15	77
Beef cattle	53.8	24.5	82
Dairy cattle	102.5	46.6	87
Chicken	0.32	0.15	73
Broiler	0.28	0.13	75
Turkey	0.89	0.40	68
Duck	0.74	0.34	
Sheep	3.8	1.7	73
Horse	44.2	20.1	60
Swine	7.1	3.2	84

Source: Meek at al.[28]

[a] Values are averages compiled from several sources by the authors.

kg ds/year and a beef cow about 1600. These are equivalent to the municipal sludge production from about 75 and 50 people, respectively. A single feed lot with 10,000 head of beef cattle would have a human population equivalent of about 500,000 people. Even the lowly chicken produces about 15 kg ds/year per head. It is little wonder that management of manures has become a major problem in some countries. The Netherlands, for example, produces a large quantity of pig manure but has little land space to absorb it. Composting should continue to be an important management alternative for manures, and there is plenty to go around.

Yard Wastes

Yard waste ranks right behind paper as the second largest component in the U.S. solid waste stream.[10] The same is true for many of the European countries. A 1986 U.S. EPA study reported that the U.S. discards 23.8 million tons of yard waste annually, representing almost 11% of the total refuse stream. Other studies report significantly higher percentages for the yard waste fraction.[10] In the Midwest and Northeast U.S. the material can make up to 17 to 20% of the residential waste stream.

Roughly 70% of yard waste consists of grass clippings. Leaves comprise about 25%, with the remaining 5% made up of brush and other miscellaneous materials.[11] The latter figures are averages and the collection rates vary significantly throughout the year. During the colorful New England autumn, for example, yard waste can be 40% of the collected solid waste. In areas with cold winters, grass clippings are typically generated from May through September. Deciduous leaves are collected between September and December and again in early spring. Prunings enter the waste stream in early spring and again in autumn. In warmer climates, grass and tree trimmings may be collected all year.

Millions of people rake leaves into piles and sometimes bag them for separate pickup. Historically, the common practice in the U.S. was to dispose of the leaves and grass along with refuse in landfills. Beginning in the 1980s, pressure began to mount to divert this relatively clean waste stream from the landfills. Many states enacted legislation banning yard wastes from municiple solid waste landfills. Composting became the preferred management practice for most of the diverted yard waste.

Leaves and grass can have a relatively high moisture content and are often produced only seasonally. Grass has a high nitrogen content and can cause increased nitrous oxide emissions if commingled with refuse at refuse-to-energy projects. For the backyard "composter" and many municipalities the quantities produced are not sufficient to justify energy recovery. As a result, energy recovery from leaves and grass is problematic at best. On the other hand, leaves and grass are relatively free of tramp materials (at least compared to municiplal solid waste), and leaves can be stored without causing putrescible odors. Composting becomes a rather ideal process for converting such organics to a usable product. For these reasons, composting for management of leaves and sometimes leaves mixed with grass exploded in popularity in the 1980s. By 1991, there were an estimated 1650 leaf and yard waste composting sites in the U.S. alone.

Meyer et al.[12] reported that a yard waste composting facility in Uzwil, Switzerland processed 1200 metric tons/year from 45,000 citizens, equivalent to about 27 kg/year per capita: this is about equal to the per capita production of raw sludge. By contrast, a suburban U.S. facility[13] reported over 10,000 tons/year of leaves, grass clippings, and woody debris from 34,000 people, equivalent to about 270 kg/year per person. The latter is about 10 times the production rate of the Uzwil residents. Obviously, the generation rate for yard wastes varies significantly from one community to another, and the generation rate can be quite high compared to sludge production.

With their low decomposition rate, leaves could be handled with "low tech" piles and windrows without forced aeration. As long as an agency had land space, leaf composting could be implemented at relatively low cost. Today, we are witnessing an evolution toward the use of "higher tech" approaches to yard waste composting. This evolution is driven by many of the same factors that drove sludge composting toward in-vessel systems in the 1970s and 80s. Many yard wastes contain grass which is more difficult to handle and more prone to odor. Even with source separation, yard waste collected from the household will always be contaminated to some extent. If the homeowner is allowed to use plastic bags, then plastic will be a major contaminant. These factors have caused leaf composting to evolve from an industry dominated by "low tech" installations to one where modern facilities are more engineered. In general, there is more pre- and postprocessing to improve product quality, better operation of the process to reduce nutrient and moisture rate limitations, and more attention to odor management.

Septage

Septage pumped from septic tank systems is becoming a more significant management problem. Regulations in the U.S. require that domestic septage be stabilized to reduce pathogens and vector attraction prior to land application. Septage generation in the U.S. is estimated to equal one third of the secondary sludge production rate from POTWs. Much of the septage generation is in rural areas served by small POTWs that cannot accept large portions of high-strength septage.

Composting of septage can present unique problems. Septage is often low in volatile solids content because of extensive anaerobic decomposition in the septic tank, and it may contain as much as 90 to 98% water. Nevertheless, Williams et al.[14] reported successful composting of septage which was dewatered and conditioned with wood shavings. Volatile solids content of the septage solids was only 27 to 33% and the wood shavings supplied additional organics to the process. Nitrogen addition in the form of urea was required to enhance decomposition of the wood shavings and increase the generation of heat.

Food and Agricultural Wastes

A variety of wastes from the food processing and agriculture industries are suitable substrates for composting. In general, if it is out of the ground, vegetative or animal in nature, and not contaminated, it can be composted. The literature contains examples of a large assortment of food and "ag" wastes that have been composted:

- Cranberry and prune wastes
- Cull potatoes (bruised, diseased, or too small to use), potato skins, and other wastes from French fry producers
- Fish wastes including lobster shells, crab scraps (the shells and viscera), scallop viscera, whole fish, and fish scraps
- Orange and citrus culls
- Apple processing residuals including apple pomace, filter cake, and biological sludge
- Wine processing wastes such as grape pomace and filter residues
- Chocolate production residuals
- Red algae, trapped fish, and other marine life from water intake screens of power stations
- Garbage wastes from food preparation facilities
- Agricultural residues including rice hulls, straws, corn cobs, cotton gin trash, and almond hulls

The above list is by no means complete but it does highlight the wide variety of vegetable and animal wastes that can be converted into useful composts.

Municipal Solid Wastes

Municipal solid waste (MSW) is all of the wastes arising from human activities that are normally solid and discarded as useless or unwanted. Production of combined residential and commercial refuse in the U.S. is estimated at about 4.3 lb/day per capita, equivalent to about 700 kg/year per capita.[10] Paper, wood, food, and yard wastes are major components of the MSW stream and are suitable composting substrates. As a result, composting as a management option for MSW has been pursued for many decades.

The history of MSW composting in the U.S. has been rocky. Facilities were implemented in the 1950s and 60s amid high hopes that this was the solution to the solid waste problem. Unfortunately, it was not. Few if any of the facilities survived because of three primary factors. First, quality of the composts was poor. Compostables in the MSW were usually commingled with other noncompostables and separation is never 100%. As a result, the product was usually contaminated with glass and plastic and there was little demand for the low-grade compost. Second, odor control was usually ignored in the plant design in the mistaken belief that aerobic composting would not produce odors. Mother nature simply does not work this way for reasons discussed in Chapter 16. Third, the compost plants were forced to compete against the economics of low cost landfills with little revenue from compost sales to help balance the books. The flush of MSW composting activity in the U.S. was essentially over by the late 1960s.

In the meantime, MSW composting did advance in other parts of the world, particularly Europe, the Middle East and South America. The practice continues to flourish in these areas probably because of better facility designs, higher costs for alternative management strategies, and specialized markets for the product. The comparative success of the European facilities led to a resurgence in MSW composting activity in the U.S. beginning in the late 1980s. Whether this renewed interest will be sufficient to sustain a prosperous industry in the U.S. is yet to be determined. As of this writing several facilities are operational with more in construction. Unfortunately, several facilities have also failed. In at least one case the failure was largely caused by a complete lack of any odor control in the original plant design. If the lessons of history are not learned, MSW composting may again fail in the U.S.

There is a growing trend in the U.S. to return to source separation of refuse materials. This can only help composting as a process. Trying to produce high quality compost from commingled refuse materials is a very frustrating task. More and more pre- and postprocessing equipment can be used, but only with higher costs and less throughput with each added process. If a composting facility is viewed as a processing plant, then the feedstocks should be free of any materials which would lower the quality of the final product. Industrial processing plants almost invariably use the highest quality feedstocks available. Why should composting be any different? In my opinion, producing a quality product demands that you use a quality feedstock. It's not that MSW composting will not work. Indeed, it has demonstrated that it can work in some countries. But the chance of success using source separated, clean feedstocks is much greater, particularly where high quality composts are already in the marketplace.

Special Wastes

Composting has even been applied on laboratory and pilot scale to hazardous industrial wastes such as TNT, petroleum sludges, and certain pesticide residues. For example, compost piles constructed of cotton gin trash and chicken manure were inoculated with concentrations of carbofuran, a carbamate insecticide, and simazine, a triazine herbicide.[15] The carbofuran was completely degraded and the simazine 98.6% degraded after 50 days of composting. In this case the compost piles served as growth media for an active biological population which also degraded the added pesticide. Based on these successful results, composting was recommended as a possible treatment system for pesticide rinseates produced by pesticide manufacturers, formulators, and applicators.

Simpkin et al.[16] applied composting technology to the bioremediation of soil contaminated with diesel, JP-4, and motor gasoline at a site near Fairbanks, Alaska. Significant temperature elevations were observed in test composters treating (1) the contaminated soil conditioned with nutrients and water and (2) the soil amended with sewage sludge. Alkylbenzenes (BTEX) were degraded from about 10 mg/kg to below the detection limit (0.006 mg/kg) in less than 35 days. Total petroleum hydrocarbons (TPH) were reduced from about 1700 mg/kg to about 114 to 132 mg/kg in 70 days. Simpkin concluded that composting has an advantage over other bioremediation technologies in that it has the potential to generate and retain heat, which can be significant in cold climates with short growing seasons.

The relatively new field of bioremediation has demonstrated that many hazardous chemicals can be biologically decomposed, including some thought to be not readily degradable. Benzene, pentachlorophenol (PCP), phthalates, light and heavy fuels, mineral spirits, coal tars, phenols, polycyclic aromatic hydrocarbons (PAHs), and even chlorinated solvents and polychloro-biphenols (PCBs) have been treated by various bioremediation technologies. Because of its enhanced microbial flora and high reaction kinetics, composting should continue to find useful applications as a bioremediation technology.

PATHS FOR RESOURCE RECOVERY

There are essentially three basic approaches to resource recovery from organic wastes: (1) use of the organics and associated nutrients either directly in the soil or after production of compost material, (2) conversion of the organics to alternative energy forms, or (3) conversion of the organics into new products such as recycling paper into new paper products. All of these paths to resource recovery have noble objectives. Compost is a proven organic supplement that, by supplying humus and nutrients to poor soils, can greatly improve crop yields and reduce irrigation requirements. Alternatively, the energy potential of organic wastes is considerable. Development of the biomass resource as an alternative, renewable fuel received much public attention following the OPEC oil embargoes of the 1970s. Interest waned when worldwide oil supplies stabilized and became plentiful in the 1980s. Interest in biomass energy returned in the 1990s because of the Persian Gulf War and long term concerns over global warming from greenhouse gases, such as carbon dioxide produced from the burning of fossil fuels. Finally, direct recycle of selected organics, such as paper, into recycled products has the obvious benefits of reducing demand on a limited resource and, in some cases, reducing energy demands. Which of these reuse possibilities will win out? To answer this, the characteristics of waste organics must be examined.

Relatively dry materials, such as municipal refuse and wood wastes, have considerable value as an energy resource. Energy can be extracted efficiently by thermal processes, such as incineration, pyrolysis, and gasification, because of the dry nature of the material. Costs of extraction may be high and certain components of a heterogeneous waste such as refuse will not be amenable to thermal processing. As moisture content increases, thermal processing becomes much less efficient. For combustion to be self-supporting, it is usually necessary for moisture levels to be less than 60 to 70%. The exact value depends on the organic characteristics and the operating parameters of the thermal process. If the waste is in the form of a liquid slurry or suspension, anaerobic digestion is the only practical energy recovery method. But what about residues remaining after anaerobic digestion, or other organic wastes too wet for efficient thermal conversion to energy? In the past, fossil fuels were often added to such wastes either to support combustion of the organics or to remove moisture. But such processes are energy intensive, and the use of fossil fuels in this manner is falling into increasing disfavor.

High-moisture organic wastes represent a rather unique management problem. Direct application to land is a popular management practice, but it is usually limited to rural areas where sufficient land is available. Composting can be particularly effective in converting wet materials to a more usable or easily disposable form. At the same time, composting can stabilize putrescible organics, destroy pathogenic organisms, and provide significant drying of the wet substrate. All these advantages are obtained with minimal outside energy input, the major energy resource being the substrate organics themselves. Furthermore, composting is a flexible process. It can be viewed as a conversion process to produce a material suitable for reuse or simply as a stabilization and drying process to provide for easier disposal. Composting is also compatible with a wide variety of feedstocks.

The question posed previously regarding resource recovery by recycling, energy recovery, or composting can now be answered. It is likely that all avenues of resource recovery will be practiced in the future as part of integrated waste management. Recycling of high quality organics, such as white paper and newsprint, to produce new paper products is probably the highest and best use of such organics. Energy recovery will find its greatest potential with dry waste organics, which cannot be recycled into new products. Wood wastes, agricultural biomass, and select fractions of refuse are examples of organics that can serve as alternative and renewable fuel sources. High-moisture materials present unique problems and are often more conducive to management by land application or composting. This is not to imply that a distinct boundary exists between wet and dry organics. Indeed it does not. Also, not all dry organics will be processed thermally because of costs, air pollution concerns, lack of available furnace capacity, and public opposition. Although each case should be evaluated on its own site-specific merits, it would seem that composting will be a favored process for many waste organics, particularly wet organic substrates and in some cases even dry organics. The quantity of suitable waste materials does not appear to be a limitation. It seems that there will be enough to keep both large and small scale "composters" happily in business for many years to come.

PROBLEMS OF WET SUBSTRATES

Composting of municipal and industrial biosolids can present unique problems because the organics may still contain 70 to 80% water. The presence of so much water in sludge can result in reduced composting temperatures and inefficient operation if moisture is not controlled. An understanding of the thermodynamics of composting is essential for designing workable sludge composting systems.

As a rule, the higher the moisture content of the organic material, the greater is the need to maintain a large void volume to ensure adequate aeration. Dewatered sludge cake is not a friable material, and it lacks the porosity of materials such as straw or refuse. Because of its plastic nature, sludge also tends to compact under its own weight, which further reduces the void volume. The high moisture content, lack of porosity, tendency to compact, and need to dry dewatered sludge during composting make biosolids composting somewhat unique and often difficult. In recent years several composting systems have been developed or modified to overcome problems associated with biosolids composting.

With wet substrates, the designer/operator will be concerned with (1) structural conditioning of the feed to achieve a friable mixture of materials and (2) energy conditioning to close the thermodynamic balance. The engineer will also consider the air supply required to dry the feed substrates and the temperature control systems needed to regulate the air supply. Protecting the composting mixture from the elements will also be a concern, because excessive rain will add more water to an already difficult problem. While the water balance and feed conditioning are primary concerns with wet substrates, there are many advantages as well. Sludges tend to be relatively homogeneous and free of trash materials. Sludges also typically contain adequate nutrients to support composting and usually very little pre- or postprocessing is required beyond dewatering.

PROBLEMS OF DRY SUBSTRATES

Having read the above, those with dry substrates may be thinking that they have avoided most of the difficult problems. Such is not the case. Dry substrates present their own unique blend of challenges. Dry substrates, such as agricultural wastes and particularly refuse and yard waste, are heterogeneous and may require source separation along with front end and back end processing systems to produce a desirable end product. These wastes often lack sufficient nutrients, particularly nitrogen, which may have to be added to avoid kinetic rate limitations. Size reduction is often required to reduce the particle size and improve the rate of composting. Finally, water usually must be added to the mixture to avoid biochemical rate limitations from lack of moisture. Because composting is a dehydrating environment, moisture addition is often necessary throughout the process.

Sometimes there are subtle differences between seemingly similar substrates. Grass and leaves are sometimes collected and composted together, but the two substrates could not be more different. Leaves are very friable and usually make a good structural mix. Leaves are characterized by relatively low rates of decomposition and usually remain fresh even if stored for short periods before pickup. However, leaves are typically low in nutrients and often require supplemental nitrogen addition and some limited shredding to reduce the particle size. Grass, on the other hand, is not friable and tends to compact on itself. Grass is characterized by a high rate of decomposition, often exceeding that of raw sewage sludge. As a result, grass is often anaerobic and odorous by the time it reaches the composting site. It cannot be stored without generating odors and must be incorporated into the composting process immediately upon receipt. Grass is also high in nitrogen and will release ammonia into the exhaust gases. Quite a difference for two substrates collected from one's backyard.

Each type of composting substrate presents its own set of challenges to the designer/operator. Knowledge of the feed characteristics and the effect of these characteristics on plant performance are essential to the proper design and operation of composting facilities. As a rule, design and operation must start with knowledge of the feed substrates.

PRODUCT QUALITY STANDARDS

Once the quality and characteristics of the substrates are defined, attention shifts to the compost quality expected from the facility. Product standards are necessary to protect public and environmental health and to assure a measure of commercial acceptability. Public health risks associated with composting facilities come from exposure to human pathogens, aerospores and vectors. Primary pathogens exist in the materials to be composted and secondary pathogens grow during the composting process. The secondary pathogens of greatest concern are spore forming fungi that can inhabit some composting environments. Pathogens of concern, conditions for heat inactivation, suggested product standards, and requirements for personnel protection are discussed in Chapters 4 and 5.

Heavy metals and trace organics contained in compost can enter the food chain through plants. Both human, animal, and plant toxicities are of concern. The most accepted approaches to regulation are to control the concentration of these constituents in the final compost product and/or restrict the uses or application rates of the compost. The U.S. EPA regulates sludge based composts under the "marketing and distribution" provisions of the Clean Water Act, Part 503. A final version of this rule was published in February 1993, and contained the metal limits presented in Table 1.6. These concentrations are based on health risk assessments of numerous exposure pathways. Sludge based compost products which meet these metal limits and also comply with pathogen quality and vector attraction standards are considered to be of exceptional quality and can be used without restrictions. The Part 503 regulations are considered significant in the U.S., because they will probably be adopted by most states and eventually extended to other composting substrates.

Dorau[17] noted that backyard composting is currently exempt from regulation in most, if not all, U.S. states, because the potential for adverse impact is minimal. Regulations regarding

Table 1.6. Metal Limits Proposed for Sewage Sludge Based Composts by the U.S. EPA under Part 503 of the Clean Water Act[a]

Element	Concentration[b]
Arsenic (As)	41
Cadmium (Cd)	39
Chromium (Cr)	1200
Copper (Cu)	1500
Lead (Pb) .	300
Mercury (Hg)	17[c]
Molybdenum (Mo)	18
Nickel (Ni)	420
Selenium (Se)	36
Zinc (Zn)	2800

[a] The U.S. EPA published a final version of this rule in February 1993, but has allowed for additional public comments; therefore, the above values should be used with caution.
[b] Units are mg/kg dry wt.
[c] Valid for all sludge uses except mushroom production.

Table 1.7. Selected U.S. State Standards for Compost Made from Municipal Solid Waste and Yard Waste

	NC	NH	FL	MN	NY
Unrestricted distribution for MSW	Y	N	N	Y	N
Element[a]					
Cadmium	10	10	15	10	10
Chromium	1000	1000		1000	1000
Copper	800	1000	450	500	1000
Lead	250	500	500	500	250
Mercury	10	10		5	10
Nickel	200	200	50	100	200
Zinc	1000	2500	900	1000	2500
PCB	2			1	1
Grades of MSW compost	2	2	5	2	2
Maximum particle size					
in.	1		1		1
cm		1		2.5	
Maximum foreign matter	6%		2%		
Stabilization	>60% VS loss	reheat or 60%	reheat or 40–60%	6 months or 60%	50 days

Source: Dorau.[17]

[a] Units are mg/kg dry wt.

composting of food processing wastes, yard wastes, and animal manures are also relatively lenient. Because of their greater potential for adverse impacts, composts made from municipal solid waste and sludges have received the most attention. Dorau reported in 1992 that six states regulated yard waste and MSW compost and that many other states were in the process of adopting rules. A summary of selected state regulations is presented in Table 1.7. Several states do not permit the use of MSW compost on crops grown for direct human consumption.

As noted above, U.S. standards are generally based on health risk assessments of exposure pathways. Limits are set below levels expected to produce an adverse effect to a reasonably exposed individual. European standards tend to be more stringent, with heavy metal limits

Table 1.8.   Voluntary Heavy Metal Standards for Composts to Bear the German RAL Label

Element	Concentration[a]
Cadmium	1.5 (later 1.0)
Chromium	100
Copper	100
Lead	150
Mercury	1.0
Nickel	50
Zinc	400

Source: Briton.[22]

[a] Units are mg/kg dry matter (adjusted to a standard 30% organic matter).

Table 1.9.   U.K. Soil Association Standards for Maximum Permissible Levels of Metals in Composts and Manures Used in Organic Farming

	Concentration[a]	
Element	Manure and Fertilizer	Seed, Potting, Blocking, and Mushroom Compost
Cadmium	10	2
Copper	400	50
Lead	250	100
Mercury	2	1
Nickel	100	50
Zinc	1000	150

Source: Bardos et al.[29]

[a] Units are mg/kg dry matter.

nearer the natural background levels in soils. The philosophy is based in part on limiting the change from natural background. The Technical Committee (TC) 223 of the European Committee for Normalization (CEN) is developing standards for "soil improvers" and "growing media". Once completed, these standards will ensure uniform product quality for composts across all members of the European Community (EC) and the European Free Trade Association (EFTA). Current examples of European and Canadian standards and guidelines are presented in Tables 1.8 to 1.10.

A comparison of metal concentrations in various composts and U.S. soils with Part 503 limits is presented in Table 1.11. MSW compost was produced at central separation/composting facilities in the U.S. "Green" waste composts were prepared in Europe by separate collection of only the compostable fraction. U.S. sludge concentrations were developed from the National Sewage Sludge Survey.[4] MSW compost generally contains higher levels of many trace elements compared to U.S. background soils but lower levels than sewage sludges. Source separation and separate collection of "green" wastes produces composts with lower metal residues than can be attained by central-separation of MSW. Ryan and Chaney[18] noted that just because lower concentrations can be achieved in "green" waste or MSW composts compared to sewage sludge does not mean that they have to be attained to make utilization of the composts on cropland a valuable practice of sustainable agriculture.

Pesticides and herbicides, either in containers or as residue on yard wastes, are generally biodegradable and decompose in the compost. Hegberg et al.[19] noted that, while published data are

Table 1.10. Quality Requirements for Compost Product Used in an
Unrestricted Manner in Ontario, Canada[a]

Element	Concentration[b]
Arsenic	10
Cadmium	3
Chromium	50
Cobalt	25
Copper	60
Lead	150
Mercury	0.15
Molybdenum	2
Nickel	60
Selenium	2
Zinc	500
PCB	0.5
Plastic	1.0% of dry wt
Total nondegradable particulate matter	2.0% of dry wt

Source: Gies[30] and Ontario Ministry of the Environment.[31]

[a] Compost must also be proven stable through testing or a 6 month curing
period. Acceptable indicators of stability include volatile solids destruction, spontaneous reheating, oxygen uptake rates, toxin production, carbon
to nitrogen ratio, seed germination and growth tests, and redox potential.
[b] Units are mg/kg dry matter.

Table 1.11. Comparison of Metal Concentrations in Various Composts, U.S.
Sludges, U.S. Soils, and the Part 503 Sludge Limits

Element	Concentration[a]					
	MSW Composts[b]		"Green" Waste Compost[c]	U.S. Sludges Geo. Mean[d]	U.S. Soils[e]	503 Sludge Limits
	Number of Samples	Geo. Mean				
As	8	2.6	—	9.9		41
Cd	72	2.0	0.5	6.9	0.18	39
Cr	66	32.6	—	118	53	1200
Cu	73	107	40	741	18.0	1500
Pb	73	169	86	134	10.6	300
Hg	31	1.1	0.17	5.2		17
Ni	66	22.7	17	42.7	16.5	420
Zn	72	418	225	1200	42.9	2800

[a] Units are mg/kg dry wt.
[b] From Epstein et al.[32]
[c] From Fricke et al.[33]
[d] From U.S. EPA.[4]
[e] From Holmgren et al.[34] and Shacklett et al.[35]

limited, most yard waste composts exhibit low levels of chlorophenoxy herbicides and organo-
phosphate pesticides. Trace concentrations of chlorinated organics, such as chlordane, DDT, and
pentachlorophenol, are sometimes observed but are usually below the tolerances for food.

There are several characteristics of compost that are important to commericial acceptance
but do not affect human health or the environment. These include color, particle size, the
presence of weed seeds and other foreign or inert material such as stones, glass shards, and
plastic, organic matter content, carbon to nitrogen ratio, salinity, pH, moisture content, and
water holding capacity. Weed seeds are a nuisance but can be effectively destroyed at the time-

temperature conditions common to composting. Golueke et al.[20] conducted germination tests of yard waste compost and noted no viable weed seeds. The Solid Waste Composting Council was organized in February 1990, to promote public acceptance and use of composting processes and products in the U.S. One of the objectives of the council was to develop model standards/guidelines for compost quality. However, Albright[21] has indicated that it has been difficult to achieve any consensus on standards for esthetic or commercial quality. Characteristics that do not effect public or environmental health will probably be left to the marketplace to dictate appropriate product quality. Albright noted an interesting example where a compost produced from brewery waste was light tan in color. Market acceptance was initially limited because the established markets were based on darker brown composts.

As discussed previously, the presence of glass, stone, and plastic in MSW compost can limit the commercial acceptance of the product. The quantities of these contaminants, and also the concentrations of heavy metals, usually decreases with increased source separation. Briton[22] has noted that in Germany the term "biowastes" now refers exclusively to kitchen and yard wastes that are collected separately from other household trash. Organic materials sifted from MSW no longer qualify. The trend in Europe is toward composting of source separated materials. Many MSW composting facilities have closed or converted to handling source separated materials primarily due to contamination of the final product.

APPROACH TO ANALYSIS

A fundamental approach is used throughout this text. Whenever possible the first principles of physics, chemistry, biology, thermodynamics, and kinetics are applied to problem analysis and solution. Particular emphasis is placed on process thermodynamics. No other science can so unify the approach to analysis. Application of thermodynamic principles is perhaps the most fundamental way of analyzing composting systems, just as it has been a fundamental tool in analyzing and coordinating physics, chemistry, and biology.

While the book contains a heavy emphasis on theory, guidance is provided to help apply theoretical concepts to the design of practical composting systems. The goal is to develop working tools for analysis of the process based on theoretically sound arguments, supported when necessary by experimental data. These tools allow one to study the relationships between process variables, determine the limits within which composting can be conducted, answer questions for which field data are lacking, and provide guidance for improved design of future systems and operations of existing ones.

Design of composting facilities is a challenge, but one that is made considerably easier by a firm grasp of the engineering principles of composting. To develop these principles, background chapters on compost systems, chemical thermodynamics, microbiology, and heat inactivation are provided (Chapters 2 through 5). From this point, the analysis centers on physical and thermodynamic aspects of composting such as feed conditioning for control of moisture, porosity, free air space, and the energy balance and aeration requirements and mechanisms (Chapters 6 through 8). Substrate degradability, process kinetics, and product stability are discussed in Chapters 9 and 10.

At this point, if the reader is not exhausted, he should have sufficient background to tackle the development of process simulation models presented in Chapter 11. These models are based on thermodynamic and kinetic principles combined to produce useful mathematical models of the compost process. The computer simulation models were developed to run on personal computers and are separately available from the author. Ordering information is contained on the inside title page of this book. Application of the computer models to various compost systems, the interplay of process variables and implications for design and operation are presented in Chapters 12 through 14.

The majority of modern facilities employ some manner of forced aeration of the composting material. Air supply is the largest term in the mass balance, so proper design of supply, distribution, and collection systems is quite important. A discussion of these important components and their design is presented in Chapter 15.

Composting can be subject to nuisance problems such as odors and dust; therefore, measures to control such conditions should be part of any process design. In particular, development of an "odor management program" must begin early in the planning stage for any new facility. Basics of odor science, odor emission rates, and treatment options are discussed in Chapter 16. Basics of meteorology and atmospheric dispersion of odors are presented in Chapter 17. Chapter 18 concludes the book with a discussion of ambient odor standards and the author's personal theorems and elements for successful odor management.

ENGLISH VS. METRIC

The magazine *Chemical Engineering* recently conducted a poll of its readers asking the question whether they preferred English or metric units.[23] The number of response letters was considerably more than for any other topic in their recent memory. Non-U.S. respondents preferred the metric system, whereas most U.S. readers voted for English units. About one third of the U.S. respondents reluctantly voted metric, some arguing that metric is more logical and that the U.S. should get in step with the rest of the world. Nevertheless, the majority of U.S. respondents favored English units (feet, pounds, Btus and degrees Fahrenheit) in part because they had developed a sense for the meaning of the numbers.

The question of metric vs. English is not as clear-cut as it might seem. The term "metric" has come to mean SI (Systeme International), which has nominally become the worldwide standard. SI uses units of meters, kilograms, joules, and degrees Kelvin. By contrast the cgs metric system uses centimeters, grams, calories, and degrees Centigrade. Many SI units are unwieldy and difficult to visualize. For example, SI units for pressure and viscosity are not convenient and only the purest practitioners express temperatures in degrees Kelvin. It seems that few of us want to accept water freezing at 273°.

The composting industry in the U.S. seems to be caught halfway between English and a form of metric. Temperatures are commonly expressed in Centigrade, and most U.S. engineers now have a good feel for such numbers. However, mass and energy balances continue to be shown in English units for the most part. Metric units were used in my first book and the overwhelming comment from U.S. practitioners was that the units were awkward for them. Academia seems to be comfortable with metric units, but the practicing engineer in the U.S. continues to conduct business in English. I didn't hear much from non-U.S. engineers, hopefully because they were happy with everything the way it was.

My goal has been to make this new book a useful tool to composters whatever their country of practice. The book is intended more for practitioners than for academia. Therefore, I decided to provide a more balanced usage of English and metric units in this work and ask that non-U.S. readers be patient when they come across an occasional Fahrenheit or Btu. Fortunately, most of the units used in composting are easily converted between systems, so non-U.S. readers shouldn't have many problems.

THEORY, PRACTICE, AND ART

It is hoped that the reader will come to appreciate the ordered complexity of what appears on the surface to be a simple process. Indeed, composting can be a very contradictory process. For example, rapid and extensive organic stabilization is desired. But this leads to rapid and extensive heat evolution, which in turn can elevate temperatures to the point where biological reaction rates become temperature limited, slowing down the process. With "energy rich" substrates, thermodynamics may limit the process kinetics. In such a case, it may be very difficult to achieve optimum values for more than one or two variable at a time. A flexible design is needed to simultaneously achieve a high level of organic decomposition, a high temperature elevation, and a high level of drying. As another example, low ambient temperatures can lead to kinetic limitations that impede development of the available energy resource. It is much like placing a large log on a fire — slow to start, but once started, hard to control. Exploring such interrelationships has been a fascinating experience, and I hope to convey a sense of this fascination to you the reader.

But composting is not theory and practice alone. One of the more fascinating aspects of composting is that it still retains elements of art. Composting is not all numbers and equations. One of the most important aspects of composting, the quality and stability of the final product, is still largely "judged" by the operator on the basis of appearance, smell, and touch. No numbers or analytical tests have been developed that can replace the operator's eye. While analytical techniques will continue to advance and provide useful input, they will never replace the operator's judgement.

Logsdon and Briton[24] wrote that "In the not-so-distant future, compost making may well be as much an artful science as wine making." The wine maker must produce a product that meets the exacting standards of the buying public. So too, the engineer must remember that he is designing a composting plant to produce a product. That product must meet quality standards for the intended use. As a minimum the compost must be safe, beneficial to plants, and attractive to the intended buyer or end user. Producing good compost requires the same level of knowledge, engineering, skill, and art required for producing good wine.

REFERENCES

1. Klass, D. L. "Wastes and Biomass as Energy Resources: An Overview," in *Clean Fuels from Biomass, Sewage, Urban Refuse, Agricultural Wastes* (Chicago, IL: Institute of Gas Technology, 1976).
2. Riley, G. A. "The Carbon Metabolism and Photosynthetic Efficiency of the Earth as a Whole," *Am. Sci.* (April 1944), p. 32.
3. U.S. EPA. 40 CFR Parts 257 and 503. "Standards for Disposal of Sewage Sludge; Proposed Rule," *Fed. Regist.* 54:23 (1989).
4. U.S. EPA. 40 CFR Part 503. "National Sewage Sludge Survey," *Fed. Regist.* 55:218 (1990).
5. Bruce, A. M., Campbell, H. W., and Balmer, P. "Developments and Trends in Sludge Processing Techniques," in *Proceedings of the 3rd International Symposium on Processing and Use of Sewage Sludge*, Brighton (Boston, MA: D. Reridel Publishing Co., 1983).
6. Powlesland, C. and Frost, R. C. "A Methodology for Undertaking BPEO Studies of Sewage Sludge Treatment and Disposal," WRC Report PRD 2305-M/1 (September 1990).
7. European Community Council Directive 91/271/EEC (May 21, 1991).
8. Wilcock, K. G. and White, M. J. D. "Sludge Management in the UK," paper by Watson Hawsley Limited, High Wycombe, England (1992).
9. Suler, D. "Composting Hazardous Industrial Wastes," *Compost Sci.* (July/August 1979).

10. Tchobanoglous, G., Theisen, H., and Elissen, R. *Solid Wastes, Engineering Principles and Management Practices* (New York: McGraw-Hill Book Co., 1977).
11. Spielmann, B. A. "A Yard Waste Primer," *Waste Age* (February 1988).
12. Meyer, M., Hofer, H., and Maire, U. "Trends in Yard Waste Composting," *BioCycle* (July 1988).
13. *Municipal Leaf Composting — A Solid Waste Recycling Program* (Kingston, PA: Royer Industries, Inc., 1973).
14. Williams, T., Gould, M., and Callihan, T. "Dewatering, Treatment, and Composting of Septage," presented at 63rd Annual Water Poll Control Conference, Washington D.C., October 1990.
15. California Agricultural Research, Inc. "Composting for Treatment of Pesticide Rinseates," prepared for Department of Health Services, State of California, (October 1988).
16. Simpkin, T. J., Walter, D., and Doesburg, J. "Treatment of Fuel Product Contaminated Soil in a Cold Climate Using Composting Technology," in *Proceedings of the 85th Annual Air & Waste Management Conference,* Kansas City, MO, Paper No. 92–27.06 (June 1992).
17. Dorau, D. A., "Solid Waste Compost Standards," in *Proceedings of the 85th Annual Air & Waste Management Conference,* Kansas City, MO, Paper No. 92–21.04 (June 1992).
18. Ryan, J. A. and Chaney, R. L. "Regulation of Municipal Sewage Sludge under the Clean Water Act Section 503: A Model for Exposure and Risk Assessment for MSW Compost," *Proceedings of the International Composting Research Symposium* (Columbus, OH: Ohio State Univeristy Press, 1992).
19. Hegberg, B. A., Hallenbeck, W. H., Brenniman, G. R., and Wadden, R. A., "Setting Standards for Yard Waste Compost," *BioCycle* 32:2 (1991).
20. Golueke, C., Diaz, L., and Gurkewitz, S. "Technical Analysis of Multicompost Products," *BioCycle* 30:55–57, (1989).
21. Albright, A. E., President, Solid Waste Composting Council, Personal communication (June 1992).
22. Briton, R. "German Composting Systems," *BioCycle* 33:6 (1992).
23. Letters to the Editor, *Chem. Eng.* (May 1991).
24. Logsden, G. "New Sense of Quality Comes to Compost," *BioCycle* 30:12 (1989).
25. "Europe: Knocking down the Trade Barriers," *USA Today* (June 28, 1991).
26. Spinosa, L. and Lotito, V. "Technical Requirements and Possibilities of Incineration," in *Sewage Sludge Treatment and Use: New Developments, Technological Aspects and Environmental Effects* (New York: Elsevier Applied Sciences, 1989).
27. Rawlinson, L., Kanak, A., and Sharp, R. "Management of Biosolids Down Under, Australia and New Zealand," in *Proceedings of the Specialty Conference Series,* Water Environment Fed. TT041, Portland, OR (July 1992).
28. Meek, B. L., Chesnin, L., Fuller, W., Miller, R., and Turner, D. "Guidelines for Manure Use and Disposal in the Western Region, USA," Bulletin 814, College of Agriculture Research Center, Washington State University, Pullman, WA (1975).
29. Bardos, R. P., Hadley, P., and Kendle, A. "Composting Guidance in the United Kingdom," *BioCycle* 33:6 (1992).
30. Gies, G. "Regulating Compost Quality in Ontario," *BioCycle* 33:2 (1992).
31. *Guidelines for the Production and Use of Aerobic Compost,* Ontario Ministry of the Environment (November 1991).
32. Epstein, E., Chaney, R. L., Henry, C., and Logan, T. J. "Trace Elements in Municipal Solid Waste Compost," *Biomass Bioenergy* (1992).
33. Fricke, K., Pertl, W., and Vogtmann, H. "Technology and Undesirable Components on Compost of Separately Collected Organic Wastes," *Agric. Ecosys. Environ.* 27: 463–469 (1989).
34. Holmgren, G. G. S., Meyer, M. W., Chaney, R. L., and Daniels, R. B. "Concentrations of Cadmium, Lead, Zinc, Copper, and Nickel in Agricultural Soils of the United States," *J. Environ. Qual.* (1992).
35. Shacklette, H. T. and Boerngen, J. G. "Element Concentrations in Soils and Other Surficial Materials of the Conterminous United States," *U.S. Geol. Survey Prof. Paper* 1270:1–105 (1984).

CHAPTER 2

Composting Systems

INTRODUCTION

This chapter provides a brief description of the available types of composting systems. It is necessary to have a basic understanding of the types of processes before proceeding too far into fundamental concepts. Therefore, this chapter introduces the general nature of composting systems, describes their basic similarities and differences, introduces terminology used in later chapters, provides an historical perspective, and transitions into the more fundamental aspects of composting presented in later chapters.

Attention is focused on fundamental concepts that underlie all composting systems and their operation. New processes and equipment will undoubtedly continue to enter the market in response to the growing interest in composting. Any text that simply describes currently available systems will be obsolete within a few years of publication. Basic principles, however, remain unchanged regardless of how new or novel the composting process.

It is also important for the student of composting to understand the history of this industry. Understanding where we are today and projecting were we may be in the future requires that we understand from where we have come. Current perspectives and future projections must spring from a platform anchored by knowledge of the past. It is equally important to study the history of composting to avoid repeating past mistakes. For example, many refuse composting facilities were closed in the 1950s and 60s because they lacked odor control facilities. It is unfortunate for the current composting industry that this problem is still needlessly recurring. In 1991, a large MSW composting facility in the U.S. operated for only a few months before being closed; one reason — no odor control. One can only wonder how such things can occur, given the knowledge of such errors in the past.

History can also teach us about the numerous devices that have been used for composting. Reactors and other equipment available today evolved from earlier, more primitive generations, many of which became extinct. Studying this evolutionary path should help in the continual forward march and growth of this industry. Therefore, this chapter is sprinkled with historical accounts and descriptions of systems, some of which have faded from use. By reviewing and understanding why certain concepts became obsolete, students of today can move ahead with new concepts and designs.

THE MODERN ERA

In his monumental book entitled *The Complete Book of Composting*, J. I. Rodale wrote in 1950 that "The origin of composting has been lost, along with thousands of other age-old practices, in the dim shadows of history. When man first discovered that manure or leaf mold from the forest was good for growing plants, is a matter of speculation. The important thing is that he did discover the benefits of organic matter, and that the knowledge spread throughout the land, and throughout the world."[1] History is peppered with accounts of composting and the importance of organic amendments to the soil. The Preface to this book presents the eloquent feelings of Victor Hugo written in 1862.

While knowledge of composting is evident from Biblical, medieval, and more current accounts, the history of the modern era of composting begins with Sir Albert Howard, a British government agronomist. Sir Howard spent the years 1905 to 1934 in India where he began to recognize that soil must be fertile to produce healthy plants and fertility meant a high percentage of humus. While stationed at the Indore Institute of Plant Industry, he developed a composting technique which has since become known as the Indore method. While the concept was simple, the Indore method represented the first organized plan for composting in the modern era. It provided the following formula:[2]

1. Place a layer of brush on the ground to provide a base for the heap.
2. Build the pile in layers, first using a 6-in. layer of "green matter" like crop wastes or leaves. Next add a 2-in. layer of manure, which in turn is covered by a light layer of topsoil and limestone (note the 3:1 volume ratio of green waste to manure).
3. Repeat the layering until the pile reaches a height of about 5 ft. Turn the pile at about 6 week intervals for about 3 months.

The Indore method was successful because the resulting piles would predictably heat up and not putrefy. The Indore method was widely adopted in the British Empire because it enabled farmers to compost crop waste that otherwise were burned. It was a significant step toward modern composting systems for at least two reasons. First, it presented a "recipe" of substrates, which gave better results than any single substrate by itself; thus, the importance of feed conditioning was recognized. Second, it recognized the importance of procedures to the overall success of the operation. The feed recipe coupled with defined procedures gave repeatable results. By 1935, the tea plantations of India and Ceylon were reportedly producing one million tons of compost a year using the Indore method.

In 1942, Rodale founded *Organic Gardening* magazine after learning of the research of Sir Albert Howard and his Indore composting method. The magazine focused on the philosophy of organic gardening and "backyard" and small-scale composting. *The Complete Book of Composting* was published by Rodale's staff in 1950. One of its contributing authors, a young man named Jerry Goldstein, saw the need for a publication that focused on large-scale uses of municipal and industrial wastes. Not long thereafter, *Compost Science* was founded under Rodale Press and published its premiere issue in Spring 1960. The magazine has since passed to JG Press and was retitled *BioCycle* in 1981. But Jerry remains the editor and publisher and under his stewardship *BioCycle* has become the flagship journal of the composting industry.

Biocycle owes much of its success to its long-time collaboration with Dr. Clarence Golueke, professor emeritus from the University of California at Berkeley and Senior Editor. Dr. Golueke began his composting work at the Sanitary Engineering Research Laboratory in Berkeley, California about 1949. This work led to several significant publications.[3] Dr. Golueke continues to be a leading force in making composting more of a science, while retaining its art, and is considered by most to be the "grandfather" of modern U.S. composting. Other prominent names in the early years of the modern era include Paul McGaughey, Harold Gotass, Lester Haug (my father), John Wiley, Eric Eweson, the Buhler brothers, Nugent Myrick, Norman Pierson, John Snell, Karl Schulze, John Jeris, Ray Regan, and many others. The names of these pioneers will become familiar as their works are referenced in this book.

The history of composting in the post World War II era is chronicled in the pages of *BioCycle*. It has witnessed the rise and fall and rise again of refuse composting in the U.S., the emergence and rapid advance of municipal sludge composting, the explosion in leaf and yard waste composting, and the many problems and solutions developed over time. The people of the industry and all of the composting systems, past and present, successful and not-so-successful, have graced its pages at one time or another. This book must necessarily focus on technologies active in the modern marketplace, but the interested reader can further explore the modern history of composting in the pages of *BioCycle*.

GENERALIZED COMPOSTING SCHEMATIC

A process diagram for a generalized composting system is presented in Figure 2.1. Feedstocks or substrates are shown entering from the left of the diagram. Strictly speaking, any material added to the system should be considered a feedstock to the process. However, conventional practice has adopted the terms "amendment" and "bulking agent" to certain types of added substrates. Both should be considered feedstocks despite the separate nomenclature.

Feed Conditioning

Wet sludge cake has never been successfully composted alone, except in small pilot-scale facilities where constant mixing was applied.[4] The high moisture content of the sludge cake saturates all of the void space. Oxygen transfer into the composting mass is effectively prevented unless high levels of agitation are used continually to expose new surfaces for oxygen transfer. By contrast, dry substrates may require water addition to produce a moist environment for the composting microbes. Nutrients may also be needed to avoid rate limitations from lack of nitrogen, phosphorus, or other trace elements. Preparation of the feedstocks to avoid the above limitations is termed "feed conditioning".

Designers and operators of composting systems have a limited number of areas over which to exercise control of the process. One area of control is the composition of the infeed mixture. The operator must adjust the proportions of the feed components to "close" the energy balance and provide porosity and perhaps add water, nutrients, or seed microbes to avoid rate limitations caused by lack of moisture, sterile feed, or low nutrient levels. Four approaches to blending feed components have been used in practice: (1) recycle of compost product and blending with the other feed components; (2) addition of organic or inorganic amendments; (3) addition of a bulking agent such as wood chips and subsequent screening of the bulking particles from the compost product; and (4) combinations of the above.

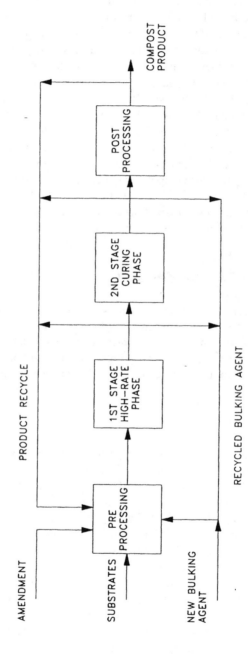

Figure 2.1.   Generalized process diagram for composting showing inputs of feed substrates, amendments, and bulking agents.

Amendments and Bulking Agents

An *amendment* is a material added to the other substrates to condition the feed mixture. Two types of amendments can be defined as follows:

1. Structural or drying amendment — an organic or inorganic material added to reduce bulk weight and increase air voids allowing for proper aeration.
2. Energy or fuel amendment — an organic material added to increase the quantity of biodegradable organics in the mixture and, thereby, increase the energy content of the mixture.

Amendments that have been used to condition wet substrates such as sludge cake include sawdust, straw, peat, rice hulls, cotton gin trash, manure, refuse fractions, yard wastes, vermiculite, and a variety other waste materials. The ideal amendment is dry, has a low bulk weight, and is relatively degradable. Compost recycling can be used to accomplish the reduction in bulk weight and in this sense can be considered an amendment. However, recycle is generally distinguished from other amendments because it does not add new material to the process. With wet substrates, compost recycle is often used along with amendment addition. The advantage of this is that compost recycle can reduce the quantity of supplemental amendments required.

A *bulking agent* is a material, organic or inorganic, of sufficient size to provide structural support and maintain air spaces within the composting matrix. *Bulking particle* may be more descriptive and is the term preferred by the author, although the term "bulking agent" is irreversibly ingrained within the industry. Bulking agents form a three-dimensional matrix of solid particles capable of self support by particle-to-particle contacts. Sludge cake can be viewed as occupying part of the void volume between particles. If the bulking agent is organic, an increase in the energy content of the mixture is a secondary benefit. Wood chips about 1 to 2 in. in size are the most commonly used bulking agent although use of other materials, such as pelleted refuse, shredded tires, peanut shells, tree trimmings, and even rocks (not recommended), has been reported.

The separate identification of amendments and bulking agents developed from the sludge composting industry, where sludge is considered the primary substrate. Unfortunately, this sometimes diverts attention away from the importance of knowing the characteristics of amendments and bulking agents added to the process. Remember, anything added to the process should be considered a feedstock, whether it is termed a substrate, amendment, or bulking agent.

Process Stages

Referring to Figure 2.1, composting systems are often divided into a first stage, high-rate phase and a second stage, curing phase. No precise definition or distinction between these two phases exists. The high-rate phase may use windrow, aerated static pile, or reactor processes. The first stage is characterized by high oxygen uptake rates, thermophilic temperatures, high biodegradable volatile solids (BVS) reductions, and higher odor potential. The second stage, curing phase may use aerated or non-aerated piles, windrows, or even enclosed reactors. The curing phase is characterized by lower temperatures, reduced oxygen uptake rates, and lower odor production potential. The curing phase provides the time required for (1) degradation of the more refractory organics, (2) overcoming the "slowing" effects imposed by kinetic rate limitations, and (3) reestablishing lower temperature microbial populations, which may be beneficial in "maturing" the compost, metabolizing phytotoxic compounds, and suppressing plant diseases.

Traditionally, the first stage of composting has been more engineered and controlled due to the need to reduce odors, supply high aeration rates, and maintain process control. The curing phase is often less engineered, less controlled, and given only token consideration in some designs. This is not acceptable from a process engineering standpoint and is not considered "good engineering practice". Both phases are integral to the design and operation of a complete composting system and both phases are necessary to produce a mature compost product.

Pre- and Postprocessing

Processing of the feedstocks prior to the first stage of composting is termed "pre" or front-end processing. Processing of the feedstocks at intermediate stages or after the curing phase is termed post- or back-end processing. Pre- and/or postprocessing may be required depending on the feedstock characteristics and desired product quality.

Biosolids Systems

Municipal sewage sludges are relatively homogeneous and free of trash materials. They are often conditioned by adding clean amendments such as sawdust. In such cases, preprocessing usually consists of storage and metering facilities for the sludge cake, sawdust, and recycle materials. Mixing by pug mill or other mechanical device is usually provided, after which the mixed materials are fed into the process. Depending on the quality of available sawdust, screening by disc screens or other equipment may be provided. Postprocessing is usually not required in this case. Postprocessing by screening is required if bulking agents are used or if a fine graded product is desired. Sludge composting facilities typically produce high quality composts that are relatively free of contaminants.

The majority of sludge composting facilities condition the wet cake with sawdust or other amendments added to the cake after dewatering. At least one facility, a silo reactor system at Bristol, Tennessee, blends sawdust amendment with the liquid sludge prior to dewatering. A 35 to 37% TS combined cake is produced compared to 20 to 22% with sludge alone. Dohoney[5] reported a more homogeneous mix with this approach compared to pug mill mixing of the components after sludge dewatering. The dewatered mixture of cake and sawdust is fed directly to the silo reactor without additional processing.

MSW, Refuse, and Garbage Systems

Composting of MSW represents the opposite extreme from municipal sludge. MSW is a very heterogeneous material, and only part of the feedstock is suitable for composting. A variety of pre- and postprocessing schemes have been used to prepare and separate compostables and remove unwanted materials from the final product. Preprocessing can be broadly divided into three categories: (1) composting of the mixed refuse fraction with minimal front-end processing, (2) front-end processing to remove undesirable materials prior to composting, and (3) composting a "clean" source separated material. Preprocessing usually involves unit processes designed to size reduce and separate the compostables. The goal of preprocessing is to produce an organically enriched, size reduced, low trash content feedstock to the composting process. Shredding by hammermill or rotary drum may be used for size reduction.

Hand sorting, trommel screening, magnetic separation of iron, induction separation of aluminum, froth separation of glass, and air classification of light and heavy fractions have all been used for preprocessing of the composting feedstock. Moisture and nutrients may also be added during preprocessing.

Postprocessing of MSW compost usually involves equipment designed to separate glass and plastics or grind the glass into small fragments. The goal of postprocessing is to improve the final product quality by removing residual unwanted materials. Screening and air classification are commonly used for this purpose. Obviously, the quality of any compost made from MSW feedstock is highly dependent on the type and extent of pre- and postprocessing. The net output of compost product using MSW feedstock can range from about 10 to 40% of the input wet weight, again depending on the type of pre- and postprocessing. Similarly, the reject or recycling stream can range from about 10 to 50%. The remainder is water and organics converted to CO_2 and H_2O, which exit the process in the exhaust gases.

Yard Waste Systems

In many respects, yard wastes represent an intermediate substrate between sludges and MSW. Yard waste is source separated and relatively clean by comparison to MSW. If the yard waste consists primarily of leaves and grass, very little preprocessing may be necessary, particularly if agitation such as windrow turning is provided for size reduction. Postprocessing by trommel screening will probably be required to produce a fine grade product and remove rocks and other unwanted debris. On the other hand, if tree trimmings are included, preprocessing for size reduction will be required. Usually this involves shredding by hammermill or tub grinder. Some trimmings, such as palm fronds, are particularly difficult to handle and may require tub grinding followed by hammermill shredding.

If yard wastes are collected in plastic bags, another level of complexity must be added to deal with this problem. Shredding will usually be required to break open the bags, followed by screening to remove as much plastic as possible before actual composting begins. Postscreening is usually required to remove residual plastic from the compost product.

Points to Remember

Three points are important when it comes to pre- and postprocessing. First, separations are never 100%, meaning that some unwanted materials, such as plastic and glass, will always find their way into the end product if they are present in the feedstock. Second, the net output of compost product will decrease as more separations are used to improve product quality. Third, no amount of pre- or postprocessing can substitute for starting with a clean feedstock in the first place. As stated previously, starting with a high quality feedstock gives the greatest assurance of producing a high quality product. This does *not* mean that marketable compost cannot be made from heterogeneous substrates such as MSW. It *does* mean that the compost quality will be lower compared to compost made from clean substrates and that more work will be needed to make the compost.

Compost as a Product

Composted material exits to the right in Figure 2.1 and is the product generated by the process. Quality of the compost product depends on the characteristics of the feed substrates, the design parameters of the high-rate and curing phases, the amount of pre- and postprocessing, and the operating conditions maintained within the system. It is important to remember that the process is designed to deliver a product of a given quality. Therefore, the desired product quality should be established early in design. In actual fact, it should probably be the very first thing considered in design.

A number of criteria can be established to define product quality. These include physical criteria such as particle size distribution, texture, color, odor, moisture content, and general appearance. Criteria can also be established to define compost stability or maturity. Specific oxygen consumption rate (mg O_2/kg volatile solids per hour), absence of phytotoxic compounds, reduction of biodegradable volatile solids (BVS) across the system, and a return to near ambient temperatures at the end of the process can be used to measure compost stability. Criteria can also be established for the chemical characteristics of the compost. This might include the nutrient content, nitrate/ammonia ratio, absence of readily degradable compounds such as starch, absence of anaerobic intermediates such as acetic acid, heavy metal content, and the effect on seed germination. The point to remember is that the compost facility is a processing plant that must be designed and operated to produce the desired product.

CLASSIFICATION OF COMPOSTING PROCESSES

Attention will now be turned to the various processes used for the high-rate and curing phases of the process blocks of Figure 2.1. A number of approaches have been used to categorize composting systems. A chemical engineering approach to classification is used here, which emphasizes reactor type, solids-flow mechanisms, bed conditions in the reactor and manner of air supply. Perhaps the most basic distinction is between systems in which the composting material is contained in a reactor and those in which it is not. Systems that use reactors are popularly termed "mechanical", "enclosed", or "in-vessel", whereas those that do not are often termed "open" systems.

The term "mechanical" is a misnomer because all modern compost systems are mechanical to some extent. Some employ mobile equipment while others use stationary reactors and conveyors. One system might be considered more mechanical than another but the basic distinction is illusive. The terms "enclosed" and "open" are also poorly defined. For example, composting material might be housed under a roofed structure and thereby be enclosed but not contained in a reactor. Would such a system be "open" if it is under a roofed structure? To avoid such confusion, the basic distinction will be between reactor and nonreactor systems. Reactor systems are those in which composting material is placed in a reactor and nonreactor systems are those in which it is not. Both types of systems can employ mechanical equipment and both may or may not be enclosed under protective housings.

U.S. practice has adopted the term "in-vessel" for reactor type systems. Much of the early literature used terms like "digester" and "fermenter". I prefer the term "reactor" because it is more in keeping with chemical engineering practice for the design of processing plants. The concept of designing and operating a compost facility as a process plant is a central theme of this book. Unfortunately, the term "in-vessel" is now solidly entrenched in the industry and there is no hope of displacing it. Therefore, the terms "reactor" and "in-vessel" are used synonymously in this text.

Nonreactor Processes

Illustrations of the basic classified nonreactor composting processes are presented in Figure 2.2. Nonreactor systems are divided between those that maintain an agitated solids bed and

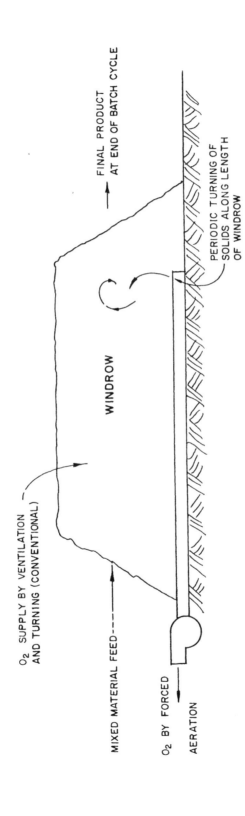

Figure 2.2A. Non-reactor, agitated solids bed (windrow system): (1) conventional — solids agitated by periodic turning without forced aeration, batch feed of solids, plug flow of solids with some dispersion from agitation; (2) forced aeration — same as conventional windrow but with provision for forced aeration.

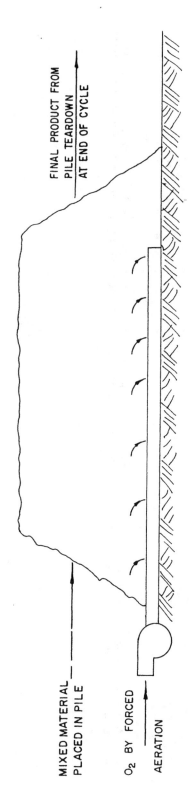

Figure 2.2B. Non-reactor, static solids bed (aerated, static pile process): no agitation or turning of static bed, batch feed of solids, no dispersion or mixing of solids while in the bed. Forced aeration.

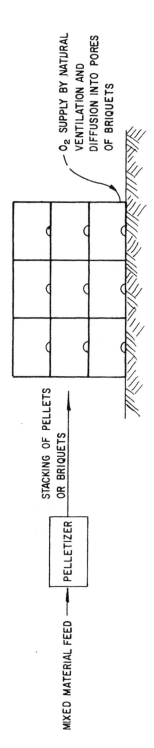

Figure 2.2C. Non-reactor, static solids bed: natural ventilation and diffusion of oxygen (example: Brikollari process).

those that employ a static bed. An agitated solids bed means that the composting mixture is disturbed or broken up in some manner during the compost cycle. This may be by periodic turning, tumbling, or other methods of agitation. There is a distinction between agitation and mixing. If a system is well mixed it means that the infeed will be dispersed throughout the bed volume as a result of agitation. In a completely mixed system no concentration gradients exist throughout the bed volume. Thus, compost withdrawn will have the same characteristics as material within the bed. For a system to be well mixed it must be agitated in some manner. However, an agitated system may be either well mixed or not. In fact, in some agitated systems very little, if any, mixing occurs as a result of the energy input.

The Windrow Process

The windrow system is the most popular example of a nonreactor, agitated solids bed system. Mixed feedstocks are placed in rows and turned periodically, usually by mechanical equipment. Height, width, and shape of the windrows vary depending on the nature of the feed material and the type of equipment used for turning. Oxygen is supplied primarily by natural ventilation resulting from the buoyancy of hot gases in the windrow, and to a lesser extent, by gas exchange during turning. In the forced aeration windrow system, oxygen transfer into the windrow is aided by forced or induced aeration from blowers. In most composting literature the term "forced" aeration is used regardless of whether aeration is forced or induced. Strictly speaking, forced aeration applies to cases where ambient air is forced into the material under positive pressure. Induced aeration applies to cases where gases are pulled from the material under negative pressure. In either case, periodic agitation by turning is used to restructure the windrow. As a result, considerable mixing can be expected along the height and width of the row, but little mixing will occur along the length.

The Static Pile Process

The aerated, static pile process is the premiere example of a nonreactor, static solids bed system. This system has seen wide application in the U.S. for wet substrates such as sludge cake. The substrate is mixed with a bulking agent, such as wood chips, and formed into a large pile. The bulking agent provides structural stability to the material and maintains air voids without the need for periodic agitation. A distribution system is used to allow either forced or induced draft aeration. No agitation or turning of the static bed occurs during the compost cycle, and the piles are formed on a batch basis. Because there is no bed agitation, no mixing occurs once the pile is formed. Some recent designs for the aerated static pile process have included provisions for tearing the pile down, delumping solids and, reforming for additional composting. If this provision is used the system can be classified as a "semi-agitated" bed system because the agitation is considerably less than that normally associated with the windrow system.

Reactor Processes

Reactor processes are first classified according to the manner of solids flow as either vertical flow reactors or horizontal flow reactors. Horizontal flow includes a number of reactor types in which the reactor is inclined slightly from the horizontal to promote solids flow. Illustrations of vertical (tower or silo) and horizontal (and inclined) flow reactors are presented in Figures 2.3 and 2.4, respectively.

Vertical Flow Processes

Vertical flow reactors are further defined according to bed conditions in the reactor. Some systems allow for agitation of solids during their transit down the reactor and are termed moving agitated bed reactors. These are usually fed on either a continuous or intermittent basis. In other systems, the composting mixture occupies the entire bed volume and is not agitated during any single pass in the bed. These are termed moving packed bed reactors and can be fed on either a continuous, intermittent, or batch basis. The moving packed bed systems shown in Figure 2.3 often allow for periodic transfer of solids from the bottom to the top of the reactor. Agitation of solids occurs as a result of this transfer, but on any single pass the bed solids remain static until they are again withdrawn from the bottom for transfer.

The vertical, moving packed bed reactor has been widely applied to composting of sludge cake amended with sawdust and other materials. Several versions of this reactor are available and include circular and rectangular reactor geometries with counter-current and co-current aeration patterns. Material depth inside the reactor is typically about 6 to 9 m. One reason for the popularity of this reactor style is the relatively low cost per unit of working volume.

Horizontal and Inclined Flow Processes

Horizontal flow reactors are divided into those that employ a rotating or rotary drum (tumbling solids bed reactor), those that use a bin structure of varying geometry and method of agitation (agitated solids bed reactors), and those that use a bin type structure but with a static solids bed (static solids bed reactor). Such systems have been applied to a wide variety of composting substrates including MSW, agricultural wastes, and sewage sludges.

Rotary (Rotating) Drums   At least three types of rotary drums can be distinguished based on the solids flow pattern within the reactor. In the dispersed flow case, material inlet and outlet are located on opposite ends of the drum. Plug flow conditions exist within the vessel except for some dispersion of material resulting from the tumbling action. This is probably the most commonly used of the drum systems and has been widely applied to MSW composting.

To prevent possible short-circuiting of material through the reactor, the drum can be compartmented into a number of cells in series. The most common mode of operation is to discharge product from the last cell. Each cell is then sequentially discharged into the next. Feed material is added to the first cell once it has been emptied. Thus, feed is added on an intermittent basis and the reactor can be viewed as a number of well-mixed cells in series. The Los Angeles County Sanitation Districts experimented with compartmented and noncompartmented drums and favored the former.[6] Reasons cited included the tendency in a noncompartmented drum to mix material uniformly rather than traveling slowly from point of entry to the discharge and the potential benefits of semicontinuous feeding with a compartmented drum.

In the last case, an attempt is made to completely mix the contents of the drum to produce a homogeneous composting mass. This requires that both feed and discharge of material occur uniformly along the length of the reactor. If feed and discharge are continuous, a considerable quantity of material will exit the reactor in less than the theoretical detention time. In such a case, further processing would be necessary to ensure a pathogen-free product. Intermittent feeding and withdrawal can avoid this problem to some extent. This reactor style is not commonly used because of the complexities of feed and withdrawal and the limited advantages of a true completely mixed system.

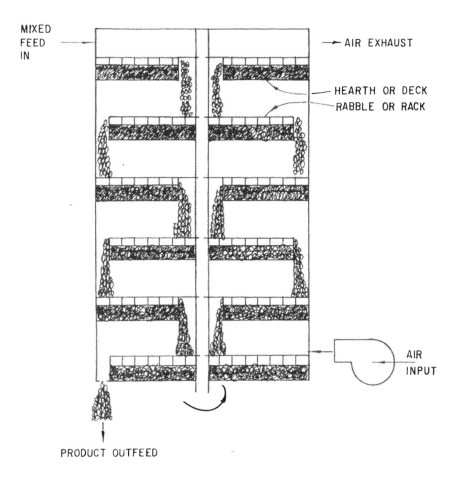

MIXED FEED IN

AIR EXHAUST

HEARTH OR DECK
RABBLE OR RACK

AIR INPUT

PRODUCT OUTFEED

Figure 2.3A. Vertical flow, agitated solids bed reactor: multiple hearths, solids are agitated during movement down the reactor, forced aeration, continuous or intermittent feeding, some mixing in reactor.

Agitated Bins   A number of bin reactors of the horizontal flow, agitated solids bed type have been developed. Both forced aeration and mechanical agitation of solids are employed, which allows considerable operational flexibility. Reactors are usually uncovered at the top and may be housed in a building for improved all-weather operation and control of nuisances. Most are operated on a once-a-day feed cycle. Both circular and rectangular shapes are currently in use, and both have been applied to a variety of substrates including MSW, yard wastes, and industrial and municipal sludges.

In the circular bin system, augers are mounted along a traveling bridge, which rotates from the center of the reactor much like a circular clarifier mechanism. Typically, the bridge takes about 2 h to complete one revolution. Material is fed along the outer periphery as the bridge rotates. Augers agitate the material within the reactor and also provide some mixing of the new feed with the already composting mass. Material is gradually transferred toward the center of the reactor, where it drops over an adjustable weir and falls down to an outlet conveyor located in a gallery below the reactor.

A number of rectangular, agitated bin systems have been developed. A variety of agitation devices are available, most of which run on rails mounted on the top of the bin walls. The

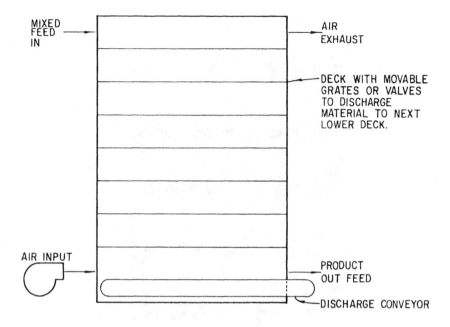

Figure 2.3B. Vertical flow, agitated solids bed reactor: multiple floors, decks, or belts.

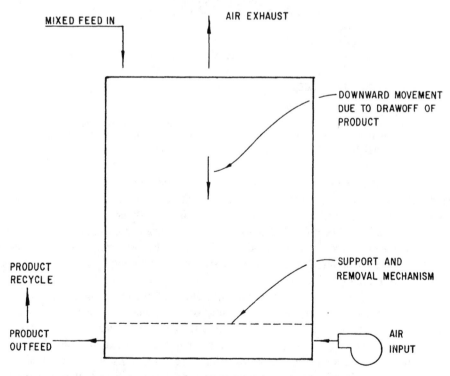

Figure 2.3C. Vertical flow, packed bed reactor (silo reactors): solids not mechanically agitated during movement down the reactor, forced aeration with counter-current or cross-current air/solids flow, plug flow of solids in reactor.

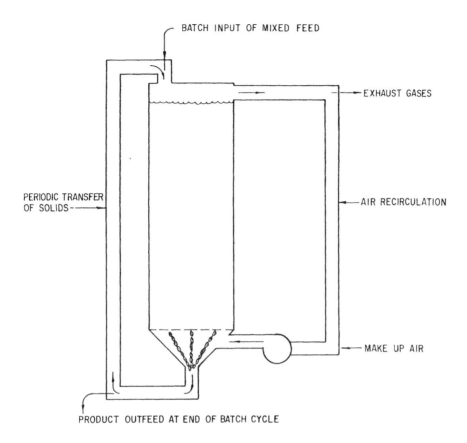

Figure 2.3D. Vertical flow, packed bed reactor.

rectangular bin systems differ primarily in the shape and size of bins and mode of operation. Nearly square bin geometries have been used with agitation provided by augers mounted on a traveling bridge. Like their circular cousins, such bins are fed on one end and material is moved along the bin by the agitation device, eventually exiting the end opposite the feed.

A number of bin systems are in use with very high length to width ratios. Among these, some equipment suppliers prefer smaller scale bins with dimensions of about 2 m wide by 2 m high. Other bins have been used with dimensions as large as 6 m wide by 3 m high and with lengths as much as 220 m. The larger bins typically operate by placing material into designated cells along the bin length. Material is turned about once a week and remains generally within the designated cell until it is ready to be removed. With the smaller bins it is more typical to agitate the material on a daily basis. In so doing, material is moved a distance down the bin with each agitation. This allows material to be fed at one end and gradually move the length of the bin, eventually exiting at the opposite end. Yet another variant uses the agitation device to lift material over the bin wall into an adjacent empty bin. Material is moved sequentially from one bin to another during each turning.

Static Bed Bins   The horizontal, static bed reactor was first developed in about 1979 from the European experience with vertical silo systems. This type of system has since been applied to a variety of feedstocks including sludge, manure, and MSW fractions. The vessel is essentially a plug flow, tubular reactor or tunnel of rectangular cross section. Reactor volume can range

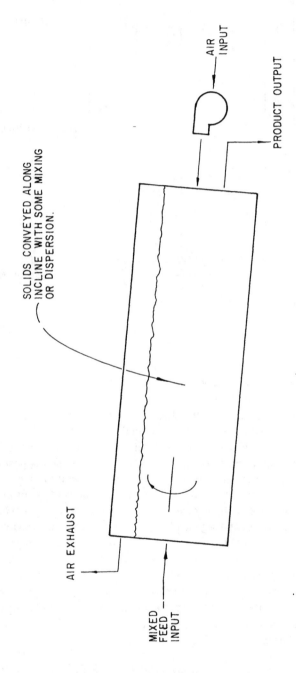

Figure 2.4A. Tumbling solids bed reactor (rotating drum or kiln) — dispersed flow: dispersion provided by constant tumbling action, reactor not segmented inside, solids agitated by nearly constant rotation of a drum and fed on a continuous or intermittent basis, forced aeration is usually provided.

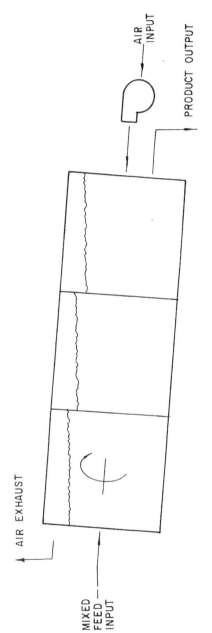

Figure 2.4B. Tumbling solids bed reactor — cells in series: solids flow is by periodic emptying and transfer of material from one cell to another, each cell is well mixed.

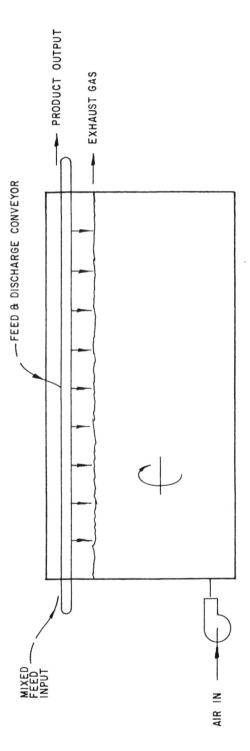

Figure 2.4C. Tumbling solids bed reactor — complete mix: uniform feed and discharge are maintained along with a high level of mixing.

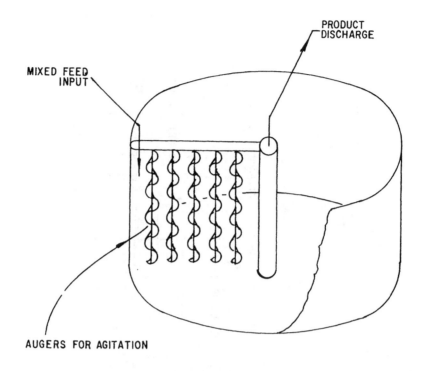

Figure 2.4D. Agitated solids bed reactors (agitated bins) — circular shape: solids agitated by mechanical turning devices and fed on a continuous, intermittent, or batch basis.

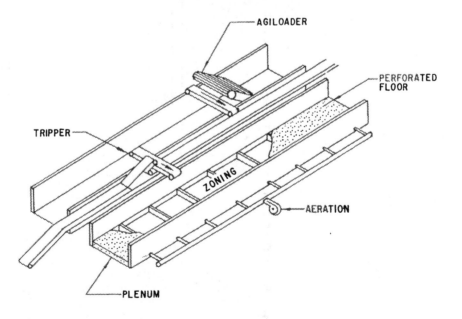

Figure 2.4E. Agitated solids bed reactor — rectangular shape.

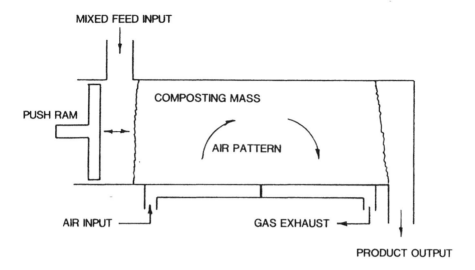

Figure 2.4F. Static solids bed reactor (tunnel shaped) — push type: solids pushed or conveyed along the reactor, solids not mechanically agitated within each reactor, but may be agitated while transferred from one reactor to another.

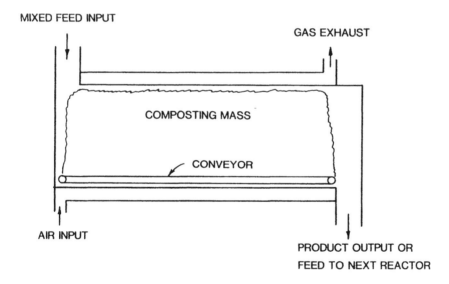

Figure 2.4G. Static solids bed reactor — conveyor type.

from 10 to 500 m³. The larger sizes are generally constructed of reinforced concrete and the smaller sizes fabricated in steel. One design uses a pusher plate with forward and backward movement which is situated at the feed end of the tunnel. When the plate is withdrawn, the daily charge of mixed materials is added to the empty zone created. When the chamber is full, the pusher plate is hydraulically moved forward, pushing the new material into the reactor and simultaneously expelling "green" compost from the far end of the reactor. The pusher plate remains in place until it is withdrawn for the next feeding cycle. Air can be supplied and gases can be removed along the length of the reactor. At least one other version of the horizontal, static bed reactor is available. It uses a type of "walking floor", commonly used in the trucking industry, to move material along the reactor.

Nonflow Processes

A number of composting technologies that use batch operated, compost "boxes" are available. Materials are loaded at the start of the cycle and typically remain in the box reactor for 7 to 14 days. Aeration is usually controlled and may alternate between positive and negative modes. Curing is usually conducted in windrows for several months.

Completing the System

The reader will note that the term "process" was used to describe the above reactor and nonreactor equipment. This was purposely done to emphasize the point that a single process does not make a complete system. For example, the rotating drums used in MSW composting facilities typically have a residence time of only 1 to 4 days. Mature compost is certainly not produced in such a short period of time. In fact, the drum serves primarily to grind down and homogenize the feedstock. Most of the organic stabilization occurs downstream of the drum, probably in static piles or windrows. Because the reactor or other equipment represents a large capital investment and is the focus of most proprietary claims, it usually receives the most attention. Composting facilities have actually been constructed (fortunately only a few) with little or no attention given to the curing phase, which is absolutely essential to producing a quality product.

Reactor based systems have been constructed in the U.S. with the intent of conducting both high-rate and curing phases within one or more reactors. Presumably, product would exit the last reactor and immediately be loaded into trucks for reuse. Unfortunately, few facilities actually operate in this manner. Some additional curing or storage almost always is necessary. Usually, this is done in windrows or static piles. Oftentimes, curing facilities have been added "after the fact" or moved to another location as a result of odors or lack of planning. There are exceptions to the above rule. The City of Hamilton, Ohio, pays to have the reactor outfeed immediately removed from the site and cured or blended with topsoil at an off-site location. Lancaster, Pennsylvania, loads directly into trucks from a silo reactor system, but it has been forced to derate the plant to operate in this manner. These examples are the exception and not the rule.

Again, the point here is that a complete *system* includes more than just the high-rate phase and more than just the reactors or nonreactors used to carry out this phase. Therefore, most composting systems use more than one type of *process* to complete the system. For example, a reactor process might be followed by a windrow or static pile process for curing to complete the overall system. The following sections focus on the process flowsheets typical of complete systems.

NONREACTOR, AGITATED SOLIDS BED SYSTEMS

Process Descriptions

The following processes are categorized as nonreactor, agitated solids bed types and either are actively used today or are significant to the history of composting.

Indore

A process developed by Sir Albert Howard in India about 1925. Material is placed in alternate layers of refuse, manure, night soil, earth, and straw or woody matter until about 5 ft high. No grinding is used. The pile is turned by hand several times during the cycle. If the pile is very large, vertical funnels or chimneys can be formed by rotating a rod into the pile or using ventilating pipes placed about 4 ft on center. Detention time is typically 120 to 180 days. The Indian Council of Agricultural Research has modified the Indore Method, and it is still widely used there under the name Bangalore process.

Highlights: Historically significant for its development of a recipe of feedstocks and procedures that gave predictable results.
Status: Active in small farm applications, particularly where labor is inexpensive.

VAM (van Maanen)

This process was developed by Mr. T. van Maanen and first used at Wijster, Netherlands, in 1931 by the company Vuilafvoer Maatschappij (VAM) to compost refuse. VAM is a version of the Indore process, but adapted for composting large quantities of refuse. Trains loaded with town refuse delivered material to the composting site. The trains moved onto one of four viaducts, each about 20 ft high and 1560 ft long. Refuse was discharged from both sides of the train cars until the viaduct was filled. Moisture was adjusted periodically by sprinkling with water and/or recirculating drainage liquor. Refuse was composted from 4 to 8 months, turned occasionally, and dug out by a grab crane at the end of the cycle. Composted materials were moved to the postprocessing building, which included screening, magnetic separation and grinding. Later versions used preshredding with a rasp (Raspel or rasping process). A second plant was started in 1955 at Mierlo, Netherlands, with a capacity of 70,000 metric tons/ yr (mtpyr) refuse. The process was conducted in remote areas because of odors and fly and rodent problems. MSW composting operations closed in 1989. VAM is now focusing its efforts on source separated organics.

Highlights: This was one of the first processes to recognize the need for moisture addition
to improve reaction rates with dry substrates.
Status: The original MSW composting system is historic.

Modern Windrow

In the windrow process, wastes are stacked in piles that are arranged in long parallel rows or windrows. In large and more "high tech" systems, the windrows are turned at regular intervals by specialized mobile equipment. In cross section, windrows may range from haystack to rectangular, trapezoidal, or triangular shape, depending largely on characteristics of the composting material and equipment used for turning. The windrow system has been used successfully for composting a wide variety of organic residue, including manures, sewage sludges, and leaves and yard wastes. It is the most commonly used process for leaf and yard waste composting. Ventilation is by natural chimney effect or forced aeration.

Windrows were once used for refuse composting in Mobile, Alabama (270 tpd), Boulder, Colorado (90 tpd), and Johnson City, Tennessee (47 tpd). Windrows are currently used in Israel and Mexico City and for later phases of other reactor composting systems. Windrow composting was first applied to sewage sludge as the primary substrate in 1972 by the Los Angeles County Sanitation Districts and the USDA at Beltsville, Maryland. Use of recycled compost and/or amendments is required for moisture control with wet substrates. Windrows are currently used for sewage sludge composting at Denver, Colorado (aerated), Los Angeles, California (nonaerated), Upper Occoquan, Virginia (aerated), Austin, Texas (nonaerated), and Hawk Ridge, Maine (aerated, raw sludge).

Highlights: Windrow composting is widely used today because it is flexible, adaptable to a variety of feedstocks, and can be implemented with relatively low capital investment. A wide variety of specialized turning equipment is available in today's market.

Status: Active.

Others

Numerous process names have been applied to systems that basically use the windrow process. Often these take the name of the front-end equipment manufacturer or the company who implemented the process. For example, Buhler-Miag, Reuter Resource Recovery, and others offer advanced windrow composting technologies for refuse under their company names.

Windrow Composting of Wet Substrates

History of Application

One of the first references to composting of sewage sludge as the primary substrate appeared in 1950. In that year, Ullrich and Smith[7] reported on experiments conducted in Austin, Texas, using digested sludge mixed with hardwood sawdust. The mixture was then windrow composted for about 11 weeks. It is interesting to note that full-scale windrow composting of digested sludge was implemented (much later) at Austin in 1987. Later experiments on sludge composting were reported by Reeves in 1959.[8] Digested sludge from the City of El Paso, Texas, was air dried for 4 to 6 months and then mixed with hardwood sawdust. The mixture was windrow composted for 2 to 3 months. Water was added and the mixture turned by a grader at 2 to 3 week intervals. The material was cured for 2 to 3 months in static, nonaerated piles.

A major adaptation that has allowed the windrow system to be more easily applied to wet substrates is the concept of recycling dry compost to blend with wet feed. The concept was pioneered for sludge composting by the Los Angeles County Sanitation District (LACSD) in 1972. The quantity of recycled material is adjusted to obtain a mixture moisture content of about 60% (40% TS). By so doing, structural integrity of the mixture is increased so that a properly shaped windrow can be maintained. Friability or porosity of mixed material is also greatly improved, which, in turn, increases the effectiveness of windrow turning. Amendments such as wood chips, sawdust, straw, or rice hulls can also be added to the mixture, either with or without dry compost recycle, and will accomplish the same purposes. Addition of amendments will also improve the energy balance because new organics are added to the process.

The windrow composting technique is believed to be the oldest "sludge only" composting system. It was first used in 1972 by the LACSD to compost approximately 100 ton ds/day

Figure 2.5. Open windrow composting of dewatered, digested sludge at the Los Angeles County Sanitation District's wastewater treatment plant in Carson, California. Sludge is amended with sawdust and recycled product. Windrows are turned about three times a week. Aeration is provided primarily by natural ventilation. Windrows shown correspond to the Step 1 windrows of Figure 2.9.

(dtpd) of digested, dewatered sludge cake.[9] The operation evolved and improved over the years and remained at the plant site until 1991, when it was moved to a more remote location because of odor concerns. A photo of the LACSD facility is presented in Figure 2.5. The Upper Occoquan Sewer Authority, Virginia, began operating a windrow composting facility in 1980 processing about 8 dtpd. The facility was the first to use advanced design concepts such as roofed coverage for all weather operation, concrete pad flooring for better equipment access and improved housekeeping, and permanent, subsurface manifolds for suction aeration.[10]

A large advanced windrow system was developed by the Denver Metropolitan Wastewater Reclamation District in the early 1980s. The facility was designed for 100 dtpd and included almost 16 acres under roof, both positive and negative aeration capability, and over 165,000 standard cubic feet per minute (scfm) of installed aeration capacity. A photo of the facility is presented in Figure 2.6. Operations began in 1986 and the facility experienced odor complaints almost immediately. A number of operational measures were taken to reduce odor emissions and the facility resumed operation at part load.

There are a number of other successful sludge windrow systems. In early 1991, San Joaquin Composting Co. began operating a windrow facility under contract to the City of Los Angeles. The facility is located in a remote part of the San Joaquin Valley and is currently composting about 100 dtpd of digested, dewatered sludge at 20% TS amended with product recycle, agricultural residues (almond wastes, cotton gin trash, and rice hulls) and some municipal yard wastes. To date, the facility has registered few odor complaints, probably because of the remoteness of the site.

The City of Austin, Texas, began operating a windrow facility composting about 10 dtpd in 1987. The sludge is first processed by anaerobic digestion followed by air drying to 30 to

Figure 2.6.   Enclosed windrow composting facility operated by the Denver Metropolitan Wastewater Reclamation District in Denver, Colorado. Substrate is dewatered, digested sludge amended with recycled product, sawdust and bark. About 16 acres of roofed coverage are provided. Note the natural ventilation evident by the condensing steam above the crown of the windrow.

40% TS in 25 acres of drying basins. Air dried cake is blended with tree trimmings, leaves, and yard waste and composted to produce a product called "Dillo Dirt". The product is either distributed to other city departments at no charge or marketed through registered vendors. Demand for Dillo Dirt often exceeds the available supply.[11] A similar concept of air drying prior to windrow composting is used at the 96th Avenue Plant in Phoenix, Arizona.

Operating Parameters

A roughly trapezoidal cross section is typically observed with sludge based windrows. The actual cross section is largely a function of the feedstock characteristics and the turning equipment used, but the dimensions of Figure 2.7 are typical. Allowing about 8 ft between the smaller windrows and 10 ft between larger ones, specific volumes of about 5000 to 5700 m^3 mixed material/hectare (2630 to 3000 yd^3/acre) can be achieved. Windrows are about 300 ft long in the Denver operation and 400 ft long in the LACSD operation.

More fibrous feedstocks can generally be formed into deeper windrows of rectangular cross section. Composting operations producing bedding material for mushroom growing often achieve rectangular windrow shapes with depths of 2.4 to 3.0 m (8 to 10 ft) and with only about 2 ft between windrows. These unique dimensions result because straw is a major component of the mix and because turning equipment special to this industry has been developed. Specific volumes as high as 20,000 m^3/hectare (10,500 yd^3/acre) can be achieved in such special cases. The 1980s saw a large increase in the number of suppliers of windrow turning equipment. A wide variety of equipment is now available in the market, spanning a wide range in price and size of windrow that can be turned.

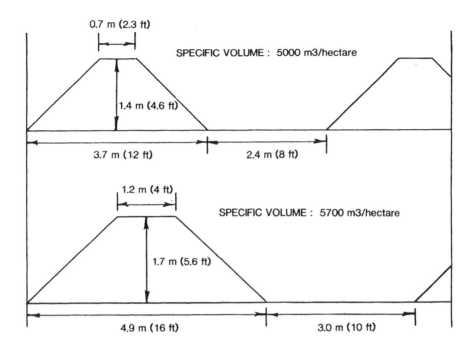

Figure 2.7. Typical windrow dimensions for a mixture of dewatered sludge amended with product recycle and sawdust to an initial 40% TS mixture. Values shown are based on measurements made at the Denver windrow facility shown in Figure 2.6.

A materials balance reported by LeBrun et al.[12] for windrow composting of digested sludge cake amended only with recycled product is shown in Figure 2.8. The balance was developed for summer weather conditions in Southern California and used a 35-day windrow detention time. It is interesting to note that even though the feed substrate was anaerobically digested, over 40% reduction of feed volatile solids (VS) was achieved. VS content of the final product was less than 40%, which is low compared to most composts because no supplemental amendments were added. The significant loss of water should also be noted.

The LACSD developed a number of improvements during the many years in which they operated the windrow process. According to Hay,[9] larger turning equipment allowed the formation of larger windrows. Increasing the speed of the rotating drums and using detachable flails gave superior grinding action and produced a more uniform, more finely divided compost and eliminated clumping problems. Research demonstrated adding external amendments such as sawdust and rice hulls is beneficial, that large windrows achieve higher temperatures and are less odorous than smaller windrows, and that larger windrows are more productive in terms of land space than small windrows.

The final version of the LACSD system used a two step process developed by Hay[9,13] as shown in Figure 2.9. Recycle windrows, which typically contained a volumetric ratio of cake, recycle, and sawdust of 1.0:0.8:0.4, were composted only between May and September. Sawdust windrows, which contained about a 1:1.2 volumetric ratio of sludge cake to sawdust, were composted throughout the year. Research demonstrated that the recycle windrows could not reliably achieve time-temperature conditions of 55°C for 15 days during the cool, rainy months of October through April. Composting in small windrows (Step 1) usually took 2 to 3 weeks and accomplished initial drying, aeration, and preliminary pasteurization. The windrows had dimensions similar to the smaller windrows of Figure 2.7. Windrows were turned

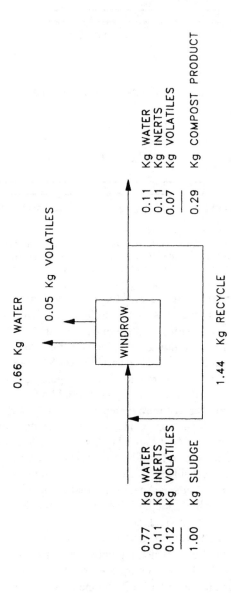

Figure 2.8. Mass balance for a windrow facility composting dewatered sludge amended only with recycled product. Recycle is adjusted to give an initial mixture typically in the range of 40 to 45% TS. From LeBrun et al.[12]

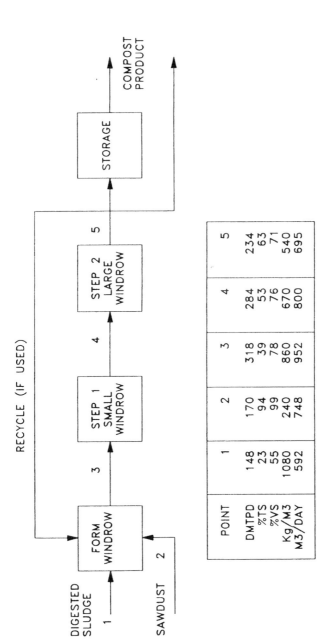

POINT	1	2	3	4	5
DMTPD	148	170	318	284	234
%TS	23	94	39	53	63
%VS	55	99	78	76	71
Kg/M3	1080	240	860	670	540
M3/DAY	592	748	952	800	695

Figure 2.9.   Process flow diagram and mass balance for a two-stage windrow facility composting digested, dewatered municipal sludge. Residence time in Step 1 is 3 to 4 weeks and in Step 2 3 to 10 weeks. Windrow shape is triangular with a 12-ft base and a 4-ft height in Step 1 and with a 16-ft base and a 6-ft height in Step 2. Adapted from Hay.[9,13]

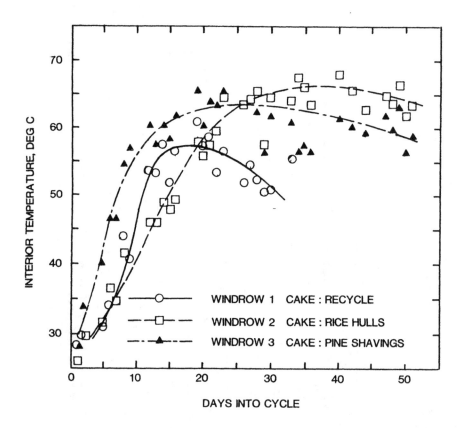

Figure 2.10.   Temperature and total solids profiles during open windrow composting of dewatered sludge conditioned with various amendments. Data were collected in May in Southern California. Windrows 2 and 3 were wetter at the start because of the low bulk weights of the amendments compared to recycled compost. Windrow cross section corresponds to the small windrows of Figure 2.7. All windrows were turned 3 times/week. From Iacoboni et al.[26]

3 to 4 times/week during Step 1. Temperatures above 55°C were usually achieved in the small windrows during warm weather but not during the cooler winter months.

When adjoining windrows completed Step 1, a front-end loader pushed together 2 or 3 adjoining windrows into a single large windrow. Additional amendment could also be added at this time depending on the temperature history during Step 1. Windrow dimensions were similar to the larger windrows of Figure 2.7. Step 2 promoted development of maximum internal temperatures (needed to thermally inactivate pathogens), dehydrated the compost further, and uniformly mixed the substrates. Large windrows were composted for about 3 weeks or longer to meet desired time/temperature/turning requirements. Large windrows were turned about 3 times/week and had lower heat losses, developed higher internal temperatures for longer periods, and were less prone to upset from rain or cool ambient temperatures. A mass balance for the sawdust windrows is presented in Figure 2.9.

Typical temperature profiles reported for windrow composting of sludge cake are presented in Figures 2.10 and 2.11. For the data of Figure 2.10, windrow dimensions were near those of the smaller windrows of Figure 2.7. The recycle windrow consisted of 1.3 m³ of compost product per metric ton of sludge cake. The sawdust and rice hull windrows did not use compost product for conditioning, but instead used 0.84 m³ of amendment per metric ton of cake. All windrows were turned 3 times/week throughout the cycle and were not combined into larger

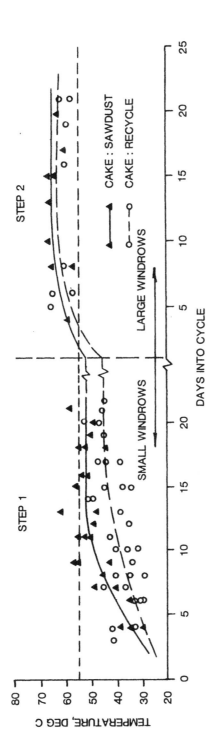

Figure 2.11. Temperature profiles for windrow composting of sludge amended with product recycle and sludge amended with sawdust. Data correspond to the 2 step process shown in Figure 2.9. From Hay.[9,13]

windrows. Temperature profiles in Figure 2.11 correspond to Hay's two step process described previously.

Windrow Composting of Drier Substrates

Windrow composting has been applied to a variety of relatively dry substrates, but in particular yard and "green" wastes. Green waste is a general term applied to a variety of natural, organic, plant materials that are often separately collected from other waste materials. Yard wastes, such as leaves and grass clippings, tree and brush trimmings, sawdust, wood chips, and even garbage are often referred to as green wastes. The greatest number of windrow facilities have been implemented for composting of leaves and leaf/grass mixtures.

Because of its large commitment to yard waste composting, the State of New Jersey published a leaf composting manual in 1985.[14] Three different levels of composting are defined in this manual. They are a minimal technology, a low-level technology, and a high-level technology.[15] Each of these is a variation on the basic windrow process.

Minimal Technology

If a large, well-isolated area is available, a very low-cost approach to leaf composting is possible. Leaves are formed into large windrows (about 12 ft high by 24 ft wide) using a front-end loader. The windrows are untended, except for a once-a-year turning with the front-end loader. Compaction of the piles can cause anaerobic zones within the windrow. Some odor can be expected prior to the first turning, with significant releases during the first turning. A minimum quarter-mile buffer is recommended with this technology. About 3 years are required to produce an acceptable compost. Small, rural communities sometimes use this approach because it is inexpensive and they are favored with abundant land areas and wide buffers.

Low-Level Technology

With "low-level" technology, more attention is paid to providing proper conditions of moisture, oxygen, and temperature within the pile. Water is usually added during the initial windrow formation to achieve about a 50% moisture content. Pile size is adjusted to help maintain aerobic conditions. Initial pile size may be reduced to about 6 ft high by 12 to 14 ft wide to enhance initial aeration. Piles may later be combined after the initial burst of microbial activity, which lasts about 1 month. Turning frequency is normally 2 to 3 times/yr. Water can be added during pile turning to readjust the moisture content. Acceptable compost can be produced in about 1.5 yr.

High-Level Technology

The authors of the New Jersey Manual (Strom and Finstein from Rutgers University) have historically been associated with the aerated, static pile system for composting of sewage sludge. Not surprising, they defined a "high-level" technology which uses forced aeration of large static piles with blowers. Water addition is necessary to adjust the moisture content. Nitrogen fertilizer may also be added to adjust the carbon/nitrogen (C/N) ratio to avoid nutrient rate limitations. After about one month of high rate processing, the blowers can be removed and mechanical turning of the piles begins. It is expected that a final compost can be produced in about 1 yr.

I am not aware of any leaf composting facility using the "high technology" approach defined by Rutgers. Little is gained by using the initial high-rate phase with static piles and blowers, particularly when mechanical turning is used for the next 11 months of a total 12 month cycle. It appears that the capital and labor costs to conduct "high-rate" composting for 1 month limited the application of this technology.

Intermediate-Level Technology

The majority of private leaf composting facilities recognized that frequent mechanical turning accomplished a number of beneficial purposes in the composting process. Leaves are "refluffed", increasing free air space and enhancing the movement of air through the windrow. Mechanical turning provides significant shredding and size reduction, which improves the attractiveness of the final product. Water can be added at frequent intervals. Finally, the marketplace has provided a variety of specialized turning machines at attractive prices. Because many composting entrepreneurs come from a farming background, they were quick to adopt equipment and practices similar to existing agricultural operations. With frequent turning (1 to 3 times/week) and moisture and nutrient adjustments, an attractive product can be produced in about 6 months.

Windrow composting with frequent agitation by specialized turning equipment developed as the most common form of leaf composting, despite not being mentioned in the original New Jersey Manual. The authors of the Manual later recognized that the composting industry was developing its own windrow technology. In response, they later added a new "intermediate-level" technology based on the windrow system.[16,17] The term "intermediate-level" is entirely arbitrary. One could argue that the windrow system is the most "advanced" technology currently available, because it produces final compost in the shortest retention time and is the most widely applied process.

Other System Improvements

Application of the windrow process to yard waste composting has seen its own evolution, as was the case with sludge composting. Most of the recent advances have resulted from the need for better odor control and more pre- and postprocessing. Composting of leaves alone can result in some odor emissions, but the problem is increased with the introduction of grass. Grass degrades quickly and is often odorous by the time it reaches the composting site. Montgomery County, Maryland experimented with windrow composting of leaf/grass mixtures at their Dickerson leaf composting operation. They found that a 1:1 ratio by volume worked, but recommended a 2:1 ratio to assure proper friability within the windrow.[18] Several other facilities, including a large yard waste composting project in Islip, New York, recommend immediate mixing of malodorous grass with 2 or 3 parts leaves by volume.[19]

Yard waste composting facilities typically use "open air" tipping areas, where the materials are received. Neutralizing and masking chemicals are sometimes sprayed into the air or on the compost to reduce odors; however, such efforts are marginally successful at best. Enclosed tipping areas are being considered in cases where grass and other malodorous compounds are received and buffer zones are limited. Enclosing the tipping area allows the building air to be collected and treated.

Yard waste composting facilities are also evolving toward increased use of pre- and postprocessing to deal with problems of contaminants, such as plastic, and size reduction of the substrates to improve the rate of degradation. The use of specially designed equipment to turn the windrows and shred the material is now common practice. Screening to improve the product quality and appearance is also becoming more common. Yard waste composting may have started as a "low tech" adventure, but the industry is rapidly seeing the need to engineer each facility to produce an attractive product with minimal odor impacts.

NONREACTOR, STATIC SOLIDS BED SYSTEMS

Process Descriptions

The following processes are classified by having beds of compost that remain largely static throughout the process. Again, each process is actively marketing today or is historical significant to the development of modern composting technology.

Aerated Static Pile (Beltsville Process)

Open piles are constructed of a mixture of dewatered sludge cake and bulking agent. Oxygen is supplied by forced aeration. Bulking agent, usually wood chips, is screened from the compost and returned for mixing with new sludge. The static pile system is widely used in the U.S. for the composting of dewatered sludges. Later versions have used other names, such as the Rutgers process. Major U.S. facilities are located at Montgomery County, Maryland, Philadelphia, Pennsylvania, Scranton, Pennsylvania, Nashville, Tennessee, Oakland, California, Columbus, Ohio, Hampton Roads, Virginia, and Washington, D.C. (Blue Plains).

Highlights: The process is generally successful with wet substrates because of the use of bulking agents. High aeration rates can be maintained because of the porous structure of the mixture.
Status: Active.

Brikollari (Caspari, Briquetting)

Ground refuse or biowaste is compressed into blocks and stacked on pallets with air spaces. The heating period lasts 5 to 6 weeks and curing from 4 to 12 weeks. Aeration is by natural diffusion and airflow through the stacks. Curing follows the initial composting and the compressed blocks can be later ground. Sludge can be added to the mixture up to about 53% moisture content. Plants were reported in Schweinfurt, West Germany, and Biel, Switzerland in 1977. The process is currently offered by Rethmann Entsorgungswirtschaft GmbH & Co of Selm, Germany.

Status: Active.

Pelleting Process

Dewatered sludge was conditioned with recycled product and extruded into pellets with a diameter of about 1 cm. Pellets were then stacked into piles for composting. Oxygen was supplied by natural ventilation through the large pore spaces between the pellets. The system was reportedly developed in Germany.

Status: Believed to be inactive.

Daneco

The Italian company Daneco, Inc. has implemented a number of composting facilities for refuse, mostly in Italy and the Middle East. Most of these use the aerated, static pile concept. Extensive pre- and postprocessing and materials recovery from the refuse is typical of the Daneco plants. A permanent ventilation system is usually used for reversible forced/induced draft aeration of the static piles through perforated channels in the concrete floor. Reversible aeration is used to minimize air channeling. Moisture can be added from an overhead spray system. After about two weeks the compost is screened and reformed in aerated static piles for four additional weeks. Moisture can also be added at this time. A 250 tpd (design) MSW facility for East Central Solid Waste Commission, Mora, Minnesota began operation in 1991. Preprocessing includes trommel screen separation, manual sorting of the oversize to recover recyclables, shredding and wet flotation to recover the compostables. Postprocessing includes shredding and screen refining.

Status: Active.

History of the Static Pile Process

The aerated static pile process was developed by the U.S. Department of Agriculture at the Agricultural Research Center at Beltsville, Maryland. In 1972–73 the Center successfully developed a windrow method using wood chips as a bulking and moisture absorbing agent for digested sludge. Difficulties were later encountered when the same windrow method was applied to raw sludge because of reported odor generation. This led the Beltsville researchers in 1975–76 to develop a new process targeted to raw sludges. The process is most often referred to as the "static pile" process and occasionally as the "Beltsville" process. Both raw and digested municipal and industrial sludges have been composted by this technique. It has enjoyed considerable popularity in the U.S., with over 76 facilities in operation as of 1990.[20] The process is most applicable to composting of wet substrates because bulking agents are used to establish porosity in the pile.

Operating Parameters

The aerated static pile process differs from the windrow process in that composting material is not turned. Aerobic conditions are maintained by forced or induced aeration through the pile. Another difference is that previously composted material is usually not recycled to produce a friable mixture or to adjust the starting moisture content. Instead, dewatered sludges are mixed with a bulking agent such as wood chips, which serves as a moisture absorbent and provides porosity to the material. The required ratio of wood chips to sludge typically ranges from 2:1 to 3:1 by volume.[21,22] Most of the experience at commercial facilities has been with wood chips, although other bulking particles have also been used successfully. As mentioned earlier, use of pelleted refuse, shredded tires, peanut shells and other materials has been reported. Obviously, both size and quantity of the bulking agent must be controlled to maintain porosity throughout the pile and assure adequate airflow without excessive blower head loss.

A generalized process diagram for the aerated static pile system is shown in Figure 2.12. Sequential steps in the formation of a pile are as follows:

1. Proportion and mix the sludge with the bulking agent either on a mixing pad or by controlled metering of the substrates from storage bins to a pug mill for mixing
2. Lay about a 1 ft base of bulking agent over the permanent aeration piping or along with temporary, perforated aeration piping
3. Place the sludge/wood chip mixture in a deep pile on the prepared bed
4. Cover the outer surface of the pile with a layer of screened or unscreened compost
5. Attach the blower to the aeration piping

At this point the pile is ready to begin operation. The blower is operated to either push air into the pile (forced draft) or pull air through the pile (induced draft). Blower operation is controlled to maintain aerobic conditions throughout the pile. An on/off sequence is sometimes used to avoid excessive cooling of the pile. Alternatively, temperature feedback control loops can be used to adjust the on/off cycle or to throttle the blower to maintain a setpoint temperature. In the induced draft mode, blower exhaust can be collected and is usually deodorized before discharge. A common practice in the past was to vent the gas through a pile of screened compost; however, more advanced deodorization techniques are now practiced in many cases (see Chapter 16).

Detention time in the aerated pile is usually about 21 days, after which the pile is dismantled. There is nothing fundamental about using 21 days. It seems to work well with most sludges, but the designer/operator should not be fearful of trying longer detention times. The mixing pile, compost cover, prepared base, and deodorization pile (if used) are usually placed in a temporary stockpile. Subsequent drying may be required before screening the mixture to separate the bulking agent. Mixture solids content is an extremely important variable which affects the ability to separate wood chips from the sludge. A minimum solids content of 50%, preferably 55% is required for efficient separation using vibrating screens and trommels.[23] The Montgomery County facility found it advantageous to reform the pile for an additional 7 days of composting. Mixing as a result of pile breakdown seemed to provide renewed temperature rise and additional drying. The Philadelphia facility constructed a separate drying building under roofed cover where the mixture is again aerated prior to screening.

Separation and reuse of bulking agent is required because of the large volumes normally used, the high cost of materials such as wood chips, and the need to improve product quality by removing the large bulking particles. Drying during the initial 21 day cycle can be enhanced by achieving high cake solids during dewatering and by maintaining high aeration rates controlled by process temperature. If wood chips or other degradable material is used as the bulking agent, some degradation and physical breakdown can be expected during composting. Eventually, size reduction will be such that some bulking material will pass through the screen with the composted sludge. Thus, continual makeup of bulking agent is necessary to balance that lost with the final product.

It is common to collect some leachate from the bottom of the piles. With induced draft aeration, a water trap must be placed on the air piping ahead of the blower to collect condensate formed when hot, moist exhaust gas is cooled. Both leachate and condensate must be collected and treated.

A cross section of a typical aerated static pile is shown in Figure 2.13. Several process modifications have been suggested to reduce the land requirement when single piles are constructed. In one modification, the "extended pile method", new piles are built on the

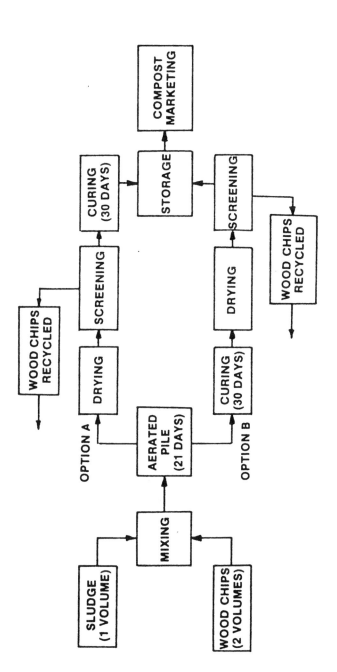

Figure 2.12. Schematic flow diagram for aerated, static pile composting of sludge cake amended with wood chips. Detention times in aeration and curing are approximate. From Willson et al.[24]

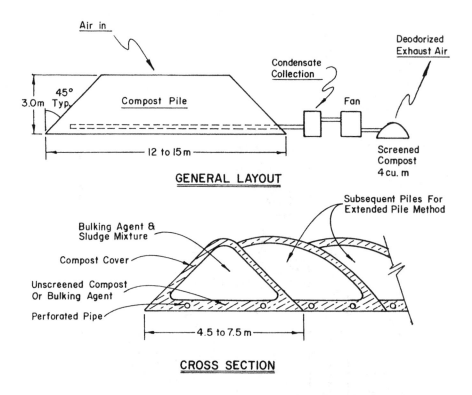

Figure 2.13.   Typical static pile dimensions for composting 40 m³ of dewatered sludge.[27]

structure of the preceding piles as shown in Figure 2.12. In another, piles as high as 18 ft were constructed using a crane and the modification was termed the "extended high pile method". Neither of these modifications represent fundamental changes and would be used simply to reduce land requirements or operational problems. The extended pile method has been adopted in many facilities, but the high pile method remains relatively unused. Static pile facilities are shown in Figures 2.14 and 2.15.

Typical temperatures recorded during composting of raw sludge with wood chips as a bulking agent are shown in Figure 2.16. In general, good temperature elevations have been observed with the aerated static pile process on raw sludge in cold and wet climates. As seen in Figure 2.16, temperatures increase rapidly during the first 3 to 5 days, hold relatively constant, and then begin to decrease after about 3 weeks. Temperature profiles with static pile composting of digested sludge blended with various amendments is shown in Figure 2.17. Good temperature elevations were achieved but air channeling was more of a problem compared to use of bulking particles such as wood chips. For both Figures 2.16 and 2.17, elevated pile temperatures remain after the "classic" 21-day cycle. Subsequent curing is required to produce a more mature product. Curing of the screened material for 30 to 60 days is typical. Screening is often conducted prior to curing to reduce the land requirement. Curing is typically conducted using aerated static piles and temporary aeration piping. Aeration of the static curing piles is necessary, particularly if extended static piles are used.

A mass balance for the enclosed, static pile system at Montgomery County, Maryland, is presented in Figure 2.18. About 400 wet tons/day (wtpd) of sludge at 25% TS are input to the process. This is blended at a ratio of about 2:1 of wood chips and sludge by volume.

Figure 2.14.   Open static pile composting facility operated as shown by the Philadelphia Water Department since 1988. Substrate is dewatered, digested sludge with wood chips as the bulking agent. Truck loading of centrifuge dewatered cake is in the foreground with static piles in the background. White piping are suction headers for induced draft ventilation of the static piles. Curing, static pile drying in a roofed facility, and trommel screening complete the process. The facility is designed for 300 dtpd annual average and 405 dtpd peak month. Compost product is used under the name Phillymulch. This is one of the largest biosolids composting operations in the world. Photo courtesy of Richard Kuchenrither, Black & Veatch.

Recent Trends

The 1980s witnessed a considerable evolution and further development of the original Beltsville static pile process. Perhaps the most significant advance was recognizing the importance of air supply rate to the process. The original Beltsville method used relatively low horsepower blowers controlled primarily by on/off timers. The very high process temperatures shown in Figure 2.16 often resulted. Gradually the industry moved toward higher aeration rates to reduce process temperatures. This had the benefit of reducing temperatures to more optimum levels, which in turned increased the rate of decomposition and heat release, which in turn supported the evaporation of more water, which in turn improved screening and overall process control.

At the same time, temperature feedback loops were implemented to control blower operation to selected setpoint temperatures. Such control loops had been used for many years in reactor composters, but had yet to be applied to windrows or static piles. The pattern of air supply changed significantly when using temperature control loops as shown in Figure 2.19. Aeration demand peaks during the first week of composting and then gradually decreases during the remainder of the cycle. This is in sharp contrast to the near constant aeration supply of the original Beltsville version.

Figure 2.15. Enclosed static pile system at the Montgomery County Composting Facility operated by the Washington Suburban Sanitary Commission near Silver Spring, Maryland. The static pile building contains two rows of extended piles. Process air comes from 56 aeration blowers that exhaust it to deodorization equipment. Material is screened after the 21-day static pile period. A separate enclosed building is used for aerated curing. All mixing, screening, and composting facilities are enclosed. The product is marketed as ComPRO. The staff at this facility have developed what is probably the most extensive system of odor controls in use at any composting facility.

It soon became apparent that larger blowers were needed to maintain setpoint temperatures. Specific aeration rate, G, defined as cubic feet of air supplied per hour per ton dry solids (cfh/ton ds) became an important process parameter. The Beltsville version typically supplied about 500 cfh/ton ds which was sufficient to maintain pile oxygen between 5 to 15%.[24] Newer designs supply an average nearer to 2000 cfh/ton ds with peak air supply rates near 5000 cfh/ton ds. High G designs are used almost exclusively today.

There has been considerable argument within the industry over the "optimum" setpoint temperature. One side favored low temperatures in the range of 45°C based on argued improvements in the health of the microbial population. The other side argued that this would not achieve the desired pasteurization and that the proposed microbial health benefits were not well documented. Most practitioners recognized that extremely high temperatures could pose kinetic rate limitations but took a more pragmatic approach to the problem. As long as the designer incorporated sufficient air supply and a workable control logic, the operator could decide for himself which temperature setpoint achieved the best results. With municipal sludges, most operators use an initial setpoint in the 55 to 65°C range to assure proper pasteurization. The operator may lower the setpoint near the end of the cycle once the goal of pathogen control has been achieved. Engineering pragmatism seems to have "won the day" judging by the recent absence of much debate over optimum temperatures, a subject that sparked heated volleys in many scientific forums in the late 1970s and 80s.

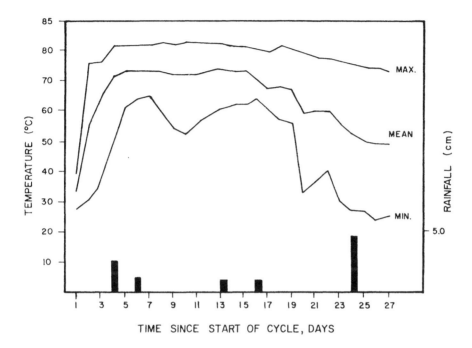

Figure 2.16. Typical temperatures recorded during aerated static pile composting of a raw sludge-wood chip mixture. Bars at the bottom indicate rainfall events. From Willson et al.[28]

REACTOR BASED SYSTEMS

The number of historic and current reactor-based systems is large indeed. It is important to study the range of reactor types to gain perspective on the current state of composting technologies. The following sections present a mixture of history, process descriptions, system flowsheets, mass balances, and equipment photos. The student should emerge with a clearer view of the breadth of application and the requirements for reactor based composting systems.

Vertical Flow, Agitated Solids Bed

Earp-Thomas

Developed by G. H. Earp-Thomas of New Jersey, a pioneer of municipal refuse composting in the U.S., the process is similar in design to a multiple hearth furnace. In it 8 to 10 decks or hearths are stacked vertically. The organic waste, after having been ground to a pulplike consistency, is introduced to the top section along with a bacterial inocula. Material is pushed outward on one deck and then inward on the next, toward openings through which it falls to the next deck. A center shaft drives plows which agitate the compost and move it downward from deck to deck. Air passes upward through the unit. Digestion time is typically 1 to 3 days and is usually followed by windrow curing. The process was used for refuse composting in Paris, France; Verone, Italy; Seoul, Korea; and Basel, Switzerland. The Paris unit used 10 hearths and was 30 ft tall and 15 to 18 ft diameter. A version of this reactor is currently used at Heidelberg, Germany, for composting source separated garbage.

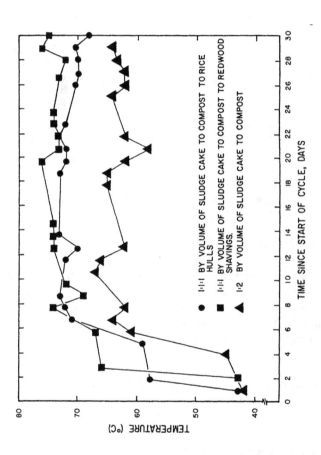

Figure 2.17. Temperature profiles during aerated static pile composting of digested sludge blended with various amendments. Each data point is the average of 10 readings made over the pile cross section. From Iacoboni et al.[26]

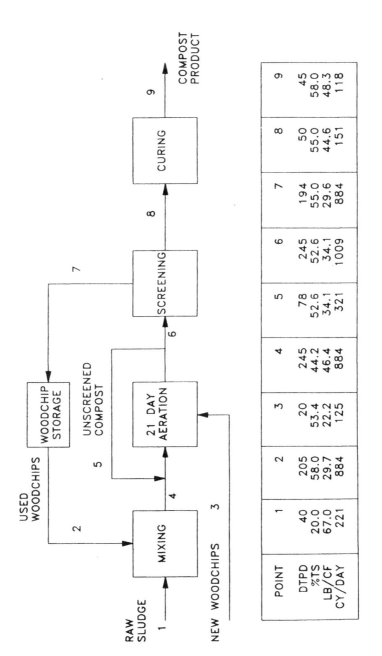

POINT	1	2	3	4	5	6	7	8	9
DTPD	40	205	20	245	78	245	194	50	45
%TS	20.0	58.0	53.4	44.2	52.6	52.6	55.0	55.0	58.0
LB/CF	67.0	29.7	22.2	46.4	34.1	34.1	29.6	44.6	48.3
CY/DAY	221	884	125	884	321	1009	884	151	118

Figure 2.18. Mass balance for the enclosed, aerated static pile process at Montgomery County, Maryland, as shown in Figure 2.15. Values shown are monthly averages for 1991. Balance was developed and provided courtesy of Chuck Murray, Manager of Sludge Operations.

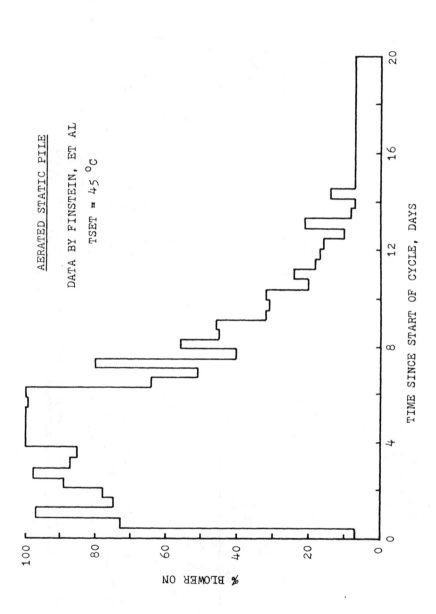

Figure 2.19.   Blower operation during static pile composting of a mostly raw 25% TS sludge using a temperature feedback control strategy with a setpoint of 45°C. From MacGregor et al.[29]

A test unit operated at Orange County Sanitation Districts, California, in 1976 under the name Thermax. Digested, dewatered sludge was conditioned with recycled product and other amendments. The unit used a 1-day residence time and was not considered successful.

Highlights: The system was often criticized as being mechanically intense, with unsupported claims of producing stable compost in only 24 h using specialized inocula.

Status: Mostly historic, but the Heidelberg plant remains in operation and Organic Waste Recyclers has an inactive unit near Portland, Oregon.

Frazer-Eweson

It was developed by Eric Eweson, a biochemist whose patents were assigned to the Frazer-Eweson Company of Rhode Island. Ground refuse was placed at the top of a vertical bin with four decks. The decks were formed by parallel grid bars spaced so that the refuse bridged across the bars. Transfer of material from one zone to a lower one was brought about by breaking the bridges using specially designed cutter arms operated from outside the digester. Air was forced upward through the bin. The detention time was typically 5 to 7 days on each deck giving a total cycle of 20 to 28 days. Final product from the lowest deck was passed through a 4 to 6 mesh screen. Problems were encountered with bridging of solids across the perforated decks. An 18 tpd facility operated on garbage in Springfield, Massachusetts, from 1954 to 1962. A pilot facility reportedly operated at the Chicago Union Stock Yards on manure.

Highlights: Recycle of exhaust gases was used to increase CO_2 content, possibly to help neutralize pH and reduce off-gas quantity and odors.

Status: Historic.

Jersey (John Thompson)

This process used a six-floor silo arrangement and was originally developed for refuse. Each floor was a series of half cylinders, open side up. Half cylinders could be rotated so that their open side was down, thus depositing compost on the floor below and agitating it in the process. Material remained on each floor for 1 day. After the 6-day period the compost was cured in sheltered windrows for 6 weeks. The first Jersey plant was constructed on Isle of Jersey, England in 1957. A 300 mtpd plant was then constructed in Bangkok, Thailand in 1959 by John Thompson Industrial Constructions, Ltd. and Composting Engineers, Ltd.

Status: Believed to be historic.

Naturizer International

The Naturizer process was developed by Norman Pierson beginning with early windrow experiments in 1947. The basic components of the technology were developed in a 4 to 5 tpd MSW experimental plant constructed at Norman, Oklahoma in the mid-1950s. The process used six digestion chambers, each designed to receive a 1-day supply of ground refuse. Each cell used about a 9-ft wide steel conveyor belt arranged to pass material vertically from one belt to another. Each belt was an insulated cell. Air passed upward through the digester, and aeration was enhanced by dropping the material daily from one cell to the next immediately below it. Total detention time through the six cells was 6 days. Sewage sludge was added as a moistening agent and nitrogen source. Commercial plants were developed in Norman, Oklahoma (1959–80), San Fernando, California (1963–64), St. Petersburg, Florida (1966–70), and Madisonville, Kentucky (early 1980s but since closed). The Norman facility operated for over 20 years and is currently used for research by Marshall Pierson, son of the original developer.

Status: Currently marketed as the Pierson/Naturizer Technology.

Riker

The process consisted of four-story bins with clamshell floors. Compost was dropped from floor to floor to provide agitation, and forced aeration was provided. Total detention time was 20 to 28 days. The reactor treated a mixture of ground garbage, corn cobs, and sludge. Problems were reported in maintaining aerobic conditions. A 4 mtpd facility operated in Williamston, Michigan (1955–62).

Status: Historic

T. A. Crane

The Crane digester consisted of three horizontal desks with two rows of horizontal ribbon screws extending the length of each deck. The slowly moving screws recirculated ground refuse from deck to deck. Air was introduced in the bottom of the cells. The refuse composted for 3 days in the reactor and was followed by curing for 7 days in a bin digester with forced aeration. A 30 tpd pilot refuse and sludge system was installed in Kobe, Japan in 1956 by the WHO under direction of Dr. John Snell.

Status: Historic.

Dambach-Schnorr "Biocell"

This system is offered by Dambach, Ltd., Gaggenau, Germany. The reactor consists of a vertical tower with 8 to 10 floors, one above the other. The bottoms of each floor are aluminum flaps, can be turned down to discharge material to the next floor. Oxygen is introduced by forced aeration. Feed consists of dewatered sludge, recycled compost, and ground bark or sawdust in proportions of 2:2:1 by volume. The composting mixture is about 1 m deep on each floor. Residence time is about 3 days on each floor, giving a total reactor time of 30 days. Facilities at Rastatt and Saarbrucken, West Germany were reported in 1977.

Status: Active.

Varro (Krupp-Varro)

The Varro reactor was originally designed for refuse that was preprocessed by shredding, magnetic iron separation, and screening. Ground refuse is placed in a six to eight deck digester and moved downward from deck to deck and stirred by plows. Two decks form one chamber with its own recirculating air supply to control the CO_2 level. Digestion time in the reactor is

2 days. Output can be sent to a granulating plant where it is compacted, milled to desired size, and screened for use as a base material for fertilizer, soil conditioner, and wallboard. A system of about 55 tpd was constructed in Brooklyn, New York, in 1971. The system is marketed by Krupp Industrie, Germany.

Status: Active.

Ebara Multiplex Paddling Fermentor (MPF)

Developed by Ebara Corp. of Tokyo, Japan, the basic design was similar to the Earp-Thomas multiple hearth digester and was primarily intended for refuse composting. Unlike the Earp-Thomas reactor, the traveling arms were equipped with rotating paddles to provide greater agitation of the material while on the hearth. Air could be supplied separately to each hearth. Exhaust gases were collected for deodorization by a biofilter or chemical scrubbing.

Status: The reactor is no longer marketed by Ebara.

NGK (NGK Insulators)

The Japanese firm NGK Insulators, Ltd. has a vertical, multi-stage, agitated bed composting facility at Aichi Prefecture, Japan, which began operation in 1989 (see Figures 2.20 and 2.21). Two reactors are provided, each with six cells arranged vertically. Material goes into the top cell of Reactor 1. The floor of each cell consists of metal plates that hold the material when in the horizontal position. The plates can be rotated to the vertical to drop material into the cell below. Material from the lowest cell of Reactor 1 is fed to the top cell of Reactor 2. The facility composts 2.1 dry metric tons/day (dmtpd) of lime and ferric conditioned, raw sludge, filter press cake. The cake is conditioned with product recycle only. Single-pass residence time is about 20 days in each reactor (3 to 4 days per cell). Total solids retention time in the system is estimated at about 70 days. The compost product can be pelletized to improve market appeal.

Status: Active.

Vertical Flow, Packed Bed

Beccari

This process was originally developed by Dr. Giovanni Beccari of Florence, Italy in the 1920s to conserve the nutritive value of manure. The reactor was a simple cuboidal structure with a loading chute on top and an unloading door on one side. Air vents with valves were attached to either supply or exclude air. The process was generally started anaerobically (air valves closed) and then converted to an aerobic condition (valves open). Cycle time was typically 20 to 40 days. In Italy and France 50 plants were reported in the 1920s and 30s. One at St. Georges, France composted sewage sludge and garbage using 40 cells each with 700 ft^3 capacity, elevated 10 ft above ground to facilitate removal of the contents. In the U.S., 5 plants were built in Florida and New York for refuse composting from 1920 to 30. Seal problems on outlet doors allowed leachate to leak, creating a nesting ground for roaches and insects.

Figure 2.20. NGK vertical, multistage, agitated bed composting system at the Aichi Prefecture, Japan. Design capacity is 2.1 dmtpd of filter press cake conditioned with product recycle. Each reactor consists of six cells in series, and the two reactors are arranged in series. An elaborate system of air supply and withdrawal ducting is evident. Product from the reactors can be pelletized prior to bagging. Photo courtesy of Yuzo Okamoto, NGK.

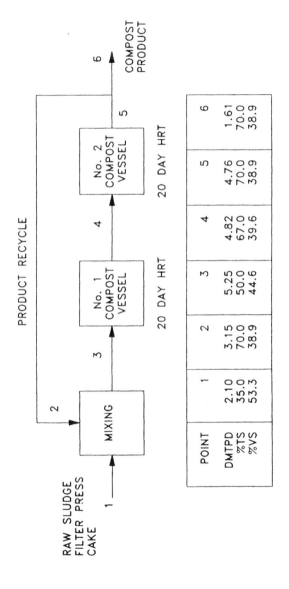

POINT	1	2	3	4	5	6
DMTPD	2.10	3.15	5.25	4.82	4.76	1.61
%TS	35.0	70.0	50.0	67.0	70.0	70.0
%VS	53.3	38.9	44.6	39.6	38.9	38.9

Figure 2.21.   Mass balance on the NGK system shown in Figure 2.20. There is a trend in Japan to rely more on product recycle for feed conditioning because of the limited supply of other amendments. Mass balance courtesy of Yuzo Okamoto, NGK.

Highlights: The Beccari process was one of the first reactor systems.
Status: Historic.

Boggiano-Pico

This is a modification of the Beccari process and is probably the first silo type system. A sealed tower was batch filled from above and emptied from below. Air was introduced from below at about 7 atm pressure. Cycle time was 35 days. A facility in Geneva, Switzerland had a 56,500 ft³ silo volume.

Highlights: One of the first silo systems.
Status: Historic.

Bordas

This is another modification of the Beccari method developed by Jean Bordas in 1931. The silo structure was designed for continual operation and was divided by a grate into upper and lower sections. Air was introduced upwards along the walls and through a central pipe. After 12 to 15 days the grate was opened to permit the top charge to fall to the lower chamber. Residence time in the lower chamber was about 8 days. Exhaust gases were passed through a scrubbing tower for absorption of ammonia by superphosphate. The resulting ammonium phosphate was then added to the final compost.

Highlights: First known attempt to recover ammonia from compost exhaust gases and
               convert it to a form that could be added back to the compost. Believed to be the
               first reactor system designed for semicontinuous feeding.
Status: Historic.

Verdier

This process was essentially the same as the Beccari process but modified to include recirculation of leachate liquor. The cycle time was about 20 days. It was once used in Cannes, France on garbage.

Highlights: Recirculation of leachate liquor.
Status: Historic.

Euramca (Roediger, Fermentechnik)

The reactor was of the tower type with a special extraction and agitation mechanism at the bottom. The reactor was batch-operated with dewatered sludge and recycled product as the feed. Once the reactor was loaded, material was frequently recycled from the bottom to the top to provide agitation and assure uniform exposure to temperature. Residence time was 6 days. After composting, the material was extruded into 2 to 3 cm diameter pellets and air-dried in windrows or an enclosed forced-air dryer. Pellets were broken before recycling to the composting reactor. In 1978 a facility was commissioned at Wutoshingen, West Germany, with a reactor volume of 50 m³. A pilot unit was tested in Southern California in 1978.

Status: Believed to be inactive.

Triga (Halbert-Triga)

Reactor is a circular, concrete tower called a "Hygiensator" that may be divided into four separate vertical compartments. Residence time is typically 4 to 12 days depending on feed. Output from one tower may be fed to a second tower in series. Vacuum pumps for each tower continually draw air up through the material. The air flow is regulated to maintain a bed temperature in the range of 70 to 80°C. A screw extractor removes and agitates material from the bottom of the reactor. Extracted material is recycled 3 to 5 times during the compost period to avoid compaction at the bottom. Over 25 European facilities have been built since 1962, many in France, for composting refuse and sludge. A plant at Saint-Palais, France composts 3.9 dmtpd of sludge amended with sawdust and pine bark. Two 260 m^3 digesters provide 4 to 6 days residence time. Output is cured for 2 months in a covered curing yard. A plant at Brasilia, Brazil, is designed for 660 tpd of refuse with 6 days residence time in the digester followed by windrow curing for 30 to 60 days. The Caju plant in Rio de Janeiro, Brazil, has 3 concrete towers, each 16 m high by 14 m diameter, designed for 1100 tpd MSW. Residence time is 4 days in the silos followed by outdoor curing for 30 to 45 days.

Status: Active.

BAV

The BAV reactor was one of the first European silo types and is manufactured by Biologische Abfallverwertungs Gesellschaft (BAV). The reactor consists of a cylindrical tower with no interior floors or other mechanisms and is termed a "bioreactor". Feed usually consists of dewatered sludge, recycled compost, and amendments such as sawdust or bark. Feed is introduced at the top and flows downward as product is removed by a screw extractor mechanism from the bottom. Air is supplied by forced aeration from the bottom. Typical residence times is 14 days, followed by maturation in curing piles for at least 6 weeks. Over 25 installations were built in Germany with reactor volumes as large as 375 m^3.

Status: Active.

Weiss Bioreaktor System Kneer (Taulman-Weiss in U.S.)

It was developed by the German scientist Franz Kneer beginning in 1972 in association with the German manufacturing firm of Gebruder Weiss, K.G. and offered by the Taulman Company in the U.S. The reactor is a completely enclosed, cylindrical silo type and is applied primarily to sludges. Output from the bioreactor is usually fed to a second cure reactor. Residence times are typically 14 days in the bioreactor and 14 to 20 days in the cure reactor. Air is supplied from a distribution grid in the bottom of each reactor and flows upward, counter-current to the flow of solids. The reactor tops are covered and exhaust gases are collected from the top of each reactor for subsequent deodorization. Over 35 facilities are reported in the U.S., Europe, and Japan. A facility in Portland, Oregon, uses four 1200 m^3 bioreactors and two 1800 m^3 cure reactors and currently composts about 25 dtpd of anaerobically digested sludge conditioned with sawdust and product recycle (see Figures 2.22 and

Figure 2.22.   Taulman-Weiss vertical flow, packed bed reactor systems at Portland, Oregon (above, under construction in 1984) and Bickenbach, Germany (below). The Portland facility includes two sawdust storage silos, four 1200 m³ bioreactors, two 1800 m³ cure reactors and is representative of a large scale reactor system. The Bickenbach facility is typical of smaller scale systems. A sawdust storage silo (right), a single 250 m³ bioreactor (center) and a single 250 m³ cure reactor (left) are provided. Total single-pass residence time at Bickenbach is about 40 days.

2.23). Other U.S. facilities include Clayton County, Georgia (3 dtpd), Dothan, Alabama (7 dtpd), East Richland County, South Carolina (5 dtpd), and Lancaster, Pennsylvania (two 1200 m³ bioreactors and two 1200 m³ cure reactors designed for 32 dtpd but operating at 18 dtpd to provide more curing and drying).

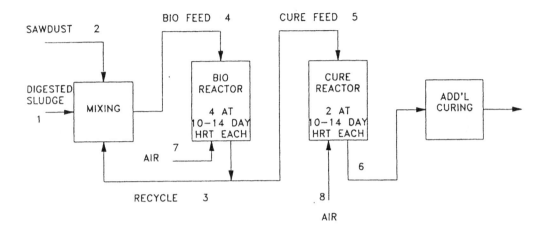

POINT	1	2	3	4	5	6	7	8
DTPD	25.0	16.6	32	73.6		22.4	302	151
%TS	25	70	42	37	42	48		
%VS	58	99						
LB/CF	42	10	32	36	32	30		
CY/DAY	176	176	176			115		
SCFM							5600	2800

Figure 2.23.   Mass balance for the vertical, packed bed reactor system at Portland, Oregon. Later designs tend toward higher air flowrates. For example, the facility at Lancaster, Pennsylvania, supplies 2300 cfm, compared to 1400 cfm at Portland, for the same size bioreactor. Portland recently retrofitted 2800 cfm blowers on each of the four bio- and two cure reactors to increase drying and improve temperature control. Mass balance based on data supplied by Dan Clark, Manager of Compost Operations.

Highlights: Generally considered a successful design. The facility in Portland, Oregon, is one of the first and largest reactor systems on sludge in the U.S., beginning operations in September 1984.

Status: Active.

Figure 2.24.   Vertical, packed bed Purac facility at Ft. Lauderdale, Florida. Rectangular silos are used for both bio- and cure reactors. This facility was designed to use yard waste for amendment. Front-end processing for the yard waste included tub grinding and hammermill shredding.

ABV Purac System

This system was developed by Purac AB, a Swedish company and member of the contracting group ABV of Malmo, Sweden. ABV is a silo type system using either cylindrical or rectangular reactors. Air is fed from the bottom of the reactor, which is pressurized and removed from the top layers of compost by a manifold withdrawal system. Exhaust air flow is larger than the amount of air supplied at the bottom. A roof without walls or a full enclosure is used at the top of the reactor. Later designs have generally avoided the walls, relying on the exhaust air withdrawal for odor control. Both bio- and cure reactors are normally provided with about 14 days residence time each. Facilities in the U.S. include Cape May, New Jersey (12 dtpd sludge), Ft. Lauderdale (32 dtpd sludge, see Figure 2.24) and Sarasota (6 dtpd sludge), Florida, and Henrico County, Virginia.

Highlights: Facility at Cape May, New Jersey, was one of the first reactor systems specifically designed for municipal sludge in the U.S., starting operations in May 1985
Status: Active.

American Bio Tech

The reactor is a rectangular shaped, silo type developed by John Laurenson in the early 1980s. A key feature of the design is the use of multiple air lances, which are mounted into manifolds above the compost and extend to just above the outfeed auger. The lances on one manifold supply air while lances on the adjacent manifold remove air. The "push" and "pull" manifolds are reversed at intervals to avoid moisture migration and plugging. The path for air travel is reduced to about 5 ft, which generally allows for increased air supply rates at lower headloss. Facilities are in Schnectady, New York (15 dtpd, startup 1987), and Hartford,

Connecticut (70 dtpd design, commissioned in 1991, but currently being retrofitted to improve odor control). The Schnectady plant has retrofitted to all positive aeration and this is also being considered at Hartford.

Highlights: Use of cross-current air: solids flow as opposed to the counter-current pattern used in most silo designs.
Status: Currently inactive.

Horizontal Flow, Rotary Drums

Dano

The Dano drum was developed by Dano Ltd. in Denmark about 1933 for composting refuse. The reactor is a dispersed flow rotating drum, termed a "bio-stabilizer", and it is slightly inclined from the horizontal, 9 to 12 ft in diameter, and up to 150 ft long. The drum is kept about half full of refuse and rotates at 0.1 to 1.0 rpm. As the cylinder rotates, the wastes move in a helical path towards the outlet and are mixed and granulated by abrasion. Typical digestion is 1 to 5 days in the drum and is usually followed by screening and windrow or static pile curing. Water and nutrients may be added to the drum and forced aeration is provided. The rotary drum has become one of the most popular reactor processes for MSW, with several hundred installations worldwide. A Dano plant in Rome, Italy once handled over 500 mtpd. A plant in Leicester, England composts refuse and sludge. U.S. installations were once located in Sacramento, California (1956–1963) and Phoenix, Arizona (1963–1965). The Dano drum system was reintroduced into the U.S. with an MSW facility in Portland, Oregon, operated by Riedel Environmental Technologies, Inc., which began accepting waste in April 1991. Unfortunately, the facility temporarily stopped receiving MSW in January 1992, due to a variety of problems, including odors and failure to produce an acceptable final product.

Highlights: The rotary drum has become a centerpiece for many refuse composting plants. In 1957 a plant at Duisburg-Huckingen, West Germany, used air recirculation to reduce the net emission of exhaust gases. Coolers and a water scrubber were used to condense water vapor in the recirculated gases.
Status: Active.

Fermascreen

The reactor is a hexagonal shaped drum, three sides of which are screens. Refuse is ground- and batch-loaded. The screens are sealed for initial composting. Aeration occurs when the drum is rotated with the screens open. Detention time was about 4 days.

Status: Historic.

Eweson (Bedminster Bioconversion Corporation)

Eric Eweson (1897–1987) was a fermentation technician active in the yeast business. He became interested in composting after reading Sir Albert Howard's opinion that modern fermentation technology could accelerate the composting process. He conferred with Sir Howard in England and developed the Eweson rotary drum based on the principles of Sir

Howard's Indore process. In the 1950s, three co-composting plants of an earlier design were operated in Spain. In the late 1960s, two prototype digesters were installed in California, one of which was used by the Los Angeles County Sanitation Districts for pilot testing of MSW composting. The first full-scale plant using the modern Eweson digester was installed at Big Sandy, Texas, in 1971. The Big Sandy facility has composted a mixture of 27 tpd refuse and 9 wtpd sludge and many other agricultural products such as brewery sludge, septage, poultry manure, sawdust, and waste paper. The facility continues in operation.

The Eweson drum was originally developed for refuse and sludge composting. The drum is a cells-in-series type. Unground refuse is placed in the rotating drum, which is 11 to 12 ft diameter, 120 to 180 ft long, and slightly inclined from horizontal. Three compartments are typically used in the drum, but as few as one and as many as six have been used in some designs. The mixture is transferred to the next compartment every 1 to 2 days, giving a total digestion time of 3 to 6 days. About 15% of material is retained in each cell upon transfer to serve as an inoculum for the incoming material. Rotation is about 1 rpm. Screened output is cured in static piles or aerated channels. Bedminster Bioconversion Corporation holds the patents and rights to the Eweson digester. Recomp, Inc., a licensee of the process, started a 100 tpd refuse facility in St. Cloud, Minnesota, in 1990 using the Eweson drum. The St. Cloud facility has retrofitted a Royer enclosed, aerated channel curing system. A Bedminster facility began operation at Pinetop, Arizona, in 1991 (8 tpd MSW and 3 wtpd sludge). A larger facility serving Sevier County, Tennessee (150 tpd MSW and 75 wtpd sludge), commenced operation and was officially accepted in January 1993.

Highlights: The Big Sandy facility has accumulated over 17 years of actual operation, making it one of the longest running plants in the U.S. and significant to the history of modern U.S. composting. See Figure 2.25.

Status: Active.

Ruthner

This rotary drum system was developed by Ruthner Industrieanlagen, Austria. The fermentation drum is generally similar to a Dano drum. A refuse/sludge co-composting facility at Siggerwiesen (near Salzburg), Austria, started operation in 1976 on 400 mtpd refuse and 120 mtpd sludge and continues in service. See Figure 2.26.

Status: Active.

Voest-Alpine

This rotary drum/static pile system was developed by Voest-Alpine of Austria. The reactor is generally similar to the Dano drum. Preprocessing usually includes hammermill shredding, magnetic separation, and trommel screening to remove oversize. Some facilities do not use a drum reactor; instead they use a rotary "homogenizing drum" of short residence time where liquid sludge or water is added and mixed. The mixture is placed in an extended static pile for about 21 days followed by static pile curing for 2 months. Postprocessing includes vibratory screens, ballistic separation, and additional magnetic separation. A plant at Allerheiligen, Austria, composts 80 mtpd MSW and 35 mtpd of liquid sludge. A facility is also located at Al Ain, UAE.

Status: Active.

Figure 2.25.   Bedminster Bioconversion system serving Sevier County, Tennessee. The facility has a design rating of 150 wtpd of MSW and 75 wtpd of biosolids cake. The mix of MSW and sludge are directly fed to 3 Eweson, multi-cell, rotary drum digesters, each 12.5 ft diameter by 180 ft long (upper photo). Residence time in the drums is 3 days. Outfeed from the drums is separated by trommel with a 1.25 inch screen size. Aerated static piles are used for curing (lower photo). Cured material is fine screened through a 0.375 inch screen. A stoner is then used to remove glass. The separated glass is crushed and returned to the digester. Static piles are designed for either upflow or downflow aeration. Note the patented "condensahoods" used to collect process gases in the upflow aeration mode. All process gases, including drum and static pile gases and hood air from the loading and discharge ends of the drums, are collected and treated by a biofilter. Note the water addition to the curing piles. Low piles in the foreground were an experiment. Six-foot deep static piles are normally used for curing. Photos courtesy of Hugh Ettinger, Bedminster Bioconversion Corporation.

Buhler Inc. (Formerly Buhler-Miag)

This composting system was first developed in 1957 by the Buhler Brothers Company in Uzwil, Switzerland, site of the foundry opened by Adolf Buhler in 1860. Since then, Buhler, Inc. and its subsidiary companies have implemented about 100 composting plants worldwide, mostly for refuse. For example, a 500 mtpd MSW plant began operation in Izmir, Turkey, in 1988. Several different composting processes are used depending on the situation. Many use a rotary drum concept and for this reason the Buhler family of processes is presented as a

Figure 2.26.   Enclosed discharge chutes and conveyors removing the outfeed from three parallel, rotary drums (above) at the Siggerwiesen co-composting facility serving Salzburg, Austria. Output from the drums is conveyed to a fermentation hall for aerated, static pile curing (below).

reactor based system. An aerated windrow process and a bin type process, termed the "Wendelin" system, are also used either alone or following the rotary drum. Extensive pre- and postprocessing are typically provided using Buhler equipment. Preprocessing often includes grinding, iron separation, screening, and fine grinding. Windrow composting is usually conducted for 3 to 4 months, using piles 1.5 to 1.8 m high. Typically, two to three turnings are used during the cycle. Reuter, Inc., commissioned a 667 tpd (5-day basis) MSW composting facility using Buhler equipment at Pembroke Pines, Florida, in 1992. The facility has 10.1 acres of enclosed space including a 6.6 acre composting "hangar" that contains 24 windrows, 720 ft long, formed over aeration channels with a design residence time of 42 days. This is considered to be a minimum requirement, and 70 to 85 days would be preferred.[25] A 165 tpd facility is also being implemented in Wright County, Minnesota.

Status: Active.

Other Drums

The number of companies offering rotary drum composters has increased significantly in recent years. These include the Alvahum Drum (Alvater & Co. of Herford, Germany), the Envital Drum (Envital GmbH of Aschaffenburg, Germany), and Lescha Drum (Lescha Maschinenfabrik GmbH & Co. of Augsburg, Germany). These systems do not differ fundamentally from those described above.

Horizontal Flow, Agitated Bins, or Open Channels

Fairfield-Hardy (Fairfield, Compost Systems Co.)

The reactor is a circular type up to 120 ft in diameter with compost typically 6 to 10 ft deep within the reactor. Vertical screws, mounted on a rotating traveling bridge, agitate the material and move it toward a center well for removal. Material is fed on a daily basis along the reactor circumference using a conveyor system mounted on the traveling bridge. Forced aeration is through distribution rings mounted in the tank bottom. Detention time is about 7 to 14 days. Equipment for this process is provided by the Fairfield Engineering Co. Systems for sludge composting are marketed by Compost Systems Co.

The original Fairfield plant was constructed at Altoona, Pennsylvania in 1951 with a capacity of 25 tpd and continued in operation until the mid-1980s. A 135 tpd facility once operated in San Juan, Puerto Rico, beginning in 1969. In 1978 a 45 tpd facility for composting garbage and sludge was constructed in Toronto, Ontario, which is planned to be retrofitted to source separated garbage (1991). A major facility operates at the Delaware Reclamation Project, Wilmington, Delaware, with a design capacity of 600 tpd of MSW heavy fraction and dewatered sludge cake (3.2:1 heavy fraction to sludge by dry weight). Four reactors are used, each with a 100-ft diameter and about a 6-ft compost depth. In 1986, the City of Plattsburgh, New York, commissioned a two reactor system designed to compost about 30 dtpd of sludge amended with sawdust. Each reactor is 112 ft in diameter and operates up to 10 ft of compost depth with a 14-day design residence time (see Figure 2.27).

Highlights: The facility at Altoona, Pennsylvania, was one of the first garbage composting facilities in the U.S. and was one of the most successful based on the length of its operation. In 1969 the digester was used to compost raw dewatered sludge conditioned with heat-dried recycle and shredded paper in a dry weight ratio of 1:1.2:0.7. Material discharged from the reactor was about 50% solids and was heat-dried to 87% before recycling. This work was instrumental in establishing the concept of sludge composting in the U.S. and is believed to be the first full-scale reactor composting of a mostly sludge mixture.

Status: Active.

Figure 2.27. Circular, agitated bed Fairfield reactor at the City of Plattsburgh, New York. The facility composts raw sewage sludge conditioned with sawdust, paper mill waste and recycle. The upper photo shows the bridge mounted augers and the circular aeration zones mounted in the floor. The lower photo shows the reactor loaded with compost. A geodesic dome covers the reactor. However, the upflow aeration has caused problems with fogging, moisture condensation and high ammonia concentrations within the dome. It is planned to retrofit a flat cover which will travel with the bridge. Process gases will be drawn off beneath the cover to improve the atmosphere within the dome.

Snell

The reactor was a rectangular tank about 8 ft deep with a porous floor equipped with air ducts for forced aeration. The tank bottom was inclined on a 6° slope. A traveling bridge with vertical paddles provided agitation and movement of material along the tank. Detention time was 5 to 8 days. A 275 tpd facility was constructed in Houston, Texas, in 1967. The facility was reported to operate successfully but local problems forced closure.

Status: Historic.

Aerotherm

Similar in concept to the Fairfield circular digester, Aerotherm was designed by John Houser, formerly manager of R&D for Fairfield Engineering Co., and incorporated some modifications to the basic Fairfield concept. No full-scale facilities were built.

Status: Historic.

Compost-A-Matic

It is bin type system developed primarily for manure composting by Farmer Automatic of America, Inc., Georgia. It uses about a 3-ft depth of manure and pit widths from 6 to 20 ft depending on the application.

Status: Active.

Metro-Waste

Metro-Waste, under the direction of Victor Brown, was the first elongated bin type system with a straightline material flow implemented in the U.S. The system was awarded five U.S. patents. Parallel bins, each about 20 ft wide, 10 ft deep, and 200 to 400 ft long were used. The design was originally intended for ground refuse. Residence time was about 7 days. An "Agiloader" moved on rails mounted on the bin walls and provided periodic agitation by turning and forward movement of the pile. This was considered one of the more successful reactor types, particularly for large quantities of refuse. A 275 tpd MSW facility operated in Houston from 1966 to 1970. Other refuse installations once operated in Largo (45 tpd) and Gainesville (135 tpd), Florida. The Houston facility demonstrated the technical feasibility of MSW composting. However, the compost market proved to be experimental and small, and the plant could not compete against the low cost, landfill tipping fees of the time.

Highlights: Generally considered one of the more successful designs in the 1950-60 era of refuse composting in the U.S.
Status: Active.

Tollemache

It is similar in design to the Metro-Waste bin system. Installations were once reported in Spain and Rhodesia in 1971.

Status: Believed to be historic.

Paygro (Compost Systems Co.)

In 1972, Carl Kipp constructed a 250 tpd manure composting facility at his feed lot near South Charleston, Ohio. The facility was based on the Metro-Waste design and remains in operation today. Sludge experiments were later conducted at this facility and the process became known as the Paygro System, marketed by Compost Systems Co. Kipp's facility was instrumental in bridging the gap between the refuse reactors of the 1950s and 60s and the new generation of sludge reactors constructed in the 1980s in the U.S. Paygro systems for sludge composting were installed in Columbus, Ohio (200 wtpd, startup 1985), Akron, Ohio (70 dtpd at 22% TS, startup 1986), and Baltimore, Maryland (210 wtpd, startup 1986). The Akron facility uses four parallel bays, each 20 ft wide, with 10-ft operating depth, and 730 ft long (see Figure 2.28). Design residence time is 21 days, followed by 6 months of static pile curing, 2 months covered and 4 months open.

Highlights: The Akron facility has retrofitted movable covers on the bins to allow collection of exhaust gases and prevent their release into the building. The 450,000 cfm odor scrubbing system will probably be the largest at any composting facility.
Status: Active.

IPS (International Process Systems)

This agitated, multibay, bin-type composting system has been offered in the U.S. by International Process Systems (IPS) since 1985. The technology is based on bin composting developments in Japan, which typically used 2 m wide bins. The first facility was installed at Lebanon, Connecticut, using a Japanese agitator that was subsequently improved and modified for computer control. The IPS bins are smaller than those used by the Metro-Waste and Paygro systems and the pattern of solid flow is different. IPS bins are typically 2 m (6 ft, 6.75 in.) inside wall to wall by 6 ft high. Material is fed at one end of each bay and is turned daily. Material moves about 10 to 12 ft along the bin with each turn. The bin length is adjusted to give the desired residence time, typically 18 to 24 days. Air is supplied from manifolds below each bin. The turning equipment is fully automatic and functions without an operator. A facility at Fairfield, Connecticut (1989) composts about 4000 wtpyr sludge mixed with 16,000 cy/yr of shredded brush, leaves, and yard wastes. Six bays, each 220 ft long, are used (see Figure 2.29 and Table 2.1). The Anheuser-Busch Breweries, New York (1989), use 12 200-ft bays to compost about 60 wtpd of sludges, using sawdust and recycled compost as amendments. Another 12-bay system was commissioned in Lockport, New York, in 1991. An

Figure 2.28. Aerated, agitated bin Paygro system at Akron, Ohio. The facility is designed for 70 dtpd of raw sludge conditioned with sawdust, wood bark, and recycle. Four parallel bins are provided, each 10 ft deep, 20 ft wide, and 730 ft long. Material is turned about once a week using the digger shown in the lower photo. Covers have been retrofitted onto the bins to allow process gases to be drawn off and avoid the moisture problems evident in the above photos.

Figure 2.29. Aerated, agitated bin IPS system at Fairfield, Connecticut. The compost turner is fully automatic and runs without an operator. The mixture is loaded at the near side of the bin and moves about 10 ft with each daily turning. Material exits the far end of the bin after about 21 days.

Table 2.1.   Mass Balances for Typical IPS Systems Composting Sludge with Yardwaste Alone and Sludge with Sawdust and Recycle

Component	Wet Wt (tons)	%TS	Dry Wt (tons)	Bulk Wt (lb/ft³)	Volume (ft³/day)
Generic 8-bay reactor with yardwastes					
Sludge	25.0	20	5.0	59.3	845
Yardwastes	26.5	55	14.5	19.3	2745
Infeed mix	51.5	38	19.5	35.5	2900
Reactor output	31.0	55	17.0	33.3	1860
Generic 8-bay reactor with saw-dust and recycle					
Sludge	25.0	20	5.0	59.3	845
Recycle	12.5	60	7.5	37.0	675
Sawdust	12.5	60	7.5	17.8	1405
Infeed mix	50.0	40	20.0	41.5	2410
Reactor output	30.0	60	18.0	37.0	1620

Source: Albright and Ferencsik.[30]

original 4-bay sludge:yard waste system at Palm Beach County, Florida, is being expanded to 36 bays. As of 1991, 13 facilities were operational or under construction in the U.S., totalling over 90 reactor bays.

Highlights: The daily agitation and forced aeration make this design applicable to a wide variety of substrates. The feed and withdrawal arrangement, the pattern of solids flow, and the turning equipment itself are operator friendly.

Status: Active.

Royer

Royer Industries, Inc. offered a multibay, bin-type system similar to the IPS system but with bays about 9.3 ft wide wall to wall. This makes the bays intermediate in size between those of the IPS and Paygro processes. Residence time is 21 days. A four-bay facility began operation in Lexington, North Carolina, about 1991, composting sewage sludge.

Status: Royer is apparently no longer offering this system, but instead is supplying turning equipment.

KOCH

The KOCH system is an agitated bed system developed by KOCH Transporttechnik GmbH of Germany. A roofed structure without side walls is usually provided. Compostables are stacked into an extended pile, about 9 ft high. Rails run the length of the pile on both sides. A beam is supported above the piles, mounted to the rails and spanning the width of the piles. A digging, turning, and transfer machine is mounted to the beam and travels in a zigzag pattern. Two facilities in Germany and one in Cascais, Portugal, are reported. A facility is planned for Riverhead, New York, that will use extensive preprocessing to separate recyclables and compostables. A drum is provided to mix separated compostables and provide water addition to 50% moisture. Two 75 by 600-ft composting troughs are provided, which will provide a 70 day residence time (at 9 ft stacking height) for the estimated 215 to 250 tpd of compostables. Postprocessing includes trommel screening, air classification, and 90 day storage.

Status: Active

Ebara Round Trip Paddling Fermentor (RPF)

This agitated bed system was designed by Ebara Corporation of Tokyo. The reactor is a rectangular-shaped, bin-type with nearly square dimensions. Material is fed at one end and exits the opposite end of the reactor. A rotating paddle suspended from a traveling house crane moves in a zigzag manner to agitate and move material toward the exit (see Figure 2.30). Forced aeration is provided through the floor. As of 1991, over 15 facilities were operational in Japan and 1 in the U.S. on a wide variety of substrates. A 235 tpd sludge/MSW facility is under design for Burlington Co., New Jersey (1993).

Figure 2.30.   Ebara fermenter, crane, and agitation paddle at the Yayoi Compost Plant, Japan (top photo). The facility is designed for 5 tpd of MSW, agricultural wastes and animal manures.[31] Bottom photo shows the paddle in operation. Photos courtesy of Carl Geary, Ebara Environmental Corporation.

Figure 2.31. Sorain Cecchini agitated bin system composting pre-processed MSW at Perugia, Italy. Material is loaded along one side of the bin length (right side in photo), moves across the bin width by action of the agitation screws, and is discharged at the opposite side. Four agitation screws are mounted on a 30 m traveling bridge. The screws move along the traveling bridge while it is stationary (from left to right in photo). Each cycle is about 15 min long. At the end of each cycle the screws retract and the bridge moves a few meters in the direction of the basin length. The bridge takes about two working shifts to run the entire 120 m basin. Photo courtesy of John Nutter, Sorain Cecchini Recovery.

Status: Active.

Sorain Cecchini

Sorain Cecchini S.p.A. is an Italian company involved in the MSW field since the early 1940s. Over a 20-year period, Sorain operated several versions of its composting technology. Early vertical reactors were eventually replaced by horizontal designs. A "fourth generation" facility was commissioned in 1987 at Perugia, Italy, and is a rectangular, bin-type system (see Figure 2.31). Digester dimensions are typically 23 m wide with the compost placed 2.0 to 2.5 m deep. Reactor length is varied depending on the MSW loading. The Perugia reactor is 120 m long and is housed in a 3900 m² building. The basin is straddled by a movable bridge crane to which agitation screws are attached. The bridge crane moves the length of the basin, agitating and moving material on a daily basis. The screws are maintained at about 45° when agitating. The agitation cycle requires 16 h in a 135 m basin. Residence time in the digester is typically 28 days. Aeration manifolds supply air from the floor of the basin. The facility at Perugia preprocesses MSW by trommel separation with the <3 in. fraction passing to a ballistic separator and then to composting. Design capacity is 200 tpd of compostables and about 171 tpd (7-day basis) was actually processed in 1988. Postprocessing includes two stages of trommel screening and air classification. The American Recovery Corporation holds rights to the Sorain technologies in the U.S. and Canada.

Status: Active.

Siloda Process (OTV, OTVD)

This is a bin-type composting system developed primarily by Jean-Pierre Levasseur of the French firm of Omnium de Traitements et de Valorisation des Dechets (OTV or OTVD). Parallel bins are typically 4 m wide and divided from the next by a 1.6- to 3 m-high wall (see Figure 2.32). The length depends on the daily rate of refuse input. A single bin is designed to receive about 2 days production of refuse. A minimum of eight bays are used. Material is turned about weekly by a specially designed paddle wheel which is about 3.5 m in diameter and spans the bin width. Blades on the paddle slice the material and a screw conveyor elevates and discharges it into the adjacent bay. The paddle wheel travels about 15 m/h. Refuse is preprocessed by shredding and screening. Forced aeration is provided through the floor of the bays. About 35 Siloda composting plants have been implemented in Europe and the Middle East. A 100 tpd refuse facility was commissioned at Prairieland County, Minnesota, in 1992.

Status: Active.

LH Resource Management Inc.

This agitated bin process is based on the developmental work of Jack Pos, professor of agricultural engineering at the University of Guelph. The process and equipment were commercialized by LH Resource Management Inc., Walton, Ontario, Canada. A 3-bay system at the Village of Hensall, Ontario, commenced operation in 1992 composting a mixture of grain screenings, cardboard, and wood waste. The grain screenings come from three local grain elevator companies which combine to make Hensall Canada's largest inland granary. Each bin is 100 ft long, 15 ft wide, and keeps a 4-ft compost depth along the length. Each channel has a design capacity of 600 ft³/day. Residence time is 12 days at Hensall. Mixed materials are loaded at one end, turned daily, and exit the opposite end of the bin. Manifolds in the floor of the channel provide aeration. Material from the reactor is cured for 1 to 2 months in large windrows turned occasionally by front-end loader. Two other facilities are currently in construction for food and yard wastes and will use a 28 day residence time.[32]

Highlights: The compost turner (biomass processor) is a mobile machine with an on-board power source and is controlled by a programmable logic controller (PLC). The turner spans the compost and the channel walls, riding on tires which run between adjacent channel walls. Flail speed and "bite" can be controlled by the PLC to vary the grinding action. Displacement along the channel can also be controlled to compensate for volume reduction and help maintain a constant material depth.

Figure 2.32.   Agitated bin OTVD composting plant at Brametot, Normandy, France (above). This plant was designed for 120 mtpd of household refuse, built in 1978, and continues in service. The Siloda paddle wheel agitates the material and transfers it to the adjacent bin about once/week. The "Prairieland" plant at Truman, Minnesota (below). The latter facility composts for 30 days in the Siloda bins followed by 60 days of aerated, static pile curing. Ten bins are provided, each 4 m wide, about 3 m high, and 41 m long. Concrete construction is used for floors, walls, and ceiling which allows for periodic washdown. The aeration duct is visible in the empty bin. Photos courtesy of Norman Schenck, OTVD, Inc.

Figure 2.33. Horizontal, static bed, push type reactor at Hamilton, Ohio. The photo shows the side of one bin and the manifold system used to supply air and remove process gases.

Status: Active.

Horizontal Flow, Static Beds

BAV Tunnel Reactor (Ashbrook-Simon-Hartley Tunnel Reactor)

The "Tunnel" reactor was first developed about 1979 based on BAV experience with vertical silo systems in Europe. The vessel is a plug flow, tubular reactor of rectangular cross section. Reactor volume can range from 10 to 500 m³. The reactor typically uses a pusher plate with forward and backward movement, which is situated at the feed end of the tunnel. The daily charge of mixed materials is added to the empty zone created when the plate is withdrawn. When the chamber is full, the pusher plate is hydraulically moved forward, pushing the new material into the reactor and simultaneously expelling composted material from the far end of the reactor. The pusher plate remains in place until it is withdrawn for the next feeding cycle. Air can be supplied and gases can be removed along the length of the reactor. The system has been applied to a variety of feedstocks including sludge, manure, and MSW fractions. Residence time is typically 15 to 28 days, followed by 30 to 60 days curing in aerated static piles. In the U.S., the system is provided by Ashbrook-Simon-Hartley, a division of Simon Waste Solutions, Inc. A facility at Hamilton, Ohio (1988 startup), uses six reactors, each 18 ft wide by 12 ft high by 63 ft long for composting about 17 dtpd of sludge conditioned with sawdust, bark, and recycled compost (see Figure 2.33). Other sludge facilities include Newburg, Oregon, and Hickory, North Carolina. A 50-dtpd, 20-tunnel system for sludge composting is under construction at Camden, New Jersey. In Baltimore, Maryland, a 700-tpd MSW facility with 14 tunnels underwent startup in March 1993. Each tunnel is 18 ft

Figure 2.34. Horizontal, static bed, conveyor type, Dynatherm system installed at Hilton Head Island, South Carolina. The steel modules were factory assembled and shipped to the job site by truck. The facility processes from 35 to 45 wet ton/week of sewage sludge amended with shredded yard waste. "Walking" floors move material along the reactors. Material is loaded into the top reactor and eventually drops to the lower reactor through an intermediate mixer, which provides agitation and remixing. Photo courtesy of Richard Ryan, Compost Systems Company.

wide by 12 ft high by 66 ft long and provides 18 days residence time for 365 tpd (5.5 day/week basis) of separated compostables.

Status: Active.

Dynatherm (Compost Systems Co.)

Compost Systems Co. of Ohio has developed a modular composting system that can be classified as a horizontal, static bed reactor. The Dynatherm reactor uses a type of "walking floor", commonly used in the trucking industry, to move material along the reactor. The moving floor provides a noncompacting method for moving material from inlet to outlet. Forced aeration is through the sliding members of the moving floor. A facility for 1.2 dtpd (5-day basis) of raw sludge amended with yard waste began operation at Hilton Head Island, South Carolina, in 1990 (see Figure 2.34). Two factory-fabricated steel modules, each 10 ft high by 12 ft wide by 44 ft long, are used. A tub grinder preshreds the yard waste. Residence time is about 22 to 23 days at average load. Field erected concrete modules can be used for larger facilities.

Status: Active.

Nonflow, Static Beds

Double T Composting System

This is a batch operated "tunnel" system developed by Thomas Thomas of Double T Equipment Manufacturing Ltd., Airdrie, Alberta, Canada. First offered in 1981, the system was originally designed for composting agricultural wastes for the mushroom industry. Over 24 tunnels are currently in operation. The smallest tunnels can be shipped by truck. The largest tunnels are about 13 ft wide and 105 ft long and are field erected. Each tunnel has a grid floor with a ventilation plenum to force air through the material. Material is batch loaded from one end using a telescoping conveyor with a swiveling head. An airspace remains above the compost after loading. A fabric net is laid on the floor prior to loading and is pulled from one end to discharge material after the compost cycle. A 5 day residence time is used for most agricultural wastes and 9 days for source separated refuse. The reactor output is cured in windrows or static piles. An 8-tunnel system near Windsor, Ontario, composts about 150 mton of agricultural wastes per tunnel per batch cycle.[33]

Highlights: Each tunnel is supplied with an automatic air recirculation system. Gases drawn from the gas space within the tunnel are cooled by cooling water or air-to-air heat exchange to remove heat and condense water vapor. Makeup air is added and the supply air rehumidified by steam or atomized water injection. The volume of blowdown gas exhausted from the recirculation system is greatly reduced compared to the case of single-pass aeration. Composting temperature is used to control the gas recirculation rate, oxygen to control the input of new air and the blowdown of recirculation gas, and relative humidity to control rehumidification. Respiration rate can by determined from the volume and oxygen content of the blowdown gas.

Status: Active.

Herhof "Box"

This is a batch operated composting system manufactured by Herhof Umwelttechnik GmbH of Niederbiel, Germany. An aerated box provides 7 to 10 days of composting. Aeration is controlled by the CO_2 content of the exhaust gas. Windrow curing typically is used after material is unloaded from the box. Each box can compost about 1000 mtpyr. Fifteen facilities were reported operating in Germany in 1992.

Status: Active.

ML Biocontainer

This is a batch-operated, box composter manufactured by MAB-Lentjes Energie und Umwelttechnik, Germany. Each box is about 22 m³ volume and is individually controlled. It provides 10 to 14 days residence time. Aeration is alternated between positive and negative modes. About 10 weeks of windrow curing is provided. Two facilities were reported to be operating in 1992.

Status: Active.

SUMMARY

Composting is often divided into a first stage, high-rate phase and a second stage, curing phase. There is no precise distinction between these two stages. In general, the first stage is characterized by high oxygen uptake rates and thermophilic temperatures. The curing phase is characterized by lower oxygen uptake rates and a decrease in process temperatures. Both phases are integral to the design and operation of a complete composting system, and both are necessary to produce a mature compost product. It is important to remember that a compost facility is a processing plant that must be designed and operated to produce the desired product.

Any material added to the composting system should be considered a substrate or feedstock to the process. An amendment is a material added to the other substrates to condition the feed mixture, either structurally or to improve its energy content. A bulking agent is a material of sufficient size that it provides structural support and maintains airspaces within the composting matrix. Anything added to the process should be considered a feedstock, whether it is termed a substrate, amendment, or bulking agent.

Composting systems can be classified according to the reactor type, solids flow mechanisms, bed conditions in the reactor, and manner of air supply. The following system can be used to classify most of the historic and current compost systems.

I.   Nonreactor systems (open)
    A.   Agitated solids bed (windrow)
        1.   Naturally ventilated
        2.   Forced aeration
    B.   Static solids bed
        1.   Forced aeration (aerated static pile)
        2.   Natural ventilation (nonaerated piles)
II.   Reactor systems (in-vessel or enclosed)
    A.   Vertical solids flow
        1.   Agitated solids bed
            a.   Multiple hearths
            b.   Multiple floors, decks, or belts
        2.   Packed bed (silo reactors)
            a.   Counter-current air:solids flow
            b.   Cross-current air:solids flow
    B.   Horizontal and inclined solids flow
        1.   Tumbling solids bed (rotary drums or kilns)
            a.   Dispersed flow
            b.   Cells in series
            c.   Completely mixed
        2.   Agitated solids bed (agitated bins or open channels)
            a.   Circular shape
            b.   Rectangular shape
        3.   Static solids bed (tunnel shaped)
            a.   Push type
            b.   Conveyor type
    C.   Nonflow (compost boxes)

Representative systems from most of the above classifications are active in today's market. Each reactor or nonreactor type has certain characteristic advantages and disadvantages relative to the other market offerings. The importance of each advantage or disadvantage depends on the particular situation. This means that (1) selection of an appropriate composting system will always be a site-specific decision, (2) there is no "right" system for all circumstances, and (3) the marketplace should remain healthy with a reasonably diverse offering of systems.

REFERENCES

1. Rodale, J. I. *The Complete Book of Composting* (Emmaus, PA: Rodale Books, Inc., 1950).
2. Howard, A. "The Manufacture of Humus by the Indore Process," *J. Royal Soc. Arts* 84:25 (1935).
3. "Reclamation of Municipal Refuse by Composting," Sanitary Engineering Research Project, University of California, Berkeley, Technical Bulletin No. 9, Series 37 (June 1953).
4. Shell, G. L. and Boyd, J. L. "Composting Dewatered Sewage Sludge," report to the U.S. Public Health Service, Bureau of Solid Waste Management (1969).
5. Dohoney, R. W., District Manager for Professional Services Group, Personal communication (December 31, 1991).
6. Compton, C. R. and Bowerman, F. R. "Composting Operation in Los Angeles County," *Compost Sci.* (Winter 1961).
7. Ullrich, A. H. and Smith, M. W. "Experiments in Composting Digested Sludge at Austin, Texas," *Sewage Ind. Wastes* 22(4):567 (1950).
8. Reeves, J. B. "Sanitary Aspects of Composted Sewage Sludge and Sawdust," *Sewage Ind. Wastes* 31(5):557 (1959).
9. Hay, J. C. and Kuchenrither, R. D. "Fundamentals and Application of Windrow Composting," *J. Environ. Eng. Div.* 116:4 (1990).
10. Robbins, M. "Solids Disposal at the Upper Occoquan Sewage Authority," in *Proceedings of the National Conference on Municipal and Industrial Sludge Composting-Materials Handling* (Rockville, MD: Hazardous Materials Control Research Institute, 1979).
11. Schoening, C. and Doersam, J. "Dillo Dirt: Austin's Solution to a Sludge Disposal Problem," presented at the 62nd Annual Conference Water Pollution Control Federation, San Francisco, CA, October 1989.
12. LeBrun, T. J., Iacoboni, M. D., and Livingston, J. R. "Overview of Compost Research Conducted by the Los Angeles County Sanitation Districts," in *Proceedings of the National Conference on Municipal and Industrial Sludge Composting* (Rockville, MD: Hazardous Materials Control Research Institute, 1980).
13. Hay, J. C. and Caballero, R. C. "Windrow Composting: Principles and Application," presented at the 1987 National Conference on Environmental Engr., ASCE, Orlando, FL, 1987.
14. Strom, P. F. and Finstein, M. S. "Leaf Composting Manual for New Jersey Municipalities," Office of Recycling, New Jersey Department of Energy and Environmental Protection (1986).
15. Strom, P. F., Flower, F. B., Liu, M. H. P., and Finstein, M. S. "Recommended Methods for Municipal Leaf Composting," *BioCycle* 27:9 (1986).
16. "Options for Municipal Leaf Composting," *BioCycle* 29:9 (1988).
17. Taylor, A. C. and Kashmanian, R. M. *Study and Assessment of Eight Yard Waste Composting Programs Across the United States,* EPA/530-SW-89-038 (April 1989).
18. Iacoboni, M., LeBrun, T., and Smith, D. L. "Composting Study," Technical Services Department, Los Angeles County Sanitation Districts (Sept.–Dec., 1977).
19. Glenn, J. "Odor Control in Yard Waste Composting," *BioCycle* 31:11 (1990).
20. Goldstein, N. and Riggle, D. "Sludge Composting Maintains Momentum," *BioCycle* 31:12 (1990).

21. Colacicco, D., Epstein, E., Willson, G. B., Parr, J. F., and Christensen, L. A. "Costs of Sludge Composting," Agricultural Research Services, USDA, ARS-NE-79 (1977).

22. Epstein, E., Willson, G. B., Burge, W. D., Mullen, D. C., and Enkiri, N. "A Forced Aeration System for Composting Wastewater Sludge," *J. Water Poll. Control Fed.* 48:4 (1976).

23. Willson, G. B. "Equipment for Composting Sewage Sludge in Windrows and in Piles," in *Proceedings of the 1977 National Conference on Composting of Municipal Residues and Sludge* (Rockville, MD: Hazardous Materials Control Research Institute, 1977).

24. Willson, G. B., Parr, J. F., Epstein, E., Marsh, P. B., Chaney, R. L., Colacicco, D., Burge, W. D., Sikora, L. J., Tester, C. F., and Hornick, S. *Manual for Composting Sewage Sludge by the Beltsville Aerated-Pile Method,* Municipal Environmental Research Lab., Cincinnati, OH, EPA-600/8-80-022 (May 1980).

25. Goldstein, J. "Start-up Times at Pembroke Pines," *BioCycle* (March 1992).

26. Iacoboni, M. D., Livingston, J. R., and LeBrun, T. J. "Windrow and Static Pile Composting of Municipal Sewage Sludges," U.S. EPA Municipal Environmental Research Laboratory, Cincinnati, OH, EPA-600/2-84-122 (July 1984).

27. Ettlich, W. F. and Lewis, A. E. "A Study of Forced Aeration Composting of Wastewater Sludge," U.S. EPA Municipal Environmental Research Laboratory, Cincinnati, OH, EPA-600/2-78-057 (1978).

28. Wilson, G. B., Epstein, E., and Parr, J. R. "Recent Advances in Compost Technology," in *Proceedings of the 3rd National Conference Sludge Management* (Rockville, MD: Disposal and Utilization, Information Transfer, 1977).

29. MacGregor, S. T., Miller, F. C., Psarianos, K. M., and Finstein, M. S. "Composting Process Control Based on Interaction Between Microbial Heat Output and Temperature," *Appl. Environ. Microbiol.* 1321 (June 1981).

30. Albright, A. E. and Ferencsik, R. International Process Systems Inc., Personal communication (1991).

31. Geary, C. H. Process Manager, Ebara Environmental Corp., Personal communication (October 1991).

32. Lee, C. President and General Manager, LH Resource Management Inc., Personal communication (April 1993).

33. Thomas, T. President, Double T Equipment Manufacturing Ltd., Personal communication (April 1993).

Chapter 3

Thermodynamic Fundamentals

INTRODUCTION

Thermodynamics is the branch of science that deals with energy and its transformations. Thermodynamics is normally associated with heat, but the subject deals not only with heat but all forms of energy. The principles of thermodynamics are well established and have been applied to physical, chemical, and biological systems. Lehninger[1] has stated that the proper study of biology, for example, should start from thermodynamic principles as the central theme that can best systematize biological facts and theories. This same statement can be made about the study of composting systems. Application of thermodynamic principles is a fundamental way of analyzing composting systems just as it has been a fundamental method for analysis of other physical, chemical, and biological processes. Because the laws of thermodynamics appear to be inviolable, application of these laws will reveal much about the limitations and expectations of composting systems.

The subject of thermodynamics should be distinguished clearly from the related subject of kinetics. Thermodynamics deals with the energy changes that accompany a process. Kinetics deals with rates or velocities of reactions and cannot be inferred from thermodynamics. For example, organic molecules in a piece of paper contain a rather substantial amount of energy. If a match is struck to the paper the energy is released at a rapid kinetic rate. If the paper is decomposed by microbial action, as in a compost pile, the energy is released at a much slower kinetic rate. The same amount of energy is released in either case but the kinetics are quite different. The subject of this Chapter is energy changes as determined by thermodynamics. Kinetic principles of composting are discussed in Chapter 10.

For most organisms life is a constant struggle or search for energy supplies needed to power the cellular machinery. Certain higher organisms, such as man, have freed themselves from the constant search for energy. Nevertheless, the human body is constantly "burning" various stored substances for energy. If these stored reserves are not replaced at periodic intervals, death is inevitable. Microorganisms, however, spend nearly their entire life cycles in search of energy sources. Despite this difference, many of the fuels used by microbes are also used by man.

From a thermodynamic standpoint, all life forms can be viewed as chemical machines which must obey the laws of energy and heat as most all other nonliving processes. Thus, thermodynamics places limits on living systems just as it does on physical systems constructed by man to extract energy from his surroundings. This discussion of thermodynamics will necessarily be brief but of sufficient depth to serve the purposes of this book. The interested reader is referred to the numerous excellent texts available on chemical thermodynamics and bioenergetics for more in-depth discussion.

Although the need for an energy supply is common to all life forms, the actual source can be markedly different. The study of different energy sources available to microbes is probably the best approach to understanding the differences between life forms. It will also lead to a better understanding of reactions that organisms mediate and the useful tasks to which they can be directed in properly engineered systems. In this sense the designer of a composting plant differs little from the biochemical engineer designing an industrial fermentation, enzyme extraction, or antibiotic production plant or the sanitary engineer designing a biological waste treatment facility. All must understand the microbes involved, reactions they mediate to obtain energy, environmental conditions required for growth and metabolism, and, in certain cases, conditions required to kill organisms such as pathogens.

HEAT AND WORK, THE CONSERVATION OF ENERGY

To begin this discussion, the terms "heat" and "work" need to be defined. Heat is energy that flows because of a temperature difference between two bodies. The basic unit of heat is the calorie, which is the energy required to raise 1 g of water 1°C. The energy required actually varies somewhat with temperature, and it is common to specify the 15°C calorie as the heat flowing into 1 g of water when its temperature increases at atmospheric pressure from 14.5 to 15.5°C The equivalent English unit is the 60°F Btu (British thermal unit), which is the quantity of heat flowing into 1 lbm (pound mass) of water when its temperature increases at atmospheric pressure from 59.5 to 60.5°F. These exact definitions are seldom necessary for environmental engineering purposes, and it is sufficient to define the specific heat of water as 1.00 cal/g °C, which is equivalent to 1.00 Btu/lb °F. The specific heat of other materials varies significantly and is usually much lower than that of water. The specific heats of ethanol, acetone, aluminum, and copper are 0.65, 0.50, 0.22, and 0.093 cal/g °C, respectively.

Note that the previous definitions specified atmospheric pressure conditions. Specific heat can be measured under both constant volume and constant pressure conditions. c_v is the specific heat at constant volume and c_p the specific heat at constant pressure. There is little difference between the two for liquids and solids. With gases, however, c_p is greater than c_v because of the added heat energy required to expand a gas against a constant pressure. Most biological processes operate under conditions of constant pressure in aqueous solutions, and the distinction is not as significant to liquid phase systems. Dry gases and water vapor are important components in a composting system, however, and the distinction will be important in later studies.

The concept of work is essential to the study of thermodynamics. Work may be mechanical, electrical, magnetic, or of other origin. Consider the mechanical system shown in Figure 3.1. Work is the application of force through a distance. If the force on the piston is constant, the incremental work, dw, in moving the piston through the distance, dl, is given as

$$dw = Fdl \qquad (3.1)$$

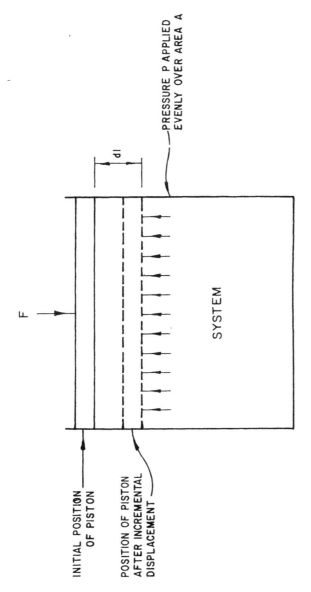

Figure 3.1.   Schematic illustration of mechanical work being performed on a system.

The force, F, is equal to the product of pressure, P, times area, A. Thus,

$$dw = PAdl = PdV \tag{3.2}$$

Since dV is positive when the volume increases, work in an expansion process is positive. Work is measured in foot-pounds, joules, or ergs.

Because heat and work are both forms of energy one should be able to equate them. James Prescott Joule verified this in a number of experimental systems between about 1840 and 1850. In each case the amount of work yielded the same amount of heat, about 4.18×10^7 ergs of work per calorie of heat. In honor of Joule, 10^7 ergs were set equal to 1 J. Thus, today we say that 4.18 J = 1 cal. In English units, 778 ft·lb is equivalent to 1 Btu.

THE FIRST LAW OF THERMODYNAMICS

Consider the system shown in Figure 3.2. If energy can be neither created or destroyed, heat energy that flows into the system, +q, must either be stored within the system, flow out of the system or appear as work done by the system. In other words, energy that flows into a system must be fully accounted for in other forms of energy. This is a statement of the law of conservation of energy which is also referred to as the first law of thermodynamics. Stated yet another way, energy can be neither created nor destroyed. The German physicist Hermann Ludwig von Helmholtz is usually credited with the first formal statement of this principle in 1847, although it had been intuitively accepted as early as the 18th century. Notice that this concept is referred to as the First Law of Thermodynamics. There is no proof of this concept. It is accepted as a first principle, a fundamental concept that describes observed phenomena and which has never been shown to be violated. We will place considerable reliance on the concept of an "energy balance" in later discussions of composting systems.

Returning to Figure 3.2, the First Law for the system shown can be stated as

$$q = \Delta E + w \tag{3.3}$$

where

E = the change in internal energy of the system
$+q$ = heat flow into the system
$+w$ = work done by the system

If the system is maintained at constant volume (isovolumetric), Equation 3.3 becomes

$$q_v = \Delta E \quad (V = \text{constant}) \tag{3.4}$$

No work of expansion or contraction can be performed in a constant volume system, and heat absorbed by the system is exactly balanced by increased internal energy within the system.

The heat per unit mass flowing into a substance can be defined as

$$dq = mcdT \tag{3.5}$$

where

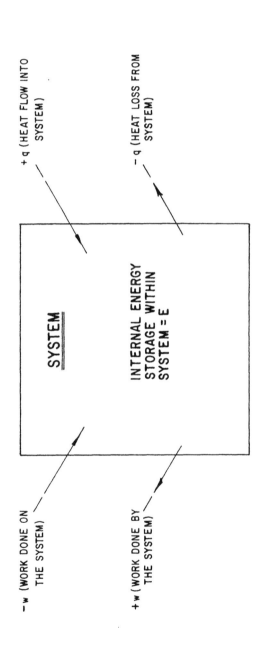

Figure 3.2.   Schematic illustration of the principle of conservation of energy, also known as the First Law of Thermodynamics. Note the sign convention on the terms for heat and work.

$$\begin{aligned}
c &= \text{the specific heat capacity} \\
m &= \text{the mass} \\
T &= \text{the temperature.}
\end{aligned}$$

Integrating this expression,

$$q = m \int_{T_1}^{T_2} c \, dT \tag{3.6}$$

The specific heat must be known as a function of T to complete the integration. Fortunately, values of specific heat for most substances remain relatively constant over small temperature ranges. If the temperature range is large, the variation of specific heat with temperature can be approximated by a polynomial, which has the advantage of being integrated easily. Usually a second or third-degree polynomial is used with the form

$$c = a + bT + cT^2 + dT^3$$

where a, b, c, and d are constants that depend on the particular gas. This complication is not necessary to the present analysis, because the change in specific heat for composting gases is not significant over the temperature range in question.

Assuming a constant-volume process with constant specific heat, Equation 3.6 integrates to

$$q_v = mc_v \Delta T \qquad (\Delta T \text{ not large}) \tag{3.7}$$

Heat flow into a substance under constant pressure can also be determined from the specific heat and temperature difference. Integrating Equation 3.6 under conditions of constant pressure gives

$$q_p = mc_p \Delta T \qquad (\Delta T \text{ not large}) \tag{3.8}$$

Equation 3.8 will be extremely useful in conducting energy balances on composting processes, because they normally operate under conditions of constant pressure.

Many chemical and most biological processes operate under conditions of constant pressure. Applying Equation 3.3 to the system in Figure 3.1, assuming constant pressure and movement of the piston over the incremental small distance, dl, gives the following:

$$dq = dE + pdV \tag{3.9}$$

Integrating and recalling that p is constant,

$$q_p = E_2 - E_1 + pV_2 - pV_1 \tag{3.10}$$

Rearranging,

$$q_p = (E_2 + pV_2) - (E_1 + pV_1) \tag{3.11}$$

Because constant pressure systems are so important in chemical and biochemical systems, the terms in parenthesis have been defined as the enthalpy, H, of the system. Thus,

$$q_p = H_2 - H_1 = \Delta H \qquad (3.12)$$

When chemical changes result in the absorption of heat, i.e., heat flows into the system, the reaction is termed endothermic and both q and ΔH are taken as positive values. Reactions that result in the evolution of heat, i.e., heat flows from the system, are termed exothermic, and q and ΔH have negative values. The chemical reactions that result during composting usually occur under conditions of constant pressure. Thus, heat release can be determined from the enthalpy changes accompanying the reactions.

It is common to measure elevations against a reference point. Thus, the elevation of a mountain is referenced to sea level or some other datum, the height of a ceiling to the floor beneath it, and so on. Similarly, measurement of enthalpy is simplified by establishing a datum against which all other values can be measured. The datum is termed a standard state and is conveniently taken as the stable state of the compound at 25°C and 1 atm pressure. For example, the standard state of oxygen is a gas. By convention, enthalpy of an element in its standard state is set equal to zero. The enthalpy of a compound is determined from the heat of reaction necessary to form the compound from its elements. For reactions in solution, the standard state corresponds to a concentration at unit activity (approximately a 1 M solution). Enthalpy change calculated under standard state conditions is usually termed $\Delta H°$. Standard enthalpy values for a few compounds of interest are presented in Table 3.1.

Both internal energy and enthalpy are properties of a system. This means that changes in enthalpy and internal energy between two equilibrium states of a system are the same for all paths followed between the two states. In other words, internal energy change and enthalpy change depend only on the beginning and end points of a process. For example, heat absorbed in an isobaric (constant pressure) process is equal to the enthalpy difference between beginning and end states regardless of path. Because enthalpy is a property of a system, its values can be tabulated. The reader is probably familiar with steam tables that contain tabulated values of enthalpy for water and water vapor as functions of temperature and pressure. Not all terms mentioned in this discussion are system properties. Work, for example, depends on the path followed between equilibrium states and is not a property of a system.

The distinction between a property of a system and other functions that depend on path may seem rather subtle. However, the distinction is quite important and has direct application to thermodynamic analysis of composting systems. The inlet and outlet conditions around a compost process often can be defined without knowledge of the path followed between end points. Fortunately, it is not necessary to know the path to determine the enthalpy change. The following example should help clarify the distinction and highlight its importance.

Example 3.1

Consider a compost process operated under constant pressure conditions in which water and vapor enter at temperature $T_1 = 15.5°C$ (60°F). The reactor is maintained at a temperature $T_2 = 71.1°C$ (160°F), and different proportions of water and vapor exit the reactor at T_2. Determine the heat required to accomplish the mass and temperature changes assuming two different thermodynamic paths as follows: (1) assume that water is vaporized at temperature T_1 and then water and vapor are heated to T_2 and (2) assume inlet water and vapor are first heated to temperature T_2 and then a portion is vaporized at temperature T_2.

Table 3.1.   Standard State Enthalpy and Free Energy of Formation at 25°C

Substance	State[a]	H°	G°
CH_4	g	−17.89	−12.14
CH_3COOH	aq	−116.74	−95.51
$C_6H_{12}O_6$	aq		−217.02
CO_2	g	−94.05	−94.26
CO_2	aq	−98.69	−92.31
H_2O	lq	−68.32	−56.69
H_2O	g	−57.80	−54.64
H_2S	aq	−9.4	−6.54
NH_4^+	aq	−31.74	−19.00
SO_4^{-2}	aq	−216.90	−177.34
H^+ (pH = 7)	aq		−9.67
NH_4Cl	c	−75.38	−48.73
Cl^-	aq	−40.02	−31.35

Source: Perry and Chilton,[2] CRC Handbook of Chemistry and Physics,[3] Lehninger,[4] and Sawyer and McCartey.[5]

[a] Aq = aqueous, g = gas, lq = liquid, c = crystal.

Inlet T = 15.5° C   Exit T = 71.7° C

0.1 g H_2O vapor → | Compost Process | → 0.6 g H_2O vapor

1.0 g H_2O liquid → | p = constant | → 0.5 g H_2O liquid

Solution

1. From steam tables the following table of enthalpy values can be constructed:

	Enthalpy (cal/g)	
Temperature (°C)	Water	Steam (Sat.)
15.5	15.59	604.28
71.1	71.09	627.89

2. For path 1 the enthalpy change can be determined as

$$\Delta H = 0.5(71.09 - 15.59) + 0.5(604.28 - 15.59) + (0.5 + 0.1)(627.89 - 604.28)$$

liquid to T_2 vaporize at T_1 all vapor to T_2

$$\Delta H = 336.26 \text{ cal}$$

3. For path 2 the enthalpy change is determined as

$$\Delta H = 1.0(71.09 - 15.59) + 0.5(627.89 - 71.09) + 0.1(627.89 - 604.28)$$

liquid to T_2 vaporize at T_2 input vapor to T_2

$$\Delta H = 336.26 \text{ cal}$$

4. The enthalpy change is the same regardless of path, and one need only select a convenient path and be consistent in any subsequent calculations. This result will have a direct bearing on energy balances conducted in later chapters.

The heat release accompanying any chemical reaction is determined as the sum of the enthalpy of products minus reactants. The equation for oxidation of glucose is

$$C_6H_{12}O_6 + 6O_2 \rightarrow 6CO_2 + 6H_2O$$

$$\Delta H_R^\circ = -673,000 \text{ cal / mole}$$

The heat of reaction ΔH°_R is negative, which means that 673 kcal of heat energy is released to the surroundings per mole (180 g) of glucose oxidized. If the reaction is conducted by microbes, some of the chemical bond energy will be captured for use by the microbes, reducing somewhat the amount that appears as heat.

For microbial reactions all reactants and products should be assumed to be in the aqueous state, since all energy yielding reactions are conducted within the cellular cytoplasm. For combustion reactions the O_2 and produced CO_2 are usually assumed to be gaseous. If the produced water remains as a vapor the enthalpy change is termed the lower heat value (LHV) of combustion. If produced water is condensed to a liquid, additional heat is released and the enthalpy change is termed the higher heat value (HHV) of combustion. Only the LHV is available for recovery from combustion reactions, whereas the HHV is available in microbial systems because the water remains as a liquid within the cell.

Every organic molecule has a characteristic heat of combustion of significant magnitude and, in each case, of negative sign. Organic molecules contain energy in the chemical bonds that form the molecule. This energy is released as the molecule is degraded into simpler compounds, such as CO_2 and H_2O, which have a lower energy content.

Heats of reaction for the aerobic oxidation of various organics are presented in Table 3.2. It is interesting to note that the heat of reaction, expressed per gram of organic, varies significantly between the three major foodstuffs: proteins, carbohydrates, and lipids (fats). Lipids contain over twice the energy per gram as proteins and carbohydrates, including both starch and cellulose. Because most organic wastes are composed of a mixture of these natural foodstuffs, the heat of combustion can be expected to vary from a low of about 2100 to about 9300 cal/g (3800 to 16,700 Btu/lb). Comparative heating values of various fuels and municipal waste products are presented in Table 3.3. The organic components in domestic sludge commonly contain about 5550 cal/g of organic matter (10,000 Btu/lb).

THE SECOND LAW OF THERMODYNAMICS

In the latter part of the 19th century there was considerable interest among the early founders of thermodynamics as to the direction in which spontaneous chemical reactions would proceed. The French chemist Pierre Berthelot proposed that reactions would proceed spontaneously in the direction of negative ΔH. In other words, only exothermic reactions were spontaneous. This seemed to make a great deal of common sense. Organic molecules would always burn to form CO_2 and H_2O, but the latter would not spontaneously combine to form organic molecules.

Table 3.2. Heats of Combustion of Some Organic Fuels[a]

Feed	Formula	Molecular Weight	ΔH°_R (cal/mol)	gCOD[b] mol	gCOD/ g organic	kcal/g	kcal/g COD
Carbohydrate							
Glucose	$C_6H_{12}O_6$	180	−673,000	192	1.067	3.74	3.51
Lactic acid	$CH_3CH(OH)COOH$	90	−326,000	96	1.067	3.62	3.40
Polysaccharides, (e.g., starch and cellulose)	$(C_6H_{10}O_5)_x$				1.185		
Lipid							
Palmitic acid	$CH_3(CH_2)_{14}COOH$	256	−2,398,000	736	2.87	9.37	3.26
Tripalmitin	$(CH_3(CH_2)_{14}COO)_3(C_3H_5)$	809	−7,510,000	2,320	2.87	9.28	3.24
Protein							
Glycine (amino acid)	$CH_2(NH_2)COOH$	75	−163,800	48	0.64	2.18	3.41
Hydrocarbons							
n-Decane	$CH_3(CH_2)_8CH_3$	142	−1,610,000	496	3.49	11.34	3.25
Methane	CH_4	16	−210,800	64	4.00	13.18	3.29

[a] Heats of combustion were obtained from *CRC Handbook of Chemistry and Physics*, 50th ed. and corrected to products of combustion as $(CO_2)_g$, $(H_2O)_{aq}$, and $(NH_a^+)_{aq}$.
[b] Chemical oxygen demand.

Table 3.3. Comparative Heating Values of Various Fuels

Fuel	Heat Value cal/gm	Heat Value Btu/lb
No. 2 oil	10,900	19,640
No. 6 oil	9,720	17,510
Natural gas	12,700	22,880
Bituminous coal	7,560	13,620
Wood (air dried)	3,060	5,510
Grease and scum	9,280	16,720
Sludge (dry, ash-free)	5,560	10,020
Digested sludge (dry solids)	2,950	5,315
Digester gas	8,560	15,420
Municipal refuse (@ 20% moisture)	2,720	4,900

Source: Burd[6] and Olexsey.[7]

Unfortunately for Berthelot, there were a few chemical reactions that did not follow this rule. For example, consider the dissolution of ammonium chloride crystals in water.

$$NH_4Cl_c \rightarrow NH_{4\,aq}^+ + Cl^-_{\,aq}$$

$$\Delta H^\circ_R = (-31.74 - 40.02) - (-75.38) = +3.68 \text{ kcal} / \text{mole}$$

When added to water the crystals dissolve completely, provided the solubility limit is not exceeded. In other words, the reaction proceeds spontaneously to the right even though the reaction is endothermic. Something was missing from Berthelot's suggestion regarding the direction of spontaneous reactions.

The missing factor was the tendency of all systems to seek a state of maximum randomness or disorder. In the above example, the solution of ammonium and chloride ions in water is more random than the highly organized structure of the original crystal. Thus, the crystal tends to dissolve even though the reaction is endothermic, driven by the tendency toward maximum randomness, entropy.

Development of mathematical formulations to describe the tendency toward randomness dominated much of theoretical thermodynamics in the 19th century. The concept of entropy was developed from a search for a function that would serve as a criterion for spontaneity of physical and chemical changes. The search led to the formulation of the Second Law of Thermodynamics, which states that all spontaneous changes in an isolated system occur with an increase in entropy or randomness. In 1876 and 1878 an American chemist, Josiah Willard Gibbs, published papers in which the two laws of thermodynamics were applied to chemical reactions. Gibbs showed that the change in enthalpy was composed of two parts: (1) a change in useful or available energy, termed free energy, ΔG (later named the Gibb's free energy) and (2) a change in entropy, ΔS. The relationship is

$$\Delta H = \Delta G + T\Delta S \quad (\text{T and p } = \text{ constant}) \tag{3.13}$$

or

$$\Delta G = \Delta H - T\Delta S \tag{3.14}$$

In any spontaneous process the change in entropy, ΔS, must always be positive. Thus, in the case of ammonium chloride the product of T ΔS is greater than the enthalpy increase, H, giving a net decrease in free energy calculated as

$$NH_4Cl_c \rightarrow NH_{4\,aq}^+ + Cl_{\,aq}^-$$

$$\Delta G_R^\circ = (-19.00 - 31.35) - (-48.73) = -1.62 \text{ kcal / mole}$$

In any spontaneous process, whether endothermic or exothermic, the entropy must increase and the free energy decrease. Final realization of this concept was one of the crowning achievements of 19th century thermodynamics.

It can be shown that ΔG gives the useful work that can be derived from a chemical reaction occurring under constant pressure and temperature conditions. Since these are exactly the conditions under which most biochemical reactions occur, the Gibb's free energy function can be used to determine the energy available to microbes from various chemical reactions. The original premise of this chapter can now be restated in a more sophisticated manner: "All life forms require a source of free energy to maintain life."

Standard states must again be established to measure changes in free energy. For convenience, standard states are defined in a manner similar to enthalpy changes. For reactions that occur in solution the standard state corresponds to a concentration at unit activity (approximately 1 M solution). In addition, hydrogen ion at unit activity (pH = 0) is assigned a standard free energy of zero. Standard free energy values for a number of compounds of interest are presented in Table 3.1.

FREE ENERGY AND EQUILIBRIUM

Consider a reaction in which "a" can proceed to "b", and vice versa:

$$a \Leftrightarrow b$$

Such a reaction is termed an equilibrium reaction. Most of the metabolic reactions mediated by microbes are composed of a series of such equilibrium reactions; thus, they are important to living systems. Let us assume that initially no "b" is present in solution. Then the reaction will proceed to the right, producing "b" and consuming "a". As soon as "b" becomes present in solution, some of it will begin to react back to the left producing "a". Eventually, a dynamic condition will be established when the rate of production of "b" is balanced by the production of "a". At that point the system is in equilibrium.

Now let us consider the free energy changes that occur in the above equilibrium. The initial movement of "a" to form "b" occurs because of a decrease in free energy for the reaction. It becomes apparent that the magnitude of the free energy change must be related to the concentration of the reactant and product. At equilibrium the rate of production of "b" is exactly balanced by production of "a". There is no tendency for the reaction to make a net movement in either direction. At this point the free energy change for the reaction must be zero. If it were negative, the reaction would move to produce more "b". If it were positive the reaction would move in the reverse direction to produce more "a."

Consider the reaction

$$aA + bB \Leftrightarrow cC + dD$$

It can be shown that the effect of concentration of the different species can be described as

$$\Delta G_R = \Delta G_R^o + RT \ln\{[C]^c[D]^d / [A]^a[B]^b\} \qquad (3.15)$$

where

ΔG_R = reaction free energy change, cal/mol
ΔG_R^o = reaction free energy change under standard state conditions cal/mol
R = universal gas constant = 1.99 cal/deg-mol
T = absolute temperature, $^\circ K$
$[A]$ = molar concentration of species A

Equation 3.15 can be used to adjust standard free energy values for the effect of concentration.

The reaction free energy ΔG_R must be zero at equilibrium, thus

$$\Delta G_R^o = -RT \ln\{[C]^c[D]^d / [A]^a[B]^b\}_{eq} \qquad (3.16)$$

The term in brackets is actually the equilibrium constant for the reaction

$$K_{eq} = \text{equilibrium constant} = \{[C]^c[D]^d / [A]^a[B]^b\}_{eq} \qquad (3.17)$$

Thus,

$$\Delta G_R^o = -RT \ln K_{eq} \qquad (3.18)$$

This equation is one of the more important results of chemical thermodynamics because it allows prediction of equilibrium constants from standard free energy changes.

REACTION RATES AND TEMPERATURE

Rates of chemical reactions are usually a function of temperature. A convenient method of expressing the effect of temperature is to determine the rate of activity at one temperature to the rate at a temperature 10°C lower. This ratio is the temperature coefficient Q_{10}. As a general rule of thumb, most chemical reaction rates about double for each 10°C rise in temperature (i.e., $Q_{10} = 2$). For biologically mediated reactions such a relationship is also observed over the limited temperature ranges suitable for living organisms. Temperatures outside this range cause enzymes responsible for mediating the desired reactions to be inactive. Diffusion-controlled reactions usually exhibit a $Q_{10} < 2$ because diffusion coefficients vary less with temperature. On the other hand, temperature coefficients for enzyme coagulation and heat inactivation of microbes are characteristically greater than 2. Therefore, the effect of temperature on chemical and biochemical reaction rates is of considerable importance in the decomposition of waste materials and in thermal inactivation kinetics.

In 1889 the Swedish chemist Svant August Arrhenius developed a mathematical relationship between temperature and the rate of reaction. Based on experimental and theoretical considerations, Arrhenius proposed the following relationship:

$$\frac{d(\ln k)}{dT} = \frac{E_a}{RT^2}$$
(3.19)

where

k = reaction rate constant
E_a = activation energy for the reaction, cal/mol
T = absolute temperature, °K

Activation energy is interpreted as the amount of energy a molecule must have to undergo a successful chemical reaction. The integrated form of Equation 3.19 becomes

$$\ln \frac{k_2}{k_1} = \frac{E_a(T_2 - T_1)}{RT_1T_2}$$
(3.20)

where k_2 and k_1 are the reaction rate constants at temperatures T_2 and T_1, respectively. Between 20 and 30°C, a Q_{10} value of 2 corresponds to an activation energy of about 12.25 kcal/mol.

There are several other commonly used forms of the Arrhenius expression based on Equation 3.20. Taking the antilog of both sides of Equation 3.20 and rearranging,

$$k_2 = [k_1 e^{(Ea/RT_1)}] e^{(-Ea/RT_2)}$$
(3.21)

Because k_1 corresponds to temperature T_1, the term in brackets can be considered a constant for a given reaction. Thus,

$$k_2 = C e^{(-Ea/RT_2)}$$
(3.22)

Most biological processes operate over a limited temperature range. This is true even in composting systems. A temperature range from 0 to 80°C, which seems fairly extreme, corresponds to a range of only 273 to 353°K. Thus, the product of T_1 and T_2 changes only slightly over the biological temperature range. It is frequently assumed that E_a/RT_2T_1 is constant over this range so that Equation 3.20 becomes

$$\ln(k_2 / k_1) = \phi(T_2 - T_1) \qquad (3.23)$$

or

$$k_2 = k_1 e^{\phi(T_2 - T_1)} \qquad (3.24)$$

Although ϕ should be reasonably constant, it has sometimes been found to vary considerably, even over small temperature ranges. It is considered good practice to state the applicable temperature range whenever a value of ϕ is given. By way of comparison, a Q_{10} of 2.0 corresponds to a ϕ of 0.069.

LIFE AND ENERGY

Strictly speaking, the thermodynamic principles outlined above apply only to "closed" systems, defined as those that do not exchange matter with their surroundings or across the system boundary. Living systems are constantly exchanging matter with their surroundings and hence are termed "open" systems. Furthermore, they seldom attain true thermodynamic equilibrium. Instead there is a continual flux of metabolic materials that may result in a dynamic "steady-state". Under such steady-state conditions the rate of formation of a component is balanced by the rate of subsequent breakdown or conversion to another component. Nevertheless, principles of equilibrium thermodynamics have been applied to living systems with much success, and we need not be overly concerned with slight deviations from theoretical accuracy. The field of irreversible thermodynamics has been developed to deal with steady-state systems such as living microbes. One useful result of irreversible thermodynamics is the realization that steady-state is characterized by achievement of the minimum possible rate of entropy production for a given substrate use rate.[8]

Another attribute of living cells is that they are highly ordered systems composed of many sophisticated molecular structures. As such, there is a local decrease in entropy within the cell. It was observed before that any spontaneous process must occur with a net increase in entropy. Therefore, the local decrease in entropy within the cell is maintained at the expense of a larger increase in entropy in the surroundings. Consider, for example, an organism using the organic molecule glucose as a food source. The end products of aerobic metabolism are CO_2 and H_2O, which are more random than the original glucose molecule. Since the glucose is taken from the media surrounding the cell and CO_2 and H_2O are discharged back into it, an increase in entropy of the surroundings has occurred. Lehninger[1,4] stated that living organisms create and maintain their essential orderliness at the expense of their environment, which they cause to become more disordered and random.

Local decreases in entropy are not unique to living systems. Energy flow into a nonliving system can often cause a local decrease in entropy even though the net entropy must increase. What is unique to living systems is the use of enzyme-catalyzed chemical reactions to effect and maintain the decreased entropy within the cell. In recognition of these factors the late Isaac Asimov[9] distinguished life and living organisms as "characterized by the ability to effect a temporary and local decrease in entropy by means of enzyme-catalyzed chemical reactions."

One final attribute of microbial systems that should be considered is their remarkable ability to exploit available sources of chemical free energy. It is rare that an organic or inorganic reaction that yields free energy is *not* used by microbes. If it is assumed that all such reactions are capable of use by microbes, one will rarely be in error. Poindexter[10] has indicated that bacteria and fungi are particularly omnivorous and as a group can use for growth every known naturally occurring organic compound. No wonder these microbes are particularly important in composting systems.

ESTIMATING HEATS OF REACTION

It is often difficult to estimate heats of reaction for organic wastes from standard enthalpy values. Usually such wastes comprise a mixture of organics of unknown composition. In such a case, standard enthalpy values are of little use except in defining the probable range of heats of reaction. However, several experimental and empirical approaches are available for determining heats of reaction.

The heating value of an organic waste can be determined by calorimetric measurements. The quantity of heat released, however, is a function of the path followed during oxidation of the sample material. One method of determining the heat of reaction for a given chemical reaction is by means of an "open calorimeter", in which pressure is maintained constant at 1 atm. Under constant pressure conditions, heat released is equal to the enthalpy change for the reaction. Another type of calorimeter is the "bomb calorimeter", in which reactions are conducted under conditions of constant volume. Thus, heat released in a bomb calorimeter would differ somewhat from that in a constant-pressure calorimeter. Fortunately, these differences are usually small for organic materials, which release considerable energy on oxidation. Methods are also available to correct bomb calorimeter results to conditions of constant pressure.[11]

The calorimeter approach is undoubtedly the most accurate way to determine heats of reaction for unknown mixtures of organics. However, calorimetric tests are not routine for most water quality laboratories, and analysis by specialty labs equipped for this purpose is usually required. As a result, a number of approximate formulas based on both theoretical and experimental approaches have been developed.

Fair et al.[12] determined fuel values of different types of vacuum-filtered sewage sludges. A bomb calorimeter was used in these experiments, but actual heat release under constant pressure conditions would probably not differ significantly. The empirical formula describing their results is

$$Q = a[P_v(100)/(100 - P_c) - b][(100 - P_c)/100] \qquad (3.25)$$

where
Q = fuel value, Btu/lb ds
a = coefficient equal to 131 for raw and digested primary sludge, 107 for raw waste activated sludge
b = coefficient equal to 10 for raw and digested primary sludge, 5 for raw waste activated sludge
P_v = percent volatile solids in sludge
P_c = percent of inorganic conditioning chemical in sludge

Table 3.4.   Representative Chemical Analysis and Heat Content of Dry Refuse and Sewage Sludges

Constituent	Refuse (wt %)	Raw Sludge (wt %)	Digested Sludge (wt %)
Carbon	33.11	37.51	24.04
Hydrogen	4.47	5.54	3.98
Oxygen	25.36	22.56	12.03
Nitrogen	0.60	1.97	2.65
Chlorine	0.41	0.33	0.17
Sulfur	0.14	0.37	0.75
Metal	11.64		
Glass, ceramics, stone	16.23		
Volatiles @ 110°C		3.66	3.01
Ash	8.04	28.06	53.37
Total	100.00	100.00	100.00
HHV, cal/gm	3280	3910	2570

Source: Loran.[15]

A formula presented by Spoehr and Milner[13] relates the heat of combustion to the degree of reduction of the organic matter. This is a rational approach because the heat of combustion has already been shown to be significantly lower for carbohydrates than for the more reduced lipids and hydrocarbons. Products of combustion are assumed to be gaseous carbon dioxide, liquid water, and nitrogen gas. The degree of reduction for any type of organic matter is

$$R = 100[2.66(C) + 7.94(H) - (O)] / 398.9 \qquad (3.26)$$

where C, H, and O are the weight percentages of carbon, hydrogen and oxygen, respectively, on an ash-free basis. The heat of combustion is

$$Q = 127R + 400 \qquad (3.27)$$

where Q is the heat of combustion, cal/g ash-free ds. Representative ultimate analyses for domestic sludge and refuse are presented in Table 3.4.

Another formula, similar to that presented by Spoehr and Milner, is called the Dulong formula and is also useful in estimating gross heating values from the feed composition:[14]

$$Q = 145.4(C) + 620[H - (O / 8)] + 41(S) \qquad (3.28)$$

where

Q =   heat of combustion, Btu/lb ash-free ds
S =   sulfur content on an ash-free basis, weight %

Equations 3.27 and 3.28 both require an ultimate analysis of the waste to determine the percentages of carbon, hydrogen, oxygen, and sulfur. This may not be a routine test in many water quality laboratories. Equation 3.25 requires analysis only of the volatile solids (VS) content, which is easily handled. However, Equation 3.25 was developed only for municipal sludge and should not be extended to other organic wastes.

A rule of thumb that is reasonably accurate for most organics is that about 3.4 ± 0.2 kcal are released per gram COD of the waste. Because the COD test is routinely practiced by many laboratories, it allows a relatively easy approximation of the heat of combustion. Heats of reaction were calculated for the foodstuffs presented in Table 3.2. When expressed in terms of kcal/g COD, the caloric values are reasonably constant even though they vary considerably when expressed per unit weight of organic.

The reason that the heat released per unit of COD is relatively constant lies in the fact that COD is a measure of electrons transferred. During aerobic oxidation, 4 mol of electrons must be transferred for each mole of substrate oxygen demand:

$$\frac{1}{4} O_2 + 1e^- = \frac{1}{2} O^{-2}$$

Thus, substrate COD is proportional to the number of electrons transferred during aerobic oxidation. Furthermore, heat of combustion per electron transferred to a methane-type bond is relatively constant at about 26.05 kcal per electron equivalent[8] or, since O_2 has four such electrons, 104.2 kcal/mol O_2. This in turn is equal to 3.26 kcal/g COD. For aerobic oxidation, therefore, the COD unit turns out to be a measure of energy release, and it is not surprising that energy release per electron transferred is relatively constant for a wide variety of substrates.

The correlation of calorific value with COD was experimentally verified by Zanoni and Mueller[16] for a number of primary, biological, and digested sewage sludges. They developed the following correlation:

$$Q = 3.81(COD) + 28.6 \tag{3.29}$$

where

$$\begin{aligned} Q \quad &= \quad \text{heat of combustion, cal/g ds} \\ COD \quad &= \quad \text{sample COD, mg COD/g ds} \end{aligned}$$

The constant 3.81 in Equation 3.29 is somewhat higher than the 3.26 factor discussed above. This may be related to formation of nitric and sulfuric acids which release heat in the bomb calorimeter but are not included in the above analysis of electron transfers.

A general rule of thumb for sewage sludges presented by Vesilind[17] and others is to expect about 5550 cal/g (10,000 Btu/lb) of dry volatile solids. This same range was verified by Zanoni and Mueller.[16] Caution should be exercised when using this rule of thumb because it is based on having a "typical" composition of proteins, carbohydrates, and fats in the sludge. This may not always be the case, particularly if there are large inputs of industrial wastes. For example, pulp and paper wastes can contain large amounts of cellulosic material which will lower the calorific content of the sludge.

Example 3.2

Given the raw sludge composition in Table 3.4, compute the heat of combustion by the techniques described above.

Solution

1. Calculate Q by Equation 3.25. The VS content is $100 - 28.06 = 71.94\%$. Assume that inorganic chemicals, such as lime and ferric chloride, are not used for conditioning so that $P_c = 0$. Using coefficients for raw primary sludge

$$Q = 131[71.94(100)/(100-0)-10][(100-0)/100] = 8114 \text{ Btu/lb ds}$$

$$Q = 4503 \text{ cal/g ds}$$

2. Calculate Q by Equation 3.27. The percentages of C, H, and O on an ash-free basis are

	% from Table 3.4	Adjusted to Ash-Free
C	37.51	57.17
H	5.54	8.44
O	22.56	34.39
	65.61	100.00

From Equation 3.26:

$$R = 100[2.66(57.17)+7.94(8.44)-34.39]/398.9 = 46.3$$

Therefore, from Equation 3.27:

$$Q = 127(46.3)+400 = 628 \text{ cal/g ash-free ds}$$

Because the sludge is 71.94% VS:

$$Q = 6280(0.7194) = 4518 \text{ cal/g ds}$$

3. Calculate Q by Equation 3.28. The percentages of C, H, O, and S on an ash-free basis are

	% from Table 3.4	Adjusted to Ash-Free
C	37.51	56.85
H	5.54	8.40
O	22.56	34.19
S	0.37	0.56
	65.98	100.00

$$Q = 145.4(56.85)+620(8.4-34.19/8)+41(0.56) =$$

$$10,847 \text{ Btu/lb ash-free ds}$$

$$Q = 10,847(0.7194) = 7803 \text{ Btu/lb ds}$$

$$Q = 4331 \text{ cal/g ds}$$

Note that sulfur can be oxidized by certain autotrophic organisms during composting. However, the contribution to the overall energy balance is small for the sulfur contents found in most sludges and other organic wastes.

4. Calculate Q based on sample COD. Average organic composition can be determined from the weight percentages and molecular weights of the components. Consider only the C, H, and O fractions:

$$C \qquad 57.17 / 12 = 4.8$$

$$H \qquad 8.44 / 1 = 8.4$$

$$O \qquad 34.39 / 16 = 2.2$$

which gives an average composition of $C_{4.8}H_{8.4}O_{2.2}$. COD can be determined by balancing the chemical equation for oxidation to carbon dioxide and water.

$$\underset{101.2}{\overset{1g}{C_{4.8}H_{8.4}O_{2.2}}} + \underset{5.8(32)}{\overset{x}{5.8O_2}} = 4.8CO_2 + 4.2H_2O$$

$$x = 5.8(32)(1 \text{ g}) / 101.2 = 1.83 \text{ g COD/g organic}$$

Assuming an average of 3.26 kcal/g COD:

$$Q = 3260(1.83) = 6292 \text{ cal} / \text{g ash–free ds}$$

$$Q = 6292(0.7194) = 4526 \text{ cal} / \text{g ds}$$

Alternatively, Q can be estimated from Equation 3.29 as

$$COD = 1.83(0.7194) = 1.32 \text{ g COD} / \text{g ds}$$

$$Q = 3.81(1320) + 28.6 = 5058 \text{ cal} / \text{g ds}$$

5. Calculate Q from the rule of thumb for VS. Consider the organic fraction to contain 5550 cal/g VS.

$$Q = 5550(0.7194) = 3993 \text{ cal} / \text{g ds}$$

6. Note that the range of values calculated by these techniques is 3993 to 5058 cal/g ds or about ±12% of the average value. If the value estimated from Equation 3.29 is excluded, the range is about ±6%, which should be sufficiently accurate for most analyses of sludge composting systems.

One final point should be made concerning heats of reaction for mixtures of organics characteristic of most natural waste products. Actual heat release during composting is determined by those organics that actually degrade during composting. Thus, in the example above, heat release will be as calculated provided all organic components are equally degradable. If all components of the mixture are not equally degradable the energy release could vary significantly. For example, if the lipid fraction were more degradable than either the protein or carbohydrate fractions, the heat release per gram of organic actually decomposed would be greater because of the higher caloric value of the lipids. The subject of degradability is discussed further in Chapter 9.

MECHANISMS OF HEAT TRANSFER

Three distinct mechanisms of heat or energy transfer can be described: conduction, convection, and radiation. Conductive transfer is defined as the transfer of heat between two points caused by a temperature difference but without any mass movement between the points. Conduction of heat into a compost particle or from one particle in contact with another are examples. The quantitative law of heat conduction was formulated in 1822 by Fourier as a generalization of his experimental investigations. Fourier's law states

$$dQ / dt = -kdAdT / dx \qquad (3.30)$$

Fourier's equation describes the amount of heat, dQ, that passes through a plane of area, dA, in time, dt, in response to a temperature gradient, dT/dx. The proportionality constant, k, is termed the thermal conductivity and usually is expressed in units of cal/(h-cm^2-°C/cm) or Btu/(h-ft^2-°F/ft). Thermal conductivity of a substance depends on the state of the substance (solid, liquid, or gas), and for a given state it will vary somewhat with temperature. Typical values of thermal conductivity in cal/(h-cm^2-°C/cm) for a number of substances are: aluminum, 1800; stainless steel, 150; concrete, 15; water, ~5.6; water-saturated wood, 3; and corkboard, 0.40. Values between 2 and 4 have been measured for compost material.

Thermal diffusivity is related to thermal conductivity by

$$a = k / \rho\, c_p \qquad (3.31)$$

where

a = thermal diffusivity, cm^2/h
ρ = mass density, g/cm^3
c_p = specific heat, cal/g-°C

Like thermal conductivity, thermal diffusivity is a property of the substance. The term $k/\rho c_p$ appears in many heat conduction problems and is related to the diffusion coefficient used in mass transfer problems.

Solutions to Equation 3.30 are beyond the scope of this book and the interested reader is referred to specialized books devoted to heat transfer. However, many useful estimates can be made for the case of one-dimensional steady-state conditions. Integrating Equation 3.30

$$q = kA\Delta T / \Delta x \qquad (3.32)$$

where

q = the rate of heat transfer
k = average thermal conductivity (assumed constant)
Δx = the length of the flow path

Convective heat transfer results from movement of mass between a high temperature zone, where heat is accepted, to a zone of lower temperature, where heat is released. Convective heat transfer occurs as a result of fluid motion in contrast to conductive heat transfer which occurs entirely by means of intermolecular energy transfers. Movement of air within a bed of composting particles is an example of convective heat transfer. Another is the evaporation of water, its transport and condensation in another part of the compost bed.

In many cases both conduction and convection operate to transfer heat. A case in point is heat transfer from the surface of a compost pile or from the walls of a compost reactor. Mass movement of ambient air across the pile will result in convective heat loss. In response to the resulting temperature difference, heat will be conducted from the interior of the pile or across the walls of the reactor. Steady-state heat transfer in such cases is often modeled by an equation of the form

$$q = UA(T_1 - T_2) \tag{3.33}$$

where

U = overall heat transfer coefficient which includes effects of both conductive and convective heat transfer, $cal/h\text{-}cm^2\text{-}°C$

A = area perpendicular to direction of heat transfer, cm^2

T_1, T_2 = temperatures at points 1 and 2, $°C$

Estimates of overall heat transfer coefficients cannot be made entirely on the basis of theoretical arguments. Empirical relationships based on experimental data are usually required. For example, the value of U for heat loss from the surface of a compost pile would likely be a function of wind speed, relative humidity, and perhaps other factors. Even so, once the surface has cooled, further heat loss is probably limited by transport of heat (either by conduction or convection) from the interior of the pile.

A final form of heat transfer can occur as a result of radiant or electromagnetic energy exchange between two bodies of unequal temperature. Radiant energy can be transmitted in a vacuum and does not depend on direct physical contact or the movement of any fluid between the bodies. Radiant energy is always being exchanged, but a net exchange in one direction results only if there is a temperature difference between the two bodies. At temperatures below 300°C practically all radiant energy is in the infrared region of the spectrum. The reader is probably familiar with infrared satellite photographs which highlight temperature differences between objects. If composting is successful, its temperature should be greater than that of the surroundings, and some radiative losses can be expected from the surface.

Radiant energy transfer between two bodies, A and B, is described by the Stefan-Boltzmann Law:

$$q = \sigma A(T_a^4 - T_b^4)F_a F_e \tag{3.34}$$

where

σ = Stefan-Boltzmann constant, $4.87 \times 10^{-8}\ kcal/(h\text{-}m^2\text{-}°K^4)$

F_a = configurational factor to account for the relative position and geometry of the bodies

F_e = emissivity factor to account for non-black body radiation

T_a, T_b = absolute temperature of bodies A and B, $°K$

Ideal radiators emit radiant energy at a rate proportional to the fourth power of the absolute temperature. As a result radiant energy losses become much more significant at higher temperatures.

Example 3.3

Determine the heat transfer rate from 1 m² of surface at a temperature of 60°C radiating to the ambient environment at 20°C. Assume an emissivity factor of 0.9 and configurational factor of 1.0. Compare this to the conductive transport of heat through a 30 cm layer of compost with a uniform temperature gradient from 60 to 20°C. Assume a k of 4.0 cal/(cm²-h-°C/cm).

Solution

1. Estimate the radiant energy transfer using Equation 3.24:

$$q = 4.87 \times 10^{-8}(1)[273 + 60)^4 - (273 + 20)^4](0.9)(1.0)$$

$$q = 216 \text{ kcal} / \text{h}$$

2. Using Equation 3.32 the conductive heat rate per m² of surface area can be estimated as

$$q = 4.0(10^4 \text{cm})(60 - 20) / 30\text{cm} = 5.3 \times 10^4 \text{ cal} / \text{h}$$

$$q = 53 \text{ kcal} / \text{h}$$

3. Using Equation 3.33, estimate a U value corresponding to the conductive heat loss rate in part 2.

$$q = UA(T_1 - T_2)$$

$$53 = U(1\text{m}^2)(60 - 20)$$

$$U = 1.3 \text{ kcal} / (\text{m}^2 - \text{h} - °\text{C})$$

4. The estimated value of U is consistent with values used for calculating losses from insulated, concrete anaerobic digesters. The latter typically range from about 0.5 to 1.5 kcal/m²-h-°C.

When a warm compost surface is exposed to cooler ambient air, such as after turning a windrow or pile, a rapid surface temperature drop should result from radiative and convective losses with the outside ambient air. Once the outer surface is cooled, further heat loss should be limited by conductive transport from the pile interior. Convective transport from air moving through the pile also occurs, but it can be accounted for from the temperature and quantity of output gases.

THERMAL PROPERTIES OF COMPOST

Before leaving this discussion of thermodynamics, the thermal properties of compost will be examined. Unfortunately, few if any tests have been reported on sludge, refuse, or yard waste composts. However, Mears et al.[18] conducted some excellent work on compost produced from swine wastes. Their results should be reasonably applicable to other composts as well. Based on the previous discussions, the thermal properties of importance include the specific heat at constant pressure and thermal conductivity.

Mears et al. used a calorimetric technique to determine specific heat values. A wide-mouth Thermos® was used as a calorimeter. To account for heat lost to the thermos body during the experiment, the Thermos® was tested with a fluid of known specific heat to determine the fluid equivalent of the Thermos® as a function of depth. Water was the fluid of choice in these studies. Material to be tested was then placed in the calorimeter and a known quantity of hot water was added. The slurry was mixed for 5 min before taking final temperature measurements. Mixing of compost and water at the same temperature did not result in any measurable rise in temperature, indicating that any heat released in the wetting of compost was insignificant. This was not true for oven-dried samples, however, which did exhibit a significant heat of wetting. Therefore, if one were to measure the specific heat of very dry materials by the method of mixtures, the heat of wetting would have to be considered.

Thermal conductivity was measured by placing the sample in a long, thin-walled cylinder constructed of material with a high thermal conductivity, in this case aluminum. Insulating plugs were placed in the ends of the cylinder to restrict heat transfer in the axial direction. The cylinder, with its sample, was then placed in a hot water bath. Subsequent heat transfer was described as occurring in the radial direction into a cylinder of infinite length. Analytical solutions to this problem have been developed. By measuring temperature differences between the center of the cylinder of compost and the hot water bath vs time, thermal conductivity can be calculated. One problem with this technique is the fact that convective currents can arise from differential temperatures, and hence densities, of gases in the void spaces between particles. This is a problem encountered in measuring thermal conductivity of any fluid, and the effects of convective heat transfer cannot be entirely avoided. Apparently, these effects were minimized to a sufficient degree in the test equipment. Measurements of thermal conductivity of known materials containing void spaces, such as sand, in the same device were within 4% of accepted values.

Using these techniques, Mears et al. determined the thermal properties of composting material composed of swine waste amended with 5% by weight straw (10% by volume). The swine waste consisted of feces, uneaten feed, bones, plastics, paper, and glass. Large pieces of bone, plastic, etc. were excluded from the analysis, but usually totaled only 2 to 10% of the sample weight. A windrow composting technique was used and representative samples collected at weekly intervals. Results are presented in Figure 3.3 and Table 3.5.

In all cases the specific heat and thermal conductivity varied linearly with moisture content. Extrapolation of the curves to 100% moisture content gave results very close to the actual specific heat of water, 1.0 cal/g-°C, and thermal conductivity, 5.62 cal/(cm²-h-°C/cm). This means that in a thermodynamic analysis the water and solid fractions can be treated separately. The water portion of compost material can be assumed to have the thermal properties of water. The solid fraction can be assumed to have thermal properties equal to the extrapolated value at zero moisture. It also means that a compost sample can be tested at a single moisture content and the entire relationship established by using this point and the value for 100% water.

From the data of Table 3.5, it appears that both specific heat and thermal conductivity of the solid fraction increase with compost age. Mears et al.[18] speculated that this may be related to an increase in the proportion of inorganic material (ash) as a result of organic decomposition. Specific heat and thermal conductivity of inorganic components are generally greater than values for the organic fraction.

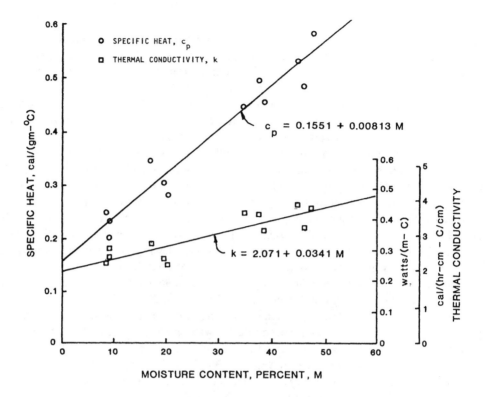

Figure 3.3.   Specific heat, c_p, and thermal conductivity, k, of compost as a function of moisture content. Compost was produced from swine waste blended with about 5% straw by weight. Data shown correspond to Sample No. 6 from Table 3.5. From Mears et al.[18]

Table 3.5.   Specific Heat and Thermal Conductivity of Swine Compost

Sample No.	Age of Windrow (days)	Zero Moisture Intercept, I	Slope of Curve, S	100% Moisture Intercept, I + 100 S
Specific Heat, cal/g-°C[a]				
1	0	0.0550	0.00940	0.9951
2	7	0.0699	0.00921	0.9909
3	14	0.0771	0.00922	0.9991
4	20	0.0834	0.00906	0.9894
5	30	0.1289	0.00834	0.9629
6	35	0.1551	0.00813	0.9831
Thermal Conductivity, cal/(h-cm²-°C/cm)[b]				
1	0	0.875	0.0498	5.854
2	7	1.086	0.0473	5.816
3	14	1.323	0.0440	5.726
4	20	1.341	0.0403	5.374
5	30	1.854	0.0383	5.682
6	35	2.071	0.0341	5.476

Source: Mears et al.[18]

[a] Specific heat of water = 1.00 cal/g-°C.
[b] Thermal conductivity of water = 5.62 cal/(h-cm²-°C/cm).

As a result of these studies it can be concluded that compost will have a low specific heat that increases with increasing moisture content. Although a function of moisture, thermal conductivity is relatively low over the entire range of moisture contents. Therefore, a large compost pile will tend to be self-insulating, and heat losses by conduction should generally be small. In the case of wet substrates, values of specific heat and thermal conductivity for the mixture will likely decrease during composting since the loss of moisture should overshadow effects of increased ash content. Finally, thermal properties are likely a function of the particular compost material in question. Given the basic organic nature of composts, however, values should not be significantly different from those reported here. The laboratory techniques used by Mears et al. to measure thermal properties are very straightforward and it is hoped that the approach will be applied to other composts in the future.

SUMMARY

Thermodynamics is the study of energy and its transformations. It provides an underlying current for the understanding of physical, chemical, and biological systems. It places distinct limits on the energy transformations within systems as small as a single microbe or as large as the universe. One of the central themes of this book is the application of thermodynamic principles to composting systems, which by their nature are composed of physical, chemical, and biological processes. No single science unifies the diverse aspects of composting as does thermodynamics.

The First and Second Laws of Thermodynamics form the foundation upon which the science is based. Both are accepted as first principles that have been repeatedly upheld by human observations. The First Law states that energy can be neither created nor destroyed. In engineering terms it is commonly referred to as the Law of Conservation of Energy. The concepts of heat, work, internal energy, and enthalpy are related to the First Law. The Second Law resulted from a search to explain the direction in which spontaneous processes would occur. This led to the realization that all spontaneous changes in an isolated system occur with an increase in entropy or randomness.

The concept of free energy was developed from the First and Second Laws. Free energy gives the useful work which can be derived from a chemical reaction that occurs under constant pressure and temperature conditions. This is extremely useful because most microbial processes occur under such conditions. Therefore, a measure exists of the useful energy available from the feed substrate being used by a microbial population.

All chemical reactions have a standard free energy change, measured with all reactants and products at unit activity (approximately a 1 M concentration). Spontaneous chemical reactions proceed in the direction of decreasing free energy. If the free energy change is zero the reaction is at equilibrium. The standard free energy change can be related to the equilibrium constant for the reaction and can be adjusted for the effect of product and reactant concentrations which differ from the standard concentration.

The effect of temperature on the rate constant for chemical reactions can be estimated from the Arrhenius equation. Various simplified forms of the Arrhenius relationship are currently used in engineering practice.

Heats of combustion vary from about 2100 to 9300 cal/g for the three major foodstuffs: proteins, carbohydrates, and lipids (fats). Lipids generally contain about twice the energy per gram as proteins or carbohydrates. Expressed on a COD basis, however, most organics have a heat of combustion of about 3.26 kcal/g COD of the organic. It is often difficult to estimate heats of reaction for organic wastes from standard thermodynamic tables because the wastes are likely to be composed of a mixture of compounds of unknown composition. Open and bomb calorimetric techniques can be used experimentally to determine the heats of combustion for such unknown materials. A number of empirical equations are also available which yield reasonably consistent results and require only routine laboratory analysis.

The water and solid fractions of a composting material can be treated as separate components from a thermodynamic standpoint. The water fraction exhibits the thermal properties of water, while the solid fraction has the properties of the dry solid. This result is significant to the development of energy balances presented in later chapters.

REFERENCES

1. Lehninger, A. L. *Bioenergetics* (New York: W. A. Benjamin, Inc., 1965).
2. Perry, R. H. and Chilton, C. H. Eds. *Chemical Engineer's Handbook,* 5th ed. (New York: McGraw-Hill Book Co., 1973).
3. *CRC Handbook of Chemistry and Physics,* 50th ed. (Cleveland, OH: CRC Press, 1970).
4. Lehninger, A. L. *Biochemistry* (New York: Worth Publishers, Inc., 1970).
5. Sawyer, C. N. and McCarty, P. L. *Chemistry for Environmental Engineering* (New York: McGraw-Hill Book Co., 1978).
6. Burd, R. S. "A Study of Sludge Handling and Disposal," Dow Chemical Co. report to the FWPCA, Department of Interior, Pub. WP-20-4 (1968).
7. Olexsey, R. A. "Thermal Degradation of Sludges," paper presented at Symposium on Pretreatment and Ultimate Disposal of Wastewater Solids, Rutgers University, NJ, May 1974.
8. Bailey, J. E. and Ollis, D. F. *Biochemical Engineering Fundamentals* (San Francisco, CA: McGraw-Hill, 1977).
9. Asimov, I. *Life and Energy* (New York: Avon Books, 1962).
10. Poindexter, J. S. *Microbiology — An Introduction to Protists* (New York: The Macmillan Co., 1971).
11. Lee, J. F. and Sears, F. W. *Thermodynamics* (Reading, MA: Addison-Wesley Publishing Co., Inc., 1963).
12. Fair, G. M., Geyer, J. C., and Okun, D. A. *Water and Wastewater Engineering, Vol. 2, Water Purification and Wastewater Treatment and Disposal* (New York: John Wiley & Sons, Inc., 1968).
13. Spoehr, H. A. and Milner, H. W. "The Chemical Composition of Chlorella; Effect of Environmental Conditions," *Plant Physiol.* 24:120 (1949).
14. Corey, R. C., Ed. *Principles and Practices of Incineration* (New York: Wiley-Interscience, 1969).
15. Loran, B. I. "Burn that Sludge," *Water Wastes Eng.* (October 1975).
16. Zanoni, A. E. and Mueller, D. L., "Calorific Value of Wastewater Plant Sludges," *J. Environ. Eng. Di., ASCE,* 108(EE1) (February 1982).
17. Vesilind, P. A. *Treatment and Disposal of Wastewater Sludges,* 2nd ed. (Ann Arbor, MI: Ann Arbor Science, 1979).
18. Mears, D. R., Singley, M. E., Ali, C., and Rupp, F. "Thermal and Physical Properties of Compost," in *Energy, Agriculture and Waste Management,* W. J. Jewell, Ed. (Ann Arbor, MI: Ann Arbor Science Publisher, Inc., 1975).

Chapter 4

Biological Fundamentals

INTRODUCTION

This chapter provides a brief introduction to the biology of organisms responsible for composting, the bacteria and fungi. The discussion should acquaint the reader with sufficient terminology to allow an understanding of the fundamental biology involved in composting. It will not be a detailed discussion from the viewpoint of the microbiologist, but it will stress the energy requirements and overall chemical reactions mediated by microbes.

The distinction of living things into plant and animal kingdoms is common to our experience. With the discovery of and experimentation with microorganisms, however, distinctions became less well defined. Microbes seemed to have characteristics of both the plant and animal kingdoms. In 1866 Ernst H. Haeckel, a German zoologist, proposed the kingdom Protista, which included bacteria, algae, protozoa, and fungi. All unicellular (single-celled) organisms and those containing multiple cells of the same type are included in this classification. Viruses, which are noncellular, were not discovered until 1882 and so escaped Haeckel's original classification. For convenience, viruses are often included in the Protist kingdom. A breakdown of the Protist kingdom into groupings convenient for our discussion is presented in Figure 4.1.

All bacteria and blue-green algae are procaryotic cell types, while other protists and all other living organisms are of the eucaryotic cell type. The distinction is based on differences in cellular anatomy. Procaryotic cells contain a single DNA strand, the nuclear substance is not enclosed within a distinct membrane, and nuclear division is less complex than in eucaryotes. Procaryotes are thought to be more primitive organisms on the evolutionary scale.

Microbes of importance in composting include bacteria and fungi. All other groups are of minor significance. It should be noted that some workers in composting include actinomycetes as a separate group distinct from bacteria and fungi. There appears to be considerable confusion on this point, because actinomycetes have been at times classified as bacteria, fungi, and even a separate phylogenetic line. Actinomycetes are filamentous in form like many of the fungi, but they have a procaryotic cell structure like the bacteria. As a class they are active

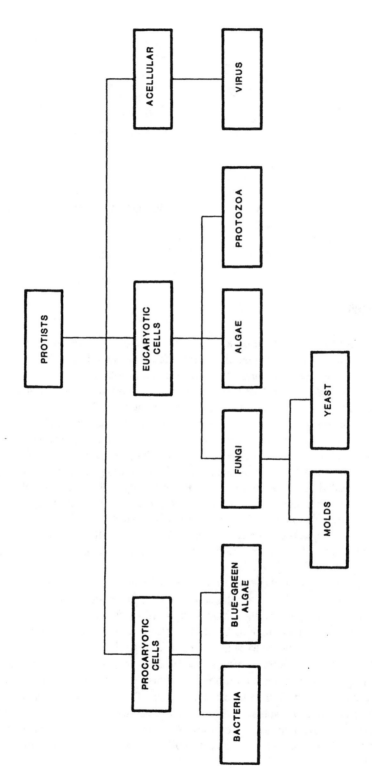

Figure 4.1.   Classification of microbes in the Protist kingdom.

in degradation of insoluble, high molecular weight organics, such as cellulose, chitin, proteins, waxes, paraffins, and rubber. As such, they should be important in composting systems. For convenience they will be considered along with bacteria, although the difficulty of classification should be recognized.

METABOLIC CLASSIFICATIONS

Metabolic distinctions between organisms provide a useful tool to understand both the effect of an organism on its surroundings and the environment necessary for proper growth of the organism. Perhaps the most basic distinction is between aerobic, anaerobic, and anoxic metabolism. Aerobic refers to respiration with oxygen. Consider the aerobic oxidation of glucose as follows:

$$\overbrace{C_6H_{12}O_{6_{aq}} + 6O_{2_g} \rightarrow 6CO_{2_{aq}}}^{-24e^-} + 6H_2O_l \qquad \Delta G_R^o = -677 \text{ kcal / mol}$$

with $+24e^-$

The reaction is of the oxidation-reduction type, because electrons are transferred from glucose and accepted by oxygen. Thus, oxygen is reduced while carbon is oxidized. In this case, oxygen is referred to as the electron acceptor. All organisms that use oxygen as an electron acceptor are termed aerobic. All macroscopic organisms, and many microscopic ones as well, are obligate aerobes. In other words, oxygen and only oxygen will serve as an electron acceptor.

Among the microbes, other compounds can be used as electron acceptors. The most notable of these are oxidized inorganic compounds of nitrogen and sulfur such as nitrate (NO_3^-), nitrite (NO_2^-), and sulfate (SO_4^{2-}). Carbon dioxide can also be used as an electron acceptor, and it is usually reduced to methane. Metabolism with these electron acceptors is termed anoxic. Many microbes can function with one or more of these electron acceptors and are termed facultative. The oxidation of glucose, using sulfate as an electron acceptor, is given as

$$\overbrace{C_6H_{12}O_{6_{aq}} + 3SO_{4_{aq}}^= + 6H_{aq}^+ \rightarrow 6CO_{2_{aq}}}^{-24e^-} + 6H_2O_l + 3H_2S_{aq} \qquad \Delta G_R^o = -107 \text{ kcal / mol}$$

with $+24e^-$ (pH = 7)

Note that considerably less energy is available compared to aerobic metabolism. Being efficient chemical factories, microbes will first use those electron acceptors that provide the greatest energy yields. The order of use is generally O_2, NO_3^-, and SO_4^{2-}. This is fortunate, because production of H_2S can be prevented by maintenance of aerobic conditions.

But what if all available electron acceptors have been used? Will electrons accumulate until the biochemical machinery grinds to a halt? The answer is no. To explain this consider the following reaction with glucose:

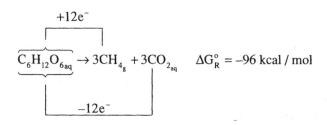

$$C_6H_{12}O_{6aq} \rightarrow 3CH_{4g} + 3CO_{2aq} \qquad \Delta G_R^\circ = -96 \text{ kcal / mol}$$

Electrons removed from CO_2 are ultimately accepted by methane. Electrons are transferred and the reaction is of the oxidation-reduction type, but the acceptor and donor of electrons originate from the same molecule. Such reactions are termed fermentations, following Louis Pasteur who called fermentations "life without air". Some confusion in nomenclature still persists between different disciplines. Most industrial microbial conversions are called fermentations even though many of the processes are aerobic. Similarly, some composting reactors are called "fermenters" even though they usually take elaborate measures to assure aerobic conditions.

Anaerobic metabolism is considerably more complex than the above fermentation reaction with glucose would indicate. Conversions are frequently mediated by a variety of organisms operating in series. In other words, the product of one organism is used as substrate by the second organism, and so on. A variety of intermediate products can be formed along the way, including low molecular weight organic acids (e.g., acetic and propionic), alcohols, and aldehydes. In sanitary engineering practice these organisms are collectively referred to as first stage or acid formers, although a variety of other intermediates can be formed. McCarty[1] has shown that with a complex starting substrate such as municipal sludges as much as 72% of the organics (measured as COD) pass through acetic acid. First-stage organisms are often important to industrial fermentation because of the end products formed. Ethanol is an obvious example of such an end product, resulting from the fermentation of sugars by selected yeast cultures.

If fermentation is allowed to go to completion, the final end products will always be methane and CO_2. At this point as much energy has been extracted from the substrate as is possible under anaerobic conditions. Microbes that convert intermediate products to methane and CO_2 are termed methane-formers and are usually strict or obligate anaerobes.

The complexity of end products that result from anaerobic metabolism is often confusing to the student. Aerobic metabolism seems somewhat simpler because end products are usually carbon dioxide and water. In fact, the biochemical pathways are remarkably similar between the two groups. To help understand the reason for such variety of first stage end products, a number of possible reactions were assembled in Table 4.1. In all cases, the starting material is glucose and the flow of electrons is balanced. There is a rather large range of free energy values between the reactions listed. In a mixed microbial population, reactions yielding more free energy predominate. Production of a single intermediate such as ethanol requires sterile substrate preparation and pure or nearly pure cultures of selected microbes, which yield the desired end product.

Table 4.1.   Possible Anaerobic Fermentations of Glucose

End Products	Reactions	ΔG_R° @ pH =7 (kcal/mol glucose)
Acetate	$C_6H_{12}O_6 \rightarrow 3CH_3COO^- + 3H^+$	−78.55
Propionate, acetate, H_2	$C_6H_{12}O_6 \rightarrow CH_3CH_2COO^- + CH_3COO^- + 2H^+ + CO_2 + H_2$	−70.84
Butyric, H_2	$C_6H_{12}O_6 \rightarrow CH_3CH_2CH_2COOH + 2CO_2 + 2H_2$	−61.7
Ethanol	$C_6H_{12}O_6 \rightarrow 2CH_3CH_2OH + 2CO_2$	−51.14
Lactate	$C_6H_{12}O_6 \rightarrow 2CH_3CH(OH)COO^- + 2H^+$	−49.5
Methanol	$C_6H_{12}O_6 + 2H_2O \rightarrow 4CH_3OH + 2CO_2$	−21.42

Table 4.2.   Metabolic Categories Based on Carbon and Energy Sources

Types of Nutrition	Principal Source of Energy	Principal Source of Carbon	Occurrence
Photoautotroph	Light	CO_2	Some bacteria, most algae, higher plants
Photoheterotroph	Light	Organics	Some algae, some bacteria
Lithoautotroph (chemoautotroph)	Inorganic oxidation-reduction reactions	CO_2	Some bacteria
Organoheterotroph[a] (chemoheterotroph)	Organic oxidation-reduction reactions	Organics[b]	Higher animals, protozoa, fungi, and most bacteria

[a] In saprophytic nutrition, nonliving (inanimate) organic material is used; in parasitic or predatory nutrition, some or all of the materials used are animate.
[b] A few bacteria classified in this group are able to use CO_2 as a carbon source although this is not an energetically favorable pathway.

Metabolic classification of organisms can be continued by considering the carbon and energy sources used by the organism. Carbon accounts for about 50% of the dry mass of most organisms and is needed to synthesize the variety of organic molecules used in the structure and machinery of the cell. Two sources of carbon are available: carbon in the form of organic molecules, used by heterotrophs, and the carbon present in carbon dioxide, used by autotrophs.

Three distinct energy sources are available to organisms: organic oxidation-reduction reactions, inorganic oxidation-reduction reactions, and the energy available in light. Organisms that use organic reactions for energy are termed organotrophs; those that use inorganic reactions are lithotrophs. Only bacteria are capable of using the energy of inorganic reactions. Organisms that use light for energy are phototrophs. By comparing the carbon and energy sources, four nutritional categories can be distinguished as shown in Table 4.2. One might suspect that with two carbon sources and three energy sources a total of six nutritional categories might be described. As far as it known today, however, all lithotrophs are also autotrophs. In other words, all lithotrophs obtain carbon from carbon dioxide. Likewise, all organotrophs will use organic carbon for synthesis. Therefore, the practical number of nutritional categories is reduced to four.

Although photoautotrophs are not significant to composting, this discussion of nutritional patterns would not be complete without a brief description of the pattern of electron flow for these organisms. Light is the ultimate source of energy for phototrophs. As such they do not depend on organic oxidation-reduction reactions in the same manner that organoheterotrophs do. However, electrons are still needed to reduce the cell carbon source, carbon dioxide, to the level needed for construction of organic molecules. These electrons are usually taken from the oxygen of water as indicated by the following simplified equation for photosynthesis:

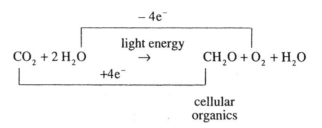

cellular
organics

Four electrons taken from oxygen in the water molecule are used to reduce the carbon dioxide. While simplified, this reaction is typical of the photosynthetic activity of algae and all higher plants.

Certain photosynthetic bacteria, notably the green and purple sulfur bacteria, are capable of extracting electrons from other sources such as sulfide, sulfur, H_2, and thiosulfate ($S_2O_3^{2-}$). A simplified equation representing the use of H_2S as an electron source is

$$-4e^-$$
$$CO_2 + 2H_2S \xrightarrow[+4e^-]{\text{light energy}} CH_2O + 2S° + H_2O$$

The reader might wonder why some bacteria are classified as photoautotrophs along with the algae and high plants. If bacteria are photosynthetic should they not then be called algae? The distinction centers on the type of chlorophyll contained in the cells. All algae, including the blue-green algae and all higher plants, contain the photosynthetic pigment chlorophyll-a, a particular type of chlorophyll, along with other light-sensitive pigments. Photosynthetic bacteria are procaryotic and do not contain chlorophyll-a. In addition, bacterial photosynthesis does not result in evolution of molecular oxygen, because electrons are obtained from sources other than water. Hence, a distinction is made between photosynthetic bacteria and all other photosynthetic organisms.

Now that these metabolic distinctions have been made, we are in a position to explore the types of reactions observed in the microbial world. Respiration using various reductants and oxidants is presented in Table 4.3. The reactions are indicative of those mediated by various groups of microbes. After studying Table 4.3 the reader should be sufficiently familiar with nutritional pathways to understand subsequent material.

BACTERIA

Bacteria are the smallest living organisms known. Bacteria are typically unicellular, but multicellular associations of individual cells are also common. They may exist in a number of morphological forms including spheres (cocci), rods (bacillus), spirals (spirillum), and a variety of intermediate forms such as comma-shaped (vibrio) and spindle-shaped (fusiform). Most bacteria reproduce by binary fission, division into two identical daughter cells. Sexual reproduction can also occur in certain cases.

Table 4.3.   General Types of Lithoautotrophic and Organoheterotrophic Metabolism

Reductant	Oxidant	Products	Carbon Source	Respiration	Metabolic Category	Type, Representative Organism
NH_4^+, NH_3	O_2	NO_2^-, H_2O	CO_2	Aerobic	Lithoautotroph	Nitrifying bacteria, *Nitrosomonas*
NO_2^-	O_2	NO_3^-	CO_2	Aerobic	Lithoautotroph	Nitrifying bacteria, *Nitrobacter*
S^{2-}, H_2S	O_2	SO_4^{2-}	CO_2	Aerobic	Lithoautotroph	Sulfur-oxidizing bacteria, *Thiobacillus, Thiothrix, Beggiatoa*
Fe^{2+}	O_2	Fe^{3+}, H_2O	CO_2	Aerobic	Lithoautotroph	Iron-oxidizing bacteria, *Ferrobacillus*
Fe^{2+}	NO_3^-	Fe^{3+}, N_2, H_2O	CO_2	Anoxic	Lithoautotroph	
H_2	O_2	H_2O	CO_2	Aerobic	Lithoautotroph	Hydrogen bacteria
H_2	CO_2	CH_4, H_2O	CO_2	Anaerobic	Lithoautotroph	CO_2-reducing bacteria, *Methanobacterium*
Organics	O_2	CO_2, H_2O	Organics	Aerobic	Organoheterotroph	Many bacteria
Organics	NO_3^-	CO_2, H_2O, N_2	Organics	Anoxic	Organoheterotroph	Denitrifying bacteria
Organics	SO_4^{2-}	CO_2, H_2O, H_2S	Organics	Anoxic	Organoheterotroph	Sulfur-reducing bacteria, *Desulfovibrio*
Organics	Organics	Many intermediates, see Table 4.1 CH_4 + CO_2 final products	Organics	Anaerobic	Organoheterotroph	Acid-forming bacteria, methane-forming bacteria

Spherical bacterial cells are usually on the order of 0.5 to 3.0 μm in diameter. Rod-shaped cells are typically 0.5 to 1.5 μm in width and 1.5 to 10 μm in length. Because of their small size the bacteria have a very high surface-to-volume ratio. This allows for rapid transfer of soluble substrates into the cell and high rates of metabolic activity. As a result, bacteria will usually predominate over other larger microbes in aqueous substrate solutions.

Accumulations of certain unicellular bacteria often occur after cell division, particularly with spherical cells. Clumps, sheets, and chains of cells can thus be formed. These accumulations are not regarded as true multicellular states because (1) they are not a regular feature of the life history of the cell and (2) metabolism and development of an individual cell is not affected by its presence in such an association.

Poindexter[2] described three types of multicellular states characteristic of certain bacteria: trichomes, microcolonies containing budding filaments, and mycelia. A trichome is a long, multicellular thread or chain of undifferentiated cells that remains intact indefinitely. In some cases a tubular sheath surrounds the chain of cells. Cells within the trichome continue to reproduce by cell division. Reproduction of the trichome itself occurs by cutting the thread into two subunits or by releasing single cells to begin a new multicellular structure. Flagellated, nonflagellated, and gliding bacteria are capable of such associations. The sulfur oxidizers *Beggiatoa* and *Thiothrix* are samples of trichome-forming, lithoautotrophic bacteria.

Certain microbes reproduce by budding (typically yeast and some bacteria). Mature cells give rise to one or more daughter cells, which are initially much smaller than the mother cell, but grow and eventually divide from the mother cell. In some bacteria a filamentous outgrowth occurs from which a bud develops at the end. More filaments can develop from the newly formed bud, eventually developing into a microcolony of cells linked by filamentous appendages (Figure 4.2).

Perhaps the most elaborate multicellular arrangement occurs among the actinomycetes. A single reproductive cell gives rise to a chain of cells which branches extensively. Eventually, a three-dimensional matrix of linked and branched filaments develops, termed mycelia. The general structure resembles that of molds, but there are several distinctive differences. Filaments of the actinomycetes are usually 1 to 5 μm in diameter, compared to the larger molds, which range around 10 to 20 μm. Furthermore, actinomycetes have a procaryotic structure, compared to the eucaryotic fungi, as well as other biochemical and sexual differences. They are, therefore, usually classified as true branching bacteria and sometimes referred to as moldlike bacteria.

The subject of classification brings up the question of taxonomic nomenclature. A sequence of taxonomic categories is employed, the idea being to group related organisms at various levels of similarity. The nomenclature commonly used in biological sciences is as follows:

Species	Organisms of one and the same kind
Genus	A group of related species
Tribe	A group of related genera
Family	A group of related tribes or genera
Order	A group of related families
Class	A group of related orders
Phylum	A group of related classes
Kingdom	A group of related phyla

In some cases additional taxa are employed, particularly where there are a large number of species.

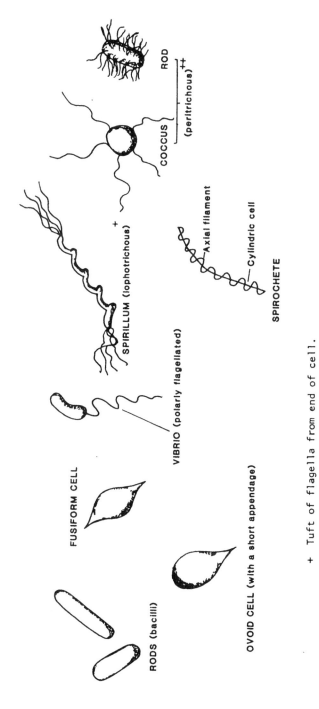

Figure 4.2A. Bacterial morphology — unicellular forms. From Poindexter.[2]

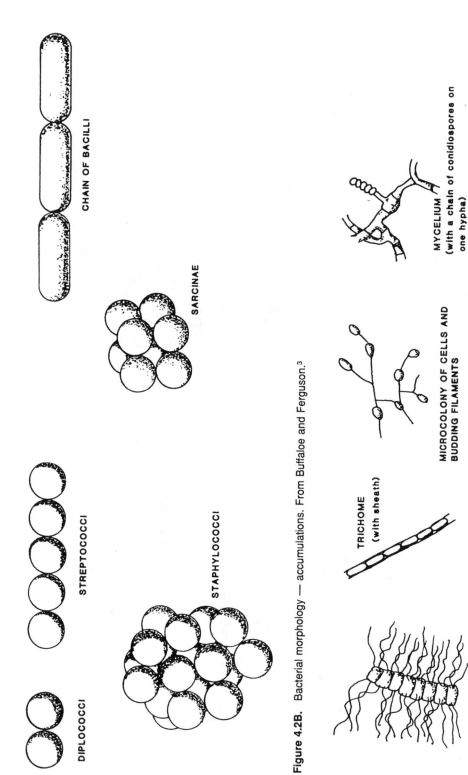

Figure 4.2B. Bacterial morphology — accumulations. From Buffaloe and Ferguson.[3]

Figure 4.2C. Bacterial morphology — multicellular forms. From Poindexter.[2]

Classification of bacteria as to genus and species is a very difficult task partly because of the extremely small size of the microbe. This means that the form (morphology) of the organism is not sufficient for complete classification. Newer classifications take into account the type of nutrition and respiration, response to chemical tests such as the gram stain, composition of the DNA molecule, sensitivity to select bacterial viruses (bacteriophage), and other factors. The most widely accepted system of classification is that in *Bergey's Manual of Determinative Bacteriology*.[4] In this system the procaryotes are divided into three classes: Class 1, Schizophyceae, contains all of the blue-green algae; Class 2, Schizomycetes, includes the entire group of "true" bacteria; and Class 3, Microtatobiotes, includes obligate intercellular parasites. Some distinctive characteristics of the orders of bacteria in Classes 2 and 3 are shown in Table 4.4.

A summary of the typical composition of bacteria, yeasts, and molds is presented in Table 4.5. The cells and their associated slime layers are about 80% water and 20% dry matter, of which about 90% is organic. Several empirical formulations for the organic fraction have been determined for mixed populations. For both aerobic and anaerobic bacteria these generally range from $C_5H_7O_2N$ to $C_5H_9O_3N$. The formulations do not vary significantly and will be useful in later analysis.

Some bacteria are capable of producing dormant forms of the cell, which are more resistant to heat, radiation, and chemical disinfection. An endospore is a thick-walled, relatively dehydrated unit formed within a bacterium and released upon cell lysis. Cysts and microcysts are formed from an entire cell that develops a thickened cell wall. Mycelial bacteria can produce spores (exospores) that can survive prolonged periods. Of the three types of dormant forms, endospores are generally more stable under adverse conditions that cysts or exospores. These forms of bacteria become significant when requirements for heat inactivation are considered (Chapter 5).

A bewildering variety of bacterial organisms can be isolated from composting material. Identification of the bacterial species present is generally an academic interest and few such studies have been conducted. It is known that mesophilic bacterial types dominate in the early stages of composting but yield to thermophilic types as temperatures increase above about 40 to 50°C. Species of the actinomycete genera, including *Micromonospora, Streptomyces,* and *Actinomyces,* can be isolated regularly from composting material. Golueke[7] indicates that actinomycetes become detectable visually in undisturbed piles near the end of the composting process. They appear as a blue-gray to light green powdery or filamentous layer in the outer 10 to 15 cm of the pile. If the compost is mixed mechanically such large colonies will not occur. For example, visual evidence of filamentous colonies is usually limited in windrow composting where mechanical action is frequent. Where mixing action is less intense, such as in the aerated static pile system, filamentous growths can become established to the point where visible colonies are formed.

Poincelot[8] analyzed the density of various microbes as a function of temperature during composting. His results are summarized in Table 4.6. Bacteria (including the actinomycetes) are usually present in larger numbers than the fungi in all temperature ranges tested. In most cases, bacterial numbers were about 100 or more times those of fungal species. This is consistent with intuitive estimates by Golueke,[7] who ascribed at least 80 to 90% of the microbial activity during composting to bacteria.

FUNGI

Fungi are of the eucaryotic cell type and are organoheterotrophic. The vast majority are saprophytic, decomposing organic matter in soil and aquatic environments. They are ubiqui-

Table 4.4.   Characteristics of the Orders of Bacteria

Order	Description	Motility	Mode of Reproduction	Representatives	Remarks
Class II — Schizomycetes					
Pseudomonadales	Many metabolic types; photosynthetic lithoautotrophs; organoheterotrophs; unicellular	Generally motile by flagella	Binary fission	*Pseudomonas, Nitrosomonas, Nitrobacter, Thiobacillus, Acetobacter*	Gram-negative, some form trichomes; widely distributed in soil and fresh- and saltwater; do not produce endospores; some characterized by the variety of metabolized substrates; includes green and purple sulfur, nitrifying, sulfur-oxidizing, sulfur-reducing bacteria
Chlamydobacteriales	"Sheathed bacteria", filaments enclosed within a sheath	Cells released from sheath called swarm cells and are motile by flagella	Usually binary fission	*Sphaerotilus*	Inhabit fresh and marine waters; *Sphaerotilus natans* associated with water heavily polluted with organics
Hypomicrobiales	"Budding bacteria", characterized by budding type of reproduction	Some motile by flagella (polar)	Budding and longitudinal fission	*Hypomicrobium*	Found in all types of aquatic environments and sewage
Eubacteriales	"True bacteria", all are organoheterotrophs, most cocci or bacillus forms; unicellular	Some motile by flagella (peritrichous)	Binary fission	*Escherichia, Syaphylococcus, Lactobacillus, Azotobacter, Rhizobium*	Ubiquitous in nature, some produce heat-resistant endospores; order includes many genera common to sanitary engineering such as *Escherichia, Enterobacter, Proteus, Salmonella, Shigella, Klebsiella*
Caryophanales	"Filamentous bacteria", disklike cells in trichomes	Motile by flagella (peritrichous)	Fission and spores	*Caryophanon*	Cells usually large, measuring 20 to 30 μm in length; cells occur in trichomes; present in water and decomposing organic material
Actinomycetales	"Moldlike bacteria", branching bacteria;	Generally nonmotile	Mostly by sporulation	*Actinomyces, Mycobacterium,*	Prevalent in decomposing organic matter; *A. israeli* causes human actinomycosis;

	some produce mycelia			*Streptomyces, Micromonospora*	*Streptomyces* are major producers of antibiotics; *M. tuberculosis* causes tuberculosis
Beggiatoales	"Gliding bacteria"	Gliding motion without flagella caused by waves of contraction	Binary fission	*Beggiatoa, Thiothrix*	Order contains both chemoautotrophs and chemoheterotrophs; filamentous with rod or coccoid cells in trichomes; often referred to as algalike because of similar motility to blue-green algae; *Beggiatoa* and *Thiothrix* are sulfide oxidizers
Myxobacterales	"Slime bacteria", often termed protozoalike due to similarities with amoeba protozoa; cells may swarm and form fruiting bodies and microcysts	Gliding motion without flagella	Binary fission and microcyst formation	*Myxococcus*	Found in soil, compost, manure, rotting wood; capable of degrading complex substrates such as cellulose and bacterial cell walls; many are predatory on other bacteria; do not form trichomes, mycelia or endospores; all are organoheterotrophs
Spirochaetales	"Spiral bacteria", elongate spiral cells	Rotary and flexing motion, no flagella	Binary fission	*Treponema pallidum* (syphillis)	Large cells from 6–500 µm long; normally occur in sewage, stagnant, fresh- or saltwater; many are harmless saprophytes while others cause diseases of man and animals such as syphilis and leptospirosis
Mycoplasmatales	Pleuropneumonialike organisms (PPLO); extremely variant morphologic form	Nonmotile	Fragmentation and formation of elementary bodies	*Mycoplasma*	Extremely small cells, smallest ranging from 0.1–0.2 µm; organisms lack cell wall; a few harmless saprophytic species are known; *M. pneumonia* causes primary atypical pneumonia
Class III — Microtatobiotes[a]					
Rickettsiales	Obligate intracellular parasites	Nonmotile	Binary fission	*Rickettsia*	Very small rods or cocci; typically parasites of arthropods such as fleas, lice, ticks; often pathogenic to man; *R. prowazekii* causes typhus; others cause rocky mountain spotted fever and Q fever; order includes the smallest living cells

Source: Bergey's Manual of Determinative Bacteriology, 7th ed.

a Virus particles are sometimes included in the Microtatobiotes as the order Virales.

Table 4.5.   Typical Composition of Organic and Inorganic Fractions of Microbes

Component[a]	Bacteria		Yeasts		Molds	
	Avg	Range	Avg	Range	Avg	Range
Organic constituents[b] (%dry wt)						
Carbon	48	46–52	48	46–52	48	45–55
Nitrogen	12.5	10–14	7.5	6–8.5	6	4–7
Oxygen		22–28				
Hydrogen		5–7				
Protein	55	50–60	40	35–45	32	25–40
Carbohydrate	9	6–15	38	30–45	49	40–55
Lipid	7	5–10	8	5–10	8	5–10
Nucleic acid	23[c]	15–25	8	5–10	5	2–8
Ash	6	4–10	6	4–10	6	4–10

	Bacteria	Fungi	Yeast
Inorganic constituents[d] (g/100 g dry wt)			
Phosphorus	2.0–3.0	0.4–4.5	0.8–2.6
Sulfur	0.2–1.0	0.1–0.5	0.01–0.24
Potassium	1.0–4.5	0.2–2.5	1.0–4.0
Magnesium	0.1–0.5	0.1–0.3	0.1–0.5
Sodium	0.5–1.0	0.02–0.5	0.01–0.1
Calcium	0.01–1.1	0.1–1.4	0.1–0.3
Iron	0.02–0.2	0.1–0.2	0.01–0.5
Copper	0.01–0.02		0.002–0.01
Manganese	0.001–0.01		0.0005–0.007
Molybdenum			0.0001–0.0002
Total ash	7–12	2–8	5–10

[a] Overall chemical composition: 80% water, bound and free; 20% dry matter, organic (90% of the dry weight), inorganic (10% of the dry weight).
[b] In part from Peppler.[5]
[c] Values this high are observed only with rapidly growing cells.
[d] From Aiba et al.[6]

tous in nature and are responsible for destruction of much of the organic matter on the earth, a largely beneficial activity that is integral to the recycling of living matter.[9] Fungi can be broadly divided between molds and yeasts. Molds are aerobic, whereas both aerobic and anaerobic metabolism is observed in yeasts. Molds tend to form filamentous structures while yeasts tend to be unicellular.

Fungi are very similar to organoheterotrophic bacteria in that they use most of the same organic substrates. They are distinguished from bacteria by their eucaryotic cell type, generally larger size, and more sophisticated methods of reproduction. Because both types of organisms are served by similar substrates, competition between them is common. Both can use solid food materials by secreting extracellular hydrolytic enzymes to dissolve the solid substrate. However, fungi are less affected by low moisture environments and can often grow on dry substrates nourished by moisture absorbed from damp atmospheres. Fungi can also extract moisture from materials that have high osmotic pressure such as syrups, jams, and pickling brines. They can withstand a broad range of pH conditions and often have a lower nitrogen requirement than bacteria. As such, fungi are common in soils and decaying vegeta-

Table 4.6.   Microbial Populations During Aerobic Composting[a]

| Microbe | No./Wet Gram Compost | | | Number of Species Identified |
	Mesophilic Initial Temp <40°C	Thermophilic 40–70°C	Mesophilic 70°C to Cooler	
Bacteria				
Mesophilic	10^8	10^6	10^{11}	6
Thermophilic	10^4	10^9	10^7	1
Actinomyces				
Thermophilic	10^4	10^8	10^5	14
Fungi[b]				
Mesophilic	10^6	10^3	10^5	18
Thermophilic	10^3	10^7	10^6	16

Source: Poncelet.[8]

[a] Composting substrate not stated but thought to be garden-type materials composted with little mechanical agitation.
[b] Actual number present is equal to or less than the stated value.

tion, where their lower moisture requirements give them a competitive advantage over bacteria. For example, it is common to observe mold development on bread where the low moisture level inhibits rapid bacterial development. As a result of these distinctive properties, molds are common inhabitants of moist, dark locations where organic matter and oxygen are available. Obviously, such conditions are found in most composting systems, and we can expect the fungi to play a prominent role in such.

McKinney[10] reported an approximate empirical formulation for the organic fraction of the fungus *Aspergillus niger* as $C_{10}H_{17}O_6N$. Assuming this formulation to be representative of other classes of fungi and comparing it with bacterial formulations given previously, the lower nitrogen content of the fungal protoplasm is immediately obvious. Thus, fungi should have a competitive edge over bacteria in nitrogen-deficient environments, because the basic nitrogen requirement is lower. This can be of practical significance because many composting substrates are cellulosic and therefore tend to be low in nitrogen.

Fungi are important industrial microbes. Yeasts are the only significant microbes used in the production of alcoholic beverages such as beer and wine. Yeasts also produce industrial alcohol and glycerol, and they are used as a leavening agent in baking. Molds are used industrially to synthesize a wide variety of valuable substances that cannot easily be made by artificial processes. These include commercial enzymes (amylases, proteases, and pectinases), antibiotics (notably penicillin), organic acids (citric and lactic), and special cheeses.

Over 80,000 species of fungi have been identified. Such a bewildering number of organisms presents major difficulties in classification. Fortunately, knowledge of detailed classification schemes is not necessary to design successful composting systems. Nevertheless, a knowledge of general groupings is useful and provides considerable insight into the variety of living habits of these interesting organisms. A partial classification of fungi to the class level is presented in Table 4.7.

Fungi are classified largely on the basis of morphology and method of reproduction. In this sense the classification scheme differs from that used for bacteria, where biochemical responses are of prime importance. Fungi are usually capable of both sexual and asexual reproduction. An exception is the class Deuteromycetes, for which sexual reproductive stages have not been observed. In either case a reproductive body termed a spore is formed. Fungal spores should not be confused with bacterial spores which serve mainly as a survival mecha-

Table 4.7.   Characteristics of the Classes of Fungi[a]

Class	Description	Representatives	Remarks
Kingdom: Protista			
Division: Mycota (fungi)			
Subdivision I: Eumycotina (true fungi)			
Class I: Phycomycetes	Primitive fungi: asexual spores contained in a sporangium; mycelia non-septated except at reproductive site	*Rhizopus* (black bread mold), *Mucor*	Most are saprophytic; class includes water molds, blights, mildews, and bread molds; some species are aquatic; terrestrial forms include *R. nigricans*, the common black bread mold; Occasionally, certain species can cause infection in man especially if large numbers of spores are inhaled
Class II: Ascomycetes	Sac fungi: sexual spores are enclosed in sacs or asci; mycelia septated; includes filamentous and non-filamentous forms (yeasts); budding is the most distinctive method of asexual reproduction in yeasts	*Aspergillus, Penicillium* (blue-green mold), *Neurospora* (red mold of bread), *Candida* (a yeast)	About 30,000 species of filamentous Ascomycetes; also includes the yeasts; important industrial uses in fermentation, baking, brewing, and biomass production; participate in production of humus and digestion of cellulose; *A. fumigatus* can cause pulmonary infection; *P. notatum* and *P.* chrysgenum prominant as sources of penicillin
Class III: Basidiomycetes	Club fungi: sexual spores are borne on basidia, a swollen clublike reproductive cell; mycelia septated		Includes macroscopic fungi such as mushrooms, puffballs, and toadstools; parasitic species such as rusts and smuts can infect plant crops; other species aid in deterioration of wooden structures and formation of humus
Class IV: Deuteromycetes	Fungi imperfecta: heterogeneous group of about 20,000 species for which a sexual stage of reproduction has not been observed	*Torula, Tricoderma, Cladosporium, Alternaria, Fusarium*	Vast majority are saprophytic, occurring in soils, decaying vegetable matter, and foods; about 50 species are opportunists, causing disease when introduced into body under proper conditions; infection of lungs and superficial skin infections most common; some are carnivorous, trapping and digesting micro-scopic animals in-cluding nematodes in soil
Subdivision II: Myxomycotina (slime fungi)			
Class I: Myxomycetes	True slime molds: in vegetative phase the microbe is a viscous mass of multi-nucleated protoplasm called a plasmodium; resemble amoebas in some respects; reproduce by spores on stalks like fungi	*Physarum, Badhamia*	Obtain nourishment by ingesting bacteria, small particulate matter, and organic materials in soils; some are parasitic on higher plants
Class II: Acrasiomycetes	Cellular slime molds: frequently occur as a pseudoplasmodia composed of many amoebalike individual cells	*Dictyostelium*	Diet consists mainly of bacteria, spores of fungi, and small piece of organic matter

[a]   Adapted from a classification scheme by Alexopoulos.[11]

A. SPORANGIOSPHORES: Spores
   inside swollen fertile
   structure called
   sporangium (limited to
   Phycomycetes).

B. CONIDIOSPORES (conidia):
   Spores supported by a
   specialized fertile
   structure, the
   conidiophore.

C. THALLOSPORES: Spores resulting
   from changes in the vegetative
   hyphae or thallus.

   1. ARTHROSPORES (oidia):
      Hyphae fragment into
      small spores with
      thickened cell walls.

   2. CHLAMYDOSPORES: Hyphae
      divide into spore-like
      cells with large food
      reserve and resistance to
      unfavorable environment.

   3. BLASTOSPORES: Produced
      by budding.

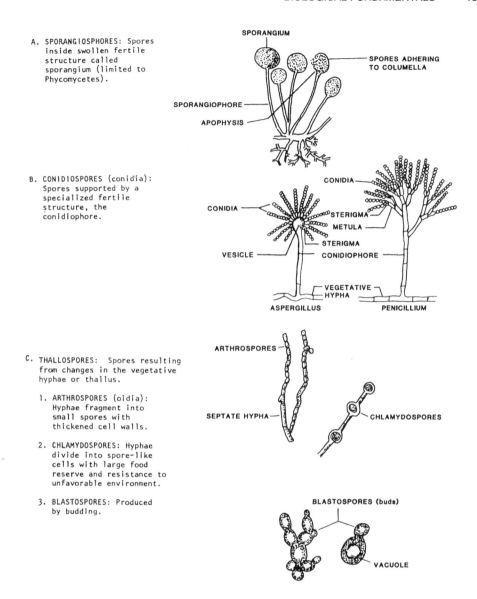

Figure 4.3.   Major types of asexual spores formed by fungal microbes. From Frobisher et al.[12]

nism. The major types of asexual and sexual spores are illustrated in Figures 4.3 and 4.4 along with much of the nomenclature common to fungal classification. Study of Figures 4.3 and 4.4 along with Table 4.7 should provide the reader with a knowledge of fungi sufficient for purposes of this text.

During the vegetative phase, virtually all filamentous fungi consist of tubular, branching filaments called hyphae. Diameter of the hyphae typically range from 10 to 50 μm. Note that this is considerably larger than the average bacterial cell, a reflection of the eucaryotic nature of the fungi. A mass of hyphae is called a mycelium. In more primitive fungi, such as the Phycomycetes, the hyphae are not septated. In other words, individual cells in the filament are not separated by a cross wall and the hyphae can be considered a single, multinuclear cell.

A. ASCOSPORES: Spores produced in a sac or ascus.

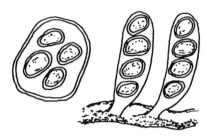

B. BASIDIOSPORES: Spores produced at the surface of a club-shaped structure, the basidium.

C. ZYGOSPORES: Spores produced by the fusion of similar-appearing gametes formed at the tips of hyphae (limited to the Phycomycetes).

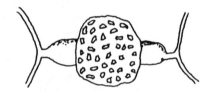

D. OOSPORES: Spores resulting from the mating of two unlike gametes (limited to the Phycomycetes).

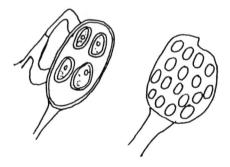

Figure 4.4. Major types of fungal sexual spores resulting from the fusion of nuclei or mating of gametes (reproductive cells). From Frobisher et al.[12]

Hyphae of other classes of fungi are generally septated, although the septum often contains pores that allow passage of material between cells.

Large numbers of fungal species can be isolated during both mesophilic and thermophilic stages of composting. Kane and Mullins[13] isolated 304 unifungal cultures from a Metro-Waste reactor system operating on municipal refuse. Of the total 304 isolates, 120 belonged to the genus *Mucor*, 97 to *Aspergillus*, 78 to *Humicola*, 6 to *Dactylamyces*, 2 to *Torula*, and 1 to *Chaetomium*. As Kane and Mullins pointed out, however, the number of isolated species in each genus may not indicate the actual number of individuals of each species in the compost.

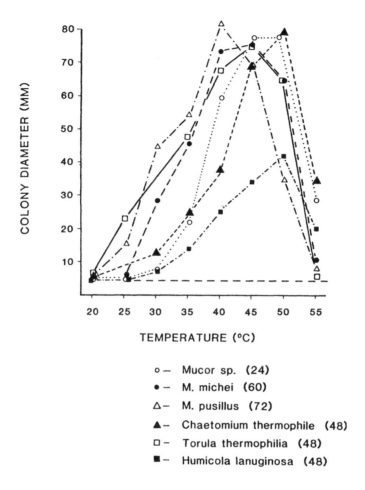

Figure 4.5.   Relation of temperature to growth of various fungi. Numbers after fungal name are the hours of growth. The fungi were isolated by Kane and Mullins[13] from a Metro-Waste reactor system composting municipal refuse.

Thus, the importance of each genus to composting is not indicated by the count of species. Thermophilic fungi were observed during all stages of the compost cycle, including the beginning of digestion under mesophilic conditions. No apparent succession of species was found.

One would expect the fungal population to be influenced by feed substrate, temperature, aeration, pH, moisture content, and perhaps the mechanical agitation applied to the compost. Fungal molds are strict aerobes and would be present only where persistent aerobic conditions are maintained. Temperature conditions during composting can often reach the point where thermophilic fungi are inactivated. Some thermophilic fungi do not appear to grow well at temperatures above about 60°C, as indicated by Figure 4.5. Acclimation of the fungal species may have allowed continued metabolic activity at temperatures above those shown in Figure 4.5. Nevertheless, fungal activity will vary depending on conditions maintained in the compost system. Kane and Mullins[13] concluded that high temperatures, acidity, and anaerobic conditions may limit fungal growth in the interior of a compost pile and restrict the role of thermophilic fungi.

OTHER PROTISTS

This discussion would not be complete without a small note of acknowledgment to the other groups of protists, i.e., algae, protozoa, and viruses. Algae include both procaryotic and eucaryotic types, are photosynthetic, contain chlorophyll-a as a photosynthetic pigment, and evolve O_2 as a by-product of photosynthesis. Because of their photosynthetic nature, algae are not significant in the decomposition of organic residues.

Protozoa are eucaryotic, organoheterotrophic microbes that exhibit a tendency toward unicellular growth with elaborate intercellular organization. They are usually large organisms, ranging in size from 10 to 100 μm. Most are motile and most reproduce by binary fission. In liquid waste treatment processes protozoa are not the primary consumers of organic material. This role is usually filled by the bacteria, which, because of their higher surface-to-volume ratio, can process substrates more rapidly. Instead, protozoa serve as scavengers of solid organic particles including bacteria and act to polish the liquid effluent. Their role in composting systems is minor.

Although the vast majority of protozoa are saprophytic or predatory, protozoan caused diseases are not rare. The causative agents of amoebic dysentery, giardiasis, malaria, African sleeping sickness, and many other human diseases are protozoan. Composting systems must be operated to assure destruction of protozoan disease agents that may be present in the starting substrate. In this regard, it should be noted that some protozoa are capable of encysting during periods unfavorable to growth. During this process the cell will produce a thickened cell wall, lose water, and maintain limited or no metabolism. Such cells are commonly called cysts and are more resistant to conditions of drought, heat, and extremes of pH. This must be considered when determining conditions for pathogen control in composting.

Viruses are acellular particles that carry genetic information for reproduction but no biochemical machinery to transcribe the information or metabolize substrates for energy. As such they are obligate parasites, using a host cell to provide the biochemical machinery they lack. Virus particles are extremely small, ranging from about 0.01 to 0.25 μm. Individual virus types are extremely host-specific, generally invading only one type of host cell. Virus particles are known to "reproduce" in cells of almost all living organisms, including other protists. As with the protozoa, the major concern with viruses is their potential for plant, animal, and human disease transmission and the conditions necessary for their destruction.

PATHOGENIC ORGANISMS

Enteric Microbes

The vast majority of protists are harmless from a disease standpoint and generally beneficial to man. However, a limited number are capable of causing human disease. These pathogenic microbes have thus attained a prominence far beyond their number. One of the major objectives of composting is to destroy pathogens that may be present in the original substrate. This section will present the types of pathogens common to municipal wastes and to waste composting systems. The following chapter will consider measures available to control pathogens.

Table 4.8 presents a summary of many of the enteric pathogenic organisms that potentially can be transmitted by water, sewage, or sludge. The table is not presented as an exhaustive list of all potentially waterborne pathogens, but it does contain those of greater importance or with a higher frequency of occurrence. Even so, the list is quite extensive. The point of Table

4.8 is not that sewage sludge, animal manure, refuse, or other substrates should be feared, only respected. Measures must be taken to assure that pathogens are controlled to avoid public health problems. Composting, of course, has long been used to destroy such pathogens.

A wide variety of pathogenic bacterial, viral, protozoan, and metazoan forms occur in sewage, hence, in sludges of sewage origin and refuse contaminated with fecal matter. However, the types of pathogens present, as well as the concentration, may vary considerably from community to community. Enteric bacterial (Enterobacteriaceae) pathogens include the causative agents for cholera, typhoid and paratyphoid fever, and various dysentery-related diseases. Numerous viruses are excreted by man including poliovirus (3 types), coxsackievirus (25 types), echovirus (25 types), reovirus (3 types), and adenovirus (33 types).[14] Even though numerous types are excreted, the only viral disease definitely known to be waterborne is hepatitis A (infectious hepatitis).

Various types of intestinal parasites have been reported in wastewater including *Entamoeba histolytica*, various nematodes such as *Ascaris lumbricoides*, hookworms, and schistosomes. The life cycle of many intestinal parasites is quite complex, but, in general, the first stage involves excretion in the feces. It should be noted that not all metazoan disease agents are of public health concern. Although they may be present in sludge, lack of a proper intermediate host often prevents transmission of the disease. Schistosomiasis is one example.

Of the parasite eggs found in sewage, those of *Ascaris* species are the most common and of major concern. Geographic factors usually influence the types of parasites expected, but the ubiquity of *Ascaris* places quantities of their eggs in almost all sewages and sludges tested. *Ascaris* was among the earliest known of human parasites. Its life cycle has been explored and well defined. The large nematode in the adult form is found in the small intestine. Females can reach a length of 20 to 35 cm and males 15 to 30 cm. Egg production of the female ascarid ranges from 200,000 to 250,000 eggs/day and is excreted in the feces of the host. The size of unsegmented eggs is 60 to 70 × 40 to 50 μm, and they are covered by a thick albuminous coat. An environment with a temperature lower than the host body, a trace of moisture, and a supply of oxygen are required for eggs to develop. If favorable conditions are realized, active embryos develop in about a week and reach an infective larval stage in 10 to 14 days. Should eggs then be ingested, larvae will hatch in the small intestine where they penetrate the mucous membrane and travel throughout the body to the liver, heart, and/or lungs. Eventually the worm settles in the small intestine and begins the cycle again.

The shell of an *Ascaris* ovum is resistant to chemicals and desiccation. Eggs will survive for weeks in a 10% formalin solution and are very resistant to chlorination. This resistance allows the ovum to survive and remain infective for years under proper conditions.

The U.S. Center for Disease Control analyzed results of over 400,000 stool specimens examined by public health laboratories throughout the U.S. for the presence of intestinal parasites.[15] *Giardia lamblia* was the most commonly identified pathogenic intestinal parasite, appearing in 3.8% of all stool specimens examined. *Entamoeba histolytica* was identified in 0.6% of the specimens. Nematode ova, arranged in the order of occurrence, were *Trichuris triciura*, 2.7% of all samples, *Ascaris lumbricoides*, 2.3%, *Enterobius vermicularis*, 1.7%, and hookworm, 0.8%. Cestodes and trematodes appeared in about 0.3 and 0.05% of the samples examined, respectively. A total of 8.4% of all samples were positive for some form of pathogenic intestinal parasite. Geographical variations were noted in most cases. For *A. lumbricoides*, for example, the positive sample frequency ranged from a high of 9.3% in Guam to a low of 0.0% in Wyoming, Arizona, and Nevada.

Because the above results are from public health laboratories, the values reflect "high risk" groups such as institutionalized patients and residents of communities with lower standards of living and poorer sanitation facilities. Results should not be interpreted as the prevalence

Table 4.8. Pathogenic Organisms that Can Potentially Be Transmitted by Water, Sewage, or Sludge

Causative Agent	Disease	Remarks
Enteric Viruses		
Poliovirus	Poliomyelitis	Exact mode of transmission not yet known; found in effluents from biological sewage treatment plants; introduction of effective vaccine made polio a rare disease in developed areas of the world in less than a decade
Virus	Infectious hepatitis type A	"Viral hepatitis" is a generic term that covers at least two distinct forms of hepatitis; hepatitis type A, known as "infectious hepatitis", is the generally accepted term for epidemic, community-acquired disease; type A virus is present in the feces and blood of infected persons who can contaminate milk, food, and water; hepatitis type B, also known as "serum hepatitis", is ordinarily transmitted by the parenteral route and is commonly associated with drug abuse and transfusion of blood and blood products
Coxsackievirus	Usually mild infections	Responsible for aseptic meningitis, pleurodynia, and infantile myocarditis; common cause of diarrhea in infants and young children
Echovirus	Usually mild infections	These viruses (enteric cytopathogenic human orphan viruses) have been associated with illnesses of aseptic meningitis, rash, diarrhea, and common respiratory diseases
Adenovirus	Respiratory infections	Many of the adenoviruses have been associated with a variety of respiratory diseases such as colds, influenzalike illnesses, bronchitis, croup, and atypical pneumonia
Reovirus		Diarrhea and respiratory diseases similar to those noted for adenovirus
Virus	Gastroenteritis and diarrhea	Causative agents not known but thought to be viral in some instances; in terms of magnitude, gastroenteritis and diarrheal disease are probably the most important viral diseases transmitted by water
Bacteria		
Coliform species	Diarrhea and internal infections, gastroenteritis	Implicated in several cases of infant diarrhea in hospitals; in rare cases known to cause cardiovascular infections
Vibrio cholerae	Cholera	Acute diarrheal disease transmitted by sewage and polluted waters; disease often terminates in death; endemic in India and SE Asia; has occurred in Europe and N. and S. America; during an epidemic in London in 1854, the causative organism was shown to be transmitted in water, making cholera the first disease for which this important fact was known; infection results from ingestion of contaminated food or drink; infective dose is near 10^9 vibrios; over 250,000 cases in major outbreaks in Peru in 1992
Salmonella (many types)	Salmonellosis	Salmonellosis may range in severity from intestinal discomfort to potentially fatal diseases such as typhoid fever (see below); food infections from salmonella are quite common; an estimated 1% of the human population may excrete salmonella at any time; three clinically distinct forms of the infection in man: enteric fevers, septicemias (rare), and acute gastroenteritis

Organism	Disease	Remarks
Salmonella typhi	Typhoid fever	Most severe of enteric fever forms of salmonellosis; occurs in all parts of the world but infrequently where good sanitation and purification of water is practiced; can survive in water for a week or more, also transmitted by milk; common in sewage and effluents in times of epidemics; enteric form of infection can be produced by other *Salmonella* species, i.e., *S. paratyphi*, which causes paratyphoid fever
Salmonella typhimurium (and others)	Salmonellosis, gastroenteritis	Most common form of salmonellosis; infection occurs from ingestion of contaminated food or drink; the disease persists for 3–5 days and is usually not severe
Shigella dysenteriae, S. sonnei, S. flexneri	Shigellosis (bacillary dysentery)	Dysentery is a clinical condition with intestinal inflamation, diarrhea, and water stools containing blood, mucus, and pus; polluted water is the main source of infection
Bacillus anthracis	Anthrax	Anthrax is a disease of sheep, cattle, horses, goats, and swine; human infection is rare but spores can be found in sewage and are resistant to treatment; in man generally appears as a disease of the skin; can be fatal if left untreated
Brucella abortus (cattle), *B. suis* (hogs), *B. melitensis* (goats)	Brucellosis	Normally transmitted from animals to man by infected milk or by contact with infected meat or placentae of infected animals; sewage also suspect; very rare in U.S. except in Midwest; hazard to slaughterhouse workers, farmers, veterinarians
Mycobacterium tuberculosis	Tuberculosis	Isolated from sewage and primary and secondary sludges; possible mode of transmission; care with sewage and sludge from sanatoria; deaths have been sharply reduced by early detection and treatment; estimated 80,000 new cases per year in U.S. with evidence of increasing rates; WHO reports death rates in Central and S. America three times greater than in North America
Leptospira interohaemorrhagiae (rats), *L. canicola* (dogs), *L. pomona* (cattle and swine)	Leptospirosis	Jaundice-like disease in man; worldwide distribution and fairly common in man; often transmitted to man by ingestion of food and drink contaminated by urine of the reservoir animal or bathing in contaminated water; can be carried by sewer rats and documented as occurring in sewer workers in England
Yersinia enterocolitica	Gastroenteritis	Role in human disease is not completely known; high levels have been reported in some sludge samples
Campylobacter jejuni	Gastroenteritis	Incidence rate is thought to approach that of Salmonellosis; commonly found in the intestines of many animals
Protozoa		
Entamoeba histolytica	Amoebiasis (amoebic dysentery)	Spread by contaminated waters and sludge used as fertilizers; also transmitted by uncooked vegetables fertilized by sewage or sludge; common in warmer countries; organism can form a cyst which is resistant to disinfection
Giardia lamblia	Giardiasis (lambliasis)	Clinical manifestations range from asymptomatic cyst passage to severe malabsorption; mean duration of the illness is often 2–3 months; in 1974 an outbreak occurred in Rome, NY, where over 5000 persons were affected; giardia cysts are not destroyed by chlorination at dosages and contact times normally employed in water treatment, but it is felt that they can be removed by coagulation, settling, and filtration
Balantidium coli	Balantidiasis	Found throughout the world, particularly in the tropics; illness similar to amoebic dysentery; some persons suffer acute dysentery, but the majority are probably carriers without symptoms; infection results from ingestion of cysts harbored in stools of man or swine; fatalities have occurred in severe infections despite treatment

Table 4.8.   (continued) Pathogenic Organisms that Can Potentially Be Transmitted by Water, Sewage, or Sludge

Causative Agent	Disease	Remarks
Isospora belli, I. hominis	Coccidiosis (isosporosis)	Infection is usually sporadic and not severe; most common in tropics and subtropics, but also reported in U.S.; infection results from ingestion of viable cysts
Cryptosporidium	Cryptosporidioses	Diarrhea like illness; outbreak from consumption of drinking water in Carrollton, GA in 1987; cyst is very resistant to chlorine
Metazoan Helminths (intestinal worms and flukes)		
Nematodes (roundworms)		
Ascaris lumbricoides	Ascariasis	A large intestinal roundworm sometimes reaching 20–40 cm in length in the intestine; the most common of the intestinal helminths of man; prevalent throughout the world and described as "one of man's most faithful and constant companions from time immemorial"; danger to man from sewage effluents and dried sludge used as fertilizer; infection occurs by ingestion of mature eggs in fecally contaminated food or drink; infection does not require an intermediate host
Ancylostoma duodenale, Necator americanus	Hookworm	*Necator* is the prevailing genus in the Western Hemisphere; formerly very prevalent in the southeast U.S.; infections developed in sewage farm workers in England; adult worms live in intestines, fastening to walls by strong mouth parts; eggs excreted in feces; subsequent larval stage may enter host through skin
Enterobius vermicularis (pinworm)	Enterobiasis	The most common cause of helminthic infection in man in the U.S.; although annoying, cure is readily effected with one of several drugs
Tricuris trichiura (whipworm)	Trichuriasis	Common parasite of man throughout the world; new infections are acquired by direct ingestion of the infective eggs passed in the feces
Strongyloides stercoralis (threadworm)	Strongyloidiasis	Prevalent in the southeastern U.S. and tropical and subtropical areas of the world; eggs secreted by the adult worm in the intestine develop into larvae which are passed in the feces; the free-living larvae can penetrate the skin of the next victim, enter the blood stream, and eventually the small intestine where maturation to the adult stage takes place
Toxocara cati (cat roundworm) *Toxocara canis* (dog roundworm)	Visceral larva migrans	Intestinal parasites found in dogs and cats; street runoff suspected as a source of eggs; recognized as a disease agent for children with pets
Ancylostoma brazilliense (cat hookworm) *A. canium* (dog hookworm)	Cutaneous larva migrans (creeping eruption)	Common infection of man in the southeastern U.S., particularly from contact with moist sandy soil contaminated with dog or cat feces; larvae invade the skin surface, usually hands or feet, and remain active for several weeks or months

Cestodes (tapeworms)		
Taenia saginata (beef tapeworm)	Tapeworm infection	Eggs very resistant; present in sewage, sludge, and sewage effluents; danger to cattle on sewage irrigated land or land manured with sludge; up to 9% of slaughtered cattle are reported to harbor cysts in Europe; man is the main host and cattle the intermediate host
Taenia solium (pork tapeworm)	Tapeworm infection (taeniasis)	Similar to above; no longer prevalent in the U.S.
Hymenolepis nana H. diminuta (dwarf tapeworm)	Tapeworm infection	Requires no intermediate host; most common tapeworm infestation in U.S., especially in the South; prevalent in the tropics and subtropics
Diphyllobothrium latum (fish tapeworm)	Tapeworm infection	Often found in Europe, Japan, and Great Lakes region of U.S.
Echinococcus granulosus, E. multi-locularis	Echinococcus (hydatid disease)	Tapeworm is found in several hosts throughout the world, including sheep, dogs, and other canines; S. America, Australia, Greece, and other Mediterranean countries are areas of heaviest human infestation; human infection results from ingestion of eggs passed in the host feces
Dipylidium caninum (dog tapeworm)	Tapeworm infection	Occasionally reported in children in Europe and the Americas who live in close association with dogs or cats; infection results from swallowing fleas or lice, which serve as the intermediate hosts
Trematodes (flukes)		
Schistosoma mansoni, S. haematobium, S. japonicum	Schistosomiasis (bilharziasis) (liver and intestinal flukes)	Eggs excreted in urine or feces of infected person; hatch on contact with water and enter snail host; emerging larvae (cercariae) leave the snail and can penetrate directly into human skin; infection may continue for years as an insidious drain on body vigor; widespread in Africa, Near East, and Orient where more than 90% of population may carry the worms; Egyptian government considers disease to be a major obstacle to country's economic progress; schistosomes are not transmitted in U.S. because the host snails are not present; in N. America a mild disease called "swimmer's itch" can occur; microbes probably killed by efficient sewage treatment
Fasciolopsis buski (giant intestinal fluke)	Fasciolopsiasis	Common parasite of man and pigs in China, Taiwan, SE Asia, and India; rare in continental U.S.; aquatic snail is the intermediate host
Fasciola hepatica (sheep liver fluke)	Fascioliasis	Infection results from ingestion of cercariae on aquatic vegetables; prevalent in sheep-raising countries, particularly where raw salads are eaten
Clonorchis sinensis (liver fluke)	Clonorchiasis	Endemic in parts of Japan, China, Formosa, and Indochina; snails and freshwater fish are intermediate host; infection results from eating such fish raw or undercooked
Paragonimus westermani (lung fluke)	Paragonimiasis	Commonly infects man throughout the Far East; snails and crabs or crayfish are intermediate hosts; infection results from eating such crustaceans, raw or pickled

of parasitic infection in the general U.S. population. Despite this limitation, the data do provide an idea of the relative prevalence of parasitic infections in a developed country. Higher incidence rates would certainly be expected in less-developed regions of the world.

Worms and eggs have a specific gravity of approximately 1.1 or greater, with settling velocities of about 3 ft/h. Tapeworm eggs are reported to have a somewhat lower settling velocity of 0.3 to 1.5 ft/h.[16,17] Therefore, most worms and eggs tend to accumulate in primary sludges and, perhaps, in waste activated sludges. Virus particles tend to be associated with particulate solids and are also concentrated in the sludge component.

Since the majority of eggs, cysts, and certain other pathogens tend to occur in sludge, their fate during subsequent sludge processing is important. Mesophilic anaerobic digestion is moderately effective against certain pathogens such as virus and protozoan cysts. Thermophilic digestion (45 to 50°C) should be more effective, but a considerable number of all pathogenic types, particularly helminth eggs, will survive. This is further assured by the fact that most modern anaerobic digesters are well mixed. Thus, a large portion of the feed passes through the digester in less than the theoretical detention time. This alone will limit pathogen destruction achievable in such a digester even if operated under thermophilic conditions. By contrast, autothermophilic aerobic digesters (ATADs) are usually operated on a fill and draw basis, with several reactors arranged in series. This allows ATADs to achieve high levels of pathogen destruction.

Air drying of sludge, such as using sand drying beds, is not particularly effective, especially against eggs and cysts that can survive desiccation. It would appear, therefore, that both raw and digested sludges, whether dewatered by mechanical or air drying techniques, are likely to contain significant numbers of pathogens. Stated another way, unless specific measures are taken to destroy pathogens, dewatered sludges produced by conventional processing can be expected to contain various concentrations of the pathogens presented in Table 4.8.

Fungal Opportunists

Although the vast majority of fungal organisms are saprophytic, about 50 species are termed opportunistic. They are capable of causing disease in man when introduced into the body under certain conditions. Frequently, such infections occur in patients already debilitated by other diseases or in weakened conditions. Fungal infection, termed a mycosis, can be divided into systemic or dermal types. Systemic infections are those that affect or pertain to the body as a whole. These affect deep tissues, usually the lungs, and therefore are regarded as serious. Dermal infections affect superficial tissues such as skin, hair, or nails. They are annoying but not generally serious. A description of the more important systemic mycotic and Actinomycetes infections and their causative agents is presented in Table 4.9.

It should be recalled that all fungal organisms, including those shown in Table 4.9, are organoheterotrophic, deriving their energy from degradation of organic matter. Even those that can cause human disease usually live naturally, as saprophytes in soil or decaying vegetation. Thus, there are innumerable opportunities for individuals to come in contact with spores of these fungi. Infection usually occurs from inhalation of large quantities of spores into the lungs or by contamination of cuts or abrasions. Under the right conditions the spores can develop, leading to fungal infection of body tissues. Such infections are rarely transmitted from person to person. Emmons[18] concluded that:

the fungi that cause systemic mycoses are normal and more or less permanent members of the microflora of soil, compost, or organic debris.... These fungi are vigorous and self-sufficient saprophytes so long as environmental conditions are favorable. They are parasites by accident.

Table 4.9.   Systemic Mycotic (Fungal) Infections

Causative Agent	Disease	Remarks
Coccidiodes immitis	Coccidioidomycosis (San Joaquin fever)	Infections results from inhalation of spores or mycelial fragments; fungus grows in soil in certain arid regions of southwest U.S. and Mexico; infections commonly pass unnoticed but can be severe in certain individuals; lung is major focus of infection; considered an occupational hazard for archaeologists in southwest U.S.
Histoplasma capsulatum	Histoplasmosis	Similar to above
Cryptococcus neoformans	Cryptococcosis	Yeastlike fungus found in soil and pigeon nests; human infection is worldwide; lungs and meninges are major focus of infection
Blastomyces dermatitides	Blastomycosis	Infections are usually severe; lungs are major focus of infection; geographically limited to central and eastern U.S. and Canada; similar infections can occur in South or Central America and Mexico caused by another microbe
Candida albicans	Candidiasis	Can be cultured from mouth, vagina, or feces of about 65% of the population; frequently a secondary invader in other types of infections; lungs, skin, intestinal tract, and internal organs can be sites of infection
Sporothrix schenkii	Sporotrichosis	Worldwide in distribution; most patients are people whose occupation brings them into contact with soil, plants, or decaying wood; infection takes place when an organism is introduced into a break in the skin
Aspergillus fumigatus (and other *Aspergillus* species)	Aspergillosis	Infection usually occurs by inhalation of large numbers of spores; lung is focus of infection and disease is not rare in man; caused by various species of *Aspergillus*, but *A. fumigatus* causes more serious infections
Micromonospora spp.	Farmer's lung	An allergic response in lung tissue caused by this *Actinomyces* species

Because fungal organisms are active in composting piles, concern has arisen over possible growth of opportunistic species during composting. To date, most concern has centered on the genus *Aspergillus*, a number of species of which can cause aspergillosis. Other fungal opportunists have not been reported to be commonly associated with composting systems. *A. fumigatus* is usually discussed in reference to aspergillosis because its infections appear to be the most common and most serious. Various members of the genus *Aspergillus*, including *A. fumigatus*, have been isolated from composting systems, particularly those with cellulosic substrates. Millner et al.[19] indicated that *A. fumigatus* has been found frequently in composting vegetative material such as grass and tree clippings, woodchip piles, refuse compost, refuse-sludge compost, sludge, and moldy hay, and it is often isolated from soil. The organism is also thermotolerant, growing over a temperature range of <20 to 50°C. The spores of *A. fumigatus* average about 3 μm diameter with a settling velocity in air of 0.03 cm/sec,[20] which means that the spores will readily disperse if released into the air. The questions, then, are To what extent are individuals already exposed to opportunistic fungi? and Do composting operations increase this exposure?

Aspergillus is a ubiquitous fungus to which every individual has daily contact throughout his entire life.[21] Solomon[22] sampled fungal spores in the air of 150 homes in winter. *Aspergil-*

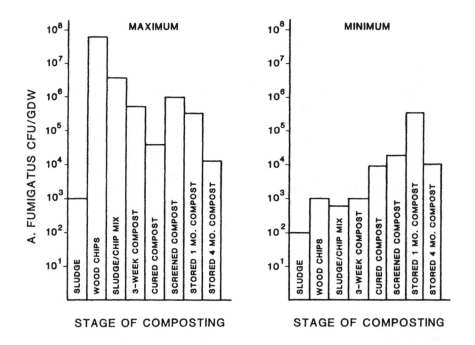

Figure 4.6.   Maximum and minimum *A. fumigatus* CFU/gm dw of substrate from each stage of the aerated static pile composting process. Numbers for stored compost represent counts from the 10-cm depth only. The composting mixture consisted of raw municipal sludge blended with wood chips. From Millner et al.[19]

lus was one of the most common fungi found. *Aspergillus*-type spores in outside air probably rarely exceed 500 colony-forming units (CFU)/m^3, with an observed range of about 10 to 10^4 CFU/m^3.[23,24] On the other hand, concentrations up to 21 million CFU/m^3 have been observed in farm buildings after shaking of moldy hay.[25] Hudson[26] reported concentrations of *A. fumigatus* in outdoor air ranging from 0 to 14 spores/m^3 with a mean of 3.2 spores/m^3.

Millner et al.[19] sampled for *A. fumigatus* throughout various stages of aerated static pile composting at Beltsville, Maryland. The composting mixture consisted of raw sludge blended with wood chips as a bulking agent. Maximum and minimum concentrations of the fungus detected in each stage of the composting process are shown in Figure 4.6. Sludge samples contained between 10^2 and 10^3 CFU/g dry weight. Wood chips appeared to be the major source of the fungus. Samples of fresh and old wood chips contained 10^3 to 2.3 × 10^5 CFU/g dw and 2.6 × 10^6 to 6.1 × 10^7 CFU/g dw of *A. fumigatus*, respectively. The abundance of *A. fumigatus* was sufficient that chips were often observed to have a gray-green coloration, attributed to dense masses of dry conidia. The authors also concluded that other bulking agents of high cellulose content, such as bagasse, carob husks, corncobs and husks, grass seed straw, cereal straws and husks, refuse, peanut hulls, and licorice root, would probably be suitable for growth of the fungus. Licorice root alone was observed to contain about 4.7 × 10^6 CFU/g dw of *A. fumigatus*. Semiquantitative studies of airspora at the composting site indicated that *A. fumigatus* constituted 75% of the total viable mycoflora captured. This decreased to only about 2% at locations 320 m to 8 km from the site. Commercial potting soils, manures, and mulches were also examined for the presence of *A. fumigatus*. Of 21 products analyzed, 5 contained no detectable *A. fumigatus*. The remainder ranged from about 10^2 to 5.7 × 10^5 CFU/g dw. Five of the products had concentrations exceeding that found in 4-month-old Beltsville compost (Figure 4.6).

Later work by Millner et al.[27] at the Blue Plains facility in Washington, D.C., demonstrated that aerosol concentrations from windblown losses from stationary piles are relatively small in comparison with those generated during mechanical movement of the piles. After the front-end loaders stopped moving compost, *A. fumigatus* concentrations at 3 and 30 m downwind of the static piles were about 33 to 1800 times less than those measured during pile movement and not significantly above background levels. In contrast, microbial aerosols emitted during pile agitation produced aerial concentrations significantly above background levels downwind from the piles. Using field measurements and Gaussian dispersion modeling, the authors estimated a maximum emission rate of 4.6×10^6 *A. fumigatus* particles/sec during the pile moving operations. Aerodynamic size fractionation of the *A. fumigatus* particles indicated that approximately 87% are breathable (≤ 7 μm), approximately 70% can pass the trachea and primary bronchi, 5% can pass the terminal bronchi, and <1% can reach the alveoli.

In a separate study, the LA/OMA project sponsored field measurements of *Aspergillus* concentrations at a windrow facility and an enclosed reactor pilot system.[28] Digested sludges were blended with recycled compost for conditioning in both cases. No cellulosic bulking agents or amendments were used. The starting compost material had a typical *Aspergillus* content of 10^3 to 10^4 CFU/g dw, similar to levels reported in the Beltsville study. If internal temperatures above 60°C were attained, *Aspergillus* levels dropped off rapidly as shown in Figure 4.7. Final concentrations <10 CFU/g dw were readily achieved. Even if temperatures greater than 60°C were not achieved in the windrow system, *Aspergillus* concentrations in the final compost ranged from about 10 to 1000 CFU/g dw, which is less than that of the original feed. *Aspergillus* was not observed growing during the compost cycle, unlike the experience at Beltsville where wood chips fostered growth of the fungus. Similar results were obtained with both the windrow and enclosed reactor systems. Therefore, growth of *Aspergillus* is probably related more to use of cellulosic bulking agents than to the type of compost system.

An aerosol sampling program was conducted to evaluate, first, those portions of the windrow composting process thought to release fungal spores into the air. Second, the compost field was monitored immediately upwind and downwind to evaluate the total emission of *Aspergillus*. Airborne concentrations measured during potentially dust-producing phases of the windrow operation are shown in Table 4.10. Samples were collected using an Anderson impingement sampler located as close as possible to the source of emission to minimize atmospheric dilution. The sampler divided collected particles into respirable and nonrespirable size fractions. Very low levels of *Aspergillus* spores were detected during these relatively dusty operations. In some cases, the Anderson sampler was covered with dust after the sampling period. The single highest observation of nonrespirable aerosols was 60 CFU/m³, and for the smaller respirable particles 42 CFU/m³. For all other samples, concentrations were <10 CFU/m³.

Aspergillus concentrations from the windrow compost field itself are presented in Table 4.11. Samples were taken upwind of the field and immediately downwind of one windrow, half the compost field, and the entire field. As expected, *Aspergillus* levels were quite low. The highest value recorded was 33 CFU/m³ downwind of the compost field, but the majority of samples were <10 CFU/m³. Considering the background levels reported earlier for normal outdoor air and the high concentrations associated with certain conventional agricultural operations, the airborne concentrations of 1 to 60 CFU/m³ observed at this windrow facility would appear to present little or no risk.

Kuchenrither[29] reported on similar studies conducted at the windrow sludge composting facility in Denver. *Aspergillus* concentrations in the composting mixture were reduced about 1 to 3 log units. This was somewhat less than observed in the LA/OMA studies and was

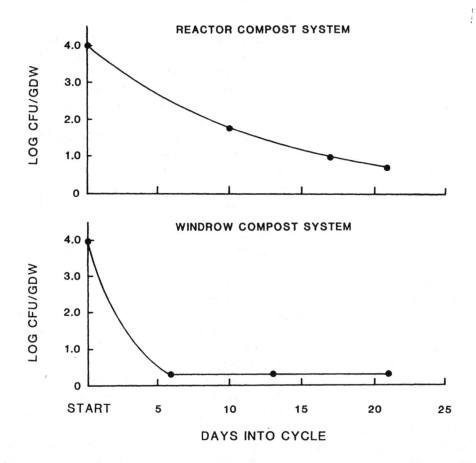

Figure 4.7.   Concentration of *A. fumigatus* during composting of digested sludge blended with recycled product. Temperatures >60°C were attained during each cycle. From LeBrun.[28]

thought to be due to the use of sawdust and wood chip amendments in the Denver operation. Concentrations adjacent to the composting turning operation were reported at 680 CFU/m³ compared to an ambient of 14 CFU/m³.

Lees and Tockman[40] reported on an 8-year sampling study conducted at the Montgomery County Composting Facility, a static pile process using sewage sludge. It was concluded that there was no significant increase in *A. fumigatus* concentrations at off-site locations during compost operations. In addition, maximum on-site concentrations were within the range of normal background concentrations. Similar studies were conducted at static pile composting sites in Westbrook, Bangor, and Old Town, Maine.[41] Aerospora levels of *A. fumigatus* were reported to be equivalent to background levels within 50 to 150 m from the composting site. Both of these studies concluded that there was no significant public health risk posed to residents in the vicinity of the composting operations.

Based on the studies discussed here, it appears that compost systems that use cellulosic substrates will be more prone to increases in airborne *Aspergillus* concentrations but any increases will be quickly dissipated. Apparently the substrate and other environmental conditions are suitable for growth of the fungus, resulting in elevated concentrations in the compost. While more data are available for the static pile process, it seems likely that the potential for elevated *Aspergillus* concentrations depends more on the use of cellulosic substrates than the

Table 4.10. Concentrations of Airborne *Aspergillus* Spores Measured Adjacent to Various Operations at a Windrow Facility Composting Digested Sludge Blended with Recycled Product

Compost Operation	Aspergillus Spores (CFU/m^3)		Sampling Location
	Respirable	Nonrespirable	
Stockpiling finished	<3	3	Downwind
windrows	2	0	Downwind
	<1	<1	Downwind
	<1	<1	Downwind
	<1	<1	Upwind
	<1	<1	Upwind
Loading finished compost	7	60	Downwind
into trucks	1	<1	Downwind
	<1	<1	Downwind
	<1	<1	Downwind
	<1	<1	Downwind
	<1	1	Downwind
Laying windrows	42	8	Downwind
	4	<1	Downwind
	4	<1	Downwind
	2	1	Downwind
	1	2	Downwind
Turning windrows	<1	<1	Downwind
	<1	<1	Downwind
	<1	<1	Downwind

Source: LeBrun.[28]

Table 4.11. Concentrations of Airborne *Aspergillus* Spores from a Windrow Compost Field (Composting Mixture Consisted of Digested Sludge Blended with Recycled Compost)

Description of Sample Location	Aspergillus Spores (CFU/m^3)	
	Respirable	Nonrespirable
Downwind from single windrow at rest — strong wind	7	19
Center of field		
Equipment working and slight wind	8	<1
	4	<1
Center of field		
No equipment working and slight wind	<1	<1
	<1	<1
Downwind from entire field with equipment working — slight wind (3 samples)	<1	<1
	1	<1
	3	1
Downwind from entire field with no equipment working		
Moderate wind	<2	6
Moderate wind	6	<2
Slight wind	31	33
Upwind from compost field	<1	<1
	<1	<1
	<1	<1

Source: LeBrun.[28]

type of composting system. Refuse, sawdust, agricultural residues, leaves, and yard wastes contain high cellulosic fractions and are probably suitable substrates for fungal opportunists. For example, refuse that has been wet pulped is known to produce large *Aspergillus* populations.

Public and Worker Protection

Many years experience with operating facilities has demonstrated that the risk from primary pathogens and fungal opportunists to workers and the general public is small. The predominant route of infection from primary pathogens is through the mouth, whereas for secondary pathogens it is by inhalation of air containing a high density of spores. People who are predisposed because of such conditions as diabetes, asthma, emphysema, tuberculosis, and positive skin tests or who may be taking medication such as corticosteroids, broad-spectrum antibiotics, or immunosuppressive drugs or who may be prone to severe allergies may be more susceptible to secondary infections.

It has been recommended that job applicants be medically screened and that those with any of the above predispositions be excluded from employment.[30] A complete physical examination was also recommended, plus inoculations for typhoid, tetanus, and polio. Pahren et al.[21] recommended that personnel working in areas with high agitation, such as mixing and screening buildings, regularly be skin-tested and/or evaluated for precipitating antibodies to *Aspergillus*. Other common sense rules regarding cleanliness should also be observed such as (1) washing hands before eating, drinking, etc. and before returning home and generally keeping hands away from the mouth, (2) storing food in a safe location, (3) providing showers and lockers at the facility, (4) providing protective clothing and changing from protective clothing and showering before going home, (5) cleaning and/or sterilizing the protective clothing, (6) minimizing fugitive dust to the extent possible by good housekeeping, water sprinkling, and process control, and (7) using face masks or respirators if the worker must be exposed to high dust operations.

Design aspects should also be considered to reduce unnecessary exposure to fungal opportunists. Pressurized cabs with filtered air supplies have been recommended for on-site equipment such as front-end loaders and windrow turners.[29] It is sometimes recommended that mixing and screening operations at static pile facilities be enclosed because the latter are major mechanisms for airborne release of spores. However, there are many static pile systems with open screening operations. It has also been recommended that surrounding regions be periodically monitored for *A. fumigatus* spores and that the proximity of planned composting operations to health care facilities be carefully considered.[31] The consensus of expert opinion appears to be that the above measures provide adequate protection for both the general public surrounding the site and workers employed at the site.

KINETICS OF MICROBIAL GROWTH

The rate at which microbes consume substrate and grow is important to all waste treatment processes including composting. Numerous models have been proposed to describe the rate of substrate use in solution. Here, the discussion will focus on the most commonly used of these models. The subject of composting kinetics is discussed again in Chapter 10.

The effect of substrate concentration on the rate of substrate use by a microbe is illustrated in Figure 4.8. Suppose a series of test tubes are inoculated with the same mass of microbes.

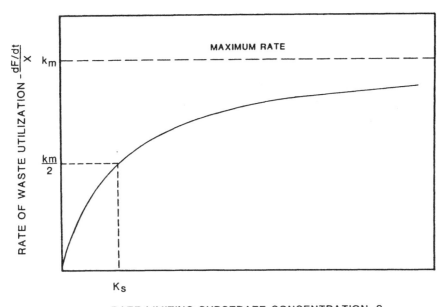

Figure 4.8.   Effect of limiting substrate concentration on the rate of substrate use.

Suppose further that the test tubes have different concentrations of a substrate that can be used by the inoculated microbes. If all other substrates, such as oxygen, are plentiful, the rate of substrate use should plot in the form shown in Figure 4.8. In 1942, Monod proposed a hyperbolic relationship of the form

$$\frac{dS}{dt} = -\frac{k_m SX}{K_s + S}$$

where

dS/dt	=	rate of substrate utilization, mass/volume-time
X	=	concentration of microbes, mass/volume
k_m	=	maximum utilization coefficient, maximum rate of substrate utilization at high substrate concentration, mass substrate/mass microbes-day
K_s	=	half-velocity coefficient, also referred to as the Michaelis-Menten coefficient, mass/volume
S	=	concentration of the rate-limiting substrate, mass/volume

The form of the equation used by Monod is pleasing in that it is similar to equations developed by Henri (1902) and Michaelis and Menten (1913) for single enzyme-substrate systems. The negative sign is used to indicate that substrate concentration decreases as a result of microbial activity.

Use of Equation 4.1 assumes that mass transfer of substrate to the cell is not rate-limiting. In other words, kinetics are controlled only by the concentration of a limiting substrate, S. It is common practice to measure substrate concentration in the immediate vicinity of the microbial cell. Thus, any subsequent mass transfer limitations across the cell wall or within the cellular protoplasm are included within the equation. Finally, the Monod model applies best to soluble substrates. Additional mass transfer limitations can be encountered with particulate substrates, a situation discussed in Chapter 10.

Recognizing these limitations, a physical interpretation of Equation 4.1 can be proposed. At high concentrations of substrate, the cellular enzyme systems become saturated with substrate and process substrate as rapidly as possible. Further increase in substrate concentration causes no further increase in the rate of substrate use. In Equation 4.1, this corresponds to the condition where $S \gg K_s$ and the form of the equation reduces to

$$\frac{dS/dt}{X} = -k_m \tag{4.2}$$

This is a zero-order reaction with regard to substrate concentration. At lower substrate concentrations cellular machinery operates at reduced rates because of the limited supply of substrate. If $S \ll K_s$, Equation 4.1 reduces to

$$\frac{dS/dt}{X} = -(k_m/K_s)S \tag{4.3}$$

which is first-order with respect to substrate concentration. Under the condition where $S = K_s$, Equation 4.1 reduces to,

$$\frac{dS/dt}{X} = -k_m/2 \tag{4.4}$$

Thus, the half-velocity coefficient corresponds to the substrate concentration where the rate of reaction is half the maximum as shown in Figure 4.8. Such a relationship has no real physical meaning in the model proposed by Monod but results simply from the form of the hyperbolic equation. With single enzyme-substrate systems, however, the coefficient does have a physical interpretation, but one that will not be explored here.

The use of substrate can be related to microbial growth through an equation of the form

$$dX/dt = Y_m(-dS/dt) - k_e X \tag{4.5}$$

where

dX/dt	=	net growth rate of microbes, mass/volume-time
Y_m	=	growth yield coefficient, mass of microbes grown/mass of substrate used
k_e	=	endogenous respiration coefficient, time^{-1} or mass of microbes respired per unit time per mass of microbes

It is assumed that the growth rate of new microbes is proportional to the rate of substrate utilization, with Y_m the proportionality coefficient. The endogenous respiration term is included to account for the observed fact that cellular mass decreases if no substrate is available. Microbes begin using stored reserves and eventually the essential protoplasm under such conditions. The rate of cell loss is generally proportional to the mass of organisms present in solution.

Substituting Equation 4.1 into Equation 4.5 gives the combined growth equation

$$\frac{dX / dt}{X} = \frac{Y_m k_m S}{K_s + S} - k_e \tag{4.6}$$

the term $(dX/dt)/X$ is often referred to as the net specific growth rate m. The product $Y_m k_m$ is termed the maximum net specific growth rate, m_m, achieved at high substrate concentrations $(S \gg K_s)$ and low endogenous respiration $(k_e = 0)$. With these terms, Equation 4.6 becomes

$$\mu = \frac{\mu_m}{K_s + S} - k_e \tag{4.7}$$

Equation 4.6 or 4.7 is the most common model used to describe microbial growth and substrate use in aqueous solutions. The model has been applied to a variety of biological processes used in sanitary and biochemical engineering practice, including the activated sludge process, oxidation towers and anaerobic digesters. Consideration of the model as applied to composting systems is deferred to Chapter 10.

Referring to Equation 4.6, four kinetic coefficients, Y_m, k_m, K_s, and k_e, must be known for the particular substrate and microbe under consideration. Values for the coefficients have been determined for a variety of substrates under both aerobic and anaerobic conditions. As might be expected, the yield coefficient is strongly influenced by free energy yield of the energy reaction. In fact, a number of predictive models have been developed based on energy available to the microbes.[32-35]

For aerobic heterotrophs, values of Y_m typically range from 0.25 to 0.5 g microbes/g substrate COD. It is common to base the yield coefficient on the COD of the substrate. As discussed in Chapter 3, substrate COD is proportional to the number of electrons transferred during aerobic metabolism. Furthermore, heat of combustion per electron transferred to a methane-type bond is reasonably constant at about 26.05 kcal/electron equivalent or, because O_2 has four such electrons, 104.2 kcal/mol O_2 (or COD). For aerobic metabolism, therefore, COD change turns out to be a measure of energy release. Thus, it is not surprising that maximum yield coefficients for aerobic metabolism fall within a reasonably small range for a variety of substrates when based on COD units.

With anaerobic metabolism, yield coefficients are significantly reduced because of the lower energy yield from fermentation reactions. Values as low as 0.04 and as high as 0.2 g microbes/g substrate COD have been determined for fatty acid and carbohydrate substrates, respectively.

The maximum utilization coefficient, k_m, represents the maximum rate at which microbes will process substrate. Typical units for k_m are g substrate COD/g microbes-day. From the previous discussion, the substrate COD is related to the number of electrons transferred. Thus, k_m can be interpreted as an electron transfer rate. McCarty[36] analyzed a number of reported values for both heterotrophic and autotrophic metabolism and for aerobic, anoxic, and anaerobic respiration. Values of k_m at 25°C appeared to vary between about 1 and 2 electron-mol/ g microbe-day (an electron-mol is one Avogadro's number of electrons), which is equivalent to 8 to 16 g COD/g microbe-day. This strongly suggests that the rate of electron transport is the limiting factor at high substrate concentrations and that the rate is reasonably constant for a wide variety of metabolic types. Considering that the basic biochemical machinery is common to all organisms, this result is not surprising.

The rate of electron transport should be a function of temperature and should obey the Arrhenius relationship discussed in Chapter 3. McCarty[36] reported values of E_a ranging from 9.8 to 18.2 kcal/mol (average 12.9) over the temperature range of 10 to 40°C. An E_a value of 12.25 kcal/mol corresponds to a doubling of the reaction rate for each 10°C rise in temperature. Temperature effects observed for microbial reaction rates are consistent with those observed for most chemical reactions.

Monod observed values of K_s on the order of 4 mg/L for glucose oxidation by dispersed aerobic growths. For pure substrates and with pure cultures of aerobic microbes, values are in the 20 mg COD/L range or lower. Higher values, usually in the range of 20 to 300 mg COD/L, have been determined with mixed cultures in activated sludge systems. This may be caused in part by added mass transport resistance into the sludge floc, which would not exist in a dispersed growth system. Resulting concentration gradients into the floc would cause the apparent value of K_s to increase.

Measured K_s values for anaerobic processes are considerably higher than for aerobic metabolism. Reasons for this are not particularly clear. Nevertheless, O'Rourke[37] found values for fatty acid fermentation ranging from about 2000 to 5000 mg COD/L. Values for K_s also appeared to be a function of temperature. Over the temperature range 20 to 35°C, the value of K_s for a mixture of fatty acid intermediates approximated

$$K_s = (1.8 \text{ g COD} / L)(1.112^{(35-T)})\qquad(4.8)$$

where

T = solution temperature°C

The endogenous respiration coefficient is used to account for the observed fact that microbial mass in a batch reactor will decrease after the initial substrate is consumed. Measured values of k_e usually range from 0.02 to 0.15/day. The meaning of the coefficient is perhaps clarified if the units are interpreted as grams of cells consumed per gram of cells per day. Endogenous respiration is the metabolic consumption of stored reserves and cellular protoplasm. Thus, the rate should be a function of temperature. McKinney[38] used the following formula to correct for temperature effects on k_e over the temperature range from 4 to 36°C:

$$k_e (\text{per hour}) = (0.02)e^{0.069(T-20)}\qquad(4.9)$$

Available data at higher temperatures is too scattered to allow development of reasonable equations.[39]

Equation 4.1 assumes that reaction rates are controlled by a single rate-limiting substrate. For most organisms a variety of substrates are involved in metabolism, any one or more of which can be potentially rate-limiting. For example, aerobic, organoheterotrophic bacteria must be supplied with both oxygen and organic matter as substrates. Depending on concentrations surrounding the cell, either substrate or both can exert a rate limitation. Such a condition has been mathematically modeled by modifying Equation 4.1 to the form

$$dS / dt = -k_m X\left[\frac{S_1}{K_{s1} + S_1}\right]\left[\frac{S_2}{K_{s2} + S_2}\right]\cdots\left[\frac{S_n}{K_{sn} + S_n}\right]\qquad(4.10)$$

where

S_n = concentration of the nth substrate which can be potentially rate limiting

K_{sn} = half-velocity coefficient for the nth substrate

If the concentration of any substrate is much greater than the corresponding K_s value, the term in brackets will approach unity. If this condition exists for all but one substrate, Equation 4.10 reduces to Equation 4.1. It should be recognized that Equation 4.10 is not based on any fundamental analysis of cellular kinetics. Rather, it is a convenient mathematical form that has been applied usefully to situations involving multiple substrates.

In summary, rates of microbial growth and substrate use can be modeled mathematically using the Monod approach based on the rate-limiting substrate(s). The model applies best to soluble substrates and dispersed growth of microbes. With particulate substrates and flocculated or fixed-film growths, additional mass transfer limitations must be considered. In composting systems the substrate is particulate by nature. To understand further the rate-limiting factors involved in composting, kinetics of cellular metabolism must be integrated with the rate of solubilization of particulate substrate and the subsequent mass transport of solubilized substrates and oxygen to the cell. These factors are considered in Chapter 10.

SUMMARY

Composting is a biological process mediated by microbes belonging to the Protist kingdom, which includes bacteria, algae, fungi, protozoa, and virus particles. Microbes can be classified into metabolic types based on the carbon and energy sources utilized by the cell. Autotrophs use carbon dioxide as a source of cell carbon, whereas heterotrophs use the carbon of organic molecules. Phototrophs obtain energy from light. Lithotrophs use the energy of inorganic chemical reactions, while organotrophs use the energy of organic chemical reactions. Most bacteria and all fungi are organoheterotrophs using organic compounds both as a source of energy and for cell carbon. They are the microbes of importance in organic composting.

Cellular respiration can be aerobic, anoxic, or anaerobic, depending on the electron acceptor used in the energy reactions. Oxygen and oxidized compounds of nitrogen and sulfur are electron acceptors for aerobic and anoxic respiration, respectively. Anaerobic fermentations are characterized by having the electron donor and acceptor originate from the same organic molecule.

Bacteria are the smallest living organisms, with representatives from all the metabolic and respiratory classifications. They have a procaryotic cell structure. Fungi are divided into molds and yeasts. Molds are usually strict aerobes while yeasts are of both aerobic and anaerobic types. Both molds and yeasts have eucaryotic cell structure and are all organoheterotrophs.

The vast majority of microbes are harmless and even helpful to man. However, a small number can cause disease and are termed pathogenic. Of these, the enteric pathogens are indigenous to the intestines and can potentially be transmitted by sludges, sewage or water. Diseases caused by enteric pathogens include the following: poliomyelitis and infectious hepatitis caused by viral agents; cholera, typhoid, and bacillary dysentery caused by bacterial agents; amoebic dysentery and giardiasis caused by protozoan agents; and roundworm, tapeworm, and fluke infestations caused by helminths.

A number of fungal organisms are termed opportunists and can cause infections in man if introduced into the body under appropriate conditions. Usually this is by inhalation of large quantities of spores. For the spores to cause infection, the host must usually be debilitated by other disease or in an otherwise weakened condition. *A. fumigatus* is the principal agent of aspergillosis and, unlike most other fungal opportunists, is thermotolerant and often isolated from composting material. Growth of *A. fumigatus* appears to be enhanced by the use of cellulosic substrates such as wood chips. However, the risk of human infections as a result of composting operations appears to be low provided common sense rules to minimize contact are observed and workers with predisposing conditions are excluded from employment.

Kinetics of microbial growth can be described in mathematical terms using the Monod model, which is similar in form to models used to describe a single enzyme-substrate complex. Rates of substrate use and microbial growth can both be described in terms of four kinetic coefficients: the yield coefficient, Y_m, maximum rate of substrate utilization, k_m, half-velocity coefficient, K_s, and the endogenous respiration coefficient k_e. The yield of organisms per unit of substrate used is a function of free energy available from the substrate. The maximum rate of substrate use is related to the electron transport rate within the cell, which is reasonably constant for a wide variety of microbes.

REFERENCES

1. McCarty, P. L. "Anaerobic Treatment of Soluble Wastes," in *Advances in Water Quality Improvement* (Austin, TX: University of Texas Press, 1968).
2. Poindexter, J. S. *Microbiology — An Introduction to Protists* (New York: The Macmillan Co., 1971).
3. Buffaloe, N. D. and Ferguson, D. V. *Microbiology* (Boston, MA: Houghton Mifflin Co., 1976).
4. Breed, R. S., Murray, R. G. E., and Smith, N. B., Eds. *Bergey's Manual of Determinative Bacteriology,* 7th ed. (Baltimore: The Williams & Wikins Co., 1957).
5. Peppler, H. J., Ed. *Microbial Technology* (New York: Reinhold Publishing Corp., 1967), p. 421.
6. Aiba, S., Humphrey, A. E., and Millis, N. F., Eds. *Biochemical Engineering,* 2nd ed. (New York: Academic Press, Inc., 1973), p. 29.
7. Golueke, C. G. *Biological Reclamation of Solid Waste* (Emmaus, PA: Rodale Press, 1977).
8. Poincelot, R. P. "The Biochemistry of Composting," in *Composting of Municipal Residues and Sludges, Proceedings of the 1977 National Conference* (Rockville, MD: Information Transfer, 1977).
9. Phaff, H. J. "Industrial Microorganisms," *Sci. Am.* 245(3) (September 1981).
10. McKinney, R. E. *Microbiology for Sanitary Engineers* (New York: McGraw-Hill Book Co., Inc., 1962).
11. Alexopoulos, C. J. *Introductory Mycology,* 2nd ed. (New York: John Wiley & Sons, Inc., 1962).
12. Frobisher, M., Hinsdill, R. D., Crabtree, K. T., and Goodheart, C. R., *Fundamentals of Microbiology* (Philadelphia, PA: W. B. Saunders Co., 1974).
13. Kane, B. E. and Mullins, J. T. "Thermophilic Fungi in a Municipal Waste Compost System," *Mycologia* 65: 1087–1100 (1973).
14. Love, G. J., Tompkins, E., and Galke, W. A. "Potential Health Impacts of Sludge Disposal on the Land," paper presented at the 2nd National Conference on Sludge Management and Disposal, Anaheim, CA (1975).
15. "Intestinal Parasite Surveillance, Annual Summary 1976," U.S. Center for Disease Control, Atlanta, GA (1977).
16. Brannen, J. P., Garst, D. M., and Langley, S., "Inactivation of Ascaris Lumbricoides Eggs by Heat, Radiation, and Thermoradiation," Sandia Laboratories, NM (1975).
17. Havellar, A. H. and Niemela, S. I. "Hygienic Constraints to the Use of Sludge in Agriculture Due to the Content of Pathogens," in *Proceedings of the 3rd International Symposium on Processing and Use of Sewage Sludge* (Lancaster, England: D. Reidel Publishing Co., 1983).

18. Emmons, C. W. "Natural Occurrence of Opportunistic Fungi," *Lab. Invest.* 11(2):1026–1032 (1962).

19. Millner, P. D., Marsh, P. B., Snowden, R. B., and Parr, J. F. "Occurrence of Aspergillus fumigatus During Composting of Sewage Sludge," *Appl. Environ. Microbiol.* 34(6) (1977).

20. Gregory, P. H. *The Microbiology of the Atmosphere* (New York: John Wiley & Sons, 1973).

21. Pahren, J., Lucas, J., and Kowal, N. E., "An Assessment of the Health Risks from the Oxon Cove Compost Piles Resulting from *Aspergillus fumigatus*," letter report to EPA Region III (1978).

22. Solomon, W. R. "Assessing Fungus Prevalence in Domestic Interiors," *J. Allergy Clin. Immunol.* 53(2):71 (1974).

23. Austwick, P. K. "Ecology of *Aspergillus fumigatus* and the Pathogenic Phycomycetes," in *Recent Progress in Microbiology* (Toronto, Canada: University of Toronto Press, 1963).

24. Austwick, D. K. "The Role of Spores in the Allergies and Mycoses of Man and Animals," in *Proceedings of the 18th Symposium, Colston Research Society* (London: Butterworths, 1966).

25. Lacey, J. and Lacey, M. E. "Spore Concentrations in the Air of Farm Buildings," *Trans. Br. Mycol. Soc.* 47:547–552 (1964).

26. Hudson, H. J. "Thermophilous and Thermotolerant Fungi in the Air-Spora at Cambridge," *Trans. Br. Mycol. Soc.* 60:596–598 (1973).

27. Millner, P. D., Bassett, D. A., and Marsh, P. B. "Dispersal of *Aspergillus fumigatus* from Sewage Sludge Compost Piles Subjected to Mechanical Agitation in Open Air," *Appl. Environ. Microbiol.* 39(5) (May 1980).

28. LeBrun, T. Memo to the LA/OMA Project on status of *Aspergillus* monitoring, Los Angeles County Sanitation Districts (1979).

29. Kuchenrither, R. Black & Veatch Consulting Engineers, Denver Composting Facility Task Memorandum No. 5, (January 1984).

30. Willson, G. B., Parr, J. F., Epstein, E., Marsh, P. B., Chaney, R. L., Colacicco, D., Burge, W. D., Sikora, L. J., Tester, C. F., and Hornick, S. *Manual for Composting Sewage Sludge by the Beltsville Aerated-Pile Method,* U.S. EPA, Municipal Environmental Research Lab., Cincinnati OH, EPA-600/8-80-022 (May 1980).

31. Marsh, P. B., Millner, P. D., and Kla, J. M. "A Guide to the Recent Literature on Aspergillosis as Caused by *Aspergillus fumigatus*," Agricultural Reviews and Manuals, USDA, ARM-NE-5 (September 1979).

32. McCarty, P. L. "Thermodynamics of Biological Synthesis and Growth," *Adv. Water Poll. Research* 2(1) (1964).

33. Servizi, J. A. and Bogan, R. H. "Free Energy as a Parameter in Biological Treatment," *J. San. Eng. Div., ASCE* 89(SA3):17 (1963).

34. Payne, W. J. "Energy Yields and Growth of Heterotrophs," *Ann. Rev. Microbiol.* 24:17 (1970).

35. Sykes, R. M. "Theoretical Heterotrophic Yields," *J. Water Poll. Control Fed.* 47(3) (1975).

36. McCarty, P. L. "Energetics and Bacterial Growth," paper presented at the 5th Rudolf Research Conference, Rutgers University, NJ, (July 1969).

37. O'Rourke, J. T. "Kinetics of Anaerobic Treatment at Reduced Temperatures," PhD Thesis, Stanford University, CA (1968).

38. McKinney, R. E. "The Value and Use of Mathematical Models for Activated Sludge Systems," in *Progress in Water Technology,* Vol. 7, No. 1, (New York: Pergamon Press, 1975).

39. Randall, C. W., Richards, J. B., and King, P. H., "Temperature Effects on Aerobic Digestion Kinetics," *J. Env. Eng. Div., ASCE* (October 1975).

40. Lees, P. S. J. and Tockman, M. S. *Evaluation of Possible Public Health Impact of WSSC Site II Sewage Sludge Composting Operations,* Johns Hopkins University School of Hygiene and Public Health, prepared for Maryland Department of Health and Mental Hygiene, Baltimore, MD (1987).

41. Energy Resources Co., Inc. (ERCO). *Monitoring of A. fumigatus Associated with Municipal Sewage Sludge Composting Operations in the State of Maine,* prepared for the Portland Water District, Portland, ME (1980).

CHAPTER 5

Kinetics of Heat Inactivation

INTRODUCTION

Many composting substrates have the potential for carrying human, animal and plant pathogens, and other undesirable biological agents such as weed seeds. Sludge from municipal sewage treatment is an obvious example of a pathogen containing substrate, but other substrates can also contain human and animal pathogens. For example, municipal refuse can contain surprisingly high concentrations of fecal coliform bacteria and other indicator organisms. Increased use of disposable diapers and the inclusion of dog and cat feces in refuse are significant sources of pathogen contamination. Yard wastes often carry plant pathogens, which are present when diseased vegetation is removed. Seeds from undesirable plants can contaminate many substrates, including municipal sludges and yard wastes. Tomato and melon seeds are ubiquitous in sewage sludges while grass and other plant seeds are prevalent in yard wastes. Fortunately, all of these biological agents are effectively controlled by maintaining elevated temperatures for minimum time periods during composting. Heat inactivation of human, animal and plant pathogens, and seeds is one of the major benefits of thermophilic composting.

Heat death of a cell results in part from thermal inactivation of its enzymes. Enzymes that may be reversibly inactivated by mild heat are irreversibly inactivated by higher temperatures. If an enzyme denatures reversibly with temperature, then at equilibrium a fraction of the enzyme will be in the active form, with the remainder in a denatured (inactive) form. Thermodynamic arguments indicate that the fraction in the active form decreases significantly over a narrow temperature range.[1] Without enzyme activity a cell cannot function and will die. Very few enzymes can withstand prolonged heat. Therefore, microbes can be expected to be very sensitive to thermal inactivation.

Processes other than heat inactivation are also available for pathogen control. Use of ionizing radiations, such as electron beam and gamma ray sources, has seen limited application on sludges. Chemical treatment of sludge, such as by lime or lime/cement addition, can also reduce pathogen levels, but much of the reduction can be attributed to the elevated temperatures that come from the heat of chemical hydration. With these minor exceptions, heat inactivation remains the most widely accepted and commonly used method for providing pathogen destruction.

This chapter reviews the kinetics of heat inactivation, discusses the parameters of importance to heat pasteurization during composting, and analyzes human health risks associated with various pathogen levels in the final compost. The focus is on destruction of human pathogens, because this assures that most animal and plant pathogens and seeds are also destroyed. It should be noted, however, that heat inactivation is not the only method of pathogen destruction in a compost system. Organisms are also destroyed or controlled by competition with other microbes, antagonistic relationships, and antibiotic or inhibiting substances produced by certain microbes.[2] Time is another factor that affects survival because of the natural die-off of pathogenic organisms in unsuitable environments. Nevertheless, temperature is the one factor that the operator can measure and control during composting. For these reasons, regulatory agencies are prone to use temperature as a measure of pathogen destruction. Therefore, this discussion will center on kinetics of heat inactivation as applied to composting systems.

REGULATORY APPROACHES

A number of regulatory approaches are used by different countries to control the quality of composts. For sludge based composts, the U.S. EPA has published performance standards for two classes of pathogen reduction.[3] Sludge based compost intended for distribution and marketing to the public must meet Class A requirements. Several alternatives are available for meeting the Class A requirements. In all cases the density of fecal coliform must be < 1000 most probable number (MPN) per gram ds or the density of *Salmonella* sp. < 3 MPN/4 g ds. In addition, either the pathogenic quality must be determined or the process monitored for time/temperature conditions. Pathogenic quality is assured if the density of enteric viruses in the compost is < 1 plaque forming unit (PFU)/4 g ds and the density of helminth ova < 1 viable ova/4 g ds. Time/temperature conditions are specified by equation, but are equivalent to 53° C for 5 days, 55° C for 2.6 days, and 70° C for 30 minutes. The minimum temperature is 50° C.

The recommended density of indicator organisms is 1000 MPN fecal coliform/g solids. Farrell based this limit on a statistical analysis of many compost samples, which showed that salmonellae were absent from all samples with <1000 fecal coliform/g. If the fecal coliform limit cannot be achieved, the facility can monitor for salmonellae. If the latter are absent, the monitoring requirement is satisfied. The use of fecal indicators as a measure of pathogen reduction applies only in cases where heat inactivation is the primary mechanism for pathogen destruction. This is because of the large compendium of laboratory and field studies that indicate that the reduction of fecal indicators to low levels satisfactorily indicates the reduction of pathogens to insignificant levels by thermal processes.[5]

The issue of potential regrowth of pathogenic bacteria, such as *Salmonella*, is addressed by requiring that stabilization be accomplished with or after the pathogen reduction. The reasons for this is that many nonpathogenic bacteria are also destroyed under the time/temperature conditions necessary to destroy the pathogens. The nonpathogenic bacteria normally act as competitors with pathogenic bacteria and help prevent regrowth. If they are absent, explosive regrowth can occur.[5] When stabilization occurs simultaneously with or shortly after pathogen reduction, explosive regrowth is prevented. Composting accomplishes both stabilization and heat inactivation within the same process, and an active bacterial population is always maintained. Therefore, the potential for regrowth is greatly reduced.

Class B requirements specify a minimum density of indicator organisms in the treated sludge. The density of indicator organisms is 2×10^6 fecal coliform/g ds. Other alternatives are also available to meet the Class B requirements. Public access and cropping restrictions are included in the Class B definition.

The above regulations are but one example of the types of management approaches being considered in various countries. For example, countries of the EEC seem to place more importance on certain target organisms. Havelaar[6] identified vegetative cells of the bacterium *Salmonella*, eggs of the tapeworm *Taenia saginata,* and the roundworm *Ascaris* as being of primary importance. The 1986 EEC Directive[55] requires that sludge be treated before it's applied to farmland surface to reduce "significantly" both the fermentability of sludge and the potential health hazards from its use. Detailed interpretation of some of the requirements is left to each EEC Member State. However, both Switzerland and the Federal Republic of Germany included requirements for disinfection in their ordinances on sewage sludge.[7] Such regulations reflect the growing concern over pathogens and the desire to minimize risk from using of sludge-based or compost products.

The establishment of acceptable levels of indicator organisms is not without some controversy. Comparing the indicator bacteria levels for EPA Class A sludge to levels in other products suggests that the proposed regulations are probably conservative. For example, milk must meet bacterial levels specified in the U.S. FDA Pasteurized Grade A Milk Ordinance[8] of 20,000 total bacteria/ml and 10 and 100 coliform/ml in packaged and bulk milk, respectively. Because 1 ml is approximately equal to 1 g for milk; the coliform limit in bulk milk is about 100/g. Therefore, the indicator organism density required for EPA Class A sludge is only 10 times that required for milk intended for human consumption.

TIME-TEMPERATURE RELATIONSHIPS

Sterilization is the process of destroying all life forms. Bacterial spores and certain cysts are usually the most resistant microbial forms to temperature, consistent with earlier remarks on endospore formation as a survival mechanism. Cysts and spores, for example, resist exposure to dry heat of over 100°C. Dry sterilization, therefore, requires a temperature of about 180°C for 2 to 3 h to assure complete destruction. Moist sterilization, on the other hand, requires only 15 min at 115°C (15 psig pressure in an autoclave or pressure cooker). Boiling at 100°C kills some cysts and spores, but induces germination in others. After a time lapse a second boiling is necessary to kill bacteria or protozoa that have emerged from the spores or cysts. Presumably, requirements for sterilization of a liquid slurry or sludge would be similar to those for moist sterilization, namely 115°C for 15 min.

The reason moist and dry heat differ in their inactivation effects is that enzyme denaturation is influenced by solvent concentration. The temperature needed to coagulate (denaturation followed by massive cross-linking of the denatured protein) egg protein albumin increases with decreasing moisture content as shown in Table 5.1. Resistance of bacterial endospores to heat is probably caused in part by their dehydrated condition.

Pasteurization implies heating to a specific temperature for a time sufficient to destroy pathogenic or undesirable organisms. In general, nonspore-forming bacteria and the vegetative cells of sporulating bacteria are destroyed in 5 to 10 min at temperatures of 60 to 70°C (moist heat). Data presented by Roedinger,[10] shown in Table 5.2, suggest that pasteurization at 70°C for 30 min destroys pathogens found in sludge. A similar compilation of time/ temperature relationships compiled by Gotaas,[11] shown in Table 5.3, also indicates that *Entamoeba histolytica* is pasteurized at relatively low temperatures. Stern[12] reported that

Table 5.1.   Effect of Hydration and Heat on Egg Albumin

Water Content (%)	Approximate Coagulation Temperature (°C)
50	56
25	76
15	96
5	149
0	165

Source: Frobisher et al.[9]

Table 5.2.   Temperature and Time for Pathogen Destruction in Sewage Sludges

Microbe	Exposure Time (min) for Destruction				
	50°C	55°C	60°C	65°C	70°C
Cysts of *Entamoeba histolytica*	5				
Eggs of *Ascaris lumbricoides*	60	7			
Brucella abortus		60		3	
Corynebacterium diptheriae		45			4
Salmonella typhi			30		4
Escherichia coli			60		5
Micrococcus pyogenes					20
Mycobacterium tuberculosis					20
Viruses					25

Source: Roediger.[10]

Table 5.3.   Temperature and Time for Pathogen Destruction in Sewage Sludges

Microbe	Destruction Time/Temperature			
	Temp (°C)	Time (min)	Temp (°C)	Time (min)
Salmonella typhosa	55–60	30	60	20
Salmonella sp.	55	60	60	15–20
Shigella sp.	55	60		
Entamoeba histolytica cysts	45	few	55	few sec
Taenia	55	few		
Trichinella spiralis larvae	55	quickly	60	few sec
Brucella abortis	62.5	3	55	60
Micrococcus pyogenes	50	10		
Streptococcus pyogenes	54	10		
Mycobacterium tuberculosis	66	15–20	67	few
Corynebacterium diphtheriae	55	45		
Necator americanus	45	50		
Ascaris lumbricoides eggs	50	60		
Escherichia coli	55	60	60	15–20

Source: Gotass.[11]

temperatures of 75°C for 1 h destroyed enteric pathogens and usually reduced coliform indicator concentrations to <1000/100 ml.

From the data presented in Tables 5.2 and 5.3 it is apparent that thermal inactivation is a time/temperature phenomenon. A high temperature for a short period of time or a lower temperature for longer duration can be equally effective. Such time/temperature relationships

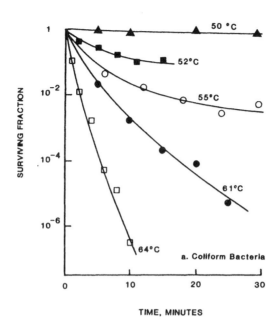

Figure 5.1A. Heat inactivation of coliform bacteria in composting sludge. From Ward and Brandon.[13]

for a variety of organisms are presented in Figures 5.1 through 5.6. Such data are more informative than tabular summaries of the kind shown in Tables 5.2 and 5.3, because they allow development of mathematical expressions or models to represent the data.

Time/temperature conditions presented in Figure 5.1 for enteric bacteria and *Ascaris* generally support the inactivation conditions presented in Tables 5.2 and 5.3 and those proposed by U.S. EPA regulations for Class A sludges. It would appear that a temperature of 60°C held for 30 min provides about 6 $\log_{10}$ reduction for both coliform and *Salmonella*. Fecal streptococcus appears to be somewhat more heat resistant, requiring about 67°C or a longer time for the same level of inactivation. A temperature of 60°C for 30 min should provide similar levels of reduction for the parasite *Ascaris lumbricoides*, considered to be one of the more heat resistant parasites.

Heat inactivation of virus is somewhat more complicated than for other classes of microbes. Time/temperature profiles developed by Ward et al.[15] for a poliovirus strain are presented in Figure 5.2 for liquid raw and digested sludge (about 5% solids) and a phosphate-buffered saline solution. Dramatically different responses to temperature are apparent in the data for raw and digested sludges. Raw sludge is more protective of the viral particle, whereas protection is significantly reduced in digested sludge. Through a series of experiments, Ward et al.[19] concluded that raw sludge contains a protective substance whose activity was overwhelmed by a virucidal agent acquired during digestion. The latter was determined to be ammonia (NH_3) in the uncharged state. As would then be expected, virucidal activity is a function of pH, increasing with higher pH. Indeed, raw sludge that was protective at pH 7 became nonprotective at pH 9 because of the shift of ammonia from the charged to the uncharged form. Ammonia was also shown to be virucidal to strains of poliovirus, coxsackievirus and echovirus. Reovirus was found to be relatively insensitive to ammonia.

Ward et al.[19] concluded that the protective agent identified previously was concentrated during the drying process because heat resistance increased in drier samples. Alternatively, it could be that resistance was increased because coagulating temperatures generally increase

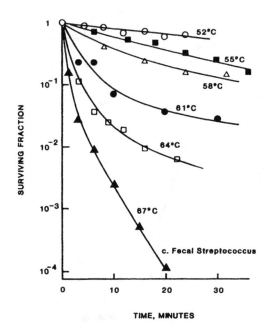

Figure 5.1B. Heat inactivation of *Salmonella enteritidis* in composting sludge. From Ward and Brandon.[13]

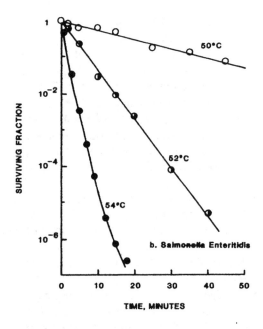

Figure 5.1C. Heat inactivation of fecal streptococcus in composting sludge. From Ward and Brandon.[13]

with solids content, as shown in Table 5.1 for egg albumin. This is probably not the case, however, because later experiments using composted sludge showed that poliovirus is inactivated more rapidly in composted sludge than in raw sludge with the same moisture content (Figure 5.4). Apparently the protective agent is concentrated as raw sludge is dewatered, but it is removed or destroyed during composting.

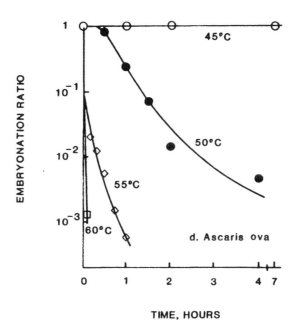

Figure 5.1D. Heat inactivation of *Ascaris* ova in composting sludge. From Sandia Laboratories.[14]

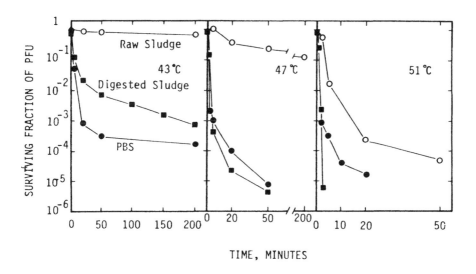

Figure 5.2.   Effect of sludge on the rate of heat inactivation of poliovirus. After a tenfold dilution into buffer (PBS), raw sludge, or anaerobically digested sludge, poliovirus strain CHAT was mixed for 15 min at room temperature and heated at the specified temperature. Samples were removed from the incubation bath at the times shown and immediately cooled in ice. After sonication in 0.1% SDS, each sample was directly assayed for total PFU on HeLa cells. From Ward et al.[15]

Various organisms have at times been suggested as the most thermal tolerant of the enteric pathogens expected in sludge. Often mentioned candidates include cysts of the protozoan *E. histolytica* and eggs of *A. lumbricoides*. Enteric bacterial pathogens are not spore-forming and are usually considered less heat tolerant than protozoan cysts or helminth eggs; however, certain vegetative bacterial cells, such as *Mycobacterium tuberculosis*, are reasonably heat

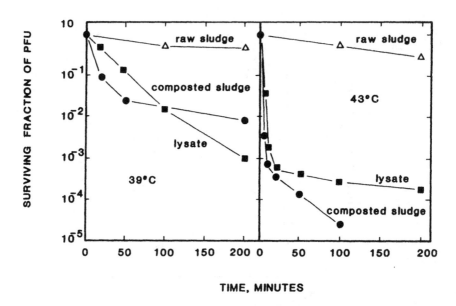

Figure 5.3.   Comparative rates of heat inactivation of poliovirus samples in dewatered raw sludge, composted sludge, and lysate of poliovirus-infected cells. Poliovirus was blended with dried raw sludge or composted sludge whose final solids contents had been adjusted to 40%. The rates of viral inactivation by heat at 39 and 43°C were then determined and compared with that in a lysate of infected HeLa cells.[14]

resistant. Kruse[18] reported the data shown in Figure 5.6, comparing temperature sensitivity of human cysts with eggs of *Ascaris*. It was concluded that conditions that disinfect *Ascaris* would also destroy protozoan cysts, and that *Ascaris* would be an appropriate indicator for determining the degree of disinfection. Indeed, *Ascaris* has been accepted widely as an indicator organism because of its thermal tolerant nature relative to other enteric pathogens.

The origin of the EPA minimum 50°C requirement is derived from results obtained by Brannen et al.[20] *Ascaris* egg densities were reduced 2 $\log_{10}$ in about 5 min at 55°C and in 60 min at 51°C, but there was essentially no destruction after 2 h at 47°C. These results are consistent with those presented in Figure 5.1D. To assure destruction, the processing temperatures should exceed this apparent threshold temperature. Thus, a minimum temperature of 50°C was selected as sufficient to cause a rapid reduction in *Ascaris* density.[3]

MODELING HEAT INACTIVATION KINETICS

First Order Decay Model

Referring to Figure 5.1, it is not uncommon to observe straight lines (or nearly so) through temperature survival data in semilog plots. Thus, inactivation kinetics are often modeled assuming first order decay as follows:

$$dn / dt = -k_d n \qquad (5.1)$$

where

n	=	viable cell population
k_d	=	thermal inactivation coefficient

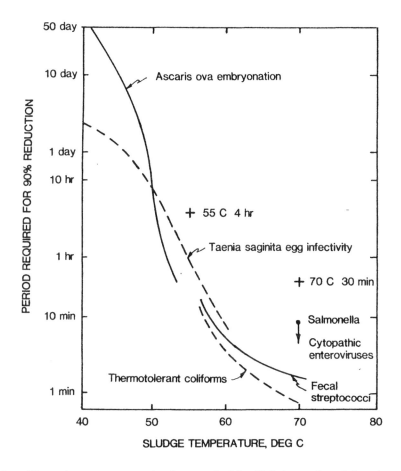

Figure 5.4.   Effect of temperature on the time required for 90% destruction of *Ascaris* ova, *Taenia saginita* eggs, fecal steptococci, and thermotolerant coliforms in sludge. Compiled from several references and presented by Bruce et al.[16]

Equation 5.1 is often referred to as "Chick's Law" after Harriet Chick[21] who reported on such exponential die-off in 1908. However, some survival curves exhibit a nonexponential or aberrant nature such that Equation 5.1 is not applicable. Nonexponential curves are discussed later.

If k_d is constant, integration of Equation 5.1 from an initial cell population, n_o, to a later population, n_t, at time, t, yields

$$n_t = n_o e^{(-k_d t)} \tag{5.2}$$

Taking the log of both sides and rearranging

$$t = [\ln(n_o / n_t)] / k_d$$

Converting to base 10 logs and considering a one log reduction in cell concentration (i.e., a reduction of 90%):

$$t_{90} = D_r = 2.303 / k_d \tag{5.3}$$

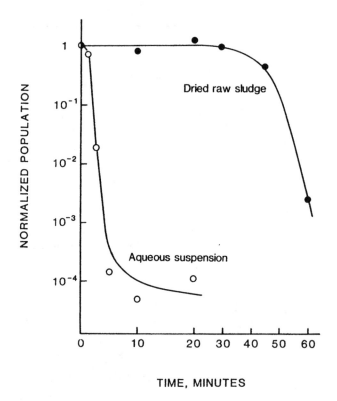

Figure 5.5.   Heat inactivation of *Aspergillus flavus* conidia in aqueous suspension and in dried raw sludge. From Brandon et al.[17]

The term D_r is the decimal reduction factor and is the time required to achieve a 10-fold reduction in cell population. Representative D_r values from the literature are presented in Table 5.4. Note that D_r for *E. histolytica* cysts is reported to be greater than that for *Ascaris* ova. This contradicts previously cited work by other authors and points out some of the variability seen in literature data.

The value of k_d is a function of temperature, the effect of which is most often modeled by the familiar Arrhenius form:

$$k_d = Ce^{(-E_d/RT_k)} \qquad (5.4)$$

where
$$T_k \quad = \quad \text{temperature, } °K$$

The range of inactivation energies, E_d, for many spores and vegetative cells is between 50 and 100 kcal/mol.[1] This implies that heat inactivation of microbes is much more sensitive to temperature than most chemical reactions. Logarithmic transformation of Equation 5.4 yields

$$\log_{10} k_d = \log_{10} C - (E_d / R)(1 / T_k) \qquad (5.5)$$

Thus, a plot of the logarithm of k_d vs. $1/T_k$ allows determination of the constant, C, and the inactivation energy, E_d.

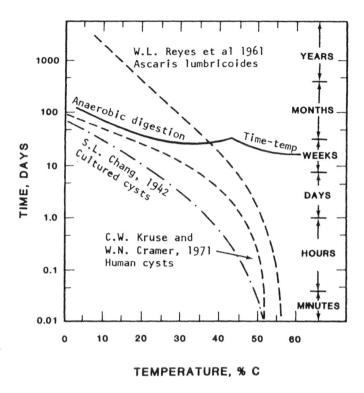

Figure 5.6.   Die-away of cultured and human cysts of *Entamoeba histolytica* and ova of *Ascaris lumbricoides* var. *suum* under anaerobic digestion conditions giving 99.9% destruction. From Kruse.[18]

Table 5.4.   Time Required for a 10-Fold Population Reduction for Various Microbes

Microbes	D_r (min)	
	55°C	60°C
Adenovirus, 12 NIAID	11	0.17
Poliovirus, type 1	1.8	1.5
Ascaris ova	—	1.3
Histolytica cysts	44	25
Salmonella[a]	80	7.5
Bacteriophage f2	267	47

Source: Burge et al.[22]

[a] Serotype Senftenburg 77W

Example 5.1

Determine values for the thermal death coefficient, k_d, and the inactivation energy, E_d, for *A. lumbricoides* from data in Figure 5.1D.

Solution

1. Since the curves in Figure 5.1D are not straight lines, some approximation will be necessary to fit the first order assumption. Fitting straight lines from the origin through the data as shown, the following table can be constructed:

Temp (°C)	n_t/n_o	Time, t (min)	k_d min^{-1}	$1/T_k$	T_c–50 °C	D_r min
50	0.001	270	0.0256	30.96×10^{-4}	0	90.0
60	0.001	45	0.154	30.49×10^{-4}	5	15.0
70	0.001	7.5	0.921	30.03×10^{-4}	10	2.5

Values of k_d were determined from Equation 5.2 as follows:

$$\ln(n_o / n_t) = k_d t$$

$$k_d = [\ln(n_o / n_t)] / t$$

For 50°C

$$k_d = [\ln(1000)] / 270 = 0.0256 \text{ min}^{-1}$$

2. Values of k_d are plotted as a function of $1/T_k$ and $(T_c - 50°C)$ in Figure 5.7. Note that a good straight-line fit is obtained, justifying use of the Arrhenius relationship. To determine E_d, values of k_d and $1/T_k$ from the curve in Figure 5.7 were applied to Equation 5.5 with the following results:

$$C = 1.81 \times 10^{49}$$

$$E_d = 75,200 \text{ cal / mol} = 75.2 \text{ kcal / mol}$$

Note that the value of E_d is consistent with the previously reported range of values for heat inactivation of microbes. Substituting these values into Equation 5.4 and recalling that R = 1.99 cal/deg-mol:

$$k_d = 1.81 \times 10^{49} e^{-(3.78 \times 10^4)/T_k}$$

This equation is somewhat awkward because of the large exponents. Using alternative forms of the Arrhenius equation, the following expression can be developed from the plot of k_d vs. T_c–50:

$$k_d = 0.025 e^{0.361(T_c - 50)}$$

3. The equation for k_d can be substituted into Equation 5.2 to give

$$\ln(n_o / n_t) = 0.025 t e^{0.361(T_c - 50)}$$

Converting to base 10 logs:

$$2.303 \log(n_o / n_t) = 0.025 t e^{0.361(T_c - 50)}$$

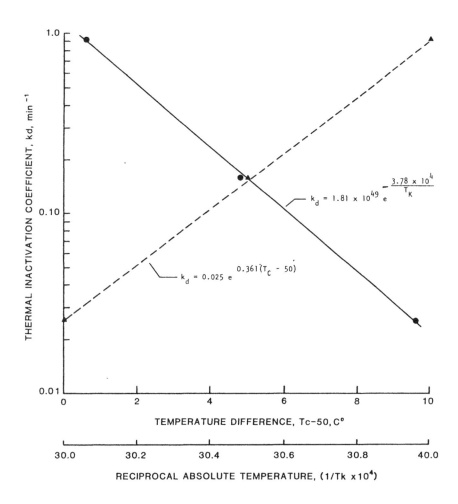

Figure 5.7.   Thermal inactivation coefficient as a function of reciprocal temperature, 1/T, and temperature difference, T_c-50. Data for *Ascaris* from Figure 5.1D.

Thus, for any combination of time, t, and temperature, T_c, the log reduction can be calculated. Using this approach the curves in Figure 5.8 were calculated. Note that over 25 $\log_{10}$ reductions can be achieved in 1 to 2 days with temperatures above 50°C. This is an astronomical number, indicating complete and total destruction of *Ascaris* for all practical purposes.

In composting as well as in many batch sterilization processes, it is common for material to be heated, held at a relatively constant temperature for a time, and then cooled. The kill resulting from such a time/temperature profile can be evaluated by first combining Equations 5.1 and 5.4:

$$dn / dt = -Cne^{-[E_d/RT_k(t)]} \tag{5.6}$$

Temperature, T_k, is a function of time and is expressed as $T_k(t)$. Separating variables and integrating from initial conditions, $n = n_o$ at $t = 0$, to final conditions, $n = n_f$ at $t = t_f$:

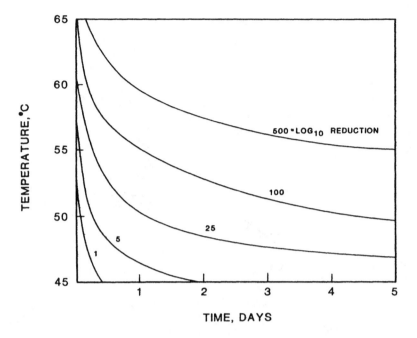

Figure 5.8.   Time/temperature conditions to achieve stated $\log_{10}$ reductions for *Ascaris* ova using equations from Example 5.1.

$$\ln(n_o / n_t) = \int_0^{t_f} Cne^{-[E_d/RT_k(t)]}dt$$

$$(5.7)$$

Before the right side of Equation 5.7 can be integrated, the temperature profile must be known. Consider first a constant temperature profile (i.e., T not a function of time). Equation 5.7 then integrates directly to

$$\ln(n_o / n_t) = Ct_f e^{-[E_d/RT_k]}$$

$$(5.8)$$

where T_k = constant. Similar integrations have been performed for a variety of other time/temperature profiles including hyperbolic, exponential, and linear increases. Bailey and Ollis[1] present further discussion of the analytic approach to solution. Regardless of the type of time/temperature profile, total kill is the sum of kills observed in the heating, constant (holding) temperature, and cooling periods. Thus,

$$\ln(n_o / n_t)_{Total} = \ln(n_o / n_t)_{heating} + \ln(n_o / n_t)_{constant} + \ln(n_o / n_t)_{cooling}$$

$$(5.9)$$

Actual time/temperature profiles observed in composting may not correspond to profiles for which analytical solutions are available. In such cases a more practical approach is to use graphical integration procedures to solve the right side of Equation 5.7. The graphical approach can be used with any profile and is not limited by our ability to describe the profile in formal mathematical terms. This method is illustrated in the following example:

Example 5.2

A time/temperature profile for a batch composting system is presented in Figure 5.9. Using a graphical integration approach, determine the log reduction of *Ascaris* ova using equations developed in Example 5.1. Assume the minimum lethal temperature to be 45°C.

Solution

1. The time/temperature profile in Figure 5.9 is approximated by constructing a number of smaller elements as shown. Note that the first element begins at 45°C, the assumed minimum lethal temperature. The width of each element is Dt. Recall that Dt approaches dt as Dt becomes small. Average temperature for an element (indicated by points on the curve) is used to determine an average k_d for that element using the equation developed in Example 5.1:

$$k_d = 0.025e^{0.361(T_c - 50)}$$

Thermal inactivation for each element is determined from Equation 5.2 as

$$\ln(n_o / n_t) = k_d \Delta t$$

The following table can be developed from the elements in Figure 5.9:

Element Number	Δt days	Avg T_c °C	k_d min^{-1}	$\ln(n_o/n_t)$
1	0.25	47.7	0.011	4
2	0.25	53.2	0.079	29
3	0.25	56.7	0.281	101
4	0.25	59.0	0.644	232
5	0.25	60.0	0.924	333
6	0.75	60.3	1.030	1112
7	0.25	60.0	0.924	333
8	0.25	58.8	0.599	216
9	0.25	57.0	0.313	113
10	0.25	54.7	0.136	49
11	0.25	52.0	0.051	19
				$\Sigma\ln(n_o/n_t) = 2541$

2. The total $\log_{10}$ reduction is then determined as

$$\log_{10}(n_o / n_t) = \frac{\ln(n_o / n_t)}{2.303} = 2541 / 2.303 = 1100$$

This is 1100 $\log_{10}$ units of reduction! In other words,

$$n_f = n_o 10^{-1100}$$

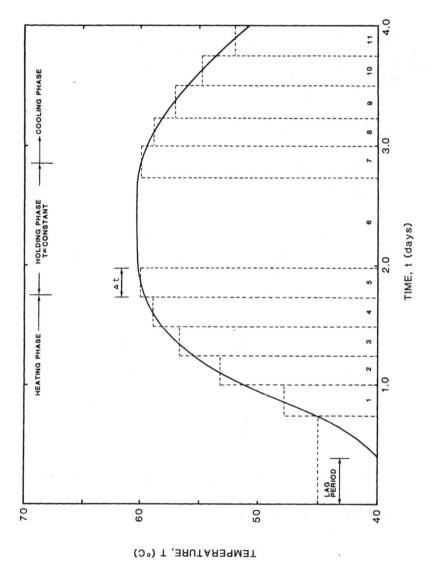

Figure 5.9.   Idealized time/temperature profile for a batch composting system (Example 5.2).

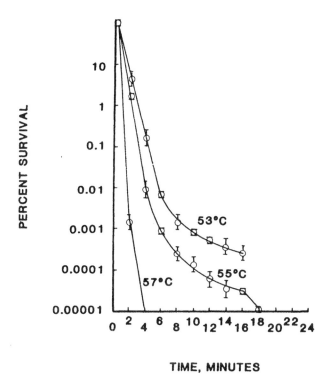

Figure 5.10,   Heat destruction of *Staphylococcus aureus* in neutral buffer. From Frobisher et al.[9]

which is essentially equal to 0. Data contained in Figure 5.1d are reasonably consistent with other data presented in Tables 5.3 and 5.4. Therefore, it must be concluded that essentially no Ascaris ova would survive the time/temperature profile of Figure 5.9. Even deviations from the first order rate model as n becomes small are not likely to change this outcome.

Nonexponential Decay Curves

In some cases the assumption of first order decay may be inappropriate to the data. Deviation from first order decay is sometimes observed as the number of microbes decreases, as illustrated in Figure 5.10 for *Staphylococcus aureus*. Similar effects can be noticed in some of the data in Figure 5.1. In explaining these effects it has often been assumed that not all cells in the culture are equally vulnerable to the temperature conditions. More vulnerable cells are assumed to be killed first (higher k_d), leaving a small number of heat-resistant cells (lower k_d). Also, as the number of organisms becomes small, the assumption of a uniform population characterized by a single value of k_d begins to fail. Whether this explanation is correct or not, temperatures can be raised to the point where such fluctuations are minimized. Note that increasing the temperature from 53 to 57°C in Figure 5.10 results in a more linear plot with nearly complete inactivation of those microbes resistant at 53°C.

Time/temperature relationships observed during heat inactivation of microbes can be generalized by the forms shown in Figure 5.11. Curve A is the first order or exponential decay form as previously discussed. Curve B shows an initial phase of first order decay followed by a subsequent decrease in the rate of inactivation. In some cases the curve can become almost asymptotic to the time axis. Curve C shows an initial lag period followed by first order decay.

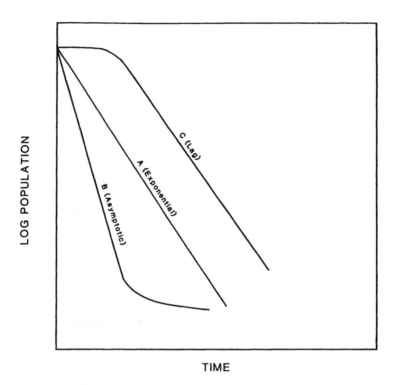

Figure 5.11. Generalized types of curves observed in studies of inactivation of microbes: (A) first order or exponential decay; (B) first order followed by asymptotic or retardant die-away; (C) initial lag or shoulder followed by first order decay.

Sigmoid or S-shaped curves are possible by combination of Curves B and C. Elements of these generalized curves can be seen in the time/temperature relationships already presented.

A disturbing question immediately comes to mind. If Curve B is applicable to some of the enteric pathogens, does this mean that there is a limit to expected inactivation at a particular temperature? Will a small but troublesome fraction of the population survive? By applying a first order model have we been guilty of bending reality to fit a mathematically pleasing equation? To answer these questions the nature and cause of the asymptotic and lag portions of Curves B and C must be explored.

Several authors have advanced "multiple target" or "multiple hit" theories to explain the death of microbes as a result of heat, chemicals, or ionizing radiation. In these theories a number of critical molecular sites, which can be inactivated by the disinfecting agent, are envisioned. For example, Moats[23] proposed that thermal injury and death could be explained by assuming that death results from inactivation of X_1 of a total of N critical sites per cell. The critical sites were thought not to be enzymes because death resulted from inactivation of only a small percentage of the total sites. Moats predicted that X_1 would vary depending on the recovery medium used for heating. This provided some explanation for the observation that some microbes are only injured by mild heat treatment and can recover if grown on enriched media. Moat's model was able to fit the lag period in the type C curve but not the asymptotic portion of the B curve. The latter was attributed to the presence of a small heat resistant population. Unfortunately, the asymptotic region is the one of most concern, as previously discussed.

Wei and Chang[24] proposed an explanation that can account for both the lag and asymptotic phases. Microbial death in a disinfection process was assumed to result from random collision between molecules of a disinfectant and the microbe. Since the disinfectant is present in extremely large numbers compared to the microbes, a Poisson probability was used to describe the collision rate with a single microbe. Microbes were also allowed to exist in clumps of various sizes, leading to a multi-Poisson distribution model. The model was tested using amoeba cysts of *Naegleria gruberi* and also applied to data on coliform and virus destruction.

The result was the realization that clumping of organisms is a major, if not the only, factor responsible for aberrant survival curves in disinfection studies. Exponential curves were obtained when cysts were treated to produce discrete particles. The lag period of type C curves results when a large percentage of the population exists in clumps. The higher the percentage and the larger the clumps, the more prominent is the shoulder. To better understand this, recall that plating techniques used to recover viable microbes cannot distinguish between a single microbe or a clump of microbes, where any from one to all of the microbes may be viable. Disinfection may proceed with first order kinetics within a clump, but as long as one organism remains viable the cell count by traditional plating techniques will remain relatively unchanged. As the number of viable cells within each clump approaches unity, exponential rate kinetics again become evident.

The asymptotic curves of type B are caused by a large percentage of the population existing as discrete particles along with a number of clumps of unusually large size. The higher the percentage of singles, the more marked is the initial exponential drop. The situation with a sigmoid inactivation curve is shown in Figure 5.12. A large majority of small aggregates produce an initial shoulder and a small number of large aggregates produce a plateau that plunges to the time axis after all individual organisms are destroyed.

Wei and Chang also showed that if the asymptotic segment is located below the 5% survival level, the value of k_d for destruction of a single microbe can be computed from the exponential segment. If the asymptotic section is above the 10% survival level, clump survival will reduce the slope of the linear segment from that which would be observed if all organisms were singles. Similar limits apply to the type C curves. Within these limits, estimating k_d and D_r values is relatively straightforward despite the aberrant nature of many curves.

An important result of this model is that the asymptotic segment will drop steeply once all organisms in the clumps have been destroyed. Furthermore, destruction of microbes within a clump appears to follow the same exponential decay coefficient observed for singles. As Burge et al.[22] pointed out, "this does much to erase from our minds the concept that a portion of a population will survive composting for some reason we have not been able to fathom."

One final point will be made before leaving this subject. Other nonexponential curves have been observed, such as the sigmoid shaped, those that curve downward with constantly increasing slope (concave viewed from the origin), those that curve downward with a constantly decreasing slope (convex) and step ladder-shaped curves. Each of these types can be explained by assuming clumps of various size and frequency of distribution. Estimating k_d and D_r values for these types is mathematically more difficult. Nevertheless, there is reasonable certainty that asymptotic sections do not represent heat resistant portions of a population.

STANDARDS OF PERFORMANCE

Stochastic models have been applied to the problem of sterilization to statistically estimate numbers of viable organisms remaining after a certain time/temperature exposure. Assuming the usual case of a large number of initial microbes ($n_o \gg 1$), application of first order kinetics described by Equation 5.1, the independence of each organism, and their exposure to the same lethal condition, the probability of at least one organism surviving is given by

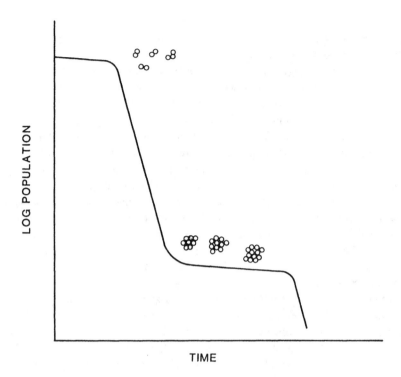

Figure 5.12.   Sigmoid type inactivation curve with aggregate sizes of microorganisms producing the shoulder and plateau parts of the curve. After complete inactivation of the large aggregates, the curve plunges to the time axis. From Burge et al.[22]

$$1 - P_o(t) = 1 - e^{-n_t} \tag{5.10}$$

where n_t is determined from Equation 5.2. $P_o(t)$ is termed the extinction probability, in other words, the probability that all organisms are inactivated. It follows then that $1 - P_o(t)$ is the probability of at least one organism surviving the lethal condition.

A statistical approach of this type is often used in industries that require pasteurization or sterilization of materials. A considerable number of surviving organisms are acceptable in certain applications. For example, viable bacterial concentration in Grade A pasteurized milk must be <20,000 cells/ml.[8] Alternatively, a fermentation industry may require that feed substrates be assured of complete sterilization in 99 out of 100 batches. In this case, economic loss associated with a lost batch of substrate must be weighed against the cost of achieving a statistically higher probability of extinction, $P_o(t)$. Perhaps the most severe requirements are found in the canning industry. A single surviving spore of *Clostridium botulinum* can lead to lethal concentrations of the exotoxin causing botulism poisoning. Therefore, assurance of virtually complete destruction is required. Design criteria often require that spore survival probability, $1 - P_o(t)$, be reduced to 10^{-12} or less.[1]

Requirements for milk pasteurization are based on conditions necessary to inactivate the most heat resistant pathogen expected to be transmitted by milk. For a long time this was considered to be *M. tuberculosis*, the causative agent of tuberculosis. *Coxiella burneti*, the causative agent of Q fever, was later discovered to be slightly more heat resistant, and time/ temperature conditions for pasteurization were adjusted to account for this. Milk is presently batch-pasteurized at 63°C for 30 min (low temperature holding or LTH) or continuously

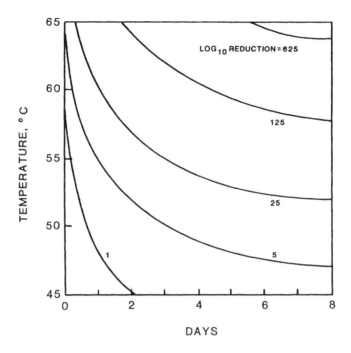

Figure 5.13.   Curves showing the time/temperature regimes necessary for inactivation of a desired number of logs of f2 bacteriophage in tryptone yeast extract medium (TYE). From Burge et al.[22]

pasteurized by heating to 71.7°C and holding for 15 to 30 sec followed by rapid cooling (high temperature short time, HTST or flash).

A somewhat similar concept has been proposed to assure satisfactory disinfection of enteric pathogens during composting. Burge et al.[22] recommended use of bacteriophage f2 as a standard organism. From Table 5.4 it can be seen that bacteriophage f2 appears to be more heat tolerant than the enteric pathogens, including viruses, bacteria, protozoan cysts, and helminth ova. Therefore, conditions designed to achieve a certain log reduction in the standard organism would assure greater destruction of the enteric pathogens. Time/temperature conditions for various log reductions of bacteriophage f2 in tryptone yeast extract medium (TYE) are presented in Figure 5.13. Comparison with similar curves developed for *Ascaris* ova in Example 5.1 confirms that bacteriophage f2 is considerably more heat resistant. Some change in the curves of Figure 5.13 can be expected as more data are developed on survival of f2 in actual composting material. However, the more thermal-tolerant nature of f2 will probably remain unaltered.

Using Figure 5.13, one would predict that 25 log units reduction would be achieved at 55°C for about 3 days or 60°C for about 0.5 days. Reduction in enteric pathogens would be considerably greater because their D_r values are lower than that for bacteriophage f2. Using this approach, health officials can establish desired log reductions based on intended use of the final product and desired safety factors. Time/temperature conditions can then be established with assurance that enteric pathogens will be controlled to even safer levels. It is expected that this approach, using either bacteriophage f2 or other standard organisms, will find increasing application because of its straightforward and scientific basis and the well-established success of similar concepts used in pasteurization of milk and other products.

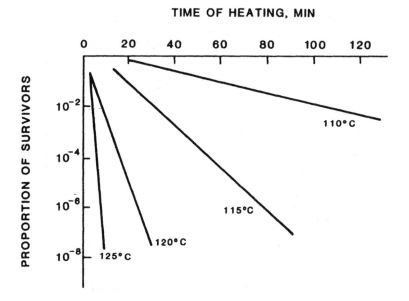

Figure 5.14.   Thermal inactivation of endospores of *Bacillus subtilis*. From Burton et al.[25]

LIMITATIONS ON MICROBIAL DESTRUCTION

Spore-Forming Bacteria

A rather dramatic reduction in response of *Ascaris* ova to a typical composting time/temperature profile was predicted in Example 5.2. However, other organisms may be more temperature resistant. Consider the data in Figure 5.14 for inactivation of *Bacillus subtilis* spores. Inactivation would not be expected until the temperature reached >100°C. The data of Figure 5.14 are typical for endospore-forming bacteria including types such as *Clostridium tetani* (tetanus) and *C. botulinum* (botulism). Thus, endospores formed by spore-forming bacteria would not be inactivated by composting. This statement does not extend to protozoan cysts and fungal spores which are much more susceptible to heat inactivation.

Fortunately, enteric bacterial pathogens are nonspore-forming. It is interesting to speculate on why this should be the case. One possibility is that the microbes found a suitable environment in the gut of man and never had to develop spore formation as a defense mechanism to assure their survival. Spore-forming pathogens may be found in sludge, yard waste, refuse, and other substrates, but probably not in levels much elevated from those naturally occurring in soils or other materials. *C. tetani*, for example, is a ubiquitous saprophytic organism commonly found in soil. The organism is dangerous to man only when introduced into wounds that become anaerobic, allowing growth of the microbe.

In summary, temperatures common to composting should be effective against enteric bacterial pathogens (as well as viral, protozoan, and metazoan forms) but not endospores produced by bacterial spore formers. Therefore, care should be exercised in the use of compost just as one would exercise care in use of soil or other natural materials.

Clumping of Solids

Referring again to Example 5.2, the outcome is predicated on the assumption that all organisms experience the time/temperature profile to an equal extent. A number of factors

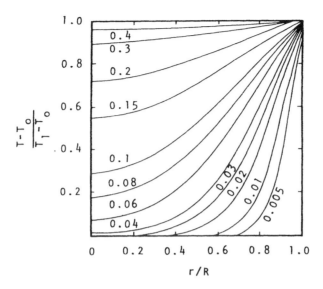

Figure 5.15. Temperature profiles within a sphere of radius, R, as a function of dimensionless time defined by Equation 5.11. Initially the sphere temperature is T_0 throughout, and the outer-surface temperature for t > 0 is T_1. From Bailey and Ollis.[1]

could prevent full realization of this assumption. For one, large particles or balls may form in the compost, these may not receive adequate oxygen, which would significantly reduce heat buildup from within the particle itself. Heat transport from surrounding compost would then be necessary to assure adequate pathogen destruction within the particle. But does this occur at a sufficient rate, and for what size particles?

The situation can be examined analytically by assuming a spherical, homogeneous ball of material heated by conduction from the outside. This is a classical problem in heat transfer for which analytical solutions are well developed. The development of these solutions will not be explored here, but their results will be used. A wealth of solutions and other data for a variety of such problems is available in Carslaw and Jaeger.[26]

Dimensionless temperature profiles within a sphere of radius R are shown in Figure 5.15 as a function of dimensionless time defined as follows:

$$\text{dimensionless time} = kt / (\rho c_p R^2) \tag{5.11}$$

where

k	=	thermal conductivity, cal/h-cm²-°C/cm
t	=	time, h
ρ	=	mass density, g/cm³
c_p	=	specific heat of the particle, cal/g-°C
R	=	radius of the spherical particle, cm

For the solution shown in Figure 5.15, it is assumed that temperature is uniform throughout the sphere at t = 0 and is equal to T_0. The outer surface temperature is equal to T_1 for all t > 0. Thus, at t > 0 heat flows into the spherical particle until the temperature becomes uniform and equal to T_1 at t = large.

Thermal properties of compost were discussed in Chapter 3. A worst-case situation is assumed with a particle of high density and specific heat and low thermal conductivity.

Table 5.5.   Estimated Heat Transfer Times into Spherical Compost Particles[a]

Particle Radius (cm)	Time to Reach $(T - T_0)/(T_1 - T_0) = 0.9$ (h)
1	0.1
10	10
20	40
50	250
100	1000

[a] Assumed conditions: ρ = 1 g/cm^3, c_p = 1 cal/g-°C, k = 3 cal/h-cm^2-°C/cm, r/R = 0.

Because compost mixtures are commonly 50% moisture, let ρ =1 g/cm^3 and c_p = 1 cal/g-°C. Thermal conductivity values between 2 and 4 cal/h-cm^2-°C/cm have been measured for compost material, and a value of 3 is assumed here.

The center of the sphere (r = 0) will heat most slowly. Therefore, if the center is adequately treated, the remainder of the particle should be as well. Consider a dimensionless temperature $(T - T_0)/(T_1 - T_0) = 0.9$ which gives an internal temperature close to the surrounding temperature T_1. From Figure 5.15, this corresponds to a dimensionless time of about 0.3 at r/R = 0. Thus,

$$kt / (\rho c_p R^2) = 0.3$$

Substituting assumed values for k, ρ, and c_p, and rearranging:

$$t = 0.1R^2 \tag{5.12}$$

Data presented in Table 5.5 were calculated from Equation 5.12.

Based on this analysis it would appear that the conductive heating time for particles of 1 to 10 cm radius would be small compared to detention times normally used in composting. If clumps larger than 20 cm radius are formed, however, heating times can be significant. This would then limit the potential kill of microbes to values less than that calculated using the approach in Example 5.2. Consideration should be given to breaking these clumps to assure full exposure of all organisms to the time/temperature profile.

Nonuniform Temperature Distributions

A second factor that may limit thermal inactivation during composting is the fact that uniform temperatures will not exist throughout the entire compost mass. Cold pockets or zones may allow pathogenic microbes to survive the composting conditions. When compost is not turned, such as in the aerated static pile process and certain reactor.systems, the engineer must assure, by proper design and operation, that cold zones are not formed. In the windrow and many reactor systems, in which compost is periodically agitated, the designer must assure a minimum statistical probability that all particles are exposed to the temperature conditions.

Compost Mixing and Turning

The case where compost is agitated at frequent intervals can be examined analytically provided a number of simplifying assumptions are made. Imagine the compost material to be composed of two zones, each of which have uniform temperatures. Temperatures in the cooler zone are sublethal and cause no organism destruction. Uniform lethal temperatures occur in the hotter zone where thermal inactivation is described by Equation 5.2. The pile is turned at time intervals of Δt. Each time the pile is turned there is sufficient mixing energy to cause a random redistribution of material after each turning. Under these conditions, thermal inactivation can be described as

$$n_t = n_o [f_l + f_h e^{(-k_d \Delta t)}]^N \qquad (5.13)$$

and

$$f_l + f_h = 1 \qquad (5.14)$$

where
n_t = number of organisms surviving
n_o = number of organisms initially present
f_l = fraction of composting material in the low temperature, sublethal zone
f_h = fraction of composting material in the high-temperature zone
Δt = time interval between compost turning
k_d = thermal death coefficient as defined by Equation 5.1
N = number of pile turnings

Equation 5.13 was solved for various assumed conditions with results shown in Figure 5.16. Two different values of f_l/f_h were evaluated. A value of $f_l/f_h = 1$ implies that half of the composting material is in the high temperature zone at any time. For $f_l/f_h = 0.25$, 80% is in the high temperature zone. Various values of $k_d \Delta t$ were assumed. Note that a value of 2.303 implies a 1 $\log_{10}$ reduction between each turning for material in the high temperature zone. A value of infinity for $k_d \Delta t$ implies that undesirable microbes in the high temperature zone are completely inactivated between each turning. For the latter case Equation 5.13 reduces to

$$n_t = n_o (f_l)^N \qquad (5.15)$$

Equation 5.15 represents the maximum destruction achievable under the particular conditions of f_l and N.

The fraction of surviving organisms shown in Figure 5.16 is considerably greater than that which would be calculated by using procedures in Example 5.2. The latter example was based on a typical time/temperature profile using k_d values representative of the enteric pathogens, and essentially complete destruction was predicted. This strongly suggests that average inactivation achieved during composting is determined as much by the ability to expose all material to the lethal temperature as by the time/temperature profile itself, provided, of course, that a reasonable value of $k_d \Delta t$ is maintained.

The amount of material in the low temperature zone should be minimized to the extent possible. Suppose that a survival fraction of 10^{-6} is determined to be adequate. If $k_d \Delta t$ is 2.303, the desired inactivation can be achieved with 11 pile turnings at $f_l = 0.2$, increasing to 23 turnings at $f_l = 0.5$. In the windrow system, f_l can be minimized by constructing larger piles to reduce the surface/volume ratio.

The assumption of random redistribution of material during turning may not be reasonable in all cases. However, many commercial turning devices impart considerable energy to the pile

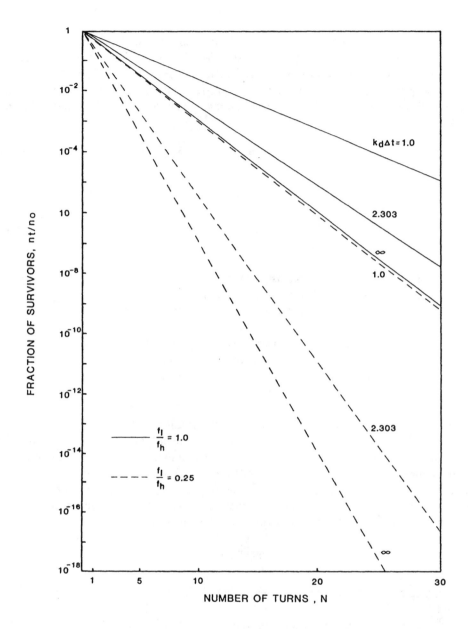

Figure 5.16.   Microbe survival as a function of the number of material turnings with $k_d\Delta t$ and f_l/f_h as parameters.

during turning, and for these the assumption is probably reasonable. On the other hand, if a front-end loader is used, energy input is lower and the assumption is probably not valid. With a front-end loader, however, the operator can apply his knowledge as to the location of the low temperature zone. The operator can then attempt to redistribute low temperature material into the high temperature zone for the next cycle. Thus, redistribution is purposely nonrandom. In other words, nonrandomized redistribution guided by information is used in one case compared to higher energy, randomized redistribution in the other. In either case, the basic conclusion remains that average thermal inactivation is determined in large measure by the ability to expose all material to the lethal temperature.

Table 5.6.   Number of Days During which Material Obtains a Temperature Equal to or Greater Than a Particular Temperature at Individual Confidence Levels[a]

	Confidence Levels, %		
Temperature (°C)	95	99	99.9
≥50	13.8	13.3	12.6
≥55	10.6	10.1	9.4
≥60	7.3	6.8	6.3
≥65	4.3	3.9	3.4
≥70	1.2	1.0	0.8

Source: Burge et al.[22]

[a] Data were developed from analysis of temperatures taken from 15 piles during the composting of raw sewage sludge and wood chips by the aerated static pile method.

The model developed here assumes that all material is redistributed after each turning. Suppose a small amount of material near the edge of the pile escaped the turning action. If this material amounted to only 0.1% of the total composting mass and if no inactivation occurred within the material, average thermal inactivation would be limited to no more than 3 $\log_{10}$ units. Thus, it is most important that all material participate in the redistribution achieved during turning. If such care is exercised, inactivation levels can be achieved as determined in Figure 5.16.

At this point it should be mentioned that average thermal inactivation levels may be difficult to measure in practice. In the above example it was assumed that 0.1% of the material was not composted. If small grab samples of the final product were analyzed, acceptable destructions might be observed in as many as 999 out of 1000 samples. The one sample, which finally did show high levels, would probably be assumed to be in error because so many previous samples were acceptable. Thus, it is important to sample all portions of the final compost pile. Even so, it is still incumbent upon the designer and operator to assure that all material is exposed to lethal temperatures.

Static Pile Systems

Static pile systems, in which compost is not turned, can be examined analytically provided temperature distributions within the piles are known. Burge et al.[22] examined 15 static piles at Beltsville, Maryland, which were composting a mixture of raw sludge and wood chips. Temperature readings were taken throughout the compost cycle from the pile center, lateral portions extending out from the center, and the region designated as the toe just beneath the outer blanket at the lower edge of the pile. The latter has traditionally been the area of lowest temperature. From the temperature history throughout the compost pile, confidence levels for achieving a particular time/temperature relationship were established for the 15 test piles as shown in Table 5.6. For example, a confidence level of 99.9% can be achieved that all material within the pile will attain 55°C for 9.4 days. Most material would see this temperature for a significantly longer period of time. Recall that 55°C for only 3 days is sufficient to achieve 25 $\log_{10}$ reductions in bacteriophage f2, which has been proposed as a standard test organism.

Mean temperature and standard deviations observed at the toes of the 15 test piles are shown in Figure 5.17. Significant reduction of bacteriophage f2 would be predicted not only on the basis of the mean temperature but also the temperature obtained by subtracting the standard deviation from the mean.

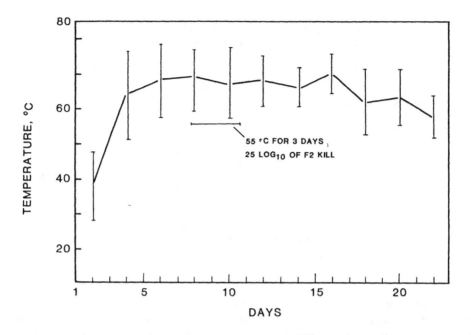

Figure 5.17. Mean temperature of the toe areas of 15 raw sludge, aerated static piles. Vertical lines show values with ± the standard error and the horizontal line shows the time/temperature regime for 25 $\log_{10}$ reduction of f2 bacteriophage. From Burge et al.[22]

It should be emphasized that confidence levels shown in Table 5.6 apply only to the 15 test piles examined by Burge et al.[22] In general, such confidence levels will vary with the composting operation. Confidence levels for a particular operation would be a function of the substrate material (i.e, raw sludge, digested sludge, manure, refuse, etc.), the type of bulking agent used, and the manner of operation of the piles. Thus, proficient design and operation are vitally important in establishing confidence that all material achieves an adequate time/temperature profile.

Short-Circuiting

The residence time distribution of solids in the composter is important to achieving a high pathogen kill. All solids must "see" at least the minimum residence time needed for pathogen destruction in the high temperature zone of the process. For batch and plug flow processes, most solids remain in the system for the theoretical residence time. By definition, however, complete mixed processes involve some short-circuiting of the feed solids. A portion of the feed passes through a complete or well-mixed reactor in less than the theoretical detention time. This is not necessarily a problem because (1) most composting systems employ residence times considerably greater than those needed for pathogen destruction, (2) true complete mixed conditions are difficult to achieve with solid materials, and (3) many reactors are fed on a "fill and draw" basis which assures a minimum residence time. On the other hand, some rotary drum systems use a short residence time, on the order of 1 day, followed by static piles. Surface material on the piles may be exposed to elevated temperatures only while in the rotary drum.

The point here is that the designer should consider the residence time distribution of all solids in the system. Extreme short-circuiting should be avoided to reduce the risk that some compost bypasses the heat inactivation zone.

Bacterial Regrowth

Another factor that may limit thermal inactivation is that certain enteric bacteria can regrow in organic materials once temperatures are reduced to sublethal levels. This phenomenon has been observed with total and fecal coliform, *Salmonella*, and fecal streptococcus bacteria grown in liquid sludges and even compost with moisture contents <40%.[13,27] Repopulation may occur through regrowth of the organisms existing in the compost at an undetectable concentration or through the growth of organisms introduced from an outside source. A likely source may be feces from salmonella-infected birds, reptiles, or other animals.[28] The most dramatic regrowth is usually observed in sterilized material recontaminated by pathogenic bacteria as shown in Figures 5.18 and 5.19. In nonsterilized material such regrowth is restricted, if not eliminated, apparently by competition with the natural microbial flora. Regrowth should not pose difficulties in well stabilized, mature compost in which a diverse natural flora has been reestablished.

The problem of bacterial regrowth was observed on a large scale in Switzerland.[30] Regulations were adopted in 1971 prescribing that sewage sludge be disinfected before application to grazing and fodder land during the vegetation season. The system of disinfection generally adopted was postpasteurization after the digestion process. By 1977 there were about 70 such facilities. Evidence of regrowth of certain bacterial pathogens began to appear. For example, Keller[31] reported enterobacteria concentrations in digested sludge of 1 to 2×10^4/g. These decreased to essentially zero just after heat pasteurization, but then exploded to values of 0.65 to 29×10^6/g after 1 to 3 days. *Salmonella* were also observed to regrow. Data of this type caused the Swiss to temporarily suspend further installations in 1977. Subsequent work demonstrated that prepasteurization ahead of the digestion process could resolve the regrowth problem. This is because the disinfected sludge is quickly reinoculated in the digester with a balanced bacterial population that inhibits regrowth of the pathogenic types. In 1981 the Swiss adopted new regulations that require heat pasteurization be conducted prior to digestion.[32] Postpasteurization is prohibited. Such systems have now been widely adopted in Switzerland. This lesson from the Swiss experience is reflected in the EPA Part 503 regulations that require that stabilization be accomplished either with or directly after the pathogen reduction step.[3]

Selna and Smith[33] observed occasional regrowth of bacteria during windrow composting of digested sludge. Regrowth of coliform organisms was the most consistently observed, particularly if windrows were moist, as during winter months. Only sporadic regrowth of *Salmonella* was observed. In a supplementary series of experiments, samples of final compost were incubated at 30°C for 30 days under controlled laboratory conditions. In interpreting the results, regrowth was defined as a concentration increase of one or more logs to a final concentration >10 most probable number (MPN)/g dry weight. Using this definition as a measure, regrowth of total coliform was observed in 25% of samples, fecal coliform in 8% of samples, and *Salmonella* in 8% of samples. The low moisture content of the final compost, estimated at 30 to 35%, along with the natural bacterial population, may have inhibited bacterial regrowth.

A larger scale regrowth experiment was conducted by Iacoboni et al.[34] using finished material from windrow composting of digested sludge cake and recycled product. In it 5 tons

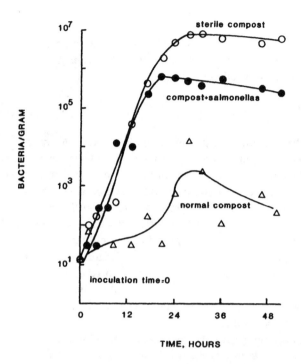

Figure 5.18. Growth of fecal streptococcus bacteria in normal composted sludge, in sterilized composted sludge, and in composted sludge saturated with *Salmonella* spp. In the last case, following sterilization of the compost, *Salmonella* spp. grew to approximately 10^8 bacteria/g. From Ward and Brandon.[13]

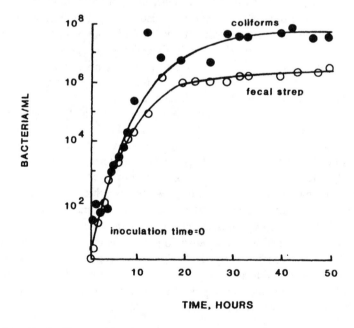

Figure 5.19. Growth of coliform and fecal streptococcus bacteria in sterile liquid digested sludge at 35°C. From Brandon.[29]

of material were rewetted to 54% TS and formed into a 4 ft high pile. Total coliform, fecal coliform, and *Salmonella* were nondetectable in the starting material. Some regrowth was noted at days 3 and 4. The maximum *Salmonella* concentration was about 5 MPN/g, which corresponded to the fecal and total coliform peaks of 4.1 and 7.7×10^4, respectively. By day 14 all values had returned to nondetectable levels below 1 MPN/g. Russ and Yanko[35] also observed a transient increase in salmonellae in compost with concentrations returning to background levels after 3 weeks.

Burge et al.[28] studied sludge composts collected from 30 U.S. facilities. Salmonellae inoculated into the composts died out, except in the compost that had been sterilized first. The authors concluded that the active, indigenous flora of composts establish a barrier to colonization by salmonellae. In the absence of competing flora, such as by sterilization, reinoculated salmonellae may grow to potentially hazardous densities. The capability of microbes from different compost temperature regimes to inhibit salmonellae growth was also determined. Material taken from a 70°C zone of a compost pile did not suppress salmonellae growth. Compost from a 55°C incubator was more suppressive, and compost from a near ambient temperature zone of a curing pile was completely suppressive. The authors concluded that salmonellae regrowth would be negligible given the diversity of the microbial population in cured compost at ambient temperature.

The studies by Burge et al.[28] and the field tests by Iacoboni et al.[34] form the basis for an important criteria for design of composting systems. Sufficient residence time should be included in the facility design for curing temperatures to return to near ambient for at least 10 to 15 days. This should assure the reestablishment of a diverse microbial flora, which will then inhibit salmonellae regrowth should the compost become reinoculated. Reestablishment of a diverse microbial population would appear to be one of the important goals of the curing stage. The diverse flora would also appear to be a major advantage of compost products because of their ability to inhibit regrowth of bacterial pathogens.

With the exception of the bacteria mentioned above, enteric virus, protozoa, and helminth pathogens are either obligate parasites or they require an intermediate host not present in compost. Human viruses, for example, can reproduce only under suitable conditions within human eucaryotic cells. Obligate parasites inactivated in one phase of the compost process cannot regrow in the next. Thus, concern over regrowth is limited to a few bacterial species and is not of concern with other classes of pathogens. The healthy and diverse microbial population characteristic of well stabilized compost should suppress the regrowth of salmonellae and other bacterial pathogens.

Reinoculation

Inoculation of the composting mixture with new pathogens after the heat inactivating phase will certainly increase the probability of pathogens surviving into the final product. For example, adding sewage sludge at the end of the process would certainly degrade the hygienic quality of the compost. This is probably obvious to most readers. However, there may be sources of reinoculation that are not so obvious. For example, Reeves[36] reported on experiments using sewage effluent for moisture addition during windrow composting of sludge amended with sawdust. Sudden and dramatic increases in the numbers of salmonellae and shigellae were observed following each effluent addition. Reeves recommended against further use of undisinfected sewage effluents because of the danger of recontamination. Other pathways for reinoculation include birds and other animals that might be attracted to the stored compost. Reasonable care and attention to management practices should prevent any significant reinoculation in most cases.

FINAL PRODUCT STANDARDS

Product Quality in Practice

Thus far, time/temperature profiles, mathematical models of heat inactivation kinetics, proposed regulatory approaches using pathogenic and indicator organisms, and potential limitations on microbial destruction have been discussed. Given all of the above, what microbial quality can be achieved in the final product in actual practice? To answer this, one must examine final product quality data developed from actual sludge composting installations.

Yanko et al.[37] studied the fate of viruses during composting of sewage sludge. The authors concluded that efficient composting can reduce virus concentrations to below their detection limit of ~0.25 PFU/g ds. They also concluded that heat inactivation is probably the significant factor influencing virus inactivation during composting. The problem lies in assuring that efficient composting consistently occurs and that all material is exposed to inactivating temperatures. Viruses were found to survive 25 days of composting if the composting mass did not achieve adequately high temperatures. Fecal coliform were also found in high concentrations if maximum composting temperatures were <50°C.

Farrell[5] concluded that a temperature of 53°C or above for sufficient time effectively eliminates pathogenic bacteria, enteric viruses, and *Ascaris* eggs. This conclusion was based in part on the work of Yanko[38] who analyzed weekly samples from both windrow and static pile sludge composting facilities for 1 year. During this time, no protozoan cysts were found, no viable helminth ova were detected, and enteric viruses were confirmed in only two samples at very low levels. Farrell[4] statistically analyzed Yanko's data and concluded that salmonellae were absent from all samples with <1000 MPN fecal coliform/g solids. These studies disclosed essentially no hazard associated with properly composted sludges from obligate parasites and viruses.

Selna and Smith[33] monitored parasitic ova during windrow composting of digested sludge. Viable *A. lumbricoides, Trichuris trichiura,* and hookworm ova were consistently present only during the first 7 to 10 days of composting. Viable ova were isolated in only 3 final compost samples and only 8 of 140 samples collected after more than 10 days of composting. This same general pattern was true for viruses, which consistently were not isolated in final compost samples. These results are particularly encouraging because the windrow system from which the samples were drawn was not the optimized system later described by Hay and Kuchenrither,[39] as discussed in Chapter 2, and did not achieve as consistent time/temperature profiles.

The question of indicator bacteria, such as total and fecal coliform, and nonobligate bacterial pathogens, such as *Salmonella,* cannot be answered from established time/temperature profiles because of the potential for regrowth. Recourse must be made to operating data from full scale facilities. Epstein et al.[40] reported fecal coliform reductions from initial concentrations of 10^6 to 10^7 CFU/g solids to 0.03 to 30 CFU/g during aerated static pile composting of both raw and digested sludges. If detected at all, *Salmonella* was usually present at levels <1 CFU/g solids. Burge et al.[28] sampled composts from 30 municipalities and found that salmonellae were nondetectable in 88% of the samples. Yanko[38] also studied sludge products collected from 24 municipalities. These samples showed random, generally low levels of salmonellae, *Yersinia*, and toxigenic *E. coli*. Relatively few salmonellae were detected in final compost from a windrow facility, but some increases occurred after the compost was blended with other materials to produce commercial soil amendments. This apparent regrowth may have resulted from blending relatively immature compost with the other organic amendments. Reestablishment of a diverse, microbial population prior to blending may have subdued the regrowth.

Hay and Kuchenrither[39] developed the data presented in Figure 5.20 on total coliform concentrations during open windrow composting of digested sludge blended with various amendments. Total coliform concentrations were reduced to 10 MPN/g solids or less by the two step windrow process. Prior to optimization of this windrow system the disinfection effectiveness during winter months was consistently less than that during summer months. For example, in August when the internal temperature and exposure time averaged 66°C and 33 days, respectively, about 85% of windrows had a median coliform count below 1 MPN/g. In February, when temperature and exposure time averaged 59°C and 50 days, respectively, none of the windrows had a median count of <1 MPN/g. Longer exposure times did not fully compensate for the lower winter temperatures. With optimization of the two step process, <6% of 49 weekly samples contained detectable salmonellae. The salmonellae were found in windrows that had been excessively cooled from heavy rainfall. No detectable salmonellae were found in the following year. During this time there were no viable *Ascaris* ova or virus detected in any samples. Based on these improvements, a quality criteria was established whereby windrows must achieve a median coliform limit of ≤10 MPN/g and/or a salmonellae limit of ≤1 MPN/g prior to release for reuse.

Strauch[41] analyzed composts produced from several European reactor systems, including a Weiss/Kneer Bioreactor, a BAV reactor, and a Schnorr Biocell system. In all cases the reactors were composting mixtures of sludge, sawdust or bark, and product recycle. Salmonellae, virus, and ascaris eggs were reliably destroyed provided that all material was exposed to inactivating temperatures. In one case salmonellae survived the reactor when the process temperatures reached only 49 to 53°C. In another test both salmonellae and virus survived because of poor temperature distribution, even though the maximum reactor temperature was 82°C. The author concluded that all of the reactor systems tested are capable of producing an hygienic product, but only if they are properly operated and monitored.

Several conclusions can be drawn from the above discussion. First, composting can provide essentially complete destruction of obligate pathogens and can reduce indicator bacteria and nonobligate bacterial pathogens to very low levels. Second, regrowth of nonobligate bacterial pathogens can be suppressed by maintenance of a diverse microbial flora. The latter is enhanced if composting temperatures are allowed to return to near ambient levels. Third, it is very important that all material be consistently exposed to the inactivating time/temperature conditions. High probability of pathogen destruction is achieved by maintaining a high statistical assurance that all material "sees" the inactivating conditions. Fourth, a quality control program is essential for assuring a statistically high probability of pathogen destruction. Fifth, covering the composting process or using reactor systems may provide higher probabilities of destruction, particularly in areas subject to cold and wet climatic conditions.

Health Risks

Actual health risk associated with various levels of pathogens in compost is difficult to assess accurately but estimates can be made. Hornick et al.[42,43] published dose response data for *Salmonella typhosa* on human volunteers which is summarized in Table 5.7. In these

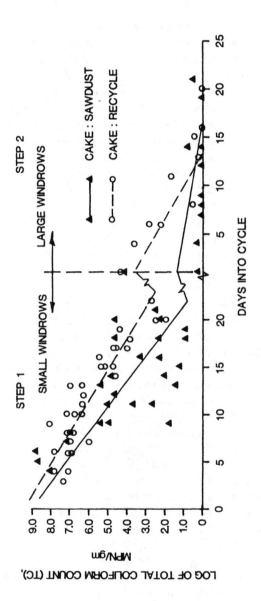

Figure 5.20. Temperature and total coliform concentrations during two step windrow composting of digested sludge. Data correspond to the temperature profiles presented in Figure 2.11. From Hay and Kuchenrither.[39]

Table 5.7.   *Salmonella typhosa* Dose Response Data on Human Volunteers

S. typhosa Strain	Challenge Dose Concentration	Number of Volunteers who Became Ill	Total Number of Volunteers Challenged
Quailes	10^3	0	14
	10^5	32	116
	10^7	16	32
	10^8	8	9
	10^9	40	42
Quailes	10^7	16	30
Zermatt	10^7	6	11
TY2V	10^7	2	6
Quailes	10^5	28	104
	10^7	15	30
	10^9	4	4

Source: Hornick et al.[42,43]

Table 5.8.   Dose of Various Species and Strains of *Salmonella* That Caused Disease in Human Volunteers

Salmonella Species/Strain	Dose at which 50% or More Respond[a]
S. meleagridis I	50,000,000
S. meleagridis II	41,000,000
S. meleagridis III	10,000,000
S. anatum I	860,000
S. anatum II	67,000,000
S. anatum III	4,700,000
S. newport	1,350,000
S. derby	15,000,000
S. bareilly	1,700,000
S. pullorum I	1,795,000,000
S. pullorum II	163,000,000
S. pullorum III	1,295,000,000
S. pullorum IV	1,280,000,000

Source: McCullough and Eisele.[45–48]

[a] Develop clinical disease.

studies, human volunteers were inoculated with various doses of *S. typhosa* and the frequency of illness was noted. *S. typhosa* tends to produce more severe symptoms than other salmonellae and is generally credited with greater virulence.[44] Data reported on other species and strains of salmonellae are presented in Table 5.8. These data suggest that large (10^4 to 10^9) numbers of *Salmonella* must be introduced into the body to achieve a high risk of illness. There is some evidence that lower doses of *Salmonella* may at times be infective, such as during actual outbreaks of the disease when the strains may be more virulent. D'Aoust and Pivnick[49] and D'Aoust[50] describe salmonellosis outbreaks that may have been initiated by infective doses as low as 10 to 100 cells.

Mechales et al.[44] conducted extensive and sophisticated statistical analyses of the data of Hornick and others. They concluded that infectivity of *S. typhosa* was similar to other *Salmonella* species and that the entire salmonellae group could be handled as a single factor. They were able to develop estimates of disease risk as a function of *Salmonella* concentration in recreational water. Partial results of their analysis are shown in Figure 5.21. It was assumed that 10 ml of water would be imbibed by a recreationist either by direct swallowing or nasal inhalation and subsequent swallowing. Thus, the concentration can be converted to a total ingested dose. This allows extrapolation of their results to the ingestion of compost. The authors were quick to point out that their analysis was based on limited data (even though the literature search was exhaustive) and that Figure 5.21 must be interpreted accordingly.

Let us assume that a health risk of 0.001% or less is desired upon ingestion or inhalation (with subsequent swallowing) of 1 g of compost solids. In other words, a risk of ill effect in 1 person out of 100,000 who each ingest 1 g. There is no firm basis for the assumption of 1 g ingested, but it does seem to be a conservative assumption given the fact that compost is not intended for direct consumption. From Figure 5.21 this level of risk corresponds to ingestion of about 0.80 MPN. Because we are considering 1 g of compost, the corresponding concentration would be 0.8 MPN/g solids. At a level of risk of 1 in 10,000 the concentration can be increased to 30 MPN/g. This analysis indicates a relatively low risk of salmonellosis from ingestion or inhalation at concentrations below 10 and even perhaps 100 MPN/g, which is readily achievable during composting. Other pathways for infection from compost products exist. Examples include contamination of garden products that are subsequently eaten raw and direct ingestion of soil/compost mixtures (Pica syndrome). However, it is not expected that these pathways would present risks significantly different from that estimated above for direct ingestion/inhalation.

Mechalas et al.[44] also investigated risks associated with ingestion of virus. Here the infective dosages seem to be markedly different than for bacterial agents. While there is considerable scatter in literature data, there is evidence that as little as 1 plaque forming unit (PFU) can cause infection, as shown in the data of Table 5.9.

Considerable variation in viral concentrations can be expected depending on the compost substrate. Raw sewage sludges probably represent a worst case with actual concentrations dependent on the incidence rate of viral disease in the population. Various sources indicate a range of about 20 to 700 PFU/100 ml in raw sewage. If the virus particles are associated with particulate solids, a range of about 600 to 25,000 PFU/g dry raw sludge solids could be expected. Another rule of thumb suggests about 15 virus particles per 10^6 coliform organisms.[44] Assuming 10^8 coliform/g of sludge, a viral concentration of 1500 PFU/g can be estimated. Farrell[5] estimated a likely range from 250 to 7,000 PFU/g solids. Obviously, the sludge treatment train must provide many logs reduction in virus to achieve an acceptable compost product.

Mechalas et al.[44] analyzed data from viral infectivity studies that listed 812 volunteer exposures. Statistical analysis yielded the dose/risk diagram for virus shown in Figure 5.22. It is obviously difficult to extrapolate dose/response data of the type shown in Table 5.9 to the low levels of risk shown in Figure 5.22. Nevertheless, it does allow estimation of the approximate level of risk associated with a given dosage.

Let us assume a desired risk level of 1:10,000 following ingestion of 1 g of compost, similar to the risk level previously assumed for *Salmonella*. From Figure 5.22 the risk corresponds to a concentration of about 7 PFU/L. Since it was assumed that 10 ml were imbibed, the viral dosage would be 0.07 PFU. Considering 1 g of compost solids, the corresponding concentration would be 0.07 PFU/g solids. In other words, if 1 g of compost at an average concentration of 0.07 PFU/g were administered to 10,000 individuals, about one case of infection would be expected. Obviously, with an average concentration of only 0.07 PFU/g, most of out 10,000 "volunteers" would not have ingested a single virus in their sample. This means that to achieve the same level of risk between virus and *Salmonella*, the virus concentration must be reduced essentially to zero. Fortunately, animal and plant viruses cannot reproduce extracellularly, and regrowth in compost is not a concern. Available data indicate that virus can be reliably reduced to below detection limits, at least with present isolation and identification methods. Reasonable detection limits for virus are currently about <0.1 to 0.25 PFU/g.

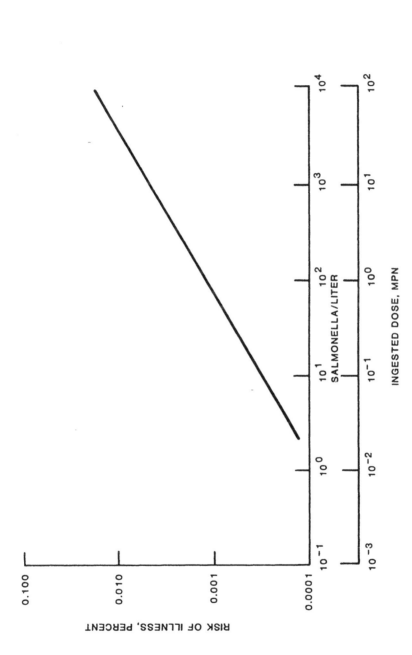

Figure 5.21.   A risk/dosage diagram for *Salmonella* developed by Mechalas et al.[44] Original work was for a water recreationist who was assumed to imbibe 10 ml of water. This assumption was used to convert the concentration axis to a total dosage to allow extrapolation of the results to inhalation or ingestion of compost.

Table 5.9.   Infection of Human Volunteers with Attenuated Poliovirus I

Dose (PFU)	Number Infected/Number Fed	% Infected
0.2	0/2	0.0
2.0	2/3	66.7
20.0	4/4	100.0
200.0	4/4	100.0

Source: Adapted from Mechalas et al.[44] Based on original data by Koprowski.[51]

The situation with protozoan and helminth disease agents is similar to that for virus. Ingestion of one protozoa or viable protozoan cyst or helminth egg is likely to produce a significant risk of infection.[52] Therefore, viable helminth ova and protozoan cysts should be below detection limits in the final compost. Reasonable detection limits for viable *Ascaris* ova are currently about <0.5 viable ova/g. Again, this level is achievable in actual composting practice.

Suggested Product Standards

Establishment of microbial standards on the final compost is an approach that regulatory agencies can use to assure an acceptably low risk to users of the material. However, setting numerical standards is at best a difficult task. The level of risk associated with proposed uses of the material must be considered along with the types of pathogens likely to be present, levels of reduction that can actually be achieved in practice, and the statistical nature of final product quality. In the previous analysis the relative risk of infection from direct ingestion of compost was considered. The number of people who would purposely ingest compost would hopefully be small. Ingestion would more likely result from inhalation of dusts, accidental hand to mouth contact, or adherence of compost to crops that are eaten raw. Thus, in a practical sense the levels of risk are probably less than calculated above, because the probability of a person actually ingesting compost must also be considered.

Establishment of standards must also recognize the fact that there is no such thing as "risk free" actions. Recovery of any resource will always involve some risk. Unfortunately, much recent environmental legislation has not recognized this fact and efforts toward resource recovery have often been delayed because of it. For example, residents near proposed composting sites often want guarantees that there will be no odors, dusts, noise, or risk of any pathogens. Of course, it is impossible to guarantee the negative. The designer can achieve insignificant risks, but never zero risks.

Addressing the concept of "risk free," Akin et al.[53] wrote:

> The concept of "acceptable risk" rather than "risk free" must be embraced as the only realistic approach for grappling with environmental health questions. It is a nebulous concept but a required one, nonetheless. However, it would be absurd to attempt to quantify the number of cases of any serious disease within a population that would be acceptable. The term "acceptable risk" has practical definition only in an economic and political sense. When a situation becomes of such little health concern that its investigation and control cannot demand sufficient priority for funding (either from the absence of documented evidence of a hazard or lack of public interest and pressure), then that risk by definition has become acceptable....When disease transmission through a single source e.g., land application of waste, occurs below the background disease transmission from all other sources, then the limits of epidemiological discernment have been reached and the health scientist is obligated to consider the risk to be at an acceptable level.

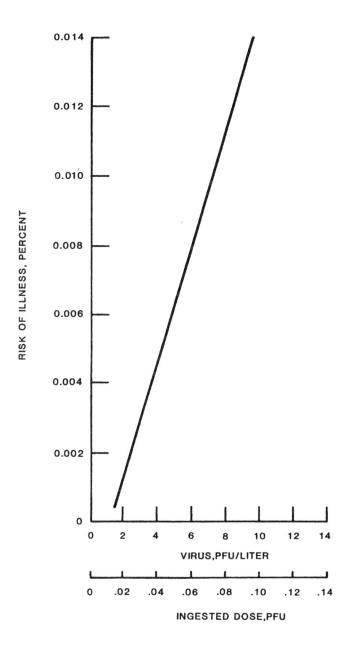

Figure 5.22. A risk/dosage diagram for virus developed by Mechalas et al.[44] for the case of a water recreationist assumed to imbibe 10 ml of water. This assumption was used to convert the concentration axis to a total dosage to allow extrapolation of the results to inhalation or ingestion of compost. The data indicate a relatively high level of risk from ingestion of small numbers of virus.

Comar[54] suggested the following guidelines in dealing with risks: (1) eliminate any risk that carries no benefit or is easily avoided; (2) eliminate any large risk (about 1 in 10,000 per year or greater) that does not carry clearly overriding benefits; (3) ignore for the time being any small risk (about 1 in 100,000 per year or less) that does not fall into category 1; and (4) actively study risks falling between these limits, with the view that the risk of taking any proposed action should be weighed against the risk of not taking that action. Comar was quick to point out that establishing such risk levels is an oversimplification of a very complex technical, social, and political problem. However, the approach should promote understanding about how to deal with risk in the real world, focus attention on actions that can effectively improve or safeguard health, and avoid squandering resources attempting to reduce small risks while leaving large ones unattended.

With the above comments in mind, the following guidance standards are proposed as being achievable with good management practices, economically reasonable, and representing an acceptably low risk to the user (all in the author's opinion of course).

1. *Virus*: No infective viruses detected by an acceptable laboratory method with a minimum detection limit of 0.1 to 0.25 PFU/g ds or less.

2. *Ascaris Ova*: No viable *Ascaris* ova detected by an acceptable laboratory method with a minimum detection limit of 0.5 viable ova/g ds or less. *Ascaris* will be considered representative of all parasites, i.e., helminth ova and protozoan cysts.

3. *Salmonella*: Median of all samples to be <1 MPN/g ds. Not more than 10% of samples to exceed 10 MPN/g ds. No sample to exceed 100 MPN/g ds. *Salmonella* will be considered representative of all bacterial pathogens capable of regrowth.

4. *Fecal Coliform*: Although the coliform group is not generally considered pathogenic, their destruction is indicative of good composting practice. Median of all samples to be <10 MPN fecal coliform/g ds. Not more than 20% of samples to exceed 1000 MPN/g ds. No sample to exceed 10,000 MPN/g ds.

Along with these guidance standards, statistical assurance must be provided that all material is exposed to the inactivating temperatures. Samples must also be collected from all regions of the windrow, pile, or reactor to assure that material does not escape the compost.

SUMMARY

Heat inactivation of microbes is a function of both temperature and length of exposure. A high temperature for a short period of time or a lower temperature for longer duration can be equally effective provided the exposure temperature is above the minimum lethal threshold. Heat inactivation kinetics are often modeled assuming first order decay. The inactivation coefficient, k_d, is a function of temperature, which can be described by the Arrhenius relationship. The inactivation energy, E_a, for many spores and vegetative cells is between 50 and 100 kcal/mol. This means that heat inactivation is strongly influenced by temperature, and a change in temperature of a few °C can cause significant changes in the rate of heat inactivation. If the value of k_d is known as a function of temperature for a particular microbe, thermal inactivation resulting from a particular time/temperature profile can be estimated.

Based on known time/temperature relationships, heat inactivation of enteric pathogens should be readily accomplished with the conditions common to composting. Temperatures of 55 to 60°C for a day or two should be sufficient to kill essentially all pathogenic virus, bacteria, protozoa (including cysts), and helminth ova to acceptably low levels. Endospores produced by spore-forming bacteria would not be inactivated under these conditions. However, enteric pathogens are not spore-forming.

A number of factors can reduce actual pathogen inactivation from that calculated by time/ temperature conditions in the composting process. These include (1) clumping or balling of solids, which can isolate material from the temperature effects; (2) nonuniform temperature distribution, which can allow pathogens to survive in colder regions; (3) short-circuiting of the feed substrate; (4) bacterial regrowth, which has been observed with coliform, *Salmonella,* and fecal streptococcus; and (5) reinoculation after the high temperature phase. When compost is not agitated, such as in the aerated static pile system and some reactor systems, it is important that uniform airflow and temperature be achieved throughout the material and that excessive clumping of solids be avoided. When compost is agitated, such as in the windrow and many reactor systems, it is important that turning frequency be sufficient to assure a minimum probability of a microbe escaping the high temperature zone. It is also important that short-circuiting be reduced to the point that all solids are exposed to high temperatures for a minimum period of time.

Analysis of available data on final product quality indicates that high product standards can be achieved with good management practices. Viruses and *Ascaris* ova can be reduced to below detection limits. *Salmonella* and total coliform can normally be reduced to levels below 1 and 10 MPN/g ds, respectively. Attainment of these standards should assure a very low risk of disease infection to users of the material.

REFERENCES

1. Bailey, J. E. and Ollis, D. F. *Biochemical Engineering Fundamentals* (San Francisco, CA: McGraw-Hill, 1977).
2. Wiley, J. S. "Pathogen Survival in Composting Municipal Wastes," *J. Water Poll. Control Fed.* 34:1 (1962).
3. "Standards for the Use and Disposal of Sewage Sludge," 40 CFR, Part 503, U.S. Federal Register, February 19, 1993.
4. Farrell, J. B. "Fecal Pathogen Control During Composting," *Proceedings of the International Composting Research Symposium* (Columbus, OH: Ohio State University Press, 1992).
5. Farrell, J. B. "Technical Support Document — Pathogen/Vector Attraction Reduction in Sewage Sludge," Office of Water Regulations and Standards, U.S. EPA, PB89-136618 (1989).
6. Havellar, A. H. "Disinfection of Sewage Sludge: A Review of Methods Applied in the European Communities," in *Proceedings of the 3rd International Symposium on Processing and Use of Sewage Sludge* (Lancaster, PA: D. Reidel Pub. Co., 1983).
7. Havellar, A. H., Oosterom, H., Notermans, S., and Van Knapen, F. "Hygienic Aspects of the Application of Sewage Sludge to Land," in *Proceedings of International Symposium on Biological Reclamation and Land Utilization of Urban Wastes,* Naples, Italy (1983).
8. U.S. FDA, PMO Grade A Pasteurized Milk Ordinance, U.S. Dept. of Health and Human Services, Public Health Service, Food and Drug Administration, U.S. Government Printing Office (1985).
9. Frosbisher, M., Hinsdill, R. D., Crabtree, K. T., and Goodheart, C.R. *Fundamentals of Microbiology* (Philadlephia: W.B. Saunders Co., 1974).
10. Roediger, H. J. "The Technique of Sewage-Sludge Pasteurization: Actual Results Obtained in Existing Plants; Economy," International Research Group on Refuse Disposal Information, Bulletin 21–31 (1964–1967).
11. Gotaas, H. B. "Composting — Sanitary Disposal and Reclamation of Organic Wastes," World Health Organization, Mono. Ser. No. 31 (1956).
12. Stern, G. "Pasteurization of Liquid Digested Sludge," in *Proceedings of the National Conference on Municipal Sludge Management* (Rockville, MD: Information Transfer, 1974).

13. Ward, R. L. and Brandon, J. R. "Effect of Heat on Pathogenic Organisms Found in Wastewater Sludge," in *Composting of Municipal Residues and Sludges, Proceedings of the 1977 National Conference* (Rockville, MD: Information Transfer, 1977).

14. "Progress Report — Beneficial Uses Program, Period Ending Sept 30, 1977," Sandia Laboratories, NM, SAND78-0242.

15. Ward, R. L., Ashley, C. S., and Moseley, R. H. "Heat Inactivation of Poliovirus in Wastewater Sludge," *Appl. Environ. Microbiol.* 32(3) (1976).

16. Bruce, A. M., Pike, E. B., and Fisher, W. J. "A Review of Treatment Options to Meet the EC Sludge Directive," *J. Inst. Water Environ. Manage.* 4(1) (February 1990).

17. Brandon, J. R., and Neuhauser, K.S. "Moisture Effects on Inactivation and Growth of Bacteria and Fungi in Sludges," Sandia Laboratories, NM, SAND78-1304 (1978).

18. Kruse, C. W. "Sludge Disinfection," in *Evaluation of Current Developments in Municipal Waste Treatment* (Technology Information Center, Energy Research and Development Administration, 1977).

19. Ward, R. L. and Ashley, C. S. "Identification of the Virucidal Agent in Wastewater Sludge," *Appl. Environ. Microbiol.* 33(4) (1977).

20. Brannen, J. P., Garst, D. M., and Langley "Inactivation of *Ascaris lumbricoides* eggs by heat, radiation, and thermoradiation," Sandia Laboratories, NM, SAND75-0163 (1975).

21. Chick, H. "Investigations of the Laws of Disinfection," *J. Hyg.* 8:92–158 (1908).

22. Burge, W., Colacicco, D., Cramer, W., and Epstein, E. "Criteria for Control of Pathogens during Sewage Sludge Composting," in *Proceedings of the National Conference on Design of Municipal Sludge Compost Facilities,* (Rockville, MD: Information Transfer, 1978).

23. Moats, W. A. "Kinetics of Thermal Death of Bacteria," *J. Bacteriol.* 105(1) (1971).

24. Wei, J. H. and Chang, S. L. "A Multi-Poisson Distribution Model for Treating Disinfection Data," in *Disinfection: Water and Wastewater,* J. D. Johnson, Ed. (Ann Arbor, MI: Ann Arbor Science Pub. Inc., 1975).

25. Burton, H. and Jayne-Williams, D. "Sterilized Milk," in *Recent Advances in Food Science, Vol. 2, Processing* (London: Butterworths & Co. Publishers Ltd., 1962).

26. Carslaw, H. S. and Jaeger, J. C. *Conduction of Heat in Solids,* 2nd ed. (Oxford: Clarendon Press, 1959).

27. Brandon, J. R. "Parasites in Soil/Sludge Systems," Sandia Laboratories, NM, SAND77-1970 (1978).

28. Burge, W. D., Millner, P. D., Enkiri, N. K., and Hussong, D. "Regrowth of *Salmonella* in Composted Sewage Sludge," report of the U.S. EPA Water Engineering Research Laboratory, Cincinnati, OH, EPA-600/S2-86/106 (March 1987).

29. Brandon, J. R. "Sandia's Sludge Irradiation Program," in *Sludge Management, Disposal and Utilization* (Rockville, MD: Information Transfer, 1976).

30. Clements, R. P. L. "Sludge Hygienization by Means of Pasteurization Prior to Digestion," in *Disinfection of Sewage Sludge: Technical, Economic and Microbiological Aspects* (London: D. Reidel Pub. Co., 1982).

31. Keller, U. "Experiences and Development of the Sludge Pasteurization in Altenrhein," in *Disinfection of Sewage Sludge: Technical, Economic and Microbiological Aspects* (London: D. Reidel Pub. Co., 1982).

32. Berger, R. "Raw Sludge Pasteurization," in *Disinfection of Sewage Sludge: Technical, Economic and Microbiological Aspects* (London: D. Reidel Pub. Co., 1982).

33. Selna, M. and Smith, D. "Pathogen Inactivation During Composting," reports of the Technical Services Deptartment, Los Angeles County Sanitation Districts (September 1975–December 1976).

34. Iacoboni, M. D., Livingston, J. R., and LeBrun, T. J. "Windrow and Static Pile Composting of Municipal Sewage Sludges," report to U.S. EPA Municipal Environment Research Laboratory, Cincinnati, OH, EPA-600/2-84-122 (July 1984).

35. Russ, C. F. and Yanko, W. A., "Factors Affecting Salmonellae Repopulation in Composted Sludge," *Appl. Environ. Microbiol.* 41:597–602 (1981).

36. Reeves, J. B. "Sanitary Aspects of Composted Sewage Sludge and Sawdust," *Sewage Ind. Wastes* 31(5):557 (May 1959).

37. Yanko, W. A., McGee, C. D., and Glass, J. S. "The Fate of Viruses During Composting of Sewage Solids — A Review of the Problem and Results of Field Studies," in *Windrow and Static Pile Composting of Municipal Sewage Sludges*, EPA-600/2-84-122 (July 1984).

38. Yanko, W. A. "Occurrence of Pathogens in Distribution and Marketing Municipal Sludges," report to the Health Effects Research Laboratory, U.S. EPA, Cincinnati, OH (1988).

39. Hay, J. C. and Kuchenrither, R. D. "Fundamentals and Application of Windrow Composting," *J. Environ. Eng., ASCE* 116(4) (July/August 1990).

40. Epstein, E., Wilson, G. B., Burge, W. D., Mullen, D. C., and Enkiri, N. "A Forced Aeration System for Composting Wastewater Sludge," *J. Water Poll. Control Fed.* 48:4 (1976).

41. Strauch, D. "Hygienic Aspects of the Composting Process," in *Disinfection of Sewage Sludge: Technical, Economic and Microbiological Aspects* (London: D. Reidel Publishing Co., 1983).

42. Hornick, R. B., Woodward, T. E., McCrumb, F. R., Snyder, M. J., Dawkins, A. T., Bulkeley, J. T., DeLaMaccora, F., and Carozza, F. A. "Study of Induced Typhoid Fever in Man. I. Evaluations of Vaccine Effectiveness," *Trans. Assoc. Am. Physicians* 79:361–367 (1966).

43. Hornick, R. B., Greisman, S. E., Woodward, T. E., Dupont, H. L., Dawkins, A. T., and Snyder, M. J. "Typhoid Fever: Pathogenesis and Immunologic Control," *N. Engl. J. Med.* 283(13):686–691 (1970).

44. Mechalas, B. J., Hekimian, K. K., Schinazi, L. A., and Dudley, R. H. "An Investigation into Recreational Water Quality — Water Quality Criteria Data Book, Vol. 4," Office of Research and Monitoring, U.S. EPA (1972).

45. McCullough, N. B. and Eisele, C.W. "Experimental Human Salmonellosis, I. Pathogenicity of Strains of *Salmonella meleagridis* and *Salmonella anatum* obtained from Spray-Dried Whole Egg," *J. Infec. Dis.* 88:278–289 (1951).

46. McCullough, N. B. and Eisele, C. W. "Experimental Human Salmonellosis, II. Immunity Studies Following Experimental Illness with *Salmonella meleagridis* and *Salmonella anatum*," *J. Immunol.* 66:595–608 (1951).

47. McCullough, N. B. and Eisele, C. W. "Experimental Human Salmonellosis, III. Pathogenicity of Strains of *Salmonella newport, Salmonella derby,* and *Salmonella bareilly* Obtained from Spray-dried Whole Egg," *J. Infect. Dis.* 89:209–213 (1951).

48. McCullough, N. B. and Eisele, C. W. "Experimental Human Salmonellosis, IV. Pathogenicity of Strains of *Salmonella pullorum* Obtained from Spray-Dried Whole Egg," *J. Infect. Dis.* 89:259–265 (1951).

49. D'Aoust, J. Y. and Pivnick, H. "Small Infectious Doses of *Salmonella*," *Lancet* 1:866 (1976).

50. D'Aoust, J. Y. "Infective Dose of *Salmonella typhimurium* in Cheddar Cheese," *Am. J. Epidem.* 122(71):720 (1985).

51. Kaprowski, H. *Am. J. Trop. Med. Hyg.* 5:440 (1956).

52. Kowal, M. E. "Health Effects of Land Application of Municipal Sludge," U.S. EPA Health Effects Research Laboratory, Research Triangle Park, NC, EPA-600/1-85-015 (1985).

53. Akin, E. W., Pahren, H. P., Jakubowski, W., and Lucas, J. B. "Health Hazards Associated with Wastewater Effluents and Sludges: Microbiological Consideration," in *Proceedings of the Conference on Risk Assessment and Health Effects of Land Application of Municipal Wastewater and Sludges* (San Antonio, TX: Center for Applied Research and Technology, University of Texas at San Antonio, 1977).

54. Comar, C. L. "Risk: A Pragmatic De Minimus Approach," *Science* 203:4378 (1979).

55. Council directive of June 12, 1986 on the protection of the environment and in particular of the soil when sewage sludge is used in agriculture, 86/278/EEC, *Official Journal of the European Communities,* No. L 181 (July 4, 1986).

CHAPTER 6

Feed Conditioning — Physical and Chemical

INTRODUCTION

Designers and operators of composting systems have a limited number of areas over which to exercise control of the process. One area of control is the composition of the feed mixture. The proportions of feed components must be adjusted to satisfy the energy balance and avoid rate limitations caused by lack of moisture, lack of free airspace, sterile feed, or low nutrient levels. The quantities of required feed components must be known to size the system and its metering, mixing, and conveying equipment. There are three aspects to feed conditioning: physical or structural conditioning, chemical conditioning, and thermodynamic or energy conditioning. The relationships between free airspace and moisture content are discussed in this Chapter. These are the most important physical factors that must be considered to assure proper structural conditioning of the feed substrates. Chemical conditioning to correct for nutrient and pH imbalances is also discussed. Energy conditioning is discussed in Chapter 8.

THE IMPORTANCE OF MOISTURE AND FREE AIRSPACE

The composting matrix is a network of solid particles that contain voids and interstices of varying size. Voids between particles are filled with air, water, or a mixture of air and water. If the voids become completely filled with water, oxygen transfer is greatly restricted and aerobic composting becomes impractical in the absence of constant agitation. If some water is removed and the voids fill with air, oxygen transfer begins and aerobic composting is possible. If too much water is removed, however, microbial kinetics will be slowed by lack of moisture and the composting activity will decline.

Decomposition of organic matter depends on the presence of moisture to support microbial activity. As Golueke[1] points out, the theoretically ideal moisture content is one that approaches 100% because under such conditions biological decomposition occurs in the absence of any moisture limitation. Autothermal thermophilic aerobic digestion (ATAD) processes are com-

Table 6.1.   Maximum Recommended Moisture Contents for Various Composting Materials

Type of Waste	Moisture Content (% of total weight)
Theoretical	100
Straw[a]	75–85
Wood (sawdust, small chips)	75–90
Rice hulls[a]	75–85
Municipal refuse	55–65
Manures	55–65
Digested or raw sludge	55–60
"Wet" wastes (grass clippings, garbage, etc.)	50–55

Source: Golueke.[1]

[a] Serves as a moisture absorbent and source of carbonaceous material. Requires addition of nitrogenous material to lower C/N ratio to a proper level.

mercially available; they achieve thermophilic temperatures during aerobic decomposition of liquid organic slurries. These processes operate without any moisture rate limitations and have occasionally been referred to as "liquid composting" processes. Such processes cannot produce a dry end product, however, and they are usually limited to situations in which the liquid slurry is land applied.

Composting is usually applied to solid or semisolid materials and the practical moisture content must be considerably <100%. If the substrate is to be placed in windrows, static piles or a reactor system, the question arises as to the maximum moisture content to begin the process. General ranges of moisture contents found suitable for various wastes are listed in Table 6.1. The values shown are related to the structural strength of the composting material. Fibrous or bulky material such as straw or wood chips can absorb relatively large quantities of water and still maintain their structural integrity and porosity. For example, McGauhey and Gotaas[2] were able to compost mixtures of vegetable trimmings and straw that had initial moisture contents as great as 85%, but 76% moisture was too great when paper was used instead of straw.

Many composting substrates, such as municipal refuse, agricultural residues, and yard wastes, begin the composting process in a relatively dry form. Even animal manures are often field-dried before composting. Many composting systems have been developed to process such dry materials, and provisions are usually made for moisture addition as composting proceeds. Furthermore, the fibrous and bulky nature of such materials allows absorption of relatively large quantities of water. Sludge and other wet organic wastes differ on both accounts. Municipal sludge seldom contains <70% water, and sludge is not a fibrous material capable of supporting such high moisture contents. If sludge cake were composted alone, nearly constant mechanical agitation would be needed to provide aeration because of the lack of free airspace in which to store oxygen.

The importance of proper moisture control was demonstrated by Senn[3] during composting of dairy manure. Composting was conducted in 2.4 m deep bins equipped with a forced aeration system mounted in the bin floor. The influence of moisture content on subsequent temperature development is shown in Figure 6.1. At 66% moisture the temperature rose to about 55°C but no higher. The bin was unloaded, dry material added, and the mix reloaded at 61% moisture. Temperature then rose rapidly to above 75°C. In a parallel test using material loaded at 60% moisture, the temperature quickly rose to above 75°C and remained for several days. The higher moisture content impeded the composting process because the excess moisture caused packing and reduced void space which prevented proper air movement throughout the material.

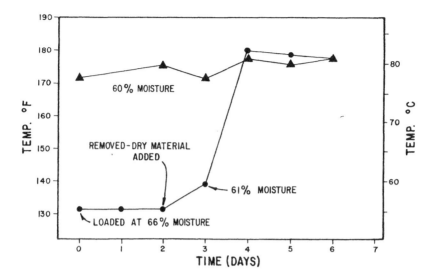

Figure 6.1.   Effect of initial moisture content on temperatures developed during composting of dairy manure in deep bins. From Senn.[3]

Achieving a balance of moisture and free airspace in the starting mixture does not guarantee that the balance is maintained once composting starts. If proper aeration rates are supplied, the thermophilic temperatures of composting will cause significant moisture removal. It is not unusual for the composting mixture to dehydrate to the point that reaction rates are seriously impacted. Water must then be added back into the process to correct the moisture imbalance.

The free airspace of a composting mixture is important in determining the quantity and movement of air through the mixture. It is generally believed that the optimum moisture content for a particular material is related to maintenance of a certain minimum free airspace. In general, more fibrous and friable materials can maintain higher moisture contents while still retaining adequate free airspace. Thus, different materials can hold different moisture levels while still maintaining the same free airspace. The term "optimum moisture" represents a tradeoff between the moisture requirements of the microbes and their simultaneous need for an adequate oxygen supply.

Maintaining proper moisture and free airspace levels is a matter of balancing numerous competing forces. Moisture levels must be high enough to assure adequate rates of biological stabilization, yet not so high that free airspace is eliminated, thus reducing the rate of oxygen transfer and in turn the rate of biological activity. In addition, it is advantageous to produce a reasonably dry compost product, one that can be stockpiled and transported economically for subsequent reuse. It may not be possible to optimize all of these factors at the same time. Tradeoffs will usually be necessary. Relationships between these competing factors are developed in the discussion of process dynamics in Chapters 11 through 14. The discussion here will focus on methods available to control moisture and free airspace in the starting mixture.

WEIGHT AND VOLUME RELATIONSHIPS

A schematic representation of the composting matrix considered as a three phase system of solids, water, and gas is shown in Figure 6.2. The figure must be understood as a schematic representation because the void and solid volumes cannot be segregated as shown. Neverthe-

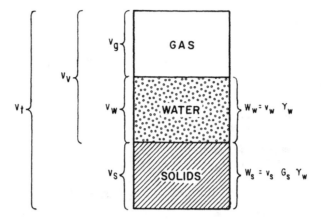

Figure 6.2.   Diagram of the compost matrix as a three phase system of solids, water, and gas.

less, the sketch facilitates understanding of the relationships between terms given in this chapter. As shown in Figure 6.2, total volume, v_t, consists of two parts, the volume of solid matter, v_s, and the volume of voids, v_v. The volume of voids is further distinguished into water volume, v_w, and gas volume, v_g.

The specific gravity of a solid is a function of the specific gravities of component parts making up the solid. If specific gravities of organic (volatile) and ash (fixed) components are known, the specific gravity of the solids as a whole can be calculated as

$$\frac{1}{G_s} = \frac{V_s}{G_v} + \frac{(1 - V_s)}{G_f} \tag{6.1}$$

where

V_s = volatile fraction of the substrate solids
G_s = specific gravity of the substrate solids
G_v = specific gravity of the organic or volatile fraction of the substrate solids
G_f = specific gravity of the ash or fixed fraction of the substrate solids

The specific gravity of the volatile solids (VS) normally is about 1.0 and that of fixed solids about 2.5.

The total volume of a composting mixture equals the sum of the volumes of water, solids, and gas contained in the mixture. The maximum bulk weight can be estimated by assuming the gas volume to be zero. Municipal and industrial sludge cakes often approach this condition. Assuming zero gas volume, total volume becomes

$$V_t = \frac{W_s}{G_s \delta_w} + \frac{W_w}{\delta_w} = \frac{W_s}{G_s \delta_w} + \frac{W_s(1 - S_s)}{S_s \delta_w}$$

$$V_t = (W_s / \delta_w)[1/G_s + (1 - S_s)/S_s] \qquad \left(V_g = 0\right) \tag{6.2}$$

where

W_s = weight of dry solids
W_w = weight of water
δ_w = unit weight of water
S_s = fractional solids content of the substrate

Let δ_s = unit bulk weight of the substrate solids = total wet weight per unit volume, and let δ_s (dry) = unit dry weight = dry weight per unit volume. Based on Equation 6.2, unit bulk and dry weights for the substrate can be calculated as

$$\delta_s = \frac{W_s}{S_s V_t} = \frac{W_s}{S_s (W_s / \delta_w)[1/G_s + (1-S_s)/S_s]}$$

$$\delta_s = \frac{\delta_w}{[(S_s / G_s) + 1 - S_s]} \qquad (V_g = 0)$$

(6.3)

$$\delta_s(\text{dry}) = S_s \delta_s$$

(6.4)

Equation 6.3 is valid as long as pore spaces within the material are completely filled with water. Wet substrates such as sludge cake usually contain little or no gas volume and generally meet the conditions of Equation 6.3. As moisture is removed, however, the gas volume begins to increase and bulk weights become less than the maximum. This is illustrated by the data of Figure 6.3 which was developed for digested sludge cake. The maximum bulk weight was calculated from Equation 6.3 using the procedures of Example 6.1. Although there is some data scatter, unit bulk weight does decrease with increasing solids content, with measured values beginning to deviate from the maximum at solids contents above ~35%. Thus, FAS becomes measurable above about 35% TS for the sludge tested for Figure 6.3. Many sludge composting systems operate with initial mixtures near 40% TS for this reason. If amendments with low bulk weights, such as sawdust, are blended with the sludge cake, FAS may be present at higher moisture contents.

Corresponding values of unit dry weight are shown in Figure 6.4. Unit dry weight is reasonably constant above solids contents of 35 to 40%. Within this range, it is likely that moisture evaporates from the solids causing little net change in the total volume. If the unit dry weight is assumed constant, Equation 6.4 becomes

$$\delta_s(\text{dry}) = \text{constant}, C = S_s \delta_s$$

$$\delta_s = C / S_s$$

(6.5)

The form of Equation 6.5 provides the basis for algorithms used to predict unit bulk weights in the simulation models of Chapters 11 through 14.

Data scatter in Figures 6.3 and 6.4 is probably caused in part by different levels of compaction and consolidation, which are functions of the methods used in handling the material. Compaction was minimized during collection of data shown in Figures 6.3 and 6.4.

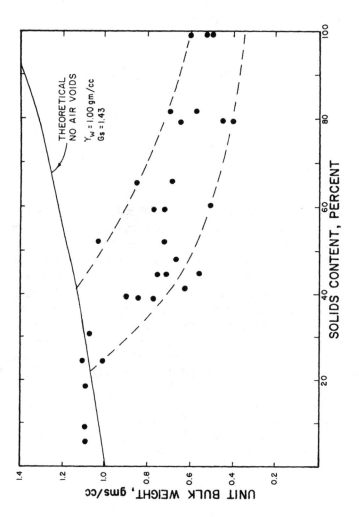

Figure 6.3. Unit bulk weight of sludge and compost as a function of solids content. Data are for sludge-only systems and do not include effects of amendment or bulking agent additions. From Haug.[4]

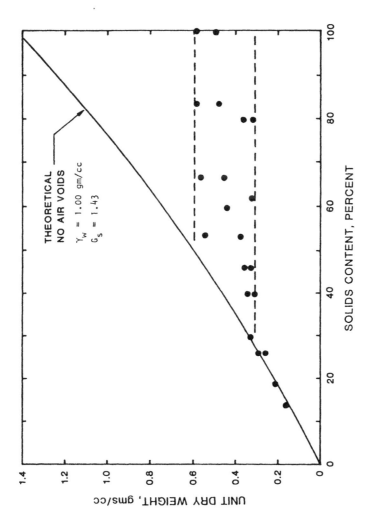

Figure 6.4.   Unit dry weight of sludge and compost as a function of solids content. Data are for sludge-only systems and do not include effects of amendment or bulking agent additions. From Haug.[4]

It should be noted that relatively high unit weights are possible if FAS is removed by compaction. Volume reductions are often observed during composting caused in part by consolidation of material. One of the functions of compost agitation or turning is to decrease the unit bulk weight and minimize effects of compaction.

It is generally advisable to measure bulk weights for the actual substrates in question. Substrates such as sawdust, wood chips, and yard wastes can vary significantly in bulk weight. For example, some dry sawdusts can have bulk weights as low as 0.15 g/cm^3 (9.4 lb/ft^3). Also, the bulk weight of mixtures of substrates is not equal to the average of the individual bulk weights. Consider the case where a small amount of sawdust is added to a very wet sludge cake. The sawdust may be completely absorbed and the bulk weight of the mixture may only be slightly reduced from that of the sludge cake. Bulk weight is also affected by the method of handling. Dry leaves generally have a very low bulk weight, which increases if they are shredded.

Example 6.1

A wet substrate with a volatile solids fraction of 0.50 is dewatered to 30% solids. Estimate the unit bulk and dry weights of the substrate assuming no measurable gas volume in the cake. Assume the specific gravities of volatile and ash fractions to be 1.0 and 2.5, respectively.

Solution

1. The specific gravity of the substrate is given by Equation 6.1:

$$\frac{1}{G_s} = \frac{0.5}{1.0} + \frac{(1-0.5)}{2.5}$$

$$G_s = 1.43$$

2. Unit bulk weight is then determined from Equation 6.3, assuming the unit weight of water as 1.00 g/cm^3:

$$\delta_s = \frac{1.00}{[(0.3/1.43)+1-0.3]} = 1.099 \text{ g/cm}^3 \quad (68.6 \text{ lb/ft}^3)$$

3. Using Equation 6.4, the unit dry weight is

$$\delta_s(\text{dry}) = (0.30)(1.099) = 0.33 \text{ g dry solids (ds)}/\text{cm}^3 \text{ or } 20.6 \text{ lb ds/ft}^3$$

4. A number of factors can cause substrates to have bulk weights below the maximum values calculated above. For example, digested sludge often contains considerable gas volume from the biogas that continues to be produced. Sludge dewatered in belt or filter presses can contain air pockets between the "pancakes" of material. Experience has shown that the bulk weights of sludge cake can be as low as 0.80 g/cm^3 (50 lb/ft^3) because of these effects.

POROSITY AND FREE AIRSPACE

One might argue that the volume of voids shown in Figure 6.2 should not include the water volume, v_w. After all, if the substrate is saturated with water it seems somewhat incorrect to say that it still has void volume. The problem is that the concept of void volume, and related ideas such as porosity, void ratio and degree of saturation, have their origin in the science of soil mechanics. Both the concepts and nomenclature have been borrowed and applied to composting systems. To change nomenclature at this point would be difficult and of dubious value to the student who must still contend with the terms as they are used in the composting literature. Instead, let us define terms as clearly as possible and overlook minor problems of nomenclature.

Volume ratios commonly used in composting are the porosity and free airspace (FAS). Porosity, n, of a composting mass is defined as the ratio of void volume to total volume:

$$n = v_v / v_t = \text{porosity} \tag{6.6}$$

$$n = (v_t - v_s) / v_t = 1 - v_s / v_t \tag{6.7}$$

Considering a unit total volume,

$$n = 1 - \frac{\delta_m S_m}{G_m \delta_w} \tag{6.8}$$

where

δ_m = unit bulk weight of the mixed material to be composted, wet weight per unit volume

S_m = fractional solids content of the mixture

G = specific gravity of the mixture solids

FAS, f, is defined as the ratio of gas volume to total volume:

$$f = v_g / v_t \tag{6.9}$$

$$f = (v_t - v_s - v_w) / v_t \tag{6.10}$$

Again consider a unit total volume:

$$f = 1 - \frac{\delta_m S_m}{G_m \delta_w} - \frac{\delta_m (1 - S_m)}{\delta_w} \tag{6.11}$$

Recall from the previous discussion that the term "optimum moisture" represents a tradeoff between the moisture requirements of the microbes and their simultaneous need for an adequate oxygen supply. The effect of moisture content on FAS is shown in Figure 6.5 for various composting substrates. Reported optimum moisture contents tend to fall in the range of FAS between 30 and 35%. Jeris and Regan[5] further examined the effect of FAS on oxygen consumption rates of mixed refuse samples. Approximately 67% moisture and 30% FAS were found to be optimum conditions as shown in Figure 6.6. About 95% of the maximum oxygen consumption rate was maintained when FAS was between about 20 and 35%. Working with garbage and sludge mixtures, Schulze[6] concluded that a minimum of ~30% FAS should be maintained.

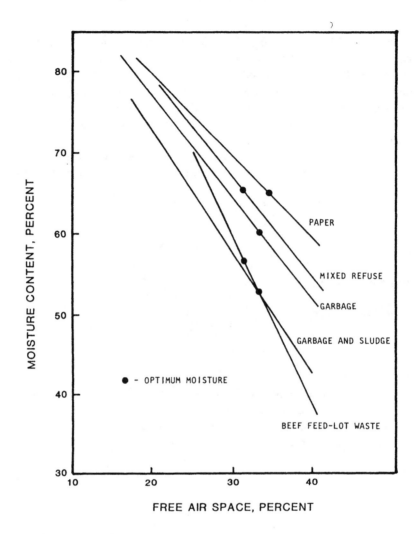

Figure 6.5.   FAS as a function of moisture content for various feed materials. Reported optimum moisture contents tend to fall within a reasonably narrow range of FAS. From Jeris and Regan.[5]

The minimum FAS is probably influenced to some extent by method of aeration. With constant tumbling or turning, a lower FAS should be acceptable because of repeated exposure of new surfaces to the oxygen. However, constant agitation is not characteristic of most compost systems. Windrow, static pile, and most reactor systems will require FAS more in keeping with previously quoted results. Therefore, a FAS of ~30% is recommended for most substrates and composting systems.

Example 6.2

An organic slurry is dewatered to 30% TS and blended with recycled compost to give a mixture solids of 45%. Assuming a mixture volatility of 0.50 and a unit bulk weight of 0.85 g/cm³, calculate the porosity and FAS of the mixed material.

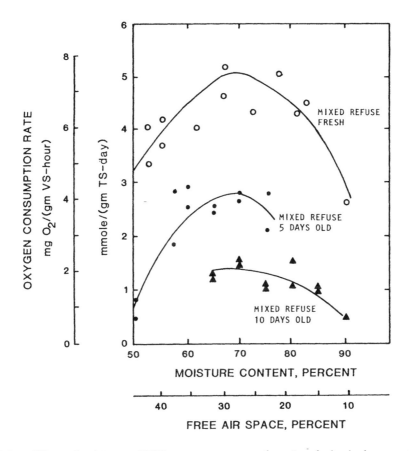

Figure 6.6.   Effects of moisture and FAS on oxygen consumption rates of mixed refuse samples. Data were developed from Warburg respirometer runs. Incubation temperatures were not specified. From Jeris and Regan.[5]

Solution

1. From Example 6.1, specific gravity of the solids, G_s, is 1.43.
2. From Equation 6.8, the porosity is calculated as

$$n = 1 - [(0.85 \times 0.45 / (1.43 \times 1.00)]$$

$$n = 0.73$$

3. Using Equation 6.11, FAS is calculated as

$$f = 1 - \frac{0.85(0.45)}{1.43(1.00)} - \frac{0.85(1 - 0.45)}{1.00}$$

$$f = 0.27$$

4. Volume and weight fractions for the substrate mixture are determined as

$$\text{Gas} \quad v_g = 0.27$$

$$\text{Water} \quad W_w = 0.85(1 - 0.45) = 0.46 \text{ g/cm}^3$$

$$v_w = 0.46/(1.00) = 0.46$$

$$\text{Solids} \quad W_s = 0.85(0.45) = 0.38 \text{ g/cm}^3$$

$$v_s = 0.38/[1.43(1.00)] = 0.27$$

$$\text{Total} \quad v_g + v_w + v_s = v_t$$

$$0.27 + 0.46 + 0.27 = 1.00$$

Note that water occupies the greatest part of the mixture volume. This emphasizes the problem of maintaining proper free airspace with wet substrates.

GENERALIZED PROCESS SCHEMATIC

A process diagram for a generalized composting system is presented in Figure 6.7. The diagram is a generalized model for performing mass balances and is applicable to all composting systems. Pre- and postprocessing stages are not shown for convenience. Inputs of substrate, amendment, and bulking agent are included in Figure 6.7. This is not meant to imply that all additives would be used at the same time. Whether one or all are used depends on the composting system under study.

The following nomenclature, first proposed by the author in 1977,[8] is used in Figure 6.7 and subsequent discussions:

X_s = wet weight of the main substrate produced per day
X_p = wet weight of compost product produced per day
X_r = wet weight of material recycle per day
X_a = wet weight of amendment added to the mixture per day
X_b = wet weight of bulking agent added to mixture per day
X_w = weight of water added to the mixture per day
X_m = wet weight of mixed feed materials entering the compost process per day
S_s = fractional solids content of the substrate
S_p = fractional solids content of compost product
S_r = fractional solids content of recycle material
S_a = fractional solids content of amendment
S_b = fractional solids content of bulking agent
S_m = fractional solids content of the feed mixture
V_s = volatile solids content of the substrate, fraction of dry solids
V_p = volatile solids content of compost product, fraction of dry solids
V_r = volatile solids content of recycle material, fraction of dry solids
V_a = volatile solids content of amendment, fraction of dry solids
V_b = volatile solids content of bulking agent, fraction of dry solids
V_m = volatile solids content of mixture, fraction of dry solids
k_s = fraction of substrate volatile solids degradable under composting

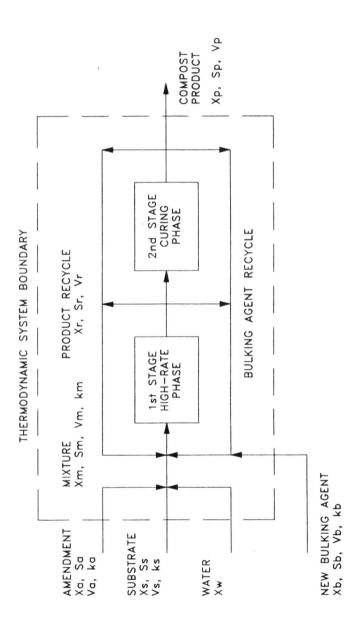

Figure 6.7.   Generalized mass balance diagram for composting showing inputs of substrate, recycled compost product, amendment, and bulking agent. Adapted from Haug.[7]

conditions
k_a = fraction of amendment volatile solids degradable under composting conditions
k_b = fraction of bulking agent volatile solids degradable under composting conditions
k_m = fraction of the mixture volatile solids degradable under composting conditions

A point should be made about the above nomenclature. All materials added to the composting process should be considered as substrates. Amendments and bulking agents are substrates because they add new materials to the process. Using separate terms for these substrates is convenient and consistent with the nomenclature used in the industry. For example, a sludge composting facility may use sawdust for feed conditioning. Sludge is considered the substrate and sawdust the amendment. There is no particular problem in this as long as one recognizes that the sawdust is also a substrate.

If multiple substrates are input to the process, the combined feed characteristics from these substrates can be determined as,

$$X_s = \sum_{j=1}^n X(j) \tag{6.12}$$

$$S_s = \left[\sum_{j=1}^n S(j)X(j)\right] / X_s \tag{6.13}$$

$$V_s = \left[\sum_{j=1}^n V(j)S(j)X(j)\right] / S_s X_s \tag{6.14}$$

$$k_s = \left[\sum_{j=1}^n k(j)V(j)S(j)X(j)\right] / V_s S_s X_s \tag{6.15}$$

where
$X(j)$ = daily wet weight of substrate j
$S(j)$ = fractional solids content of substrate j
$V(j)$ = fractional VS content of substrate j
$k(j)$ = fractional degradability of substrate j
n = the number of substrates

The following discussion assumes that, where multiple substrates are involved, Equations 6.12 through 6.15 have been used to convert them into an equivalent single substrate with characteristics X_s, S_s, V_s, and k_s.

CONDITIONING OF WET SUBSTRATES

A number of approaches are available to overcome the problem of high moisture content in the feed substrate. First, dry compost product can be recycled to condition the starting mixture. Second, dry amendments, such as sawdust or yard wastes, can be added to the wet substrate either with or without compost product recycle. Third, bulking agents, such as wood chips, can be added to maintain structural integrity. Fourth, the wet substrate can be air or heat dried to decrease the moisture content before composting. Fifth, constant agitation can be provided by mechanical means. Shell and Boyd[9] successfully used the latter approach on a bench scale, but even then found that compost recycle was advantageous. The use of constant agitation has never been practical on a full scale. Therefore, the following discussion addresses the use of recycled compost, amendments, bulking agents, and drying for feed conditioning.

Conditioning with Compost Recycle

Compost product can be recycled and used to condition the feed substrate. The quantity of recycled material required to adjust the mixture moisture content can be determined from the process diagram of Figure 6.7. Total dry solids produced per day is $S_s X_s$. Similarly, dry solids in recycled compost product is $S_r X_r$. Assuming no amendment or bulking agent addition, a mass balance on total wet weight gives

$$X_s + X_r = X_m \tag{6.16}$$

Similarly, a mass balance on dry solids gives

$$S_s X_s + S_r X_r = S_m X_m \tag{6.17}$$

Substituting Equation 6.17 into 6.16:

$$S_s X_s + S_r X_r = S_m (X_s + X_r)$$

$$X_r = X_s \frac{(S_m - S_s)}{(S_r - S_m)} \tag{6.18}$$

Let R_w be defined as the wet weight recycle ratio defined as

$$R_w = X_r / X_s \tag{6.19}$$

Substituting into Equation 6.18 gives

$$R_w = \frac{(S_m - S_s)}{(S_r - S_m)} \tag{6.20}$$

Let R_d be defined as the dry weight recycle ratio defined as

$$R_d = S_r X_r / S_s X_s \tag{6.21}$$

Substituting the expression for R_d into Equation 6.17 and rearranging gives

$$R_d = \frac{[(S_m / S_s) - 1]}{[1 - (S_m / S_r)]} \tag{6.22}$$

Equations 6.20 and 6.22 can be used to calculate the required recycle ratios given the substrate solids, S_s, compost recycle solids, S_r, and desired solids content in the mixture, S_m.

Example 6.3

10 dry metric tpd (dmtpd) of substrate at 20% TS is to be reactor composted using recycled compost for feed conditioning. The desired mixture solids is 40% and recycled compost is 60% TS. Calculate the required recycle ratios and the total weight of mixture processed daily.

Solution

1. From Equation 6.20, the wet weight recycle ratio, R_w is

$$R_w = (0.40 - 0.20) / (0.60 - 0.40) = 1.00$$

2. The dry weight recycle ratio, R_d is determined from Equation 6.22:

$$R_d = \frac{[(0.40 / 0.20) - 1]}{[1 - (0.40 / 0.60)]} = 3.0$$

3. The daily weight to be processed can be determined from either R_d or R_w as follows:

$$\text{total weight} = \text{substrate weight} + \text{recycle weight}$$

Based on R_d: total wt $= 10 / 0.20 + 10(3.00) / 0.60 = 100$ wmtpd

Based on R_w: total wt $= 10 / 0.20 + 10(1.0) / 0.20 = 100$ wmtpd

4. Using procedures from Example 6.2, the mixture FAS should be ~20%, sufficient to avoid significant rate limitations.

A bar diagram of the substrate, recycle, and mixed materials from Example 6.3 is shown in Figure 6.8 based on a unit dry weight of substrate. Some new terms are introduced in this diagram. WAT is the water component, BVS the biodegradable volatile solids component, NBVS the nonbiodegradable volatile solids component, and ASH the ash or inert component. Equations used to calculate these component fractions of the substrate are as follows:

$$BVS = K_s V_s S_s X_s \tag{6.23}$$

$$NBVS = (1 - K_s) V_s S_s X_s \tag{6.24}$$

$$ASH = (1 - V_s) S_s X_s \tag{6.25}$$

$$WAT = X_s - S_s X_s \tag{6.26}$$

Similar equations can be developed for amendment, bulking agent, and recycle components. Further development and use of these Equations is presented in Chapter 11.

Example 6.4

Based on the results of Example 6.3, develop the bar diagram shown in Figure 6.8. Assume 1 kg of substrate solids. The substrate has an organic content of 75% ($V_s = 0.75$) and degradability of 50% ($k_s = 0.50$).

Solution

1. The wet weight of substrate is determined as

$$X_s = 1/0.20 = 5 \text{ kg}$$

2. The 5 kg of substrate contains the following components based on Equations 6.23 to 6.26:

$$BVS = 0.5(0.75)(0.20)(5) \qquad = 0.375 \text{ kg}$$

$$NBVS = (1-0.50)(0.75)(0.20)(5) = 0.375 \text{ kg}$$

$$ASH = (1-0.75)(0.20)(5) \qquad = 0.250 \text{ kg}$$

$$WAT = 5-(0.20)(5) \qquad = 4.000 \text{ kg}$$

The left bar in Figure 6.8 represents the substrate with the above component fractions.

3. From Example 6.3, $R_w = 1.0$; therefore, $X_r = X_s R_w = 5(1) = 5$ kg. All BVS are assumed to be degraded during composting so that none appear in the compost recycle, i.e., $k_r = 0$. The NBVS and ASH components are assumed to act conservatively in the process; therefore, their distribution in the recycle reflects their distribution in the feed substrate.

$$BVS = \qquad\qquad\qquad\qquad\qquad 0.00$$

$$NBVS = [0.375/(0.375+0.250)](0.60)(5) \quad = 1.80$$

$$ASH = [0.250/(0.375+0.250)](0.60)(5) \quad = 1.20$$

$$WAT = 5-(0.60)(5) \qquad\qquad\qquad = 2.00$$

The middle bar in Figure 6.8 represents the compost recycle with the above component fractions.

4. The right bar in Figure 6.8 represents the combined mixture of substrate and recycle. Each fraction in the mixture bar is the sum of substrate and recycle contributions. Note that the mixture solids content is 40%, which is the desired condition from Example 6.3.

Based on Examples 6.3 and 6.4, it would appear that the mixture is properly conditioned. Indeed, from the structural standpoint of moisture and free airspace the mixture is acceptable. However, mixture conditions must also be checked from the standpoint of the energy balance. The latter is considered in Chapter 8.

Equations 6.20 and 6.22 were solved for various substrate and compost product solids contents, assuming a desired mixture solids of 40% ($S_m = 0.40$). Resultant total and dry weight recycle ratios are presented in Figures 6.9 and 6.10, respectively. Referring to Figure 6.9, R_w

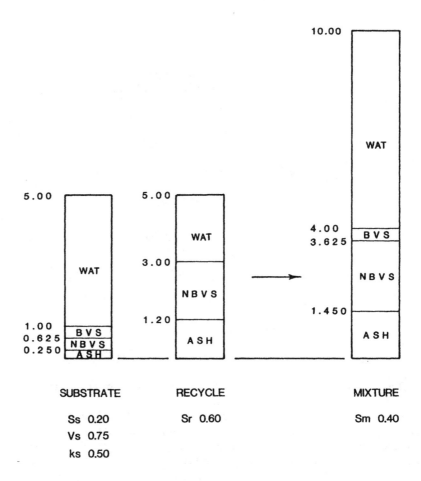

Figure 6.8.   Feed conditioning of substrate using recycled compost for the conditions of Example 6.4.

is zero at a substrate solids of 40%. This reflects the fact that a 40% TS mixture was assumed necessary to begin the process. R_w increases as substrate solids decrease below 40%, with the slope depending on the solids content of the recycled product. Obviously, the drier the recycled product, the lower the required recycle ratio. Equation 6.20 can be solved for the case where $S_s = 0$, in other words, no solids in the substrate (i.e., pure water). This is a mathematical anomaly that results from the assumption of a fixed solids content for the recycle product. Valid application of Equations 6.20 and 6.22 depends on the ability to achieve the assumed compost product solids, a subject of discussion in later chapters.

The total quantity of mixed material that must be handled daily is a function of the initial substrate solids. For example, assume a substrate solids of 30% and a compost product of 70% solids. From Figure 6.9, R_w for this case is 0.33. Assuming 100 ton ds/day of substrate, X_s is 333 tpd and X_r 110 tpd, giving a mixture of 443 tpd. If substrate solids are reduced to 20%, R_w increases to 0.67, X_s to 500 tpd, X_r to 335 tpd, and the total quantity of mixed material, X_m, to 835 tpd. Thus, reducing substrate solids from 30 to 20% results in a near doubling of the total weight of mixed materials to be handled each day.

The recycle ratio based on dry weight, R_d, is presented in Figure 6.10. As S_s approaches zero, Equation 6.22 tends toward infinity and the boundary problems associated with Equation 6.20 are not encountered. Based on Figure 6.10 the dry weight of recycled compost can, in

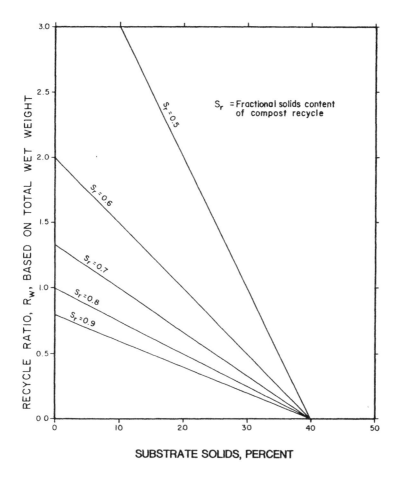

Figure 6.9.   Effect of substrate solids content on the wet weight recycle ratio needed to achieve a 40% mixture solids content. From Haug.[8,10]

certain cases, exceed the quantity of substrate solids. Again, assuming a substrate of 30% solids and a compost product of 70% solids, R_d from Figure 6.10 is 0.78. In other words, 0.78 g ds of compost product must be recycled for each 1.0 g ds of substrate to achieve 40% solids in the mixture. If substrate solids are reduced to 20%, R_d increases to 2.34. Thus, decreasing the substrate solids from 30 to 20% increases the mixture dry weight from 1.78 to 3.34 g and increases the dry weight percentage of recycled compost in the mixture from 44 to 70%. Also, as previously mentioned, the daily mixture wet weight nearly doubles. This is not meant to imply that a substrate with 20% solids cannot be composted. It merely points out the importance of substrate solids in determining both the quantity of material to be handled each day and the relative percentage of compost product in the mixture.

Equation 6.20 was again solved, in this case to determine the effect of mixture solids content on the recycle ratio. Substrate solids were assumed to be constant at 25%, with recycle solids ranging from 50 to 90%. Results are presented in Figure 6.11. Mixture solids content has a significant effect on the required recycle ratio, particularly at lower recycle solids contents. As recycle becomes drier, the effect of mixture solids becomes less pronounced. Assuming recycle at 70% solids, R_w increases from 0.29 to 1.25 as S_m increases from 35 to 50% solids. In practice it is desirable to minimize the mixture solids, S_m to reduce the weight

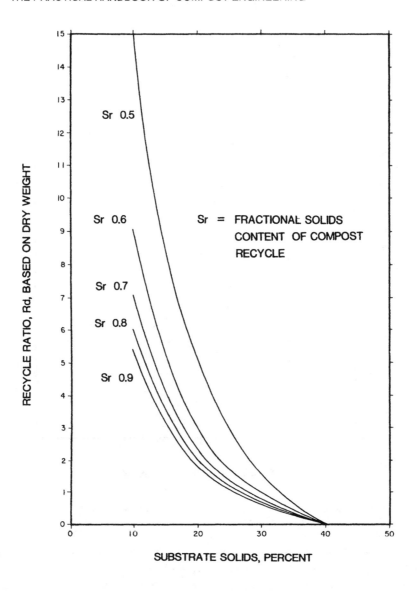

Figure 6.10. Effect of substrate solids content on the dry weight recycle ratio needed to achieve a 40% mixture solids content. From Haug.[8,10]

and volume of material processed daily. However, care must be taken to avoid decreasing mixture solids to the point that free airspace is reduced below the optimum range. For purposes of planning and design, it is better to err on the side of higher mixture solids and let operations "discover" if lower mixture solids can actually be used in practice.

Conditioning with Amendments

Organic or inorganic amendments can be added to condition the composting mixture either with or without use of product recycle. Using the process diagram of Figure 6.7 and assuming no bulking agent or recycle addition, the following equation can be derived for the required daily weight of amendment X_a:

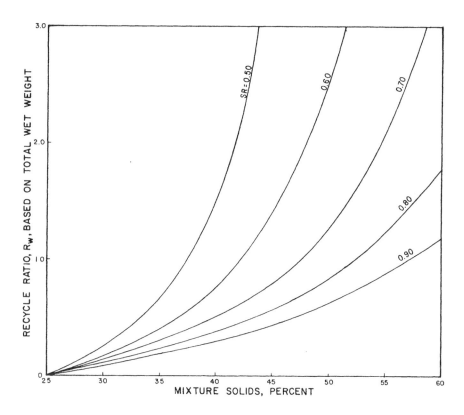

Figure 6.11. Effect of mixture solids content on the wet weight recycle ratio for various recycle solids contents. Substrate solids, S_s, assumed to be 25%.

$$X_a = X_s \frac{(S_m - S_s)}{(S_a - S_m)} \qquad (X_r = 0) \tag{6.27}$$

Again, let R_w be defined as the wet weight ratio and R_d as the dry weight ratio. Then,

$$R_w = \frac{X_a}{X_s} = \frac{(S_m - S_s)}{(S_a - S_m)} \tag{6.28}$$

and

$$R_d = \frac{S_a X_a}{S_s X_s} = \frac{[(S_m / S_s) - 1]}{[1 - (S_m / S_a)]} \tag{6.29}$$

Example 6.5

The 10 dmtpd of substrate from Examples 6.3 and 6.4 is to be conditioned using sawdust as a structural amendment. The sawdust is 65% TS, 95% VS, with an estimated 20% degradability of the VS. Desired mixture solids is 40%. Determine the quantity of required amendment and develop a bar diagram showing the substrate, amendment, and mixture conditions.

Solution

1. The quantity of required amendment is determine from Equation 6.27 as

$$X_s = 10 / 0.20 = 50 \text{ tpd}$$

$$X_a = 50(0.40 - 0.20) / (0.65 - 0.40)$$

$$X_a = 40 \text{ tpd}$$

2. From Equations 6.28 and 6.29, the wet and dry mixture ratios are

$$R_w = 40 / 50 = 0.80$$

$$R_d = 40(0.65) / 50(0.20) = 2.60$$

3. Four kg of sawdust are required for each 5 kg of substrate. The WAT, BVS, NBVS, and ASH components of the 4 kg of amendment are determined from Equations 6.23 to 6.26 as

$$BVS = 0.20(0.95)(0.65)(4) = 0.494 \text{ kg}$$

$$NBVS = (1 - 0.20)(0.95)(0.65)(4) = 1.976 \text{ kg}$$

$$ASH = (1 - 0.95)(0.65)(4) = 0.130 \text{ kg}$$

$$WAT = (1 - 0.65)(4) = 1.400 \text{ kg}$$

4. A bar diagram of the sludge, sawdust and mixture components is presented in Figure 6.12. Again, the mixture is properly conditioned from a structural standpoint, but the energy balance must be checked using procedures in Chapter 8.

The majority of sludge composting facilities condition the wet cake with sawdust or other amendments added to the cake after dewatering. At least one facility, a silo reactor system at Bristol, Tennessee, blends sawdust with the liquid sludge prior to dewatering. Sludge and sawdust at a 1:1 to 1.2:1 ratio by dry weight are metered and blended in a liquid mix tank before belt press dewatering. Dohoney[11] reported better water drainage, reduced variability in feed characteristics, reduced polymer consumption, better cake release, and no decrease in liquid loading to the press. A 35 to 37% TS cake is produced compared to 20 to 22% without sawdust addition. The dewatered mixture is then directly fed to the silo reactor.

Conditioning with Recycle and Amendments

Using amendments for structural conditioning without product recycle can result in consumption of considerable quantities of amendment. This can be expensive, depending on local availability of suitable materials. Also, the quantity of final product is increased compared to systems that rely all or in part on compost recycle for conditioning. Whether this is an advantage or not depends on the expected market for the final product. A compromise between use of product recycle alone and use of amendment alone is possible. If recycle is used for part of the structural conditioning, smaller quantities of amendment can be used for the balance.

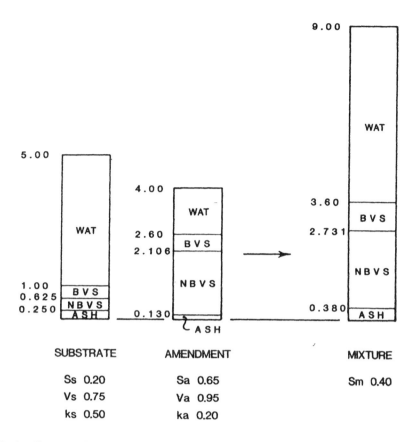

Figure 6.12. Feed conditioning of substrate using amendment for the conditions of Example 6.5.

Approaches to Solution

If both recycle and amendment are added to the infeed mixture, the mass balances for total and dry weight are

$$X_s + X_a + X_r = X_m \tag{6.30}$$

$$S_s X_s + S_a X_a + S_r X_r = S_m X_m \tag{6.31}$$

The mass balance cannot be uniquely solved because there are three unknowns (X_r, X_a, and X_m) and only two mass equations (wet weight or water and dry weight balances). A unique solution is possible if a third equation can be developed. Three approaches to a third equation are possible: (1) a mass balance on volatile solids, (2) assuming a value for X_a or X_r, and (3) an energy balance on the system. The first two approaches are considered here. The third approach of using the system energy balance to develop a third equation is considered in Chapter 8.

Referring to the system diagram in Figure 6.7, a volatile solids balance gives a third equation as

$$V_s S_s X_s + V_a S_a X_a + V_r S_r X_r = V_m S_m X_m \tag{6.32}$$

Combining Equations 6.30, 6.31, and 6.32 and rearranging gives the following solutions:

$$X_a = \frac{X_s\left[S_r(S_s/S_m - 1)(V_r - V_m) + S_s(S_r/S_m - 1)(V_m - V_s)\right]}{S_r(1 - S_a/S_m)(V_r - V_m) - S_a(S_r/S_m - 1)(V_m - V_a)} \tag{6.33}$$

$$X_r = \frac{S_s X_s(V_m - V_s) + S_a X_a(V_m - V_a)}{S_r(V_r - V_m)} \tag{6.34}$$

The procedure for a problem of this type is to assume all input variables and consecutively solve Equations 6.33, 6.34, and 6.30 for X_a, X_r, and X_m, respectively.

Alternative to the above approach, a third equation can be developed by assuming a value for either the amendment, X_a or recycle, X_r, i.e., X_a = constant or X_r = constant. If X_r is assumed, Equations 6.30 and 6.31 can be solved for X_a as follows:

$$X_a = \frac{X_s(S_m - S_a) + X_r(S_m - S_r)}{(S_a - S_m)} \tag{6.35}$$

If X_a is assumed, the solution for X_r is

$$X_r = \frac{X_s(S_m - S_a) + X_a(S_m - S_a)}{(S_r - S_m)} \tag{6.36}$$

Example 6.6

Ten dtpd of substrate at 20% solids is combined with 20 wtpd of sawdust at 65% solids. These substrates are further conditioned with recycle compost at 60% TS. Other characteristics are the same as in previous examples. Determine the quantity of recycle required to achieve a 40% TS mixture.

Solution

1. The total weight of substrate = 10/0.20 = 50 tpd.

2. The quantity of required recycle is determined from Equation 6.36 as

$$X_r = \frac{50(0.40 - 0.20) + 20(0.40 - 0.65)}{(0.60 - 0.40)}$$

$$X_r = 25 \text{ tpd}$$

3. The quantity of mixed materials is determined from Equation 6.30.

$$X_m = 50 + 20 + 25 = 95 \text{ tpd}$$

4. Referring to Example 6.3, the recycle is reduced from 50 to 25 tpd and the mixture weight reduced from 100 to 95 tpd by use of the combination of sawdust and recycle. Referring to Example 6.5, amendment use is reduced from 40 to 25 tpd.

5. A bar diagram showing the mixture and individual components is presented in Figure 6.13.

Particle Size Distribution

The particle size distribution of the amendment can be important to the structural conditioning of wet substrates. If the amendment is too fine, it may not provide the expected increase in free air space even though the mixture solids content is acceptable. As an example, some sludge composting facilities have noted that very fine sawdust, termed wood "flour", tends to produce a muddy consistency in the mix. The saturated condition can produce anaerobic conditions due to lack of free air space. As a result, several agencies have revised their amendment specifications to define the desired particle size distribution. Lancaster, Pennsylvania, specifies sawdust with a minimum 65% TS, with 95% passing a 12.5 mm screen opening but <50% passing a 2.23 mm opening. Cape May, New Jersey, specifies a coarser sawdust between 50 and 70% TS, with 95% passing a 12.5 mm screen opening but <20% passing a 4.75 mm opening.[12] The limit on fines is designed to avoid problems with wood flour. The limit on larger sized particles is applied to avoid the need to screen the final compost. The maximum particle size should be ≤10 mm if the compost product is intended for general horticultural use or as a top dressing on lawns and gardens.

The particle size distribution is also important to energy conditioning, which is discussed in Chapter 8. Generally, the biodegradability of amendments such as sawdust is favored by small particle size. Too many fines may be undesirable from a structural standpoint, but very attractive for energy conditioning. The demands for structural and energy conditioning should both be considered when developing amendment specifications.

Moisture Control by Drying

Air Drying

In lieu of using compost recycle or amendments to provide all of the structural conditioning, some wet substrates can be dried to further reduce the moisture content before composting. Air drying is used at both Phoenix, Arizona, and Austin, Texas, to reduce the moisture content of digested sludges before windrow composting. Liquid digested sludge is sent to decanting/ evaporation lagoons. Free water is decanted as supernatant from the liquid surface. Evaporation then continues to remove water until a cake is formed. Mobile turning equipment is used during the evaporation period to prevent a crust from forming, which would reduce the surface evaporation rate. Cake is generally removed from the drying beds when it reaches ~30+% solids and is conditioned with recycle or other amendments before windrow composting. Little if any composting occurs during the drying process because of the limited aeration. Therefore, biodegradable solids are largely conserved for use in the composting process.

Air drying of wet substrates prior to composting is a form of conditioning, just as adding recycle or other amendments. Removal of water by air drying effectively increases the solids content of the feed substrate. From the previous discussions, this will in turn reduce the amount of recycle or amendment needed for supplemental conditioning. Obviously, air drying is only effective in arid regions where evaporation exceeds precipitation and where relatively large land areas are available. If the climate is characterized by wet and dry seasons, the composting process will experience peak loads, because sludge will essentially be stored in the wet season and subsequently dried during the dry season. Finally, air drying is usually restricted to relatively stable substrates, such as digested sludge and some manures, because of the odor potential associated with putrescible substrates, such as raw sludge. Given these constraints, air drying is an extremely effective and often low cost method of structural

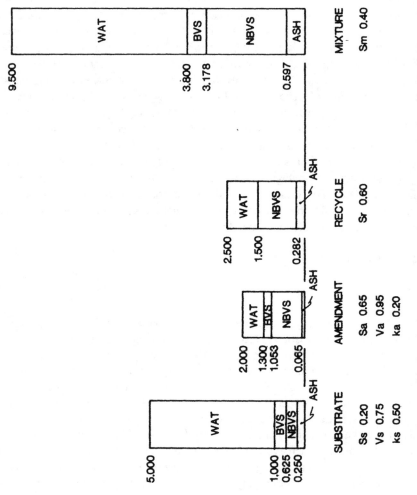

Figure 6.13. Feed conditioning of substrate using amendment and recycle for the conditions of Example 6.6.

conditioning. Because air drying removes water from the process, it is also a form of energy conditioning.

The time required for evaporative drying can be estimated by considering a layer of wet substrate of initial depth, d_{si}, and unit surface area. The substrate is to be dried from an initial solids content, S_{si}, to a final solids content, S_{sf}. Assuming that the substrate remains homogeneous throughout its depth as water evaporates from the surface, the following equation can be developed:

$$t_a = \frac{d_{si}\delta_{si}}{(E-P)\delta_w}\left(1-S_{si}/S_{sf}\right) \qquad (E > P) \qquad (6.37)$$

where

t_a = time required for air drying, days
d_{si} = initial depth of the wet substrate, cm
δ_{si} = initial bulk weight of the substrate, g/cm³
δ_w = bulk weight of water, g/cm³
E = average evaporation rate over the drying period, cm/day
P = average precipitation rate over the drying period, cm/day
S_{si} = initial solids content of the wet substrate, fraction
S_{sf} = final solids content of the dried substrate, fraction

The area required for drying can then be determined knowing the daily weight of sludge cake.

$$\text{area} = \frac{\text{volume}/\text{day}}{\text{depth}}\left(t_a\right)$$

$$\text{area} = \frac{X_s/\delta_{si}}{d_{si}}\left(t_a\right) \qquad (E > P)$$

Substituting Equation 6.37 for t_a and applying unit conversions:

$$A = \frac{X_s(1-S_{si}/S_{sf})}{(E-P)\delta_w(10^5)} \qquad (E > P) \qquad (6.38)$$

where

A = required drying area, hectares
X_s = substrate wet weight, kg/day

Should precipitation exceed evaporation during any period, no drying will result. Thus, Equations 6.37 and 6.38 apply to periods of the year when evaporation exceeds precipitation or to situations where the drying area is covered to prevent rainfall from reaching the substrate.

The most readily available information on evaporation rates is usually pan evaporation data recorded at a weather station near the proposed facility. However, evaporation rates from a wet organic substrate may not necessarily equal the pan evaporation rate. This would obviously be true if a dry crust were allowed to form, restricting passage of water to the surface. It is important that the substrate be turned frequently to reduce the likelihood of such a rate limitation. Even so, evaporation rates per unit surface area are likely to be less than pan evaporation rates, particularly as the cake moisture content decreases. On the other hand, surface area of cake per unit of ground surface is probably greater than unity because of the irregular surface of the dry cake. In the absence of actual field data and for purposes of estimation, pan evaporation rates can be assumed over the ground surface area occupied by the wet substrate.

Equations 6.37 and 6.38 represent a rather simplified approach to analysis because both evaporation and precipitation are assumed constant over time. Obviously, this is a simplified model of the actual case. In real situations, both the time required for drying and the land area will vary dynamically throughout the year. Simulation models have been developed to predict the drying time and area requirements in response to changing weather conditions.[14]

Heat Drying

Another approach to increasing the solids content of wet substrates is to heat dry either all or a portion of the substrate before composting. This is sometimes referred to as "thermal dewatering". Consider a substrate at 25% TS. If the substrate is thermally dewatered to 40% solids, about 2.5 g of water are removed per gram of solids. Thermal dewatering can accomplish the same feed conditioning benefits as air drying, namely biodegradable solids are conserved and excess water is removed. In addition, the heat drying process is not subject to outside climatic conditions. The downside is the fuel demand of the dryer and its capital cost.

A factor that may limit the application of heat drying prior to composting is that a dry product can be produced directly from the dryer with only modest additional energy input. For example, further drying of the above substrate to 90% solids requires removal of only 1.4 g additional water. Essentially complete drying can be obtained with only modest increases in the energy investment over that required for thermal drying. The number of required dryers is increased significantly, however. Today, there are no known applications of thermal dewatering prior to composting, but its possible use in selected situations should not be discounted.

Another potential application of heat drying is to dry the compost product and/or recycled compost to better assure a proper moisture content. If compost product is dried from 50 to 70% solids, about 0.57 g additional water must be removed per gram of compost solids, significantly less than that calculated for complete drying of the above substrate. The moisture removal that can be achieved during composting is determined by the thermodynamic balance of the system, which is discussed in Chapter 8. The compost process can remove significant quantities of water using the heat of biological decomposition. Heat drying can then be used to "polish" the composted product to the desired moisture content. This can be an attractive concept because energy requirements for heat drying can be significantly reduced without placing the total drying burden on the compost process or on the uncertainties of open-air drying. Postdrying was used at the Fairfield Hardy composter at Altoona, Pennsylvania, during composting of sludge cake and paper.[13] A materials balance for that facility is shown in Figure 6.14. Postdrying facilities were also designed into the Delaware Reclamation Project which composts MSW heavy fraction (see Chapter 2). However, the heat dryers have seen only limited service.

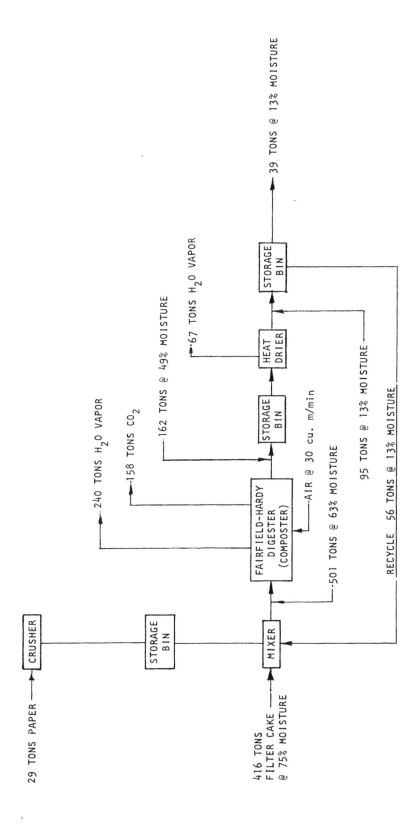

Figure 6.14.   Process flow diagram and materials balance during composting of raw sludge cake at the Fairfield Hardy digester in Altoona, Pennsylvania. The system included a pelletizer ahead of the heat dryer which was not used in the study. At the start of the study, shredded paper was the only amendment used at a ratio of 4.0 tons ds cake to 1.6 tons ds paper. Later in the study, compost product that had been heat dried to 13% moisture was used as the sole amendment.

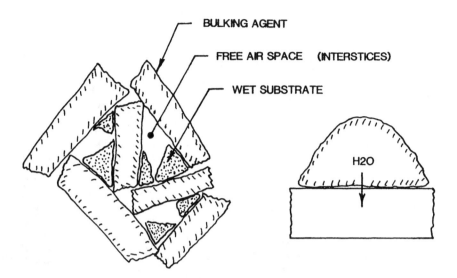

Figure 6.15.   Schematic illustration of substrate/bulking agent mixture showing water absorption from wet substrate into bulking particles.

Conditioning with Bulking Agents

Functions of a Bulking Agent

As discussed previously, the free airspace of wet substrates, such as sludge cake and wet manures, can reasonably be assumed to be zero. The function of a bulking agent is to provide structural support for the wet substrate, to provide free airspace within the voids between particles, to increase the size of pore spaces, and to allow easier air movement through the mixture. A bulking agent can be viewed as a three-dimensional matrix of solid particles capable of self-support. The void volume and size of the pore spaces is determined by the shape and size of the bulking particles. Conceptually, the wet substrate can be viewed as occupying part of the void volume in the bulking agent, as shown schematically in Figure 6.15.

If the substrate is assumed to occupy the void spaces, limits must exist on the ratio of bulking agent to substrate. If too little bulking agent is added the individual bulking particles will not be in contact with each other. Instead they will be immersed in the substrate and no practical increase in FAS or pore size will result. On the other hand, addition of bulking agent beyond that required to assure adequate FAS will increase the quantity of material to be handled daily, increase the consumption of bulking agent (assuming it to be degradable), and result in greater land requirements and higher costs.

Numerous bulking agents have been used in practice, including wood chips, straw, pelleted refuse, shredded tires, rice hulls, peanut shells, and other materials. Most of these are composed of cellulosic material and are degradable to some extent during composting. Decomposition will use up a portion of the bulking agent and size reduction will allow an additional fraction eventually to pass through the screening process with the composted sludge. Thus, continual makeup of bulking agent is necessary to balance that which is degraded and that which becomes part of the final product. Certain bulking agents, such as shredded tires and plastic materials, are extremely resistant to microbial decomposition and are probably not affected by the composting process. Hence, they should act almost as conservative substances. Assuming that the bulking agent can be screened efficiently from the compost mixture and reused, little or no makeup would be required in such a case.

Bulking agents can also be classified as to their moisture-absorbing characteristics. Most natural cellulosic materials are porous and capable of significant moisture absorption. Nonporous materials, such as plastics, can be assumed to be nonabsorptive for all practical purposes. Similarly, a porous bulking material that is saturated with water is not capable of further absorption unless it is dried before reuse.

The moisture-absorptive capacity of the bulking agent is important in determining the quantity required for structural conditioning. Assume the bulking agent to be dry and porous before mixture with the wet substrate. Once mixed, moisture will be drawn from the substrate into the bulking agent. Thus, a volume of water will penetrate the bulking agent, leaving an equivalently greater void volume. If the bulking agent is nonporous or water-saturated, a greater quantity of bulking agent will be required to produce the same FAS.

The quantity of moisture that can be absorbed by a bulking agent such as wood chips is considerable. For example, assume 1 m^3 of dry wood chips with a bulk weight of 288 kg/m^3 (18 lb/ft^3) is saturated with water to a final moisture content of 60%. In such a case, the 288 kg of dry chips will absorb 434 kg of water, i.e., (288/0.4) − 288, or 0.43 m^3. Thus, a volume of moisture greater than 40% of the wood chip volume can be absorbed. Considering that the porosity of randomly piled wood chips is probably on the order of 40%, the importance of moisture absorption can easily be seen.

Development of a Conceptual Model

A mass balance diagram for the addition of bulking agent to wet substrate is presented in Figure 6.16. The required ratio of bulking agent to substrate can be determined from analysis of the mass balance diagram provided certain assumptions are made. First, the substrate is semifluid in nature and occupies the interstices of the mixture. Second, sufficient bulking agent must be added so that contacts between the bulking particles provide structural support for the mixture. In general, this condition is satisfied as long as the substrate/bulking agent mixture remains porous. If insufficient bulking agent is added, bulking particles will be suspended in the substrate and will not provide structural support. Third, moisture absorption by the bulking agent is limited to a maximum moisture content (i.e., minimum solids content). Fourth, moisture release from the substrate is limited to a maximum solids content. Fifth, if other amendments or recycled compost are added to the mixture, the quantities added are sufficiently small to not violate the first assumption. Sixth, the individual bulking particles are solid with no internal FAS. Any free airspace that might exist in the cellular matrix of a wood chip or other porous particle is of little value to the substrate in the void space between particles, at least in terms of oxygen transfer to the substrate.

Moisture relationships with absorption limited by the bulking agent (bulking agent limited) are shown in Figure 6.17, and the case for moisture absorption limited by the substrate (substrate limited) is shown in Figure 6.18. Nomenclature used in Figures 6.17 and 6.18 and in the subsequent discussion, which has not been previously defined, is as follows:

S_{sm} = fractional solids content of substrate in the substrate/bulking agent mixture after moisture absorption

S_{sm}^m = maximum fractional solids content of substrate achievable by absorption of moisture from substrate to bulking agent

S_{bm} = fractional solids content of bulking agent in substrate/bulking agent mixture after moisture absorption

S_{bm}^m = minimum fractional solids content of bulking agent achievable by absorption of moisture from substrate to bulking agent

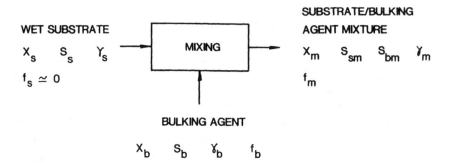

Figure 6.16.   Mass balance diagram for bulking agent addition to a wet substrate.

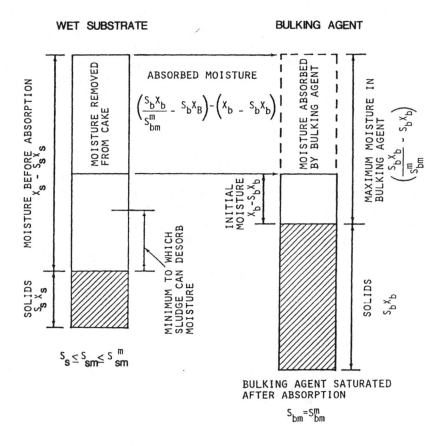

Figure 6.17.   Moisture relationships for a substrate/bulking agent mixture with water absorption limited by the bulking agent. From Haug.[4]

f_b = FAS within the interstices of a bulking agent before substrate addition

f_s = FAS within the interstices of a wet substrate, usually assumed to be zero

f_m = FAS within the interstices of a substrate/bulking agent mixture

δ_b = unit bulk weight of bulking agent, wet weight per volume

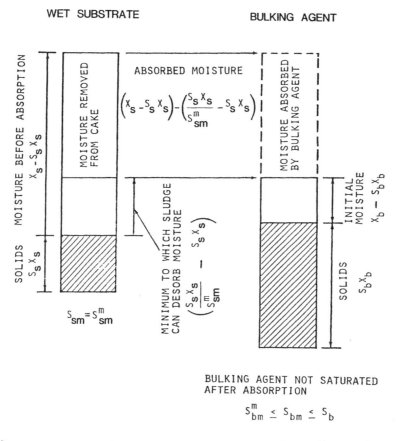

Figure 6.18. Moisture relationships for a substrate/bulking agent mixture with water absorption limited by the substrate. From Haug.[4]

For the case where moisture transfer is limited by absorption into the bulking agent (bulking agent limited), the following equation describes the volume of substrate in the void spaces after moisture absorption:

(initial substrate volume) − (absorbed moisture volume) =

(mixture volume) − (bulking particle volume) − (mixture FAS)   (6.39)

Substituting terms from Figure 6.17 into Equation 6.39:

$$(X_s / \delta_s) - (1/\delta_w)\left\{\left[(S_b X_b / S_{bm}^m) - S_b X_b\right] - (X_b - S_b X_b)\right\} =$$

$$(X_m / \delta_m) - (X_b / \delta_b)(1 - f_b) - (X_m / \delta_m)f_m \qquad (6.40)$$

Let the volume of bulking agent to substrate be defined as the volumetric mixing ratio, M_{bs}, as follows:

$$M_{bs} = \text{volumetric mixing ratio} = \frac{X_b / \delta_b}{X_s / \delta_s} \qquad (6.41)$$

If the substrate is very fluid it will occupy only the interstices of the mixture and will not add significantly to the total mixture volume. As the substrate solids increase, however, the structural strength also increases. The substrate then becomes capable of partially supporting its own weight. Therefore, as substrate solids increase the substrate begins to add to total mixture volume. To account for this, let M_{mb} be defined as the volume ratio of mixed materials to bulking agent:

$$M_{mb} = \frac{\text{volume of mixture}}{\text{volume of bulking agent}} = \frac{X_m / \delta_m}{X_b / \delta_b}$$

$$M_{mb} = f(S_s, M_{bs}) \qquad (6.42)$$

M_{mb} equals 1.0 if mixture volume equals the bulking agent volume, i.e., the substrate occupies only void space in the initial bulking agent. M_{mb} will be greater than 1.0 if the mixture volume is greater than that of the initial bulking agent.

The value of M_{mb} should be a function of substrate solids and the volumetric mixing ratio, M_{bc}. The effect of substrate solids is illustrated in Figure 6.19 for blends of wood chips and sludge cake. FAS was not measured at the different test points in these studies. Some of the variation in M_{bs} may have been caused by differences in FAS. Nevertheless, the tendency is for M_{bs} to decrease as substrate solids increase. This results from the increasing structural strength of the substrate, which then increases M_{mb} above 1.0. Exact relationships between these variables have not been determined experimentally, and additional work is needed in this area. Limited data[16] suggest an M_{mb} of about 1.2 for a 20% TS sludge cake.

Substituting Equations 6.41 and 6.42 into Equation 6.40 and rearranging:

$$1 / M_{bs} = (\delta_b / \delta_w)[S_b / S_{bm}^m - 1] + M_{mb}(1 - f_m) - (1 - f_b) \qquad (6.43)$$

Equation 6.43 can be used to estimate the volumetric mixing ratio, M_{bs}, required to achieved a desired mixture free airspace, f_m, when the quantity of moisture removed from the substrate is limited by that which the bulking agent can absorb (bulking agent limited).

Because Equation 6.43 assumes that moisture absorption is limited by the bulking agent, the solids content of the bulking agent in the mixture, S_{bm}, must equal the minimum solids content, S_{bm}^m. The solids content of the wet substrate, on the other hand, will increase to a level between the initial substrate solids, S_s, and the maximum solids, S_{sm}^m. The actual value of S_{sm} can be determined as

$$S_{sm} = \frac{\text{wt solids}}{\text{wt solids} + \text{wt initial water} - \text{wt water absorbed}}$$

Substituting terms from Figure 6.17:

$$S_{sm} = \frac{S_s X_s}{S_s X_s + (X_s - S_s X_s) - \{[(S_b X_b / S_{bm}^m) - S_b X_b] - (X_b - S_b X_b)\}}$$

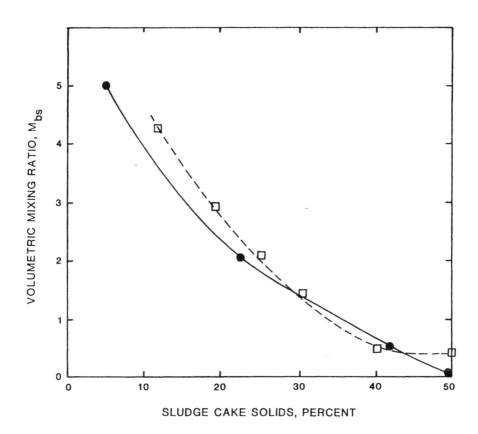

Figure 6.19. Effect of sludge cake solids on the volumetric mix ratio of woodchips to sludge. Solid line represents Willson's data[15] and dashed line represents that of Epstein.[16] FAS was not measured in these tests, so values may have varied at the different test points.

Substituting Equation 6.41 and rearranging:

$$S_{sm} = \frac{S_s}{1 + M_{bs}(\delta_b / \delta_s)[1 - (S_b / S_{bm}^m)]} \qquad S_s \leq S_{sm} \leq S_{sm}^m \qquad (6.44)$$

The wet substrate cannot give up unlimited quantities of water. Thus, a situation can exist where moisture transfer is limited by the amount that the substrate is capable of yielding (substrate limited), a situation illustrated in Figure 6.18. Substituting appropriate terms from Figure 6.18 into Equation 6.39 yields

$$(X_s / \delta_s) - (1/\delta_w)\{(X_s - S_s X_s) - [(S_s X_s / S_{sm}^m) - S_s X_s]\} =$$
$$(X_m / \delta_m) - (X_b / \delta_b)(1 - f_b) - (X_m / \delta_m)f_m \qquad (6.45)$$

Substituting Equations 6.41 and 6.42 into Equation 6.45 and rearranging:

$$M_{bs} = \frac{1 - (\delta_s / \delta_w)[1 - (S_s / S_{sm}^s)]}{M_{mb}(1 - f_m) - (1 - f_b)} \qquad (6.46)$$

With moisture absorption limited by the substrate, the substrate solids content, S_{sm}, will equal the maximum substrate solids, S_{sm}^m. The bulking agent solids content, S_{bm}, will decrease to a value less than the initial solids content, S_b, but greater than the minimum solids content with maximum moisture absorption, S_{bm}^m. The value of S_{bm} can be estimated as

$$S_{bm} = \frac{\text{wt of bulking agent solids}}{\text{wt solids} + \text{wt initial water} + \text{wt water absorbed}}$$

Substituting terms from Figure 6.18:

$$S_{bm} = \frac{S_b X_b}{S_b X_b + (X_b - S_b X_b) + \{(X_s - S_s X_s) - [(S_s X_s / S_{sm}^m) - S_s X_s]\}}$$

Substituting Equation 6.41 and rearranging:

$$S_{bm} = \frac{S_b}{1 + (1/M_{bs})(\delta_s / \delta_b)[1 - (S_s / S_{sm}^m)]} \qquad S_{bm}^m \leq S_{bm} \leq S_b \qquad (6.47)$$

For a porous bulking agent the bulk weight, δ_b, is a function of the moisture content and, hence, the solids content, S_b. Because both terms appear in Equation 6.43, the relationship between the two must be known. Wood chips are the most commonly used bulking agent. Available data suggest a bulk weight of about 0.288 g/cm³ for oven-dried wood chips. Bulk weight at other moisture contents can be estimated as

$$\delta_b = \frac{0.288 \text{ g/cm}^3}{S_b} \qquad (6.48)$$

For a wet substrate the bulk weight, δ_s, and solids content, S_s, both appear in Equation 6.46. The relationship between these two can be determined from Equations 6.1 and 6.3.

Because either of the above cases may be limiting, the procedure for calculating the required M_{bs} for a given situation is as follows:

1. Calculate M_{bs} and the corresponding S_{sm} from Equations 6.43 and 6.44, respectively.
2. Calculate M_{bs} and the corresponding S_{bm} from Equations 6.46 and 6.47, respectively.
3. Determine the required mixing ratio as the greater of the two calculated values. If M_{bs} calculated by Equation 6.43 is greater, moisture transfer is limited by the amount that can be absorbed by the bulking agent, therefore, $S_{bm} = S_{bm}^m$. If M_{bs} calculated by Equation 6.46 is greater, absorption is limited by the amount released from the wet substrate, therefore, $S_{sm} = S_{sm}^m$. Equations 6.43 and 6.46 will calculate to the same M_{bs} if $S_{sm} = S_{sm}^m$ and $S_{bm} = S_{bm}^m$.

Example 6.7

A wood chip bulking agent is to be used to provide airspace for composting 40 dmtpd of raw sludge cake at 25% TS and 70% VS. The bulking agent has a porosity of 35% and a moisture content of 40%

(60% TS). The minimum FAS to be maintained in the mixture is 15%. It is expected that the sludge can further dewater to a maximum of 35% TS ($S_{sm}^m = 0.35$) and the bulking agent absorb water to a maximum of 60% moisture ($S_{bm}^m = 0.40$). M_{mb} is estimated at 1.10. Determine the required volumetric mixing ratio, the moisture contents of the cake and bulking agent in the mixture, and the total volume and weight of mixed materials to be processed daily.

Solution

1. Specific gravity of sludge solids from Equation 6.1 is

$$1/G_s = (0.70/1.0) + (1 - 0.7)/2.5$$

$$G_s = 1.22$$

2. Bulk weight of the sludge solids from Equation 6.3 is

$$\delta_s = \frac{1.00}{[(0.25/1.22) + 1.0 - 0.25]}$$

$$\delta_s = 1.047$$

3. Estimate the bulk weight of wood chips from Equation 6.48 as,

$$\delta_b = 0.288/0.60 = 0.48 \text{ g/cm}^3$$

4. Calculate M_{bs} by Equation 6.43:

$$1/M_{bs} = (0.48/1.00)[(0.60/0.40) - 1] + 1.10(1 - 0.15) - (1 - 0.35)$$

$$M_{bs} = 1.91$$

5. Calculate M_{bs} by Equation 6.46:

$$M_{bs} = \frac{1 - (1.047/1.00)[1 - (0.25/0.35)]}{1.10(1 - 0.15) - (1 - 0.35)}$$

$$M_{bs} = 2.46$$

The required M_{bs} is 2.46 and moisture absorption is limited by the sludge cake. Therefore,

$$S_{sm} = S_{sm}^m = 0.35$$

6. S_{bm} is then given by Equation 6.47:

$$S_{bm} = \frac{0.60}{1 + (1/2.46)(1.047/0.48)[1 - (0.25/0.35)]}$$

$$S_{bm} = 0.479$$

Note that $S_{bm}^m \le S_{bm} \le S_b$ because $0.40 < 0.479 < 0.60$

7. Calculate the total daily volume of mixed material as follows:

$$\text{volume sludge cake} = 40(1000 \text{ kg / mt})(1 / 0.25)(1 \text{ m}^3 / 1047 \text{ kg}) = 153 \text{ m}^3 / \text{day}$$

$$\text{volume of bulking agent} = 2.46(153) = 376 \text{ m}^3 / \text{day}$$

$$\text{total volume} = 376(1.10) = 414 \text{ m}^3 / \text{day}$$

8. Calculate the total daily weight of mixed materials as

$$\text{wt sludge cake} = 40(1 / 0.25) = 160 \text{ mtpd}$$

$$\text{wt bulking agent} = 376 \text{ m}^3/\text{day } (480 \text{ kg/m}^3)(1 \text{ mt} / 1000 \text{ kg}) = 180 \text{ mtpd}$$

$$\text{total wt} = 160 + 180 = 340 \text{ mtpd}$$

Example 6.8

If a nonporous bulking agent is substituted for the wood chips in Example 6.7, calculate the required volumetric mixing ratio if all other conditions remain the same.

Solution

1. Because no moisture is absorbed by the bulking agent the following conditions hold:

$$S_{bm}^m = S_{bm} = S_b = 1.0 \text{ for the nonporous case}$$

$$S_{sm} = S_s$$

2. Calculate M_{bs} from Equation 6.43:

$$1 / M_{bs} = (0.48 / 1.00)[1.0 / 1.0) - 1] + 1.10(1 - 0.15) - (1 - 0.35)$$

$$M_{bs} = 3.51$$

3. M_{bs} calculated from Equation 6.46 will remain at 2.46 as in Example 6.7. Therefore, the required M_{bs} is 3.51 as calculated above.

4. Note that use of a nonporous bulking agent, or one that is already saturated with water, significantly increases the required mixing ratio over that calculated in Example 6.7. This in turn would increase the daily volume of material to be handled and the land area required for composting.

Example 6.9

The following data[16] represent the mass flow of materials for a mixture of sludge cake and wood chips to be composted by the aerated static pile technique. Estimate the FAS of the mixture.

	Bulk wt kg/m³	wet tpd	% TS
Raw sludge	1068	600	20.0
New chips	297	58	67.1
Recycled chips	415	444	62.8
Mixed blend	726	1102	39.7

Solution

1. Calculate the average solids content and bulk weight of the new and recycled wood chips and sludge:

$$S_b = \frac{58(0.671) + 444(0.628)}{58 + 444} = 0.633$$

$$\delta_b = \frac{58(297) + 444(415)}{58 + 444} = 401 \text{ kg} / \text{m}^3$$

$$\delta_s = 1068 \text{ kg} / \text{m}^3 \text{ from the above table}$$

2. Calculate the volumetric mixing ratio as

$$M_{bs} = \frac{\text{volume bulking agent}}{\text{volume sludge}}$$

$$M_{bs} = \frac{444(1000) / 415 + 58(1000) / 297}{600(1000) / 1068} = 2.25$$

3. Assume S_{bm}^m to be 0.40, S_{sm}^m as 0.40, and f_b as 0.40. First consider the case where moisture transfer is limited by that which the bulking agent can absorb and estimate f_m from Equation 6.43.

$$1 / 2.25 = (0.401 / 1.00)[(0.633 / 0.40) - 1] + 1.10(1 - f_m) - (1 - 0.40)$$

$$f_m = 0.263$$

The corresponding S_{sm} is determined from Equation 6.44:

$$S_{sm} = \frac{0.20}{1 + 2.25(0.401 / 1.068)[1 - (0.633 / 0.40)]} = 0.394$$

$$S_{sm} = 0.394 < 0.40; \quad S_{sm} < S_{sm}^m \text{ which is OK}$$

4. Now consider the case where moisture transfer is limited by that which can be released by the sludge cake. Estimate f_m from Equation 6.46.

$$2.25 = \frac{1 - (1.068 / 1.00)[1 - (0.20 / 0.40)]}{1.10(1 - f_m) - (1 - 0.40)}$$

$$f_m = 0.266$$

The corresponding value of S_{bm} is determined from Equation 6.47.

$$S_{bm} = \frac{0.633}{1 + (1 / 2.25)(1.068 / 0.401)[1 - (0.20 / 0.40)]} = 0.397$$

$$S_{bm} = 0.397 < 0.40;\ S_{bm} < S_{bm}^m \text{ which is not OK}$$

5. The condition $S_{bm} < S_{bm}^m$ cannot exist, so the conditions as determined in part 3 are correct and

$$f_m = 0.263$$

$$S_{sm} = 0.394$$

$$S_{bm} = 0.40$$

Other Model Results

Examples 6.7 and 6.8 highlight the importance of moisture absorption by the bulking agent in determining the required volumetric mix ratio. The same calculation procedures were used to prepare the results shown in Figure 6.20. The bulking agent was assumed to be wood chips with a porosity of 0.40, capable of absorbing water to a maximum 60% moisture, $S_{bm}^m = 0.40$. The wet substrate was assumed to release moisture to a maximum solids content of 40%, S_{sm}^m = 0.40. Specific gravity of the substrate solids was assumed to be 1.43 and 1.00 for water. For convenience, M_{mb} was assumed to be 1.0 for all calculations.

Analysis of published data for the aerated static pile process suggests a typical FAS of about 20% in the mixed materials. Referring to Figure 6.20, therefore, consider a mixture, f_m, of 0.20. Reading upward, the first intercepted curve is for a cake solids of 10%. All of the solid curves, which represent different bulking agent solids contents, lie above the line for $S_s = 0.10$. Under these conditions, moisture transfer is limited by the absorptive capacity of the bulking agent (solid lines). If $S_b = 0.80$, for example, the required M_{bs} is about 1.8, increasing to M_{bs} = 2.9 if $S_b = 0.50$. If substrate solids increase to 20%, M_{bs} is determined by moisture release from the substrate if $S_b > 0.60$. In other words, the dashed curve for $S_s = 0.20$ lies above the solid curves for $S_b \geq 0.60$ at $f_m = 0.20$. In this case, drying the bulking agent beyond 60% solids does not decrease the required M_{bs} of ~2.3. If $S_b = <0.60$, M_{bs} is determined from the solid curves. Considering an S_b of 0.40, for example, M_{bs} is ~5.0, decreasing to ~2.9 at $S_b = 0.50$.

As seen above, moisture content of the bulking agent can have a significant influence on the total quantity of mixed material to be processed each day. Drying the bulking agent before recycling can be considered as a way to reduce the required M_{bs}. The facility shown in Figure 6.21 has implemented a storage silo with forced aeration to provide further drying of screened wood chips prior to recycle. This should be an effective strategy to reduce the required M_{bs} within certain limits. For the assumptions of Figure 6.20, drying beyond an S_b of ~0.60 with

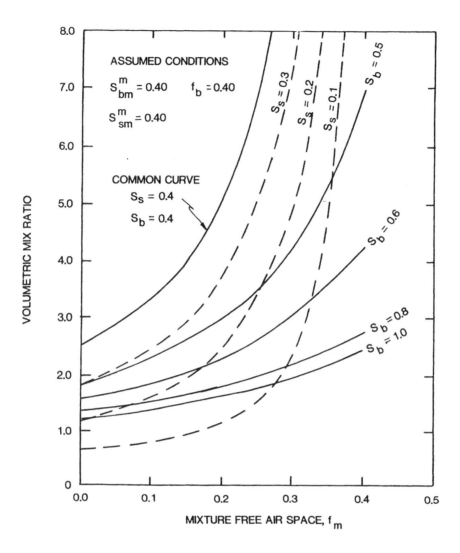

Figure 6.20. Volumetric mixing ratio for bulking agent/substrate mixtures as a function of mixture FAS for various values of substrate solids, S_s, and bulking agent solids, S_b. From Haug.[4]

a substrate of 20% solids may no longer reduce M_{bs}, because moisture removal from the substrate becomes limiting at about that point.

Remember that the curves in Figure 6.20 are based on assumed values for a number of variables. Although the values assumed are reasonable, variations should be anticipated depending on conditions specific to a particular problem. The examples discussed above are intended to illustrate the factors involved in determining the volumetric mixing ratio. If better input data are available for a specific case, the procedures of Example 6.7 can be used to develop curves similar to those presented in Figure 6.20.

Even though assumed values have been used in the previous analysis, it is interesting to note that the general range of M_{bs} values is in good agreement with values reported in practice. With wood chips as a bulking agent and sludge cake solids of ~20%, volumetric mix ratios of 1.0 to 4.0 have been reported.[18-20] Based on the model results in Figure 6.20, M_{bs} will likely range between 1.5 and 3.0 if the bulking agent is capable of some moisture absorption and

Figure 6.21.   Vertical, aerated silo used to store and provide supplemental drying for screened wood chips. Site is the Montgomery County Composting Facility operated by the Washington Suburban Sanitary Commission, Maryland. The facility composts raw sludge with wood chips as a bulking agent using the aerated static pile process.

substrate solids are between 15 and 30%. The general agreement is encouraging and suggests that the model is a reasonable representation of the physical events that occur when using bulking agents for control of FAS.

CONDITIONING OF DRY SUBSTRATES

Dry substrates are generally friable with available free airspace and usually do not need the extensive structural conditioning required for wet substrates. Bulking agents are seldom required for dry substrates. Preprocessing for size reduction or separation of undesirable materials may be required and may have some structural benefits. The present discussion will focus on the need for water addition to adjust dry substrates to a proper range of moisture content.

Referring to Figure 6.7 and assuming water is added to the mixture at a rate, X_w, the mass balance on total weight becomes

$$X_s + X_a + X_r + X_w = X_m \qquad (6.49)$$

The mass balance on dry weight remains the same as Equation 6.31:

$$S_s X_s + S_a X_a + S_r X_r = S_m X_m \qquad (6.31)$$

S_m in Equation 6.31 is interpreted as the desired solids content after moisture addition. Combining and solving for X_w:

$$X_w = \frac{X_s(S_s - S_m) + X_a(S_a - S_m) + X_r(S_r - S_m)}{S_m} \qquad (X_w > 0) \qquad (6.50)$$

If $X_w < 0$ the mixture does not require water addition and the actual mixture solids are less than the desired S_m. In such a case, the actual S_m can be calculated from simultaneous solution of Equations 6.30 and 6.31.

Example 6.10

50 tpd of air dried manure at 60% TS is blended with 25 tpd of yard waste at 55% TS and 25 tpd of compost recycle at 65% TS. Calculate the water required to bring the initial mix to 50% moisture.

Solution

1. The desired solids content in the initial mix, S_m, is $1 - 0.50 = 0.50$.

2. The required water rate, X_w, to achieve $S_m = 0.50$ is determined from Equation 6.50:

$$X_w = \frac{50(0.60 - 0.50) + 25(0.55 - 0.50) + 25(0.65 - 0.50)}{0.50}$$

$$X_w = 20 \text{ tpd water} = 20(2000) / (8.34 \text{ lb} / \text{gal}) = 4795 \text{ gal} / \text{day}$$

3. Without water addition the mixture would have an S_m determined from Equations 6.30 and 6.31 as

$$X_m = 50 + 25 + 25 = 100 \text{ tpd}$$

$$S_m(100) = 50(0.60) + 25(0.55) + 25(0.65)$$

$$S_m = 0.60 = 60\% \text{ TS or } 40\% \text{ moisture}$$

4. As will be seen in later chapters, rate limitations become significant at moisture contents below 40 to 50%. Thus, it is important to consider water addition whenever dry substrates are used for composting.

Adjusting the initial mixture to a proper moisture content does not guarantee that the moisture content stays within an acceptable range throughout the composting process. Composting tends to be a dehydrating environment. Supplemental water addition may be required during the composting cycle to compensate for moisture lost in the exhaust gases. Determining the required rate of supplemental water addition is a considerably more difficult problem because it depends on both process kinetics and thermodynamics. This subject is revisited in Chapters 11 and 14 as part of the analysis of simulation models.

CHEMICAL CONDITIONING

Some substrates may require additional conditioning beyond adjustment of free airspace and moisture. Cellulosic rich substrates, such as some yard wastes and MSW fractions, may lack the nutrients necessary to sustain rapid microbial growth rates. Other substrates may have extremely low or high pH levels which will also impede microbial growth. Feed conditioning may be necessary to remove these potential limitations to microbial growth.

Nutrients

As discussed in Chapter 4, a number of inorganic nutrients are required to support microbial synthesis in biological systems. Nitrogen has received the most attention in composting systems, because higher concentrations are required than for any other inorganic nutrient. The composting industry generally uses the carbon/nitrogen ratio (C/N) as the measure of a proper nitrogen balance. The C/N ratio has a fundamental significance because nitrogen is necessary to support cellular synthesis and carbon makes up the largest fraction of organic molecules in the cell.

During active aerobic metabolism, microbes use about 15 to 30 parts of carbon for each part of nitrogen, i.e., C/N = 15 to 30. To understand this ratio consider a microbe using a starch or cellulosic substrate as an energy source. Glucose is the base monomer of both starch and cellulose, and the energy reaction can be written as

$$x\left[C_6H_{12}O_6 + 6O_2 \rightarrow 6CO_2 + 6H_2O + energy\right]$$

$$x(180)$$

As discussed in Chapters 3 and 4, this reaction yields free energy that the cell can apply to drive the synthesis of new microbial mass. Assuming an average cell formulation of $C_5H_7O_2N$ and ammonia as the source of cell N, the synthesis reaction is

$$5CO_2 + 2H_2O + NH_3 + energy \rightarrow C_5H_7O_2N$$

$$113$$

The maximum yield of cells is limited by thermodynamics to ~0.4 g cells/g glucose. The actual yield is usually lower, particularly in systems with long residence times where endogenous respiration is significant. Let us assume a net yield in the range of 0.1 to 0.2 g cells/g glucose as typical for composting of relatively degradable substrates. Using 0.1 as the yield coefficient, the moles of energy reaction required per mole of synthesis reaction can be determined as

$$1/[x(180)] = 0.1/113$$

$$x = 6.3, \text{ say 6 moles / mol}$$

Multiplying the energy reaction by 6 and adding to the synthesis reaction gives the net metabolic reaction:

$$6C_6H_{12}O_6 + 31O_2 + NH_3 \rightarrow C_5H_7O_2N + 31CO_2 + 34H_2O$$

36 mol of carbon are used for each 1 mol of N. Thus, the required C/N ratio is 36(12)/1(14) = 30.9. If the cell yield were higher, say 0.2 g cells/g glucose, the required C/N ratio decreases to ~15. Hence, rapid composting is favored by maintaining a C/N ratio of ~30 or less.

The theoretical arguments are supported by actual field experiments. McGaughey and Gotass[2] composted mixed refuse materials with initial C/N ratios from 20 to 78. An initial C/N between 30 and 35 was recommended as optimum for rapid composting of the refuse materials. Below this range, rapid composting was accompanied by increasing losses of excess nitrogen from ammonia volatilization. Above this range, composting time increased with increasing C/N ratio. The composting activity accomplished in 12 days with a C/N ratio of 20 required about 21 days with a C/N of 78.

The effect of low nitrogen concentration is due to growth limitations imposed by lack of this particular nutrient. With growth limited, the overall process kinetics become limited in turn. Referring back to Equation 4.10, nitrogen can be a limiting substrate as well as oxygen or another solubilized organic substrate. Problems associated with high C/N ratios can be alleviated by removing a portion of the carbonaceous material (high C/N ratio substrates) and/ or by adding nitrogen. If the C/N ratio is ≤15 to 30, nitrogen is present in excess and no rate limitation is imposed. This is often misunderstood in the composting literature. It is not necessary that the mixed substrates have a C/N ratio within the range of 15 to 30. It is only required that C/N be <15 to 30 to remove nitrogen from being rate limiting. Below a C/N of 15 to 30, excess nitrogen will be available and some will likely be lost as volatilized ammonia in the exhaust gases. Kayhanian and Tchabanoglous[22] noted some problems with ammonia toxicity during anaerobic composting of MSW fractions. However, this problem is unique to the anaerobic environment and probably would not be a problem with aerobic composting.

Nitrogen (N), phosphorus (P), potash (K), and C/N ratios for various organic substrates are presented in Table 6.2. Many composting substrates have high nitrogen contents including night soil, manures, sludges, and grass clippings. Others are quite low in nitrogen and these tend to be cellulosic materials such as sawdust, straw, leaves, and the paper fractions of refuse. It is often advantageous to blend high nitrogen substrates with low nitrogen substrates. Ammonia released from the nitrogen rich substrate can be captured for synthesis by microbes using the nitrogen poor substrate. Thus, more of the nutritive value is conserved and less ammonia is released to the atmosphere. If a nitrogen rich substrate is not available, supplemental nitrogen is usually added in the form of urea, ammonium sulfate, or ammonium chloride. Ammonium phosphate can also be used if a phosphorus deficiency is suspected or if it is desired to fortify the final product.

Determining the C/N ratio requires knowledge of the chemical composition of the substrate. Total Kjeldahl nitrogen can be determined by most wastewater laboratories, however, the necessary test for carbon may not be a routine practice. New Zealand researchers[24] have suggested the following relationship between carbon content and ash fraction:

$$\% \text{ carbon} = (100 - \% \text{ ash}) / 1.8 \qquad (6.51)$$

The ash fraction can be analyzed by most labs. Equation 6.51 provided results accurate to within 2 to 10% during the New Zealand field studies and should be sufficiently accurate for most practical purposes.

One caution with using C/N ratios should be noted. It is assumed that both the carbon and nitrogen sources are relatively degradable. If the nitrogen were present in a nondegradable organic molecule, the nitrogen would not be available regardless of the C/N ratio. Fortunately, most of the nitrogen in naturally occurring substrates is present in protein molecules, which are relatively degradable. Alternatively, a relatively nondegradable carbon source would not experience rate limitations even with a high C/N ratio. This is because there would be little decomposition and therefore little synthesis to be supported. Kayhanian and Tchobanoglous[22] recommended that the C/N ratio should be based on biodegradable carbon, particularly for

Table 6.2.   Nutrient Content of Various Composting Substrates

Material	Nitrogen as N (% dw)	Phosphorus as P_2O_5 (% dw)	Potash as K (% dw)	C/N ratio
Night soil	5.5–6.5			6–10
Urine	15–18			0.8
Blood meal	10–14	1–5		3.0
Animal tankage				4.1
Cow manure	1.7–2	1.0	2.0	18
Poultry manure	5–6.3	1.9	1.2	15
Sheep manure	3.8			
Pig manure	3.8			
Horse manure	1.2–2.3	1.0	1.6	25
Raw sewage sludge	4–7			11
Digested sewage sludge	2–4	1.5	0.2	
Activated sludge	5			6
Grass clippings (green)	2.4–6			12–15
Grass clippings and weeds	2.0	1.1	2.0	
Mixed grasses	2.4			19
Nonlegume vegetable wastes	2.5–4			11–12
Bone meal	2.0	23		
Coffee grounds	2.1	0.3	0.3	
Cottonseed meal	6.6	2.0–3.0	1.0–2.0	
Eggshells	1.2	0.4	0.1	
Fish scraps	2.0–7.5	1.5–6.0		
Garbage	2.0–2.9	1.1–1.3	0.8–2.2	
Meat scraps	5–7			
Potato tops	1.5			25
Seaweed	1.7	0.8	4.9	
Salt marsh hay	1.1	0.3	0.8	
Straw, wheat	0.3–0.5			128–150
Straw, oats	1.1			48
Leaves, fresh	0.5–1.0	0.1–0.2	0.4–0.7	41
Sawdust	0.1			200–500
Food wastes	3.2			15.6
Mixed paper	0.19			227
Yard wastes	1.95			22.8
Wood ashes		0.1–2.0	4.0–10	

Source: Adapted from Golueke,[1] Poincelot,[23] and Kayhanian and Tchobanoglous.[22]

MSW feedstocks. Based on their pilot plant testing, almost all of the nitrogen in the organic material was available. Assuming that all nitrogen is available, the C/N ratio should be based on the nitrogen content of the total organic mass and the carbon content of the biodegradable organic mass. Biodegradable fraction was determined using Equation 9.1 based on the liqnin content of each substrate.

Estimates of the C/N ratio for various MSW fractions based on biodegradable and total carbon are presented in Table 6.3. C/N ratio based on biodegradable carbon is always lower than the value based on total carbon. Kayhanian and Tchobanoglous also noted that the C/N ratio in the organic fraction of MSW will change as new mandated recycling programs are initiated. For example, recycle of the paper fraction will remove this high C/N component from the MSW, resulting in a lower C/N ratio in the remaining material. They determined that 70% recycle of paper would reduce the biodegradable C/N ratio of from about 34 to 17 based on the MSW characteristics at Folsom, California. This would cause a significant increase in ammonia release if the remaining feedstocks are composted. The lesson here is that feedstock characteristics should be carefully reviewed to determine the effect of seasonal changes and changes resulting from societal activities such as recycling.

Table 6.3. C/N Ratios for MSW Fractions Based on Biodegradable and Total Organic Fractions

Component	C/N Ratio Based On	
	Biodegradable Organic Carbon[a]	Total Organic Carbon[b]
Food wastes	12.4	15.6
Mixed paper	143.1	227.1
Yard wastes	14.5	22.8
Mixed waste	34.4	59.9

Source: Kayhanian and Tchobanoglous.[22]

[a] Biodegradable organic carbon is obtained by correcting the total dry weight carbon value for the ash content and then multiplying by the biodegradable fraction determined from Equation 9.1.

[b] Total organic carbon is obtained by correcting the total dry weight carbon value for the ash content.

Example 6.11

Estimate the C/N ratio for raw sewage sludge.

Solution

1. Assume a chemical composition of $C_{10}H_{19}O_3N$ for raw sludge:

$$C / N = 10(12) / 1(14) = 8.6$$

2. Using the New Zealand formula, Equation 6.51, and assuming an ash content of 25%, the percentage carbon is estimated as

$$\% \text{ carbon} = (100 - 25) / 1.8 = 41.7\% \text{ of TS}$$

If nitrogen is 5% of TS from Table 6.2:

$$C / N = 41.7 / 5 = 8.3$$

3. Based on the above, municipal sludges should have more than sufficient nitrogen to satisfy the growth requirement. In fact, considerable volatilization of ammonia is usually observed because of the excess of nitrogen.

Example 6.12

Your firm has been retained to design an MSW composting facility. Refuse is to be received and preprocessed to recover 100 tpd of an "organic rich" fraction which will then be composted. The organics are expected to have a formulation of $C_{99}H_{148}O_{59}N$, contain 25% inert or ash fraction, and have 30% moisture content. Determine the quantity of nitrogen which must be added to the refuse fraction to adjust the feed mixture to a C/N ratio of 25. The nitrogen is to be added as either ammonium sulfate $(NH_4)_2SO_4$ or in the form of raw sludge cake with characteristics per Example 6.11. Determine the quantities of each required.

Solution

1. The C/N ratio of the refuse organics is

$$C / N = 99(12) / 14 = 85$$

Additional nitrogen will be needed to achieve the desired 25 C/N in the feed substrate.
2. The additional nitrogen can be estimated as follows:

The molecular weight of the organics is

Carbon	99(12)	=	1188
Hydrogen	148(1)	=	148
Oxygen	56(16)	=	896
Nitrogen	1(14)	=	14
Total		=	2246

The daily weight of refuse organics is

$$100(1 - 0.25)(1 - 0.30) = 52.5 \text{ dtpd of organic solids}$$

Nitrogen content in the feed substrate is

$$52.5(14) / 2246 = 0.33 \text{ tpd nitrogen}$$

Carbon content in the feed substrate is

$$52.5(1188) / 2246 = 27.8 \text{ tpd carbon}$$

The nitrogen required at a C/N ratio of 25 can then be estimated as

$$N = C / 25 = 27.8 / 25 = 1.11 \text{ tpd nitrogen}$$

The required additional nitrogen is then

$$1.11 - 0.33 = 0.78 \text{ tpd additional N}$$

3. If the 0.78 tpd of N is added as ammonium sulfate, the feed requirement is determined as follows:

The molecular weight of ammonium sulfate is

Nitrogen	2(14)	=	28
Hydrogen	8(1)	=	8
Sulfur	1(32)	=	32
Oxygen	4(16)	=	64
Total		=	132

The weight of required ammonium sulfate is then determined as

$$0.78(132) / 28 = 3.7 \text{ tpd}$$

4. If raw sludge is used as the nitrogen source, the quantity required to achieve a combined mixture C/N of 25 can be determined as follows.

Molecular weight of the raw sludge is

Carbon	10(12)	=	120
Hydrogen	19(1)	=	19
Oxygen	3(16)	=	48
Nitrogen	1(14)	=	14
Total		=	201

Let x be the required dtpd of sludge organics. x can then be determined from the desired C/N in the combined mixture:

$$C/N = 25 = \frac{52.5(1188/2246) + x(120/201)}{52.5(14/2246) + x(14/201)}$$

$$x = 17.1 \text{ dtpd of sludge organics}$$

Accounting for the ash and water fractions, the wet weight of sludge cake required can then be determined as

$$17.1/[1-0.25)(0.25)] = 91.2 \text{ wtpd sludge cake}$$

5. As can be seen by this example, the requirement for additional nitrogen can be significant if the substrate is nitrogen poor. Combining substrates with low and high nitrogen contents can be advantageous and can avoid the need to add outside sources of nitrogen.

Hydrogen Ion Concentration

Both hydrogen ion (H^+) and hydroxide ion (OH^-) are very toxic to microorganisms. For example, most bacteria will not survive in a pH 3 solution, but at pH 3 the H^+ concentration is only about 1 mg/L. Similarly, most bacteria begin to succumb at high pH levels, above 10.5, with significant kills above ~11.5.[21] The high and low pH causes a change in the ionization state of various protein components such as amine and carboxyl groups. This in turn causes changes in the physical structure of the protein and hence a loss of enzymatic activity.

Extremes of low or high pH are most often encountered with municipal sludges. Most raw sludges have a pH in the range of 5 to 6.5, while digested sludges are well buffered in the range of 7 to 8 pH. High pH values above 10 are encountered when lime is used for coagulation or precipitation in the wastewater treatment process or as a conditioning chemical during dewatering. Similarly, low pH values may be encountered when ferric chloride or alum are used for sludge conditioning. It is more common in modern practice to encounter high pH sludges because lime conditioning and lime stabilization are widespread practices. Under low or high pH conditions, an initial lag in the composting rate can be expected.

Composting has a rather unique ability to buffer both high and low pHs back to a neutral range as composting proceeds. This is because both a weak acid (CO_2) and a weak base (NH_3) are produced as a result of microbial decomposition. CO_2 is an end product in all organic decomposition, while NH_3 is an end product of protein decomposition. The CO_2 effectively neutralizes high pH conditions (excess OH^-) and the NH_3 low pH conditions (excess H^+).

Thus, it is common to observe a gradual neutralization of low or high pH as composting proceeds.

Using a well-mixed reactor, Shell and Boyd[9] composted digested sludge conditioned with 4.5% ferric chloride and 15.5% lime based on dry weight. The filter cake pH averaged 11.0 while pH of the compost product averaged 6.5. Obviously, sufficient CO_2 was produced to neutralize the lime content. It was also noted in this study that sludge conditioned with lime and ferric produced compost that mixed and handled better than sludge conditioned with organic polymers. With batch processes a lag period is usually observed until the persistent microbial activity neutralizes the high or low pH. The batch, static pile process has been applied to many lime conditioned sludges. A lag period of a few days to one week is typically observed before sufficient CO_2 is evolved to neutralize the excess lime.

Davis et al.[25] studied the pH changes of lime/ferric conditioned and polymer conditioned sludge cakes during composting and subsequent use of the composts in plant growth experiments. Initial pH of the lime/ferric cake was about 12.1 and about 8.4 for the polymer cake. Both sludge types were composted by both windrow and static pile processes using subtropical yard wastes for amendment. A decreasing pH trend was observed during the composting and curing periods. All final composts had pH levels between 7.7 and 7.9. The composts were then used alone and blended with pine bark and sand to produce a potting soil for the growth experiments. During an 8-week growth period the composts alone showed a continued decrease in pH to a range of 6.6 to 7.2. Similar levels were observed with the potting blend. There were no observed pH increases from the initial lime in the sludge. The authors concluded that (1) the conditioning method for dewatering did not affect growth of short-term crops, (2) once neutralized the high pH should remain neutralized, and (3) lime conditioning should not limit marketability even on calcareous soils.

For material not conditioned with lime, it is common in the early stages of batch composting to observe a decrease in pH from CO_2 and organic acid production. pH as low as 4.5 to 5.0 may at times be observed. However, the pH sag is rapidly overcome as organic acids are further decomposed and temperature rises. Protein decomposition will contribute NH_3 which will also tend to neutralize the acids. Increased rates of aeration will tend to decrease CO_2 levels in the compost, which in turn will tend to increase pH. pH usually stabilizes in the near-neutral range of values. McGaughey and Gotaas[2] examined the effect of calcium carbonate addition to reduce the normal pH sag observed during windrow composting of refuse. Although a slight reduction in the pH sag was observed, it was concluded that the practice was of limited value to process kinetics and that pH control was not necessary in the composting of garbage and municipal refuse.

Based on the above, it is not common to adjust the pH conditions of the feed substrates for composting. However, some cautions should be noted to this general rule. High lime dosages, such as those that occur with lime stabilization processes, could impose an extensive lag period on a batch composting process. If an extended lag period is undesirable, a number of approaches can be used to condition the starting substrate. First, acid can be added, but this is usually difficult to properly control. Second, CO_2 rich gases from the actively composting material can be recycled and passed through the starting substrate to neutralize the excess lime. This is usually quite effective and avoids the problem of accidentally producing low pH conditions from oversupply of acid. The latter can occur when strong, mineral acids (sulfuric, hydrochloric, and phosphoric) are used for neutralization. Third, a well-mixed process can be used so that the new substrate is mixed with actively composting material, which is already neutralized.

One other caution is noted for substrates that are both low in nitrogen and low in pH. For example, an industrial sludge could be mostly cellulosic with low pH from residual acids used in the process. If this were the case, neutralization of the low pH may not occur because ammonia is not available from the substrate. The only recourse in this case would be to neutralize the substrate pH. Lime, calcium carbonate, sodium hydroxide, sodium bicarbonate, or ammonia could be used for this purpose.

Example 6.13

Assume $C_{10}H_{19}O_3N$ as the chemical composition of the organic fraction of a raw sludge. The sludge is 30% ash by weight and is conditioned with 10% lime as $Ca(OH)_2$ based on dry weight. Estimate the fraction of organics that must degrade to produce sufficient CO_2 to neutralize the lime. Neglect any neutralization reactions that may occur during initial lime addition to the sludge.

Solution

1. Oxidation of the organics and neutralization by the produced CO_2 can be represented as

$$\overset{y}{C_{10}H_{19}O_3N} + 12.5O_2 = \overset{x}{10CO_2} + NH_3 + 8H_2O$$
$$\underset{201}{\phantom{C_{10}H_{19}O_3N}} \qquad \underset{10(44)}{}$$

$$\overset{x}{CO_2} + \overset{0.10}{Ca(OH)_2} = CaCO_3 + H_2O$$
$$\underset{44}{} \quad \underset{74}{}$$

Consider 0.10 g lime/g dry solids. The CO_2 requirement is then given by

$$x/44 = 0.10/74$$
$$x = 0.0595 \text{ g } CO_2$$

To supply this CO_2 the following organic decomposition is required:

$$y/201 = 0.0595/[10(44)]$$
$$y = 0.027 \text{ g}$$

2. The fraction of substrate organics that must decompose can be estimated as

$$0.027/0.70 = 0.0386 \text{ g/g VS} = 3.9\%$$

Thus, only a small fraction of the substrate must decompose to neutralize the initial high pH. Lime doses as high as 30 to 40% are used for stabilization. Even these large amounts would not overly tax the neutralizing capacity of the composting process. Rate limitations in the early stages of composting might occur until sufficient organic decomposition is realized to lower the pH.

3. The reader should note that a portion of $Ca(OH)_2$ added to sludge will react with CO_2 and bicarbonate alkalinity naturally present to form $CaCO_3$. Thus, the above calculations are conservative and tend to overestimate the required organic decomposition.

Example 6.14

Assume the sludge in Example 6.13 is conditioned by adding 2% $FeCl_3$ based on dry weight. Estimate the fraction of organics that must degrade to produce sufficient ammonia to neutralize acids produced from the addition of ferric chloride. Neglect any neutralization reactions that may occur during initial ferric addition to the sludge.

1. Production of acid from addition of ferric chloride can be described as

$$0.02$$

$$FeCl_3 + 3H_2O = 3H^+ + 3Cl^- + Fe(OH)_3$$

$$162.5$$

2. Oxidation of organics and neutralization by produced NH_3 can be represented as

$$y \qquad\qquad\qquad x$$

$$C_{10}H_{19}O_3N + 12.5O_2 = 10CO_2 + NH_3 + 8H_2O$$

$$201 \qquad\qquad\qquad 17$$

$$x$$

$$3NH_3 + 3H^+ + 3Cl^- = 3NH_4^+ + 3Cl^-$$

$$3(17)$$

3. Consider 0.02 g $FeCl_3$/g ds. The NH_3 requirement is then given as

$$x/[3(17)] = 0.02/162.5$$

$$x = 0.0063 \text{ g } NH_3$$

To supply this NH_3 the following organic decomposition is required:

$$y/201 = 0.0063/17$$

$$y = 0.0745 \text{ g}$$

4. The fraction of substrate organics that must decompose is then estimated as

$$0.0745/0.70 = 0.106 \text{ or about } 10.6\%$$

The organic decomposition required to neutralize low pH conditions will depend largely on the degradable protein content of the feed substrate, because this is the primary form of organic nitrogen. Substrates low in degradable protein will produce less ammonia and have a reduced neutralizing capacity against low pH conditions.

SUMMARY

Conditioning of the feed substrates is important to successful operation of the composting process. Feed conditioning is a process by which potential process limitations imposed by lack of moisture, free airspace, nutrients, or energy are selectively removed. Three types of conditioning can be defined: physical or structural, chemical, and energy or thermodynamic conditioning. Physical or structural conditioning is defined as conditioning of the substrates to remove potential rate limitations caused by lack of moisture and/or free airspace. Chemical conditioning is defined as conditioning of the substrates to remove potential rate limitations caused by lack of nutrients or other chemical imbalances. Thermodynamic conditioning is defined as conditioning of the feed mixture to assure that sufficient energy is available to drive the process.

Moisture is essential to maintain microbial activity. Lack of moisture can impose severe rate limitations on the process. Moisture content should generally be as high as possible without saturating the substrate and removing all available free airspace. Free airspace (FAS) is important in maintaining aerobic conditions within a composting material. There is considerable evidence that the optimum moisture content for a particular substrate is related to maintenance of a certain minimum FAS. Reported optimum moisture contents tend to fall in the range of FAS between about 20 and 35%.

Wet substrates, such as municipal and industrial sludges and wet manures, often contain little or no FAS. Because of the high moisture content, these materials generally lack the structural strength to maintain a properly shaped windrow or pile and the friability to assure aerobic conditions in the composting material. It is necessary to reduce the bulk weight and increase the structural strength of these materials to produce free airspace within the composting matrix. Four techniques have been demonstrated in commercial practice to deal with the problem of high moisture content:

1.   recycle of dry, previously composted material
2.   addition of dry amendments either with or without product recycle
3.   use of bulking agents such as wood chips
4.   drying the wet substrate before composting to reduce the moisture content

Recycle of compost product has been used in the windrow and many reactor systems. In some cases it is the only conditioning material added to the wet substrate; however, it is more common to combine product recycle with supplemental amendments, such as sawdust. Bulking agents are usually used in the aerated static pile process. Predrying has been used with digested sewage sludges and may be combined with the use of product recycle. A fifth technique, the use of nearly constant agitation to achieve aeration, has been successful only in bench scale applications and is not considered a feasible alternative for full scale practice.

Product recycle and amendments are effective in physical (structural) conditioning because they (1) decrease the bulk weight of the wet substrate and (2) increase the structural properties of the mixture to the point that it can be piled without consolidating or losing its free airspace. Bulk weights for most wet substrates begin to decrease from the theoretical maximum at solids contents between about 30 to 40%. Using product recycle and/or amendments to adjust mixture solids contents to the range of 35 to 45% is the common practice with wet substrates.

By comparison, bulking agents provide structural conditioning by maintaining a three-dimensional matrix of particles whose strength is provided by particle-to-particle contacts. The wet substrate occupies part of the void spaces between bulking particles. Sufficient bulking agent must be added to assure that the wet substrate volume does not exceed the available void volume. Use of dry porous bulking agents is advantageous because moisture is absorbed from the wet substrate, thus increasing the FAS. Use of a nonporous bulking agent, or one that is already saturated with moisture, increases the required quantity of bulking agent because no moisture is absorbed from the substrate.

Dry substrates usually do not require the structural conditioning provided by product recycle, amendments, or bulking agents. Recycle and amendments may be used for other than structural reasons, however, such as for energy conditioning or microbial inoculation of the dry substrate. Water addition is important to the conditioning of dry substrates to remove rate limitations from lack of moisture. Generally, the moisture content of the initial mixture should be at least 50%. Composting tends to be a dehydrating environment. Supplemental water addition may be required during the composting cycle to compensate for moisture lost in the exhaust gases.

Nutrient additions may be required for some substrates, particularly high cellulosic ones that lack sufficient nitrogen. The C/N ratio of the feed mixture should normally be less than 30 to assure that nitrogen does not become rate limiting. Blending with other high-nitrogen substrates or addition of supplemental chemical nitrogen can be used to condition a low-nitrogen substrate. With some substrates the potential exists that other macronutrients, such as phosphorus and potassium, or micronutrients, such as iron and magnesium, could be present in low enough concentrations to be rate limiting. Fortunately, such cases are rare and limitations can easily be avoided by adding a small quantity of a nutrient rich substrate such as sewage sludge, garbage, or manure.

Very low or very high pH levels can impose rate limitations to microbial activity. High pH levels are commonly encountered with municipal sludges because of the practice of lime conditioning. Fortunately, composting has a rather unique ability to buffer both high and low pHs back to a neutral range as composting proceeds. This ability to buffer extremes of pH is caused by the fact that both carbon dioxide (a weak acid) and ammonia (a weak base) are released as a result of organic decomposition. These compounds will tend to neutralize extremes of low or high pH. Therefore, pH adjustment of the starting substrates is usually not required and is not a common practice. Two cautions to this general rule are noted. The first applies to sludges conditioned with very high lime dosages which may exhibit an excessively long lag period before neutralization. The second applies to low pH, low nitrogen substrates which may lack sufficient protein to supply the ammonia needed to neutralize the acidic conditions.

REFERENCES

1. Golueke, C. G. *Biological Reclamation of Solid Waste* (Emmaus, PA: Rodale Press, 1977).
2. McGaughey, P. H. and Gotaas, H. G. "Stabilization of Municipal Refuse by Composting," *Trans. Am. Soc. Civil Eng.* Paper No. 2767.
3. Senn, C. L. "Dairy Waste Management Project — Final Report," University of California Agricultural Extension Service, Public Health Foundation of Los Angeles County (1971).

4. Haug, R. T. "Composting Wet Organic Sludges — A Problem of Moisture Control," in *Proceedings of the National Conference on Design of Municipal Sludge Compost Facilities* (Rockville, MD: Information Transfer, 1978).

5. Jeris, J. S. and Regan, R. W. "Controlling Environmental Parameters for Optimum Composting, Part II: Moisture, Free Air Space and Recycle," *Compost Sci.* (March/April 1973).

6. Schultz, K. L. "Continuous Thermophilic Composting," *Compost Sci.* (Spring 1962).

7. Haug, R. T. "Composting Process Design Criteria, Part I: Feed Conditioning," *Biocycle* 38, (August 1986).

8. Haug, R. T. and Haug, L. A. "Sludge Composting — A Discussion of Engineering Principles," *Compost Sci.* (November/December 1977, January/February 1978).

9. Shell, G. L. and Boyd, J. L. "Composting Dewatered Sewage Sludge," Public Health Service Publication No. 1936, U.S. Department of Health, Education, and Welfare (1969).

10. Haug, R. T. "Engineering Principles of Sludge Composting," *J. Water Poll. Control Fed.* 51(8) (1979).

11. Dohoney, R. W. District Manager for Professional Services Group, personal communication (December 31, 1991).

12. "Guidelines for Controlling Sewage Sludge Composting Odors," U.S. EPA, Washington D.C. (in press).

13. Houser, J. E. "Test Report for the Reduction of Organic Matter in Municipal Waste Water Under Aerobic Conditions in the Thermophilic Phase by the Fairfield Digester System at Altoona, PA," Fairfield Engineering Co., Marion, OH, unpublished.

14. Haug, R. T. "Quantitative Analysis of Sludge Management Projects," a report of the LA/OMA Project, Whittier, CA (1979).

15. Willson, G. B. "Equipment for Composting Sewage Sludge in Windrows and in Piles," in *Proceedings of the 1977 National Conference on Composting of Municipal Residues and Sludges* (Rockville, MD: Information Transfer, 1977).

16. Epstein, E. "Bulking Materials,"in *Proceedings of the National Conference on Municipal and Industrial Sludge Composting — Materials Handling* (Rockville, MD: Information Transfer, 1979).

17. "Washington Suburban Sanitary Commission Compost Facility — Site 2 Proposed Design Criteria and Plans," Toups and Loiederman Engineers, (1978).

18. Epstein, E., Willson, G. B., Burge, W. D., Mullen, D. C., and Enkiri, N. "A Forced Aeration System for Composting Wastewater Sludge," *J. Water Poll. Control Fed.* 48:4 (1976).

19. Willson, G. B., Parr, J. F., Epstein, E., Marsh, P. B., Chaney, R. L., Colacicco, D., Burge, W. D., Sikora, L. J., Tester, C. F., and Hornick, S. "Manual for Composting Sewage Sludge by the Beltsville Aerated-Pile Method," U.S. EPA Municipal Environmental Research Laboratory, Cincinnati, OH, EPA-600/8-80-022 (May 1980).

20. Williams, T., Hazlewood, B., Epstein, E., and Frye, W. "The City of Nashville's Sludge Composting Program — Implementing a 40 Dry Ton per Day Operation," presented at 62nd Ann. Conf. Water Poll. Control Fed., San Francisco, CA (October 1989).

21. Farrell, J. B., Smith, J. E., Hathaway, S. W., and Dean, R. B., "Lime Stabilization of Primary Sludges," *J. Water Poll. Control Fed.* 46(1) (1974).

22. Kayhanian, M. and Tchobanoglous, G. "Computation of C/N Ratios for Various Organic Fractions," *BioCycle* 33(5) (May 1992).

23. Poincelot, R. P. "A Scientific Examination of the Principles and Practice of Composting," *Compost Sci.* (Summer 1974).

24. "Second Interim Report of the Inter-departmental Committee on Utilization of Organic Wastes," *New Zealand Eng.* (November/December 1951).

25. Davis, P. A., Fitzpatrick, G. E., Goscicki, E., Wohlgemuth, G., and Svenson, S. E. "Effects of Sludge Processing Parameters on Finished Compost Quality and Long-term Marketability," presented at the WPCF 64th Annual Conf., Toronto, paper no. AC91-018-002 (October 1991).

CHAPTER 7

Aeration Requirements

INTRODUCTION

Air must be supplied to a composting material for three basic purposes. First, air must be supplied to satisfy the oxygen demand from organic decomposition (stoichiometric demand). Second, air must be supplied to remove water from wet substrates to provide drying (drying demand). As air is heated by the composting material, it picks up moisture and thus dries the remaining material. Drying is an important benefit to be achieved during composting, particularly with wet substrates. Third, air must be supplied to remove heat generated by organic decomposition to control process temperatures (heat removal demand). If not controlled, process temperatures can reach such high levels that biological activity is actually impeded. The aeration rate can be used to control the rate of heat removal and thereby adjust the system temperature.

The aeration system is an important feature of all modern composting systems. The ability to control aeration is one of the key points of process control and is an important consideration in process selection. Simplified procedures for calculating aeration demands for various process conditions and feed substrates are presented in this Chapter. More exact estimates can be made using the simulation models presented in Chapters 11 to 14. Fundamentals of aeration manifold design are presented in Chapter 15.

The reader may have noted that this chapter on aeration is presented between Chapters 6 and 8, both of which deal with feed conditioning. The reason for this is that the requirements for energy conditioning covered in Chapter 8 depend on the aeration supply. Therefore, a short recess from feed conditioning is necessary to first discuss the aeration demands imposed by composting.

STOICHIOMETRIC DEMANDS

The stoichiometric oxygen requirement can be determined from the chemical composition of the organic solids and the extent of degradation during composting. Representative chemi-

Table 7.1.   General Chemical Compositions for the Organic Fraction of Various Organic Materials

Waste Component	Typical Chemical Composition	Reference
Carbohydrate	$(C_6H_{10}O_5)x$	
Protein	$C_{16}H_{24}O_5N_4$	
Fat and Oil	$C_{50}H_{90}O_6$	
Sludge		
Primary	$C_{22}H_{39}O_{10}N$	
Combined	$C_{10}H_{19}O_3N$	McCarty[1]
Refuse (total organic fraction)	$C_{64}H_{104}O_{37}N$	Haug et al.[2]
	$C_{99}H_{148}O_{59}N$	Corey[3]
Wood	$C_{295}H_{420}O_{186}N$	Corey[3]
Grass	$C_{23}H_{38}O_{17}N$	Corey[3]
Garbage	$C_{16}H_{27}O_8N$	Corey[3]
Food wastes	$C_{18}H_{26}O_{10}N$	Kayhanian and Tchobanoglous[4]
Mixed paper	$C_{266}H_{434}O_{210}N$	Kayhanian and Tchobanoglous[4]
Yard wastes	$C_{27}H_{38}O_{16}N$	Kayhanian and Tchobanoglous[4]
Bacteria	$C_5H_7O_2N$	
Fungi	$C_{10}H_{17}O_6N$	

cal compositions of various organics and organic mixtures are presented in Table 7.1. Using municipal sludge as an example and assuming an average composition from Table 7.1 of $C_{10}H_{19}O_3N$, the carbonaceous oxygen demand can be determined as

$$\underset{201}{\overset{1}{C_{10}H_{19}O_3N}} + \underset{12.5\,(32)}{\overset{x}{12.5\,O_2}} \rightarrow 10\,CO_2 + 8\,H_2O + NH_3 \tag{7.1}$$

Based on the assumed substrate composition, the oxygen demand can be estimated as

$$x = 12.5(32)/201 = 1.99 \text{ g O}_2 \text{ g substrate BVS}$$

Ammonia released as a result of organic decomposition can be oxidized according to the following nitrification reaction:

$$\underset{2(32)}{\overset{y}{NH_3}} + 2O_2 \rightarrow NO_3^- + H_2O + H^+ \tag{7.2}$$

Assuming all ammonia is oxidized, the maximum nitrification demand for the above substrate can be estimated as

$$y = 2(32)/201 = 0.32 \text{ g O}_2/\text{g substrate BVS}$$

The nitrification demand is significantly less than that required for organic oxidation. Also, most of the produced ammonia will not be oxidized. Some will be used directly for cell synthesis, but the largest fraction will probably be lost from the compost by volatilization. Therefore, oxygen demands for nitrification generally are not considered when developing aeration demands for composting.

Many other composting substrates, sawdust for example, are largely cellulosic in nature. Their decomposition reaction can be estimated by assuming a carbohydrate composition:

$$\overset{1}{\underset{162}{C_6H_{10}O_5}} + \overset{x}{\underset{6(32)}{6O_2}} \rightarrow 6CO_2 + 5H_2O \tag{7.3}$$

The oxygen demand for cellulosic substrates can be estimated as

$$x = 6(32)/162 = 1.185 \text{ g } O_2/\text{g BVS}$$

The stoichiometric demand for oxygen varies from a low of about 1.0 g O_2/g organic for highly oxygenated substrates such as starch and cellulose to a high of about 4.0 g O_2/g organic for saturated hydrocarbons (refer to Table 3.2). A value of 2.0 g O_2/g organic is typical for many complex substrates, such as sewage sludge, which are composed of mixtures of protein, carbohydrate, and fats.

For cases where a mixture of substrates are input into the process, the total oxygen demand can be determined from the assumed composition of the individual fractions and their weighted average oxygen demands. For example, refer back to Figure 6.13 where both sludge and a sawdust amendment are the input substrates. Assuming that Equation 7.1 applies to the sludge BVS component and Equation 7.3 to the sawdust BVS, the total oxygen demand can be estimated as

$$[(0.375)(1.990) + (0.247)(1.185)]/(0.375 + 0.247)$$

$$= 1.67 \text{ g } O_2/\text{g feed BVS}$$

Air contains 23.2% oxygen by weight. Therefore, the weight requirement of air for the above mixture of sludge and sawdust is determined as

$$1.67/0.232 = 7.20 \text{ g air/g feed BVS}$$

This is a total *quantity* of air which must be supplied over the composting cycle to meet the stoichiometric demand. The quantity of air supply (i.e., lbs, kg, ft^3, or m^3) must be distinguished from the *rate* at which the air is supplied (i.e., kg/min or m^3/min). The rate of air supply is considered in later sections of this Chapter.

It is usually more convenient to express the stoichiometric demand on the basis of total dry weight of the feed substrates. This requires that we know the biodegradable fraction of the volatile solids. Substrate degradability is the subject of Chapter 9. For the present purposes we will assume reasonable values for degradability coefficients and leave the more in-depth discussion to Chapter 9.

Example 7.1

100 metric tons of sludge cake are to be composted. The sludge cake is conditioned with sawdust amendment and product recycle with the proportions shown in Figure 6.13 (see also Figure 7.3). The sludge has the composition given by Equation 7.1 and the sawdust given by Equation 7.3. Determine the quantity of air that must be supplied to meet the stoichiometric demand.

Solution

1. From Figure 6.13 the following values apply:

$$S_s = 0.20 \qquad S_a = 0.65$$
$$V_s = 0.75 \qquad V_a = 0.95$$
$$k_s = 0.50 \qquad k_a = 0.20$$

Also from Figure 6.13, the 100 metric tons of sludge are blended with $100(2/5) = 40$ metric tons of sawdust.

2. The stoichiometric demand for the sludge cake is determined as

$$100(0.20)(0.75)(0.50)(1.99) = 14.93 \text{ metric ton of } O_2$$

$$14.93(1000) = 14,930 \text{ kg } O_2$$

3. The stoichiometric demand for the sawdust amendment is determined as

$$40(0.65)(0.95)(0.20)(1.185) = 5.85 \text{ metric ton of } O_2$$

$$5.85(1000) = 5,850 \text{ kg } O_2$$

4. The total stoichiometric demand is the sum of the individual substrate demands:

$$14,930 + 5,850 = 20,780 \text{ kg } O_2$$

5. The quantity of air required to meet the stoichiometric demand is determined as

$$20,780 / 0.232 = 89,570 \text{ kg air}$$

$$89,570 / 1000 = 89.57 \text{ metric ton air}$$

6. Air at 25°C and 1 atm pressure has a specific weight of 1.20 kg/m^3 (0.075 lb/ft^3). The volume of air required to meet the stoichiometric demand is determined as

$$89,570 / (1.20) = 74,640 \text{ m}^3$$

7. The stoichiometric air quantity on a dry weight basis can be determined as

$$89.57 / [100(0.20) + 40(0.65)] = 1.95 \text{ kg air / kg ds mixed substrates}$$

The above example points out the importance of air supply to the composting process. The weight of air required to just meet the stoichiometric demand is almost twice the dry weight of the feed substrates. Air is usually the largest term in the mass balance for any composting system.

In any practical composting system, it is necessary to supply an excess of air over the stoichiometric requirement to assure fully aerobic conditions. The excess air ratio (EAR) is defined as the ratio of actual air supplied to that required to just satisfy the stoichiometric demand.

Another approach to estimating oxygen requirements during aerobic composting is based on the composition of feed substrate and final product. This approach is useful if bench-scale composting studies can be conducted to estimate final product composition. Degradability of the feed substrate is thereby included in the analysis. Rich[5] suggested the following stoichiometric equation:

$$C_aH_bO_cN_d + 0.5(ny + 2s + r - c)O_2$$

$$= nC_wH_xO_yN_z + sCO_2 + rH_2O + (d - nz)NH_3$$

$$(7.4)$$

where

$$r = 0.5[b - nx - 3(d - nz)]$$
$$s = a - nw$$

The terms $C_aH_bO_cN_d$ and $C_wH_xO_yN_z$ represent the compositions of feed substrate and final product, respectively. An elemental analysis is required for evaluation of the subscripts in these terms. Use of this approach is illustrated in the following two examples adapted from Rich.[5]

Example 7.2

Bench scale tests of aerobic composting were conducted on a substrate with starting composition $C_{31}H_{50}O_{26}N$. Tests indicated that 1000 kg ds of the substrate were reduced to 200 kg ds at the completion of composting. Final product composition was determined to be $C_{11}H_{14}O_4N$. Determine the stoichiometric oxygen requirement per 1000 kg of feed.

Solution

1. Molecular weight of the substrate is

Carbon	31(12)	=	372
Hydrogen	50(1)	=	50
Oxygen	26(16)	=	416
Nitrogen	1(14)	=	14
Total		=	852

kg-mol of organics entering the process = 1000/852 = 1.173

2. Molecular weight of the compost product is

Carbon	11(12)	=	132
Hydrogen	14(1)	=	14
Oxygen	4(16)	=	64
Nitrogen	1(14)	=	14
Total		=	224

The kg-mol of compost leaving the process per kg-mol entering the process = $n = 200/(1.173)(224)$ = 0.76

3. The following values apply:

a	=	31	w	=	11
b	=	50	x	=	14
c	=	26	y	=	4
d	=	1	z	=	1

Using these values, r and s can be determined as

$$r = 0.5\{50 - 0.76(14) - 3[1 - 0.76(1)]\} = 19.33$$

$$s = 31 - 0.76(11) = 22.64$$

4. From Equation 7.4, the quantity of oxygen required by the process is

$$W = 0.5[0.76(4) + 2(22.64) + 19.33 - 26](1.173)(32) = 783 \text{ kg}$$

5. Check with a materials balance using Equation 7.4:

In	
Substrate	1000
Oxygen	783
Total In	1783 kg
Out	
Compost	200
Carbon dioxide 1.173(22.64)(44)	1170
Water 1.173(19.33)(18)	408
Ammonia [1 − 0.76(1)](1.173)(17)	5
Total Out	1783 kg

6. The air required is 783/0.232 = 3375 kg. This is equivalent to 3375/1000 = 3.375 kg air/kg ds substrate. As was the case in Example 7.1, the stoichiometric air demand is several times the dry weight of feed substrates.

Example 7.3

Estimate the energy released as heat in the bench tests described in Example 7.2.

Solution

1. The percentage elemental composition of organics in the substrate and compost are

In
%C = 31(12)/852   =   43.6
%H = 50(1)/852   =   5.9
%O = 26(16)/852   =   48.9
Out
%C = 11(12)/224   =   59.0
%H = 14(1)/224   =   6.3
%O = 4(16)/224   =   28.6

2. From Equation 3.26, the R values of the substrate and compost are

$$\text{Substrate} \ \ R = 100[2.66(43.6) + 7.94(5.9) - 48.9] / 398.9 = 28.6$$

$$\text{Compost} \ \ R = 100[2.66(59.0) + 7.94(6.3) - 28.6] / 398.9 = 44.8$$

3. From Equation 3.27, the unit heats of combustion of the organics are

$$\text{Substrate} \ \ Q = 127(28.6) + 400 = 4030 \ \text{cal/g VS}$$

$$\text{Compost} \ \ Q = 127(44.8) + 400 = 6100 \ \text{cal/g VS}$$

4. Total energy released as heat is

Energy in: 1000(1000)4030 = 4030×10^6 cal
Energy out: 200(1000)6100 = 1220×10^6 cal
_____
Energy release: 2810×10^6 cal

AIR DEMAND FOR MOISTURE REMOVAL

Determination of the air requirement for drying requires further analysis of the composting process. Referring to the mass balance diagram shown in Figure 6.7 and neglecting any amendment or bulking agent addition, the quantity of water to be evaporated daily, W, is given by

$$W = (X_s - S_s X_s) - (X_p - S_p X_p) \tag{7.5}$$

Considering the ash fraction to be conservative, a mass balance on the inert fraction yields

$$(1 - V_s)S_s X_s = (1 - V_p)S_p X_p \tag{7.6}$$

Solving Equation 7.6 for X_p, substituting into Equation 7.5, and rearranging gives

$$W = [(1 - S_s) / S_s] - [(1 - V_s) / (1 - V_p)][(1 - S_p) / S_p] \tag{7.7}$$

Equation 7.7 was solved for various substrate and compost solids contents. Results and other assumptions are presented in Figure 7.1. Conditions appropriate to a digested sludge cake were used to obtain the results shown in Figure 7.1. However, similar analysis for other substrate

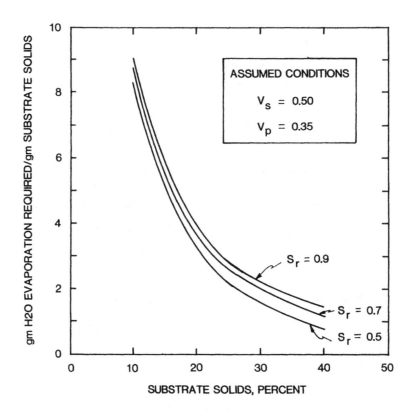

Figure 7.1.   Effect of substrate solids and compost product solids on required moisture evaporation. From Haug and Haug.[6]

conditions results in very little change from the values indicated. The quantity of moisture to be evaporated is determined primarily by the input solids at substrate solids below ~30%. Above 30%, both feed solids and the final compost solids are important in determining the quantity of moisture to be evaporated. Rates of biological activity begin to decrease below moisture levels of about 40 to 50%.[7] Therefore, drying to a moisture level below ~30% with the heat supplied by composting is difficult.

The quantity of water vapor that can be carried in saturated air at different temperatures is shown in Figure 7.2. The curve was constructed from standard psychometric charts and steam tables[8] for a total atmospheric pressure of 760 mm Hg. The quantity of moisture in saturated air increases exponentially with increasing air temperature. Gas that leaves a wet composting material will be near saturation and about the same temperature as the composting material. If thermophilic temperatures are maintained, considerable moisture will be removed with the exhaust gases.

If the temperature difference between inlet and outlet air is greater than ~25°C (45°F), relative humidity of the inlet air will have a minor effect in determining the overall moisture removal. This means that drying can occur even in climates with high ambient humidity in the feed air.

The specific humidity, w, is defined as the weight of water carried per unit weight of dry air, i.e., g water/g dry air. Values of specific humidity can be read from Figure 7.2. However, the following algorithm for calculating specific humidity is useful and is incorporated into

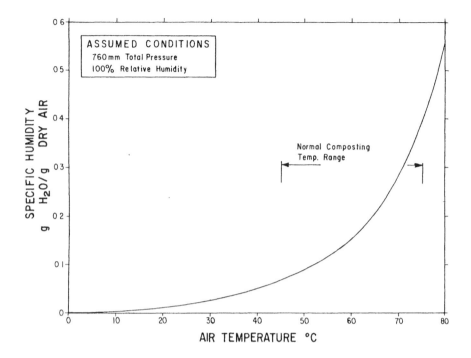

Figure 7.2.   Specific humidity as a function of air temperature under conditions of 100% relative humidity and 760 mm Hg total pressure.

later simulation models. Saturation vapor pressure (PVS) is a function of temperature and can be mathematically modeled by the Antoine equation:[9]

$$\log_{10} PVS = a / T_a + b \tag{7.8}$$

where

PVS	=	saturation water vapor pressure, mm Hg
a	=	constant equal to −2238 for water
b	=	constant equal to 8.896 for water
T_a	=	absolute temperature, °K

The actual water vapor pressure (PV) is determined as

$$PV = RHAIR(PVS) \tag{7.9}$$

where

PV	=	actual water vapor pressure, mm Hg
RHAIR	=	relative humidity, fraction of saturation vapor pressure

Specific humidity can be determined from the water vapor pressure and the molecular weights of water (18.015) and dry air (28.96):

$$W = (18.015 / 28.96)[PV / (PAIR - PV)] \tag{7.10}$$

where   w = specific humidity, g water/g dry air
        PAIR = atmosphere pressure, mm Hg

Example 7.4

A wet organic substrate at 25% TS is to be composted and dried to a final solids content of 65% TS. The substrate has a VS content of 75% with the chemical composition given by Equation 7.1. The compost is expected to be 45% VS. Ambient air to the process is 20°C with 75% relative humidity and total pressure of 760 mm Hg. Exit gas temperature is 55°C saturated. Determine the weight and volume of air and the excess air ratio (EAR) required to remove the required moisture.

Solution

1. The requirement for water evaporation can be determined from Equation 7.7:

$$W = [(1 - 0.25) / 0.25] - [(1 - 0.75) / (1 - 0.45)][(1 - 0.65) / 0.65]$$

$$W = 2.76 \text{ g water/g dry substrate}$$

2. Determine the moisture-carrying capacity of the exit gas as follows. From Equation 7.8 the saturation vapor pressure is determined as

$$\log_{10} PVS = -2238 / (273 + 55) + 8.896 = 2.073$$

$$PVS = 118.3 \text{ mm Hg}$$

Because the exit gas is saturated, the relative humidity is 1.00. The actual vapor pressure is then determined from Equation 7.9 as

$$PV = (1.00)118.3 = 118.3 \text{ mm Hg}$$

The specific humidity is determined from Equation 7.10 as

$$w = (18.015 / 28.96)[118.3 / (760 - 118.30)]$$

$$w = 0.1147 \text{ g water/g dry air}$$

Note that this value corresponds with that from Figure 7.2 for a temperature of 55°C.

3. The specific humidity of the inlet air at 20°C and 75% relative humidity is determined in the same manner:

$$\log_{10} PVS = -2238 / (273 + 20) + 8.896 = 1.258$$

$$PVS = 18.10 \text{ mm Hg}$$

$$PV = (0.75)18.10 = 13.6 \text{ mm Hg}$$

$$w = (18.015 / 28.96)[13.6 / (760 - 13.6)]$$

$$w = 0.0113 \text{ g water/g dry air}$$

The above value cannot be read from Figure 7.2 because the latter figure applies only to saturated conditions.

4. Net moisture removed with the exhaust gases is

$$0.1147 - 0.0113 = 0.1034 \text{ g water/g dry inlet air}$$

Note the relatively small moisture contribution of the inlet air if the temperature difference is about 20°C or greater.

5. The weight of required air is determined as

$$2.76/0.1034 = 26.7 \text{ g dry air/g ds substrate}$$

Note that the air required for drying is significantly greater than that required for biological oxidation as determined in Example 7.1.

6. From Example 7.1, the specific weight of air at standard temperature and pressure (STP) is 1.20 g/L. Therefore,

$$26.7/1.2 = 22.25 \text{ l dry air/g ds substrate}$$

Again, the air weight and volume are total quantities that must be supplied during the composting cycle to remove a given quantity of water.

7. The EAR is calculated as follows: from Equation 7.1 the stoichiometric demand is 1.99 g O_2/g BVS. One g ds of substrate contains 0.75 g VS. Assuming the 0.25 g ash to be conservative, the final compost will contain $0.25/(1 - 0.45) - 0.25 = 0.205$ g VS. Therefore, about $(0.75 - 0.205)/0.75 = 72.7\%$ of the VS are degradable. Therefore, the stoichiometric demand can be converted to units of g air/g ds as follows:

$$1.99(1/0.232)(0.75)(0.727) = 4.68 \text{ g air/g ds}$$

The EAR is determined as

$$EAR = 26.7 / 4.68 = 5.7$$

Another approach to determining the quantity of water that must be removed is to use the bar diagrams introduced in Chapter 6. Referring back to Figure 6.13 for the case of wet substrate conditioned with amendment and recycle, the bars for substrate and amendment are repeated in Figure 7.3. The right bar represents the compost product assuming 60% TS and loss of all BVS during the composting process. Recycle is not shown in Figure 7.3 because it is within the system boundary and does not add new water to the process. For this case, 3.581 g water must be removed per g ds of the substrate. Using the same inlet air exhaust gas conditions from Example 7.4, about 34.6 g air/g ds substrate are required for drying. The bar diagrams provide visualization of the problem and are particularly useful when multiple substrates or amendments are involved.

Using the procedures of Example 7.4, the quantity of air required for moisture removal was determined as a function of substrate solids content. Inlet air was assumed to be 20°C and 100% relative humidity. Exit gas temperatures were assumed to be 40°C or greater. Thus, inlet

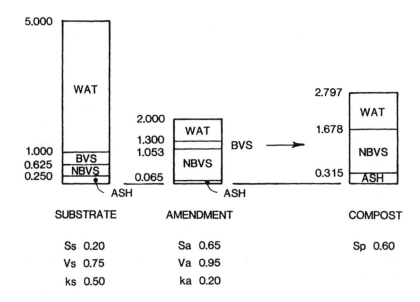

Figure 7.3.   Bar diagrams from Figure 6.13 for the case of a wet substrate at 20% TS conditioned with amendment and recycle. Recycle is not shown because it is within the system boundary and does not contribute new water to the system.

air conditions are not critical to the problem. A final compost moisture content of 30% was assumed. Results and other assumptions are presented in Figure 7.4. The typical range of stoichiometric air requirements for complex substrates such as digested and raw municipal sludge is also shown. With wet substrate, the air required for moisture removal is significantly greater than the stoichiometric demand for biological oxidation and is influenced largely by the substrate solids content and exit gas temperature. At substrate solids of 20%, the air requirement for drying can be 10 to 30 times that for biological oxidation. With dryer substrates, the air requirements for drying and biological oxidation become more equivalent. For example, at 40% TS substrate and 70°C exit temperature the two air requirements become essentially equivalent. This may be somewhat misleading, however, because water addition may be required in the latter case to prevent moisture limitations during composting. This will add to the evaporative burden and increase the required air supply over that shown in Figure 7.4.

Because the stoichiometric demand is generally less than the demand for drying, control of the air supply becomes an important factor in operation of the composting process. The extent of drying can be regulated by control of the air supply. It will be shown in Chapter 8 that composting can be divided into two distinct thermodynamic regions: one in which the energy supply is sufficient for both composting and drying, and one in which energy is sufficient for composting with only limited drying. The air supply can be used to control the thermodynamic region of operation. Thus, the air supply is one of the operator's most important tools for process control.

AIR DEMAND FOR HEAT REMOVAL

Rates of biochemical reactions generally increase exponentially with temperature. However, process temperatures can elevate to the point of thermal inactivation of the microbial population. Temperature then becomes rate limiting. To control process temperatures in a

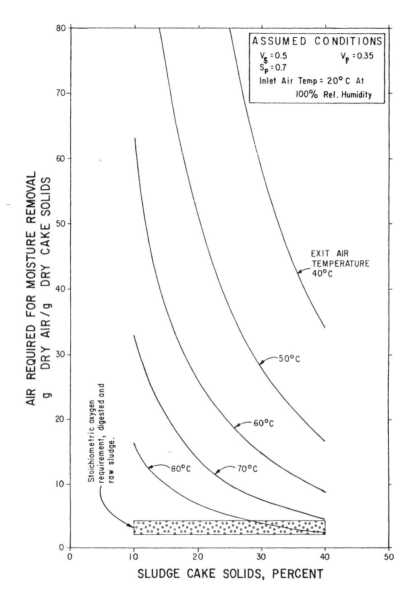

Figure 7.4.   Effect of substrate solids and exit air temperature on air requirements for moisture removal to a compost solids content of 70%. From Haug and Haug.[6]

more optimum range for the microbial population, air supply can be increased to remove more heat.

Referring to Chapter 3, the heat of combustion per electron transferred to a methane-type bond is essentially constant at ~26.05 kcal per electron transferred or, since O_2 accepts four electrons, 104.2 kcal/mol O_2. This equals 3260 cal/g O_2 or 5870 Btu/lb O_2. Thus, heat release can be estimated from the stoichiometric demand for oxygen.

Example 7.5

Using the oxygen demands determined by Equations 7.1 and 7.3 and the mixture of substrate and amendment shown in Figure 7.3, estimate the heat release for 1 g ds of feed substrate.

Solution

1. From Equation 7.1 the oxygen demand of the substrate is 1.99 g O_2/g BVS. From Figure 7.3, 1 g ds of substrate contains $1.000 - 0.625 = 0.375$ g BVS. Oxygen demand and the corresponding heat release for the substrate is

$$0.375(1.99) = 0.746 \text{ g } O_2 / \text{g ds}$$

$$0.746(3260) = 2430 \text{ cal/g ds}$$

2. From Equation 7.3, the oxygen demand of the amendment is 1.185 g O_2/g BVS. From Figure 7.3, $1.300 - 1.053 = 0.247$ g BVS are contained in the amendment. Oxygen demand and heat release from the amendment is

$$0.247(1.185) = 0.293 \text{ g } O_2/\text{g ds}$$

$$0.293(3260) = 955 \text{ cal/g of substrate ds}$$

3. Total heat release is the sum of the above,

$$2430 + 955 = 3385 \text{ cal/g of substrate ds}$$

Heat released during composting will be removed primarily by the hot, moist exhaust gases leaving the process. Some heat will be removed in the output solids and some lost to the surroundings. However, these are relatively small compared to that contained in the exhaust gases. Heat released by organic decomposition will heat the incoming dry air and water vapor to the exit temperature, supply the heat of vaporization, and heat the evaporated water vapor to the exit gas temperature. Considering only these major heat demands and neglecting other minor losses allows a closed solution to the problem of determining the air demand for heat removal.

Example 7.6

Estimate the air demand to remove the heat generated in Example 7.5. Assume the gas conditions of Example 7.4, namely ambient air at 20°C and 75% relative humidity and exhaust gases at 55°C saturated.

Solution

1. Let x = the lb dry air required to maintain the compost temperature at 55°C. Then,

Heat of vaporization at 20°C	=	$(x)(0.1034)(585.8)$ ·
Heat inlet water vapor and evaporated water vapor to 55°C	=	$(x)(0.1034 + 0.0113)(0.44)(55 - 20)$
Heat dry air to 55°C	=	$(x)(0.24)(55 - 20)$

2. For the above equations, 585.8 is the heat of vaporization at 20°C and 0.44 and 0.24 are the specific heats of water vapor and air in cal/g-°C, respectively.

3. Heat release from Example 7.5 is 3385 cal/g ds. At steady state, heat release must balance heat demand. Therefore,

$$(x)(60.57 + 1.77 + 8.40) = 3385$$

$$x = 47.9 \text{ g air/g of substrate ds}$$

4. From Example 7.1 the stoichiometric air demand is 1.95 g air/g substrate ds. The EAR is determined as

$$EAR = 47.9 / 1.95 = 24.6$$

For the examples based on Figure 7.3, the air quantity required for heat removal is greater than that required either for stoichiometry or for moisture removal. Therefore, in this case the requirement for heat removal governs. The reader should note that this conclusion is valid only for the above example. Requirements for moisture removal could govern under other process conditions, such as a wetter feed substrate. Because the heat of vaporization is generally the largest term in the energy balance, moisture removal and heat removal are related and the requirements for each should not differ by much for wet substrates. With dry substrates the air demand for heat removal will be greater than that for moisture removal. Requirements for moisture and heat removal are significantly greater than that required by stoichiometry and will usually be the governing conditions.

Composting has often been described as a problem of materials handling. If so, one should be aware that input air and exhaust gases are usually the dominant terms of the mass balance. In some cases the weight of air moved through the process can be 30 to 50 times that of the dry weight of substrates. The significance of the air movement is sometimes overlooked because the gases are not visible. The reader is cautioned to always remember that materials handling includes both solids and gases.

AERATION RATES

The air demands estimated above for stoichiometry, moisture and heat removal represent total *quantities* which must be supplied over the composting cycle. Converting these quantities to actual *rates* of air supply requires knowledge of the type of composting system and the kinetics of microbial oxidation.

Batch Processes

For a batch process, such as the windrow and aerated static pile, the total air *quantity* can be converted to an aeration *rate* by considering the time duration of the process. Assume a batch process with a total air demand based on heat removal from Example 7.6 of 47.9 g air/g of substrate ds. Assume that this quantity is supplied at a constant rate over a 25-day composting cycle. Then the *average* rate of aeration can be estimated as

$$(47.9)(10^6) / [1.2(25)(24)(10^3)]$$

$$= 66.5 \text{ cmh / dmt (cu. meters per hour per dry metric ton)}$$

$$= 2135 \text{ cfh / dt (cu. feet per hour per dry ton)}$$

where 1.20 g/L is the density of air at standard conditions.

A number of factors can cause the *peak* rate of aeration to exceed the average rate calculated above. First, the rate of organic oxidation, and therefore the rate of heat release, will vary throughout the composting cycle. Second, the type of aeration control may increase the peaking factor. An on/off aeration control logic is used in some systems. Obviously, the total air demand can be supplied only during the "on" sequence. Temperature feedback logic is popular with many composting systems and can also result in high peaking factors.

The peak rate of air supply can be estimated from data developed by Wiley,[10] Jeris and Regan,[11] Schulze,[12,13] and Snell.[14] These researchers observed that the rate of oxygen consumption is a function of temperature. Peak rates of about 4 to 14 mg O_2/g VS-h were observed in the temperature range of 45 to 65°C (see Figure 10.17). For the feed mixture presented in Figure 7.3, this is equivalent to 0.0035 to 0.0121 g O_2/h per g ds of mixed substrate and amendment. Using 3260 cal/g O_2 consumed, the heat release rate can be estimated as

$$(0.0035 \text{ to } 0.0120)(3260) = 11 \text{ to } 39 \text{ cal/h per g of feed solids}$$

$$20 \text{ to } 70 \text{ Btu/h per lb of feed solids}$$

Using the same assumptions and calculation procedures for heat removal presented in Example 7.6, the aeration rate required to maintain 55°C is about

$$(x)(60.57 + 1.77 + 8.40) = 11 \text{ to } 39$$

$$x = 0.15 \text{ to } 0.55 \text{ g air/h per g feed solids}$$

Assuming an air density of 1.20 g/L, the peak aeration rate becomes

$$(0.15 \text{ to } 0.55)(10^3)/1.20 = 125 \text{ to } 460 \text{ cmh/metric ton of feed solids}$$

$$4000 \text{ to } 14,700 \text{ cfh/ton of feed solids}$$

Using simulation models of the aerated pile process operated with temperature feedback control logic, Haug[15] predicted peak aeration rates of 3800 to 4800 cfh/dt of sludge for raw sludge blended with wood chips. Murray and Thompson[16] reported that peak aeration rates of 4000 to 5000 cfh/dt were sometimes insufficient to keep process temperatures below 60°C when composting a raw sludge/wood chip mixture. Williams and North[17] reported a peak aeration rate approaching 10,000 cfh/dt during static pile composting of a raw sludge/wood chip mixture. Both the theoretical and measured values compare favorably with the estimate based on oxygen consumption data. The subject of peak aeration rates is addressed again in Chapter 14.

It should be noted that the peak aeration rate may be maintained for only a short period until the peak demand has passed. In the model presented by Haug,[15] the required aeration rate exceeded 4000 cfh/dt for about 2 days, 3000 cfh/dt for about 4 days, and 2000 cfh/dt for about 8 days. Thus, there is a time duration associated with any peak demand. We might speak of the instantaneous peak, the 1-h peak, or the 24-h peak. The oxygen consumption data presented above is probably on the order of 1-h peak data, whereas most of the reported field data represent 24-h peak demands.

If the aeration system cannot meet the peak demand, process temperatures will exceed the desired set point. The designer must make a tradeoff between aeration system capacity, and hence capital cost, and the needs for process temperature control. It may be more cost-effective to size the system for less than the peak demand and accept process temperatures above setpoint for a short period. The simulation models in Chapters 11 through 14 are useful to develop statistical information on which to base such decisions.

Aeration systems for batch processes take many forms. Some aerated pile and windrow systems use a single blower connected to a single pile. In this case, the peak demand is carried by the blower. Other systems use a central blower facility connected to many individual piles. In this case, the peaking factor can be "shaved" because it is unlikely that all piles simultaneously would be at the point of maximum oxygen consumption.

Minimum aeration rates should also be considered because the system must be capable of operating over the turndown range from peak to minimum rates. For an individual pile or windrow the minimum average rate usually occurs at the very beginning or end of the batch cycle and may be as low as 200 to 500 cfh/dt.

Continuous Processes

Many reactor systems operate on a continuous or semicontinuous feeding schedule. Typically, substrate is conditioned and loaded into the reactor on a daily basis. Reactor residence time is usually greater than 10 days, so the daily feeding schedule can be modeled as a continuous process. For a continuous feed process, the quantity of required air can be converted to an *average* rate of aeration by considering a daily unit weight of feed substrate. Again using the air quantity for temperature control determined in Example 7.6, the aeration rate becomes

$$47.9 \text{ g/day of air per g/day of substrate ds}$$

This is equivalent to 1660 cmh/dmtpd of substrate (m^3 per hour per dry metric ton per day) or 53,000 cfh/dtpd (ft^3 per hour per dry ton per day). This is typical of the range of values predicted from simulation modeling of continuous processes as presented in Chapter 13.

With continuous feeding, the *peak* aeration demand should not exceed the average demand over the composting cycle, i.e., (peaking factor = 1). With semicontinuous feeding, some peaking might occur between feed periods. However, the 15 to 20% contingency normally applied to blower design capacity and the usual practice of providing standby capacity appear adequate to handle most above average aeration demands.

The designer should consider that feed quantity and composition will vary throughout the life of the project. The process should be examined under all expected conditions to determine whether seasonal or periodic peaking factors should be applied. Similarly, minimum flow conditions should be considered. This is particularly important if the process will operate below design loadings in early years. There is a tendency to focus attention only on average or peak design loadings. However, the full range of operation should be considered to avoid control problems during minimum turndown conditions.

CONTROLLING THE AERATION RATE

Once minimum, average, and peak aeration rates have been determined, the designer must select a control system to regulate the aeration rate. There are a number of control strategies which have been used in practice. These can be summarized as follows:

1. **Uncontrolled** — Windrow aeration by natural ventilation is an example of this approach. Such an approach is not applicable to aerated pile or reactor systems.
2. **Manual** — Throttling valves can be used to manually control the aeration rate. This is a common practice where lateral ducts are tied to a manifold which in turn is connected to a central blower system. Throttling valves can also be used with a single blower/manifold system. Flow through each lateral or single blower is manually adjusted by the throttling valve throughout the composting cycle.
3. **On/Off Sequencing by Timer** — With a single blower design, the average rate of air supply can be controlled by regulating the blower "on-time" by means of a timer control. This approach is often used with the static pile system and some reactor systems. Timer control can be used in conjunction with throttling valves to allow adjustment of both on-time and flowrate.
4. **Feedback Control Based on O_2 or CO_2 Content** — Oxygen or carbon dioxide probes inserted into the composting mixture or the exhaust gas ducts can generate a control signal used to modulate air supply and maintain a setpoint O_2 or CO_2 content. O_2 or CO_2 is the *controlled* variable and aeration rate the *manipulated* variable. This control strategy is only usable with dry substrates and where temperature control is not used because the aeration supply is then near the stoichiometric demand. With wet substrates or where temperature control is used, the aeration demands for moisture and heat removal are so high that the O_2 content often differs only slightly from that of ambient air (21% by volume).
5. **Feedback Control Based on Temperature** — Thermocouples inserted into the composting mixture can generate a control signal used to modulate air flowrate or blower on-time to maintain a setpoint temperature. Temperature is the *controlled* variable and aeration rate the *manipulated* variable. This control approach is used in many static pile and most reactor systems and is considered well proven.
6. **Air Flowrate Control** — If the daily quantity of feed material remains reasonably constant, it is possible for the operator to control air flowrate to a fixed value by either manual adjustment of fan dampers or by a setpoint flowrate control loop. This approach is used in some reactor systems where conditions are not expected to change rapidly.

Where a control signal, temperature for example, is used for regulation, the air supply can be adjusted by a number of techniques. The on-time can be controlled in a timer system. Throttling of inlet guide vanes or inlet dampers is commonly used with fixed speed aeration blowers. Outlet dampers can also be used, but the possibility of fan surging at low flow conditions should be considered. A variety of variable speed drives can also be used to adjust the aeration rate. Speed control becomes more cost-effective in large systems where the power savings can offset the added capital cost. Again, it is important to consider minimum, average, and peak aeration rates to define the range of system operation. Flow measuring devices and throttling valves must operate over the entire range of expected conditions.

MECHANISMS OF AERATION

For the aerated static pile and most reactor compost systems, the mechanism for supplying oxygen is obvious: air exchange is provided by forced ventilation using blowers to pull or push air through the material. In other systems, particularly the non-aerated windrow process, the mechanisms of aeration are not so obvious. On first impulse one is tempted to assume that windrow turning is the primary method for oxygen replenishment within the free airspace. As will be seen, however, this does not appear to be the principal mechanism. The purpose of this section is to explore the factors responsible for supplying oxygen to pile or windrow systems that do not use forced aeration.

Exchange Volumes

For any substrate the air volume required to supply the stoichiometric oxygen demand can be estimated using the procedures of Example 7.1. Once the total air volume is estimated the number of exchange volumes can be calculated provided the free airspace in the mixture is known. The number of exchange volumes is defined as the number of times the mixture free airspace must be renewed with air to supply the stoichiometric oxygen demand. Knowledge of the number of required exchanges can provide insight into the mechanisms responsible for aeration.

Example 7.7

A yard waste consisting of grass and leaves is to be windrow composted. No recycle is used. The mixed yard wastes are estimated to be 80% VS with a degradability of 50%. Initial moisture content in the windrow is adjusted to 50%. If unit weight of the mixture is 0.65 g/cm³, estimate the number of volume exchanges required to satisfy the stoichiometric oxygen demand during composting.

Solution

1. Specific gravity of the mixture can be calculated from Equation 6.1:

$$1 / G_s = (0.80 / 1.00) + [(1 - 0.80) / 2.5]$$

$$G_s = 1.14$$

2. Determine the mixture FAS from Equation 6.11:

$$f = 1 - [0.65(0.50) / (1.13)(1.00)] - [0.65(1 - 0.50) / 1.00]$$

$$f = 0.387$$

3. The stoichiometric air requirement for composting can be estimated assuming the chemical composition for grass from Table 8.1:

$$\underset{600}{\overset{1}{C_{23}H_{38}O_{17}N}} + 23.25O_2 = \underset{744}{\overset{x}{23CO_2}} + 17.5H_2O + NH_3$$

$$x = 744 / 600 = 1.24 \text{ g } O_2/\text{g substrate BVS}$$

$$x = 1.24(0.80)(0.50) / (0.23) = 2.16 \text{ g air/g ds}$$

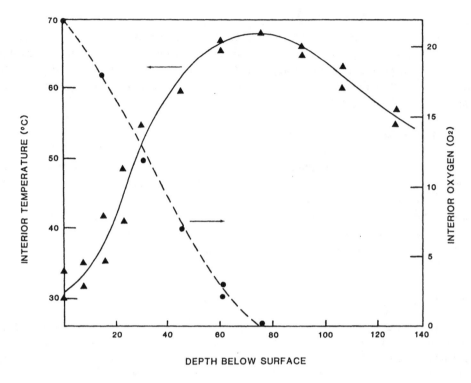

Figure 7.5.   Profiles of oxygen concentration (%) and temperature with depth (cm) into a windrow composed of digested sludge cake and recycled compost. Measurements were made 5 h after windrow turning. From Iacoboni et al.[19]

4. The volume of air to be supplied can be estimated assuming air density at 1.20 g/L:

$$x = 2.16 / 1.20 = 1.80 \ \text{l/g ds}$$

5. Each cc of mixture contains 0.65(0.50) = 0.325 g ds and contains 0.387 cc of free airspace. The number of void volume exchanges can be estimated as

$$\frac{0.325(1.80)(1000)}{0.387} = 1510 \ \text{exchanges}$$

Suppose it is assumed that each turning of the windrow completely renews air contained in the voids and that no other aerating mechanisms are at work. With these assumptions, the number of windrow turnings must equal the number of required exchange volumes. Therein lies the problem. The number of turnings normally used in windrow composting is nowhere near the exchange volumes estimated in Example 7.7. Furthermore, no combination of reasonable values for the variables used in the analysis seems likely to reduce significantly the required number of exchange volumes.

Smith[18] and Iacoboni et al.[19] observed that oxygen concentrations in the interiors of non-aerated sludge windrows often decreased to undetectable levels within a few minutes to several hours after turning. A profile of oxygen content with depth into such a windrow is shown in Figure 7.5. The data were recorded 5 h after turning. Based on the windrow dimensions it was estimated that about 50% of the windrow volume was aerobic, while the rest was either slightly aerobic or anaerobic. Maier et al.[20] observed a similar phenomenon in windrows composed of finely ground refuse (particle size <1.25 cm) and reported oxygen depletion in the bottom interior 4 to 5 h after turning. These observations seem reasonable in light of the analysis of required exchange volumes.

In discussing the turning frequency required for windrow composting of ground refuse, Golueke[7] indicated that the windrow needed to be turned every other day for a total of only 4-5 turns. By the fourth or fifth turn the material should be so advanced toward stabilization as to require no further turning. More frequent turning would be beneficial if the material were very wet or compacted. Based on experience gained during windrow composting studies at the University of California, McGauhey and Gotaas[21] and later Golueke[22] recommended the following turning schedule as adequate for municipal refuse (2.5 to 5 cm probable particle size):

Moisture Content (%)	
>70	Turn daily until the moisture content is reduced to <70%
60–70	Turn at 2-day intervals (number of required turns is about five)
40–60	Turn at 3-day intervals (approximate number of turns is four)
<40	Add moisture

The question that remains is how such minimal turning as described above can result in the aerobic conditions desired during composting. Obviously, exchanging air contained in the FAS by windrow turning is entirely inadequate to supply the air quantities required. Other aeration mechanisms must be operating in the windrow process.

Molecular Diffusion

Assuming no forced aeration, there are two ways oxygen can be supplied through pore spaces into the interior of the windrow between turnings: (1) by molecular diffusion and (2) by mass movement of air through the pores in response to an energy gradient. Molecular diffusion results from constant and random collisions between molecules of a fluid. As a result of such collisions, there is a tendency for molecules to move from a zone of high concentration to a zone of lower concentration. If oxygen is depleted within the windrow, a net movement of oxygen from the surrounding air into the windrow will occur as a result of molecular diffusion. Similarly, carbon dioxide and water vapor will diffuse from the interior of the windrow where it is produced to the outside air where the concentration is lower.

The problem with molecular diffusion, however, is that the process is extremely slow compared to rates at which oxygen is required for composting. Shell[23] and Snell[14] studied the diffusion of oxygen through composting refuse and concluded that diffusion alone supplied only a small percentage of the oxygen needed during composting. Furthermore, the rate of diffusion decreased as moisture content of the compost increased, a definite problem with wet substrates. Only when refuse was placed in thin layers (5 to 10 cm) did molecular diffusion become significant. Even in this case the rate of oxygen diffusion was estimated to be <5% of the maximum rate of oxygen demand.

Natural Draft Ventilation

Because molecular diffusion is not a practical transport mechanism, it appears that only a mass flow of air in response to an energy gradient can supply the required oxygen. Remember, forced ventilation or aeration by means of mechanical blowers is not considered here. The question then is what forces are operating within a composting windrow or pile to produce a mass flow of air?

Referring to Figures 2.5 and 2.6, it is common to observe steam continually issuing from the top of windrows even if they are turned only infrequently. This implies a mass flow of water vapor and other exhaust gases from the pile interior. The movement is in response to the high temperature of gases within the windrow, analogous to hot flue gases drafting up a chimney. This effect is termed natural draft ventilation.

The density of dry and saturated air as a function of temperature is shown in Figure 7.6. As temperature increases, the density of dry air decreases. However, the degree of saturation can also affect the density, particularly at higher temperatures, because the saturation vapor pressure increases exponentially with temperature (Figure 7.2). The effect of water vapor on density is explained by the fact that the molecular weight of water is considerably less than that of the oxygen and nitrogen it displaces. Carbon dioxide produced from organic decomposition has the opposite effect because its molecular weight is greater than either oxygen or nitrogen.

The density difference between warm moist air within the interior of the windrow or pile and colder, less moist ambient air produces an upward buoyant force that induces a natural ventilation of the windrow. A schematic illustration of the natural draft ventilation process is presented in Figure 7.7. The concept of natural drafting is easily understood. However, the question is whether the rate of natural ventilation is sufficient to satisfy the actual oxygen demands.

Haug[24] developed hydraulic models to simulate the process of natural ventilation. Based on results of these models and observations of actual composting systems, it was concluded that natural drafting is the likely driving force for most oxygen transfer in a conventional (no forced aeration) windrow or pile system. The ventilation rate was found to be a function of the density difference between inside and outside gases, the particle size of the compost, and the free air space of the pile. It was also suggested that the ventilation rate is independent of pile height, all other factors being constant. As the height increases, the rate of natural ventilation should remain relatively constant whereas the rate of oxygen consumption per unit of ground area increases with height. Thus, there is a limit to the height of pile beyond which adequate oxygen cannot be supplied by natural ventilation. The Haug[24] model suggested that natural draft could supply adequate oxygen with pile heights from 1 to 3 meters, particle sizes about 0.1 cm and above, and a temperature difference of about 40°C. Maier et al.[20] observed that aeration of windrows composed of finely ground refuse was improved when the windrow was placed on elevated racks so that air could permeate from below and produce a natural upward draft. The natural draft was reported to successfully aerate windrows placed about 1.2 m deep.

With regard to aeration, the function of windrow turning is to assure that adequate free airspace is maintained. Ground refuse and grass, for example, are noted for their tendency to consolidate during composting. If left unchecked, FAS can decrease to the point where ventilation becomes inadequate. Periodic turning decreases the unit bulk weight of the mixture, increases free airspace and assures the highest possible ventilation rate for the particular particle sizes in the mixture. While turning itself does not supply adequate air quantities by the mechanism of volume exchange, the mechanical agitation produces conditions within the pile that enhance natural draft ventilation.

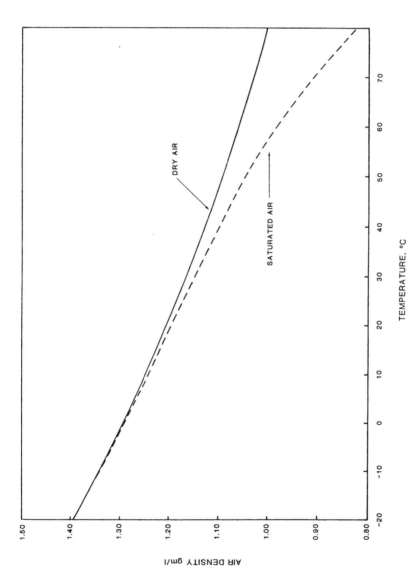

Figure 7.6.   Density of dry and saturated air as a function of dry bulb air temperature at 760 mm Hg pressure.

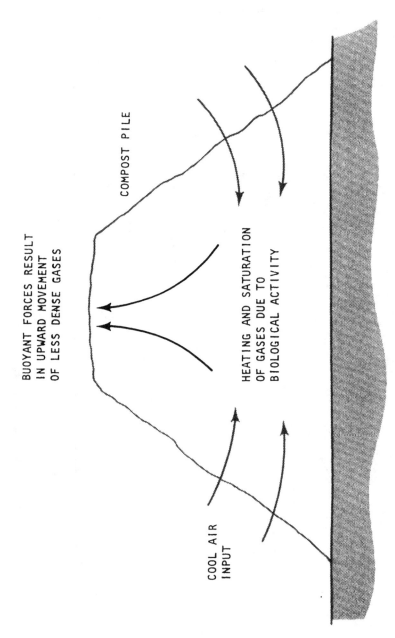

Figure 7.7.   Schematic illustration of natural draft ventilation induced by buoyant forces acting on hot, moist air produced during active composting.

SUMMARY

Air is required during aerobic composting for three purposes: (1) supplying oxygen for biological decomposition (stoichiometric demand), (2) removing moisture from the composting mass (drying demand), and (3) removing heat to control process temperatures (heat removal demand). The stoichiometric demand depends on the chemical composition of the organics. Values ranging from about 1.2 to 2.0 g O_2/g BVS are typical for most composting substrates, but values as high as 4.0 g O_2/g BVS are possible with highly saturated organics.

The moisture carrying capacity of air increases exponentially with increasing air temperature. Relative humidity of the inlet air is not a significant variable affecting drying if the exhaust gas temperature is 25°C or more above that of the inlet air.

The air requirements for drying and process temperature control are usually much greater than the requirement for biological oxidation. Either drying or temperature control can govern the air requirement depending on process conditions. With very wet substrates the requirement for moisture removal will tend to govern, whereas the requirement for heat removal will become dominate with dryer substrates. Only with relatively dry substrates and high exit air temperatures do the air requirements for drying and biological oxidation become equivalent. With substrates solids of 20%, the air requirement for drying can be as much as 10 to 30 times that for biological oxidation. The different air requirements for stoichiometry and moisture and heat removal allow control over the extent of drying by control of the air supply. Control over the extent of drying has a significant influence on the thermodynamic balance achieved during composting.

For a batch process, the quantity of air required can be converted to an average aeration rate by considering the time duration of the process. Peaking factors are significant in batch processes. Peak aeration rates can exceed the average rate by a factor of 3 to 5 or more. Peaking factors are less significant with continuous feed processes. Procedures presented in this chapter allow the designer to estimate the quantity of air required and the minimum, average, and peak aeration rates for various process conditions and feed substrates.

A number of control strategies are available to regulate air supply to the process. These range from simple manual control systems to more sophisticated feedback control loops using temperature, oxygen or carbon dioxide as the controlled variables and air supply as the manipulated variable. Aeration supply and control systems must be designed to operate over the full range from minimum to peak flow conditions.

Forced ventilation by means of blowers or fans is the major aerating mechanism in the aerated static pile and most reactor systems. In nonaerated windrow or pile systems, however, the mechanisms of aeration are not as obvious. Required air volumes to supply the stoichiometric oxygen requirements are significantly greater than can be accounted for by periodic mechanical turning, even if practiced on a daily or more frequent basis. Theoretical modeling and field observations suggest that natural draft ventilation is the most significant aeration mechanism in the non-aerated windrow process.

Natural ventilation occurs as a result of the density difference between warm, moist gases contained within the windrow and cooler, less moist, ambient air. The rate of ventilation is a function of the density difference between inside and outside gases, particle size, and free air space of the composting mixture. Natural ventilation is enhanced by increasing the density difference (i.e., higher internal temperatures), increasing the free air space (i.e., reduced bulk weight and moisture content), and increasing the particle size. The role of mechanical turning is to increase the mixture free air space to assure the highest possible ventilation rate for the particular composting mixture.

REFERENCES

1. McCarty, P. L. "Anaerobic Processes," paper presented at the Birmingham Short Course on Design Aspects of Biological Treatment, International Association of Water Pollution Research, England, (1974).
2. Haug, R. T., Tortorici, L. D., and Raksit, S. K. *Sludge Processing and Disposal — A State of the Art Review* (Whittier, CA: LA/OMA Project, 1977).
3. Corey, R. C., Ed. *Principles and Practices of Incineration* (New York: Wiley-Interscience, 1969).
4. Kayhanian, M. and Tchobanoglous, G. "Computation of C/N Ratios for Various Organic Fractions," *BioCycle* 33(5) (May 1992).
5. Rich, L. G. *Unit Processes of Sanitary Engineering* (New York: John Wiley & Sons, Inc., 1963).
6. Haug, R. T. and Haug, L. A. "Sludge Composting — A Discussion of Engineering Principles," *Compost Sci.* (November/December 1977 and January/February 1978).
7. Golueke, C. G. *Biological Reclamation of Solid Waste* (Emmaus, PA: Rodale Press, 1977).
8. Lee, J. F. and Sears, F. W. *Thermodynamics* (Reading, MA: Addison-Wesley Publishing Co., Inc., 1963).
9. Weber, J. H. "Vapor Pressure vs Temperature," in *Microcomputer Programs for Chemical Engineers* (New York: McGraw-Hill Publications Co., 1984).
10. Wiley, J. S. "Studies of High-Rate Composting of Garbage and Refuse," *10th Annual Purdue Industrial Waste Conference* (West Lafayette, IN: Purdue University Press, 1955).
11. Jeris, J. S. and Regan, R. W. "Controlling Environmental Parameters for Optimum Composting, Part I: Experimental Procedures and Temperature," *Compost Sci.* (January/February 1973).
12. Schulze, K. L. "Continuous Thermophilic Composting," *Compost Sci.* (Spring 1962).
13. Schulze, K. L. "Rate of Oxygen Consumption and Respiratory Quotients During the Aerobic Decomposition of a Synthetic Garbage," *Compost Sci.* (Spring 1960).
14. Snell, J. R. "Some Engineering Aspects of High-Rate Composting," *J. San. Eng. Div., ASCE* Paper 1178 (1957).
15. Haug, R. T. "Modeling of Compost Process Dynamics," *Proceedings of the National Conference on Composting of Municipal and Industrial Sludges* (Silver Spring, MD: Hazardous Materials Control Research Institute, May 1982).
16. Murray, C. M. and Thompson, J. L. "Strategies for Aerated Pile Systems," *Biocycle, J. Waste Recycling* 27(6) (July 1986).
17. Williams, T. O. and North, O. "Performance Testing of an Innovative Sludge Composting Aeration System in Nashville, TN," WPCF 64th Annual Conference, Paper No. AC91-018-003, Toronto (October 1991).
18. Smith, D. Los Angeles County Sanitation Districts, personal communication (1977).
19. Iacoboni, M., LeBrun, T., and Livingston, J. "Composting Study," Technical Services Department, Los Angeles County Sanitation Districts (1979).
20. Maier, P. P., Williams, E. R., and Mallison, G. F. "Composting Municipal Refuse by the Aeration Bin Process," *Proceedings of the 12th Purdue Industrial Waste Conference* (West Lafayette, IN: Purdue University Press, 1957).
21. McGauhey, P. H. and Gotaas, H. G. "Stabilization of Municipal Refuse by Composting," *Trans. ASCE* Paper No. 2767.
22. Golueke, C. G. *Composting — A Study of the Process and Its Principles* (Emmaus, PA: Rodale Press, 1972).
23. Shell, B. J., "The Mechanism of Oxygen Transfer Through a Composting Material," PhD Thesis, Civil and Sanitary Engineering Department, Michigan State University, East Lansing: (1955).
24. Haug, R. T. *Compost Engineering, Principles and Practice* (Lancaster, PA: Technomics Publishing Co., 1980).

CHAPTER 8

Feed Conditioning — Energy

INTRODUCTION

The first law of thermodynamics states that energy can neither be created nor destroyed. Thus, energy that flows into a system must be fully accounted for either as energy stored within the system or energy that flows out of the system. In a composting process the major energy input is the organic molecules of the feed substrates. As these molecules are broken down by biological activity, energy is either transformed into new organic molecules within the microorganisms or is released as heat to the surroundings. Thus, it is energy release from organic decomposition that drives the composting process, causes temperature elevation and produces the drying that is so desirable with wet substrates. Indeed, it is the opportunity to capture a portion of this energy that causes microbes to decompose the organics in the first place.

This chapter examines the material and energy balances that govern the composting process and the feed conditioning necessary to provide sufficient energy to drive the process. The latter is termed "energy" conditioning. This chapter will show how composting can be divided into distinct thermodynamic regions: one in which sufficient energy is available for both composting and evaporative drying and another in which energy is sufficient only for composting with limited drying. It is essential that the feed substrates contain sufficient energy to accomplish the thermodynamic objectives of the process.

A simplified approach to material and energy balances is presented in this chapter. Closed form analytical solutions can be developed based on certain simplifying assumptions. These solutions can then be used to determine whether the feed mixture is properly conditioned from an energy standpoint. More detailed energy balances are pursued in Chapters 11 through 14.

HEAT AND MATERIAL BALANCES

Heat liberated from the decomposition of organics increases the temperature of solids, water, and air in the composting mixture. The released energy also drives the evaporation of water, which is carried from the process in the exhaust gases. Because the compost is at a higher temperature than the surroundings, heat losses will occur from exposed surfaces of the compost. These losses are mitigated to some extent by the insulating effect of the compost, which limits conduction of heat. Losses will also occur as windrows or piles are mechanically turned. Under equilibrium conditions, the process temperature will rise to a point where energy inputs are balanced by outputs. However, maximum composting temperatures are generally limited to about 75 to 85°C because rates of biological activity become seriously reduced by thermal inactivation at such temperatures. The highest temperatures observed by this author are 88 to 90°C reported at the circular, agitated bed system in Plattsburgh, New York.

A schematic diagram of the composting process showing the thermodynamic system boundary and major process inputs and outputs is presented in Figure 8.1. Major inputs are substrate, other amendments, and process air, and its associated water vapor. Major outputs are the compost product, dry exhaust gases, and exit water vapor. The heat inputs and outputs associated with these materials are included in the analysis. Heat loss to the surroundings is not included, but it is usually a small fraction of the other heat outputs. Note also that compost recycle and bulking agent recycle are not shown. These material flows are internal to the system boundary and, therefore, do not affect the overall system balance. They are important to the balances on internal stages within the system boundary and are discussed further in Chapter 11.

Some new mass terms are introduced in this section. The terms BVS, NBVS, ASH, and WAT were defined in Chapter 6. Additional mass and energy terms are defined as follows (use of these terms is presented in the following examples):

Mass Terms

NBVSO	=	nonbiodegradable volatile solids in the compost product
ASHO	=	ash component in the compost product
WATSO	=	water component in the compost product
DAIRI	=	dry air input to the process
WATVI	=	water vapor associated with the air input
WATP	=	water produced during organic decomposition
DGASO	=	dry gas output from the process
WATVO	=	water vapor associated with the gas output

Energy Terms

HSI	=	sensible heat in the substrate solids
HWI	=	sensible heat in the water component of the substrate
HDAIRI	=	sensible heat in the input dry air
HSWVI	=	sensible heat in the water vapor associated with the input air
HORG	=	heat release by biological oxidation of the substrate
HTOTI	=	total input energy
HSO	=	sensible heat in the output product solids
HWO	=	sensible heat in the output water contained in the compost product
HDGASO	=	sensible heat in the output dry gases
HSWVO	=	sensible heat in the water vapor associated with the output gases
HLWVO	=	latent heat difference between the output water vapor and the input water vapor
HTOTO	=	total output energy
CPWAT	=	specific heat of water, taken as 1.00 cal/g-°C or Btu/lb-°F
CPSOL	=	specific heat of solids, taken as 0.25 cal/g-°C or Btu/lb-°F
CPGAS	=	specific heat of dry gases, taken as 0.24 cal/g-°C or Btu/lb-°F

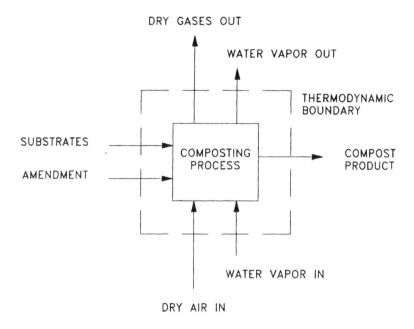

DRY GASES OUT

WATER VAPOR OUT

THERMODYNAMIC
BOUNDARY

SUBSTRATES

AMENDMENT

COMPOSTING
PROCESS

COMPOST
PRODUCT

WATER VAPOR IN

DRY AIR IN

Figure 8.1.   Thermodynamic system boundary and major material inputs and outputs to the composting process.

CPWATV	=	specific heat of water vapor, taken as 0.44 cal/g-°C or Btu/lb-°F
H_s	=	higher heat of combustion of the substrate, cal/g or Btu/lb of organics oxidized
H_a	=	higher heat of combustion of the amendment, cal/g or Btu/lb of organics oxidized

Composting of Wet Substrates

With wet substrates it is usually desirable to evaporate water during composting to produce a relatively dry product. The evaporative burden with wet substrates is considerably greater than with dry substrates. Therefore, the energy available from wet substrates is very critical because of the large energy burden placed on the system to support evaporation. The mass and energy balances for composting with drying are explored in the following examples.

Example 8.1

A wet substrate at 20% TS is to be composted. Compost recycle is used for conditioning with no other amendment or bulking agent added. The substrate is 80% VS ($V_s = 0.80$) with an estimated degradability of 60% ($k_s = 0.60$) and a composition of $C_{10}H_{19}O_3N$. A 60% TS content is desired in the compost product. Inlet air is 20°C and 75% relative humidity. The compost process will operate at 55°C. Assume the process gases are exhausted at 55°C saturated. Develop a system mass balance for these conditions based on a substrate feed rate of 5 kg/day (1 kg/day of substrate solids).

Solution

1. Using Equations 6.23 to 6.26 the components of the feed substrate are

$$
\begin{array}{llll}
\text{BVS} & = & (0.60)(0.80)(0.20)(5) & = & 0.480 \text{ kg/day} \\
\text{NBVS} & = & (1-0.60)(0.80)(0.20)(5) & = & 0.320 \\
\text{ASH} & = & (1-0.80)(0.20)(5) & = & 0.200 \\
\text{WAT} & = & 5-(0.20)(5) & = & 4.000 \\
\end{array}
$$

Substrate feed   =   5.000 kg/day

2. The weight of compost product can be determined by assuming that all BVS are lost during the composting process and that the NBVS and ASH fractions are conservative. Therefore,

$$
\begin{array}{llll}
\text{NBVSO} & = & \text{NBVS} & = & 0.320 \text{ kg/day} \\
\text{ASHO} & = & \text{ASH} & = & 0.200 \\
\text{Product solids} & & & = & 0.520 \text{ kg/day}
\end{array}
$$

The product solids are assumed to be 60% TS. The water component is

$$WATSO = (0.520 / 0.60) - 0.520 = 0.347 \text{ kg} / \text{day}$$

3. Water produced during composting WATP can be determined from the chemical formula for the substrate organics and assuming that all BVS are decomposed:

$$0.480$$

$$C_{10}H_{19}O_3N + 12.5\,O_2 = 10\,CO_2 + NH_3 + 8\,H_2O$$

$$201 \quad 12.5(32) \quad 10(44) \quad 17 \quad 8(18)$$

$$WATP = 8(18)(0.480) / 201 = 0.344 \text{ kg} / \text{day}$$

4. A mass balance on water gives

$$\text{water in} + \text{water produced} = \text{water out}$$
$$WAT + WATVI + WATP = WATVO + WATSO$$

Rearranging:

$$WATVO - WATVI = WAT + WATP - WATSO$$
$$WATVO - WATVI = 4.000 + 0.344 - 0.347 = 3.997 \text{ kg} / \text{day}$$

5. Using Equations 7.8 to 7.10 as shown in Example 7-4, the inlet and outlet gases have the following specific humidities:

$$\text{inlet air} = 0.0113 \text{ g water/g dry air}$$

$$\text{outlet gas} = 0.1147 \text{ g water/g dry gas}$$

6. Air required for moisture removal can be determined as follows. A water balance on the inlet and outlet gases gives the following equations:

$$\text{WATVO} = \text{DGASO} (0.1147)$$

$$\text{WATVI} = \text{DAIRI} (0.0113)$$

$$\text{DGASO}(0.1147) - \text{DAIRI}(0.0113) = 3.997$$

The quantity of dry exit gas DGASO will equal the inlet air, less oxygen consumed, plus the carbon dioxide and ammonia formed:

$$\text{DGASO} = \text{DAIRI} - \frac{0.480(12.5)(32)}{201} + \frac{0.480(10)(44)}{201} + \frac{0.480(17)}{201}$$

Solving the above simultaneous equations:

$$\text{DAIRI} = 38.501 \text{ kg/day}$$
$$\text{DGASO} = 38.638 \text{ kg/day}$$

7. The water vapor inputs and outputs are determined as

$$\text{WATVI} = 38.501(0.0113) = 0.435 \text{ kg/day}$$
$$\text{WATVO} = 38.638(0.1147) = 4.432 \text{ kg/day}$$

8. Check the total mass balance as follows:

In		Out	
BVS	0.480	NBVSO	0.320
NBVS	0.320	ASHO	0.200
ASH	0.200	WATSO	0.347
WAT	4.000	WATVO	4.432
WATVI	0.435	DGASO	38.638
DAIRI	38.501		
Total	43.936	Total	43.937

The total mass balance checks within the limits of roundoff errors. The water balance can be checked as follows:

In		Out	
WAT	4.000	WATSO	0.347
WATVI	0.435	WATVO	4.432
WATP	0.344		
Total	4.779	Total	4.779

The mass balance for the conditions of Example 8.1 is presented in Figure 8.2. Sufficient water is removed to produce a relatively dry compost product. Therefore, the conditions of Figure 8.2 can be described as "composting with drying". The question that now arises is whether sufficient energy is supplied by the feed substrate to accomplish both the composting and drying shown in Figure 8.2. If the substrate is "energy poor" then the conditions of Figure

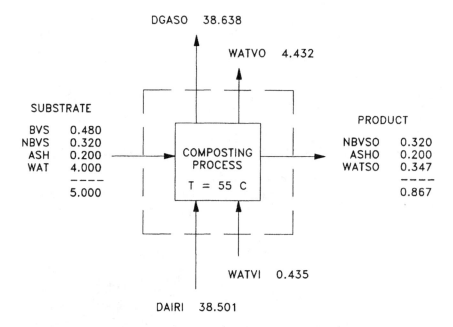

Figure 8.2.   Mass balance for the conditions of Example 8.1. The case is for a wet substrate with significant demands for drying. All units are kg/day and the balance is based on 1 kg/day of substrate solids.

8.2 may not be achievable. If this were the case and if the operator maintained the air supply as shown in Figure 8.2, process temperatures would fall. This reduces the evaporative heat loss, and the system will rebalance itself but at a lower process temperature. Alternatively, the operator could reduce the air supply to reduce the evaporative and other heat losses in the exhaust gas. Process temperatures could continue to be maintained, but drying would be limited. This illustrates the importance of knowing whether the feed substrate is "energy rich" or "energy poor" relative to the energy demands on the process.

The energy balance corresponding to Figure 8.2 is developed in the following example. Energy terms must be related to a datum or reference point for measurement. For convenience, a reference temperature TREF of 0°C (32°F) is used throughout this text. Further discussion of the reference datum and the assumed pathways for heat release is presented in Chapter 11.

Example 8.2.

Develop the energy balance for the conditions of Example 8.1. Assume substrate and air are input at 20°C and compost product and exhaust gases exit at 55°C.

Solution

1. The input energy terms HSI, HWI, HDAIRI, and HSWVI are determined using the form of Equation 3.8 as follows:

$$
\begin{aligned}
\text{HSI} &= 1.0(0.25)(20 - 0) &&= 5.0 \text{ kcal/day} \\
\text{HWI} &= 4.0(1.00)(20 - 0) &&= 80.0 \\
\text{HDAIRI} &= 38.501(0.24)(20 - 0) &&= 184.8
\end{aligned}
$$

$$\text{HSWVI} \quad = \quad 0.435(0.44)(20-0) \quad = \quad 3.8$$

The heat release during composting is determined by assuming that all BVS are decomposed with a unit heat release, H_s, of 5550 cal/g BVS:

$$
\begin{array}{lll}
\text{HORG} & = 0.480(5550) & = \underline{2664.0} \\
\text{HTOTI} & & = 2937.6 \text{ kcal/day}
\end{array}
$$

2. The output energy terms HSO, HWO, and HDGASO are determined as follows:

$$
\begin{array}{llll}
\text{HSO} & = & 0.520(0.25)(55-0) & = & 7.2 \text{ kcal/day} \\
\text{HWO} & = & 0.347(1.00)(55-0) & = & 19.1 \\
\text{HDGASO} & = 38.638(0.24)(55-0) & = & 510.0
\end{array}
$$

Referring back to Example 3.1, a path must be defined to determine the sensible and latent heats in the exhaust water vapor. A convenient path is to assume that input water vapor and liquid water, which is evaporated, are first heated to the composting temperature (55°C in this case) and then the water is evaporated at the composting temperature. Sensible heat demand for this pathway is determined as

$$
\begin{array}{lll}
\text{HSWVO} = 0.435(0.44)(55\text{--}0)+(4.432\text{--}0.435)(1.00)(55\text{--}0) & = & 230.4 \\
\text{HSWVO} & = & 230.4
\end{array}
$$

The latent heat of vaporization at 55°C is determined from steam tables to be 565.7 cal/g. The latent heat of the exhaust water vapor is

$$
\begin{array}{lll}
\text{HLWVO} & = (4.432\text{--}0.435)565.7 & = \underline{2261.1} \\
\text{HTOTO} & & = 3027.8 \text{ kcal/day}
\end{array}
$$

The total of energy output terms HTOTO is slightly greater than the total of input energy terms (HTOTI) in Example 8.2. This means that the mass balance in Figure 8.2 must be adjusted slightly to reduce HTOTO to balance that supplied by the substrate. As a practical matter, the air input could be reduced slightly to reduce the evaporative burden. This in turn would slightly reduce the output solids content. The solution point is achieved when both the mass and energy balances are closed simultaneously. Obviously, a trial and error approach is required to simultaneously close both the mass and energy balances. The reader should not despair at the prospect of doing trial and error solutions. The computer programs presented in Chapter 11 can manage such tedious calculations for the reader.

As an alternative to producing a wetter product, the substrate in Example 8.1 could be "energy conditioned" to increase the energy supply and close the balance. The most common approach to energy conditioning is to add other degradable, relatively dry substrates to increase the energy resources without significantly adding to the water burden. This approach to closing the energy balance is discussed later in this chapter.

Several lessons are apparent from the results of Examples 8.1 and 8.2. First, air supply is the dominant term in the mass balance, accounting for nearly 88% of the total mass of input materials. Second, the heat release from organic oxidation (HORG) is the major energy input term. Sensible heats contributed by input solids, water, and air are minor by comparison. Third, the latent heat of vaporization (HLWVO) is the dominant energy output term. With the wet substrate of Examples 8.1 and 8.2, latent heat accounts for about 75% of the total output energy. Sensible heat in the output dry gases (HDGASO) is the next largest term, accounting for about 17% of the total. Together HLWVO and HDGASO account for over 90% of the total energy demand.

Composting of Dry Substrates

The requirement for moisture removal is not as severe with dry substrates as it is with wet substrates. The reduced demand for water evaporation means that the air supply can be reduced toward the stoichiometric demand. Heat removal will be less in this case and very high process temperatures may result. The mass and energy balances associated with the composting of dry substrates are explored in the following examples.

Example 8.3

The company producing the wet substrate of Example 8.1 is considering the installation of additional dewatering equipment to increase the solids content to 45%. Develop the new mass balance for this situation assuming all other conditions remain the same.

Solution

1. To maintain consistency with Example 8.1, again assume 1 kg of substrate solids. The total weight of feed substrate is then $1/0.45 = 2.222$ kg. Using Equations 6.23 to 6.26 the components of the feed substrate are

BVS	=	$(0.60)(0.80)(0.45)(2.222)$	=	0.480 kg/day
NBVS	=	$(1 - 0.60)(0.80)(0.45)(2.222)$	=	0.320
ASH	=	$(1 - 0.80)(0.45)(2.222)$	=	0.200
WAT	=	$2.222 - (0.45)(2.222)$	=	1.222
Substrate feed			=	2.222 kg/day

2. The weight of compost product remains the same as in Example 8.1:

NBVSO	=	0.320 kg/day
ASHO	=	0.200
WATSO	=	0.347
Compost product	=	0.520 kg/day

3. Assuming that all BVS decompose, the water produced during composting will remain the same as Example 8.1. Therefore, WATP = 0.344 kg/day.

4. The water balance becomes

$$WATVO - WATVI = WAT - WATSO + WATP$$
$$WATVO - WATVI = 1.222 + 0.344 - 0.347 = 1.219 \text{ kg/day}$$

5. The inlet and outlet gases have the same specific humidities determined in Example 8.1. A water balance on the inlet and outlet gases gives the following equations:

$$\text{WATVO} = \text{DGASO} (0.1147)$$
$$\text{WATVI} = \text{DAIRI} (0.0113)$$

$$\text{DGASO}(0.1147) - \text{DAIRI}(0.0113) = 1.219$$

The quantity of dry exit gases (DGASO) will equal the inlet air, less oxygen consumed, plus the carbon dioxide and ammonia formed:

$$\text{DGASO} = \text{DAIRI} - \frac{0.480(12.5)(32)}{201} + \frac{0.480(10)(44)}{201} + \frac{0.480(17)}{201}$$

Solving the above simultaneous equations:

$$\text{DAIRI} = 11.634 \text{ kg/day}$$
$$\text{DGASO} = 11.771 \text{ kg/day}$$

6. The water vapor inputs and outputs are determined as

$$\text{WATVI} = 11.634(0.0113) = 0.131 \text{ kg/day}$$
$$\text{WATVO} = 11.771(0.1147) = 1.350 \text{ kg/day}$$

7. Check the total mass balance as follows:

In		Out	
BVS	0.480	NBVSO	0.320
NBVS	0.320	ASHO	0.200
ASH	0.200	WATSO	0.347
WAT	1.222	WATVO	1.350
WATVI	0.131	DGASO	11.771
DAIRI	11.634		
Total	13.987	Total	13.988

The water balance can be checked as follows:

In		Out	
WAT	1.222	WATSO	0.347
WATVI	0.131	WATVO	1.350
WATP	0.344		
Total	1.697	Total	1.697

7. The stoichiometric oxygen demand can be determined from the balanced oxidation reaction presented in Example 8.1:

$$\text{Oxygen demand} = 0.480(12.5)(32) / 201 = 0.955 \text{ kg O}_2 / \text{day}$$

The stoichiometric air demand is determined by recalling that air is 23.2% oxygen by weight:

$$\text{Air demand} = \frac{0.955}{0.232} = 4.116 \text{ kg air} / \text{day}$$

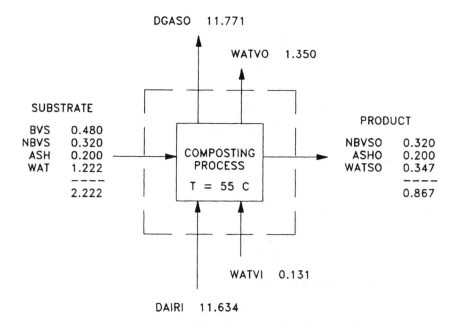

Figure 8.3.   Mass balance for the conditions of Example 8.3. The case is for a relatively dry substrate with limited demands for moisture removal. All units are kg/day and the balance is based on 1 kg/day of substrate solids.

8. Recall that the excess air ratio EAR is defined as the ratio of actual air supply to the stoichiometric demand. The EAR for this case is

$$EAR = 11.634 / 4.116 = 2.83$$

Even though the substrate in Example 8.3 is relatively dry, the air requirement for moisture removal is several times greater than the stoichiometric demand. However, the EAR of 2.83 is about as low as would normally be maintained in any composting system. The average oxygen content in the exhaust gas would be only ~13.6% with an EAR of 2.83. Further reducing the oxygen content would increase the possibility of oxygen rate limitations. Therefore, the mass balance presented in Example 8.3 should be reasonably representative of "composting with limited drying".

The mass balance for the conditions of Example 8.3 is presented in Figure 8.3. The demand for evaporative drying is greatly reduced compared to the wet substrate of Example 8.1. Nevertheless, air supply remains the dominant mass term accounting for over 83% of the total mass input. The energy balance for the conditions of Figure 8.3 is considered in Example 8.4.

Example 8.4

Develop the energy balance for the conditions of Example 8.3. Other conditions remain the same as for Example 8.2.

Solution

1. The input energy terms HSI, HWI, HDAIRI, and HSWVI are determined as

HSI	=	1.0(0.25)(20 – 0)	=	5.0 kcal/day
HWI	=	1.222(1.00)(20 – 0)	=	24.4
HDAIRI	=	11.634(0.24)(20 – 0)	=	55.8
HSWVI	=	0.131(0.44)(20 – 0)	=	1.2

Heat release is the same as for Example 8.2

HORG	=	0.480(5550)	=	2664.0
HTOTI			=	2750.4 kcal/day

2. The output energy terms are determined as follows:

HSO	=	0.520(0.25)(55 – 0)	=	7.2 kcal/day
HWO	=	0.347(1.00)(55 – 0)	=	19.1
HDGASO	=	11.771(0.24)(55 – 0)	=	155.4
HSWVO	=	0.131(0.44)(55 – 0) +		
		(1.350–0.131)(1.00)(55 – 0)	=	70.2
HLWVO	=	(1.350–0.131)565.7	=	689.6
HTOTO			=	941.5 kcal/day

The energy balance for the case of a relatively dry substrate differs markedly from that for a wet substrate. Total energy demand (HTOTO) is much less than that available from the substrate. No further energy conditioning is required for the substrate of Examples 8.3 and 8.4. Because heat release is so much greater than heat demand, changes in the conditions of Figure 8.3 would be necessary to reach a solution point where both the mass and energy balances are closed. Process temperatures are likely to exceed the assumed 55°C because so much additional heat is available. The higher process temperatures will cause an increase in the evaporative loss of water and result in a dryer product. The combination of high temperature and lower moisture content could impose kinetic limitations to the extent that the assumption of complete BVS loss is no longer valid. Methods to control the process to avoid such limitations and close both mass and energy balances are discussed in Chapters 11 to 14.

Several lessons are apparent from the results of Examples 8.3 and 8.4. First, air supply remains the dominant term in the mass balance even for relatively dry substrates. Second, the energy balance is not as critical as with wet substrates because the evaporative burden is greatly reduced. Third, despite the reduction in the evaporative burden the latent heat of vaporization (HLWVO) remains the dominant energy demand. With the relatively dry substrate of Examples 8.3 and 8.4, the latent heat accounts for ~73% of the total energy demand.

ENERGY FACTORS

If energy supply is less than demand, the feed mixture must be energy conditioned to remove the thermodynamic limitation. Several factors and rules of thumb were developed by Haug and Haug[1] and Haug[2] to judge the need for energy conditioning, but without the need to conduct mass and energy balances of the type presented in Examples 8.1 through 8.4.

The Water Ratio, W

The assumptions of Examples 8.1 and 8.2 are reasonably typical of conditions expected with composting and drying of wet substrates. A rule of thumb termed the water ratio, W, can be developed from the results of Examples 8.1 and 8.2. W is defined as the ratio of water to degradable organics. Defining W on the basis of water content is a rational approach because evaporation represents most of the total energy demand. For the conditions of Example 8.1, produced water (WATP) nearly equals that in the compost product (WATSO). Therefore, water input with the substrate WAT (4.0 kg) is nearly equal to that evaporated. Substrate BVS was 0.48 kg. Using these values the W ratio is about 4.0/0.48 = 8.3.

Considering the assumptions involved, a W value of about 8 to 10 g H_2O/g BVS can be used to judge the thermodynamic characteristics of the composting process. Because water evaporation is the major energy use, and as long as moisture in the substrate is the major water input, the factor should apply equally to windrow, aerated pile, and reactor systems. If W < 8, sufficient energy should be available for temperature elevation and water evaporation. If W > 10, the substrate alone may not provide sufficient energy. Lower process temperatures or less drying can be expected.

To determine W for a particular substrate, consider the process diagram of Figure 6.7. The weight of water in feed substrate and amendments is given by

$$\text{weight water} = (X_s - S_s X_s) + (X_a - S_a X_a) \tag{8.1}$$

The quantity of degradable organics BVS in the mixture is given by

$$BVS = k_s V_s S_s X_s + k_a V_a S_a X_a \tag{8.2}$$

Combining Equations 8.1 and 8.2:

$$W = \frac{\text{weight of water}}{\text{weight degradable organics}}$$

$$W = \frac{(X_s - S_s X_s) + (X_a - S_a X_a)}{k_s V_s S_s X_s + k_a V_a S_a X_a} \tag{8.3}$$

The rationale for use of total infeed water in Equation 8.3, instead of the amount actually evaporated, is that most of the infeed water is evaporated with wet substrates and the calculations are thus simplified. It should be remembered that W is intended only as a tool or rule of thumb to judge the thermodynamic characteristics of the composting process. Thus, some compromise in technical accuracy is justified to give a tool that can be applied easily. W is not a substitute for a complete energy balance.

The Energy Ratio, E

One limitation with the W ratio is the assumption that all substrate organics have the same heat value. Of course, substrates can vary significantly in heat content depending on their composition. This limitation can be removed by including the heat content, H, in the analysis. Again referring to Examples 8.1 and 8.2, the heat released from the 0.48 Kg of BVS at 5550 kcal/kg is 2664 kcal. Dividing this value by the 4.0 Kg of substrate water gives a value of 666 cal/g H_2O. The latter is termed E or the energy ratio defined as follows:

$$E = \frac{\text{heat released}}{\text{weight of water}}$$

Substituting terms from Equations 8.1 and 8.2:

$$E = \frac{(k_s V_s S_s X_s)H_s + (k_a V_a S_a X_a)H_a}{(X_s - S_s X_s) + (X_a - S_a X_a)}$$ (8.4)

Considering the assumptions involved, an E ratio of about 700 cal/g water (1260 Btu/lb) can be used as a measure of the energy content of the substrate. If E is greater than 700 cal/g sufficient energy should be available for both composting and drying. If E is <600 cal/g, drying may have to be reduced to maintain process temperatures. Like its cousin the W ratio, the E ratio is a rule of thumb that compromises some technical accuracy to provide a usable tool.

Example 8.5

A municipal agency produces both raw and digested biosolids at 20% TS. The raw sludge is 80% VS with a 65% degradability. The digested sludge is 55% VS with 45% degradability. The heat value for both sludge types is 5550 cal/g VS. Estimate the W and E ratios for both sludges.

Solution

1. Assume 1 g of dry solids for each sludge type, equal to 1/0.2 = 5 g wet weight. W and E for the raw sludge are determined from Equations 8.3 and 8.4:

$$W = [5 - (0.20)(5)] / [0.65(0.80)(0.20)(5)]$$

$$W = 7.69 \qquad\qquad (W < 8)$$

$$E = [0.65(0.80)(0.20)(5)](5500) / [5 - (0.20)(5)]$$

$$E = 720 \text{ cal/g} \qquad\qquad (E > 700)$$

The raw sludge meets both rules of thumb.

2. W and E for the digested sludge are similarly calculated:

$$W = [5 - (0.20)(5)] / [0.45(0.55)(0.20)(5)]$$

$$W = 16.2 \qquad\qquad (W > 8)$$

$$E = [0.45(0.55)(0.20)(5)](5500) / [5 - (0.20)(5)]$$

$$E = 343 \text{ cal/g} \qquad\qquad (E < 700)$$

The digested sludge does not meet either of the rules of thumb.

3. These two sludge types are markedly different in their energy contents. The digested biosolids would require further energy conditioning to accomplish both composting and drying. The raw biosolids, however, should have sufficient energy for both composting and drying without supplemental energy amendments.

CONTROLLING THE ENERGY BALANCE

If a substrate or mixture of substrates does not contain sufficient energy to drive the composting process, further conditioning is required to control the energy balance. In such a case, the energy budget can be controlled by one or more of the following approaches: (1) limiting drying during composting, (2) reducing the substrate water content, and (3) adding supplemental energy amendments.

Control by Limiting Drying

One approach to controlling the energy balance is to reduce the evaporation of water and, thereby, reduce the total heat demand. This can be accomplished by controlling and reducing the air supply to limit evaporation. The problem with this approach is that reducing the drying potential removes one of the major advantages of composting, the production of a relatively dry product. With wet substrates the drying usually must be reduced to the point that the final product is still very wet. This can be seen by examining the digested sludge case in Example 8.5. The digested sludge solids contain about 1373 cal/g. Using 700 cal/g of water, about 1373/700 = 1.96 g water evaporation could be supported. The final product would contain about 4.0 g from the original substrate plus about 0.18 g produced water minus the 1.96 g evaporated. This gives about 2.22 g water in a final compost quantity of $[1 - 0.45(0.55)(1)] = 0.75$ gms. The solids content of the product would be about $0.75/(2.22 + 0.75) = 0.25$ or ~25% TS to close the energy balance. With such a high moisture content, the final product would be unsuitable for reuse in most cases.

Control of the energy balance by limiting drying is a viable approach for wet substrates only if provision is made to adjust the final product moisture content by one or both of the following methods. First, moisture removal by air drying can be achieved in arid climates with open systems, such as the windrow process. In such cases, a dry product can sometimes be achieved, even though the starting sludge may not contain sufficient energy for complete moisture removal. Alternatively, heat drying could be used to dry the compost product. Second, a dry, stable structural amendment can be added to produce an acceptable final product. Some sawdusts meet these criteria, for example. While the amendment may add to the energy balance, it also serves to condition the final compost to an acceptable moisture content.

Improved Dewatering

A second approach to closing the energy balance is to reduce the water content of the substrate. With municipal and industrial sludges this means achieving higher cake solids from dewatering. This approach is examined in the following example.

Example 8.6

The municipal agency in Example 8.5 plans to install improved dewatering equipment for the digested biosolids. A 35% cake is expected. Determine the W and E ratios assuming that all other conditions remain the same.

Solution

1. Assuming 1 g of solids, the cake wet weight is $1/0.35 = 2.86$ g. The W and E ratios become

$$W = [2.86 - (0.35)(2.86)]/[0.45(0.55)(0.35)(2.86)]$$

$$W = 7.5 \qquad\qquad (W < 8)$$

$$E = [0.45(0.55)(0.35)(2.86)](5500)/[2.86 - (0.35)(2.86)]$$

$$E = 740 \text{ cal/g} \qquad\qquad (E > 700)$$

2. The 35% TS digested sludge cake satisfies the rules of thumb and should contain sufficient energy for both composting and drying.

With wet substrates such as municipal biosolids, maximizing cake solids from dewatering is one of the most effective approaches to controlling the energy balance. Dewatering and composting are integrally related. Success in sludge composting depends heavily on the ability to achieve dry cake solids. Both the designer and operator should consider dewatering as part of the feed conditioning process.

Product Recycle and the Energy Budget

The use of product recycle for structural conditioning was discussed in Chapter 6. Product recycle is a proven method of providing structural conditioning of wet substrates. However, recycle of compost product does not add new BVS to the system. Material recycle is an internal loop within the system boundary and therefore does not add to the system energy balance. Any BVS remaining in the recycle are contributed by the original substrate or amendment. Since the latter are already included in the balances presented in this chapter, including them again would cause a double accounting. Therefore, product recycle is useful for structural conditioning but not energy conditioning.

Amendment Addition

Case studies using amendment alone and amendment with product recycle for purposes of structural conditioning were presented in the bar diagrams of Figures 6.12 and 6.13. The use of amendment alone can provide effective structural conditioning, but at the cost of using considerable amounts of amendment. Product recycle can also provide effective structural conditioning, but it does not contribute new organics to the system. A compromise between use of product recycle alone (potential energy problem) and use of amendment alone (quantity problem) is possible. If recycle is used for most of the structural conditioning, then smaller quantities of amendment can be added for control of the energy balance. In other words, amendment is added for energy conditioning and recycle added as necessary for further structural conditioning.

Equations for estimating the quantity of amendment required for energy conditioning can be developed based on certain simplifying assumptions. Based on the above examples, the latent heat of vaporization usually represents about 70 to 80% of the total energy demand. The system energy demand can be estimated with sufficient accuracy by multiplying the pounds of water evaporated by a factor termed H_e. The latter is an estimate of the system energy demand expressed as cal/g of water evaporated. H_e is typically in the range of 700 to 850 cal/g water evaporated (1260 to 1530 Btu/lb).

Referring to the process diagram of Figure 6.7, there are three unknowns in the problem: X_r, X_a, and X_m. Three equations are required for solution and can be developed from balances on water, solids, and energy across the system boundary in Figure 6.7.

Water balance
water in $=$ $(X_s - S_s X_s) + (X_a - S_a X_a)$
water out $=$ $(X_p - S_p X_p) = X_p (1 - S_p)$
water evap $=$ water in − water out

$$\text{water evap} = (X_s - S_s X_s) + (X_a - S_a X_a) - X_p (1 - S_p) \tag{8.5}$$

Note that Equation 8.5 does not include the water of formation WATP, which is relatively small and neglected in this analysis.

Solids balance
solids remaining $=$ solids in − solids degraded

$$S_p X_p = S_s X_s + S_a X_a - (k_s V_s S_s X_s) - (k_a V_a S_a X_a) \tag{8.6}$$

Heat balance
heat in $=$ heat out

$$(k_s V_s S_s X_s) H_s + (k_a V_a S_a X_a) H_a = (\text{water evap})(H_e) \tag{8.7}$$

Substituting Equation 8.6 into 8.5 for X_p, substituting Equation 8.5 into 8.7, and rearranging gives the following equation for X_a:

$$X_a = \frac{H_e[(X_s - S_s X_s) - (S_s X_s - k_s V_s S_s X_s)(1/S_p - 1)] - H_s(k_s V_s S_s X_s)}{H_a(k_a V_a S_a) + H_e[(S_a/S_p - 1) - (k_a V_a)(S_a/S_p - S_a)]} \tag{8.8}$$

Once X_a is calculated from Equation 8.8, the mixture moisture content resulting from sludge and amendment alone can be determined as

$$S_m = (S_s X_s + S_a X_a)/(X_s + X_a) \tag{8.9}$$

Let SMMIN = the minimum acceptable mixture solids content, usually between about 0.35 and 0.45. Then,

$$\text{if } S_m \geq \text{SMMIN, then } X_r = 0$$
$$\text{if } S_m < \text{SMMIN, then recycle must be added}$$

The quantity of recycle to be added if $S_m < \text{SMMIN}$ is determined as

$$\text{SMMIN} = (S_s X_s + S_a X_a + S_r X_r) / (X_s + X_a + X_r)$$

solving for X_r:

$$X_r = \frac{\text{SMMIN}(X_s + X_a) - S_s X_s - S_a X_a}{(S_r - \text{SMMIN})} \qquad (\text{for } X_r > 0) \qquad (8.10)$$

Example 8.7

10 dtpd of the digested sludge cake in Example 8.5 is to be composted to produce a 60% TS product. The sludge will be conditioned with amendment and recycled product. Sawdust is the energy amendment and is to be added in sufficient quantity to close the energy balance. Recycled product will provide additional structural conditioning as necessary to reach a minimum mixture solids content of 40%. The sawdust is 70% TS, 95% VS, and 40% degradable. H_e is estimated at 1500 Btu/lb water evaporated. H_s is 10,000 Btu/lb of sludge organics and H_a 6500 Btu/lb of sawdust organics. Determine the required quantities of amendment and product recycle.

Solution

1. X_s is 10/0.20 = 50 tpd.

2. X_a is determined from Equation 8.8 as

$$X_a = \frac{1500\{[50 - 0.20(50)] - [0.20(50) - 0.45(0.55)(0.20)(50)]}{6500[0.40(0.95)(0.70)] + 1500\{[(0.70/0.60) - 1] - }$$
$$\frac{[(1/0.60) - 1]\} - 10000[0.45(0.55)(0.20)(50)]}{[0.40(0.95)[(0.70/0.60) - 0.70]]\}}$$

$$X_a = 16.2 \text{ tpd}$$

3. Using Equation 8.9 the mixture solids content from the substrate and amendment is

$$S_m = \frac{0.20(50) + 0.70(16.2)}{50 + 16.2} = 0.322 = 32.2\% \text{ TS}$$

4. Because S_m is less than the desired SMMIN of 0.40 (40% TS) recycle of product is necessary to provide additional structural conditioning. The required X_r is determined from Equation 8.10:

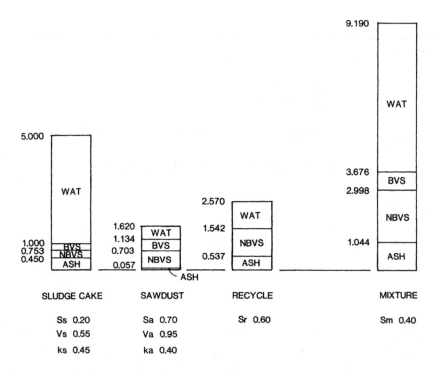

Figure 8.4.   Bar diagram of the substrate/amendment/recycle mixture for conditions of Example 8.7. The mixture is properly conditioned from both standpoints of energy and structure.

$$X_r = \frac{0.40(50 + 16.2) - 0.20(50) - 0.70(16.2)}{(0.60 - 0.40)}$$

$$X_r = 25.7 \text{ tpd}$$

5. A bar diagram showing the mixture and individual components is presented in Figure 8.4. The mixture is properly conditioned in terms of both structure and energy. The mixture should contain sufficient energy resources to accomplish the objective of composting with drying.

Use of amendment for energy conditioning and product recycle for additional structural conditioning represents an optimal approach to adjusting the composition of the infeed mixture.[3] Amendment use is reduced to the minimum required to satisfy the energy balance. Free air space is then adjusted as necessary using product recycle, which is available at no cost to the operator.

SUMMARY

Energy released by organic decomposition is the driving force for organic stabilization, temperature elevation, and moisture evaporation, all of which are desirable aspects of composting. Therefore, application of thermodynamic principles is a fundamental method for analyzing composting systems. Composting can be divided into two thermodynamic regions based on the energy and moisture contents of the substrate: one in which sufficient energy is available for both composting and drying and another in which energy is sufficient for composting with only limited drying. "Energy" conditioning of the feed mixture is a key point of process control for the system designer and operator.

The energy balance with dry substrates differs uniquely from that for wet substrates. The added evaporative burden with wet substrates imposes additional energy demands that must be provided by the substrates. With wet substrates, particular attention must be paid to assure proper energy conditioning. With dry substrates, an excess of energy supply can result in high process temperatures and over drying, both of which must be corrected by plant operations.

Air supplied to the process and exhaust gases removed from the process are the dominant terms in the mass balance. This is true even with relatively dry substrates for which the air supply is reduced to an excess air ratio of only 2 to 3. Heat released from organic oxidation is the major energy input term. The sensible heat contributed by input solids, water, and air is minor by comparison. The latent heat of vaporization is the dominant output energy term, even with relatively dry substrates where the evaporative burden is reduced. The latent heat usually accounts for about 75% of the total output energy.

Two rules of thumb are presented to determine the requirements for energy conditioning without the need to conduct complete mass and energy balances. Each is based on the fact that the latent heat of vaporization is the major energy output. The water ratio, W, is defined as the ratio of substrate water to degradable organics. The energy ratio, E, is defined as the heat released per unit weight of substrate water. If $W < 8$ or $E > 700$ cal/g sufficient energy should be available for both composting and moisture evaporation. At $W > 10$ or $E < 600$ cal/g, composting temperatures can be maintained only if drying is limited by control of the air supply. Both W and E are rules of thumb that compromise some technical accuracy to provide a tool that is easy to use.

If a substrate or mixture of substrates does not contain sufficient energy to drive the composting process, further conditioning is required to control the energy balance. In such a case, the energy budget can be controlled by one or more of the following approaches: (1) limiting drying during composting by controlling and reducing the air supply, (2) reducing the substrate water content by improved dewatering, and (3) adding supplemental energy amendments. If the energy balance is controlled by reducing the air supply, provision must be made to either dry the final product or add moisture absorbing amendments.

Adding supplemental energy amendments is a favored approach to closing the energy balance. Use of degradable amendments for energy conditioning and product recycle for additional structural conditioning represents an optimal approach to adjusting the composition of the infeed mixture. Amendment use is reduced to the minimum required to satisfy the energy balance. Free air space can then be adjusted as necessary using product recycle, which is available at no cost to the operator.

With municipal biosolids, the cake solids from dewatering is the critical parameter that determines the thermodynamic region of operation. Using typical values for degradability, the W and E ratios should be in a range where energy is sufficient for both composting and drying, with cake solids above about 30 to 35% for digested biosolids and 20 to 25% for raw biosolids.

The dewatering of wet substrates such as biosolids is integrally related to the composting process. The two unit processes should not be viewed separately because success in composting depends heavily on the ability to achieve dry cake solids. The solids content produced during dewatering is probably the single most important variable in determining the successful composting of biosolids. Implementation of any composting system for biosolids cake should be coordinated with design of the dewatering process to obtain cake solids with sufficient energy to drive the composting process. If this cannot be accomplished by dewatering alone, the addition of supplemental energy amendments is the favored approach to "energy condition" the feed mixture.

REFERENCES

1. Haug, R. T. and Haug, L.A. "Sludge Composting: A Discussion of Engineering Principles, Parts I and II," *Compost Sci.* 18(6), 19(1) (November/December 1977 and January/February 1978).
2. Haug, R. T. "Engineering Principles of Sludge Composting," *J. Water Poll. Contr. Fed.* 5(8):2189 (August 1979).
3. Haug, R. T. "Composting Process Design Criteria, Part I — Feed Conditioning," *Biocycle* 27(7) (August 1986).

CHAPTER 9

Substrate Biodegradability

INTRODUCTION

Substrate biodegradability was used in previous equations to determine the amount of substrate energy available to drive the composting process, the requirement for additional energy amendments, the stoichiometric oxygen demand, the air demand to remove heat from the process, and the final product mass. There is no factor more significant to the design and successful operation of a composting system.

In other areas of sanitary engineering practice, the design of biological treatment processes is almost always done in conjunction with laboratory testing to determine the degradability of the subject waste. The use of BOD_5, BOD_L, COD, and other tests to determine waste degradability and associated first order, rate constants is a standard practice for biological process design. Unfortunately, this practice is less common with the design of composting systems. Expected heat and materials balances are often projected largely on the basis of assumed degradability, which is usually based on previous studies or literature values. As a result there is a tendency to repeat the use of an assumed degradability whether or not it is correct or applicable to the current situation.

Design of new composting systems is usually based on experiences with previous systems. Because composting is a relatively "forgiving" process, this practice has worked successfully in most cases. However, other biological treatment processes, such as activated sludge for sewage treatment, are also well established. Nevertheless, engineering firms would generally consider it imprudent to design a sewage treatment plant without knowledge of the influent sewage characteristics. The same should be true for the substrates used in a composting system.

The purposes of this chapter are to review available data on substrate biodegradability, discuss alternative techniques for measuring degradability, present experimental results using respirometric techniques, and review a case study where laboratory testing of substrate degradability helped in solving a rather severe operating problem.

Table 9.1.   Degradability of Certain Organic Substrates

Substrate	Degradability (% of VS)	Reference
Refuse (total organic fraction)	43–54	Gossett and McCarty,[1] Golueke,[2] Pfeffer and Liebman[3]
Garden debris	66	Klein[4]
Chicken manure	68	Klein[4]
Steer manure	28	Klein[4]
Garbage	66	Klein[4]

BACKGROUND DATA

Degradability coefficients for the various feed components were first defined in Chapter 6 as follows:

k_s = fraction of substrate volatile solids degradable under composting conditions
k_a = fraction of amendment volatile solids degradable under composting conditions
k_b = fraction of bulking agent volatile solids degradable under composting conditions

The coefficients are not defined in terms of total degradable organics, but in terms of the degradable fraction expected under composting conditions. Given enough time all organics could be expected to degrade, but this is not practical under the normal conditions of composting. Therefore, a functional definition is required.

The composition and degradability of some common organic substrates are presented in Tables 9.1 and 9.2. Many of the degradability measurements reported in the literature were made under anaerobic conditions for two primary reasons. First, much of the work was conducted to support development of anaerobic biogasification as an alternative energy resource. Second, nutritional studies on feed substrates for ruminate animals are usually conducted in a manner simulating the anaerobic rumen. While some differences may be expected in aerobic composting environments, the general range of values should be applicable.

In many cases the organic substrates and amendments used in composting systems are derived from plant materials. Wood, sawdust, tree trimmings, yard wastes, refuse, paper, and agricultural wastes are examples of biomass substrates directly derived from plant material. The various animal manures, paunch, and even municipal biosolids are indirectly derived from plant materials. These substrates have been termed "once used" biomass because they are composed largely of undigested plant or animal materials. Thus, the degradability of plant materials is of central importance when considering amendment additions.

Cellulose and hemicellulose are the major structural molecules used by plants. Each is a polymeric material composed primarily of simple sugar subunits and glucose in the case of cellulose, and glucose, xylose, galactose, and others in the case of hemicellulose. Cellulose comprises more than 50% of the total organic carbon (TOC) in the biosphere and is the major molecule used for providing structural support in the plant kingdom. Wood is often 45% cellulose and cotton nearly 95% cellulose. The cellulose contained in wood and other sources contains lignin and other binders in a very complex structure. Lignin is a complex, aromatic, three-dimensional polymer that retards decomposition of the cellulose. The term lignin is a collective term for a whole series of similar, large polymeric molecules. The retardation is thought to be primarily physical because the presence of lignin between cellulose fibrils decreases the available surface area and prevents ready access to the cellulose by invading microbes and their enzymes.[1,5]

Table 9.2.   Composition and Degradability of Municipal Refuse and its Components

Component	Percent of Total Dry Wt Organics	Degradability (%)
Cellulose		
Kraft processed	40	90
Mechanical pulp	15	50
Hemicellulose	10	70
Other sugars	10	70
Lignin	10	0
Lipids	8	50
Protein	4	50
Other (plastics, etc.)	3	0

Source: Gossett and McCarty.[1]

Lignocellulosic materials (woody biomass) have a common basic structure, but vary greatly in chemical composition. The cellulose is arranged in fibrils, a number of which are bundled together and surrounded by hemicellulose to form microfibrils. The latter are again surrounded by hemicellulose and then lignin. Lee[6] presented typical compositions for woody matter as ranging from 30 to 60% cellulose, 10 to 30% hemicellulose (polyoses), and 10 to 20% lignin. Cellulose provides strength and flexibility to the wood. Lignin cements the cellulose fibers together and protects against biological and chemical attack. Hemicellulose helps bonds lignin to the cellulose fibrils.

The degradability of lignocellulosic substrates during composting depends in large measure on the previous processing. A fairly high degree of decomposition is observed in paper products that have been chemically treated to remove lignin (Kraft and sulfite pulping for delignification). This includes most paper and cardboard products, except newsprint. Golueke[2] reported about 90% decomposition of cellulose in Kraft pulp during anaerobic digestion at 37°C for 30 days. This same value was reported by Gossett and McCarty[1] in Table 9.2. Newsprint is mechanically pulped or beaten. Mechanical pulping does not remove lignin but disrupts the structure, providing increased surface area and accessibility to the cellulose fibers. Mechanical pulp should have a digestibility somewhere between that of Kraft pulp and native wood. Golueke[2] observed that cellulose in newsprint was 50% decomposed during anaerobic digestion at 37°C for 30 days, whereas native wood showed little, if any, degradation of the cellulose.

Regin and Jeris[7] conducted continuous, aerobic composting studies on mixed refuse samples. A high degree of experimental control was maintained in the laboratory composters to achieve maximum composting rates. The mixed refuse was prepared by controlled mixing of kitchen and market wastes with newsprint. The "synthetic" refuse was then ground with a garden shredder so that no particles exceeded 0.64 cm (0.25 in.) in size. Ammonium chloride and phosphorus were added to supplement the available nutrients and moisture was adjusted to ~67%. The refuse typically contained 60 to 70% paper. The lab composters were run for about 10 to 30 days to establish steady state conditions. The hydraulic residence times (mass turnover time) was about 4 to 5 days. Volatile matter was reduced an average 45% during 24 experimental runs. The components of the total volatile matter decomposed as follows: lipids 86%, other volatile matter (mostly carbohydrates) 65%, fiber 30%, and protein 22%. This gives an average degradability coefficient for typical refuse of ~0.45, which is consistent with the range reported in Table 9.1.

Youngberg and Vough[8] studied the digestibility of Oregon grass straws. Digestibility was measured with an *in vitro* technique using rumen fluid from an animal conditioned on grass straw. The grass samples were milled to pass a 20-mesh screen prior to analysis. Results are presented in Table 9.3. In general, grasses with higher cell wall fractions exhibited lower

Table 9.3.   Digestibility of Oregon Grass Straws Based on *In Vitro* Testing with Rumen Fluid

Grass Species	Number of Samples	Digestibility (% of dry matter)	
		Range	Mean
Tall fescue	7	44.1–53.8	48.8
Ryegrass, perennial turf-type	13	42.8–55.9	48.2
Bluegrass	6	40.1–53.9	46.7
Bentgrass	10	37.9–50.7	43.0
Ryegrass, perennial forage-type	12	39.7–48.3	42.9
Ryegrass, annual	11	34.1–41.5	36.8
Fescue, chewings and red	6	27.3–38.9	34.9
Orchardgrass	14	28.2–42.0	34.7

Source: Youngberg and Vough.[8]

digestibility. This is consistent with forage models which suggest that the cell contents are essentially 100% digestible whereas the cell wall digestibility is low. The reported degradability coefficients for grasses ranged from about 0.35 to 0.49.

Klass et al.[9] studied the anaerobic decomposition of grass clippings collected from a residential lawn. The grass was predominately Kentucky bluegrass with some red fescue. Using batch experiments, the volatile solids reduction in a 45-day period ranged from 37.5 to 53.4%. The temperature of digestion varied from 30 to 60°C in these tests. These results are consistent with those reported in Table 9.3.

Klass et al.[9] also studied the effect of particle size on the decomposition of the grass clippings. Three particle sizes were studied: coarse grass 60 to 100 mm in length as received, medium grass passing a No. 10 sieve (2 mm openings) and retained on a No. 16 sieve (1.18 mm openings), and fine grass passing a No. 30 sieve (0.6 mm openings). Over a 45-day period there was no significant difference in digestibility for the different particle sizes. Volatile solids reduction ranged from 52 to 54% at 35°C for the three grass sizes. However, the rate of decomposition was significantly affected by particle size with the fine grass decomposing more rapidly during the initial time of the test runs. This was expected because of the higher surface area of the fine grass.

Poincelot[10] studied the decompostion of leaves amended with municipal sludge, industrial wastes, and other additives. Cellulose decomposition over a 200-day period ranged from 23 to 54%. The lowest decompositions (23%) were noted when leaves were blended with paper fiber. Leaves alone showed 31% cellulose loss. Leaves amended with sludges or inorganic nutrients showed over 40% cellulose loss. Cellulose breakdown was concluded to be the limiting step in the composting process. Poincelot concluded that sugar and starch are most readily utilized by the microbes. Lipids, or fats, are also readily degradable. Cellulose and hemicellulose are intermediate in their resistance to degradation, while lignin is the most resistant. The high content of cellulose and lignin were given as the reason for the relatively slow breakdown of wood and paper.

Allison[11] reported on the decomposition of wood and bark species when incubated in soil for 60 days. Results of these studies are presented in Table 9.4. There appears to be a rather dramatic difference in the decomposition of various wood species in the soil. Many of the softwood and softwood barks exhibited relatively low decomposition after 60 days in the soil. The decomposition for all softwood species averaged 12% and for softwood barks 8.2%. Hardwoods and hardwood barks showed significantly higher biodegradability. The decompo-

Table 9.4.   Decomposition of Woods and Barks Incubated in Soil for 60 Days[a,b]

Tree Species	% Wood Carbon Released as CO_2	% Bark Carbon Released as CO_2
Softwoods		
California incense cedar	4.9	5.0
Red cedar	3.9	18.2
Cypress	3.6	4.5
Redwood	5.3	2.1
Western larch	11.7	4.8
Eastern hemlock	3.4	4.2
Red fir	8.0	7.5
White fir	11.8	7.9
Douglas fir	8.4	7.8
Engleman spruce	15.1	16.1
White pine	9.5	3.0
Shortleaf pine	51.0	4.1
Loblolly pine	8.6	3.5
Slash pine	15.5	5.5
Longleaf pine	13.8	9.3
Ponderosa pine	10.2	11.1
Western white pine	22.2	13.8
Lodgepole pine	12.9	23.2
Sugar pine	8.0	3.8
Average for all softwoods	12.0	8.2
Hardwoods		
Black oak	46.5	25.3
White oak	49.1	26.4
Red oak	43.3	32.2
Post oak	42.1	26.2
Hickory	48.1	9.6
Red gum	48.9	20.0
Yellow poplar	44.3	42.0
Chestnut	38.5	27.6
Black walnut	44.7	11.4
Average for all hardwoods	45.1	24.5
Averages, all species	22.6	13.4

Source: Allison.[11]

[a] Nitrogen was added to the soil mixture to prevent nutrient rate limitations.
[b] Comparative studies showed a release of 54.6% of the carbon of wheat straw in 60 days.

sition of all hardwood species averaged 45.1% and for hardwood barks 24.5%. It would appear that there is a significant difference between softwoods and hardwoods, and even between various species within the softwood and hardwood families. For example, among the softwood species Shortleaf pine showed 5 times the degradability of Loblolly pine.

Many sludge composting facilities add sawdust and/or bark to the feed mixture. Often the selection of amendment is based only on the physical characteristics such as moisture content and particle size distribution. This author has yet to review a sludge composting design where the type of amendment is identified beyond the generic term "sawdust". Allison's data,[11] along with other data presented later in this chapter, suggest that the wood species must be considered because the biodegradability between species can differ by over a factor of 10. This is particularly true if the sawdust or bark is added as an "energy" amendment, because the available heat release could differ by the same factor of 10.

Chandler et al.[12] studied the decomposition of organic substrates during long-term, batch anaerobic fermentation at 35°C. The fermentation periods, lignin content and volatile solids

Table 9.5.   Degradability of Selected Substrates during Anaerobic Digestion at 35°C

Substrate	Fermentation Period (days)	Lignin$_s$[a] (% of VS)	Degradability (% of VS)
Wheat straw	120	8.9	55.4
Corn stalks	120	3.9	77.2
Corn leaves	120	3.8	71.8
Cattails	120	8.5	59.3
Treated kelp	120	6.0	62.0
Water hyacinth	120	8.7	58.8
Corn meal	90	2.0	84.9
Newsprint	90	20.9	28.1
Elephant manure	120	10.4	52.5
Chicken manure	120	3.4	75.6
Pig manure	120	2.2	72.7
Cow manure R_1[b]	120	8.1	58.8
Cow manure R_2[b]	120	7.9	57.5
Cow manure R_3[b]	120	10.1	52.8

Source: Chandler et al.[12]

[a] Lignin values were determined by the 72% sulfuric acid extraction method and are designated as lignin$_s$.
[b] Dairy cow manures from animals housed in the same barn but fed high (R_1), medium (R_2), and low (R_3) energy rations.

reductions are presented in Table 9.5. The degradability coefficient for corn meal was as high as 0.85, for pig and chicken manure over 0.70, and for cow manures over 0.50. Newsprint was much less degradable with a coefficient <0.30. Obviously, there is a considerable range in the degradabilities determined by these studies. The researchers also noted significant differences in the rate of fermentation between substrates. For example, chicken manure produced 90% of its cumulative gas production in only 16 days, compared to 70 days for wheat straw. The kinetics of substrate decomposition are discussed further in Chapter 10.

Chandler used procedures developed by Van Soest[13] to partition the substrates of Table 9.5 into detergent soluble fractions, including cell soluble constituents, hemicellulose, cellulose, and lignin. A strong correlation between volatile solids destruction and lignin content was determined as shown in Figure 9.1. The authors concluded that lignin content was the predominant factor in determining the extent of substrate degradation. The following equation was determined by Chandler to provide the best predictive model for substrate biodegradability.

$$B = 0.830 - (0.028)X \qquad (9.1)$$

where

B = biodegradable fraction of the volatile solids
X = lignin content, % of VS

The biodegradable fraction, B, is the same as the degradability coefficient, k_s. The laboratory techniques necessary to determine lignin content can be conducted at many nutrition laboratories with minimal time and cost. Chandler's predictive model can be very useful when long term biodegradability studies cannot be conducted.

Equation 9.1 suggests that a substrate containing no lignin would only achieve a maximum degradability of 83%. The reason for this is that the decomposition of the substrate organics

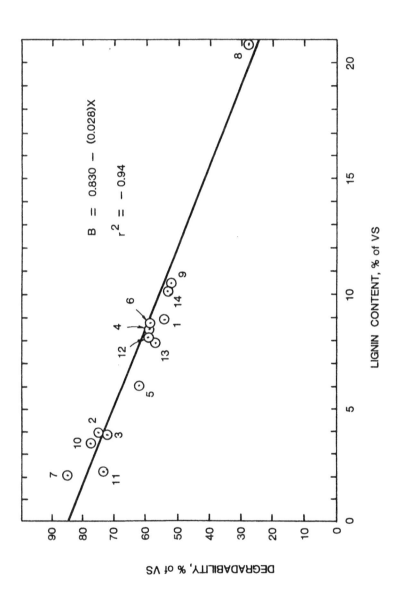

Figure 9.1.   VS destruction efficiency during anaerobic fermentation as a function of VS lignin content. Data points and numbers correspond to the substrates from Table 9.5. From Chandler et al.[12]

Table 9.6.   Estimated Biodegradable Fraction of Selected Organic Waste
Components Based on Lignin Content

Component	Lignin Content (% of VS)	Biodegradable Fraction (% of VS[a])
Food wastes	0.4	81.9
Newsprint	21.9	21.7
Office paper	0.4	81.9
Mixed paper[b]	5.8	66.7
Yard wastes	4.1	71.5

Source: Kayhanian and Tchobanoglous.[14]

[a] Computed from Equation 9.1.
[b] 25% newspaper, 75% office paper.

is coupled with production of bacterial by-products, some of which themselves are not readily degradable. The production of bacterial by-products limits the maximum volatile solids destruction that can be achieved to a probable upper limit of 80 to 90%.

Kayhanian and Tchobanoglous[14] determined the lignin content of selected MSW fractions and used Equation 9.1 to estimate the biodegradable fraction. Their results are presented in Table 9.6. As expected there was a considerable range in predicted biodegradability for the different fractions, from a low of 21.6% for newsprint to a high of 81.9% for food wastes.

With municipal biosolids, a reduction in volatile solids content from about 50 to 55% to about 35 to 40% was reported by Horvath[15] during windrow composting of primary digested sludge without other amendments. Digestion was conducted under well-mixed, mesophilic conditions for about 20 days. Based on the range of VS contents, a degradability coefficient for digested primary sludge can be estimated from 0.33 to 0.56, with an average value ~0.45. In other words, about 33 to 56% of the organics in digested primary sludge should degrade during subsequent composting. Of course, this will depend on conditions maintained during both digestion and subsequent composting.

Raw municipal biosolids usually have a volatile solids content in the range of 70 to 80%. It is reasonable to assume that the organic content after composting would be similar to that for digested, composted sludge, about 0.35 to 0.40. Based on this expected change in volatility, the degradability coefficient for raw sludge can be estimated as ranging from 0.72 to 0.82. By way of comparison, about 50% VS destruction is observed during anaerobic digestion of raw sludge. Because further decomposition is observed when digested sludge is composted, the estimated range in the degradability coefficient for raw sludge appears reasonable. Andrews and Kambhu[16] also assumed a degradability coefficient of 0.70 when studying aerobic, thermophilic digestion of raw domestic sludges.

A well-stabilized sample of digested, windrow-composted sludge was rewetted to a moisture content of 60% and incubated under controlled aerobic conditions at 49°C (120°F). After 18 days, the VS content decreased from 0.325 to 0.29, corresponding to a further degradability of ~15%. A VS content of 30% is probably about the lowest achievable with municipal biosolids and reasonable composting conditions.

MEASURING DEGRADABILITY

A number of techniques are available for either measuring or estimating substrate degradability. To date, these techniques have not found widespread application in the composting industry. It is hoped that their use will gain increased favor because of the importance of substrate degradability to process design and operation.

Substrate Composition

Agronomists routinely determine the chemical composition of substrates to determine their nutritional value as animal feed. Digestibility of the substrate is a key factor to agronomists in determining the energy available from the substrate. Acid detergent fiber, neutral detergent fiber, protein, cellulose, hemicellulose and lignin are often determined in such testing. From these values an estimate of the digestibility and nutritive value of the substrate can be made. This approach to determining substrate degradability can be extended to the composting industry.

The Chandler formula given by Equation 9.1 allows estimation of substrate degradability from measurement of the lignin and VS content. The sequence of sample analysis is presented in Figure 9.2. A portion of the sample is dried and ashed in a muffle furnace to determine the TS and VS fractions. A second portion of dried sample is refluxed with acid detergent (AD), cetyltrimethyl ammonium bromide in 1N sulfuric acid. Cell soluble (CS) and hemicellulose (HC) are solubilized, yielding an acid detergent fiber (ADF) residue. The ADF is extracted with 72% sulfuric acid to solubilize cellulose (CE), leaving the lignin and ash fraction. The latter is fired in a muffle furnace and the lignin determined by weight difference. Many laboratories are capable of performing the lignin determination outlined in Figure 9.2.

One caution is noted regarding Equation 9.1. The correlation in Equation 9.1 is between lignin content and anaerobic degradability or fermentability. While the equation should be reasonably applicable to aerobic composting conditions, this has yet to be verified in actual testing.

Mass Balance Approach

If pilot studies can be conducted or if there is an operating facility using similar substrates, a mass balance can be conducted to determine the degradability of the mixed substrates. A limitation with this technique is that only the total degradability of all mixed materials can be determined. If more than one substrate is used, individual degradabilities cannot be determined. Two approaches to analysis are discussed below.

Total Mass Loss

Referring to Figure 6.7, a balance on total solids into and out of the process gives

solids in = solids out + volatile solids lost

Substituting terms from Figure 6.7:

$$S_m X_m = S_p X_p + VS \text{ lost}$$

$$VS \text{ lost} = S_m X_m - S_p X_p$$

(9.2)

The degradability coefficient for the mixture, k_m, can be defined as VS lost from the process divided by the VS input to the process or

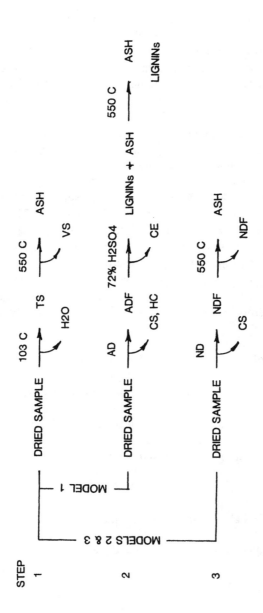

Figure 9.2.   Sequence of analyses used by Chandler et al.[12] based on the techniques of Van Soest[13] to determined substrate compositions. Models II and III were rejected in favor of Model I which is given as Equation 9.1. (AD = acid detergent, ADF = acid detergent fiber, CE = cellulose, CS = cell soluble, HC = hemicellulose, ND = neutral detergent, NDF = neutral detergent fiber).

$$k_m = \frac{VS \text{ lost}}{VS \text{ in}} \qquad (9.3)$$

Substituting Equation 9.2 into 9.3 and recalling that VS input is $V_m S_m X_m$ gives

$$k_m = \frac{S_m X_m - S_p X_p}{V_m S_m X_m} \qquad (9.4)$$

Equation 9.4 can be used to determine the average degradability of the mixed substrates into the process.

Conservation of Ash

One difficulty with using Equation 9.4 is that it requires accurate measurement of the weights of infeed and outfeed materials. Some composting facilities are not equipped for this. Another approach, which avoids the need to measure input and output weights, relies on the assumption that the ash or inert fraction should behave as a conservative substance through the process. In other words, inerts entering the process should equal inerts leaving the process at steady state. Bar diagrams representing the feed mixture and compost product are presented in Figure 9.3. Note that the ash components of the mixture and product are the same.

Terminology used in Figure 9.3 and the following analysis is as follows:

VS_m	=	volatile solids component of the feed mixture, fraction of TS
$VS_m\%$	=	mixture volatile solids content, % of TS
ASH_m	=	ash component of the feed mixture, fraction of TS
$ASH_m\%$	=	mixture ash content, % of TS
VS_p	=	volatile solids component of the product, fraction of TS
$VS_p\%$	=	product volatile solids content, % of TS
ASH_p	=	ash component of the product, fraction of TS
$ASH_p\%$	=	product ash content, % of TS

The percentage ash in the feed mixture $ASH_m\%$ can be determined as

$$ASH_m\% = \frac{ASH_m(100)}{ASH_m + VS_m} \qquad (9.5)$$

Rearranging:

$$VS_m = \frac{ASH_m(100)}{ASH_m\%} - ASH_m \qquad (9.6)$$

Similarly, $ASH_p\%$ and VS_p can be determined as

$$ASH_p\% = \frac{ASH_p(100)}{ASH_p + VS_p} \qquad (9.7)$$

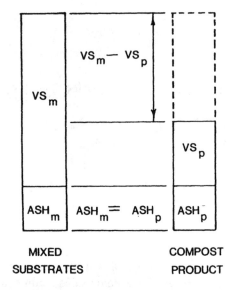

Figure 9.3.   Generalized bar diagram showing VS and ASH components for substrate mixture and compost product. ASH fraction is conservative.

$$VS_p = \frac{ASH_p(100)}{ASH_p\%} - ASH_p \qquad (9.8)$$

From Figure 9.3, the degradability coefficient, k_m, can be defined as

$$k_m = \frac{VS_m - VS_p}{VS_m} \qquad (9.9)$$

Substituting Equations 9.6 and 9.8 into Equation 9.9 and rearranging gives

$$k_m = \frac{(ASH_p\% - ASH_m\%)100}{ASH_p\%(100 - ASH_m\%)} \qquad (9.10)$$

Equation 9.10 is similar in form to one originally presented by Schulze.[17]

A similar equation based on volatile solids can be developed by recalling that ASH and VS percentages are related as follows:

$$ASH_m\% = 100 - VS_m\% \qquad (9.11)$$

$$ASH_p\% = 100 - VS_p\% \qquad (9.12)$$

Substituting Equations 9.11 and 9.12 into Equation 9.10 and rearranging:

$$k_m = \frac{(VS_m\% - VS_p\%)100}{VS_m\%(100 - VS_p\%)} \qquad (9.13)$$

Example 9.1

10 dtpd of feed substrates are composted to yield 5 dtpd of product. The feed mixture is 80% VS and the product 60% VS. Determine the degradability coefficient, k_m, for the feed mixture.

1. Using the mass loss approach, the following values apply:

$$S_m X_m = 10 \text{ dtpd}$$

$$V_m S_m X_m = (0.80)10 = 8 \text{ dtpd}$$

$$S_p X_p = 5 \text{ dtpd}$$

Substituting into Equation 9.4:

$$k_m = \frac{10 - 5}{8} = 0.625$$

2. Ash input with the feed substrates is

$$(1 - V_m)S_m X_m = (1 - 0.80)10 = 2 \text{ dtpd}$$

Assuming the ash is conservative, 2 dtpd of ash should be contained in the 5 dtpd of product, giving a product ash content of 2/5 = 40%. This is equivalent to 60% VS which checks with the original problem statement.

3. Assuming ash as conservative, k_m is determined from Equation 9.13 as

$$k_m = \frac{(80 - 60)100}{80(100 - 60)} = 0.625$$

4. Both the mass loss and ash conservative approaches check perfectly in this problem. The reader should note that it is seldom this easy to close mass balances in full scale facilities.

Respirometry

The most common method for determining degradability for aerobic systems is to measure oxygen consumption under conditions that avoid rate limitations from lack of nutrients, oxygen, moisture, imbalanced pH, or inadequate seed microbes. The pattern of oxygen consumption with time can be used to determine the rate constant for decomposition of the substrate. Because oxygen is consumed as a result of metabolic respiration, equipment used in such testing is referred to as a "respirometer".

BOD and COD Testing

The standard BOD bottle is perhaps the simplest form of respirometer. The bottle is initially charged with an oxygen saturated solution of water, sample, and seed microbes and then incubated under constant temperature conditions. Biological decomposition of the substrate consumes oxygen from the solution. The oxygen concentration remaining in solution is measured with time to determine the respiratory consumption. The standard BOD test is designed primarily for liquid samples of small volume. Also, the oxygen resource is limited to that which can be dissolved in the dilution water. This limits the quantity of sample that can be analyzed and makes the technique difficult to apply to solid samples. Within the constraints of these limitations, the BOD technique has been successfully applied to homogeneous substrates such as sewage sludge.

By comparison, the COD measures the equivalent oxygen demand resulting from complete breakdown of the organics using oxidative chemicals. The COD is a measure of ultimate oxygen demand because it does not depend on the organics being biodegradable. The use of BOD and COD data to determine degradability is illustrated in the following example based on actual laboratory testing.

Example 9.2

A composite sample of raw, combined sludge cake from the City of Plattsburgh, New York was collected. The cake tested at 19% TS and 83% VS. A known quantity of cake was diluted and homogenized to produce a liquid slurry. Known volumes were pipetted into BOD bottles. Standard BOD testing techniques were then used and the samples incubated at 20°C for various time periods. An inhibitor was added to suppress nitrification. Results of the BOD testing are given below:

Days	BOD (mg O_2/g VS)
5	190
10	320
20	540
30	640
40	970
50	930
60	1000

A plot of this data is included in Figure 9.7. Replicate COD tests using standard reflux techniques were also conducted on the cake sample. Average COD was determined to be 1.65 g O_2/g VS. Estimate the sludge degradability.

Solution

1. Ultimate BOD (BOD_L) is often interpreted to mean the BOD determined at day 20 (BOD_{20}). Referring to the above table, however, the BOD_{20} is only ~55% of that exerted by day 40. Using 20-day BOD would give a completely incorrect determination of degradability for this sludge.

2. The rate of oxygen consumption slows significantly beyond day 40. Based on this data it is reasonable to assume a BOD_L of about 1000 mg O_2/g VS as representative of the ultimate demand.

3. The oxygen demand from complete oxidation of all organics is well modeled by the COD test. Therefore, the degradability coefficient for the sludge cake, k_s, can be determined as the ratio of BOD_L/COD:

$$k_s = \frac{1000}{1650} = 0.61$$

Constant Volume Respirometers

A wide variety of other techniques have been employed in the study of chemical and biological reactions. The type of respirometer that has met with the widest use is called the "Warburg" instrument.[18] It is a standard apparatus in many biological laboratories. The Warburg is based on the principle that at constant temperature and constant gas volume any changes in the amount of a gas can be measured by changes in its pressure. A manometer is attached to the reaction vessel to measure the pressure change. Hence, the Warburg is often described as a manometric, constant volume respirometer. While the Warburg overcomes some of the problems of the BOD bottle, it is still limited to relatively small, homogeneous samples. The maximum reaction flask that can be attached to the Warburg unit is about 500 ml. Also, the instrument is not common to sanitary or treatment plant labs and is relatively expensive. Regan and Jeris[7] and Pressel and Bidlingmaier[19] successfully used the Warburg instrument in studies on the composting rates of refuse.

A number of respirometers have been developed that use electrolysis of water to replenish oxygen consumed in the reaction vessel. These are generally referred to as electrolytic, constant volume respirometers. Consumption of oxygen within the reaction vessel, coupled with absorption of produced carbon dioxide, results in a pressure drop within the reactor. The reduced pressure initiates operation of an electrolytic cell which dissociates water into hydrogen and oxygen. The cell is designed to direct the produced oxygen into the reactor while venting the hydrogen to atmosphere. The quantity of oxygen produced is determined from measurement of the applied current and its duration. A number of available commercial offerings are suitable for large sample sizes. The complexity of electrolytic respirometers can range from "homebuilt" models using microamp meters and timers to units with complex electronics and even direct computer hookups for automatic data analysis. These units are generally well suited for analysis of compost substrates, but they are also relatively expensive. Also, the addition of electronics and computers does not increase the inherent accuracy of the unit. Usui et al.[20] and Arthur[21] reported on the use of electrolytic respirometers for determining rates of decompostion for composting substrates and compost product.

Willson and Dalmat[22] developed a simple, constant volume respirometer for use in measuring the stability of compost products. A schematic of their apparatus is presented in Figure 9.4. In the Willson and Dalmat respirometer 400 g of compost sample is placed in a 1-l erlenmeyer flask. A second 1-l flask is filled with 400 g of water. The two flasks are equilibrated at room temperature or in a constant temperature bath. A vial containing 6 ml of 2 N sodium hydroxide is inserted into the flask with the compost sample. The reaction flask is connected to the low pressure side of a differential pressure gauge and the balancing flask connected to the high pressure side. The balancing flask eliminates the need to compensate for changes in barometric pressure, temperature, and water vapor pressure.

The gas volume in both flasks of the Willson respirometer should be the same. The gas volume in the reaction flask can be measured by placing a 400 g sample in the flask and measuring the amount of water added to fill the flask. However, the specific gravity of compost is sufficiently close to one that the gas volume can usually be determined as the total flask volume less 400 ml. The 400 g of water added to the balancing flask is used to bring its void volume to that of the reaction flask. A test is usually run for 3 h. The amount of oxygen consumed is calculated from the gas laws using the change in partial pressure of the oxygen and the volume of gas in the flask and tubing. The Willson respirometer is a simple apparatus that has proven to give reliable results when measuring the stability of compost products. However, the oxygen resources within the reaction flask are limited. This generally restricts

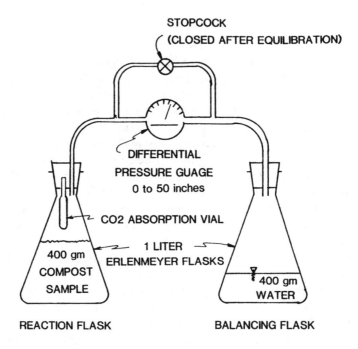

STOPCOCK
(CLOSED AFTER EQUILIBRATION)

DIFFERENTIAL
PRESSURE GUAGE
0 to 50 inches

CO2 ABSORPTION VIAL

1 LITER
ERLENMEYER FLASKS

400 gm
COMPOST
SAMPLE

400 gm
WATER

REACTION FLASK   BALANCING FLASK

Figure 9.4.   Schematic diagram of the constant volume respirometer developed by Willson and Dalmat[22] for testing of compost product stability.

its application to short-term tests on more stable materials. For this reason, it is not well suited to long-term degradability studies.

The Constant Pressure Respirometer

A constant pressure respirometer is based on the principle that at constant temperature and constant pressure any change in the amount of a gas can be measured by changes in its volume. Measuring volume change is relatively simple and allows development of an apparatus that can be constructed largely from normal laboratory supplies.

A diagram of a constant pressure respirometer developed by Haug and Ellsworth[23] to study the long-term decomposition of composting substrates is presented in Figure 9.5. A 2-l erlenmeyer flask is used as the reaction vessel. The large reaction vessel was favored because it increased the sample size that could be used. The substrate to be tested is suspended in about 1.5 l of solution within the flask. Use of an aqueous slurry allows easy mixing of the sample during testing and eliminates the potential for rate limitations due to lack of moisture. A magnetic stirrer is used to agitate the suspension and enhance mixing and oxygen transfer across the liquid surface. The reaction flask is wrapped in aluminum foil to discourage algal growth and associated oxygen production.

A 1- or 2-l erlenmeyer flask is connected to the reaction flask by Tygon tubing. This flask contains a solution of sodium or potassium hydroxide and is used to absorb carbon dioxide released during decomposition of the substrate. The hydroxide solution is agitated by means of a magnetic stirrer to enhance absorption. Two tubing connections to the reaction flask are shown. It was found that this produced a pumping action and improved transfer of CO_2 from the reaction flask.

The reaction flask is connected to a calibrated glass tube which is used to measure the volume of consumed oxygen. About 450 ml of gas volume can be measured by the calibration

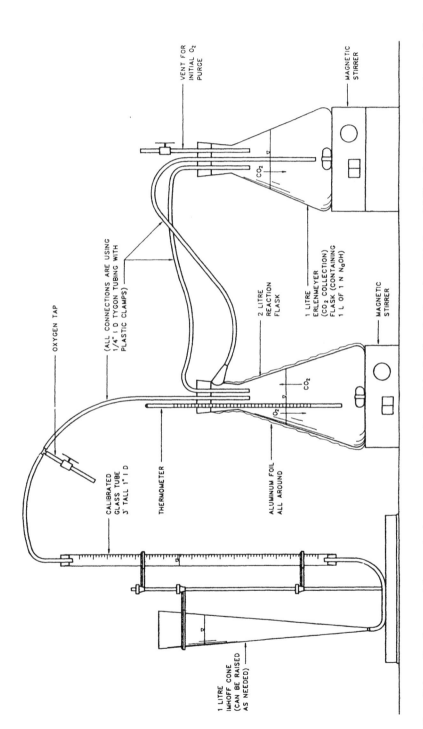

Figure 9.5.   Schematic diagram of the constant pressure respirometer used for testing the long-term biodegradability of composting substrates by Haug and Ellsworth.[23]

tube. The tube is connected to a water reservoir and acts like a "U" tube manometer. An Imhoff cone is shown as the water reservoir, but any appropriate vessel can be used. As oxygen is consumed during a test, liquid level in the calibration tube rises. The height of the Imhoff cone is adjusted so that the liquid level in the Imhoff cone is slightly below that in the calibration tube. This keeps the reaction flask under a slightly negative pressure so that any air leakage is into the reactor. This assures that any measured oxygen uptake is from microbial metabolism and not possible leakage. In actual practice, leakage has not proven to be a problem.

Readings are usually taken daily, depending on the rate of gas consumption. To take a reading the Imhoff cone is raised or lowered until liquid levels in the cone and calibration tube are equal. This assures that the reactor pressure is 1 atm when all readings are made, hence the term constant pressure respirometer. The gas volume is then recorded. Barometric pressure of the atmosphere and temperature in the reaction flask are also recorded to later correct the readings to standard conditions.

Following addition of the substrate suspension to the reaction flask and the hydroxide solution to the CO_2 absorption flask, both flasks are sealed with their rubber stoppers. The vent lines are then opened and the gas space is completely purged with pure oxygen. Once purged, the vent lines are closed and the experiment begins. Standard safety precautions should be taken when connecting the respirometer to the oxygen supply. A regulator or other device should be used to assure that the system cannot be over pressurized.

Eventually, oxygen consumption can exceed the available measuring volume in the calibration tube. Before this happens, the vent lines are reopened, the system refilled with pure oxygen, the liquid level in the calibration tube reset to zero, and the system resealed. This procedure is repeated as many times as necessary during the test. In the beginning it may be required every few days, depending on the size and degradability of the substrate sample. Refilling and resetting usually become less frequent in the later stages of the experiment.

In the initial experiments, air and not oxygen was used in the gas space. Each time the system was opened and reset to zero the consumed oxygen was replaced by air. Because air is only 21% oxygen by volume, the oxygen content in the gas space was depleted each time the system was reset. Eventually, oxygen transfer into the reaction flask became rate limiting. Once this was observed, air was replaced by oxygen and this resolved further problems. Using pure oxygen also increases the transfer rate into solution and the resulting solution concentration. This assures that oxygen transfer is not rate limiting. Spot checks of oxygen content can be made when the system is reset to zero. Oxygen is usually above 15 mg/l in solution, which is sufficient to avoid rate limitations. Air can be used in place of oxygen but it requires that the reactor volume be completely purged each time the system is reset to avoid depleting the oxygen partial pressure.

Reaction Flask Solution   The solution used in the reaction flask should contain sufficient macro and micro nutrients to avoid rate limitations from lack of nutrients. The solution should also be buffered to a near neutral pH, again to avoid rate limitations. However, the solution should not contain components that could exert an oxygen demand. The principle concern here is the use of NH_4Cl or other ammonium salts that can be an energy source for nitrifying bacteria. Such compounds are used in the Standard Methods[24] dilution water for BOD testing, but are usually not a concern in the standard 5-day test. They should be avoided here because the longer run times offer greater potential for nitrification.

After some experimentation the following "recipe" has proved workable and appears to avoid obvious rate limitations:

1. Blend a weighed sample to 500 ml of deionized water (DW) in a Waring Blender. 20 to 50 g samples (wet weight) have been used in testing to date. Note that the blender should be used only to mix the sample but not grind it to a particle size smaller than that expected in the composting process.

2. Add 0.1 g finished compost as a seed. The choice of seed is probably not critical. Finished compost was selected because it should contain microbes acclimated to the composting environment and the substrates under test. Seed has not been used with municipal sludge samples, which should already be adequately seeded.

3. Add 1.8 g KNO_3 as a nitrogen nutrient. Note that nitrate does not contribute to nitrogenous oxygen demand. Use of KNO_3 is optional with high nitrogen substrates.

4. Add 5 ml each of the $CaCl_2$, $FeCl_3$, and $MgSO_4$ nutrient solutions made up per Standard Methods BOD test procedures. This is about three times the amount used in BOD testing because of the higher concentration of substrate.

5. Add 12 ml of phosphate buffer solution made up per Standard Methods for BOD testing. This is about eight times the concentration used in BOD testing. Do not add NH_4Cl to the phosphate buffer as specified by Standard Methods.

6. Add 1.5 g $NaHCO_3$ or baking soda for added pH buffer control.

7. Add TCMP nitrification inhibitor made up per Standard Methods.[24] Use five times amount specified for BOD testing. See also the discussion on nitrification below.

8. Add DW to bring the volume to 1.5 l.

9. Add the 1.5 l solution to the reaction flask, seal, and purge with oxygen.

10. Measure COD, total solids, volatile solids, and heat value on a separate portion of the substrate sample. An ultimate analysis for C, H, O, N, and S content is recommended.

11. At the completion of the test, filter the reaction flask solution to recover the suspended solids. Rinse the recovered solids with DW to remove salts. Dry the solids and weigh.

Buffer Capacity   Oxidation of a complex substrates, such as sewage sludge, can be represented as

$$C_5H_{19}O_3N + 12.5O_2 \rightarrow 10CO_2 + NH_3 + 8H_2O$$

$$201 \quad 12.5(32) \quad 10(44) \quad 17 \quad 8(18)$$

(9.14)

Suppose 20 g of the substrate are added to the reaction flask. Assuming 20% TS, 80% VS, and 50% degradability, about 20(0.2)(0.8)(0.5) = 1.6 g will be decomposed according to Equation 9.14. This will produce 1.6(17/201) = 0.135 g NH_3. The NH_3 can be oxidized by nitrifying bacteria as

$$NH_3 + 2O_2 \rightarrow NO_3^- + H^+ + H_2O$$

(9.15)

Hydrogen ion produced from Equation 9.15 will be neutralized as

$$H^+ + HCO_3^- \rightarrow CO_2 + H_2O$$

$$61 \quad 44$$

(9.16)

Thus, the 0.135 g NH_3 produced by Equation 9.14 requires $0.135(61/17) = 0.48$ g HCO_3^- for neutralization, equivalent to 0.66 g $NaHCO_3$. The 1.5 gms sodium bicarbonate added in Step 6 above is used to neutralize the acid that would be produced from nitrification of produced ammonia. Nitrification represents the most significant acid producing reaction. The 1.5 g should provide a sufficient excess buffering capacity in addition to the phosphate buffer added in Step 5.

Nutrient Availability   The major nutrients of concern are the macro nutrients nitrogen and phosphorus, and the micro nutrients such as iron, calcium, and magnesium. The micro nutrients are satisfied by adding BOD nutrient solutions in Step 4 and sodium bicarbonate in Step 6. Reagent grade sodium bicarbonate and commercial baking soda both contain an extensive list of micro nutrients. Baking soda, being less refined, contains more of the micro nutrients. Phosphorus is supplied by the phosphate buffer added in Step 5.

Nitrogen is required for synthesis of microbial cell matter. If not present in adequate amounts, synthesis will be limited and overall reaction rates reduced from their maximum values in the absence of nutrient limitation. With nitrogen "rich" substrates, such as sewage sludge, grass, and most food wastes, nitrogen should be available in adequate amounts without supplemental addition. On the other hand, cellulosic materials including tree trimmings, leaves, paper, and some MSW fractions, can be nitrogen "poor".

The worst case situation will occur with cellulosic material, which can be represented by the oxidation of glucose, $C_6H_{12}O_6$. Using $C_5H_7O_2N$ to represent synthesized cells and a cell yield of 0.2 g cells/g glucose, the overall metabolic reaction can be approximated as

$$3C_6H_{12}O_6 + 11O_2 + H^+ + NO_3^- \rightarrow C_5H_7O_2N + 13CO_2 + 15H_2O$$

$$3(180) \qquad\qquad 62 \qquad 113 \qquad 13(44) \qquad\qquad\qquad (9.17)$$

Assuming 50 g of substrate added to the reaction flask, 60% TS, 90% VS, and 30% degradability, about $50(0.6)(0.9)(0.3) = 8.1$ g of glucose will be lost. Based on Equation 9.17, this requires $[8.1(62)]/[3(180)] = 0.93$ g NO_3^-, equivalent to $[0.93(101)]/62 = 1.5$ g KNO_3. 1.8 g KNO_3 is added in Step 3 to provide an excess.

The above calculations should be reviewed for each experiment to assure that adequate nitrogen is available. The cell yield used above is less than the maximum thermodynamically possible for aerobic, heterotrophic microbes. A lower cell yield is used because endogenous respiration becomes significant in long duration testing. This reduces the net synthesis and, therefore, the net requirement for nitrogen.

Carbon Dioxide Absorption Solution CO_2 produced by Equations 9.14 and 9.16 will desorb from the reaction flask solution, disperse through the gas phase, and eventually be absorbed in the hydroxide solution of the absorption flask. The chemical reaction in the absorption flask with NaOH is

$$2NaOH + CO_2 \rightarrow Na_2CO_3 + H_2O$$

$$2(40) \qquad 44 \qquad\qquad\qquad\qquad\qquad\qquad (9.18)$$

Referring to the above example using a 20 g sludge sample, 1.6 g of substrate oxidation was estimated. CO_2 produced per Equations 9.14 and 9.16 is $1.6\{[(10)(44)/201]+44/201\} = 3.85$ g.

Using Equation 9.18, about $3.85[2(40)]/44 = 7.0$ g NaOH is required for neutralization. Using the example of a 50 g cellulose sample, CO_2 production can be estimated from Equation 9.17 as $8.1[13(44)]/3(180) = 8.6$ g. Again using Equation 9.18, about 16 g NaOH is required for neutralization. To assure that excess hydroxide is available for all possible reactions, 1-l of 1 N NaOH, which contains 40 g NaOH, is used.

Readers familiar with the Warburg apparatus may be surprised at the quantity of hydroxide required for carbon dioxide absorption. A typical Warburg experiment uses only a few milliliters of hydroxide solution. The rather large quantity required for the apparatus of Figure 9.4 is due to the much larger size of sample.

Nitrification Effects   Substrate degradability is determined by measuring the carbonaceous oxygen demand. If ammonia released per Equation 9.14 is oxidized per Equation 9.15, the nitrogenous oxygen demand will interfere with interpretation of the results. Thus, it is desirable to limit nitrification, and an inhibitor is added in Step 7 for this purpose.

TCMP (2-chloro-6-trichloromethyl-pyridine) is commonly used to inhibit nitrification in the BOD test. The inhibitor is effective for the 5-day period normally used for BOD testing. However, there is a question about its effectiveness during long duration testing with the respirometer. A modest attempt was made to determine the long-term effectiveness of the inhibitor. Standard BOD bottle tests, with and without nitrification inhibitor, were conducted using municipal sludge. The tests were conducted for time periods up to 60 days. Measured BOD with inhibitor was less than the sample COD, even after 60 days. Measured BOD without inhibitor exceeded sample COD after about 30 days, indicating that nitrogenous demand was exerted. While this does not prove the effectiveness of the inhibitor for larger samples and longer duration, it does indicate that it can be at least partially effective. Therefore, the addition of inhibitor was included in the reactor solution.

The potential effect of nitrification can also be assessed by measuring the nitrate concentration in the reaction solution at the end of the experiment. However, this approach is only effective with substrates that are nitrogen "rich". In this case, the potassium nitrate would not be added in Step 3. Any nitrate found at the end of the experiment can then be assumed to result from nitrification and the nitrogenous oxygen demand estimated from Equation 9.15. With nitrogen "poor" substrates, KNO_3 must be added in Step 3. In this case, however, the effects of nitrification should be minimal.

Incubation Conditions   Strict scientific procedure would specify that the experiment be conducted at constant temperature and that a control unit be run in parallel. The control would contain all equipment, chemicals, and seed, but not the substrate sample itself. Standard Methods testing for BOD and normal Warburg testing both specify such procedures.[18,24] Any volume change in the control, as a result of oxygen consumption, temperature or barometric pressure change or other factors, is used to correct readings taken on the reaction flask. The need for strict adherence with such procedures is reduced with the respirometer of Figure 9.5 because the large sample size reduces the effect of such factors. Total oxygen consumption usually varies from about 2 to 15 l with the apparatus of Figure 9.5. With such large total gas consumption, the effect of control and other corrections becomes relatively minor.

Incubation can be conducted at constant temperature if an incubator or constant temperature bath is available. Alternatively, incubation can be conducted at room temperature in a laboratory environment. Laboratories are usually controlled so that temperature fluctuations rarely exceeded 5°C. Reaction flask temperature and barometric pressure are recorded daily so that measured gas volumes can be corrected to standard conditions. To apply these corrections, it is necessary to measure the total gas volume within the apparatus, including reaction flask, absorption flask, calibration tube, and connecting tubing, at the beginning of the experiment.

Reaction rates measured by this testing will correspond to the average temperature maintained during the experiment. This will usually be 20 to 25°C if the experiment is conducted at normal lab temperatures. To be useful in simulation modeling of the compost process, these rate constants must be corrected to actual composting temperatures. Relationships are available for this purpose. Nevertheless, there are advantages to conducting the experiment at the elevated temperatures more common to composting. The temperature correction to the rate constant is reduced, the duration of the experiment should be reduced, and nitrification effects can be eliminated.

Regarding duration, some experiments have been conducted for over 800 days. Such long experiments are not always necessary, but most tests have been conducted for a minimum of 60 to 90 days. Higher temperature incubation may reduce the time required.

Regarding nitrification, Haug[25] and Haug and McCarty[26] reported the effect of temperature on selected cultures of nitrifying microbes shown in Figure 9.6. Using a Warburg respirometer, the rate of nitrification was a maximum in the range of 25 to 30°C, but ceased completely at a temperature near 40°C. Incubation at a temperature of 45°C would appear to be near optimum. The temperature is high enough to eliminate nitrification, but not so high to raise concerns about inhibiting the normal composting microbes.

Calculating Degradability   Two approaches to calculating degradability are possible based on the respirometer results. First, the ultimate oxygen demand can be projected from the curve of cumulative oxygen uptake vs time. Substrate degradability can then be determined by dividing the projected ultimate demand by the total substrate oxygen demand as measured by COD, heat value, or ultimate analysis. Second, solids remaining in the reaction flask can be filtered, dried, and weighed. Degradability can then be determined using Equation 9.4. The disadvantage of the latter approach is that solubilized organics will pass through the filter and not be included in the analysis. This disadvantage is mitigated by the fact that soluble organics should normally not be present in significant concentrations given the long duration of testing. The first approach is favored with the second approach used for confirmation. The manner of analysis is discussed further in the next section.

CASE STUDY

The City of Plattsburgh, New York, operates and maintains the Clinton County Composting Facility, an in-vessel system for raw biosolids composting shown in Figure 2.27. The compost plant was dedicated in April 1986. A number of mechanical problems with the ventilation and aeration systems surfaced almost immediately. The system also experienced operating difficulties in cold weather. Thermophilic temperatures were difficult to maintain, drying was limited, and the process failed repeatedly in the winter. These symptoms were evidence of a "stressed" energy balance. This pattern repeated itself over several consecutive winters.

Like most "real world" problems, a number of factors contributed to the operating difficulties. Dewatered cake solids content tended to decrease in winter months which placed an increased water load on the plant. Cake solids were sometimes <18% TS. In addition, degradability of the sludge and sawdust then used as an "energy" amendment came into question. The sludge is raw, combined primary, and waste activated sludge. However, the treatment plant then received ~70% of its influent load from a local pulp and paper industry.

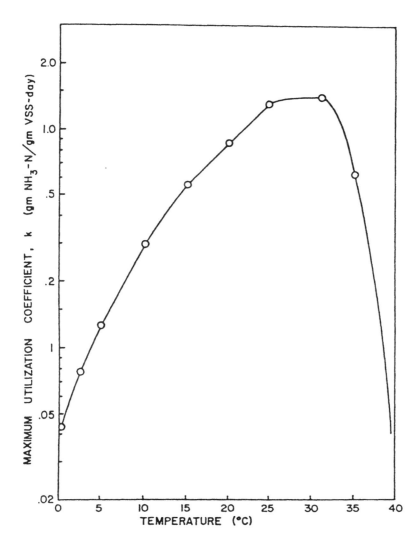

Figure 9.6.   Maximum utilization coefficient (reaction rate constant) as a function of temperature for nitrifying biofilm. Data were developed using a Warburg respirometer. From Haug.[25]

Therefore, there was reason to question the use of "textbook" values for degradability. Laboratory investigations were initiated to measure substrate biodegradability and determine the effect of this parameter on the energy balance. The laboratory work was conducted at the Plattsburgh Water Pollution Control Plant laboratory using the constant pressure respirometer of Figure 9.5.

Results obtained with the respirometer using Plattsburgh sludge cake are presented in Figure 9.7. Two respirometer runs along with data from long term BOD testing are presented. Oxygen uptake rates were relatively high at the beginning of the test (days 1 to 20) and then decreased gradually. Very little additional uptake was noted between days 50 and 60 for Run A, indicating that most degradable organics had been consumed. Run B continued to consume oxygen beyond day 60 and did not completely plateau until after day 200. Nevertheless, the oxygen consumption at day 60 was about 0.8 g O_2/g VS for both tests. COD measured at the beginning of testing was 1.65 g O_2/g VS. Therefore, the sludge degradability was estimated

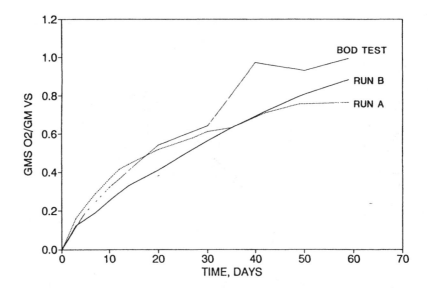

Figure 9.7.   Respirometer results with raw sludge cake from the City of Plattsburgh, New York. From Haug and Ellsworth.[27]

at about 0.8/1.65 = 48%. This was lower than expected and lower than "typical" values for raw sludge. Weighing of the solids remaining from Run B suggested a 66% degradability after a total 242 days of incubation. This suggests that any value of degradability should be keyed to a time frame, i.e., 48% at 60 days or 66% at 242 days for the sludge of Figure 9.7.

Long-term BOD testing was conducted in parallel with the respirometer tests. BOD test results are also presented in Figure 9.7. The BOD data are consistent with those from the respirometer. It is interesting to note that the Standard Methods test for ultimate BOD is normally run for 20 days. Only about 50 to 65% of the total oxygen consumption was realized by day 20. This highlights the need for longer duration testing for the substrates common to composting because of their lower reaction rate constants.

Results of respirometer testing on amendments used at the Plattsburgh facility are presented in Figure 9.8. White pine sawdust and product recycle were used for feed conditioning during the early phases of plant operation. The initial rate of oxygen consumption for the white pine sawdust was quite high as shown by the slope of the curve in Figure 9.8 between days 0 and 10. This initially high rate indicates that a fraction of the sawdust is readily biodegradable. A plateau was reached between about day 30 and day 60, after which there was another surge in activity from days 50 to 90. Cumulative oxygen consumption at day 90 was about 0.14 g O_2/g VS. COD of the sawdust was measured at 1.25 g O_2/g VS. Degradability for the pine sawdust was estimated at only 0.14/1.25 = 11.2% (90 day). This was considerably less than expected and suggested that the pine sawdust was not contributing significantly to the system energy balance.

Tests of other possible amendments were undertaken to identify a more degradable replacement for the pine sawdust. A potential amendment termed "pressate" was produced by a local pulp and paper industry. The pressate showed considerably more oxygen uptake as shown in Figure 9.8. The pressate exhibited a noticeable "lag" period for about the first 15 days of testing before significant oxygen consumption commenced. This highlights the importance of longer term testing. Oxygen consumption reached about 0.48 g O_2/g VS at day 200. COD of the pressate was 1.3 g O_2/g VS. Degradability at day 200 was estimated as 0.48/1.3 = 37% (200

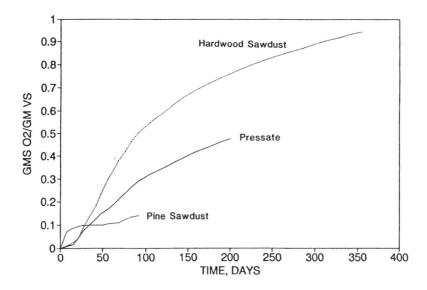

Figure 9.8.   Respirometer results with various substrates used as amendments at the sludge composting facility at Plattsburgh, New York. From Haug and Ellsworth.[27]

day). Multiple regression analysis, using a first order rate equation, suggested an ultimate oxygen demand of 0.57 g O_2/g VS. Ultimate degradability for the pressate was estimated as 0.57/1.3 = 44% (ultimate).

A locally available hardwood sawdust (species unidentified) was also tested. Referring to Figure 9.8, an initial lag phase was noted for about the first 20 days of testing. Following the lag phase, oxygen consumption proceeded vigorously and continued for over 350 days of testing. Oxygen consumption at day 350 was 0.94 g O_2/g VS. COD was 1.2 g O_2/g VS, giving a degradability of 0.94/1.2 = 78% (350 day). The hardwood sawdust was significantly more degradable than the softwood pine first used in the composting facility. The difference in degradability between these two wood species is consistent with data presented previously in Table 9.4 which suggests that the hardwoods are generally more degradable. The ultimate degradability of the hardwood is also consistent with the upper limit of 80 to 90% discussed previously.

These results were coupled with simulation modeling (see Chapters 11 through 14) and together they explained the problems observed during winter operation. The energy balance was indeed marginal to the point that process failure occurred when the system was stressed by cold weather or incidents of low cake solids. Based on these results, the City of Plattsburgh undertook a program to increase cake solids and switched to a more degradable amendment, a combination of hardwood bark and pressate. At the same time much of the pulp and paper waste was removed from the sewerage system. Dewatering improvements increased the average cake solids to levels consistently above 27% TS and reduced the frequency of lower cake solids. The combination of these actions resolved the problems with the energy balance. VS reduction across the system, which was barely measurable before, increased to over 50%. After many years of troubled operation, the facility was able to stabilize its operation and compost sales exceed the available supply.

As an interesting aside to the above story, converting the energy balance from an "energy poor" to an "energy rich" condition produced another set of problems. Process temperatures increased to levels that sometimes exceeded 80°C (176°F). In response the operators at tempted to reduce process temperatures by increasing the air supply. This worked but then caused excessive drying of the material in the reactor. Dust became a very serious problem and there was concern over potential fire hazards. Supplemental water addition became necessary to increase the water content, avoid moisture rate limitations, and control dusting. As the reader might imagine, adding supplemental water was somewhat of a "hard sell" to the plant operators who had battled excessive water and a poor energy balance for so many years.

The Plattsburgh story points out some of the interesting problems of process control during composting. It is sometimes difficult to balance the seemingly conflicting demands imposed by the process. Temperatures should be high enough to raise reaction rates and destroy pathogens and seeds, but not so high that reaction rates are decreased by thermal inactivation. Drying is very desirable with most substrates, but not so much drying that reaction rates drop and dusting becomes a problem. Design and operating procedures to balance these competing demands are discussed further in Chapters 11 through 14.

SUMMARY

Substrate biodegradability is an important parameter that should be determined during design of any composting facility. Degradability of the feed substrates determines (1) the quantity of available heat, which is the major energy input driving the composting process, (2) the mass balance including the quantity of final product and the air supply needed for heat removal, and (3) the stoichiometric oxygen demand. Because biodegradability can vary widely and because the energy balance is so important to the process, actual degradability should be measured whenever possible.

Available data suggest that the degradability of substrates common to composting can vary widely. The degradability of lignocellulosic substrates during composting depends in large measure on their previous processing. A fairly high degree of decomposition is observed in paper products that have been chemically delignified, compared to those that are only mechanically pulped. For example, Kraft processed paper can have a degradability coefficient, k_s, approaching 0.9, compared to 0.1 to 0.3 for newsprint which is mechanically pulped. Typical values for the degradability coefficient, k_s, for municipal refuse are about 0.45. Grass straws tend to range from 0.35 to 0.50. Hardwood sawdusts and barks show significantly higher degradabilities on average compared to softwood species. However, even among softwood species, degradability can vary by a factor of 5. The wood species and its degradability must be considered during facility design, because the degradability between species can differ by over a factor of 10. This is particularly important if sawdust or bark is added as an "energy" amendment, because the available heat release could differ by the same factor of 10.

Substrate degradability appears to strongly correlate with lignin content for a wide range of substrates. An upper limit on degradability of 0.80 to 0.90 is suggested because the decomposition of the substrate organics is coupled with production of bacterial by-products. Some of these by-products are themselves not readily degradable.

A number of approaches are available to determine substrate degradability. First, literature reviews can be conducted. Data presented in this chapter is a good start but should not be considered an exhaustive review. Second, mass balances can be conducted at full scale or pilot facilities composting similar substrates. Third, compositional analysis of the substrate can be conducted to determine its lignin and other fractions. Lignin content appears to correlate with biodegradability. Fourth, direct laboratory testing of degradability can be conducted. A number of laboratory approaches have been used including batch and continuous anaerobic digestion, fermentations with rumen fluid, and aerobic respirometric techniques.

A constant pressure, aerobic respirometer can be assembled from generally available laboratory supplies. The respirometer can measure substrate degradability with large size samples. Experience with the apparatus has demonstrated that it can provide useful data at relatively low capital cost. Procedures to avoid rate limitations have been refined and a test methodology is presented. It is hoped that development of relatively simple procedures and equipment will make measurement of substrate degradability a more common practice for both design and operation of composting facilities.

Application of the respirometer to sludges and cellulosic amendments confirms that biodegradability can vary widely, particularly between wood species. Test durations over 500 days were required with some substrates to realize 90% of their ultimate oxygen demand.

REFERENCES

1. Gossett, J. M. and McCarty, P. L. "Heat Treatment of Refuse for Increasing Anaerobic Biodegradability," Civil Engineering Tech. Report No. 192, Stanford University, Stanford, CA (1975).
2. Golueke, C. G. "Bioconversion of Energy Studies at the University of California (Berkeley)," *Proceedings of the Bioconversion Energy Research Conference* (Amherst, MA: University of Massachusetts, 1973).
3. Pfeffer, J. T. and Liebman, J. C. "Biological Conversion of Organic Refuse to Methane, Annual Progress Report," NSF Grant No. GI-39191 (1974).
4. Klein, S. A. "Anaerobic Digestion of Solid Waste," *Compost Sci.* (January/February 1972).
5. Poincelot, R. P. and Day, P. R. "Rates of Cellulose Decomposition During the Composting of Leaves Combined with Several Municipal and Industrial Wastes and Other Additives," *Compost Sci.* (May/June 1973).
6. Lee, J. M. *Biochemical Engineering* (Englewood Cliffs, NJ: Prentice Hall, 1992).
7. Regan, R. and Jeris, J. S. "Cellulose Degradation in Composting," report to the U.S. Public Health Service by the Civil Engineering Department, Manhattan College, Research Grant EC 00161 (January 1971).
8. Youngberg, H. and Vough, L. "A Study of the Nutritive Value of Oregon Grass Straws," unpublished.
9. Klass, D. L., Ghosh, S., and Conrad, J. R. "The Conversion of Grass to Fuel Gas for Captive Use," in *Symposium on Papers Clean Fuels from Biomass, Sewage, Urban Refuse, Agricultural Wastes,* Orlando, FL, sponsored by Institute of Gas Technology (1976).
10. Poincelot, R. P. "A Scientific Examination of the Principles and Practice of Composting," *Compost Sci.* (Summer 1974).
11. Allison, *Organic Matter and Crop Management Problems,* (1965).
12. Chandler, J. A., Jewell, W. J., Gossett, J. M., Van Soest, P. J., and Robertson, J. B. "Predicting Methane Fermentation Biodegradability," *Biotechnology and Bioengineering Symposium No. 10* (New York: John Wiley & Sons, Inc., 1980).
13. Van Soest, P. J. *J. Animal Sci.* 26(1):119–128 (1967).
14. Kayhanian, M. and Tchobanoglous, G. "Computation of C/N Ratios for Various Organic Fractions," *BioCycle* 33(5) (May 1992).
15. Horvath, R. W. "Operating and Design Criteria for Windrow Composting of Sludge," in *Proceedings of the National Conference on Design of Municipal Sludge Composting Facilities* (Rockville, MD: Information Transfer, 1978).
16. Andrews, J. F. and Kambhu, K. "Thermophilic Aerobic Digestion of Organic Solid Wastes," report to the U.S. EPA, Office of Research and Development by Environmental Systems Engineering Department, Clemson University, EPA-670/2-73-061 (August 1973).
17. Schulze, K. L. "Aerobic Decomposition of Organic Waste Materials," Final Report, Project RG-4180 (C5R1), National Institute of Health (1961).
18. Umbreit, W., Burris, R., and Stauffer, J. *Manometric Techniques* (Minnesota: Burgess Publishing Co., 1964).

19. Pressel, F. and Bidlingmaier, W. "Analyzing Decay Rate of Compost," *BioCycle* 22(5) (September/October 1981).

20. Usui, T., Shoji, A., and Yusa, M. "Ripeness Index of Wastewater Sludge Compost," *BioCycle* 24(1) (January/February 1983).

21. Arthur, R. M. personal communication (November 1985).

22. Willson, G. B. and Dalmat, D. "Measuring Compost Stability," *BioCycle* 27(7) (August 1986).

23. Haug, R. T. and Ellsworth, W. F. "Measuring Compost Substrate Degradability," *BioCycle* 32(1) (January 1991).

24. "Standard Methods for the Examination of Water and Wastewater," 17th Edition, APHA, AWWA, WPCF, New York (1989).

25. Haug, R. T. "Nitrification with the Submerged Filter," PhD Dissertation, Department of Civil Engineering, Stanford University, (August 1971).

26. Haug, R. T. and McCarty, P. L. "Nitrification with Submerged Filters," *J. Water Poll. Contr. Fed.* 44(11) (November 1972).

27. Haug, R. T. and Ellsworth, W. F. "Degradability of Composting Substrates," *BioCycle* in press.

CHAPTER 10

Process Kinetics and Product Stability

INTRODUCTION

Kinetics is the study of rates or velocities of reactions. Kinetics must be distinguished clearly from the related subject of thermodynamics. The latter deals with energy changes accompanying chemical reactions but does not reveal how fast these reactions occur. For example, glucose oxidation is exothermic, but in the form of cellulose its biological oxidation proceeds slowly at best. If one should ignite the cellulose, such as by setting fire to this paper, oxidation proceeds rapidly indeed. The same total energy is released whether the paper is burned or biologically oxidized to carbon dioxide and water. However, the kinetics of the two cases are markedly different.

Substrate biodegradability was studied in Chapter 9 and is important in determining the thermodynamic limits of the system. The mass of biodegradable organic times its heat content determines the *quantity* of energy available to drive the process. This chapter deals with the *rate* at which this energy is released, in other words, the kinetics of the system. This is a subject of vital interest to the design engineer, who must determine the type of reactor and curing systems and the detention times required to achieve a given degree of organic stabilization. Measured rates of decomposition for composting substrates are reviewed along with the factors most responsible for limiting the rates of reaction.

Because kinetics deals with time, the concepts of hydraulic retention time, HRT, and solids residence time, SRT, are introduced. These parameters are important design criteria in achieving a desired product quality. The subject of product stability is also addressed in this chapter. It is included here because a product is "stable" when its rate of decomposition has been reduced to a low level, i.e., low kinetic rate.

MICROBIAL KINETICS

Rates of decomposition vary widely depending on the organic substrate. Chandler et al.[1] studied the anaerobic fermentation of substrates and found significant differences in the rate of fermentation between substrates. For example, the time required to evolve 90% of the cumulative methane production was 16 days for chicken manure, 33 days for cow manure, and 70 days for wheat straw. The decomposition rate for chicken manure was over four times as fast as that for wheat straw. If all other factors are equal, a composting reactor designed to achieve 90% BVS reduction would require over four times the residence time with wheat straw compared to chicken manure. This should highlight the importance of kinetic rates to facility design.

The discussion will return to field and laboratory measurements made on actual compost substrates, but first the theoretical concepts that govern the kinetics of microbe-substrate systems will be presented. These concepts will provide insight into the manner in which mass transport can be linked to chemical and biochemical reaction kinetics.

Classification of Microbe-Substrate Systems

Microbe-substrate systems are generally divided into two distinct types, homogeneous and heterogeneous. The systems are summarized in Table 10.1. In a homogeneous system, microbes are dispersed in an aqueous solution containing a soluble substrate. The mass of microorganisms is completely dispersed throughout the reactor volume. Concentration gradients of substrate between cells are minimized, and each cell sees virtually the same concentration if the system is well mixed. Such conditions are approximated in many industrial fermentations and waste treatment processes. The activated sludge process is an example of a nearly homogeneous system. It is not completely homogeneous, however, because flocculation causes some degree of separation between substrate and microbes. In addition, concentration gradients may exist inside the floc particle so that each organism is not surrounded by the same substrate concentration. Nevertheless, each floc particle is randomly dispersed in the fluid phase, and from this standpoint the activated sludge process can be viewed as a nearly homogeneous system.

Homogeneous systems are traditionally modeled using the Monod kinetics developed in Chapter 4.[2] The assumption is usually made that mass transport of substrate to the cell is not limiting, so that Equation 4.1 can be applied directly. Other authors have examined the effects of substrate diffusion through suspended flocs of microbes.[3,4]

A heterogeneous system is one in which either the microbes or substrate are separated from the fluid phase containing the other component. Two distinct types of heterogeneous systems are possible. The type most common to liquid waste treatment is a system in which microbes are separated from the fluid phase containing the substrate. The trickling filter, oxidation tower, rotating biological contactor, submerged filter, anaerobic filter, and anaerobic fluidized bed are examples of reactors that employ heterogeneous conditions. Each of these reactors contains an inert medium used to support the growth of microbes. Biological film constitutes one phase of the system, while liquid containing the substrate constitutes the other. A definite interface exists between the microorganisms and the liquid phase. Substrate must move across this interface before it can be used by the biological film. A concentration gradient must exist between the microbial film and the bulk liquid to assure a mass flow of substrate into the film. In addition to the waste treatment reactors described above, several industrial processes employ enzymes isolated from living organisms and immobilized on or within solid supports. Production of high-fructose syrups from corn starch is an example.[3]

A second type of heterogeneous system is one in which the substrate is insoluble and present in a particulate or solid form. Two subcategories of this system can be described: (1) the solid substrate is suspended in a bulk fluid phase, and (2) the aqueous phase is limited to

Table 10.1. Description of Biological Systems Normally used in Biochemical and Environmental Engineering Practice

System	Description
Homogeneous systems	Individual microbes are uniformly dispersed in a solution of soluble substrate
	The model is also applied to cases of flocculated microbes in a solution of soluble and fine particulate solids
	The activated sludge process treating industrial or municipal wastewater is typical of this system
Heterogeneous systems	
Attached microbial growth	Microbes are separated from the aqueous substrate usually by attachment to a solid surface
Falling film	Microbes are attached to a surface which is washed by a falling aqueous film containing the substrate
	The trickling filter, oxidation tower, and rotating biological contactor are examples
Submerged film	Microbes are attached to a surface with the void spaces filled with fluid containing the substrate
	The submerged filter, anaerobic filter and fluidized bed reactor are examples
Solid substrates	
Aqueous solution	Solid or insoluble substrate is immersed in an aqueous phase containing microbes, some of which attach to the substrate surface
	Industrial growth of microbes on insoluble hydrocarbon substrates and anaerobic digestion of sludge solids can be classified in this category
	The latter is also analyzed using homogeneous kinetics
Limited moisture	Moisture required for microbial growth is limited to that associated with the solid organic substrate
	Composting of organic residues and decomposition of organic solids in soils are examples

bound water associated with the solid substrate. In either case, microbes must attach to the substrate surface. Hydrolysis of chemical components making up the solid substrate is then necessary before the cell can absorb the solubilized substrate through its cellular membrane. In the limited moisture case, available water is limited to that associated with the solid substrate. Thus, an additional limitation can occur if water levels decrease to a point where they reduce the microbial reaction rates.

Most composting substrates consist of solid organic matter with moisture limited to that bound with the substrate. Thus, composting can be described as a heterogeneous system with solid substrate and limited moisture. Kinetics developed for solid substrate-microbe systems should apply reasonably well to the case of composting.

Heterogeneous Systems — Solid Substrate

The sequence of events involved in metabolizing solid substrates can be conceptually described as follows:

1. release of extracellular hydrolytic enzymes by the cell and transport of the enzymes to the surface of the substrate
2. hydrolysis of substrate molecules into lower molecular weight, soluble fractions
3. diffusion transport of solubilized substrate molecules to the cell

4. diffusion transport of substrate into the microbial cell, floc, or mycelia
5. bulk transport of oxygen (usually in air) through the voids between particles
6. transport of oxygen across the gas-liquid interface and the laminar boundary layers on either side of such an interface
7. diffusion transport of oxygen through the liquid layers bound to the solid substrate
8. diffusion transport of oxygen into the microbial cell, floc, or mycelia
9. aerobic metabolism of the substrate and oxygen within the microbial cell

A rather complicated sequence of events is necessary before substrate can be composted successfully. Because the above events are arranged in series, any one of the events could limit the overall process kinetics. Several of the more important processes are discussed further to gain insight into the rate limitations most common in composting systems.

Kinetics of Solubilization

Consider a hydrolytic enzyme that adsorbs to an active site on a solid substrate surface. The following equilibrium can be described:

$$E + A \underset{k_2}{\overset{k_1}{\leftrightarrow}} EA \overset{k_3}{\leftrightarrow} P + E \tag{10.1}$$

where A is a vacant site on the substrate surface, E is a free hydrolytic enzyme in solution that can adsorb to the surface with a reaction rate constant, k_1, and desorb with rate constant, k_2, and EA is the enzyme substrate complex that can either desorb to the original constituents with a rate constant, k_2, or react irreversibly to yield the original enzyme, E, and the desired product, P, with a rate constant, k_3. If a_o is the total number of adsorption sites per unit volume, then

$$a_o \leftrightarrow a + (ea) \tag{10.2}$$

where
a = number of free sites per unit volume
(ea) = number of sites with an adsorbed enzyme
e = number of free enzymes per unit volume of the reaction mixture

The change in concentration of the enzyme substrate complex can be described as

$$\frac{d(ea)}{dt} = k_1(e)(a) - k_2(ea) - k_3(ea) \tag{10.3}$$

Under steady-state conditions d(ea)/dt will equal zero. Thus,

$$k_1(e)(a) = k_2(ea) + k_3(ea) \tag{10.4}$$

Solving for a:

$$a = \frac{(ea)}{e}\left(\frac{k_2 + k_3}{k_1}\right) = \frac{(ea)}{e} K_a \tag{10.5}$$

Substituting Equation 10.5 into Equation 10.2 and rearranging:

$$(ea) = \frac{a_o(e)}{K_a + e} \tag{10.6}$$

The rate of product formation, v, is given in Equation 10.3 as $k_3(ea)$. Thus,

$$v = k_3(ea) = \frac{k_3(a_o)e}{K_a + e} \tag{10.7}$$

If e is the concentration of free enzyme, it can be related to the total concentration at the start of the experiment, e_o, by,

$$e_o = e + (ea) \tag{10.8}$$

In certain enzyme systems it is often the case that e_o is much greater than a_o. Equation 10.8 then becomes

$$e_o = e \tag{10.9}$$

Substituting Equation 10.9 for e, Equation 10.7 becomes

$$v = \frac{(k_3 a_o e_o)}{(K_a + e_o)} \tag{10.10}$$

Equation 10.10 is similar in form to Equation 4.1 developed for the case of a homogeneous system. In the present case, however, the rate of reaction reaches a maximum value at high enzyme concentrations. Physically, this corresponds to essentially complete adsorption of enzyme on the available substrate surface. Equation 10.10 has been used to describe kinetics of solid substrate-enzyme systems with reasonable success. Data on enzyme-solid substrate reactions are presented in Figures 10.1 and 10.2. The latter is a Lineweaver-Burke double reciprocal plot ($1/v$ vs $1/e_o$). From Equation 10.10 the reciprocal plot should be linear, which is the case in Figure 10.2.

The form of Equation 10.10 probably can be adapted to the case of microbes growing on a solid substrate. The concentration of extracellular enzymes, e_o, is likely a function of the mass concentration of microbes, X. Also, the total number of absorption sites, a_o, is likely related to the available surface area per unit volume, A_v. Substituting these terms into Equation 10.10 and dropping unnecessary subscripts:

$$v = -\frac{dS}{dt} = \frac{kA_v X}{K_x + X} \tag{10.11}$$

where
$\quad$ dS/dt $\ =\ $ rate of hydrolysis of solid substrate
$\quad\quad$ k $\ =\ $ maximum rate of solid substrate hydrolysis that occurs at high microbial concentration

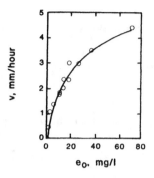

Figure 10.1. Dependence of the rate of disappearance of solid substrate (thiogel) on the concentration e_o of enzyme in solution. From Tsuk and Oster.[5]

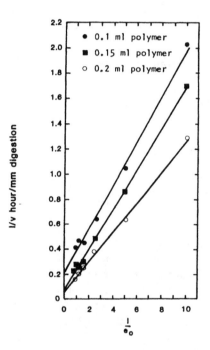

Figure 10.2. Double reciprocal plot for digestion of an insoluble substrate (poly-B-hydroxybutyrate particles) by an enzyme (depolymerase of *P. lemoignei*) in solution. From McLaren and Packer.[6]

K_x = half velocity coefficient equal to the microbial concentration
    where $ds/dt = k/2$

The general form of Equation 10.11 is graphically illustrated in Figure 10.3. In the limiting cases where X is very high $(X \gg K_x)$ and where X is very low $(X \ll K_x)$, Equation 10.11 is approximated by the following discontinuous functions:

$$\frac{dS}{dt} = -kA_v \qquad (X \gg K_x) \qquad (10.12)$$

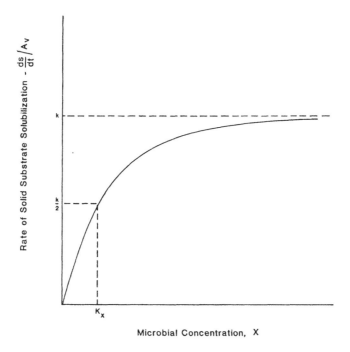

Figure 10.3. Rate of substrate hydrolysis as a function of the microbial concentration for a heterogeneous system with solid substrate.

$$\frac{dS}{dt} = -\frac{k}{K_x} A_v X \qquad (X \ll K_x) \qquad (10.13)$$

Equation 10.12 is a zero order reaction with respect to microbial concentration while Equation 10.13 is first order.

Actual composting processes are not as "ideal" as pure enzyme systems, and reaction kinetics have not been developed to the same extent as in biochemical engineering. However, the form of Equation 10.11 can be used to explain several phenomena observed in actual composting operations. Jeris and Regan[7] observed considerably lower oxygen consumption rates when composting newsprint compared to mixed refuse. Also, the effect of mechanical and chemical pulping in increasing the rate of degradation of wood and its products was discussed in Chapter 9. Differing rates of decomposition can probably be interpreted as differences in the value of kA_v. Thus, a substrate such as natural wood fiber, which is resistant to enzyme attack, would have a lower kA_v value relative to a substrate more amenable to solubilization by hydrolytic enzymes. This can be interpreted as a lower number of available enzyme binding sites or a lower number of successful enzyme reactions in a more resistant substrate.

The composting of most substrates is characterized by an initial period of high oxygen uptake followed by a longer period of low oxygen uptake. Complex substrates, such as refuse and sludges, are composed of a mixture of organics of varying kA_v values. During early stages, substrates with high values of kA_v are decomposing and the microbial population is increasing. Eventually the concentration of these "high rate" substrates is exhausted. However, substrates with low kA_v values continue to decompose at lower rates for a longer period of time.

Table 10.2. Comparison of Thermophilic Actinomycete in Various Seeding Materials

Material	Organism Count (number of colonies per gram of material)
Commercial inoculum	15.8×10^7
Rich soil	13.4×10^7
Poor soil	1.6×10^7
Horse manure	15.0×10^7

Source: Adapted from McGauhey and Gotaas.[10]

Klass et al.[8] studied the effect of particle size on the anaerobic decomposition of grasses. Fine graded grass, which passed a No. 30 sieve having 0.6 mm openings, decomposed faster than coarser grades. This was attributed to the higher surface area, A_v, of the fine grass which promoted faster solubilization of the solid substrate. Total gas production was about the same for all size fractions, only the rates of decomposition were effected by particle size. This observation is consistent with the form of Equation 10.11.

The Effect of Microbial Concentration

Referring to Equation 10.11, increasing the mass concentration of microbes, X, should increase the rate of solubilization as long as $X < K_x$. If the concentration increase much beyond K_x, however, the rate will approach a maximum value. This may have practical applications with many composting substrates.

The concentration of microbes necessary to avoid rate limitations has been a subject of controversy for many years. There is no question that most organic wastes will decompose through activity of the indigenous microbial flora. However, this does not assure that the microbial concentration is not limiting, particularly in the early stages of composting. In fact, lag periods are often observed at the start of batch composting operations, although the lag could also be caused by other factors such as oxygen availability, low starting temperatures, or poor feed conditioning. Certainly if the waste material is sterile, seeding with microbes should increase the kinetics according to Equation 10.11.

Golueke[9] distinguished between "minute" inoculation and "mass" inoculation with microbes. "Minute" inoculation referred to the introduction of a relatively minute quantity of microbes into a large quantity of substrate. The comparative number of thermophilic actinomycete isolated in various seeding materials is shown in Table 10.2. Consider a commercial inoculum consisting of 1 l of a 10^6 bacteria/ml suspension added to 1 ton of a substrate such as refuse. Compared to the background number of microbes for the substrates of Table 10.2, it seems inconceivable that such a small commercial inoculum could significantly increase the mass concentration of microbes unless the starting substrate is sterile. Use of such small additions of inoculum to increase the rates of reaction has generally been discounted. Results of a comparative study of inoculated and uninoculated composting material is shown in Figure 10.4. The striking similarity between the temperature curves indicates that the inoculum had minimal effect. Golueke[9] concluded that if the addition of "minute" inoculum contributed anything to facilitate the compost process it was so minute as to be undetectable. Even if the feed material were sterile it would seem that rich soil or horse manure would be as effective as a commercial inoculum.

Mass inoculation refers to addition of large quantities of microbial culture. This is generally accomplished by recycling compost product or using a completely or partially mixed reactor.

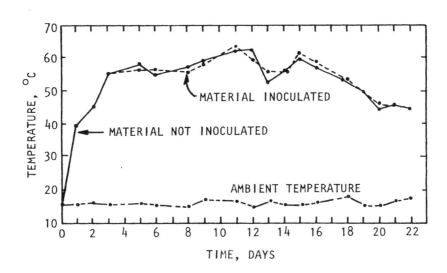

Figure 10.4.   Temperature curves showing similarity between "minute" inoculated and uninoculated aerobically composting material. Feed material was composed of shredded vegetable trimmings and paper. Inoculum was purported to be a culture high in thermophilic actinomycete. From McGauhey and Gotass.[10]

In the latter case the infeed is continually inoculated with the microbial population developed in the reactor. Using continuous composters operated on refuse materials, Regan and Jeris[11] examined the effect of seed recycle rates of 25, 50, and 75%, equivalent to R_d values of 1.3, 2.0, and 4.0, respectively. Referring to Figure 10.5 at 50 to 53°C, the composting rates with 25 and 50% seed were 10 and 90%, respectively, of the maximum rates with 75% seed. During these tests other environmental factors such as moisture, temperature, FAS, and aeration were held under conditions previously determined to be optimum. Therefore, the observed effect of recycle can probably be ascribed to the increased microbial concentration and not to other environmental factors influenced by product recycle.

In deference to the above, the literature contains conflicting reports on the utility of mass inoculation by compost recycle. Golueke[9] reported no significant acceleration of windrow refuse composting through product recycle. McGauhey and Gotass[10] examined the effect of product recycle, addition of soil, and addition of up to 30% horse manure to refuse. They concluded that none had any measurable effect on the rate of composting or the composition of the final product. On the other hand, product recycle is generally considered beneficial in sludge composting systems. However, the effect may be related more to improved structural conditioning than to increased microbial concentrations. Senn[12] observed that recycle of at least 10% of finished product during composting of raw manure greatly facilitated production of a relatively odorless material that did not attract or produce fly larvae. Without product recycle, similar composting temperatures developed, but the final product was odorous and developed fly larvae on rewetting. The APWA[13] also suggested that 1 to 10% recycle is beneficial to continuous reactor systems using refuse.

From the above discussion it is evident that the literature is unclear regarding the mass concentration of microbes required to avoid rate limitations. Indeed, values of K_x are likely a function of the type of substrate and should increase as the number of active sites per unit volume increases. Nevertheless, once the active sites on a substrate are saturated with enzymes ($e_o \gg K_a$ in Equation 10.10 or $X \gg K_x$ in Equation 10.11) the rate of solubilization should become constant. However, there are a number of other ways to further increase the rate of

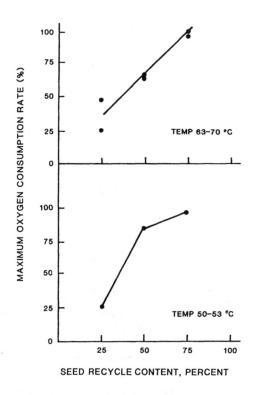

Figure 10.5. Effect of product recycle on the oxygen consumption rate measured during bench-scale composting of refuse. From Regan and Jeris.[11]

reaction. The value of k in Equation 10.11 should be a function of temperature and moisture content. The operator can provide proper conditions for each and also avoid other rate limitations such as low oxygen concentration. Beyond this, however, there is little else that can be done to improve the rate of solubilization except by altering the substrate molecule itself. The effect of chemical and mechanical pulping on cellulose degradation was discussed previously. Both pulping techniques serve to increase the susceptibility of the wood structure to enzyme attack. Grinding the substrate to increase the number of sites per unit volume should also increase the rate of reaction. For example, Pfeffer and Liebman[14] noted that smaller particle sizes resulted in faster rates of reaction during anaerobic digestion of refuse slurries.

Chemical techniques are also available to solubilize certain substrates and make the remaining solids more amenable to biological attack. Alkaline and enzymatic treatment to hydrolyze cellulosic material has been used to produce fermentable sugars. Gossett and McCarty[15] noted that heat treatment of refuse improved its degradability under anaerobic conditions. However, economic and practical considerations will likely limit the application of such techniques to composting.

The Role of Curing

As discussed above, solubilization of the solid substrate is probably a significant rate controlling mechanism during composting. Consider the composting of raw and digested sludges. During anaerobic digestion of municipal sludge ~50% of the solid organic substrate that enters the digester is solubilized and fermented. The solubilization reactions apparently proceed fast enough to prevent this step from limiting the overall reaction sequence.[3] Indeed, most kinetic models of anaerobic digestion assume that the subsequent fermentation to methane and carbon dioxide is rate-controlling. The point is that the outfeed from the digester has not been solubilized and has resisted hydrolytic attack in the anaerobic reactor. Such resistance is likely maintained in the aerobic composting environment and, therefore, the rate of hydrolysis is likely a serious rate-controlling step with digested sludge. Raw sludge, on the other hand, contains the resistant fraction characteristic of digested sludge as well as the fraction more conducive to hydrolysis. Therefore, rates of reaction may not be limited by solubilization in the early stages of composting, but they may become so in later stages after the readily hydrolyzed fraction has been degraded.

All composting substrates are likely to contain a fraction of solids resistant to hydrolytic enzymes, such as cellulose fiber and certain proteins. Furthermore, microbes synthesized on the feed organic will themselves contain resistant fractions, such as the cell wall structure. If these organic structures become saturated with microbes $(X \gg K_x)$ the rate of hydrolysis will become constant at $dS/dt = -kA_v$ (Equation 10.12). If the product of kA_v is small for the particular substrate, solubilization will require a considerable period of time. There is little the engineer can do to increase the rate other than to assure all environmental conditions are optimum to keep kA_v as large as possible. No particular reactor design or special inoculum of microbes is likely to reduce the time required as long as $X > K_x$. Therefore, the function of the curing phase is to allow time for the more resistant reactions to occur. There would appear to be little that can be done to shorten the time required for these resistant reactions to complete themselves.

Kinetics in the Aqueous Phase

Once the solid substrate has been solubilized, individual molecules can be transported by diffusion to the cell. The substrate is then transported across the cell wall and is biochemically metabolized by the cell. Diffusion resistance across the cell wall, any internal diffusion resistances, and the actual kinetics of metabolism are incorporated into the Monod kinetic model described in Chapter 4. Given the close proximity between microbes and substrate in the heterogeneous composting system, diffusion resistance through solution is likely low. Therefore, microbial kinetic rates defined by the Monod expression (Equation 4.1) probably govern:

$$\frac{dS}{dt} = -\frac{k_m SX}{K_s + S}$$

$$(4.1)$$

In the aqueous phase the rate of substrate use is a linear function of the microbial concentration, X, but nonlinear with substrate concentration. As explained in Chapter 4, zero order kinetics result when $S \gg K_s$ as a result of saturation of the microbial metabolic system to the point where substrate is processed at the maximum possible rate. This is exactly opposite to the kinetics of the heterogeneous system as represented by Equation 10.11. For the heterogeneous system, the kinetic rate is a linear function of the number of active sites, A_v, and nonlinear with the microbial concentration.

Whang and Meeneghan[16] applied the Monod equation to data developed from batch composting of cattle manure. The authors concluded that their results were reasonably modeled by the Monod expression. Using their data and assuming reasonable values for the manure characteristics, the product of $k_m X$ ranged from 0.005 to 0.014 day^{-1}. Operating conditions were not specified so the temperatures corresponding to these rate constants are not known. The development of similar rate constants from respirometric data will be discussed later in this chapter.

Kinetics of Oxygen Transport

Along with the organic substrate, oxygen must be available to complete aerobic metabolism. To supply this oxygen, air must first be supplied to the airspace within the composting mixture. Oxygen will then transport to the gas/liquid interface, diffuse across the interface, and then diffuse through the liquid phase to the microbes. Consumption of the oxygen by the microbial population produces a concentration gradient causing further diffusion from the airspace. Conversely, metabolic end products such as CO_2, H_2O, and NH_3 will be at elevated concentrations in the liquid phase and will diffuse toward the airspace and ultimately be removed with the gas flow.

Mass transfer across a gas/liquid interface is widely analyzed using the two-film model developed by Lewis and Whitman in 1924. An illustration of the idealized system is presented in Figure 10.6. Two laminar films are envisioned adjacent to the interface and provide resistance to mass transport of the gas molecules. Applying Fick's Law of molecular diffusion,

$$\frac{dF}{dt} = -AD_l \left(\frac{dS}{dz} \right)_l = -AD_g \left(\frac{dS}{dz} \right)_g \qquad (10.14)$$

where

dF/dt = mass rate of substrate transfer, mass/time
A = surface area, length^2
D_l = diffusion coefficient in the liquid phase, length^2/time
D_g = diffusion coefficient in the gas phase, length^2/time
dS/dz = substrate gradient in a direction perpendicular to the surface layer, mass/volume/length

Assuming linear concentration gradients as shown in Figure 10.6, Equation 10.14 becomes

$$\frac{dF}{dt} = -AD_l \left(\frac{C_i - C_l}{\delta_l} \right) = -AD_g \left(\frac{P_g - P_i}{\delta_g} \right) \qquad (10.15)$$

Gas phase diffusion coefficients are typically greater than those in the liquid phase by a factor of ~10^4. For slightly soluble gases, therefore, it has been shown that essentially all resistance to mass transfer lies on the liquid film side.[3] Under these conditions oxygen transport in the liquid phase becomes rate limiting and any resistance in the gas phase can be neglected.

As oxygen diffuses through the aqueous phase it will be consumed to support the biochemical oxidation of the substrate. This produces a concentration gradient allowing further mass transport into the composting matrix. Because oxygen is in the aqueous phase, its rate of consumption should be governed by the Monod kinetic model. However, application of such a model is difficult without better information on the active microbial concentration, X. Instead, a more simplified approach can be used to provide an order of magnitude estimate of

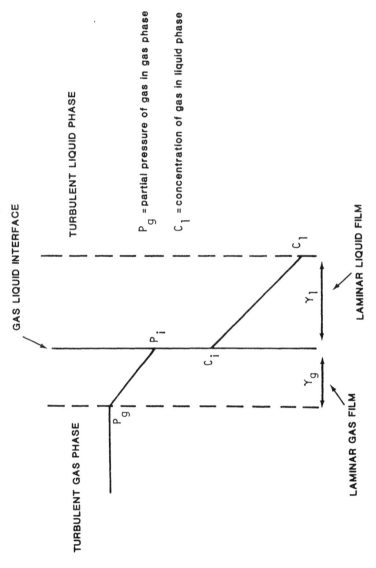

Figure 10.6.   Idealized two-film model for mass transfer from the gas to the liquid phase.

the mass transfer rate for oxygen. The calculated rates can then be compared with measured oxygen consumption rates to indicate the conditions under which oxygen supply can be rate-limiting.

The diffusion coefficient is an important parameter in mass transport. It determines the rate at which substrate materials diffuse into the matrix and the rate at which metabolized end products diffuse out of that matrix. Flow conditions within a composting particle are likely to be very quiescent. Therefore, mixing by turbulent or eddy diffusion should be negligible. The actual diffusion coefficient should approach the value of the molecular diffusion coefficient, the limiting value imposed solely by molecular motion of the diffusing materials.

Experimentally determined values for the molecular diffusion coefficient of nonelectrolytes such as oxygen can be found in Reid and Sherwood[17] and Perry and Chilton.[18] Values for dilute solutions can also be estimated from the Wilke-Chang Equation:[18]

$$D = \frac{7.4 \times 10^{-8} T_k (FM)^{0.5}}{n(V_o)^{0.6}} \qquad (10.16)$$

where

$$
\begin{aligned}
D &= \text{diffusion coefficient, cm}^2/\text{sec} \\
n &= \text{solvent viscosity, cP} \\
F &= \text{association factor for solvent, 2.6 for water} \\
M &= \text{solvent molecular weight} \\
V_o &= \text{solute molal volume at normal boiling point, 25.6 cm}^3/\text{g-mole} \\
&\quad \text{for oxygen} \\
T_k &= \text{temperature, °K}
\end{aligned}
$$

Equation 10.16 can be used to estimate the molecular diffusion coefficient for small molecules in low-molecular-weight solvents, usually to better than 10 to 15% accuracy.

The actual diffusion coefficient in a composting particle and its water layer may be less than the molecular diffusion coefficient through water alone. Blocking of the diffusing molecules by particulate matter or by changes in the viscosity of the fluid itself can reduce the diffusion coefficient. If molecules are forced to diffuse around particulate matter, such as a bacterial cell, the diffusion coefficient decreases because of the increased path length. If viscosity of the fluid increases, as might occur if the bacteria secreted a slime layer or extracellular polysaccharide, the resistance to passing molecules would increase and the diffusion coefficient decrease.

A substantial amount of work has been conducted to determine diffusion coefficients through biological materials. Most of the measured values for biological materials approximate the values found for water.[19-25] Atkinson et al.,[26] working with glucose oxidation by a fixed film growth, assumed that the diffusion coefficient within the film was equal to the molecular diffusion coefficient for glucose. The assumption worked well under their experimental conditions. Mueller[4] experimentally determined the oxygen diffusivity through pure culture flocs of *Zoogloea ramigera*. Oxygen diffusion values ranged from 0.1 to 2 times that of the molecular diffusion coefficient, primarily because of difficulty in measuring the surface area of the floc particles. However, when the nominal diameter of the floc particle was used to determine the surface area, experimental values were reasonably close to the molecular diffusion coefficient. In detailed experiments by Williamson and McCarty[27,28] diffusion coefficients were measured for NH_4^+, NO_2^-, NO_3^-, and O_2 through nitrifying films. Values ranged from 80 to 100% of corresponding values through water. It was concluded that a value of 80 to 90% of the molecular diffusion coefficient in water would be a reasonable estimate of the actual coefficient in the biofilm.

The simplified model shown in Figure 10.7 can be used to estimate oxygen diffusion rates through a water-saturated matrix of solid substrate and microbes. Diffusion is assumed to occur from both sides of the particle. The oxygen concentration decreases linearly from a saturation value at the outer interface to zero at the particle midpoint. Obviously, such a model is a greatly simplified version of the actual composting matrix. However, it is unlikely that a more complex model would yield improved results because of the numerous ill-defined factors and unknown kinetic coefficients. The simple model of Figure 10.7 is used in the following example to determine the importance of oxygen diffusion as a rate-controlling mechanism.

Example 10.1

Using the simplified model of Figure 10.7a, estimate the oxygen flux at 60°C for a particle thickness of 0.05 cm. Assume a saturation oxygen concentration of 6 mg/L at the gas-particle interface. Estimate the time required to supply the stoichiometric oxygen quantity assuming the matrix to be 50% TS (S_m = 0.5), with 50% VS content (V_m = 0.5) and degradability of 50% (k_m = 0.5).

Solution

1. Estimate the liquid phase diffusion coefficient D_l for oxygen using the Wilke-Chang correlation. Assume the viscosity of water at 60°C to be 0.45 cP:

$$D_l = \frac{7.4 \times 10^{-8}(273+60)}{0.45} \left\{ \frac{[2.6(18)]^{0.5}}{25.6^{0.6}} \right\}$$

$$D_l = 4.62 \text{ cm}^2 / \text{day}$$

2. Estimate the specific gravity and bulk weight of the particle using Equations 6.1 and 6.3:

$$\frac{1}{G_s} = \frac{0.50}{1.0} + \frac{1-0.5}{2.5}$$

$$G_s = 1.43$$

$$\delta_s = \frac{1.0}{\dfrac{0.5}{1.43} + 1 - 0.5} = 1.18 \text{ g/cm}^3$$

3. Determine the flux across the interface. The gradient of oxygen is given by

$$\frac{dS}{dz} = -\left(\frac{6 \text{ mg} / L}{0.025 \text{ cm}} \right) = -240 \text{ mg} / L - \text{cm}$$

Using Equation 10.15 and considering 1 cm² of area on each side of the particle:

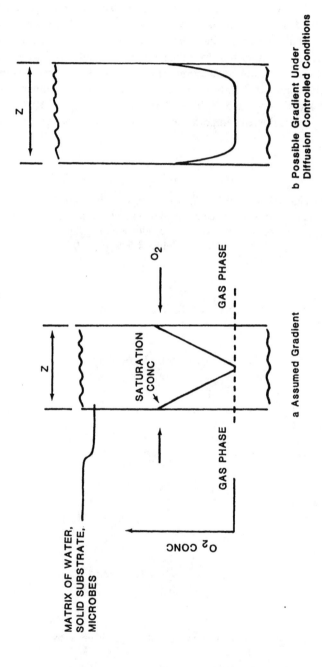

Figure 10.7.   Simplified model of gas transfer used to estimate oxygen transport rates through a saturated matrix of solid substrate and microbes.

$$\frac{dF}{dt} = -(2.0 \text{ cm}^2)(4.62 \text{ cm}^2 \text{ / day})(-240 \text{ mg / L} - \text{cm}) \frac{1 \text{ L}}{1000 \text{ cm}^3}$$

$$\frac{dF}{dt} = 2.22 \text{ mg O}_2 \text{ / day}$$

4. Each cm² of particle surface will contain 1.0(0.05) = 0.05 cm³ of volume. The matrix mass in this volume is 1.18(0.05) = 0.059 g. The mass of volatile solids VS is

$$VS = 0.059(V_m)(S_m)$$

$$VS = 0.059(0.5)(0.5)$$

$$VS = 0.0148 \text{ gm}$$

The oxygen flux can then be expressed as

$$\frac{dF}{dt} = \frac{2.22}{0.0148} = 150 \text{ mg O}_2/\text{g VS} - \text{day}$$

5. The time required to satisfy the stoichiometric demand can be estimated assuming about 2 g O_2/g VS (see Equation 7.1) and 50% degradability:

$$\text{stoichiometric demand} = 0.0148(0.5)(2.0) = 0.0148 \text{ g O}_2$$

$$\text{time required} = \frac{0.0148(1000)}{2.22} = 6.6 \text{ days}$$

Oxygen flux and time requirements were calculated for various particle sizes using the same approach and assumptions as in Example 10.1. Results are presented in Figure 10.8 along with the observed range of oxygen demands discussed later in this chapter. Oxygen flux is shown to decrease as the particle thickness increases. This obviously leads to an increase in the time required to supply the stoichiometric oxygen requirement. Conditions imposed by the model may be overly severe because the diffusion gradient decreases as particle size increases. In actual fact the gradient may remain constant, leading to the concentration profile shown in Figure 10.7b. Such a profile would be expected when process kinetics are diffusion controlled. Under such conditions the flux may no longer be a function of particle thickness. However, time required to satisfy the oxygen demand would still increase with increasing particle thickness, but at a lower rate than shown in Figure 10.8.

Despite the modeling uncertainties, it appears that diffusion transport of oxygen can match the oxygen consumption rate if the particle size is sufficiently small. Particle size on the order of 1.0 cm appears to result in large diffusion resistances which would dominate the process kinetics. Particle sizes of about 0.10 cm and lower appear to be small enough that diffusion supplied oxygen can balance the observed rates of oxygen demand. If particle thickness decreases below about 0.05 cm, oxygen diffusion would have a negligible effect on process kinetics. Golueke[9] noted that complete aeration of all particles "would involve reducing all particles to a size less than a millimeter or two, because by its very dimensions a particle any

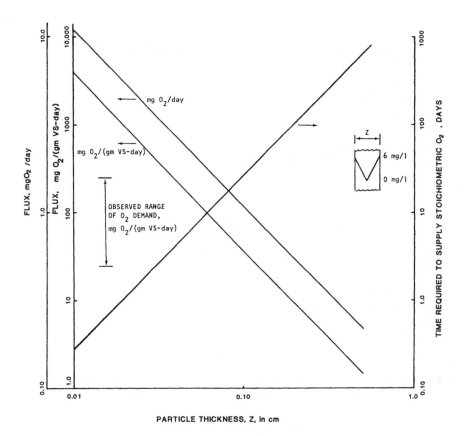

Figure 10.8.   Effect of particle thickness on oxygen flux and time needed to satisfy the stoichiometric oxygen requirement. Values were calculated using the simplified model of Figure 10.7a and the procedure and assumed values of Example 10.1.

larger could be anaerobic in its interior." Although no data were supplied to support this statement, the general agreement with the simplified model presented here is interesting.

Several other observations can be drawn from the model results. The fact that oxygen is present in the pore spaces between composting particles does not mean that oxygen diffusion is not rate-limiting. Oxygen is required *in* the matrix of water, substrate, and microbes. The fact that oxygen is present in the pore space does not necessarily mean that the flux rate of oxygen into the particles is not rate-limiting. It is common practice to use oxygen probes to measure the oxygen content of gases in the composting material. This is a useful practice but it only assures that oxygen is present in the pore space. It should not be interpreted to mean that all oxygen rate limitations have been removed. Indeed, process kinetics could still be controlled by diffusion transport within the compost particles.

The fact that diffusion transport is enhanced by smaller particle size leads to an unavoidable paradox for systems that rely on natural draft ventilation for aeration. Referring to Chapter 7, it was concluded that (1) natural draft was a primary ventilation mechanism in nonaerated windrow composting systems and (2) that natural ventilation is enhanced by larger particle sizes that increase the dimensions of the void spaces. Ventilation is enhanced by larger particle size whereas diffusion transport is not, an interesting paradox. A balance between these competing effects is necessary to assure that diffusion transport is not overly enhanced at the sake of oxygen supply, or vice versa.

Finally, it should be noted that the model assumed a particle with a continuous matrix of water/substrate/microbes and no void spaces within the particle itself. With wet substrates, such as sludges, wet manures, and wet sawdust, the water will fill many of the voids and pores within the substrate particles. The model assumptions would seem to be reasonable in this case. As composting proceeds, however, water will be removed from the micropores and voids within the substrate. Evaporation or absorption of excess water by recycled product, bulking agent, or amendment could be responsible for the moisture removal. Other substrates, such as dry wood fiber, will contain a significant volume of open micropores within the particle. As previously noted, diffusion coefficients through a gaseous phase are about four orders of magnitude greater than through the liquid phase. Therefore, diffusion through small gas pores within the substrate particle itself should occur rapidly. Also, the interfacial area for transfer into the liquid phase will be greatly increased by the micropores. Thus, diffusion transport of oxygen will be a more controlling factor with wet substrates and assume less significance with dry substrates or as moisture is removed from wet ones.

Potential Rate Limitations

The previous sections examined a number of possible rate limitations, including solubilization of the solid substrate and mass transport of both oxygen and the solubilized substrate to the cell. The effect of moisture content on the mass transport of oxygen was also discussed. However, even if oxygen and a solubilized and degradable substrate are available to the cell, there are a number of other factors that can limit microbial kinetics. Microbial reaction rates can be limited by at least the following:

- lack of degradable organics
- very low or high process temperatures
- low moisture conditions
- lack of free air space
- low oxygen content
- imbalanced pH conditions
- lack of inorganic nutrients
- lack of microbes (sterile substrate)
- the presence of toxic substances

Structural conditioning to adjust FAS and chemical conditioning to remove nutrient and pH rate limitations were discussed in Chapter 6. Aeration needs to reduce oxygen limitations were discussed in Chapter 7. Energy conditioning to assure an adequate supply of degradable organics was discussed in Chapters 8 and 9. The use of product recycle for seeding was discussed previously in this chapter and can be used to reduce rate limitations from a lack of sufficient microbial mass. Most composting substrates do not contain toxic substances in sufficient quantity to limit process kinetics. Therefore, the present discussion will address the effects of moisture and process temperature.

Moisture

The effect of excessive moisture on reducing FAS and limiting the mass transport of oxygen has already been discussed. The opposite case, lack of sufficient moisture, can also limit the reaction kinetics. A number of factors probably combine to account for this effect.

First, most biological reactions during composting are mediated by bacteria that require an aqueous environment. Second, mass transport limitations for soluble components may be encountered under low moisture conditions. In general, moisture contents should be maintained as high as possible without violating requirements for minimum FAS.

Literature data on the effect of moisture content on oxygen consumption rates for various composting materials are presented in Figure 10.9. The trend toward decreasing reaction rates at low moisture content is clearly evident. Below 20% moisture very little, if any, biological activity occurs. From that point rates of oxygen uptake increase in a more or less linear fashion to maximum values that begin at about 50 to 70% moisture. Rates begin to decrease again at high moisture contents, undoubtedly from loss of FAS. This points out the difficulty in isolating effects of moisture alone in such experiments because of the relationships between moisture, bulk weight, and FAS. This probably accounts for much of the data scatter in Figure 10.9, as well as the slightly different trends observed for the various composting materials.

Composting tends to be a drying environment. Even if the initial mixture is conditioned to a proper moisture content, supplemental water addition may be necessary at periodic intervals to avoid rate limitations. This is particularly true with dry substrates, but it may also be true with wet substrates that are "energy rich". The provision for supplemental water addition should be considered during design of any composting system.

Process Temperatures

All organisms have a temperature range over which biochemical functions can be maintained. Within certain limits the rate of these biochemical reactions about doubles for each 10°C rise in temperature. However, excessively low or high process temperatures can limit the overall process kinetics.

Microbial rates of reaction are markedly reduced as the temperature is decreased from 20 to 0°C. If the starting substrates are very cold or frozen and the input air is also cold, a significant lag period may be observed before thermophilic temperatures are developed. In extreme cases, the pile may fail to show any significant temperature rise. This phenomenon is occasionally observed in actual operations in cold climates. When all feed temperatures are low (i.e., near 0°C), newly formed, batch operated piles have been observed to sit with little temperature development. Recourse in such situations has often been to apply additional insulating layers to the pile and reduce the air flowrate. These measures would be effective if the problem were one of process thermodynamics. Unfortunately, it is not, and the above procedures will have only a marginal effect.

The problem with low substrate temperatures is primarily a kinetic problem. Reaction rates are so slow that the rate of heat generation is less than the rate of heat loss to the cold surroundings. The energy is still available within the substrate, but it is not released in significant amounts because of the low rate of biological activity. The pile sits and appears to do nothing. The author has observed composting operations where frozen substrates are piled and then aerated with sub-zero air with the expectation that the pile will elevate in temperature. There is a very good chance that it will not. One must remember that freezing is a method of preservation because it so greatly reduces microbial reaction rates. In effect, the energy to perform the composting task is thermodynamically present, but the spark is slow to ignite the process.

The best approach to overcome low temperature kinetic limitations in cold climates is to assure that any heat in the feed substrates is conserved to give as high a starting temperature as possible. If the substrates are frozen and the air supply is subzero it will be difficult to "ignite" the process. Therefore, efforts should be made to avoid loosing heat from the substrate. For example, sewage sludges should be stored adiabatically or processed as soon as

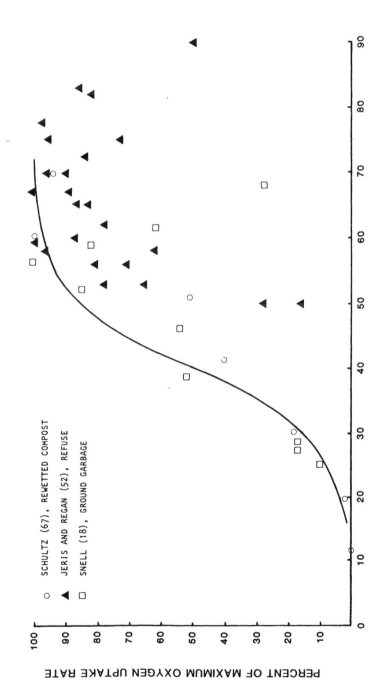

Figure 10.9.   Effect of low and high moisture contents on oxygen consumption rates measured for various composting materials.

possible to avoid losing temperature. If this is not sufficient it may be necessary to heat the substrates or preheat the air supply during the initial lag period until the process "ignites". Once started, further heating should be unnecessary. Air preheat could be supplied from conventional fuel sources or by heat exchange with hot exhaust gases.

The potential for cold weather effects on process kinetics highlights a distinction between batch and continuous feed, well-mixed processes. Infeed temperature in a well-mixed system has very little impact on process kinetics because it is the reactor temperature that governs. As long as reactor temperatures are thermophilic, the cold feed will be mixed with the hot reactor contents and low temperature kinetic problems are avoided. This is a significant advantage for compost operations in very cold climates.

If extremely cold temperatures can impose kinetic limitations, so too can very high process temperatures. In a mixed population of microbes, a natural transition from mesophilic to thermophilic temperatures can be expected. Reaction kinetics will tend to increase exponentially with temperature. However, there is an upper limit to the exponential increase with temperature. At some point even the thermophilic microbes cannot overcome the effects of thermal denaturation of their enzymes. At this point, rates of reaction will decrease with further increases in temperature. The temperature corresponding to the maximum reaction rate has been the subject of much debate. Different investigators have reported optimum values during composting ranging from as low as 40 to as high as 70°C. It should be noted, however, that nature's microbes are extremely resilient. Clark[29] reported isolating extreme thermophiles from deep sea hydrothermal vents with temperature optima of 85°C and maximum growth temperatures as high as 110°C.

The point to remember here is that extremely high temperatures do not necessarily imply a high rate of reaction. This is a common mistake because one is inclined to measure composting effectiveness by high temperature elevation. Should temperatures develop to the range of 75 to 85°C, rates of reaction are likely depressed because of the high temperature. Process kinetics would be improved by reducing the temperature. To accomplish this, generated heat must be removed at a greater rate. Increasing the aeration rate will increase the rate of heat loss and is the most effective approach.

The problem of high process temperatures is not unique to composting. Many industrial fermentations also face the problem of high heat generation rates. Because industrial fermentations are usually conducted in aqueous solution, cooling tubes can be placed in the fermenter to remove the excess heat. Unfortunately, this is not practical in a composting system.

RATES OF BIODEGRADATION

Background Data

Oxygen consumption rates during composting have been investigated extensively by numerous researchers using a variety of experimental procedures and feed materials. Both batch and continuous composters have been used. Most data have been developed using controlled, laboratory or bench scale composters where various factors, such as temperature, pH, moisture, and free airspace, can be held reasonably constant. Many studies have focused on garbage and refuse materials, probably because of the universal concern over their proper management. Results of some of the more notable studies will be presented here.

Among the more thorough studies of oxygen uptake rate were those conducted by Schulze.[30-33] His experiments are a good example of the protocols required for such studies. Continuous composters were operated to achieve steady state conditions. Each composter consisted of a 55-gal rotating drum that was normally filled to about two thirds of its volume. Feed materials consisted of various garbage/sludge mixtures. Feed material was added every 1 to 2 days and the composter was rotated for about 5 min before and after each feeding. Thus, conditions expected in a semicontinuous feed, well-mixed reactor were simulated. Data were collected after establishment of steady state conditions within the reactor.

Schulze examined a number of feed mixtures composed of one or more of the following: (1) shredded garbage consisting primarily of table scraps; (2) dewatered, digested sewage sludge containing 70 to 80% moisture conditioned with about 4% $FeCl_3$ and 12% lime on a dry-weight basis; (3) air-dried digested sludge; (4) air-dried compost; (5) shredded wastepaper, mostly newsprint; and (6) vermiculite, an expanded mica consisting of particles 0.6 to 1.3 cm in size. The vermiculite served as a bulking agent in the same way that wood chips are used in the static pile process. Data on the average composition of the substrate mixtures are presented in Table 10.3.

During initial testing it was found that plain ground garbage, a mixture of garbage and dewatered sludge, or dewatered sludge cake by itself, would not successfully compost. Material in the drum became too dense and formed large balls that impeded oxygen transfer. Because the drum was rotated only intermittently, the tumbling action was apparently insufficient to supply oxygen to the wet material. Schulze then conditioned the wet substrates using the dry amendments noted above to reduce the mixture moisture content to 50 to 60%. For all mixtures listed, the FAS ranged from about 40 to 60%. It is felt that bulk weight measurements recorded by Schultz may have been low, particularly for dewatered sludge cake. Therefore, the values of FAS shown in Table 10.3 may be higher than those actually present in the drum reactor.

Schulze concluded that ground food waste and dewatered sludge cake are too high in bulk weight and moisture to compost as such and that "those materials have to be mixed with a dry and bulky component such as refuse, waste paper, corncobs, wood shaving, rice hulls, etc. in order to obtain a suitable moisture content and bulkweight." This was a prophetic statement made in 1962, a decade before the successful sludge composting operations by the Los Angeles County Sanitation Districts, which began in 1972 and pioneered the use of product recycle for conditioning, and by the U.S. Department of Agriculture at Beltsville, Maryland, which began in 1973 and pioneered the use of bulking agents.

Average operating data observed during the test periods are presented in Table 10.4 and Figures 10.10 to 10.12. Residence times varied between 7 and 18 days and volatile solids, VS, destructions from 37 to 45%. Detention times were estimated from the total weight of ash in the reactor divided by the daily input of ash. Moisture content of the outfeed approximated that of the infeed, indicating that air supply was adjusted primarily to the stoichiometric demand rather than that required for drying. Temperatures remained constant in the thermophilic range between 43 and 68°C.

Rates of oxygen consumption were determined throughout run 2c which used a mixture of garbage, sludge cake, and vermiculite. Results are shown in Figure 10.13. Oxygen consumption was a function of temperature and followed the relationship

$$w_{O_2} = 0.11(1.066)^T \tag{10.17}$$

where

w_{O_2} = rate of oxygen consumption, mg O_2/g VS-h

T = temperature, °C

Table 10.3. Average Analytical Data for Raw Materials and Mixtures Used by Schulze [30]

Item	Moisture (% fresh weight)	Ash (% dry weight)	pH	Wet Bulk Weight (g/L)	Specific Gravity (g/cm³)	Dry Bulk Density (g/cm³)	Porosity (%)	Free Airspace[a] (%)
Ground garbage	63	10	5.9	740	1.064	0.237	77.7	27.4
Moist sludge cake	72	50	8.2	660	1.43	0.185	87.1	39.6
Dry sludge cake	6.0	50	8.4	390	1.43	0.367	74.3	72.0
Vermiculite	1.0	100	7.5	90	2.5	0.09	96.4	96.3
Dry compost	10	60	8.0	290	1.563	0.261	83.3	80.4
Shredded paper	8.0	8.0	5.0	25	1.0	0.023	97.7	97.5
Mixture A[b]	47.8	31.3	6.7	642	1.23	0.353	72.8	42.1
Mixture B[c]	56.5	49.4	5.9	410	1.42	0.178	87.4	64.3
Mixture C[d]	50.5	22.1[e]	6.2	410	1.154	0.203	82.4	61.7
Mixture D[f]	60.0	17.3[e]	6.0	410	1.152	0.164	85.8	61.2
Mixture D₁[g]	57.0	23.2	6.0	410	1.153	0.177	84.7	61.3

[a] Recalculated from original data according to Equation 7.10.
[b] Mixture A = 20 lb garbage, 10 lb air-dry sludge cake.
[c] Mixture B = 20 lb garbage, 10 lb moist sludge cake, 3 lb vermiculite.
[d] Mixture C = 20 lb garbage, 5 lb air-dry sludge cake, 4 lb paper.
[e] Computed from average data for components.
[f] Mixture D = 20 lb garbage, 10 lb moist sludge cake, 5 lb paper.
[g] Mixture D₁ = 20 lb garbage, 10 lb moist sludge cake, 5 lb paper, 2.5 lb air-dry compost.

Table 10.4. Average Operating Data Observed During Continuous Thermophilic Composting

	Run No. 1	Run No. 2	Run No. 2c	Run No. 3b
Feed mixture	A	B	B	D, D_1
Feed cycle, days	1	2	1	2
Test duration, days		52	75	167
Data collected, days[a]	11	34	23	35
Residence time, days	8.9	12.7	7.0	18.3
Red. in vol. mat., %	36.8	43.1	42.4	45.2
Moisture, %				
In	47.8	56.5	58.6	57.0
Out	51.3	55.2	58.4	56.9
pH				
In	6.7	5.9	5.6	6.0
Out	7.6	7.8	6.6	8.1
Wet weight, g/L				
In	642	410	404	412
Out	657	567	587	611
Free airspace, %[b]				
In	42.1	64.0	64.8	61.1
Out	42.4	52.8	50.6	44.3
Air supply, m^3/kg VS-day	0.28	0.55	0.73	0.31
Temperature range range, °C	43–64	53–68	59–68	62–68

Source: Schulze.[30]

[a] Indicates time over which data were collected. For Run 2b, e.g., data collection began on day 18 and was completed on day 52.
[b] Recalculated from original data according to Equation 6.11.

The rate of oxygen consumption continued to increase with temperatures up to 68°C, which was the maximum observed in run 2c. Experience has shown, however, that the oxygen consumption rate would be expected to decrease at higher temperatures.

The pH of the reactor contents was consistently above 7, except for run 2c as shown in Figure 10.12. In this case about 25% of the volatile matter in the reactor was replaced with each daily feeding, the highest feed rate attempted. Apparently, the high loading rate shifted the process into the acidic range, which indicates that the reactor output was not as close to a finished compost as material produced at lower feed rates.

Material removed from the drums was stored in open bins and for several days developed temperatures near 40°C. After 2 to 3 weeks of storage, temperatures generally decreased to ambient and the compost reportedly had a pleasant greenhouse odor. Odor from the sewage sludge was reported to be absent. Thus, a curing phase was required even after continuous thermophilic composting at residence times from 7 to 18 days.

Jeris and Regan[7,11,34,35] conducted an interesting set of experiments using both batch and continuous composters fed with mixed refuse, newsprint, and compost produced from mixed refuse at the circular, agitated bed reactor system at Altoona, Pennsylvania (see Chapter 2). The batch experiments were conducted using the Warburg respirometer and shaker flasks. The continuous composters were similar in concept to those used by Schultz, but differed in that process temperatures were artificially maintained constant by heating rods compared to the self-heating and, therefore, somewhat variable conditions used by Schultz. The mixed refuse was a simulated mix of cafeteria and supermarket wastes (meat and fish scraps, lettuce, and other vegetable trimmings) and newsprint. The mixture contained 60 to 70% paper and was considered a high cellulosic substrate. Because of the high paper content, nitrogen and phosphorus were added to provide supplemental nutrients. Tap water was added to adjust the

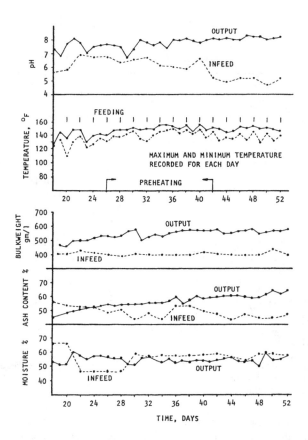

Figure 10.10. Average operating data recorded by Schulze[30] during continuous composting of a garbage/sludge mixture. Feeding was on a 2-day cycle. Data correspond to Run No. 2b, using Mixture B as identified in Table 10.3. Air supply was approximately 6.5 mg O_2/g VS-h with a detention time based on an ash balance of 12.7 days. The data indicate that reasonably steady state conditions were achieved.

moisture content. It was assumed that the tap water supplied any necessary trace elements. The materials were shredded to a particle size no greater than 0.25 inch.

Oxygen consumption rates measured on the various substrates are presented in Figure 10.14. The compost and newsprint showed oxygen consumption rates about one order of magnitude less than freshly mixed refuse. This illustrates that compost may be stable, but it is certainly not inert. It will continue to decompose, but at greatly reduced rates. The low rates observed for newsprint are caused by the structural resistance of mechanically pulped wood products. Jeris and Regan[11] developed best-fit equations for their data over the temperature range from 35 to 70°C. These were adjusted for units and are as follows:

Newsprint

$$w_{O_2} = -0.00147(T^2) + 0.1413(T) - 2.907 \qquad (10.18)$$

Composted mixed refuse

$$w_{O_2} = -0.00133(T^2) + 0.1013(T) - 1.587 \qquad (10.19)$$

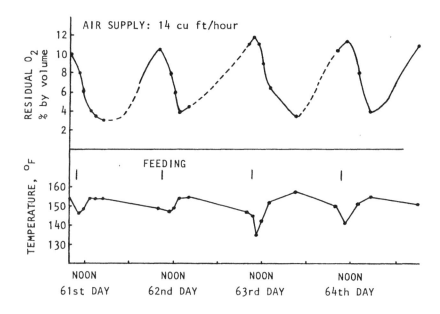

Figure 10.11. Relationship between feed cycle, temperature, and residual oxygen during continuous composting as determined by Schulze.[30] Data correspond to Run No. 2c, Mixture B, as defined in Table 10.3.

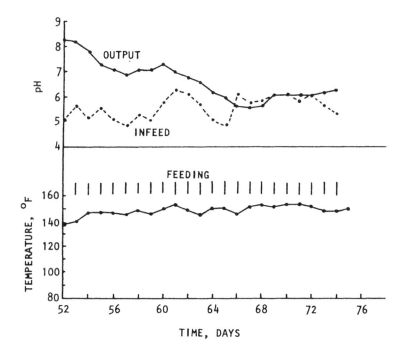

Figure 10.12. Effect of high organic loading on product pH. Data correspond to Run No. 2c, Mixture B, as defined in Table 10.3. The daily feed schedule reduced the detention time to 7 days. Apparently, at the high loading rate the rate of organic acid production exceeded the rate of acid consumption and the pH shifted from the slightly alkaline range to the acid range. From Schulze.[30]

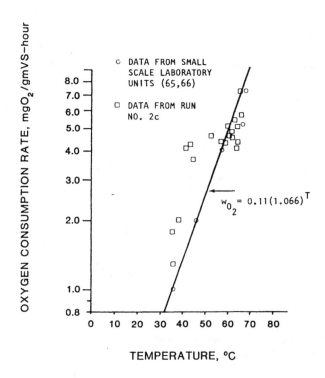

Figure 10.13. Relationship between temperature and oxygen consumption rate observed by Schulze[30] during continuous composting experiments with garbage and digested dewatered sludge cake.

Fresh mixed refuse

$$w_{O_2} = -0.0387(T^2) + 4.560(T) - 117.9 \qquad (10.20)$$

Jeris and Regan determined optimum temperatures for each substrate by differentiating Equations 10.18 to 10.20. Optimum temperatures were 38°C for the composted mixed refuse, 48°C for the newsprint, and 59°C for the fresh mixed refuse. When the process temperature was maintained within 5 to 10°C of the optimum value for each material, 95% of the maximum respiration rate was obtained. The lower temperature optima for the newsprint and compost compared to the fresh mixed refuse was attributed to the higher fractions of cellulose remaining in these materials. The authors felt that this confirmed literature reports[36] that the predominate number of species of microbes capable of utilizing cellulose operate most efficiently within the 45 to 50°C range. This may also explain the lower temperature optima for the fresh mixed refuse compared to the mixtures of garbage and sludge used by Schulze which were lower in cellulose content.

Use of a continuous composter is necessary to determine temperature effects under steady state conditions. However, many composting systems are operated on a batch basis and true steady state conditions are never achieved. Numerous studies of oxygen consumption have been conducted using batch reactors. An example of the type of data derived from such studies is shown in Figure 10.15 based on the work of Snell.[37] Small, bench scale composters were used with temperature and moisture conditions held constant. Moisture control became difficult at higher temperatures and the samples became dried, which likely introduced a

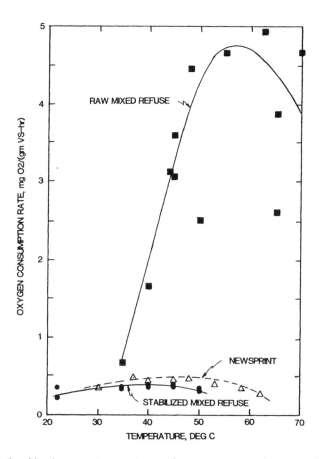

Figure 10.14. Relationships between temperature and oxygen consumption rates observed by Regan and Jeris[11] during continuous composting experiments with mixed refuse, newsprint, and compost made from mixed refuse. All substrates were high in cellulose content.

moisture rate limitation. No seeding or reseeding was practiced and the feed material consisted of ground garbage. Because of the batch nature of the process, a characteristic lag period was observed at the start of composting. As long as 8 days was required to achieve maximum rates of oxygen uptake. The maximum rate would hold for several days and then begin to decrease as the more readily degradable feed components were exhausted.

A temperature curve typical of a batch windrow system is shown in Figure 10.16. The windrow consisted of a large mass of aerobically composting refuse. The difference between temperature curves for batch and continuous reactors can be seen by comparing Figures 10.10 and 10.16. If the feed schedule is continuous or semicontinuous, and if the reactor contents are well mixed, feed material will be quickly inoculated with the mixture of microbes developed for the particular steady state conditions. The material will also quickly be brought to conditions of temperature, pH, moisture, and FAS established in the reactor. Thus, the lag period common to batch systems can be reduced or eliminated with continuous composters. Notice in Figure 10.16 that the temperature curve does not hesitate at the transition from mesophilic to thermophilic temperatures. In smaller masses of compost there is sometimes a temporary temperature plateau as the mesophilic population declines and the thermophilic population develops. In large piles where heat loss from the interior is slow, the effect is dampened.

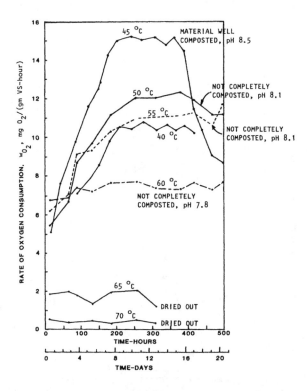

Figure 10.15. Rate of oxygen consumption for various temperatures using a batch composter with ground garbage. Note the somewhat linear increase in oxygen uptake rate to a maximum value for each temperature. Peak values are plotted in Figure 10.17. From Snell.[37]

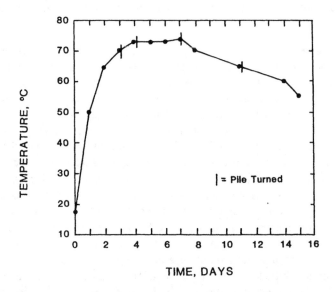

Figure 10.16. Temperature curve typical of a large mass of refuse material during aerobic composting in a batch operated windrow system. From McGauhey and Gotaas.[10]

Wiley and Pearce[38] and Wiley[39-41] studied the rates of CO_2 evolution from samples of mixed garbage and refuse with a paper fraction estimated to be at least 50%. Samples were ground to a size <5/8 in. Batch composters with a 1.5 ft³ working volume were used. The composters were insulated and no external heating or cooling was applied. Air was supplied at a constant rate to each composter and the exhaust gas analyzed to determine the rate of CO_2 evolution. Thus, the substrate self-heated to a point where heat production was balanced by heat loss. Over 245 data points relating temperature and CO_2 production rate were developed, which spanned the temperature range from 15 to 70°C. A best fit equation developed from their CO_2 data can be converted to an equivalent O_2 consumption rate by assuming 1 mol O_2/mol CO_2. The latter ratio is applicable to a largely cellulosic substrate. With these assumptions, the data of Wiley can be represented as

$$w_{O_2} = 0.3432(1.0429)^T \qquad (10.21)$$

A summary of some of the available data on oxygen uptake rates is presented in Figure 10.17. Despite the variety of procedures and feed materials, the data are reasonably consistent. All studies show an increase in the rate of oxygen uptake with increasing temperature. Some of the studies show maximum rates at intermediate temperatures of 40 to 60°C, while others show consistently increasing rates to ~70°C. All studies show decreasing rates at temperatures higher than the optimum temperature. As Regan and Jeris[11] suggest, the differences in temperature optima may in part be due to differences in the cellulosic content of the substrate.

In any composting system, the aeration system must be capable of meeting the maximum oxygen consumption rate demanded by the microbial population. From Figure 10.17, a maximum rate in the range of 4 to 14 mg O_2/g VS-h would appear to be sufficient in all but the most extreme cases. This equates to about 460 to 1620 ft³/ton VS-h (14 to 50 m³/metric ton VS-h). These values represent *peak* stoichiometric demands. Higher aeration rates would be required to remove the heat released from consumption of the oxygen (see Chapter 7).

First Order Reaction Rates

The shape of the oxygen consumption curves from the long term respirometric tests presented in Figures 9.7 and 9.8 strongly suggest a first order rate equation of the form

$$\frac{d(BVS)}{dt} = -k_d(BVS) \qquad (10.22)$$

where

BVS	=	the quantity of biodegradable volatile solids, usually kg or lbs
t	=	time, days
k_d	=	rate constant, g BVS/(g BVS-day) or day^{-1}

According to Equation 10.22, the rate of BVS oxidation, d(BVS)/dt, is a function of the quantity of remaining BVS. The negative sign indicates that the quantity of BVS decreases with time.

The assumption of first order kinetics has worked well in describing numerous processes involving biological oxidation. Included among these is the familiar first order expression for organic oxidation in such wide ranging systems as biological oxygen demand (BOD) bottles

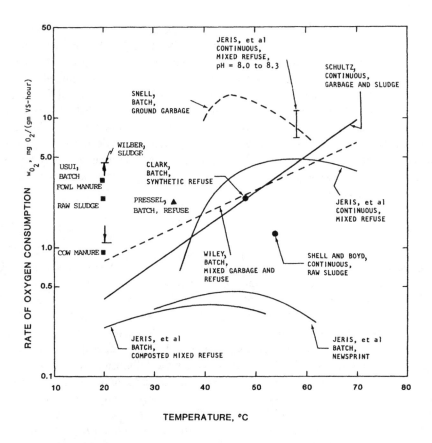

Figure 10.17. Observed oxygen consumption rates for various composting mixtures and reactor types as a function of temperature. Each curve represents the best fit of observed data.

and streams. The approach has also worked well in describing more complex processes such as activated sludge, trickling filters, submerged filters for nitrification, anaerobic filters, and aerobic digesters.

Nolan[42] studied the decomposition of papermill sludges and other substrates in the soil to determine their reaction rate constants. The substrate/soil mixtures were incubated under controlled conditions at 25°C and optimum moisture. CO_2 free air was passed through the incubator and CO_2 evolution observed. Test runs were usually 30 to 60 days duration. Nolan used a first order rate equation to model the data, but divided the substrate VS content into "fast" and "slow" fractions each with their own first order rate constants. It was found that in some cases this provided a better fit to the experimental data. Complex substrates contain a mixture of organics, some of which are likely to degrade faster than others. Therefore, dividing the substrate into faster and slower fractions would seem to be a reasonable assumption.

The first order rate constants and the fast and slow fractions developed by Nolan are presented in Table 10.5. It is interesting to note that the rate constants for grass often exceed those for digested and raw sludges. This is the reason that grass must be carefully handled to avoid odor conditions. It is also interesting to note that crude oil has a relatively high rate constant for decomposition. It is only necessary that moisture, nutrients, microbes, and a support matrix such as soil or compost be provided to increase the oil/water interface area.

The use of respirometers to measure substrate degradability was discussed in Chapter 9. Oxygen uptake curves for several substrates were presented in Figures 9.7 and 9.8. First order

Table 10.5. First Order Rate Constants (base e) at 25°C for Various Substrates Incubated with Soil

Substrate	k_d (day^{-1})		Percent	
	Fast	Slow	Fast	Slow
Digested Sludge				
(Average)[a]	.0282	.0037	34	66
(Standard Deviation)	.0169	.0029	12	12
Limed Raw Sludge				
(Average)[b]	.0293	.0045	32	68
(Standard Deviation)	.0164	.0075	10	10
Primary papermill sludge				
(Average)[c]		.0033	0	100
(Standard Deviation)		.0006		
Primary papermill sludge	.0333	.0060	26	74
Kraft papermill sludge				
(Average)[d]		.0015	0	100
(Standard Deviation)		.0002		
Sawdust	.0100	.0016	20	80
Crude oil	.0170	.0110	35	65
Wheat straw		.0029	0	100
Wood bark		.0004	0	100
Bermuda grass	.0383	.0132	40	60
Rye grass	.0699	.0172	28	72

Source: Adapted from Nolan.[42]

[a] Average of 31 60-day composite samples collected from the same facility over a 5-year period.
[b] Average of 17 60-day composite samples collected from the same facility over a 5-year period.
[c] Average of 4 samples.
[d] Average of 4 samples.

rate constants can be developed from such data using procedures described by Haug and Ellsworth.[43] An analysis for the raw sludge sample identified as Run A in Figure 9.7 is presented in Figure 10.18. Regression analysis on the data projected an ultimate oxygen uptake of 0.778 g O_2/g BVS. Degradability was estimated at 45.8% based on a sample COD of 1.70 g COD/g VS. The best fit of experimental data resulted when the BVS was divided into a 19% "fast" fraction with rate constant 0.15 day^{-1} and an 81% "slow" fraction with rate constant 0.05 day^{-1}. Projected oxygen uptake curves for the fast and slow fractions and the summation of the two are presented in Figure 10.18. The summation curve corresponds closely to the experimental data.

A similar analysis for a hardwood sawdust, shown in Figure 9.8, is presented in Figure 10.19. A significant lag period was observed with the hardwood substrate, probably due to the need to acclimate the seed microbes. In actual practice such lag periods can usually be avoided by using product recycle to supply an acclimated seed. Therefore, the lag period is removed as shown in Figure 10.19. Regression analysis projects an ultimate degradability of 78.4%. In this case the data is modeled to acceptable accuracy by use of a single rate constant of 0.0081 day^{-1}. A summary of fast and slow fractions and associated rate constants developed by Haug and Ellsworth[43] is presented in Table 10.6.

The oxygen uptake rates summarized by Equation 10.17 can be used to predict reaction rate constants as a function of temperature. The feed material used in deriving Equation 10.17 consisted of a mixture of ground garbage and dewatered, digested sludge cake. Oxygen

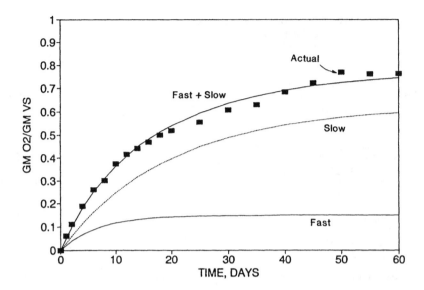

Figure 10.18. Oxygen uptake data for raw sludge compared to a first order reaction rate model. The data were developed using the constant pressure respirometer of Figure 9.5. The data were modeled assuming 19% of the BVS with a rate constant of 0.15 day^{-1} and 81% at 0.05 day^{-1}. From Haug and Ellsworth.[45]

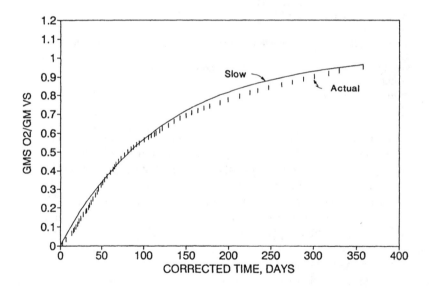

Figure 10.19. Oxygen uptake data for hardwood sawdust compared to a first order reaction rate model. The data were developed using the constant pressure respirometer of Figure 9.5. The data were modeled assuming one rate constant at 0.0081 day^{-1}. From Haug and Ellsworth.[43]

equivalent of the organics was not presented but was estimated from the original data to be about 1.5 g O_2/g VS oxidized. Using this factor, Equation 10.17 can be converted to a rate constant with the following form:

$$k_d = 0.00632(1.066)^{T-20} \text{ (g BVS oxidized/g VS} - \text{day)} \tag{10.23}$$

Table 10.6. First Order Rate Constants (base e) at 20 to 25°C and Substrate Degradabilities Determined by Long Term Respirometry

Substrate	Test Duration (days)	Ultimate Degradability[a] (%)	k_d, day^{-1} Fast	k_d, day^{-1} Slow	Percent Fast	Percent Slow
Raw sludge[b]	60	45.8	0.15	0.05	19	81
Raw sludge	242	66	0.015	0.004	40	60
Pulpmill sludge	200	44.1	—	0.0095	0	100
Sawdust, pine softwood	90	11.2	0.15	0.02	71	29
Sawdust, hardwood	368	78.4	—	0.0081	0	100

Source: Haug and Ellsworth.[43]

[a] Ultimate degradability determined by projection of the oxygen uptake data.
[b] Raw primary and secondary sludge from Plattsburgh, NY.

Examination of Schulze's original data indicates a reduction in volatile matter ranging from about 37 to 45%. Assuming the degradable fraction to be 50% of the total, Equation 10.23 becomes

$$k_d = 0.0126(1.066)^{T-20} \text{ (g BVS oxidized/g BVS} - \text{day)} \tag{10.24}$$

Units on the rate constant in Equation 10.24 are now appropriate to the present analysis. At 20°C the rate constant per Equation 10.24 is 0.0126 day^{-1} (base e), which is consistent with the range of values reported in Tables 10.5 and 10.6. Equation 10.24 can be applied up to about 68°C, the maximum temperature tested by Schultz.

Equation 10.21 was developed from Wiley's experiments conducted on garbage and refuse with a high paper fraction. Using similar assumptions as above, an equation for the reaction rate constant can be developed as

$$k_d = 0.0254(1.0429)^{T-20} \text{ (g BVS oxidized/g BVS} - \text{day)} \tag{10.25}$$

The range of rate constants predicted from Equations 10.24 and 10.25 are consistent with the values presented in Tables 10.5 and 10.6. Equation 10.25 predicts a higher rate constant than Equation 10.24 at 20°C. However, the rate constants at 50°C become essentially equivalent because of the larger temperature coefficient in Equation 10.24.

Several observations can be made from the above data and equations. First, the first order kinetic model appears to accurately describe the decomposition of many composting substrates, even in long term respirometer studies spanning over 800 days duration. Second, the accuracy of fit can sometimes be improved by dividing the substrate into fractions with different rate constants. Third, the rate constants for composting substrates are typically less than for other liquid wastes. For example, the first order rate constant for the BOD test on municipal sewage is typically taken to be about 0.23 day^{-1} (base e) at 20°C and may be considerably higher for simple substrates such as glucose. By comparison, the rate constant for many composting substrates is on the order of 0.01 day^{-1} at 20°C. This rather remarkable difference in reaction rates is due to two factors; the nature of the organic matter and the ability of the organisms to utilize the organic matter. Organic matter that exists in true solution is generally readily available, whereas solid substrates must await hydrolytic action before they can diffuse into the bacterial cells. This suggests that the rate of hydrolysis is the controlling factor in composting kinetics. Also, composting substrates commonly contain lignin and other compounds that are resistant to microbial attack. These factors combine to make the rate constants for solid substrates considerably less than for the more soluble substrates common to wastewater treatment. As a result, the residence times required to produce a stable compost product are considerably longer than for most other biological treatment processes.

HYDRAULIC AND SOLIDS RETENTION TIMES

The rate of biochemical reaction determines the speed at which composting can proceed. The same decomposition can be achieved by a fast reaction rate operating over a short time or a slower reaction rate operating over a longer time. Because kinetics deals with rates of reaction, the concept of time is important to the design and operation of composting systems. In this section, the concepts of hydraulic retention time, HRT, and solids residence time, SRT, are developed and applied to the composting process.

Distinction Between HRT and SRT

For liquid phase systems, two detention times can be defined, one based on liquid retention time and one based on solids residence time. Detention time based on liquid retention time is usually termed "hydraulic retention time". Detention time based on the average residence time of solids in the system is usually termed "solids residence time". For a reactor without recycle of solids, the HRT and SRT are equivalent. If solids are recycled, however, the SRT will be greater than the HRT.

The distinction between SRT and HRT is an important concept. For homogeneous, liquid phase systems, the efficiency of BVS decomposition is determined by the system SRT. The minimum HRT is determined primarily by time constraints imposed by oxygen transfer and the ability to maintain the required microbial concentration. The same concepts can be applied to solid phase composting systems. The extent of microbial decomposition is determined by the system SRT, whereas reactor or process stability is largely determined by HRT.

Concepts of HRT and SRT for liquid phase systems with recycle are presented in Figure 10.20. The single pass hydraulic retention time is defined as

$$HRT = \frac{V}{Q+q} \tag{10.26}$$

where

$$
\begin{aligned}
HRT &= \text{single pass, hydraulic retention time} \\
V &= \text{volume of reactor or system} \\
Q &= \text{volumetric flowrate of material, excluding recycle} \\
q &= \text{volumetric flowrate of recycle material}
\end{aligned}
$$

SRT for a liquid phase system is defined as

$$SRT = \frac{V}{Q} \tag{10.27}$$

A. SINGLE REACTOR WITH RECYCLE

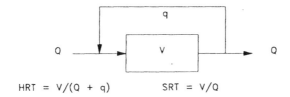

$$HRT = V/(Q + q) \qquad SRT = V/Q$$

B. TWO REACTORS WITH INTERMEDIATE RECYCLE

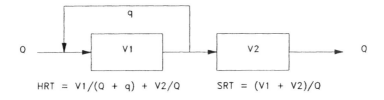

$$HRT = V1/(Q + q) + V2/Q \qquad SRT = (V1 + V2)/Q$$

C. TWO REACTORS WITH PRODUCT RECYCLE

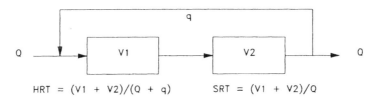

$$HRT = (V1 + V2)/(Q + q) \qquad SRT = (V1 + V2)/Q$$

Figure 10.20. Concepts of hydraulic retention time (HRT) and solids residence time (SRT) for liquid phase systems with recycle.

where
     SRT = mean solids residence time

For liquid phase systems, the volumetric flowrate remains essentially constant across the reactor, simplifying the calculation of average residence time. The equations for HRT and SRT presented in Figure 10.20 apply to liquid phase systems, but can be adapted to apply to composting systems.

Concepts of HRT and SRT can be defined for composting systems analogous to the concepts used for liquid phase systems. HRT for a composting system is defined as follows: *HRT is the single pass, mean residence time of the mixed materials including recycle.* This definition is equivalent to Equation 10.26 as applied to liquid phase systems. SRT for a composting system is defined as follows: *SRT is the mean residence time of the feed solids excluding recycle.* This definition is equivalent to Equation 10.27.

The volume of mixed materials entering a composting reactor does not remain constant with time. Moisture is lost by evaporation and BVS solids are lost by microbial decomposition. As a result, mixture volume usually decreases across the system. Recall that mass is conservative but volume is not. To account for the volumetric changes that occur during composting, the basic equations for HRT and SRT can be modified to

$$HRT = \frac{V}{\left[\dfrac{(Q+q)_{in} + (Q+q)_{out}}{2}\right]} \tag{10.28}$$

$$SRT = \frac{V}{\left[\dfrac{Q_{in} + Q_{out}}{2}\right]} \tag{10.29}$$

A number of alternative mathematical formulations to Equation 10.29 can be developed to estimate SRT from the general definition given above. Gossett[44] used an equation of the form

$$SRT = HRT\left(\frac{W_1}{W_2}\right) \tag{10.30}$$

where

W_1 = weight of reactor outfeed
W_2 = weight of reactor outfeed less the weight of recycle

The important points here are the definitions of HRT and SRT and the distinction between the two. The mathematical approach used for estimation is developed further in Chapter 11.

Design Criteria

HRT is an important parameter that affects composting temperature, output solids content, and the stabilization of BVS. These factors are quantitatively discussed in Chapter 12. In summary, process stability is favored by longer detention times. A minimum design HRT of 10 to 20 days is reasonable for continuous, well-mixed reactor systems, because the process is stable over a wide range of conditions and high output solids contents and BVS reductions are possible. Longer detention times produce smaller improvements in these parameters for a comparable increase in reactor volume. Many reactor systems are designed for an HRT of 14 to 28 days which is consistent with the above discussion. Batch processes may require longer detention times because of the lag phase encountered at the start of composting.

SRT is the most important factor in determining the stability of the compost product. Again, the effect of SRT on product stability and quality is quantitatively discussed in Chapters 12 to 14. The minimum system SRT for design is a function at least the following:

1. the types of substrates and amendments used and their corresponding reaction rate constants
2. the extent of process control incorporated into the design and the processes used for the high-rate and curing phases
3. the extent to which kinetic rate limitations are avoided
4. the end use of the product

The operating records of a number of composting systems, using a variety of substrates, were reviewed. These data, together with the simulation model results in Chapters 13 and 14 and the discussion on product stability presented later in this Chapter, suggest that a minimum system SRT of about 60 to 180 days is required to produce compost with sufficient stability and maturity to avoid reheat and phytotoxic effects. This assumes that the feed materials are properly conditioned to close the energy balance and reduce kinetic limitations. It further assumes a reasonable level of process control to prevent excessive rate limitations during the process.

Design Approach

The minimum system SRT, SRT_{min}, is a key process criteria that should be established at the beginning of the design phase. System SRT should also be monitored by the operator of an existing facility to guide operational practices and measure system performance. SRT_{min} can be established by a review of similar facilities and/or by simulation modeling to determine the required SRT to produce a given product quality. The latter approach is discussed in Chapters 11 through 14.

Once SRT_{min} is established, the next step is to define the desired HRT for the first stage, high rate phase, HRT_r. If a reactor (in-vessel) system is used for the first stage, the minimum reactor HRT should be in the range of 12 to 20 days. A minimum HRT of 20 to 30 days is typically used with the aerated static pile and windrow processes. The proper ratio of mixture components for structural, chemical, and energy conditioning is then determined using the procedures of Chapters 6 and 8. The required volume of reactor, REACVOL, aerated piles, or windrows can then be determined using Equation 10.28.

Once the volume of the first stage is known, the corresponding SRT in the first stage, SRT_r, can be determined from Equation 10.29 or 10.30. The difference between the minimum system SRT and the SRT of the first stage equals the additional SRT, which must be provided by the second stage, curing phase, SRT_c. In other words, $SRT_c = SRT_{min} - SRT_r$. Once SRT_c is determined, Equation 10.29 can be used to determine the volume required for the curing phase, CUREVOL, and Equation 10.28 to determine the curing HRT, HRT_c.

This design approach is developed further in Chapters 11 to 14. The procedure allows the designer to integrate the first stage, high rate phase and the second stage, curing phase into an overall system design. For example, decreasing the size of the first stage process will increase the size of the second stage process required to achieve the same SRT_{min}. This approach to the problem of design allows for optimization of individual processes, based on site specific constraints, while maintaining minimum system requirements.

PRODUCT STABILITY

Definition and Purpose

The term "stabilization" refers to the oxidation of organic matter or its conversion to a more refractory form. When an organic substrate is oxidized by microbes, a portion of the released energy is captured and used to support the synthesis of new cell material from the substrate. When the microbe dies the cell material becomes food for other microbes and a further transformation to CO_2, H_2O, and cell matter occurs. Each time this process is repeated a portion of the remaining organic matter is very resistant to microbial attack. This is commonly called humus. As composting proceeds, the readily degradable organics in the substrate are oxidized and gradually replaced by increasingly less degradable humus materials. The more stable compounds that remain at the end of composting are still degradable, but at a much reduced rate compared to the original feed substrate.

The question that arises is, how much stabilization is enough? There is no precise answer to this question. For example, anaerobically digested sludge is generally considered stable. In the liquid or cake form, digested sludge can be applied to land in controlled amounts without producing nuisance conditions. On the other hand, if digested sludge cake is allowed to sit in open piles, septic and odorous conditions can often develop. As another example, raw sludge can be heat dried and stored for long periods without producing odorous conditions. The low moisture content limits the rate of biological activity resulting in a "pseudostable" material. In the dried state, raw sludges are even bagged and sold by nurseries, perhaps the most demanding market in terms of product quality. When the material is rewetted, normal rates of biological activity will resume. By this time, however, the material should be incorporated into the soil, thus reducing the nuisance potential.

Another aspect of stabilization is the effect of the organics on plant growth. Numerous researchers have observed that "immature" composts can contain metabolites that are toxic to plants (phytotoxicity). Zucconi et al.[45] noted that the introduction of decomposing organic matter in the soil may damage existing plant roots and inhibit growth. If the organic matter has a high C/N ratio and decomposes rapidly, it can rob the soil of the nitrogen needed to support plant growth. If the organic matter has a low C/N ratio, released ammonia can be phytotoxic. Metabolic products of biodegradation, particularly organic acids such as acetic acid, can also exhibit a toxic effect on plant growth. Zucconi et al.[46] composted MSW fractions using mechanically agitated windrows. Samples collected throughout the process were tested for phytotoxicity. Toxicity was associated with the initial 3 to 4 weeks of composting. Following this stage, toxicity rapidly decreased, although it had not completely disappeared after 60 days of composting.

The purpose of composting is to produce an organic soil amendment that is beneficial to plants. All compost facilities should be designed and operated to produce a stable product that is beneficial to plants and not phytotoxic. Tuttle[47] has noted the truism that "if it kills plants, it's *not* compost." This does not necessarily mean that the compost must be fully cured on site. Wheeler[48] reported that the City of Hamilton, Ohio pays a soil blender to remove uncured, sludge compost immediately upon exit from their reactor system. Subsequent storage and dilution with soil materials apparently produces an acceptable product.

We come back to the original question, how much stabilization is enough before the product can be termed a compost? Certainly, there is no such thing as complete stabilization as long as any organic matter remains, because it will continue to decompose at some rate. In a strict sense, complete stabilization would require the oxidation of all organic matter to CO_2 and H_2O. However, complete stabilization is not desirable because the value of compost as a soil amendment depends in part on its organic content. The following working definition of compost is suggested: *Compost is an organic soil conditioner that has been stabilized to a humus-like product, that is free of viable human and plant pathogens and plant seeds, that does not attract insects or vectors, that can be handled and stored without nuisance, and that is beneficial to the growth of plants.* Such a definition is not precise, but it is practical. Stabilization must be sufficient to reduce the nuisance potential and phytotoxic metabolites, but not so complete that organics are unnecessarily lost from the final product. Stabilization is sufficient when the rate of oxygen consumption is reduced to the point that anaerobic or odorous conditions are not produced to such an extent that they interfere with storage or end use of the product and phytotoxic compounds have been metabolized.

Measuring Stability

A number of approaches have been used to measure the degree of stabilization and to judge the condition of the compost product. Included among these are:

1. decline in temperature at the end of batch composting or curing
2. a low level of self-heating (reheat) in the final product
3. organic content of the compost as measured by VS content, chemical oxygen demand (COD), carbon content, ash content, or C/N ratio
4. oxygen uptake rate
5. the effect on seed germination and plant growth
6. the presence of particular constituents such as nitrate and the absence of others such as ammonia, sulfides, organic acids, and starch
7. lack of attraction of insects or lack of development of insect larvae in the final product
8. characteristic changes in odor during composting and odor producing potential of the final product upon rewetting
9. rise in the redox potential
10. experience of the operator

Some of these approaches are qualitative while others are at least semiquantitative. Most can provide the operator a comparative tool by which to judge the compost product.

Temperature Decline

A temperature decline back to near ambient conditions at the end of composting is a good indication that the process is nearing completion. The proviso is that the temperature drop is not caused by thermal kill, oxygen shortage, low moisture, lack of FAS, or lack of sufficient pile insulation. This approach is based on the fact that the rate of heat production is proportional to the rate of organic oxidation, which decreases after the more degradable material is decomposed. A declining temperature is inevitable as the energy balance adjusts to the decreasing rate of heat input. Conversely, composting material that is still in a thermophilic condition is probably not stable and not yet mature.

Reheat Potential

Reheating potential of the product is another useful but comparative tool. In a method proposed by Niese,[49] the sample is adjusted to optimum conditions and placed in a Dewar flask. The flask is placed in a special incubator equipped with temperature control circuits to maintain a constant temperature difference between the incubator and the Dewar flask. Niese observed temperatures >70°C with raw refuse, 40 to 60°C with partly stable refuse, and <30°C with fully stable material. Willson and Dalmat[50] noted that the decline in temperature in windrows and piles is actually a large scale variation of Niese's lab test, but lacks its precision due to variable heat loss parameters. One difficulty with this approach is that apparent "stability" may result in part from other rate limitations such as low moisture content or lack of FAS. Willson tried the technique but judged the results as too erratic to be useful.

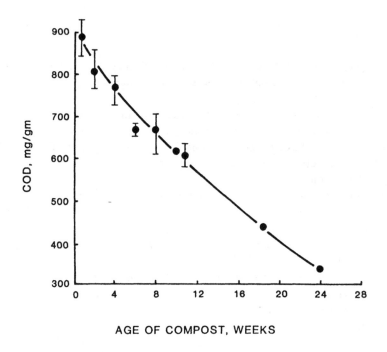

Figure 10.21. Effect of compost age on the COD of samples collected during windrow composting of refuse. From Lossin.[51]

Organic Content

The organic content of a compost will depend largely on the organic characteristics of the feed substrate. Therefore, organic content can provide a measure of stability for a particular feed, but should not be used to compare composts produced from different feed substrates. A number of chemical tests have been used to either directly or indirectly measure the organic fraction, including VS, ash or carbon contents, COD, and C/N ratio. The COD of composting samples reported by Lossin[51] during windrow composting of refuse is presented in Figure 10.21. The compost was considered stable and ready for use after eight weeks, which corresponded to a COD <700 mg/g compared to about 900 mg/g for fresh refuse. If the refuse is assumed to be primarily cellulose, the final COD corresponds to a VS content of ~70% (1 g glucose = 1.07 g COD). By contrast, compost produced from a windrow composed of digested sludge conditioned solely with product recycle usually has a VS content of only 30 to 40%. This emphasizes the point that organic content can be a useful measure when applied to a particular feed substrate, but should not be used to compare composts produced from different starting substrates. It is interesting to note from Figure 10.21 that the VS content after 24 weeks was about 30 to 40%.

Chemical Characteristics

Particular constituents in compost are often characteristic of the level of stability. Ammonia is usually present in the early stages of composting as organic nitrogen is decomposed. The ammonia concentration is eventually reduced through volatilization or oxidation to the nitrate form. Thus, the presence of nitrate and absence of ammonia are indicative of a stabilized condition. Lossin[52] reported on the use of starch content as a measure of stability. The rationale

Table 10.7. Acetic Acid Concentrations during Windrow Curing of Refuse Compost from the Dano facility at Ghent, Belgium

Windrow Curing Time (days)[a]	Acetic Acid in Extract (ppm)	
	Turned Windrow[b]	Nonturned
1	25,980	18,500
5	25,570	20,630
9	21,400	27,800
13	14,600	15,500
36	12,400	14,800
68	400	690
120	0	0

Source: DeVleeschauwer et al.[53]

[a] The curing time followed 3 days in a Dano drum.
[b] One windrow was turned frequently by means of a bulldozer. The other was left undisturbed — neither windrow was force ventilated.

rests on the assumption that most feed substrates contain a measurable quantity of starch. Because starch is easily broken down and metabolized, it should be degraded before the compost is considered acceptable. The test procedure involves formation of a colored starch-iodine complex in an acid extract from the compost material. Lossin reported that well-stabilized compost never gave a positive starch reaction with this test.

DeVleeschauwer et al.[53] studied the phytotoxicity of compost produced from town refuse at a Dano facility in Ghent, Belgium. The refuse spent about 3 days in the Dano stabilizer followed by windrow curing. Phytotoxicity was correlated with the organic acid content of the composting material. Acetic, propionic, isobutyric, butyric, and isovaleric acids were quantified by extraction into 50 ml of water added to 10 gms of compost and acidified to pH 2. Phytotoxicity was measured by bioassays with cress seed. Acetic acid was determined to be the primary organic acid present and was responsible for most of the phytotoxic effect. Phytotoxic effects began when the acetic acid concentration in the extract was above 300 ppm. Above ~2000 ppm, no seed germination was observed. Acetic acid concentrations determined during the composting cycle are presented in Table 10.7. Very high concentrations were observed during the first month of composting. It took 120 days of windrow curing to reduce acetic acid concentrations below the phytotoxic level. No seed germination was noted until the day 120 samples. The authors concluded that at least 4 months composting would be required before the product could be used safely in horticulture or agriculture.

Plant Bioassays

Perhaps the most direct method to determine whether the compost product has been cured sufficiently from the standpoint of phytotoxicity is to determine its effect on plant growth. Potting studies could be conducted but these require considerable time before the results are known. Several researchers have worked on ways of reducing the time required for testing. Most have used the germination rate of a test seed, usually cress seed *Lepidium sativum,* because of its rapid response. Spohn[54] described a procedure that measured the germination and growth of cress seed planted in beds of compost that required 5 days.

Zucconi et al.[45,46] developed a 24-h bioassay test that measured germination and root elongation in an aqueous extract from the compost. In the water extract 6 to 8 cress seeds are incubated for 24 h in the dark at 27°C, 10 to 15 replicate samples are recommended. The water extract is obtained by adjusting the sample moisture content to 60% followed by pressure filtering. The filtrate is sterilized by millipore filtration prior to incubation with the cress seeds. Both germination fraction and root growth are measured and expressed as a percentage of the control. A "germination index" was defined as the product of germination percentage and root growth (both expressed as percentages of control). If the germination index is above 50 to 60%, olive plants showed no injury, although they required some time for adapting to the organic matter.

Zucconi used the germination index to determine toxicity during windrow composting of sludge/MSW mixtures. Windrows were turned twice per week. The germination index during the first 3 to 4 weeks of composting ranged from 0 to 50%, indicative of phytotoxicity. Much of the toxicity was removed after 60 days of composting. In a second series of tests, aerated-static piles were formed from sludge/MSW mixtures. One pile was aerated by induced draft fans (vacuum) and the second by forced draft fans (pressure ventilated). The disappearance of toxicity was rapid with both systems. Toxicity was reduced to acceptable levels in about 2 weeks, i.e., germination index >50%. Zucconi concluded that phytotoxicity is associated with the initial stages of decomposition. It is a transient condition, and production ceases and the toxins themselves are inactivated in the succeeding decomposition stages. The time required to remove the phytotoxicity appears to be related to the aerobic condition of the material. The windrows were judged to be poorly aerated compared to the static piles and required a longer time to reach a germination index above 50%.

Oxygen Uptake Rate

The oxygen respiration rate is a fundamental parameter by which to measure stability because it is proportional to the rate of organic decomposition. A number of respirometric techniques have been used to measure the oxygen uptake rate of composts. Chrometzka[55] used a Warburg apparatus and determined that under optimum conditions the uptake rate of fresh substrate was about 30 times that of mature compost. Pressel and Bidlingmaier[56] used the Warburg and measured oxygen uptake rates of 120 and 42 mg O_2/kg ds-h for 8 and 10 month old composts, respectively, made from refuse. These rates were about 5% or less of the uptake rate of the starting substrate. Usui ct al.[57] used an electrolytic respirometer and determined uptake rates from 40 to 127 mg O_2/kg ds-h in cured compost made from raw sludge in a horizontal, agitated bed reactor.

Willson and Dalmat[50] used the constant volume respirometer described in Figure 9.4 to measure oxygen uptake rates at three static pile facilities using raw sludge as substrate and wood chips for bulking. The effect of curing time on respiration rate is presented in Figure 10.22. At Site A the compost was cured for 30 days in aerated piles and then transferred to unaerated piles. At Sites B and C the compost was cured in large unaerated piles. The rate of stabilization in aerated piles was 6 to 8 times that of the unaerated piles. A 75% decrease in respiration rate was observed during the 30 days of aerobic curing. The curing time to achieve a respiration rate of 20 mg O_2/kg ds-h (ln = 3 in Figure 10.22) was about 30 days for aerated curing and over 200 days with unaerated curing.

Willson concluded that a respiration rate of 100 mg/kg ds-h would be acceptable for most field applications but suggested that a rate nearer 20 mg/kg ds-h would be desirable for horticultural uses with sensitive plants. It should be noted that the respiration rates after the first stage, static pile phase were about 150 to 250 mg/kg ds-h. The compost at this point is clearly not stable. The second stage, curing phase is essential to achieving a stable compost product that can be stored and used without phytotoxic effects. Aeration during curing should be considered because of the more rapid reduction in respiration rates.

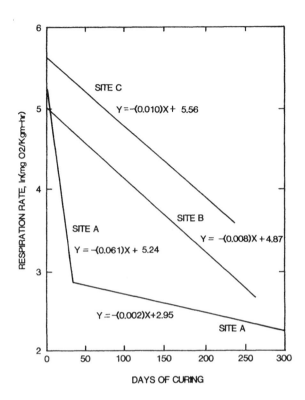

Figure 10.22. Natural logarithms of the respiration rates during curing at 3 static pile sludge composting facilities. Time 0 corresponds to the end of the static pile phase, which was probably about 28 days long. Site A used aerated curing for 30 days, followed by unaerated curing for the remainder of the time. Sites B and C used only unaerated curing. Respiration rates were measured using the constant volume respirometer from Figure 9.4. From Willson and Dalmat.[50]

Other Parameters

All of the above parameters are useful characteristics for predicting the plant response to a compost. In their work on respiration rates Willson and Dalmat[50] noted that other compost characteristics are also important to plant response. The respiration rate should be used in conjunction with other quality characteristics such as nutrient content, heavy metal content, pathogen content, C/N ratio, salt content, and pH. Also, there is no substitute for the experience of the operator. An experienced operator can usually tell the condition of a compost by its appearance, color, odor, texture, particle size, and other physical characteristics. It may not be the most scientific approach, but it is workable and usually accurate.

SUMMARY

Kinetics deals with rates of reactions and must be distinguished from thermodynamics, which deals with the energy changes that accompany physical and chemical reactions. Thermodynamics can predict the energy change from a chemical or biochemical reaction, but kinetics predicts how fast it will occur.

The composting system can be described as a heterogeneous system with solid substrate and limited moisture. Microbes either surround or are immersed in the solid substrate.

Moisture is limited to that associated with the solid substrate. In such a case, a rather complex sequence of events is necessary for metabolism of the substrate. Solubilization of substrate by exocellular enzymes, mass transport of oxygen to the cell, and utilization of the solubilized substrate and oxygen by the cell are likely the predominant rate-controlling steps during aerobic composting. Kinetic models can be developed for each of these steps. A model of substrate solubilization suggests that the kinetics for this step are a function of the mass concentration of microbes, X, at concentrations of X less than a constant, K_x. The rate becomes independent of X at higher concentrations where $X > K_x$. Under the latter conditions the rate of solubilization becomes constant at a given temperature. This means that for materials resistant to solubilization, longer time periods are required to allow solubilization and decomposition. This appears to be the function of the curing stage commonly used in composting systems.

Particle thickness on the order of 1 cm can result in large oxygen mass transport limitations which would dominate and slow the process kinetics. Particle thickness of about 0.1 cm or less is small enough that diffusion supplied oxygen can match the peak rates of oxygen demand. The fact that oxygen is present in the pore spaces of a composting material does not by itself assure that oxygen diffusion is not rate-limiting, because oxygen must still be transported within the composting particle. The moisture content of individual composting particles is significant because diffusion coefficients through a gaseous phase are much greater than through the liquid phase. Thus, diffusion transport of oxygen may be a controlling factor when particle moisture contents are high and assume less significance as moisture is removed from the particle.

A number of other factors can limit microbial reaction rates during composting. These include lack of degradable organics, very low or high process temperatures, low moisture conditions, lack of free air space, low oxygen content, imbalanced pH conditions, lack of inorganic nutrients, lack of microbes (sterile substrate), and the presence of toxic substances. The effect of these potential rate limitations can be controlled by assuring proper structural, chemical, and energy conditioning of the feed mixture, supplying adequate air to the process, and using some product recycle to seed the feed mixture if it is suspected that microbial concentrations are low.

Rates of oxygen consumption have been measured in both batch and continuous composting reactors on a variety of feed materials. In general, rates of oxygen utilization increase exponentially with temperature up to a peak value. Further temperature increases result in decreased rates of reaction from thermal inactivation. Temperature optima from 40 to 70°C have been reported. Data suggest that cellulosic substrates may have a temperature optima in the 45 to 50°C range because of a larger number of microbial species capable of utilizing cellulose in this temperature range. Maximum rates of oxygen consumption are likely to range from 4 to 14 mg O_2/g VS-h. These values represent *peak* stoichiometric demands.

Organic decomposition during composting can be modeled as a first order reaction. The accuracy of the model can sometimes be improved by dividing the substrate into fractions with different first order rate constants. The rate constants for most composting substrates are typically lower than for more solubilized substrates in the liquid phase. This suggests that the rate of hydrolysis may be the controlling factor in composting kinetics. Because of the slower kinetic rates, the residence times required to produce stable composts are considerably longer than for most other biological treatment processes.

Concepts of hydraulic retention time (HRT) and solids residence time (SRT) can be defined for composting systems analogous to the concepts used for liquid phase systems. HRT for a composting system is defined as the single pass, mean residence time of the mixed materials including recycle. SRT for a composting system is defined as the mean residence time of the feed solids, excluding recycle. The recommended design approach is to establish the minimum system SRT necessary to produce a given product quality. Volumes of the high rate and curing phases can then be adjusted as necessary to achieve the minimum system SRT. This approach to design allows for optimization of individual processes, based on site specific constraints, while maintaining minimum system requirements.

The purpose of composting is to produce an organic soil amendment that is beneficial to plants. All compost facilities should be designed and operated to produce a stable, mature compost. To be defined as compost, the organic substrates must be stabilized to a humuslike product that is free of viable human and plant pathogens, plant seeds, and insect eggs, that can be handled and stored without nuisance, and that is beneficial to the growth of plants. A number of test procedures have been developed to measure the degree of stabilization. Data from these tests indicate that curing is essential to produce a stable product. System SRTs from 60 to over 180 days are generally necessary to stabilize organics and reduce phytotoxic effects in high rate composting systems. Longer SRTs may be required in less controlled processes.

REFERENCES

1. Chandler, J. A., Jewell, W. J., Gossett, J. M., Van Soest, P. J., and Robertson, J. B. "Predicting Methane Fermentation Biodegradability," in *Biotechnology and Bioengineering Symposium No. 10* (New York: John Wiley & Sons, Inc., 1980).
2. Lawrence, A. W. and McCarty, P. L. "A Unified Basis for Biological Treatment, Design and Operation," *J. San. Eng. Div., ASCE*, 96(SA3) (1970).
3. Bailey, J. E. and Ollis, D. F. *Biochemical Engineering Fundamentals* (San Francisco: McGraw-Hill, 1977).
4. Mueller, J. A. "Oxygen Diffusion through a Pure Culture Floc of *Zoogloea ramigera*," PhD Thesis, University of Wisconsin (1966).
5. Tsuk, A. G. and Oster, G. "Determination of Enzyme Activity by a Linear Measurement," *Nature* 190:721 (1961).
6. McLaren, A. D. and Packer, L. "Some Aspects of Enzyme Reactions in Heterogeneous Systems," *Adv. Enzymol. Rel. Sub. Biochem.* 33:245 (1970).
7. Jeris, J. S. and Regan, R. W. "Controlling Environmental Parameters for Optimum Composting, Part I: Experimental Procedures and Temperature," *Compost Sci.* (January/February 1973).
8. Klass, D. L., Ghosh, S., and Conrad, J. R. "The Conversion of Grass to Fuel Gas for Captive Use," *Symposium Papers Clean Fuels from Biomass, Sewage, Urban Refuse, Agricultural Wastes*, Orlando, FL, sponsored by Institute of Gas Technology (1976).
9. Golueke, C. G. *Biological Reclamation of Solid Waste* (Emmaus, PA: Rodale Press, 1977).
10. McGauhey, P. H. and Gotass, H. G. "Stabilization of Municipal Refuse by Composting," *Trans. Am. Soc. Civil Eng.* Paper No. 2767.
11. Regan, R. W. and Jeris, J. S. "Cellulose Degradation in Composting," report to the U.S. Public Health Service, Research Grant EC 00161, Civil Engineering Department, Manhattan College (January 1971).
12. Senn, C. L. "Dairy Waste Management Project — Final Report," University of California Agricultural Extension Service, Public Health Foundation of Los Angeles County (1971).
13. American Public Works Association. *Municipal Refuse Composting,* 2nd ed. (Chicago, IL: Public Administration Service, 1966).
14. Pfeffer, J. T. and Liebman, J. C. "Biological Conversion of Organic Refuse to Methane, Annual Progress Report," NSF Grant No. GI-39191 (1974).
15. Gossett, J. M. and McCarty, P. L. "Heat Treatment of Refuse for Increasing Anaerobic Biodegradability," Civil Engineering Technical Report No. 192, (Stanford, CA: Stanford University 1975).

16. Whang, D. S. and Meenaghan, G. F. "Kinetic Model of Composting Process," *Compost Sci./Land Utilization* 21(3) (May/June 1980).

17. Reid, R. C. and Sherwood, T. K. *The Properties of Gases and Liquids* (New York: McGraw-Hill Book Co., 1958).

18. Perry, R. H. and Chilton, C. H., Eds. *Chemical Engineer's Handbook,* 5th ed. (New York: McGraw-Hill Book Co., 1973).

19. Krough, A. "The Rate of Diffusion of Gases Through Animal Tissues, with Some Remarks on the Coefficient of Invasion," *J. Physiol.* 52:391–408 (1919).

20. Greven, K. "Uber die Bedeutung der Zellmembranen fur die O_2-Diffusion im Gewebe," *Pflueger's Arch. Gesamte Physiol.* 273:353–366 (1961).

21. Longmuir, I. S. and Bourke, A. "The Measurement of the Diffusion of Oxygen Through Respiring Tissue," *Biochem. J.* 76:225–229 (1960).

22. Thews, G. "Oxygen Diffusion in the Brain," *Pfluegers Arch. Gesamte Physiol.* 271:197–226 (1960).

23. Greenwood, D. J. and Goodman, D. "Effect of Shape on Oxygen Diffusion and Aerobic Respiration in Soil Aggregates," *J. Sci. Food Agric.* 15:781–790 (1964).

24. Wise, D. L. "The Determination of the Diffusion Coefficients of Ten Slightly Soluble Gases in Water and a Study of the Solution Rate of Small Stationary Bubbles," PhD Thesis, University of Pittsburgh (1963).

25. *International Critical Tables* (New York: McGraw-Hill Book Co., 1929).

26. Atkinson, D., Swilley, E. L., Busch, A. W., and Williams, D. A. "Kinetics, Mass Transfer, and Organism Growth in a Biological Film Reactor," *Trans. Inst. Chem. Eng.* 45:T257 (1967).

27. Williamson, K. and McCarty, P. L. "A Model of Substrate Utilization by Bacterial Films," *J. Water Poll. Control Fed.* 48(1) (1976).

28. Williamson, K. and McCarty, P. L. "Verification Studies of the Biofilm Model for Bacterial Substrate Utilization," *J. Water Poll. Control Fed.* 48(2) (1976).

29. Clark, D. S. "Sea-Floor Bacteria Research May Bring Diverse Benefits," *R&D Magazine* (March 1992).

30. Schulze, K. L. "Continuous Thermophilic Composting," *Compost Sci.* (Spring 1962).

31. Schulze, K. L. "Rate of Oxygen Consumption and Respiratory Quotients during the Aerobic Decomposition of a Synthetic Garbage,"in *Proceedings of the 13th Annual Purdue Industrial Waste Conference* (Lafayette, IN: Purdue University 1958).

32. Schulze, K. L. "Rate of Oxygen Consumption and Respiratory Quotients during the Aerobic Decomposition of a Synthetic Garbage," *Compost Sci.* (Spring 1960).

33. Schulze, K. L. "Relationship Between Moisture Content and Activity of Finished Compost," *Compost Sci.* (Summer 1961).

34. Jeris, J. S. and Regan, R. W. "Controlling Environmental Parameters for Optimum Composting, Part II: Moisture, Free Air Space and Recycle," *Compost Sci.* (March/April 1973).

35. Jeris, J. S. and Regan, R. W. "Controlling Environmental Parameters for Optimum Composting, Part III," *Compost Sci.* (May/June 1973).

36. Siu, R. *Microbial Decomposition of Cellulose* (New York: Rheinhold, 1951).

37. Snell, J. R. "Some Engineering Aspects of High-Rate Composting," *J. San. Eng. Div., ASCE* Paper 1178 (1957).

38. Wiley, J. S. and Pearce, G. W. "A Preliminary Study of High-Rate Composting," *Trans. Am. Soc. Civil Eng.,* Paper No. 2895, 1009–1034, (December 1955).

39. Wiley, J. S. "Studies on High-Rate Composting of Garbage and Refuse," in *Proceedings of the 10th Annual Purdue Industrial Waste Conference* (Lafayette, IN: Purdue University 1955).

40. Wiley, J. S. "Progress Report on High-Rate Composting Studies," in *Proceedings of the 11th Annual Purdue Industrial Waste Conference* (Lafayette, IN: Purdue University 1956).

41. Wiley, J. S. "II. Progress Report on High-Rate Composting Studies," in *Proceedings of the 12th Annual Purdue Industrial Waste Conference* (Lafayette, IN: Purdue University 1957).

42. Nolan, W. E. personal communication (April 7, 1989).

43. Haug, R. T. and Ellsworth, W. F. "Degradability of Composting Substrates," *BioCycle* in press.
44. Gossett, R., personal communication, Compost Systems Co. (May 1986).
45. Zucconi, F., Forte, M., Pera, A., and de Bertoldi, M. "Evaluating Toxicity of Immature Compost," *BioCycle* 22(2) (March/April 1981).
46. Zucconi, F., Forte, M., Monaco, A., and de Bertoldi, M. "Biological Evaluation of Compost Maturity," *BioCycle* 22(4) (July/August 1981).
47. Tuttle, R., personal communication (October 25, 1991).
48. Wheeler, M. L., personal communication, Wastewater Superintendent, City of Hamilton, OH (March 1992).
49. Niese, G. "Experiments to Determine the Degree of Decomposition of Refuse by Its Self-Heating Capability," in *International Research Group on Refuse Disposal Information Bulletin 17* (1963).
50. Willson, G. B. and Dalmat D. "Measuring Compost Stability," *BioCycle* 27(7) (August 1986).
51. Lossin, R. D. "Compost Studies, Part III: Disposing of Animal Wastes, Measurement of the Chemical Oxygen Demand of Compost," *Compost Sci.* (March/April 1971).
52. Lossin, R. D. "Compost Studies, Part I," *Compost Sci.* (November/December 1970).
53. DeVleeschauwer, D., Verdonck, O., and Van Assche, P. "Phytotoxicity of Refuse Compost," *BioCycle* 22(1) (January/February 1981).
54. Spohn, E. "How Ripe is Compost," *Compost Sci.* 10(3) (1969).
55. Chrometzka, P. "Determination of the Oxygen Requirements of Maturing Composts," International Research Group on Refuse Disposal, Information Bulletin 33 (1968).
56. Pressel, F. and Bidlingmaier, W. "Analyzing Decay Rate of Compost," *BioCycle* 22(5) (September/October 1981).
57. Usui, T., Shoji, A., and Yusa, M. "Ripeness Index of Wastewater Sludge Compost," *BioCycle* 24(1) (January/February 1983).
58. Clark, C. S., Buckingham, C. O., Bone, D. H., and Clark, R. H. "Laboratory Scale Composting Techniques," *J. Env. Eng. Div., ASCE* 5:893–906 (1977).

CHAPTER 11

Process Dynamics I —
Development of Simulation Models

INTRODUCTION

Physical, chemical, and biological aspects of composting have been discussed in some depth in the preceding chapters. Mass and energy balances, aeration, moisture control, free airspace, thermodynamics, and kinetics have all been discussed. Although relationships between these factors have been stressed, it is often difficult to synthesize such a large volume of material. Understanding the subject would be easier were it tied together into a more unified and understandable package.

Earlier chapters have revealed much about the thermodynamic barriers and constraints that govern the composting process. However, composting is very much a dynamic process governed by the laws of chemical and biochemical kinetics. Thermodynamics can reveal much about the process, but it cannot answer questions that involve process kinetics. For example, what composting time is necessary to produce a stable end product? What is gained by longer or shorter reaction times? Can lag periods at the onset of composting be explained by process kinetics? Would a complete mix, continuous feed reactor offer process advantages compared to batch processes? To answer such questions it is necessary to integrate the intrinsic rate equations with conservation principles to produce a dynamic model of the process.

Dynamic models have been applied to numerous biological and chemical processes used in sanitary and chemical engineering practices. In addition to answering questions such as those posed above, dynamic models are a useful engineering tool to improve understanding of a process and the relationships between process variables. Such models can also guide engineering designs, supported of course by experimental and operational data from existing facilities. The simulation model also forms a bridge between laboratory or pilot data and the design of a full scale system. The laboratory can provide information on the organic content, composition, degradability, heat of combustion, and dewaterability of the waste under study. The simulation model can then use these results in a predictive mode to guide the engineering

385

design. This approach is commonly used in other areas of engineering practice and it is an approach to which the practice of composting is advancing.

The dynamic models developed in this chapter are available for use on the personal computer. Information on ordering the modeling software is presented on the inside title page of this book. These models allow the reader to quickly reduce the theory presented in this Chapter to actual practice as presented in Chapters 12 to 14. However, it is not necessary that the reader have these models (or even be interested in modeling) to comprehend the material presented here. Results presented in Chapters 12 to 14, while derived from the simulation models, do not require a computer background to be understood. Readers not interested in the details of computer modeling are urged to read at least the next section of this chapter because it presents the basic approach to analysis. This will be very helpful in understanding later chapters.

APPROACH TO THE PROBLEM

It is important that the entire composting process be included within the envelope of the simulation model. The initial "high rate" stages traditionally receive the most attention. Much of the capital cost and most proprietary designs focus on the early stages of the composting process. However, the later curing stages are also important and should not be excluded from any simulation model. In fact, the industry has often been guilty of paying attention only to the early stages, sometimes ignoring the critical role played out in the later stages. The dynamic models presented here encompass the entire process from the initial feed mixture to the final product.

Nomenclature used in the simulation models is presented in Appendix B. The nomenclature is, for the most part, self-explanatory and will become familiar to the reader after a little study.

Thermodynamic Boundaries

A schematic of a generalized composting system as used in the simulation model is presented in Figure 11.1. The system is broken into a number of stages. All feed substrates are introduced into stage 1. Output from stage 1 becomes the input to stage 2, and so on. Output from the final stage is the end product of the process. The entire system is surrounded by a thermodynamic system boundary. All mass and heat flows across the boundary must be balanced. In other words, mass into the boundary must equal mass out of the boundary. Similarly, heat which flows into the boundary or is released within the boundary must equal heat that flows out of the boundary.

Each stage of the process is surrounded by a thermodynamic stage boundary. All mass and heat flows across the individual stage boundaries must balance. This fact allows us to solve for process conditions within each stage. The sum of heat and materials flows for the individual stages must equal that for the entire system.

Substrates and Stages

Multiple substrates can be input to the process, up to a total of 10 different substrates. Substrate characteristics are contained in an array of values, each element in the array identified by the letter J. J equals 1 for substrate 1, 2 for substrate 2, and so on. For example, the feed rate for each substrate is identified as X(J). SUBNUM identifies the total number of

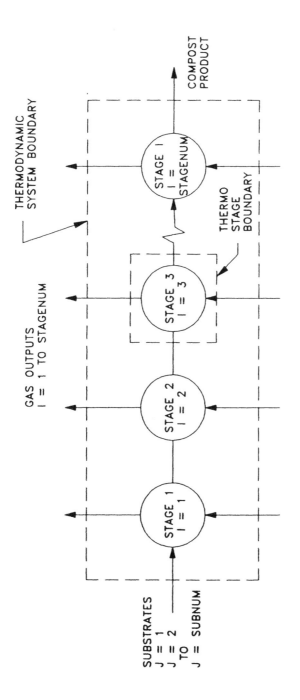

Figure 11.1.   Schematic diagram of a generalized composting process with multiple substrate inputs, stages arranged in series, and thermodynamic stage and system boundaries.

substrates input to the process. Thus, feed rate for the last substrate in the array is X(SUBNUM). Each substrate is identified by name in an array SUBNAME(J). The name can be up to 10 characters long.

The process is divided into a number of stages arranged in series, up to a maximum of 50 stages. Data for each stage are arranged in an array, each element identified by the letter I. I equals 1 for stage 1, and so on. Each stage is assumed to be completely mixed. The HRT for each stage is input to the model as HRT(I). Air flowrate to each stage is input as QAIR(I). STAGENUM identifies the total number of stages used to model the process.

HRT does not have to be the same for each stage. It is common to use shorter HRTs for the initial stages in the simulation model because of the higher rates of reaction. Conversely, longer HRTs can be used for the later curing stages where rates of reaction are reduced.

Model Types

Two basic model types are developed here. The first model is named CMSYS41B and applies to systems in which all feed components are homogeneously mixed and flow through the process without subsequent separation. Most reactor systems and most windrow processes satisfy this assumption. For example, sludge is often mixed with amendments such as sawdust and product recycle. Similarly, MSW fractions may be mixed together or yard wastes and leaves commingled with other wastes prior to composting. In either case, the mixture is then introduced to the composting process and is never separated back into its individual components. At some point in the composting cycle a portion of the mixed material may be recycled back to the feed mixture. The recycle and also the final product are homogeneous mixtures of organics remaining from the starting substrates.

A schematic diagram of the homogeneous system is presented in Figure 11.2. All substrates are input to stage 1. If recycle is used, it is also input to stage 1. Recycle can occur from any stage of the process, identified as STAGER. Output from the last stage, identified as STAGENUM, is the final product from the process. Air flowrate to each stage, QAIR(I), can either be fixed or adjusted by the program to maintain a desired temperature within the stage. Water can also be added to any stage, WATADD(I), either in fixed amounts or as required to maintain a desired moisture content within the stage.

The second model is named CMSYS52B and applies to systems that use bulking particles to maintain free air space and that utilize screening to return most of the bulking particles back to the system. The aerated, static pile process is an example of such a system and is most applicable to wet substrates such as municipal and industrial sludges and some food processing wastes. Because the bulking particles are easily identified within the mix, the static pile process is termed a heterogeneous system. Bulking particles are mixed with the substrates and introduced to stage 1. At some point in the composting cycle, the mixture is screened to recover the bulking particles which are then returned to the beginning of the process.

A schematic diagram of a heterogeneous system is presented in Figure 11.3. Again, all substrates are input to stage 1. New bulking particles are input as the last substrate, X(SUBNUM). Screening of the bulking agents follows the stage identified as STAGESCRN. Screened bulking particles, and other substrate fractions that may be incidently attached to the particles, are recycled to stage 1. Screened product becomes the input to stage STAGESCRN + 1. Recycle can occur from any stage downstream of the screening process and is again identified as STAGER. Air flowrate and water addition to each stage can either be fixed or allowed to vary to maintain desired temperature and moisture conditions within the stage.

Either metric or English units can be specified with both CMSYS41B and CMSYS52B. For clarity, only English units and associated equations are presented here.

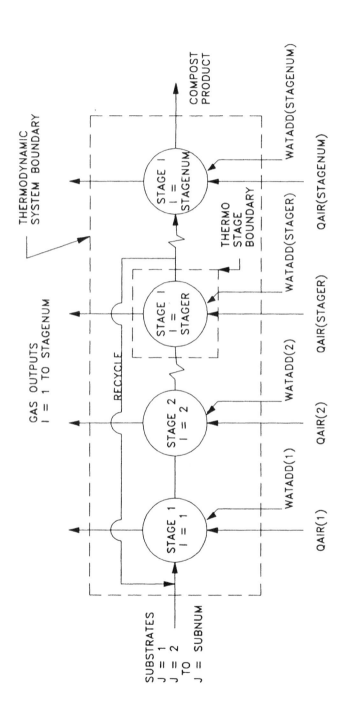

Figure 11.2. Schematic diagram of a system composting homogeneous substrates. A portion of the mixture is recycled downstream of stage STAGER and returned for conditioning of the feed to stage 1.

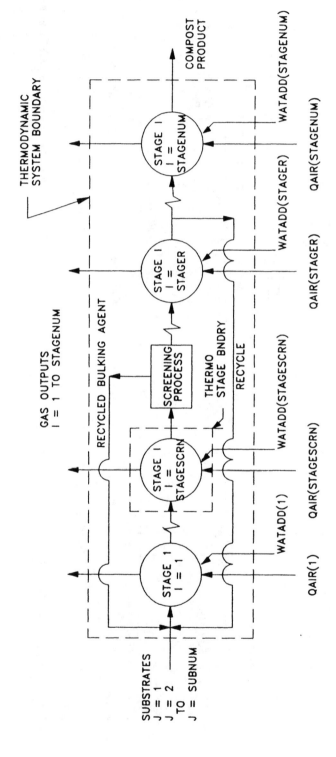

Figure 11.3.   Schematic diagram of a system composting heterogeneous substrates. Bulking agent is screened downstream of stage STAGESCRN and returned for conditioning of the feed to stage 1. A portion of the mixture can also be recycled from the output of stage STAGER and is also returned to the feed mixture to stage 1.

The CFCM Stage

Each stage of the process is modeled as a continuous feed, complete mix reactor. Most compost systems are fed on a daily basis. Considering that controlled composting usually requires at least 60 and sometimes 180 days or more residence time to produce a highly stable end product, the assumption of continuous feed is a reasonable approximation of the actual feed cycle.

Schematic diagrams showing mass and energy terms for a single stage of the process are shown in Figures 11.4 and 11.5. The number of mass and energy terms is considerably greater than that allowed by previous mass balance approaches used in this text. However, there are other factors that complicate the analysis more than the number of energy terms. The most important of these is the fact that as reactor temperature changes, so does the rate of decomposition. If the rate constant increases, for example, volatile solids, VS, in the outfeed will decrease, which alters the mass balance. Also, an increase in the rate constant means more biological heat production in the reactor. This means that a trial-and-error solution to the problem is required. The number of calculations required is somewhat staggering and computer solutions are required. Fortunately, the advent of personal computers has brought massive computing power to almost everyone. The personal computer, combined with the expert models presented here, brings the power of dynamic process modeling to the reader.

The solution point for any stage includes the steady state mass and energy balance as well as the temperature maintained in the stage. The approach to obtaining such a solution is illustrated in Figure 11.6. The sum of heat input terms, HTOTI, will characteristically curve upward with increasing temperature because of the exponential increase in biological activity with temperature. Increasing temperature will eventually result in thermal inactivation, a corresponding decrease in the rate constant and a reduction in total energy input. The sum of output energy terms, HTOTO, will generally curve upward as shown because of the exponential increase in water vapor pressure and latent heat loss as a function of temperature. Solution of the problem occurs when total energy input balances total energy output.

Three different output energy curves are shown in Figure 11.6. Two of the curves, A and B, have solution points, while curve C does not. A number of factors could account for the higher output energies of curve B as compared to curve A. Included among these would be higher heat loss rates to the surroundings or a higher aeration rate that increases latent heat losses. One might also imagine circumstances in which output energy is greater than input energy even at the beginning of the compost cycle (curve C). Extremely high aeration rates with cold ambient air could cause such a condition. In this case, compost material would tend to cool from its original input temperature. No solution point would exist except perhaps at a temperature below the inlet temperature of the feed materials.

The procedure for obtaining the solution point for a particular stage is as follows: (1) assume a stage temperature; (2) determine the rate constants for organic decomposition and calculate the mass balance about the stage; (3) sum HTOTI and HTOTO for the assumed temperature; and (4) compare HTOTI with HTOTO and estimate a new temperature based on the difference between the two. Because each stage of the process is affected by conditions in the surrounding stages, the solution for a single stage cannot be found independent of the other stages. To account for this, the solution algorithm moves sequentially from stage to stage. Once the last stage is reached, the process is again repeated beginning at stage 1 and using the new stage temperatures and other conditions calculated from the previous iteration. The process continues until both the mass and energy balances for each stage are closed to within a small limit of error. For the case of the energy balance, this means that the difference

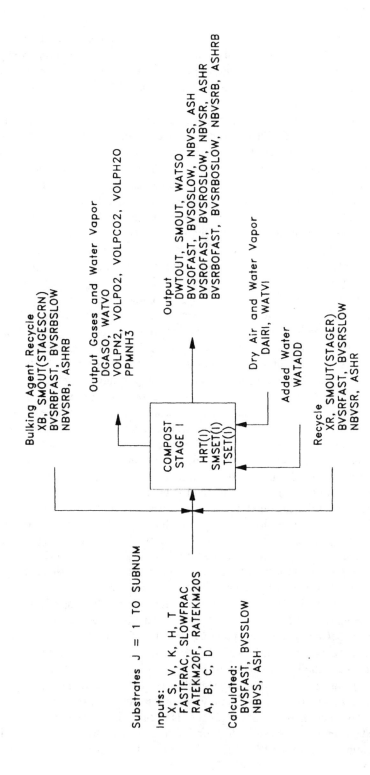

Figure 11.4.   Mass and related terms for one stage of the composting system.

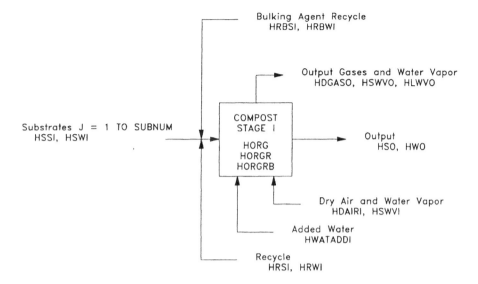

Figure 11.5. Energy and related terms for one stage of the composting system.

between HTOTI and HTOTO for each stage must be small. The procedure is well suited to computer analysis.

Batch Processes

The windrow and static pile systems are examples of batch operated processes. Each static pile is mixed at time zero and is relatively unaffected by other piles. The pile completes the compost cycle and does not receive addition substrate at any time in the process. Batch processes can also be modeled as a number of complete-mix stages arranged in series.

The reader may be somewhat reluctant to accept the fact that a batch process can be modeled as a series of complete mix stages. However, this modeling technique is commonly applied to fluid systems. In such a case, it can readily be shown that both a plug flow reactor with continuous feed and a batch process can be modeled as a large number of complete mix stages arranged in series. In a plug flow reactor with continuous feed the reactor itself is divided into a series of stages. For a batch process the entire pile contents are considered and the process divided into a series of time steps. The proof of this is relatively straight-forward, and the interested reader is referred to the large number of textbooks on reactor analysis in the chemical and sanitary engineering literature.

KINETIC ANALYSIS

From previous discussions it is obvious that the subject of composting kinetics is complex. Reaction rates can be limited by solubilization of substrate, mass transport of oxygen, utilization of the solubilized substrate and oxygen, and a variety of other factors such as lack of moisture or nutrients. A complete mathematical description of composting kinetics based entirely on first principles is probably not practical at this time. Therefore, reliance will be made on empirical expressions developed from experimental data. Although one might wish for more complete information, the data at hand must suffice.

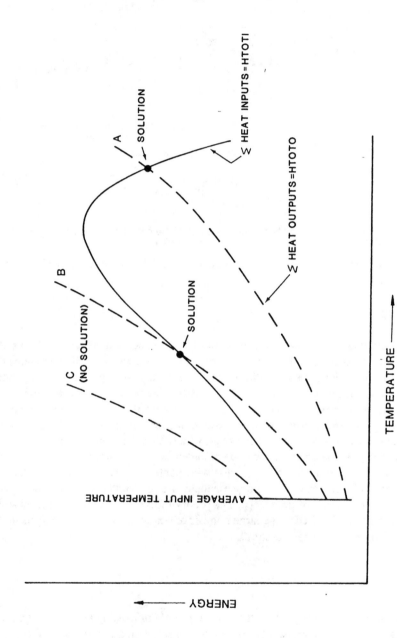

Figure 11.6.   Schematic illustration of the characteristics of energy input and output curves and solution points.

First Order Assumption

Oxidation of biodegradable volatile solids, BVS, is assumed to be first order with respect to the quantity of BVS as follows:

$$\frac{d(BVS)}{dt} = -k_d(BVS) \tag{11.1}$$

where

    BVS = biodegradable volatile solids, lb
    t = time, days
    k_d = rate constant, day^{-1}

As shown in Chapter 10 the first order model works well in describing the decomposition of composting substrates in controlled respirometer experiments. For many substrates the curve of oxygen uptake is best modeled by assuming faster and slower fractions for the organics. Equation 11.1 can then be written as

$$\frac{d(BVSFAST)}{dt} = -RATEKF * BVSFAST \tag{11.2}$$

$$\frac{d(BVSSLOW)}{dt} = -RATEKS * BVSSLOW \tag{11.3}$$

where

    BVSFAST = BVS characterized by a higher rate constant
    BVSSLOW = BVS characterized by a slower rate constant
    RATEKF = rate constant for the faster fraction
    RATEKS = rate constant for the slower fraction

Nomenclature in Equations 11.2 and 11.3 is that used in the computer simulation models.

The general form of the first order expression is shown in Figure 11.7. Total VS is divided between biodegradable (BVS) and nonbiodegradable volatile solids, NBVS. This same distinction has often been assumed in other biological waste treatment systems. It is not so much a distinction between degradable and absolutely nondegradable fractions, but rather a distinction between more degradable materials and those that are very slow to degrade. The latter are often referred to as refractory compounds. There is often no sharp dividing line between the two, particularly in composting systems where long detention times are typical.

The Effect of Temperature

The reaction rate constant is a function of temperature. Experimental data shown in Figure 10.17 and previous theoretical discussions clearly suggest an exponential relationship. The rate should about double for each 10°C temperature rise. As temperature continues to increase, however, heat inactivation will become more pronounced and the rate constant will begin to decrease. As discussed in Chapter 10, the temperature corresponding to the maximum rate constant has been the subject of considerable debate within the industry. A significant

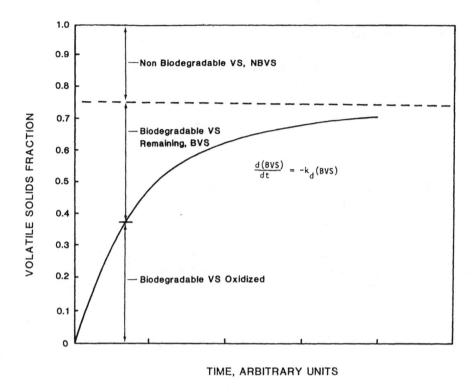

Figure 11.7. General form of the first order expression for oxidation of BVS. Note the distinction between degradable and nondegradable fractions of the total volatile solids.

advantage of computer simulations is that algorithms can be changed easily to use different temperature optima.

Andrews and Kambhu[1] used the following expression to describe effects of temperature on rate constants during aerobic digestion of liquid wastes:

$$k_d = k_{dRl} * [C_1^{(T-TRl)} - C_2^{(T-TR2)}] \qquad (11.4)$$

where

$$
\begin{aligned}
K_{dRl} &= \text{rate constant at temperature TR1, per day} \\
C_1, C_2 &= \text{temperature coefficients} \\
TR1, TR2 &= \text{reference temperatures} \\
T &= \text{substrate temperature}
\end{aligned}
$$

The form of Equation 11.4 is similar to that of Equation 3.24 which is based on fundamental thermodynamic principles. Using the data of Schulze from Figure 10.13 and assuming a temperature optimum near 70°C, the following equations are used to adjust the rate constants from a reference temperature of 20°C

$$RATEKMF(J,I) = RATEKM20F(J) * [1.066^{(TNEWC-20)} - 1.21^{(TNEWC-60)}] \qquad (11.5)$$

$$RATEKMS(J,I) = RATEKM20S(J) * [1.066^{(TNEWC-20)} - 1.21^{(TNEWC-60)}] \qquad (11.6)$$

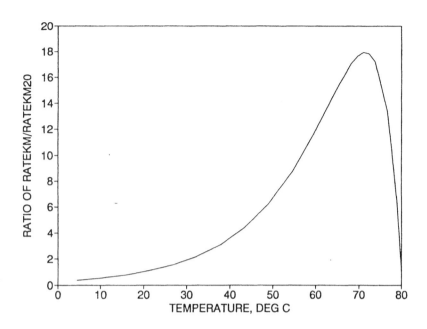

Figure 11.8. The assumed effect of temperature on the ratio of reaction rate constants at process temperature and 20°C.

where

RATEKM20F(J)	=	maximum rate constant for the fast fraction of substrate J at 20°C
RATEKM20S(J)	=	maximum rate constant for the slow fraction of substrate J at 20°C
RATEKMF(J,I)	=	rate constant for the fast fraction of substrate J in stage I at temperature TNEWC
RATEKMS(J,I)	=	rate constant for the slow fraction of substrate J in stage I at temperature TNEWC
TNEWC	=	substrate temperature in stage I, °C

The script "M" is added to the rate constant term to indicate that this is the maximum rate constant uncorrected for effects of moisture, FAS, or oxygen content. Equations 11.5 and 11.6 are plotted in Figure 11.8. The exponential dependence on temperature continues to about 65°C. The temperature coefficient Q_{10} (see Chapter 3) is about 1.9 in this range. Beyond 65°C the effects of heat inactivation become pronounced. The maximum occurs at about 70°C. Thermal inactivation is quite rapid above 70°C consistent with the inactivation energies for vegetative cells discussed in Chapter 5.

Referring back to Figure 10.7, optimum temperatures as low as 45°C have been reported. The data of Schulze,[2] using continuous composters under steady state conditions, and Wiley and Pearce,[3] using batch composters, were weighted heavily in formulating Equations 11.5 and 11.6. The latter data indicate an exponential relationship to temperatures as high as 70°C. Equations 11.5 and 11.6 are also consistent with the observation that composting temperatures rarely exceed 80°C, a point at which the rate of heat inactivation must be overwhelming even for thermophiles.

Regan and Jeris[4] have suggested that the temperature optima may depend on the substrate in question. Largely cellulosic substrates, such as paper, newsprint, and sawdust, may have a

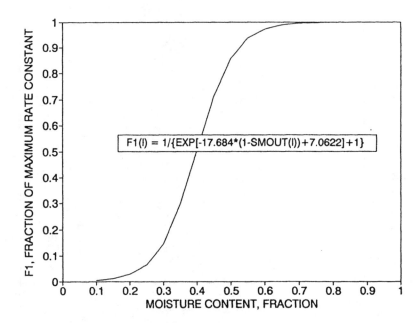

Figure 11.9. Assumed effect of mixture moisture content on the reaction rate constant.

lower optima compared to more heterogeneous mixtures of organics, such as sewage sludge. On the other hand, oxygen consumption rates shown in Figure 10.17 are reasonably similar over a wide range of organics. Although the limitations of Equations 11.5 and 11.6 are recognized, experimental data suggest that it is reasonably applicable to a wide variety of solid substrates.

The Effect of Moisture

Experimental data on the effect of moisture content on oxygen consumption rates (and hence rate of BVS oxidation) were presented in Figure 10.9. Based on the trend of experimental data, the logistics curve in Figure 11.9 was constructed and used to predict the effect of mixture moisture content on the rate constant. Because there is considerable scatter in the experimental data, a great number of possible curves might be constructed. An S-shaped logistics curve was selected since its shape is theoretically pleasing and generally consistent with the experimental data. The calculation procedure is to first determine the rate constants from Equations 11.5 and 11.6 for the assumed conditions of temperature, and then correct the rate constant by the moisture adjustment factor given as

$$F1(I) = \frac{1}{e^{[-17.684*[1-SMOUT(I)]+7.0622]} + 1} \qquad (11.7)$$

where

$F1(I)$ = factor to adjust the rate constant for the effect of moisture content, fraction

$SMOUT(I)$ = solids content of the mixture in stage I, fraction

The Effect of Free Airspace

As discussed in Chapter 6, excessive moisture or compaction can reduce FAS to the point where oxygen storage and transport through the void spaces is reduced. Reaction rates can then become oxygen limited. This effect is observed in the data of Figure 10.9, where oxygen consumption rates were actually reduced at high moisture levels. The effect results not from excessive moisture, but from a lack of FAS. Considering the general trend of data in Figure 10.9 and recalling that optimum FAS is generally about 30%, the curve in Figure 11.10 was constructed and used to represent the effect of FAS on the reaction rate constant. Although data are limited the selected curve is consistent with available data, and the general shape of the curve is theoretically reasonable.

Application of the correction factor, F2, requires that the bulk weight be known to calculate FAS. For a wet substrate such as sludge cake, which has essentially no void volume, the specific gravity, GAMMAM, and bulk weight, BULKWT, are estimated as

$$GAMMAM = \frac{1}{[(S/GS) + 1 - S]} \qquad (6.3)$$

$$BULKWT = 62.4 * GAMMAM \qquad (11.8)$$

where
  S = fractional solids content of the substrate
  GS = specific gravity of the substrate solids, calculated from Equation 6.1

For lighter weight substrates and mixtures of substrates that have a significant void volume, GAMMAM is calculated as,

$$GAMMAM = \frac{COEFFM}{SM} \qquad (11.9)$$

where
  COEFFM = bulk weight coefficient for the substrate or mixture
  SM = fractional solids content of the substrate or mixture of substrates

COEFFM typically ranges from 0.15 to 0.4 for substrates such as sawdust, paper, and MSW fractions and mixtures of substrates such as sludge and sawdust. Together, Equations 6.3 and 11.9 are consistent with the trend of data presented in Figures 6.3 and 6.4. Equations 6.3 and 11.8 are used if the substrate name SUBNAME(J) is identified as SLUDGE. Equation 11.9 is used for all other substrate names.

Knowing the bulk weight and mixture solids content, FAS for stage I is calculated using the form of Equation 6.11 as

$$FAS(I) = 1 - \left[\frac{GAMMAM * SMOUT(I)}{GS} \right] - \{GAMMAM * [1 - SMOUT(I)]\} \qquad (11.10)$$

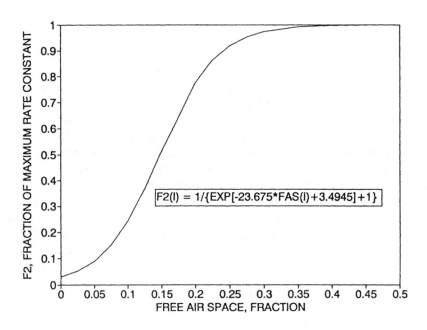

Figure 11.10. Assumed effect of mixture FAS on the reaction rate constant.

The correction factor F2, is then calculated from the logistics curve in Figure 11.10 as follows:

$$F2(I) = \frac{1}{e^{[-23.675*FAS(I)+3.4945]}+1} \qquad (11.11)$$

where

F2(I)   =   factor to adjust the rate constant for the effect of FAS in stag⸱
            I, fraction

The Effect of Oxygen Content

Even though the effect of FAS has been accounted for, it does not by itself ensure that oxygen is not a rate-limiting substrate. The kinetics of oxygen transport were discussed in some detail in Chapter 10. The key points of that discussion are as follows: (1) transport is probably limited by diffusion through the particle matrix of solids and water; (2) even though oxygen partial pressure in the FAS is high, it does not necessarily mean that oxygen transport into the particle is not rate-limiting; and (3) whether the mass flux of oxygen exceeds the oxygen demand depends on the size of the particle. On the latter point, particle thickness on the order of 1.0 cm would appear to present large diffusion resistances that would tend to dominate the process kinetics.

This leads to a rather complex picture of oxygen supply because diffusion transport and particle size should both be considered. Such a sophisticated model is beyond the present state of the art. Consider again the compilation of measured oxygen uptake rates shown in Figure 10.17. Different substrates with different particle sizes were used in these experiments. Can it be stated that oxygen was not rate-limiting in some cases? Actually, the answer to this question can be obtained indirectly by noting the rather significant effect of temperature on oxygen uptake. If oxygen diffusion was controlling, the influence of temperature would be

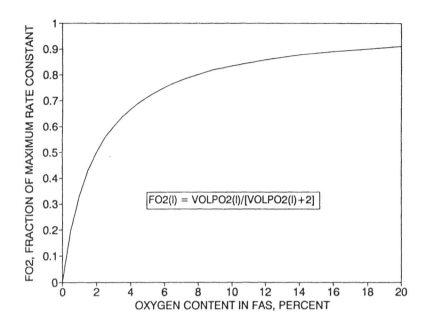

Figure 11.11. Assumed effect of oxygen content on the reaction rate constant.

much less significant. Refer back to Equation 10.16 to see the effect of temperature on the diffusion coefficient. Therefore, the rate of oxygen diffusion probably did not control process kinetics for most of the data in Figure 10.17.

Insight on the effect of oxygen content can be developed by again reviewing the data of Schulze.[2] He used continuous composters that were fed intermittently. Airflow through the composters was adjusted so that residual oxygen concentration in the exhaust gas remained between 5 and 10%. Because the composters were well mixed, oxygen levels in the FAS between composting particles should have been in the same range. Thus, there is some assurance that oxygen levels of 5% in the FAS do not impose severe oxygen limitations, provided, of course, that particle sizes are similar to those used by Schulze.

Because the effect of particle size is difficult to model, a more simplified approach was adopted. The effect of oxygen concentration in the free air space was assumed to follow a Monod-type expression as shown in Figure 11.11:

$$FO2(I) = \frac{VOLPO2(I)}{VOLPO2(I) + 2} \tag{11.12}$$

where

$FO2(I)$ = factor to adjust the rate constant for the effect of oxygen content in stage I, fraction

$VOLPO2(I)$ = volume percent of oxygen in the exhaust gas from stage I

The half velocity coefficient is assumed to be 2.0% oxygen by volume in the FAS. This assures that oxygen effects are minimal at concentrations above ~5%. Also, the rate of organic decomposition reduces to zero if the oxygen concentration is zero, a boundary condition known to be true for aerobic metabolism. An inherent assumption in this approach is that particle sizes are sufficiently small to avoid oxygen transport limitations.

Net Rate Coefficient

Once the maximum rate coefficients are determined from Equations 11.5 and 11.6, the factors F1, F2, and FO2 are applied to determine the actual rate constant under conditions of moisture, FAS, and oxygen content in the composting matrix of stage I. The actual rate constants for the fast and slow fractions for substrate J in stage I are then determined as

$$RATEKF(J,I) = RATEKMF(J,I) * F1(I) * F2(I) * FO2(I) \qquad (11.13)$$

$$RATEKS(J,I) = RATEKMS(J,I) * F1(I) * F2(I) * FO2(I) \qquad (11.14)$$

Other Rate Limitations

The above algorithms account for the effects of degradable and nondegradable organic fractions, faster and slower decomposing fractions of a substrate, temperature, moisture, FAS, and oxygen content on the overall rate of reaction. This is a reasonably sophisticated model and includes the factors that are most likely to limit kinetic rates. However, there are other potential rate limitations that are not included in the present model. For example, the model assumes that the feed mixture is not sterile and that it contains sufficient microbes to avoid the rate limitations that would result from an inadequate microbial population. The common use of recycle should assure that the feed substrate is adequately seeded. Second, it is assumed that the substrate is free of toxic compounds that would inhibit microbial metabolism. Third, rate limitations from excessively high or low pH are not considered. The latter can cause a lag period in the early stages of batch processes, but the effect usually lasts less than a week. Fourth, required nutrients, such as nitrogen and phosphorus, are assumed to be present in adequate amounts.

SOLIDS BALANCES

Mass balances are required on all components entering and leaving the system boundary and the individual stage boundaries, including organic and inorganic solids, water, and gases. Each of these components will be discussed separately.

Substrate Solids

For each feed substrate the wet weight (X), solids fraction (S), volatile solids fraction (V), and degradability coefficient (K) are input to the model. Based on these inputs, the BVS (both fast and slow fractions), NBVS, and ash (ASH) components of the substrate are determined. For substrate J, the components input to stage 1 are calculated as

$$BVS(J,1) = K(J) * V(J) * S(J) * X(J) \qquad (11.15)$$

$$BVSFAST(J,1) = BVS(J) * FASTFRAC(J) \qquad (11.16)$$

$$BVSSLOW(J,1) = BVS(J) * SLOWFRAC(J) \qquad (11.17)$$

$$NBVS(J,1) = [1 - K(J)] * V(J) * S(J) * X(J) \qquad (11.18)$$

$$ASH(J,1) = [1 - V(J)] * S(J) * X(J) \qquad (11.19)$$

The total dry weight of substrate solids (DWTS) and associated water (WATS) input to stage 1 are determined as

$$DWTS = \sum_{J=1}^{SUBNUM} [S(J) * X(J)] \tag{11.20}$$

$$WATS = \sum_{J=1}^{SUBNUM} [X(J) - S(J) * X(J)] \tag{11.21}$$

Organic decomposition in a particular stage is determined from a mass balance about that stage. Recall that a complete mix reactor is assumed for each stage. Considering the degradable fractions, a mass balance on BVS gives

Rate of storage = input − output + sources − sinks

$$VOL * \frac{d(BVS)}{dt} = BVS - BVSO + 0 - [RATEK * (BVS) * HRT]$$

where

    VOL = volume of stage I
    HRT = single pass residence time in stage I
    BVSO = output BVS from stage I

Assuming steady state conditions where d(BVS)/dt = 0, the above expression can be rearranged to

$$BVSO = \frac{BVS}{1 + RATEK * HRT} \tag{11.22}$$

Considering the faster and slower fractions for BVS, and generalizing to substrate J in stage I, Equation 11.22 becomes

$$BVSOFAST(J,I) = \frac{BVSFAST(J,I)}{1 + RATEKF(J,I) * HRT(I)} \tag{11.23}$$

$$BVSOSLOW(J,I) = \frac{BVSSLOW(J,I)}{1 + RATEKS(J,I) * HRT(I)} \tag{11.24}$$

The loss of BVS for substrate J across stage I is determined as

$$DBVS(J) = BVSFAST(J,I) + BVSSLOW(J,I) -$$
$$BVSOFAST(J,I) - BVSOSLOW(J,I) \tag{11.25}$$

ASH and NBVS fractions are assumed to be conservative. Therefore, the input to stage I + 1 equals the output from stage I.

Recycled Solids

Recycle of mixed solids can occur from any stage of the process. The stage from which recycle is drawn is identified as STAGER in the model. Recycled solids are assumed to have the same composition as the mixed outfeed from stage STAGER and are always returned to stage 1. Depending on the model, the wet weight of recycle (XR) can either be fixed or varied to maintain a setpoint solids content in the feed mixture. The latter is identified as SMMIN. XR required to achieve the input value of SMMIN is calculated as

$$SUMX = \sum_{J=1}^{SUBNUM} X(J) \tag{11.26}$$

$$SUMSX = \sum_{J=1}^{SUBNUM} [S(J) * X(J)] \tag{11.27}$$

$$XR = \frac{SMMIN * SUMX - SUMSX}{SMOUT(STAGER) - SMMIN} \tag{11.28}$$

where
SMOUT(STAGER)   =   fractional solids content in mixture out of stage STAGER

If XR is varied to maintain the solids content of the feed mixture, problems can arise if drying becomes limited. If SM(STAGER) approaches SMMIN, Equation 11.28 tends toward infinity. To prevent this condition, minimum and maximum values for XR are input to the program. The latter are defined as XRMIN and XRMAX, respectively. XR cannot fall below XRMIN or above XRMAX. If the program should encounter either of these two constraints, mixture solids cannot be maintained at the desired SMMIN. The program will recalculate the actual mixture solids in this case. For cases where bulking agents are used for control of FAS, recycled solids (if used) are held constant at the input value and the limits of XRMIN and XRMAX do no apply.

The dry weight of solids recycled (DWTR) and associated water (WATR) are determined as

$$DWTR = XR * SMOUT(STAGER) \tag{11.29}$$

$$WATR = XR - XR * SMOUT(STAGER) \tag{11.30}$$

Recycled material will contain biodegradable solids that did not degrade on their original pass through the system. These solids will continue to degrade as recycled materials again pass through the process. Therefore, degradable fractions in the recycle must be accounted for in the overall balance. Biodegradable solids in the recycle are identified as BVSR. Again, both faster and slower fractions are tracked and identified as BVSRFAST and BVSRSLOW. Ash and nonbiodegradable fractions are identified as ASHR and NBVSR. Because the composition of recycle is the same as the output from stage STAGER, the quantity of the recycle components can be determined from the ratio of dry weight recycled to the dry output from stage STAGER.

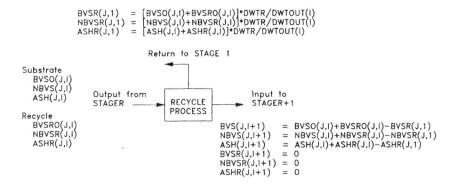

$$
\begin{aligned}
\text{BVSR(J,1)} &= \left[\text{BVSO(J,I)+BVSRO(J,I)}\right]\text{*DWTR/DWTOUT(I)} \\
\text{NBVSR(J,1)} &= \left[\text{NBVS(J,I)+NBVSR(J,I)}\right]\text{*DWTR/DWTOUT(I)} \\
\text{ASHR(J,1)} &= \left[\text{ASH(J,I)+ASHR(J,I)}\right]\text{*DWTR/DWTOUT(I)}
\end{aligned}
$$

Return to STAGE 1

Substrate
BVSO(J,I)
NBVS(J,I)
ASH(J,I)

Recycle
BVSRO(J,I)
NBVSR(J,I)
ASHR(J,I)

Output from STAGER ⟶ RECYCLE PROCESS ⟶ Input to STAGER+1

$$
\begin{aligned}
\text{BVS(J,I+1)} &= \text{BVSO(J,I)+BVSRO(J,I)−BVSR(J,1)} \\
\text{NBVS(J,I+1)} &= \text{NBVS(J,I)+NBVSR(J,I)−NBVSR(J,1)} \\
\text{ASH(J,I+1)} &= \text{ASH(J,I)+ASHR(J,I)−ASHR(J,1)} \\
\text{BVSR(J,I+1)} &= 0 \\
\text{NBVSR(J,I+1)} &= 0 \\
\text{ASHR(J,I+1)} &= 0
\end{aligned}
$$

Figure 11.12. Schematic illustration of the recycle process and the mass balance equations for the feed substrates and recycle. The diagram applies to feed substrates J = 1 to SUBNUM with I = STAGER. For clarity the fast and slow fractions of BVS are not shown.

The recycling process is illustrated in Figure 11.12. Mixed output from stage STAGER is input to the recycling process. Substrate from the original feed and substrate returned with the recycle material are included in the balance. Equations used to model the recycling process are shown in Figure 11.12. For purposes of clarity, the faster and slower BVS fractions are shown as a single component. The equations for each component are developed from the following mass balance considerations:

output from stage STAGER = input to recycle process

input to recycle process = return to stage 1 + input to stage STAGER + 1

$$\text{return to stage } 1 = \frac{\text{DWTR}}{\text{DWTOUT(STAGER)}} * \text{input}$$

By definition, no recycle components continue to the stages downstream of STAGER. In other words, BVSR, NBVSR, and AHSR are all zero for stages STAGER + 1 and above.

Recycle components are tracked through the process in the same way as the original feed substrates. Recycled biodegradable solids in the output from stage I are determined as

$$\text{BVSROFAST(J,I)} = \frac{\text{BVSRFAST(J,I)}}{1 + \text{RATEKF(J,I)} * \text{HRT(I)}} \tag{11.31}$$

$$\text{BVSROSLOW(J,I)} = \frac{\text{BVSRSLOW(J,I)}}{1 + \text{RATEKS(J,I)} * \text{HRT(I)}} \tag{11.32}$$

The loss of BVSR for substrate J across stage I is determined as

$$
\begin{aligned}
\text{DBVSR(J)} = \text{BVSRFAST(J,I)} + \text{BVSRSLOW(J,I)} - \\
\text{BVSROFAST(J,I)} - \text{BVSROSLOW(J,I)}
\end{aligned}
\tag{11.33}
$$

ASHR and NBVSR fractions are assumed to be conservative throughout the system. There-fore, the input to stage I + 1 equals the output from stage I.

Bulking Agent Solids

If bulking particles (agents) are added to the composting mix, they must be tracked through the process using procedures similar to those for recycled solids. However, normal composting practice is to screen the bulking agent from the composting mixture and then return the screened particles for mixing with the feed substrates. New bulking agent may also be added to the starting mix. The purpose of screening is twofold: first, recover the bulking particles so that they can be reused and second, separate the substrates from the bulking particles so that substrates can continue on through the process.

As was the case with recycled solids, recycled bulking agent will contain biodegradable solids that did not degrade on their original pass through the system. If the screening is efficient, most of the recycled solids will consist of clean bulking agent with very little other substrates attached to the particles. However, a portion of original feed substrates will always be present. This is particularly true if the mixture is wet. Under these conditions, a good separation is not achieved and more of the original substrates are returned with the bulking particles.

Solids recycled with the bulking agent will continue to degrade as the material passes again through the process. Therefore, the degradable fractions in the recycled bulking agent must be accounted for in the overall balance. Biodegradable solids in the recycled bulking agent are identified as BVSRB. Again, both faster and slower fractions are identified as BVSRBFAST and BVSRBSLOW. Ash and nonbiodegradable fractions are identified as ASHRB and NBVSRB. Screening of all material occurs after the stage identified as STAGESCRN. New bulking agent is entered as the last substrate, i.e., X(SUBNUM).

The Screening Process

The screening efficiency, SCRNEFF, is defined as the ratio of the substrate that separates from the bulking agent and passes through the screening process divided by the feed substrate:

$$\text{SCRNEFF} = \frac{\text{wt of substrate passing screen}}{\text{wt of feed substrate to screen}} \tag{11.34}$$

Screening efficiency is known to be adversely effected by moisture. Exact correlations between moisture and screening efficiency have not been developed. The type of screening equipment probably has some effect on the correlation, but all separation equipment is adversely affected by moisture. The industry consensus seems to be that the solids content must be at least 50%, and preferably 55%, to effect good separation. The correlation shown in Figure 11.13 is used to model SCRNEFF with the following formula:

$$\text{SCRNEFF} = \frac{1}{e^{[-17.4734*\text{SMOUT(I)}+7.863]} + 1} \tag{11.35}$$

At 50% solids SCRNEFF = 0.7 and increases to over 0.9 at 60% solids. At 40% solids SCRNEFF is only 0.3. These values are consistent with the general range of values observed in the field.

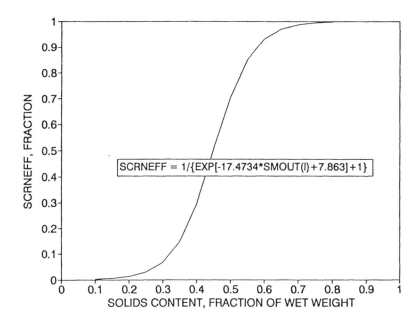

Figure 11.13. Assumed effect of mixture solids content on the efficiency of screening bulking particles from the remaining substrates.

The screening process is illustrated in Figure 11.14 for the feed substrates exclusive of new bulking agent (i.e., substrates J = 1 to SUBNUM − 1). Mixed output from stage STAGESCRN is input to the screening process. Substrate from the original feed, substrate returned with any recycle material, and substrate recycled with returned bulking agent are included in the balance. Equations used to model the screening process are also shown in Figure 11.14. Again for the purpose of clarity, the faster and slower BVS fractions are shown as a single BVS component. An assumption inherent in Figure 11.14 is that all substrates other than the bulking agent itself have a particle size smaller than the screen size. This is usually the case, particularly with sludge substrates. Equations for each component were developed from the following mass balance considerations:

output from stage STAGESCRN = input to screening

input to screening = return to stage 1 + input to stage STAGESCRN + 1

return to stage 1 = (1 − SCRNEFF) * input

The number of terms and equations in Figure 11.14 may seem a little overwhelming on first glance. However, the equations for an individual component are relatively straightforward. The reader should focus on the concepts embodied in Figure 11.14 and rely on the computer to keep track of the large number of equations.

The situation for bulking agent is somewhat more complicated than for the other substrates. Mechanical and biological breakdown of the bulking particles can occur during composting. Biological breakdown is accounted for in the BVS terms. However, breakdown due to mixing, turning, and other mechanical actions is believed to be responsible for most of the loss of bulking particles such as wood chips. To account for mechanical breakdown, a new term, MECHLOSS, is introduced. MECHLOSS is defined as the fractional loss of bulking agent per

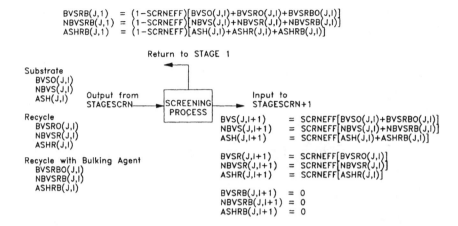

$$BVSRB(J,1) = (1-SCRNEFF)[BVSO(J,I)+BVSRO(J,I)+BVSRBO(J,I)]$$
$$NBVSRB(J,1) = (1-SCRNEFF)[NBVS(J,I)+NBVSR(J,I)+NBVSRB(J,I)]$$
$$ASHRB(J,1) = (1-SCRNEFF)[ASH(J,I)+ASHR(J,I)+ASHRB(J,I)]$$

Return to STAGE 1

Substrate
BVSO(J,I)
NBVS(J,I)
ASH(J,I)

Output from
STAGESCRN

SCREENING PROCESS

Input to
STAGESCRN+1

Recycle
BVSRO(J,I)
NBVSR(J,I)
ASHR(J,I)

$$BVS(J,I+1) = SCRNEFF[BVSO(J,I)+BVSRBO(J,I)]$$
$$NBVS(J,I+1) = SCRNEFF[NBVS(J,I)+NBVSRB(J,I)]$$
$$ASH(J,I+1) = SCRNEFF[ASH(J,I)+ASHRB(J,I)]$$

Recycle with Bulking Agent
BVSRBO(J,I)
NBVSRB(J,I)
ASHRB(J,I)

$$BVSR(J,I+1) = SCRNEFF[BVSRO(J,I)]$$
$$NBVSR(J,I+1) = SCRNEFF[NBVSR(J,I)]$$
$$ASHR(J,I+1) = SCRNEFF[ASHR(J,I)]$$

$$BVSRB(J,I+1) = 0$$
$$NBVSRB(J,I+1) = 0$$
$$ASHRB(J,I+1) = 0$$

Figure 11.14. Schematic illustration of the screening process and the mass balance equations for substrate components from the original feed, those recycled with the bulking agent, and those recycled from any product recycle. Diagram applies to feed substrates J = 1 to SUBNUM − 1 with I = STAGESCRN. For clarity the fast and slow fractions of BVS are not shown.

pass through the system. For example, a MECHLOSS = 0.1 means that on each pass through the process 10% of the bulking agent will be mechanically broken down to a size that can pass through the screen.

A diagram of the screening process for bulking agent (J = SUBNUM) is illustrated in Figure 11.15. Equations for each component are again developed by considering a mass balance about the screening process as follows:

output from stage STAGESCRN = input to screening

input to screening = return to stage 1 + input to stage STAGESCRN + 1

return to stage 1 = unbroken bulking agent at 100% capture +
(1 − SCRNEFF) times broken bulking agent +
(1 − SCRNEFF) times recycled bulking agent

The above balance assumes that bulking particles that are not broken down by mechanical means are screened at 100% efficiency. For a typical case using nominal 1 to 2 inch wood chips and a screen with say 0.5 inch openings, the above should be a reasonably good model of the actual process. If the screen size is very large or the bulking agent is irregularly shaped, the model will be less applicable. Bulking agent that is mechanically broken is assumed to be screened with the same efficiency as other substrates. In other words, a fraction of the broken bulking agent will be retained with the recycled bulking agent and the remaining fraction passed through the screen. The latter increases with increased screening efficiency.

The dry weight of solids in the recycle bulking agent, DWTRB, and associated water, WATRB, are determined as

$$DWTRB = \sum_{J=1}^{SUBNUM} [BVSRBFAST(J,1)+BVSRBSLOW(J,1)+ ASHRB(J,1)+NBVSRB(J,1)] \qquad (11.36)$$

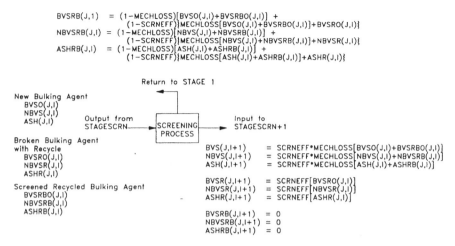

$$
\begin{aligned}
\text{BVSRB(J,1)} &= (1-\text{MECHLOSS})[\text{BVSO(J,I)}+\text{BVSRBO(J,I)}] + \\
&\quad (1-\text{SCRNEFF})\}\text{MECHLOSS}[\text{BVSO(J,I)}+\text{BVSRBO(J,I)}]+\text{BVSRO(J,I)}\} \\
\text{NBVSRB(J,I)} &= (1-\text{MECHLOSS})[\text{NBVS(J,I)}+\text{NBVSRB(J,I)}] + \\
&\quad (1-\text{SCRNEFF})\}\text{MECHLOSS}[\text{NBVS(J,I)}+\text{NBVSRB(J,I)}]+\text{NBVSR(J,I)}\} \\
\text{ASHRB(J,I)} &= (1-\text{MECHLOSS})[\text{ASH(J,I)}+\text{ASHRB(J,I)}] + \\
&\quad (1-\text{SCRNEFF})\}\text{MECHLOSS}[\text{ASH(J,I)}+\text{ASHRB(J,I)}]+\text{ASHR(J,I)}\}
\end{aligned}
$$

Figure 11.15. Schematic illustration of the screening process and the mass balance equations for the bulking agent. Diagram applies to the bulking agent identified as J = SUBNUM and with I = STAGESCRN. For clarity the fast and slow fractions of BVS are not shown.

$$
\text{WATRB} = \frac{\text{DWTRB}}{\text{SMOUT(STAGESCRN)}} - \text{DWTRB} \tag{11.37}
$$

By definition, no recycled bulking agent components continue to stages downstream of STAGESCRN. In other words, BVSRB, NBVSRB, and ASHRB are all zero for stages STAGESCRN + 1 and above.

Recycled bulking agent is tracked through the process in the same way as the original feed substrates. Biodegradable solids from recycled bulking agent in the output from stage I are determined as

$$
\text{BVSRBOFAST(J,I)} = \frac{\text{BVSRBFAST(J,I)}}{1+\text{RATEKF(J,I)} * \text{HRT(I)}} \tag{11.38}
$$

$$
\text{BVSRBOSLOW(J,I)} = \frac{\text{BVSRBSLOW(J,I)}}{1+\text{RATEKS(J,I)} * \text{HRT(I)}} \tag{11.39}
$$

The loss of BVSRB for substrate J across stage I is determined as

$$
\begin{aligned}
\text{DBVSRB(J)} = \text{BVSRBFAST(J,I)} + \text{BVSRBSLOW(J,I)} - \\
\text{BVSRBOFAST(J,I)} - \text{BVSRBOSLOW(J,I)}
\end{aligned} \tag{11.40}
$$

ASHRB and NBVSRB fractions are assumed to be conservative. Therefore, the input to stage I + 1 equals the output from stage I.

If recycle is used along with a bulking agent, the recycle process must occur downstream of the screening process. In other words, STAGER must be greater than STAGESCRN. Recycling ahead of the screening process does not make sense from a process standpoint and, therefore, has been excluded from the model.

New Bulking Agent

The demand for new bulking agent at the start of the process will depend primarily on mechanical losses and biodegradation within the system. Maintaining a proper volume ratio of bulking agent to other substrates is critical to the process. The physics of using bulking particles and typical volume ratios to maintain FAS were discussed in Chapter 6. For the simulation model, the desired volume ratio is input as VOLRATIO and is defined as

$$VOLRATIO = \frac{\text{volume of new} + \text{recycled bulking particles}}{\text{volume of all nonbulking feed substrates}} \tag{11.41}$$

The quantity of new bulking agent is input as X(SUBNUM). This value is interpreted as a minimum and is set equal to XBMIN. If additional bulking agent is required to achieve the desired VOLRATIO, the quantity of new bulking agent is recalculated. First, the volume of all nonbulking agent substrates, VOLUMESUBS, is determined as

$$VOLUMESUBS = \sum_{J=1}^{SUBNUM-1} VOLUME(J) \tag{11.42}$$

where VOLUME(J) is the volume of substrate J calculated as

$$VOLUME(J) = \frac{X(J)}{BULKWT(J)} \tag{11.43}$$

The volume of recycled bulking agent is determined from the wet weight of recycled bulking particles as

$$SUBNUMXRB = \frac{[BVSRBFAST(SUBNUM,1) + BVSRBSLOW(SUBNUM,1) +}{} $$
$$\frac{NBVSRB(SUBNUM,1) + ASHRB(SUBNUM,1)]}{SMOUT(STAGESCRN)} \tag{11.44}$$

$$BULKWTXRB = \frac{62.4 * COEFFM}{SMOUT(STAGESCRN)} \tag{11.45}$$

$$VOLSUBNUMXRB = \frac{SUBNUMXRB}{BULKWTXRB} \tag{11.46}$$

where
SUBNUMXRB   =   wet weight of recycled bulking agent, lb/day
BULKWTXRB   =   bulk weight of recycled bulking agent, lb/ft³
VOLSUBNUMXRB =   volume of recycled bulking agent, ft³/day

Note that the above equations include only the actual bulking agent component. Any substrates recycled with the bulking agent are not included in the volume calculation. This is because only actual bulking particles are included in the numerator of Equation 11.41. The required volume of new bulking agent is then determined as

$$VOLUME(SUBNUM) = (VOLUMESUBS * VOLRATIO)_j - VOLSUBNUMXRB \quad (11.47)$$

The wet weight of new bulking agent, X(SUBNUM), to achieve the desired VOLRATIO is then determined from

$$X(SUBNUM) = VOLUME(SUBNUM) * BULKWT(SUBNUM) \quad (11.48)$$

If X(SUBNUM) calculated by Equation 11.48 is less than XBMIN, then the value of X(SUBNUM) is set equal to XBMIN. In this case the actual volume ratio will be greater than the input value VOLRATIO. New values for BVSFAST(SUBNUM,1), BVSSLOW(SUBNUM,1), NBVS(SUBNUM,1), and ASH(SUBNUM,1) are determined using Equations 11.15 to 11.19 based on the new value of X(SUBNUM).

If STAGESCRN = 0, which indicates that the system does not include a screening process, then X(SUBNUM) is not recalculated to achieve a desired VOLRATIO. Instead, the program fixes the input value of X(SUBNUM), as long as it is greater than XBMIN, and does not readjust it to achieve the desired VOLRATIO. The reason for this is illustrated in Examples 14.3 and 14.4 of Chapter 14.

Output Solids

Dry weight output from any stage I, DWTOUT(I), is determined by summing all component fractions for substrate, recycled material, and any bulking agent. Recall that ASH and NBVS fractions are assumed to be conservative. The following equation applies to the output from any stage I:

$$
\begin{aligned}
DWTOUT(I) = \sum_{J=1}^{SUBNUM} & [BVSOFAST(J,I) + BVSOSLOW(J,I) + ASH(J,I) + NBVS(J,I) + \\
& BVSROFAST(J,I) + BVSROSLOW(J,I) + \\
& ASHR(J,I) + NBVSR(J,I) + \\
& BVSRBOFAST(J,I) + BVSRBOSLOW(J,I) + \\
& ASHRB(J,I) + NVBVSRB(J,I)]
\end{aligned}
\quad (11.49)
$$

GAS BALANCES

Stoichiometry of Solids Oxidation

Carbon, hydrogen, oxygen, and nitrogen composition for each substrate is input to the model according to the formula $C_aH_bO_cN_d$. This formula applies to the organic fraction of the substrate and does not include the ASH component. Decomposition of biodegradable organics in the substrate is assumed to occur according to the formula:

$$C_aH_bO_cN_d + \frac{2a+(b-3d)}{(2-c)(2)}O_2 \rightarrow aCO_2 + \frac{b-3d}{2}H_2O + dNH_3 \qquad (11.50)$$

As an example of the use of this formula, sewage sludge is often assumed to have an overall formula of $C_{10}H_{19}O_3N$. The oxidation will proceed as

$$C_{10}H_{19}O_3N + 12.5\,O_2 \rightarrow 10\,CO_2 + 8\,H_2O + NH_3$$

Similarly, cellulosic materials can be modeled with stoichiometry represented by the oxidation of glucose:

$$C_6H_{12}O_6 + 6\,O_2 \rightarrow 6\,CO_2 + 6\,H_2O$$

Molar ratios for both formulas can easily be derived from the generalized Equation 11.50.

Based on the assumed stoichiometry and knowing the organic loss for each component, stoichiometric factors for oxygen consumed and carbon dioxide, water, and ammonia produced are calculated as shown in Table 11.1. If recycle is used in the mixture, the recycle will contain some nondegraded organics from the original feed substrates. Each substrate in the recycle is tracked individually and the corresponding stoichiometric formula is applied to each substrate. The same is true for substrates recycled with screened bulking agent.

Input Dry Air and Water Vapor

The aeration rate into each stage, QAIR, is input to the model as the actual flowrate corresponding to the ambient temperature, TAIR, and atmospheric pressure, PAIR. TAIR is input to the model, while PAIR is calculated from the site elevation. Based on data for the U.S. Standard Atmosphere the following equations were developed:

$$PAIR = 760.2 - (0.02478 * ELEV) \quad ELEV \leq 5000 \text{ ft} \qquad (11.51)$$

$$PAIR = 742.8 - (0.0213 * ELEV) \quad ELEV > 5000 \text{ ft} \qquad (11.52)$$

where
    PAIR = atmospheric pressure, mm Hg
    ELEV = site elevation, ft above sea level

The mass of input dry air and water vapor is then calculated from the equation of state for an ideal gas:

$$PV = mRT_a \qquad (11.53)$$

where

Table 11.1. Expressions Used to Calculate Oxygen Consumption and Carbon Dioxide, Water, and Ammonia Production for the Feed Substrates J = 1 to SUBNUM

Factor	Expression	Equation
Molar carbon	A(J)	
Molar hydrogen	B(J)	
Molar oxygen	C(J)	
Molar nitrogen	D(J)	
Molecular weight	MOLWT(J)	$A(J) \cdot 12.01115 + B(J) \cdot 1.00797 + C(J) \cdot 15.9994 + D(J) \cdot 14.0067$
O_2 consumed	WTO2FACTOR(J)	$\{[2 \cdot A(J) + [[B(J) - 3 \cdot D(J)]/2] - C(J)]/2\} \cdot 31.9988/MOLWT(J)$
CO_2 produced	WTCO2FACTOR(J)	$A(J) \cdot 44.00995/MOLWT(J)$
H_2O produced	WATPFACTOR(J)	$\{[B(J) - 3 \cdot D(J)]/2\} \cdot 18.01534/MOLWT(J)$
NH_3 produced	WTNH3FACTOR(J)	$D(J) \cdot 17.03061/MOLWT(J)$
Change in dry gas	DGASFACTOR(J)	$WTCO_2FACTOR(J) + WTNH_3FACTOR(J) - WTO_2FACTOR(J)$

$$
\begin{aligned}
P &= \text{absolute pressure, mm Hg} \\
V &= \text{volume, ft}^3 \\
m &= \text{molar mass, lb-mol} \\
R &= \text{universal gas constant, 555 mm-ft}^3/\text{lb mol-deg R} \\
T_a &= \text{absolute temperature, deg Rankin}
\end{aligned}
$$

Rearranging Equation 11.53 and converting volume to flowrate gives

$$\frac{m}{t} = \frac{P * QAIR * 1440}{R(460 + TAIR)} \tag{11.54}$$

where

$$
\begin{aligned}
m/t &= \text{mass flowrate, lb-mol/day} \\
QAIR &= \text{flowrate of ambient air to stage I, acfm} \\
TAIR &= \text{ambient air temperature, } °F
\end{aligned}
$$

The moles of gas can be converted to actual weight by multiplying Equation 11.54 by the gas molecular weight. For dry air the average molecular weight is 28.96 lb/lb-mol based on the U.S. Standard Atmosphere.

The weight of water vapor in both inlet air and outlet gases must be known to complete the mass and energy balances. This is a particularly important term because heat of vaporization is a major component of the energy balance. Water vapor pressure is a function of temperature and can be approximated by an equation of the form

$$\log_{10} PVS = \frac{a}{T_a} + b \tag{11.55}$$

where

$$
\begin{aligned}
PVS &= \text{saturation water vapor pressure, mm Hg} \\
a &= \text{constant equal to } -2238 \text{ for water} \\
b &= \text{constant equal to } 8.896 \text{ for water}
\end{aligned}
$$

Relative humidity in the feed air to each stage, RHAIR(I), is input to the model. Knowing the relative humidity, the actual water vapor pressure is determined as

$$PV(I) = RHAIR(I) * PVS \tag{11.56}$$

where

PV = actual water vapor pressure to stage I, mm Hg

$RHAIR$ = relative humidity, fraction of saturation vapor pressure, to stage I

Equation 11.54 is then applied to determine the weight flow of dry air and water vapor:

$$DAIRI(I) = \frac{[PAIR - PV(I)] * QAIR(I) * 1440 * 28.96}{555 * [460 + TAIR(I)]} \tag{11.57}$$

$$WATVI(I) = \frac{PV(I) * QAIR(I) * 1440 * 18.015}{555 * [460 + TAIR(I)]} \tag{11.58}$$

$$TGASI(I) = DAIRI(I) + WATVI(I) \tag{11.59}$$

where

$DAIRI$ = input dry air to stage I, lb/day

$WATVI$ = input water vapor to stage I, lb/day

$TGASI$ = sum of input dry air and water vapor to stage I, lb/day

Setpoint Temperature Control

The air flowrate to any stage can either be fixed at a constant rate or varied by the simulation model as necessary to maintain a setpoint temperature within that stage. Setpoint temperature for stage I is termed TSET(I). If a setpoint temperature is defined for stage I, the initial flowrate, QAIR(I), is assumed to be a minimum value defined as QAIRMIN(I). Air flowrate is not allowed to drop below QAIRMIN(I) even if the setpoint temperature cannot be maintained as a result. Use of setpoint temperatures is particularly useful because it simulates a composting process operated with temperature feedback control of the aeration supply. The latter type of control is often used in reactor, static pile, and windrow processes. It also reduces the need to accurately know the proper range of air supply before running the model.

Output Gases and Water Vapor

Output dry gas, DGASO, is determined from the input gas quantity, the amount of organic decomposition, and the reaction stoichiometry. The change in dry gas quantity equals the production of carbon dioxide and ammonia minus the consumption of oxygen. Using appropriate terms from Table 11.1, the following equation for DGASO results:

$$DGASO(I) = DAIRI(I) + \sum_{J=1}^{SUBNUM} \{DGASFACTOR(J) * [DBVS(J) + DBVSR(J) + DBVSRB(J)]\} \tag{11.60}$$

Output water vapor (WATVO) is determined by first using Equation 11.55 to predict the saturation water vapor pressure at the exit gas temperature. Exit gas temperature is assumed to be the same as that in the composting reactor or pile. It is reasonable to assume that output gases are saturated if the compost is reasonably moist, say greater than 50% moisture. At some point, however, the gases will no longer exit under saturated conditions. Suppose the composting material were only 20 to 30% moisture. It is hard to imagine saturated exit gases under such dry conditions. Transfer of moisture to the gas phase during composting is a difficult mass transfer problem because of the complexity of the physical system. To simplify the situation, the partial pressure of water vapor in the exit gas is assumed to be a function of solids content and is calculated as

$$PVO(I) = PV(I) + \{[PVSO - PV(I)] * F1(I)\} \tag{11.61}$$

where

PVO = actual water vapor pressure in exit gas, mm Hg
PVSO = saturation vapor pressure in exit gas corresponding to the exit gas temperature, mm Hg
F1 = moisture adjustment factor

Note that the adjustment factor, F1, is the same as that used to adjust the reaction rate constant for the effect of moisture. The shape of the curve seems reasonable for the present purposes. However, it does not imply any fundamental relationship between the two. Equation 11.61 provides a reasonable approximation of the exit water vapor pressure. However, the heat of vaporization is such a significant term in the energy balance that additional work is certainly warranted to better define the exit gas condition.

Having estimated the exit water vapor pressure, the output water vapor, WATVO, from stage I can be calculated by applying the form of Equations 11.57 and 11.58 to the outlet conditions and combining to give

$$WATVO(I) = DGASO(I) * \left(\frac{18.015}{28.96}\right) * \left[\frac{PVO(I)}{PAIR - PVO(I)}\right] \tag{11.62}$$

Output Gas Composition

The composition of output gases is computed in a rather straightforward manner. Referring to Table 11.1, the weight flowrates of oxygen, carbon dioxide, and nitrogen in the output gas become

$$\text{wt } O_2 = (0.2314) * DAIRI(I) - \sum_{J=1}^{SUBNUM} \{WTO2FACTOR(J) * [DBVS(J) + DBVSR(J) + DBVSRB(J)]\} \tag{11.63}$$

$$\text{wt } CO_2 = (0.000501) * DAIRI(I) + \sum_{J=1}^{SUBNUM} \{WTCO2FACTOR(J) * [DBVS(J) + DBVSR(J) + DBVSRB(J)]\} \tag{11.64}$$

$$\text{wt N}_2 = (0.7681) * \text{DAIRI(I)} \tag{11.65}$$

In developing these equations, air is assumed to be 23.14% oxygen, 0.0501% carbon dioxide, and 76.81% nitrogen by weight. Nitrogen content of the atmosphere is actually about 75.52% by weight. It is conventional practice to include other minor components of the atmosphere, such as argon, with the nitrogen, which slightly raises its apparent weight percentage. It is also common practice to express gas composition in terms of volume percent. This conversion is made by dividing the above weights by their respective molecular weights as follows:

$$\text{VOLO2} = \frac{\text{wt O}_2}{31.9988}$$

$$\text{VOLCO2} = \frac{\text{wt CO}_2}{44.00995}$$

$$\text{VOLN2} = \frac{\text{wt N}_2}{28.0134}$$

$$\text{VOLH2O} = \frac{\text{WATVO(I)}}{18.01534}$$

Total gas volume, VOLTOT, then becomes

$$\text{VOLTOT} = \text{VOLN2} + \text{VOLO2} + \text{VOLCO2} + \text{VOLH2O} \tag{11.66}$$

The volume percentage for each component in the exit gas from stage I is determined as

$$\text{VOLPO2(I)} = \left(\frac{\text{VOLO2}}{\text{VOLTOT}} \right) * 100 \tag{11.67}$$

$$\text{VOLPCO2(I)} = \left(\frac{\text{VOLCO2}}{\text{VOLTOT}} \right) * 100 \tag{11.68}$$

$$\text{VOLPN2(I)} = \left(\frac{\text{VOLN2}}{\text{VOLTOT}} \right) * 100 \tag{11.69}$$

$$\text{VOLPH2O(I)} = \left(\frac{\text{VOLH2O}}{\text{VOLTOT}} \right) * 100 \tag{11.70}$$

Output Ammonia Concentration

Ammonia is an end product resulting from decomposition of nitrogen containing organics per Equation 11.50. Ammonia that is released can either remain in solution in the ionized form, NH_4^+, be converted to nitrite and nitrate by nitrification, be used for synthesis of new cell material, or be released to the gas phase. Modeling all possible pathways is a difficult task. An attempt is made here to estimate the potential for ammonia emissions in the vapor by considering ammonia produced and ammonia used for synthesis.

Ammonia released as a result of biological oxidation, WTNH3REL, is estimated as

$$WTNH3REL = \sum_{J=1}^{SUBNUM} \{WTNH3FACTOR(J)*[DBVS(J) + DBVSR(J) + DBVSRB(J)]\} \tag{11.71}$$

Ammonia required for cell synthesis, WTNH3SYN, depends on the cell yield, CELLYIELD, and cell nitrogen content, CELLN. To simplify the analysis, CELLYIELD is assumed to be 0.20 wt cells/wt organic decomposed regardless of the organic. This is not strictly correct but is probably justified for the present simplified model. The value of 0.2 is reasonable for a process of long cell residence time, and certainly composting qualifies here. CELLN is assumed to be 0.12 (12%) based on an average cell formulation of $C_5H_7O_2N$. Ammonia used for synthesis is then estimated as

$$WTNH3SYN = CELLYIELD * CELLN *$$
$$\sum_{J=1}^{SUBNUM} [DBVS(J) + DBVSR(J) + DBVSRB(J)] \tag{11.72}$$

The weight and volume of ammonia released to the gas phase become

$$WTNH3 = WTNH3REL - WTNH3SYN \tag{11.73}$$

$$VOLNH3 = \frac{WTNH3}{17.03061} \tag{11.74}$$

Even with high nitrogen substrates, the concentrations of ammonia in the gas phase are typically <0.1%. Therefore, is it more convenient to express ammonia in parts per million, PPMNH3, as follows:

$$PPMNH3(I) = \frac{VOLNH3 * 10^6}{VOLTOT} \tag{11.75}$$

It is important to remember that ammonia concentrations calculated by this procedure should be considered as a high estimate. All released ammonia, except that used for synthesis, is assumed to report to the gas phase. In actual fact, some ammonia can be retained in the compost in the ionized form and perhaps even nitrified.

WATER BALANCE

Having completed balances for solids and gases, the only remaining balance is that for water. Water vapor in inlet and outlet gases has already been accounted in the gas balance.

Here the subjects of water in the feed materials, produced water, added water, and water contained in the output solids are considered.

Feed Water

Water will be contained in the feed substrates, recycled substrates, and recycled bulking agent. Total water in the solids to stage 1 is termed WATSI(1) and is calculated as

$$WATSI(1) = WATS + WATR + WATRB \qquad (11.76)$$

Water input to subsequent stages is equal to the output from the previous stage less any water in recycled solids or recycled bulking agent. Water output in the solids from stage I is termed WATSO(I). Water input to stage I + 1 is determined as

$$WATSI(I+1) = WATSO(I) \text{ for } I \gtrless STAGER \text{ and } \gtrless STAGESCRN \qquad (11.77)$$

$$WATSI(I+1) = WATSO(I) - WATRB \text{ for } I = STAGESCRN \qquad (11.78)$$

$$WATSI(I+1) = WATSO(I) - WATR \text{ for } I = STAGER \qquad (11.79)$$

Produced Water

Water produced from organic decomposition, WATP, is determined from the assumed stoichiometric equations. Referring to Table 11.1, the following expression for produced water in stage I can be developed:

$$WATP(I) = \sum_{J=1}^{SUBNUM} \{WATPFACTOR(J) * [DBVS(J) + DBVSR(J) + DBVSRB(J)]\} \qquad (11.80)$$

where
   WATP   =   water produced by biological oxidation in stage I, lb/day

Added Water

Composting is generally a dehydrating environment because moisture is carried away in the hot process gases. This loss of water is usually greater than that produced by organic decomposition as defined above in Equation 11.80. With wet substrates such as sludge cake, the addition of supplemental water is not a routine practice. However, with dry substrates, such as yard wastes, leaves, and refuse fractions, water addition is often a normal part of process operation. The reader is cautioned that water addition sometimes may be required even with sludge cake. If the sludge cake is "energy rich" sufficient water can be removed to produce a dry, dusty end product. Not only does the rate of composting decrease under such conditions, the product can become a fire hazard.

Water can be added to any stage of the process. WATADD(I) is defined as the daily rate of supplemental water input to stage I in lb/day. WATADD(I) can either be held at a constant rate or varied by the simulation model as necessary to maintain a setpoint solids content within that stage. The setpoint solids content for stage I is termed SMSET(I). If a setpoint solids content is defined for stage I, WATADD(I) is assumed to be a minimum value defined as WATADDMIN(I). Water addition is not allowed to drop below WATADDMIN(I) even if the setpoint solids content cannot be maintained as a result. Use of setpoint solids contents is particularly useful because it reduces the need to know what water addition is required prior to running the simulation model. It also assures that the composting environment is maintained at proper levels of moisture.

Water with Output Solids

Water contained in the output solids, WATSO, is determined by applying the basic mass balance equation at steady state as follows:

$$output = input + sources - sinks$$

Adding the previously defined terms, the following equation results for stage 1:

$$WATSO(1) = WATS + WATR + WATRB + WATVI(1)$$
$$+WATADD(1) + WATP(1) - WATVO(1) \tag{11.81}$$

and for subsequent stages (I > 1):

$$WATSO(I) = WATSI(I) + WATVI(I) +$$
$$WATADD(I) + WATP(I) - WATVO(I) \tag{11.82}$$

Having determined WATSO from Equation 11.82 and DWTOUT from Equation 11.49, the output solids content from any stage can now be determined as

$$SMOUT(I) = \frac{DWTOUT(I)}{WATSO(I) + DWTOUT(I)} \tag{11.83}$$

ENERGY INPUTS

Following an iteration to complete the mass balance on stage I, the corresponding energy balance is determined for stage I. If the energy balance is not closed, a new temperature is selected for the next iteration. The calculation process then continues to stage I + 1 where mass and energy balances are completed and a new temperature selected as necessary. This process continues until the last stage of the system. If a single energy balance remains unclosed, the calculations begin again at stage 1.

Because many components, including solids and gases, enter the composting system, there are likely to be differences in feed temperatures. Therefore, a reference temperature, TREF, is established against which all energy inputs and outputs are measured. This does not affect the final energy balance, it merely provides a convenient datum against which all sensible heat values are measured. TREF is set at 32°F (0°C) for the simulation models presented in this book.

Solids, Water, and Gases

Sensible heat of solids, water, and gas components is determined from the basic heat equation,

$$q_p = mc_p \Delta T = \Delta H \tag{3.8}$$

or

$$q_p = mc_p (T - TREF) \tag{11.84}$$

Recall that heat flow in isobaric (constant pressure) processes is equal to the enthalpy change. The value of specific heat at constant pressure, c_p, varies depending on the component. The following values in units of Btu/lb-°F or cal/g-°C are assumed in the simulation models: water, 1.00; solids, 0.25; dry gases, 0.24; and water vapor, 0.44. Values for water, dry gas (air), and water vapor are available in handbooks. The values chosen are applicable over the temperature range common to composting. Specific heat for composting solids is based on the work of Mears et al.[5]

The use of constant specific heats for the solid and gas components is a reasonable assumption over the temperature range normal to composting. Having studied Chapter 3, the reader may recall that specific heats are actually a function of temperature. The variation of c_p with temperature is often approximated by a second or third degree polynomial. However, the temperature range necessary to affect a significant change in c_p is generally quite large compared to the temperature ranges for composting. Therefore, this complication is not necessary in the present models.

It is also assumed that the specific heat for the mixture of output dry gases is equal to that for air. The exact specific heat for a mixture of gases can be determined from the mole fraction of the individual gases. Again, this complication is not necessary in the present models. The large N_2 component of the inlet air acts as a conservative substance throughout the process and minimizes any significant changes in mixture specific heat.

Using Equation 11.84, the following equations for the sensible heat of the substrate solids and water input to stage 1 can be developed:

$$HSSI(J) = S(J) * X(J) * CPSOL * [T(J) - TREF] \tag{11.85}$$

$$HSWI(J) = [X(J) - S(J) * X(J)] * CPWAT * [T(J) - TREF] \tag{11.86}$$

where
  HSSI = sensible heat in the feed solids for substrate J, Btu/day
  HSWI = sensible heat in the feed water for substrate J, Btu/day
  CPSOL = specific heat of solids, Btu/lb-°F
  CPWAT = specific heat of water, Btu/lb-°F
  T = input temperature for substrate J, °F

The sensible heat in product recycle and recycled bulking agents input to stage 1 are determined as

$$HRSI = DWTR * CPSOL * [TNEW(STAGER) - TREF] \qquad (11.87)$$

$$HRWI = WATR * CPWAT * [TNEW(STAGER) - TREF] \qquad (11.88)$$

$$HRBSI = DWTRB * CPSOL * [TNEW(STAGESCRN) - TREF] \qquad (11.89)$$

$$HRBWI = WATRB * CPWAT * [TNEW(STAGESCRN) - TREF] \qquad (11.90)$$

where
  HRSI = sensible heat in recycled product solids, Btu/day
  HRWI = sensible heat in the water fraction of recycled product, Btu/day
  HRBSI = sensible heat in recycled bulking agent, Btu/day
  HRBWI = sensible heat in the water fraction of recycled bulking agent, Btu/day
  TNEW = temperature in stage I, °F

Total sensible heat in the feed to stage 1 is determined as

$$HSI(1) = \left[\sum_{J=1}^{SUBNUM} HSSI(J) \right] + HRSI + HRBSI \qquad (11.91)$$

$$HWI(1) = \left[\sum_{J=1}^{SUBNUM} HSWI(J) \right] + HRWI + HRBWI \qquad (11.92)$$

where
  HSI = sensible heat in the total solids feed, Btu/day
  HWI = sensible heat in the water fraction of feed solids, Btu/day

Sensible heat in the combined feed to subsequent stages is determined as

$$HSI(I) = DWTIN(I) * CPSOL * [TNEW(I) - TREF] \qquad (11.93)$$

$$HWI(I) = WATSI(I) * CPWAT * [TNEW(I) - TREF] \qquad (11.94)$$

Sensible heats in dry air, water vapor, and any added water input to stage I are determined as

$$HDAIRI(I) = DAIRI(I) * CPGAS * [TAIR(I) - TREF] \qquad (11.95)$$

$$HSWVI(I) = WATVI(I) * CPWATV * [TAIR(I) - TREF] \qquad (11.96)$$

$$HWATADDI(I) = WATADD(I) * CPWAT * [TWATADD - TREF] \qquad (11.97)$$

where

$$
\begin{array}{rcl}
HDAIRI & = & \text{sensible heat content of infeed dry air to stage I} \\
HSWVI & = & \text{sensible heat content of input water vapor to stage I} \\
HWATADD & = & \text{sensible heat content of added water to stage I} \\
CPGAS & = & \text{specific heat of dry gases, Btu/lb-°F} \\
CPWATV & = & \text{specific heat of water vapor, Btu/lb-°F}
\end{array}
$$

Organic Decomposition

Biological heat production is estimated from the change in BVS and heat of combustion for each substrate. Heat release for substrate J in stage I is determined as

$$HORG(J,I) = DBVS(J) * H(J) \qquad (11.98)$$

where

$$
\begin{array}{rcl}
HORG & = & \text{heat release by biological oxidation, Btu/day} \\
H & = & \text{higher heat of combustion for substrate J,} \\
& & \text{Btu/lb of organic oxidized}
\end{array}
$$

Heat is also released from substrate contained in a recycle stream or substrate recycled with bulking agents. For substrate J in stage I, these are determined as,

$$HORGR(J,I) = DBVSR(J) * H(J) \qquad (11.99)$$

$$HORGRB(J,I) = DBVSRB(J) * H(J) \qquad (11.100)$$

where

$$
\begin{array}{rcl}
HORGR & = & \text{heat released by biological oxidation of substrate in the} \\
& & \text{recycle, Btu/day} \\
HORGRB & = & \text{heat released by biological oxidation of substrate contained in} \\
& & \text{recycled bulking agent, Btu/day}
\end{array}
$$

The heat of combustion for a particular substrate can be determined from bomb calorimeter data, COD data, or any of the approximation techniques discussed in Chapter 3. The heat of combustion includes only organic oxidation. Oxidation of inorganic substrates, such as nitrification of ammonium to nitrate, is not considered in the energy balance.

The reader should note that total heat release from biological oxidation cannot exceed the heat available in the original feed substrates. Recycle of substrates, either by direct product recycle or by recycle of bulking agents, does not add new organics to the system. In other words, recycle lines are internal to the thermodynamic system boundary and, therefore, do not contribute new organics to the system. However, a recycle line can cross an individual stage boundary and therefore affect conditions within that stage. This is the reason for tracking both product recycle and bulking agent recycle. While new biodegradable organics are not added to the system, nondegraded organics may remain from the original substrate and can contribute to heat release in a particular stage of the system.

Total Input Energy

Total energy input to stage I is the summation of all input terms to that stage, or

$$
\begin{aligned}
HTOTI(I) = {} & HSI(I) + HWI(I) + HWATADDI(I) + \\
& HDAIRI(I) + HSWVI(I) + HORG(J) + \\
& HORGR(J) + HORGRB(J)
\end{aligned} \tag{11.101}
$$

where

      HTOTI = total input energy to stage I, Btu/day

ENERGY OUTPUTS

Solids, Water, and Gases

Energy contained in output solids and water from stage I are computed from Equation 11.84 using the DWTOUT and WATSO components as determined from the mass balance:

$$
HSO(I) = DWTOUT(I) * CPSOL * [TNEW(I) - TREF] \tag{11.102}
$$

$$
HWO(I) = WATSO(I) * CPWAT * [TNEW(I) - TREF] \tag{11.103}
$$

where

      HSO = sensible heat in the output solids, Btu/day
      HWO = sensible heat in the output water, Btu/day

Sensible heat in the output gases from stage I is determined as

$$
HDGASO(I) = DGASO(I) * CPGAS * [TNEW(I) - TREF] \tag{11.104}
$$

$$
\begin{aligned}
HSWVO(I) = {} & WATVI(I) * CPWATV * [TNEW(I) - TREF] + \\
& [WATVO(I) - WATVI(I)] * CPWAT * [TNEW(I) - TREF]
\end{aligned} \tag{11.105}
$$

where

      HDGASO = sensible heat in the output dry gases, Btu/day
      HSWVO = sensible heat in the output water vapor, Btu/day

All components, including solids, water, gases, and vapor, are assumed to exit the process at the same temperature, TNEW.

The rationale behind Equation 11.105 may not be readily apparent to the reader. Recall that enthalpy is a property of a system and is not a function of the path followed in moving between two equilibrium states, i.e., inlet to outlet conditions. In these calculations it is assumed that entering water vapor is heated to the exit temperature TNEW, which accounts for the first half of Equation 11.105 and is the reason for using the specific heat of water vapor CPWATV. It is further assumed that entering liquid water is first heated to TNEW. If a portion of the water is vaporized, it does so at temperature TNEW. Thus, CPWAT is used in the second half of Equation 11.105 instead of CPWATV even though the water exits as a vapor. Other paths could be assumed, but the change in enthalpy remains the same regardless of path (refer to Example 3.1). This is fortunate because it is unlikely that all entering water follows the same thermodynamic path in moving from inlet to outlet conditions. The path chosen here is convenient and easily visualized.

Latent Heat

The latent heat of water vaporized in stage I is determined from the water vapor content of the inlet and outlet gases as follows:

$$HLWVO(I) = [WATVO(I) - WATVI(I)] * DELHFG \qquad (11.106)$$

where
   HLWVO = latent heat of output water vapor minus latent heat of the input water vapor, Btu/day
   DELHFG = enthalpy change from liquid to vapor at temperature TNEW, Btu/lb

Given the previous assumptions regarding thermodynamic path, it is necessary that the heat of vaporization in Equation 11.106 be evaluated at temperature TNEW, the assumed temperature of vaporization. DELHFG is a function of temperature, and over the temperature range from 32 to 212°F it varies by about 10%. The heat of vaporization is such a large energy output that a 10% variation is larger than many of the other energy terms. Therefore, the effect of temperature on DELHFG must be considered. Based on data from the 1967 ASME Steam Tables the following equation can be developed for the heat of vaporization at 1 atm over the temperature range from 32 to 212°F,

$$DELHFG = 1093.7 - (0.5683) * TNEW(I) \qquad (11.107)$$

Total Output Energy

Total energy output from stage I is the sum of individual output terms for that stage as follows:

$$HTOTO(I) = HSO(I) + HWO(I) + HDGASO(I) +$$
$$HSWVO(I) + HLWVO(I) \qquad (11.108)$$

where
   HTOTO = total output energy from stage I, Btu/day

PROGRAM CLOSURE

As discussed previously, the iteration procedure is to solve the mass and energy balance for stage I based on the assumed temperature, TNEW(I), for that stage. If the energy balance is not closed to sufficient accuracy, a new temperature is assumed for the next iteration. The program then moves sequentially to the next stage I + 1, solves the mass and energy balance, estimates a new temperature, TNEW(I + 1), and moves to the next stage. When this procedure is completed on the last stage, STAGE(STAGENUM), the program returns to stage 1 and repeats the iteration process.

After each iteration through all stages the program checks whether the balances are closed within acceptable limits. Four conditions must be satisfied for the solution to be judged complete. First, the temperature difference between successive iterations for each stage must be less than 0.1°F,

$$\text{ABS}[\text{TNEW}(I) - \text{THOLD}(I)] \leq 0.1^\circ \text{F} \tag{11.109}$$

where

$$
\begin{aligned}
\text{THOLD} &= \text{temperature in stage I for iteration k} \\
\text{TNEW} &= \text{temperature in stage I for iteration k + 1} \\
\text{ABS} &= \text{absolute value}
\end{aligned}
$$

Second, energy output must balance energy input at the solution point as illustrated in Figure 11.6. The equation used to establish closure on the energy balance is

$$\text{ABS}\left[\frac{\text{HTOTI}(I) - \text{HTOTO}(I)}{\text{HTOTI}(I)}\right] \leq 0.00025 \tag{11.110}$$

An error of ≤0.025% means that energy is balanced to 2.5 parts in 10,000. This is a very high level of accuracy and would be unattainable by hand calculations. Equation 11.110 must be satisfied for all stages of the process. Conditions 1 and 2 assure that mass and energy balances for all stages are satisfied to acceptable accuracy. Solution closure for all individual stages assures that the system balance is also closed.

Two other conditions must also be satisfied if setpoint temperatures, TSET, and setpoint moisture contents, SMSET, are assumed for any stages of the process. For each stage for which a TSET(I) or SMSET(I) is assumed, the following conditions must be satisfied:

$$\text{ABS}[\text{SMSET}(I) - \text{SMOUT}(I)] \leq 0.005 \tag{11.111}$$

$$\text{ABS}[\text{TSET}(I) - \text{TNEW}(I)] \leq 0.5^\circ \text{F} \tag{11.112}$$

Equations 11.111 and 11.112 assure that setpoint conditions are achieved to reasonable accuracy. If SMSET conditions are not satisfied for stage I, the quantity of added water WATADD(I) is adjusted and the program continues. Similarly, if TSET conditions are not satisfied for stage I, the quantity of added air QAIR(I) is adjusted and the program continues.

Readers with experience in programming are probably familiar with the fact that closure algorithms can sometimes refuse to close. Computers are perfectly capable of iterating forever under such circumstances. To prevent this from happening, the maximum number of iterations is limited to 2000. Experience has shown that this is usually sufficient to obtain an accurate solution even if all four of the above conditions are not satisfied. This usually occurs when the program is very close to solution, but with each iteration there are small jumps around the solution point without further closure. Considerable effort has been made to develop the algorithms for selecting new values for TNEW, WATADD, and QAIR so that the program moves toward a solution within the established limits.

VOLUME RELATIONSHIPS

Mixture Volumes

The volume of feed components into and from each stage is required to calculate the individual stage volumes, STAGEVOL. For stage 1 the input volume of all components is determined as

$$VOLUMEI(1) = \frac{SUMX + XR + XRB}{BULKWTI(1)} \tag{11.113}$$

where
$\quad$ VOLUMEI(1) $\quad$ = volume of infeed mixture to stage 1, ft³/day
$\quad$ BULKWTI(1) $\quad$ = bulk weight of the infeed mixture to stage 1, lb/ft³

BULKWTI(1) is determined as

$$BULKWTI(1) = \frac{62.4 * COEFFM}{SM} \tag{11.114}$$

where
$\quad$ COEFFM $\quad$ = bulk weight coefficient for the mixed materials
$\quad$ SM $\quad$ = fractional solids content of the mixed feed to stage 1

For all following stages (I > 1), the input volume is determined as

$$VOLUMEI(I) = \frac{\left[\dfrac{DWTIN(I)}{SMOUT(I-1)}\right]}{BULKWTI(I)} \tag{11.115}$$

For Equation 11.115, DWTIN to stage I is divided by the solids content from the previous stage SMOUT(I – 1) to give the wet weight into stage I. The latter is divided by infeed bulk weight to determine the corresponding volume. BULKWTI to stage I is determined as

$$BULKWTI(I) = \frac{62.4 * COEFFM}{SMOUT(I-1)} \tag{11.116}$$

The output volume of mixed components from all stages is determined as

$$VOLUMEO(I) = \frac{\left[\dfrac{DWTOUT(I)}{SMOUT(I)}\right]}{BULKWTO(I)} \tag{11.117}$$

where
   VOLUMEO(I) = volume of the outfeed mixture from stage I, ft³/day
   BULKWTO(I) = bulk weight of the outfeed mixture from stage I, lb/ft³

Stage Volumes

The volume of an individual stage is based on the average of infeed and outfeed mixture volumes and the HRT within that stage. Average mixture volume is calculated as

$$AVGVOL = \frac{VOLUMEI(I) + VOLUMEO(I)}{2} \qquad (11.118)$$

where
   AVGVOL = average volume of infeed and outfeed mixture, ft3/day

The required volume for stage I is then determined as

$$STAGEVOL(I) = AVGVOL * HRT(I) \qquad (11.119)$$

where
   STAGEVOL = volume of stage I, ft³

Net Substrate Volumes

The net volume of feed substrates, excluding product recycle and bulking agent recycle, is estimated for later use in calculating solids residence times. The net infeed volume of substrates is termed NETVOLUMEI. For stage 1 the calculation is straightforward as

$$NETVOLUMEI(1) = \frac{SUMX}{BULKWTI(1)} \qquad (11.120)$$

where
   NETVOLUMEI(1) = infeed volume of substrates to stage 1, ft³/day

For later stages it is necessary to sum the BVS, NBVS, and ASH fractions for the substrates. Net infeed volume to stage I is then determined as

NETVOLUMEI(I) =

$$\left\{ \sum_{J=1}^{SUBNUM} \frac{[BVSFAST(J,I) + BVSSLOW(J,I) + NBVS(J,I) + ASH(J,I)]}{SMOUT(I-1)} \right\} / BULKWTI(I) \qquad (11.121)$$

where
   NETVOLUMEI(I) = infeed volume of substrates to stage I, ft³/day.

The net outfeed volume of substrates is termed NETVOLUMEO and is calculated as follows for all stages:

NETVOLUMEO(I) =

$$\left\{ \sum_{J=1}^{\text{SUBNUM}} \frac{[\text{BVSOFAST}(J,I) + \text{BVSOSLOW}(J,I) + \text{NBVS}(J,I) + \text{ASH}(J,I)]}{\text{SMOUT}(I-1)} \right\} / \text{BULKWTO}(I) \qquad (11.122)$$

where

NETVOLUMEO(I) = outfeed volume of substrates from stage I, ft³/day.

SYSTEM AND STAGE RESIDENCE TIMES

Following closure of the mass and energy balances for each stage of the process, the program moves to calculation of those parameters that depend on the closed solution. One of these parameters is the solids residence time, which is discussed below along with hydraulic retention time.

Hydraulic Retention Time

Hydraulic retention time, HRT, is defined as the single pass, mean residence time of the mixed materials, including product recycle and bulking agent recycle. HRT(I) is an input condition for each stage I. Total system HRT is the summation of HRTs for the individual stages and is calculated as

$$\text{TOTHRT} = \sum_{I=1}^{\text{STAGENUM}} \text{HRT}(I) \qquad (11.123)$$

where

TOTHRT = total system HRT, days.

SRT for Homogeneous Mixtures

Solids residence time, SRT, is defined as the mean residence time of the feed substrates within the process. For a system without recycle, HRT and SRT are equivalent. If solids are recycled, however, SRT will be greater than HRT. The following calculation procedures are used for estimating stage and system SRTs in the simulation models. The procedures are first applied to compost systems in which all materials move through the process in a homogeneous manner without screening, in other words, systems that do not use bulking particles. For such systems the SRT is the same for any substrate.

SRT by Volume

The first approach to estimating SRT is based on using component volumes. The net daily throughput of substrate components in stage I is

$$NETVOL = \frac{NETVOLUMEI(I) + NETVOLUMEO(I)}{2} \qquad (11.124)$$

where

NETVOL   =   average of infeed and outfeed volume of
substrate components to stage I, ft^3/day

SRT by volume for stage I is then estimated by dividing the stage volume by the net volume of daily throughput:

$$SRTVOL(I) = \frac{STAGEVOL(I)}{NETVOL} \qquad (11.125)$$

where

SRTVOL(I)   =   SRT for stage I estimated by volume calculation, days

Total system SRT by volume, TOTSRTVOL, is the sum of SRT's for the individual stages:

$$TOTSRTVOL = \sum_{I=1}^{STAGENUM} SRTVOL(I) \qquad (11.126)$$

When taken together, Equations 11.125 and 11.126 state that the system SRT is the total system volume divided by the net throughput of substrates. These equations are analogous to similar expressions used for liquid systems. In the latter case, however, the approach is to determine the total system biomass, which is then divided by the daily mass of biomass wastage to give the system SRT. Because volume is conservative in liquid systems, the SRT can also be estimated as the total volume of biomass in the reactor divided by the daily volume of wastage taken directly from the reactor. However, volume is not conservative with semi-solid systems such as composting. Therefore, an estimate of net volume across each stage is required to improve the accuracy of estimate.

SRT by Dry Weight

Another approach to estimating SRT in stage I is to multiply the HRT in stage I by the ratio of mixture dry weight to substrate dry weight. Simply stated, if the mixture throughput is 2 lb/day of which 1 lb/day is feed substrate, then the SRT in that stage will be about twice the HRT. This is because substrate passes twice through the stage on average. Average mixture dry weight, MIXDRYWTI, into stage I is estimated as

$$MIXDRYWTI = \sum_{J=1}^{SUBNUM} [BVSFAST(J,I) + BVSSLOW(J,I) + NBVS(J,I) +$$
$$ASH(J,I) + BVSRFAST(J,I) + BVSRSLOW(J,I) + \qquad (11.127)$$
$$NBVSR(J,I) + ASHR(J,I)]$$

Average mixture dry weight out of stage I, MIXDRYWTO, is determined as

$$\text{MIXDRYWTO} = \sum_{J=1}^{\text{SUBNUM}} [\text{BVSOFAST}(J,I) + \text{BVSOSLOW}(J,I) + \text{NBVS}(J,I) + \\ \text{ASH}(J,I) + \text{BVSROFAST}(J,I) + \text{BVSROSLOW}(J,I) + \\ \text{NBVSR}(J,I) + \text{ASHR}(J,I)] \tag{11.128}$$

The dry weight of feed substrates into and out of stage I, SUBDRYWTI and SUBDRYWTO, are estimated by

$$\text{SUBDRYWTI} = \sum_{J=1}^{\text{SUBNUM}} [\text{BVSFAST}(J,I) + \text{BVSSLOW}(J,I) + \\ \text{NBVS}(J,I) + \text{ASH}(J,I)] \tag{11.129}$$

$$\text{SUBDRYWTO} = \sum_{J=1}^{\text{SUBNUM}} [\text{BVSOFAST}(J,I) + \text{BVSOSLOW}(J,I) + \\ \text{NBVS}(J,I) + \text{ASH}(J,I)] \tag{11.130}$$

SRT by dry weight across stage I, SRTDWT, is then estimated as

$$\text{SRTDWT}(I) = \text{HRT}(I) * \frac{\text{MIXDRYWTI} + \text{MIXDRYWTO}}{\text{SUBDRYWTI} + \text{SUBDRYWTO}} \tag{11.131}$$

Total system SRT by dry weight, TOTSRTDWT, is determined as

$$\text{TOTSRTDWT} = \sum_{I=1}^{\text{STAGENUM}} \text{SRTDWT}(I) \tag{11.132}$$

In general, estimates of SRT by volume and by dry weight are reasonably consistent. Neither approach appears to be decidedly superior. For this reason both calculation procedures are included in the program.

SRT for Heterogeneous Mixtures

Recall again that, for a system which does not use bulking agents, the SRT is the same for all substrates. However, this is not the case when bulking agents are used. Because they are screened and returned to the process, SRT for the bulking agent can be many times that of the other substrates. SRT by volume, SRTVOL, is calculated in the same manner as described above for nonbulking systems using Equations 11.125 and 11.126. New bulking agent addition, X(SUBNUM), is included in this analysis. Therefore, the resulting SRT represents an estimate of the average SRT of substrates and bulking agent. Consider the case where the bulking agent suffered no mechanical breakdown and was completely screened on each pass. In this case the bulking agent itself would remain in the system essentially forever and the calculated SRT would have little physical significance. This situation aside, the volume calculation gives a reasonable estimate of average SRT for all substrates, including the bulking agent.

SRT for Substrates

The SRT for nonbulking agent substrates ($J = 1$ TO SUBNUM $- 1$) is estimated using dry weight calculations similar to those of Equations 11.127 through 11.132. Average mixture dry weight for the nonbulking agent substrates into stage I is estimated as

$$
\begin{aligned}
\text{NBADRYWT} = \sum_{J=1}^{\text{SUBNUM} - 1} & [\text{BVSFAST}(J,I) + \text{BVSSLOW}(J,I) + \text{NBVS}(J,I) + \text{ASH}(J,I) + \\
& \text{BVSRFAST}(J,I) + \text{BVSRSLOW}(J,I) + \\
& \text{NBVSR}(J,I) + \text{ASHR}(J,I) + \\
& \text{BVSRBFAST}(J,I) + \text{BVSRBSLOW}(J,I) + \\
& \text{NBVSRB}(J,I) + \text{ASHRB}(J,I)]
\end{aligned}
\tag{11.133}
$$

where

NBADRYWTI = dry weight of nonbulking agent substrates input to stage I, lb/day

Average mixture dry weight for nonbulking agent substrates out of stage I is estimated as

$$
\begin{aligned}
\text{NBADRYWO} = \sum_{J=1}^{\text{SUBNUM} - 1} & [\text{BVSOFAST}(J,I) + \text{BVSOSLOW}(J,I) + \text{NBVS}(J,I) + \text{ASH}(J,I) + \\
& \text{BVSROFAST}(J,I) + \text{BVSROSLOW}(J,I) + \\
& \text{NBVSR}(J,I) + \text{ASHR}(J,I) + \\
& \text{BVSRBOFAST}(J,I) + \text{BVSRBOSLOW}(J,I) + \\
& \text{NBVSRB}(J,I) + \text{ASHRB}(J,I)]
\end{aligned}
\tag{11.134}
$$

where

NBADRYWTO = dry weight of nonbulking agent substrates in the output from stage I, lb/day

The dry weights of nonbulking feed substrates into and out of stage I, SUBDRYWTI and SUBDRYWTO, are estimated as

$$SUBDRYWI = \sum_{J=1}^{SUBNUM-1} [BVSFAST(J,I) + BVSSLOW(J,I) + NBVS(J,I) + ASH(J,I)] \tag{11.135}$$

$$SUBDRYWO = \sum_{J=1}^{SUBNUM-1} [BVSOFAST(J,I) + BVSOSLOW(J,I) + NBVS(J,I) + ASH(J,I)] \tag{11.136}$$

SRT for the nonbulking substrates in stage I is then estimated as

$$SRTSUB(I) = HRT(I) * \frac{NBADRYWTI + NBADRYWTO}{SUBDRYWTI + SUBDRYWTO} \tag{11.137}$$

where

$SRTSUB(I)$ = SRT in stage I for the nonbulking feed substrates, days

Total system SRT for the non-bulking substrates, TOTSRTSUB, is then determined as,

$$TOTSRTSUB = \sum_{I=1}^{STAGENUM} SRTSUB(I) \tag{11.138}$$

SRT for the Bulking Agent

SRT for the bulking agent itself, SRTBA, can be determined by using Equations 11.133 to 11.136, but with J = SUBNUM. SRT for all stages from 1 to STAGESCRN is then determined by application of Equation 11.137. SRT for stages STAGESCRN + 1 to STAGENUM is the same as that calculated above for the other substrates. Therefore, total system SRT for the bulking agent is determined as

$$TOTSRTBA = \sum_{I=1}^{STAGESCRN} SRTBA(I) + \sum_{I=STAGESCRN+1}^{STAGENUM} SRTSUB(I) \tag{11.139}$$

where

$SRTBA(I)$ = SRT in stage I for the bulking agent, days
$TOTSRTBA$ = total system SRT for the bulking agent, days

SPECIFIC OXYGEN CONSUMPTION RATES

Oxygen consumption rate is an important criterion for determining compost product stability. Specific oxygen consumption rate is defined as the mg O_2 consumed per kg of volatile solids per hour (mg O_2/kg VS-h). Temperature is standardized at 25°C and rate limitations from lack of moisture or lack of free air space are assumed to be absent. The rate equations can be used to estimate specific oxygen consumption for the starting substrates, the mixtures into each stage, and the final compost product from the last stage STAGENUM.

Specific O_2 Consumption for Substrates

The first step in calculating specific consumption rates for the feed substrates is to use Equations 11.5 and 11.6 to determine the rate constant at 25°C for each substrate:

$$RATE25F(J) = RATEKM20F(J) * (1.066^5 - 1.21^{-35})$$ (11.140)

$$RATE25S(J) = RATEKM20S(J) * (1.066^5 - 1.21^{-35})$$ (11.141)

where
      RATE25F = rate constant for the faster fraction for substrate J at 25°C, per day
      RATE25S = rate constant for the slower fraction for substrate J at 25°C, per day

The daily loss of fast and slow BVS for substrate J input to stage 1 is determined by application of Equations 11.23 and 11.24 with HRT equal to 1 day:

$$DBVSFAST(J) = BVSFAST(J,1) - \left[\frac{BVSFAST(J,1)}{1 + RATE25F(J)} \right]$$ (11.142)

$$DBVSSLOW(J) = BVSSLOW(J,1) - \left[\frac{BVSSLOW(J,1)}{1 + RATE25S(J)} \right]$$ (11.143)

where
      DBVSFAST = loss of fast BVS for substrate J, lb/day
      DBVSSLOW = loss of slow BVS for substrate J, lb/day

The hourly consumption of oxygen for substrate J is determined as

$$WTO2 = [DBVSFAST(J) + DBVSSLOW(J)] * \frac{WTO2FACTOR(J)}{24}$$ (11.144)

Specific oxygen consumption for substrate J input to stage 1 then becomes

$$RATEO2SUB(J) = \frac{WTO2 * 10^6}{BVSFAST(J,1) + BVSSLOW(J,1) + NBVS(J,1)}$$ (11.145)

where
      RATEO2SUB = specific oxygen consumption rate for substrate J, mg O_2/Kg VS-h

Specific oxygen consumption for the combined feeds is termed RATEO2FEED and is defined as the specific consumption rate that would result if all feed substrates were mixed together as a single feed source. Any possible diluting effects of product or bulking agent recycle are not considered. RATEO2FEED is determined by summing the WTO2 for each feed substrate and applying Equation 11.145 using the summation of BVS and NBVS for all substrates in the denominator.

Specific O_2 Consumption for Stage Feeds

Specific oxygen consumption is also determined for the combined feed to each stage and the outfeed from the last stage STAGENUM (equivalent to the input to the imaginary stage STAGENUM + 1). The calculation procedure for the combined feeds to each stage is similar to that above for the feed substrates, but it includes the effects of any product or bulking agent recycle.

The daily loss of fast and slow BVS for substrate J input to stage I is determined as

$$DBVSFAST(J) = BVSFAST(J,I) - \left[\frac{BVSFAST(J,I)}{1 + RATE25F(J)} \right] \qquad (11.146)$$

$$DBVSSLOW(J) = BVSSLOW(J,I) - \left[\frac{BVSSLOW(J,I)}{1 + RATE25S(J)} \right] \qquad (11.147)$$

The comparable equations for product and bulking agent recycle are as follows:

$$DBVSRFAST(J) = BVSRFAST(J,I) - \left[\frac{BVSRFAST(J,I)}{1 + RATE25F(J)} \right] \qquad (11.148)$$

$$DBVSRSLOW(J) = BVSRSLOW(J,I) - \left[\frac{BVSRSLOW(J,I)}{1 + RATE25S(J)} \right] \qquad (11.149)$$

$$DBVSRBFAST(J) = BVSRBFAST(J,I) - \left[\frac{BVSRBFAST(J,I)}{1 + RATE25F(J)} \right] \qquad (11.150)$$

$$DBVSRBSLOW(J) = BVSRBSLOW(J,I) - \left[\frac{BVSRBSLOW(J,I)}{1 + RATE25S(J)} \right] \qquad (11.151)$$

Hourly consumption of oxygen for the combined feed to stage I is determined as

$$WTO2 = \sum_{J=1}^{SUBNUM} \{ [DBVSFAST(J) + DBVSSLOW(J) + DBVSRFAST(J) +$$
$$DBVSRSLOW(J) + DBVSRBFAST(J) + \qquad (11.152)$$
$$DBVSRBSLOW(J)] * WTO2FACTOR(J)/24 \}$$

Specific oxygen consumption for the combined feed to stage I, RATEO2, then becomes

$$
\begin{aligned}
RATEO2(I) = (WTO2 * 10^6) / \sum_{J=1}^{SUBNUM} [& BVSFAST(J,I) + BVSSLOW(J,I) + NBVS(J,I) + \\
& BVSRFAST(J,I) + BVSRSLOW(J,I) + \\
& NBVSR(J,I) + BVSRBFAST(J,I) + \\
& BVSRBSLOW(J,I) + NBVSRB(J,I)]
\end{aligned}
\tag{11.153}
$$

where
RATEO2(I) = specific oxygen consumption rate for the combined mixture into stage I, mg O_2/kg VS-h

Probably the most important of all the oxygen consumption rates is that in the final compost product from the last stage. Again, recall that the output from the last stage is equivalent to the input to an imaginary stage STAGENUM + 1. Therefore, the product consumption rate is equal to RATEO2(STAGENUM + 1) as calculated above.

SPECIFIC OXYGEN SUPPLY

Specific oxygen supply is defined as the weight of oxygen supplied per weight of biodegradable volatile solids in the feed substrates. This parameter has proven to be very useful in judging the overall quantity of air supplied to the composting process. Specific oxygen supply to the system is calculated as

$$
SPECO2SUP = \frac{\displaystyle\sum_{I=1}^{STAGENUM} [DAIRI(I) * 0.2314]}{\displaystyle\sum_{J=1}^{SUBNUM} [BVSFAST(J,1) + BVSSLOW(J,1)]}
\tag{11.154}
$$

where
SPECO2SUP = specific oxygen supply with units taken as g O_2/g feed BVS

Air is 23.14% oxygen by weight. Therefore, the specific supply of air is directly proportional to specific oxygen supply. Specific air supply can be determined by dividing specific oxygen supply by 0.2314.

Assuming that most of the feed BVS are consumed during composting, the demoninator of Equation 11.154 is proportional to the heat released in the process. A quantity of air or oxygen must be supplied to carry away this heat. Therefore, there is an upper range to specific oxygen or air supply beyond which more heat is withdrawn from the process than supplied. Therefore, this is a useful parameter to measure the capacity of the aeration system. The upper range of acceptable values for specific supply will be discussed in later chapters.

SUMMARY

Mathematical models have been developed to study the dynamics of composting processes. These simulation models allow integration of the thermodynamic and kinetic principles that govern the composting process. The approach, assumptions, and equations used for simulation modeling are described in this chapter.

Two basic model types are developed. The first applies to systems in which all feed components are homogeneously mixed and flow through the process without subsequent separation. This model applies to most compost systems that do not use bulking agents (bulking particles). The second model applies to systems that use bulking agents to maintain free air space and screen the mixture to return most of the bulking particles back to the system.

For each of the two model types, the composting system is divided into a number of stages arranged in series. Each stage of the system is assumed to be completely mixed. Different substrates can be input to stage 1 of the process. The mixture output from one stage becomes the input to the next stage, and so on. Each stage has a defined HRT. Air flowrate to each stage can either be fixed or adjusted by the program to maintain a setpoint temperature within the stage. Water can be added to any stage either in a fixed amount or the water addition can be adjusted by the program to maintain a setpoint solids content within the stage.

Product recycle can be assumed from any stage of the process. Product recycle is returned to stage 1 for mixing with the feed substrates. If bulking agents are used, they can be screened from the output of any specified stage. Screened bulking agent is returned to stage 1 for mixing with the feed substrates. If both product recycle and bulking agents are used within the same process, the product recycle must occur downstream of the bulking agent screening.

Thermodynamic boundaries are drawn around the entire system, which includes all stages, and the individual stages themselves. Input and output mass terms and input and output energy terms must balance across the thermodynamic stage boundaries and the thermodynamic system boundary. All thermodynamic boundaries must be balanced within acceptable accuracy at the solution point. The temperature difference between iterations and the difference between actual and setpoint conditions for temperature and moisture are also considered in determining the solution point.

It is assumed that oxidation of BVS is first order with respect to the quantity of remaining BVS. Both "faster" and "slower" BVS fractions can be assumed for each substrate. Effects of temperature, moisture content, FAS, and oxygen content on the first order rate constant are considered in the model. The computational procedures allow for more sophisticated mass and energy balances that include the effects of process kinetics.

REFERENCES

1. Andrews, J. F. and Kambhu, K. "Thermophilic Aerobic Digestion of Organic Solid Wastes," Office of Research and Development, U.S. EPA NTIS PB-222-396 (1973).
2. Schulze, K. L. "Continuous Thermophilic Composting," *Compost Sci.* (Spring 1962).
3. Wiley, J. S. and Pearce, G. W. "A Preliminary Study of High-Rate Composting," *Trans. Am. Soc. Civil Engr.* Paper 1178 (1957).
4. Regan, R. and Jeris, J. S. "Cellulose Degradation in Composting," report to U.S. Public Health Service from the Civil Engineering Department, Manhattan College, Research Grant EC 00161 (January 1971).

5. Mears, D. R., Singley, M. E., Ali, C., and Rupp, F. "Thermal and Physical Properties of Compost," in *Energy, Agriculture and Waste Management,* W.J. Jewell, Ed. (Ann Arbor, MI: Ann Arbor Science Publisher, Inc., 1975).

CHAPTER 12

Process Dynamics II — The CFCM Stage

INTRODUCTION

The previous chapter described an approach to mathematical analysis of the composting process. Mass and energy balances were related to process kinetics to produce a dynamic model of the system. The purposes of this Chapter are to introduce the reader to the computer simulations models, begin applying the models to actual composting situations, present results of the simulations and discuss their implications for design and operation of actual compost systems. It is not necessary that the reader have the computer software or be versed in "computereze" to understand the material presented here.

The applications begin with a one stage system to introduce the concepts and demonstrate some of the important process variables that affect the continuous feed complete mix (CFCM), stage. The feed substrates are sludge cake in one case and mixed compostables in another to span the range of substrates common to modern practice. Model complexity is then increased slightly by considering two stage systems. In this way the reader is introduced gradually to the subject and, hopefully, avoids being overloaded.

Readers of the first edition of this book may recall that the corresponding chapters were filled with graphs and charts presenting results of simulation modeling. Personal computers were not readily available at that time and any results had to be presented in the text. There was no way for the reader to explore on his own. With today's personal computers and the expert software now available, the reader is free to conduct his or her own simulations. Therefore, much of the previous data has been omitted and the emphasis directed toward demonstrating the power and versatility of the computer programs as tools for engineering design and operational simulation.

BASIC ASSUMPTIONS

The CFCM stage simulates a process where material is fed on a continuous basis and the contents are completely mixed. Neither condition is easily achieved in actual practice with solid substrates. However, intermittent feeding at regular intervals is commonly practiced and model results should be reasonably applicable to such conditions. For example, many systems are fed on a daily basis. If the CFCM stage has an HRT of more than 4 to 5 days, the error introduced by assuming continuous feed is small. Complete mixing is a difficult condition to attain even in aqueous systems let alone one with a solid substrate. However, systems that provide periodic agitation usually can be assumed to be reasonably "well mixed". Systems that do not provide agitation, such as the static pile process, can be broken into a series of stages to simulate the plug flow condition.

All the simulations presented in this chapter assume a feed substrate load of 10 dry tons per day (dtpd). HRT for the stage is 10 days unless otherwise noted. All infeed temperatures are 68°F (20°C). Inlet air humidity is 50%. For Case 1, biosolids cake is assumed to be raw sludge at 30% TS. For Case 2, mixed compostables composed of food waste, yard waste, and sawdust are assumed with a total dry weight of 10 dtpd. The minimum mixture solids content to stage 1 is 40% TS in all cases.

DYNAMICS OF A SINGLE STAGE SYSTEM

Case 1 — Biosolids Cake

The Effect of Air Supply

Example 12.1

10 dtpd of raw biosolids cake at 30% TS is input to a single CFCM stage. Product from the stage is recycled and mixed with the sludge cake for conditioning. Minimum mixture solids content is 40% TS. HRT is 10 days. Evaluate the operation of such a system.

Solution

1. The program DATIN41B is included in the software package and is used to create data files for CMSYS41B. Load DATIN41B and follow the menu to create a data file named DATA1201. Input the following data:

MDY	=	(enter current value)
IDEN	=	EXAMPLE 12.1
ME	=	E (for English units)
SUBNUM	=	1
STAGENUM	=	1
STAGER	=	1
SUBNAME	=	SLUDGE
X	=	66666.7
S	=	0.3
V	=	0.8
K	=	0.65
FASTFRAC	=	0.4
SLOWFRAC	=	0.6
H	=	9500
T	=	68

RATEKM20F	=	0.015
RATEKM20S	=	0.005
COEFF	=	0
A	=	10
B	=	19
C	=	3
D	=	1
HRT	=	10
QAIR	=	4000
TAIR	=	68
RHAIR	=	0.5
WATADD	=	0
SMSET	=	1
TSET	=	200
XRMIN	=	20000
XRMAX	=	160000
SMMIN	=	0.4
COEFFM	=	0.3
ELEV	=	0
TWATADD	=	68

The above values are typical for raw sludge solids and are consistent with the basic assumptions defined previously for this chapter.

2. Exit DATIN41B, load CMSYS41B, and run using DATA1201 when prompted for the data file name. The program should close to solution about iteration 39. The complete program output for this case is presented in Appendix C.

3. Conduct additional runs to determine the effect of QAIR. First, reload DATIN41B and follow the menu to copy data file DATA1201 under a new file name, for example DATA1202. Continue to follow the menu to modify DATA1202 by changing the value of QAIR to 2000 acfm. Rerun CMSYS41B using DATA1202 when prompted for the data filename. Repeat this procedure for various QAIR values from 1000 to 10,000 acfm.

Predicted stage temperature, output solids content, and BVS reduction are presented as a function of air supply flowrate in Figure 12.1. The reaction rate constant for the fast fraction, RATEKF, and outlet O_2 and CO_2 concentrations are presented in Figure 12.2. Specific oxygen consumption for the product from stage 1 is presented in Figure 12.3. Finally, stage volume and system SRT are presented in Figure 12.4. In all cases, air flowrate is normalized to the input tonnage and expressed as acfm/dtpd. For example, the 4000 acfm used in Example 12.1 is normalized to 400 acfm/dtpd in Figure 12.1. Stage temperature is strongly influenced by the air supply rate, decreasing with increasing air supply. However, output solids content, SMOUT, actually increases with air supply to a maximum value of ~61% TS at 300 to 400 acfm/dtpd. SMOUT then begins to decrease at higher QAIR values. The curve for BVS reduction parallels that for SMOUT. The same trend is observed with the reaction rate constant in Figure 12.2.

The process dynamics presented in Figures 12.1 to 12.4 can be explained by considering the effect of temperature on the reaction rate constant. With a low air supply, the stage operates at very high temperatures to the point where significant thermal limitations are imposed. In

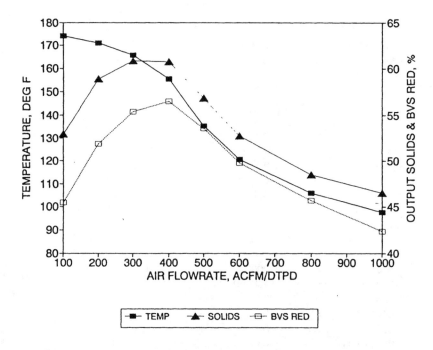

Figure 12.1.   Single stage system, biosolids cake — effect of air flowrate on stage temperature, output solids content, and BVS reduction for Example 12.1.

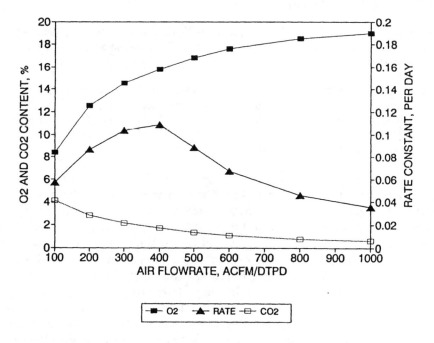

Figure 12.2.   Single stage system, biosolids cake — effect of air flowrate on oxygen and carbon dioxide concentrations and the reaction rate constant, RATEKF, for Example 12.1.

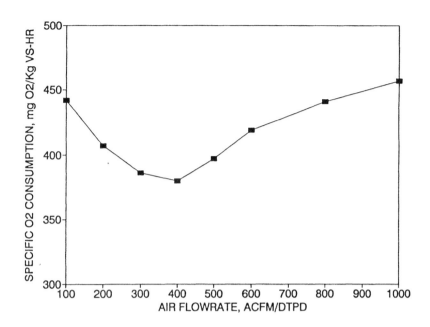

Figure 12.3. Single stage system, biosolids cake — effect of air flowrate on specific O_2 consumption of the stage product for Example 12.1.

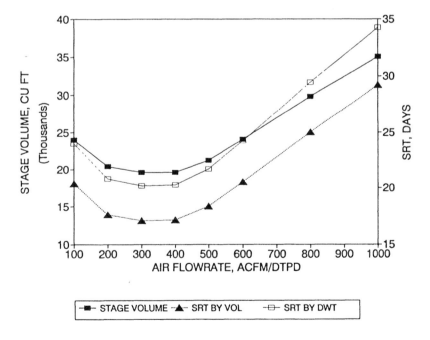

Figure 12.4. Single stage system, biosolids cake — effect of air flowrate on stage volume and SRTs for Example 12.1.

this region an increase in air supply removes more heat which reduces the stage temperature. The reduced temperature results in an increase in the reaction rate constant. The increased rate constant means that more BVS are oxidized. This in turn releases more heat to drive the "evaporative engine" and increases the output solids content. The increase in BVS reduction improves the stability of the product as measured by specific oxygen uptake in Figure 12.3. The lesson here is that a compost process actually can operate with too high a temperature. Reaction rates decrease and stability of the output product suffers as a result.

Output solids content continues to increase up to an air supply of 300 to 400 acfm/dtpd. A plateau region is eventually reached with maximum output solids ~61% and maximum BVS reduction about 57%. Output solids begin to decrease at higher air supply in response to the decreasing stage temperatures. Considering that infeed mixture solids to the stage is 40%, considerable drying is achieved in the system. The pattern of outfeed solids content answers one frequently asked question — should the air supply be increased to produce a drier product? Stated another way, if air supply is increased will the decrease in reactor temperature and hence saturation vapor pressure be offset of the increased mass flowrate of air? Apparently at low air rates, up to about 400 acfm/dtpd, the advantage is with increased air supply. Beyond about 500 acfm/dtpd, however, the effect of decreased stage temperature becomes dominant and outfeed solids content decreases.

O_2 and CO_2 content of the exhaust gas is shown in Figure 12.2. At 100 acfm/dtpd, the lowest air supply examined, exhaust gases still contain over 8% O_2 by volume. Oxygen content increases rapidly to about 16% at 400 acfm/dtpd. These concentrations are sufficiently high to avoid significant rate limitations. CO_2 content decreases as O_2 content increases because of the greater dilution afforded by the higher air flowrates. CO_2 is about 2% by volume at 400 acfm/dtpd.

Stability of the stage outfeed is presented in Figure 12.3. Specific oxygen consumption, RATEO2, decreases to a minimum value of about 380 mg O_2/g VS-h. The minimum value corresponds with the point of maximum BVS reduction and maximum reaction rate constant. Specific oxygen consumption then increases with continued increase in air supply. The point here is that a stable end product is produced by oxidizing the BVS fraction. Therefore, the system must be operated under conditions favorable to BVS oxidation. On one hand, temperatures should not be too high to avoid thermal limitations on the rate constant. On the other hand, moisture removal should not progress to the point where lack of moisture limits the rate of reaction.

Specific oxygen consumption should be in the range of 20 to 100 mg O_2/g VS-h for high quality composts. Obviously, the compost produced from this single stage system has not been sufficiently stabilized. Further composting would be necessary to reduce the specific oxygen consumption to an acceptable range. Later sections will build on the results presented here to show the total system requirements to produce the desired end product.

Referring to Figure 12.4, the stage volume required to achieve a 10 day HRT varies from a low of about 19,550 ft^3 at 300 to 400 acfm/dtpd air supply to over 35,000 ft^3 at 1000 acfm/dtpd. The reason for this range is that the solids content of product recycle reaches a high of about 61% in the range of 300 to 400 acfm/dtpd. With a high solids content, less recycle is needed to meet the minimum infeed mixture solids content of 40%. This reduces the volume and weight of materials entering the process, which in turn reduces the stage volume required to achieve the desired HRT. At 1000 acfm/dtpd the product is only 46.5% TS, which requires a much higher quantity of recycle to meet the same 40% TS input condition.

SRT for the one stage system is also shown in Figure 12.4. SRTs determined by both volume and dry weight calculations are shown. Both calculation procedures give the same pattern of results with SRTDWT consistently higher than SRTVOL by about 3 to 4 days.

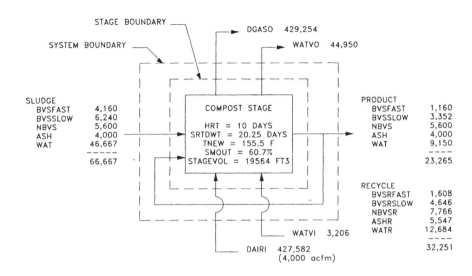

Figure 12.5. Single stage system, biosolids cake — mass balance for the case of 400 acfm/dtpd air flowrate from Example 12.1.

SRTDWT is about 20 days over the range of 300 to 400 acfm/dtpd, increasing to almost 35 days at 1000 acfm/dtpd. Again, this results from the lower product recycle rate required in the air supply range of 300 to 400 acfm/dtpd.

Mass and Energy Balances

A mass balance for solids, water, and gas components is shown in Figure 12.5. Conditions correspond to those used in Example 12.1 and as shown in Figures 12.1 to 12.4 for an air supply of 400 acfm/dtpd. The reader can refer to Appendix C to verify the mass balance. Both system and stage boundaries are shown in Figure 12.5. Because recycle crosses the stage boundary but not the system boundary, the mass balance for the two cases differs by the mass of the recycle component. This can be verified by referring to the mass balance summary in Appendix C. The reader can also verify that "mass in" equals "mass out" across both system and stage boundaries.

BVS reduction is about 57% across the system, which accounts for the rather large contribution of BVS in the recycle component. Recall that recycle does not contribute new organics to the system. In other words, recycle does not cross the system boundary. However, the use of recycle allows additional opportunity for biological oxidation of the substrate organics and, therefore, the effect must be accounted for. The combined effects of BVS loss and moisture removal result in a 65% loss in wet weight from starting substrate to stage product. Even though 57% of the BVS are lost, volatile solids content is only reduced from 80% in the starting substrate to about 71.7% in the stage product. This illustrates the problem of using VS measurements across a compost facility. Often the percentage VS reduction seems small even though actual BVS loss is sufficient to drive the process.

Considerable moisture is removed in the hot gases that exit the process. Solids content in the stage is nearly 61% TS, which is sufficiently high to induce a significant moisture rate limitation. The moisture adjustment factor, F1, is about 0.47. This means that the reaction rate constant is only 47% of that which would occur in the absence of any moisture limitation. The problem is that the starting sludge is "energy rich" and generates significant heat output. A high air flowrate is required to remove the generated heat from the process. The high flowrate of hot gas removes water vapor to the point that process kinetics become significantly reduced by the lack of water.

Particular note should be made of the gas components of the mass balance. Gases supplied to and removed from the stage are the largest components of the balance by a considerable margin. Dry air input is nearly 20 times the dry weight of feed substrate! Composting has often been described as a "problem in materials handling". If so, the major material handled is air and exhaust gas. Unfortunately, since one cannot usually see the gas components attention is more often focused on solids handling. If you remember one thing from this text, remember that gas is the major component handled at a composting plant, followed by solids and water. One should also note the quantity of water vapor in the exhaust gas. It is nearly equal to that supplied by the feed sludge. This water vapor has a tendency to condense, sometimes in the wrong places. Care must be taken when handling the process gas to allow for condensation and its removal.

A bar graph showing energy inputs and outputs is presented in Figure 12.6. The balance shown is for the system boundary. Both system and stage balances are presented in the computer printout of Appendix C. Again, because recycle crosses the stage boundary but not the system boundary, the energy balance for the two cases will differ by the sensible heat contained in the recycle. Total energy input is about 61.6 MBtu/day (million Btu/day) and is dominated by organic decomposition, provided by both the feed substrate and recycle components. All other energy inputs are relatively minor by comparison. Organic substrate in the recycle accounts for about 38% of BVS input to the stage.

Total output energy equals total input energy with closure to within 0.023%. As expected, energy output is dominated by the latent heat of water vaporization, which accounts for about 68% of total output energy. Sensible heat in the hot exhaust gases and water vapor are the next highest energy outputs. Sensible heat in the output solids and water is relatively minor. Solids, water, and gases are all assumed to exit at the stage temperature. Therefore, the high mass flowrate of gases accounts for their higher sensible heat loss. The latent heat of vaporization can be expected to remain a major output energy. Other energy outputs will vary in importance depending on the particular operating conditions within the system.

The Effect of HRT

The above calculations are based on an HRT of 10 days in the compost stage. Intuitively, the results should be significantly influenced by the assumed HRT. This section will examine the effect of this important variable.

Example 12.2

Assuming the sludge characteristics and stage parameters from Example 12.1, determine the effect of changing the stage HRT to 5 and 20 days.

Solution

1. Load DATIN41B and follow the menu to modify data file DATA1201 as follows:

$$HRT = 5 \text{ days}$$

2. Modify other data files created for Example 12.1 to an HRT of 5 days. Exit DATIN41B, load CMSYS41B, and run the program for each of the data files.

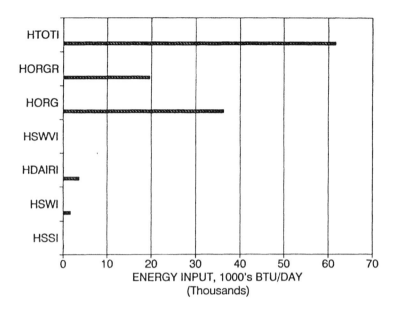

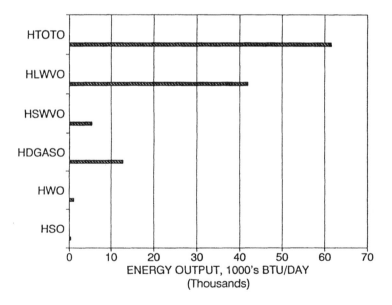

Figure 12.6.   Single stage system, biosolids cake — energy input and output terms corresponding to the mass balance of Figure 12.5.

3. Repeat steps 1 and 2 for each data file modifying the HRT to 20 days.

The effects of HRT on stage temperature, output solids content, and BVS reduction as a function of air supply are presented in Figures 12.7 to 12.9. Increasing stage HRT has the effect of increasing the range of air supply over which adequate temperatures can be maintained. In this respect, process stability and the "window of operability" are both increased.

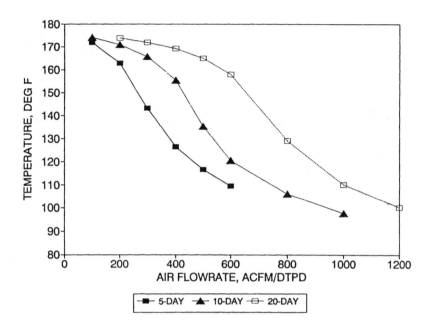

Figure 12.7. Single stage system, biosolids cake — effect of air flowrate and stage HRT on process temperature for Example 12.2.

The concept of a window of operability is important. The window is the range of process parameters within which the process will operate in a stable manner. If the window is large, the operator has considerable latitude within which to adjust process parameters. Also, the process can withstand more deviations and transients without extending itself outside the window. If the window is small, the process can still operate, but its stability and resistance to adverse conditions are reduced. Referring to Figure 12.7, maximum air supply to maintain a temperature above 130°F (54.4°C) is about 400 acfm/dtpd at a 5 day HRT. This increases to about 800 acfm/dtpd with a 20 day HRT. The "window" of acceptable air supply is about doubled with an HRT of 20 days compared to 5 days.

Referring to Figure 12.8, output solids content increases with increasing HRT for a given air supply rate. Again, the window of operability increases with increased HRT. For example, achieving 55% output solids with a 5 day HRT is limited to a range of air supply from only about 150 to 250 acfm/dtpd. With a 20 day HRT the operating band is increased eightfold to about 200 to 1000 acfm/dtpd. A similar pattern is observed for BVS reduction as shown in Figure 12.9. BVS reduction increases with increasing HRT for a given air supply rate. The band of acceptable BVS reduction widens significantly with increasing HRT.

One advantage of a shorter HRT is the reduced stage volume required and the associated capital cost savings. However, this savings is offset by the smaller range of operating parameters, i.e., specific air supply, the reduction in maximum output solids and lower BVS stabilization. Clearly, the tradeoff is between capital cost savings and operational performance. It should also be noted that requirements for subsequent curing stages will be influenced by the HRT of previous stages. If the goal is to achieve a product of given stability, then BVS not oxidized in early stage(s) must be stabilized in later curing stages. Therefore, design of the early "high rate" stages cannot be separated from design of the later "curing" stages. The compost plant must be viewed as a total system. This will be discussed further in Chapter 13.

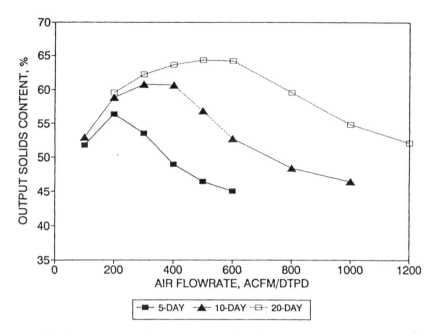

Figure 12.8.   Single stage system, biosolids cake — effect of air flowrate and stage HRT on output solids content for Example 12.2.

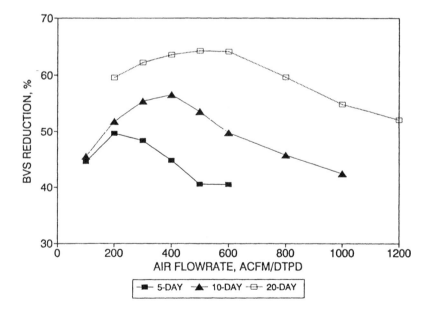

Figure 12.9.   Single stage system, biosolids cake — effect of air flowrate and stage HRT on BVS reduction for Example 12.2.

With regard to total system design, it is interesting to note that BVS reduction does not exceed 64% even with a 20 day HRT. Specific oxygen uptake is predicted to remain above about 330 mg O_2/kg VS-h, which is still too high for most compost uses. Therefore, additional stabilization is needed.

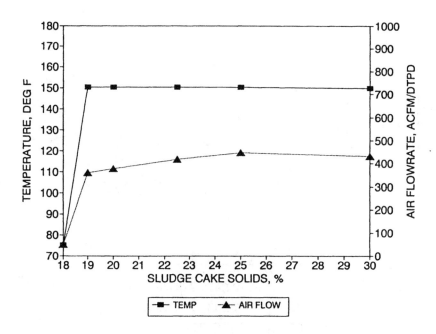

Figure 12.10. Single stage system, biosolids cake — effect of biosolids cake solids on stage temperature and air flowrate needed to maintain setpoint temperature for Example 12.2.

Reactor Response to Feed Cake Solids

Previous chapters demonstrated that the solids content from dewatering is perhaps the most important variable in determining the system energy balance when composting wet sludges. The simulation model will now be used to explore the response of the reactor system to feed cake solids. Conditions of Example 12.1 are again used in the analysis. Dry tonnage remains at 10 dtpd and HRT returns to 10 days. Solids content varies from 18 to 30% and the value of X adjusted accordingly. Also, a setpoint temperature is established at 150°F (65.6°C). The simulation model then adjusts the air flowrate as necessary to maintain the setpoint. XRMAX is increased to 500,000 lb/day to allow for the increased recycle needed as output solids content decreases.

The effects of cake solids on stage temperature and air flowrate are presented in Figure 12.10. Setpoint temperatures are maintained at cake solids above 19%. Below this, the reactor fails completely. With 18% cake, a reactor temperature of only 75°F (23.9°C) is predicted. This illustrates the significant effect of cake solids on reactor performance. The air flowrate to maintain setpoint temperature is about 360 acfm/dtpd with 19% TS cake, increasing slightly to about 450 acfm/dtpd at 25% TS cake.

The recycle rate, XR, needed to maintain the assumed 40% feed mixture solids content is presented in Figure 12.11 along with the output solids content from the reactor. With 30% TS cake, output solids content is 60% and only 33,000 lb/day of recycle is needed. As cake solids decrease, so too does the output solids content. With 20% TS cake, output solids is reduced to about 47%. Recycle increases to almost 300,000 lb/day to maintain a 40% TS feed mixture. While reactor conditions with 20% TS cake are theoretically possible, they are rapidly approaching the limits of practicality. Reactor recycle is almost three times the weight of sludge substrate. With 19% cake, output solids content is predicted to be only slightly above 40% TS. Obviously, this poses an extreme problem in maintaining a 40% mixture.

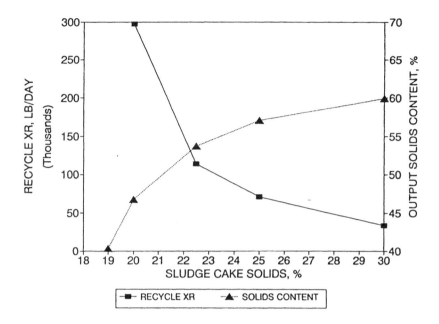

Figure 12.11. Single stage system, biosolids cake — effect of cake solids on output solids content and recycle rate, XR, for Example 12.2.

The lesson from Figures 12.10 and 12.11 is that sludge cake solids is extremely important in determining reactor performance. As a general rule, raw cake solids below 25% should sound a warning that the energy balance may be strained. There are generally two courses of action in this case. First, cake solids can be increased by improving dewatering performance. Second, an energy amendment can be added to the feed mixture to increase the energy resources.

The simulation model predictions regarding the effect of cake solids have been verified by application to the reactor compost system at the City of Plattsburgh, New York. The latter system suffered process failures during three successive winters. Respirometer studies verified that the energy resources were marginal. Degradability and rate constants were developed from the respirometer data. The simulation models then confirmed that the combination of winter temperatures and low cake solids were responsible for the repeated failures. The city moved to increase cake solids and changed to a more degradable amendment. This combination of actions resolved the problem.

Case 2 — Mixed Compostables

The Effect of Air Supply

Example 12.3

A food processing plant produces a waste product that resembles mixed garbage and is 35% TS. The plant wishes to convert this waste to a reusable product by composting. The plant is considering amending the food waste with yard waste and sawdust which are both locally available. The yard waste is a mixture of grass and tree trimmings. The plant has retained your firm, Compost R'Us, to evaluate this concept. Your first evaluation is for a single stage system with no product recycle. Because you expect considerable drying you plan to add 10,000 lb/day of supplemental water.

Solution

1. Load DATIN41B and follow the menu to create a data file named DATA1210. Input the following data:

MDY	=	(enter current value)		
IDEN	=	EXAMPLE 12.3		
ME	=	E (for English units)		
SUBNUM	=	3		
STAGENUM	=	1		
STAGER	=	0		
SUBNAME	=	FOOD WASTE	YARD WASTE	SAWDUST
X	=	22857	12000	10000
S	=	0.35	0.5	0.6
V	=	0.95	0.9	0.97
K	=	0.60	0.6	0.35
FASTFRAC	=	0.7	0.6	0.6
SLOWFRAC	=	0.3	0.4	0.4
H	=	9500	7500	7500
T	=	68	68	68
RATEKM20F	=	0.05	0.01	0.01
RATEKM20S	=	0.005	0.005	0.005
COEFF	=	0.26	0.30	0.25
A	=	16	23	295
B	=	27	38	420
C	=	8	17	186
D	=	1	1	1
HRT	=	-10		
QAIR	=	3000		
TAIR	=	68		
RHAIR	=	0.5		
WATADD	=	10000		
SMSET	=	1		
TSET	=	200		
XRMIN	=	20000		
XRMAX	=	160000		
SMMIN	=	0.4		
COEFFM	=	0.3		
ELEV	=	0		
TWATADD	=	68		

The above values are typical for the substrates under consideration and are consistent with the basic assumptions defined previously for this chapter. Note that XRMIN, XRMAX, and SMMIN are not important to the input because STAGER = 0.

2. Exit DATIN41B, load CMSYS41B, and run using DATA1210 when prompted for the data file name. The program should close to solution about iteration 76. The complete program output for this case is presented in Appendix D.

3. Conduct additional runs to determine the effect of QAIR. First, reload DATIN41B and follow the menu to copy data file DATA1210 under a new file name, for example DATA1211. Continue to

follow the menu to modify DATA1211 to QAIR = 2000 acfm. Rerun CMSYS41B inputting DATA1211 when prompted for the data filename. Repeat this procedure for various QAIR values from 500 to 6000 acfm.

Predicted stage temperature, output solids content, and BVS reduction are presented as a function of air supply flowrate in Figure 12.12. The reaction rate constant for the fast fraction, RATEKF, and outlet O_2 and CO_2 concentrations are presented in Figure 12.13. Specific oxygen consumption for the product from stage 1 is presented in Figure 12.14. Similar to the Case 1 study, stage temperature is strongly influenced by the air supply rate, decreasing with increasing air supply. Output solids content, SMOUT, increases with air supply to a maximum value of about 62.4% TS at 300 acfm/dtpd. SMOUT then begins to decrease at higher QAIR values. The curve for BVS reduction parallels that for SMOUT. The same trend is observed with the reaction rate constant in Figure 12.13. Maximum BVS reduction is 45.6%.

The process dynamics presented in Figures 12.12 to 12.14 are again explained by considering the effect of temperature on the reaction rate constant. With a low air supply, reaction rates are limited by the very high stage temperatures. In this region an increase in air supply removes more heat which reduces the stage temperature. Reduced temperature results in an increase in the reaction rate constant. The increased rate constant means that more BVS are oxidized. The point of maximum BVS reduction and maximum output solids both coincide with the point of maximum rate constant, i.e., 300 acfm/dtpd. This occurs at an intermediate temperature, one that is low enough to avoid thermal death to the microbes yet high enough to take advantage of the exponential phase of the temperature curve. Further increases in air supply remove more heat and reduce the operating temperature. This in turn reduces the rate constant. BVS reduction and output solids content are also reduced as a result.

Stability of the stage outfeed is presented in Figure 12.14. Specific oxygen consumption decreases to a minimum value of about 400 mg O_2/g VS-h. Again, the minimum value corresponds with the point of maximum BVS reduction and maximum reaction rate constant. Specific oxygen consumption then increases with continued increase in air supply. Similar to the Case 1 results with sludge cake, compost produced from this single stage system with mixed compostables has not been sufficiently stabilized. Further composting is necessary to reduce the specific oxygen consumption to an acceptable range. Later sections will build on the results presented here to show the total system conditions required to produce the desired end product.

Stage volume and SRT as a function of air supply rate are presented in Figure 12.15. The relationships are significantly different compared to the Case 1 study. Because there is no recycle, system SRT equals HRT (10 days in this case) and is independent of air supply rate. Also, SRT is the same whether calculated by volume (SRTVOL) or by dry weight (SRTDWT). The stage volume required to achieve an HRT of 10 days varies only slightly with air supply rate. The lowest stage volume is about 9500 ft³ at 300 acfm/dtpd and the highest about 10,000 ft³ at 600 acfm/dtpd. This slight variation is due to the loss of BVS in the stage. Recall that STAGEVOL is calculated from Equation 11.119 and is based on the average of infeed and outfeed volumes. Outfeed volume varies slightly depending on BVS loss within the stage.

The differences between a system with recycle and one without are important to consider. Differences in SRT can be very significant. For the Case 1 study, SRTDWT was about 20 days and BVS reduction almost 57% with an air supply rate of 400 acfm/dtpd. Maximum BVS reduction for Case 2 is only about 45% with an air supply rate of 300 acfm/dtpd. However, SRTDWT is only 10 days compared to the 20 days for Case 1. BVS reduction is primarily a function of system SRT. Therefore, the Case 1 and 2 studies are not directly comparable when

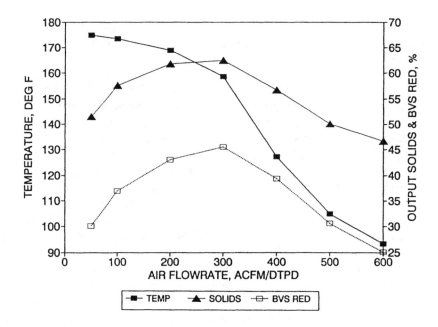

Figure 12.12. Single stage system, mixed compostables — effect of air flowrate on stage temperature, output solids content and BVS reduction for Example 12.3.

it comes to BVS reduction. While HRT for both systems is 10 days, SRT is quite different because product recycle is used in Case 1. To the extent possible, alternative systems should be evaluated at comparable SRTs.

The increased SRT for Case 1 does not come without a price, namely the larger STAGEVOL required to achieve the assumed 10 day HRT. At the air supply rates noted above, STAGEVOL is about 19,600 ft³ for Case 1, compared to only 9500 ft³ for Case 2. It is interesting to note that using two reactors in series for Case 2 would give about the same total stage volumes and system SRT as Case 1. This would be a more appropriate comparison between the two cases. Using two stages in series is the subject of the next section.

Mass and Energy Balances

A mass balance for solids, water, and gaseous components is presented in Figure 12.16. Conditions correspond to those used in Example 12.3 and shown in Figures 12.12 to 12.14 for an air supply of 300 acfm/dtpd. The mass balance is based on the computer printout presented in Appendix D. System and stage boundaries are the same for this case because there is only one stage and product recycle is not used.

Mixed compostables enter the process at a combined mixture solids content of 44.6% TS. This does not include the 10,000 lb/day of supplemental water addition which is added directly to the stage. An SM of 44.6% should be more than sufficient to avoid rate limitation from lack of FAS. Indeed, FAS is predicted to be about 53% within the stage and the FAS adjustment factor, F2, is almost 1.0. However, even with 10,000 lb/day of supplemental water addition the stage is predicted to operate at 62.5% TS. This is sufficiently dry to place a rather significant limit on the reaction rate. The moisture adjustment factor, F1, is predicted to be about 0.40. In other words, the reaction rate is only 40% of what it would be in the absence of the moisture limitation.

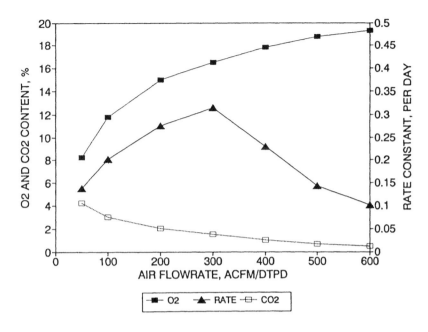

Figure 12.13. Single stage system, mixed compostables — effect of air flowrate on oxygen and carbon dioxide concentrations and the reaction rate constant, RATEKF, for Example 12.3.

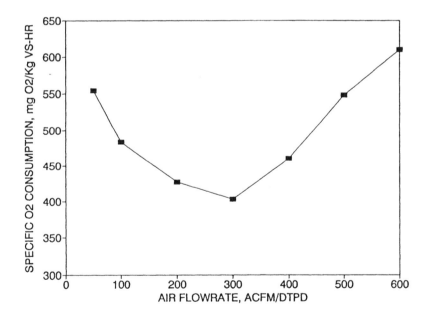

Figure 12.14. Single stage system, mixed compostables — effect of air flowrate on specific O_2 consumption of the stage product for Example 12.3.

BVS reduction is only about 46%, in part because of the moisture limitation placed on the rate constant. However, it is also important to recall that system SRT is only 10 days because no product recycle is used. Even in the absence of any rate limitations, a fully stable product could not be achieved with such a short system SRT.

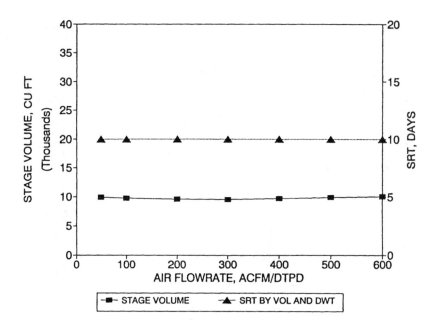

Figure 12.15. Single stage system, mixed compostables — effect of air flowrate on stage volume and SRT for Example 12.3.

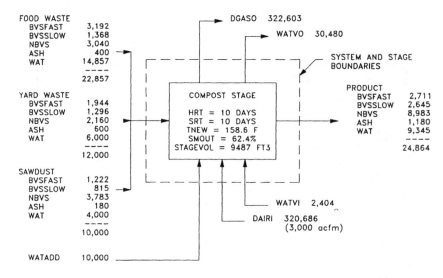

Figure 12.16. Single stage system, mixed compostables — mass balance for the case of 300 acfm/dtpd air flowrate for Example 12.3.

The combined effects of BVS loss and moisture removal result in a 45% loss in wet weight from starting substrates to stage product. The loss is about 55% if the supplemental water addition is included with the infeed substrates. Volatile solids content is reduced from about 94% in the starting substrate to about 92% in the stage product. Again, this illustrates the problem of using VS measurements across a compost facility. Often the percentage VS reduction seems small even though actual BVS loss is sufficient to drive the process. A two percentage point reduction in VS content would be hard to actually measure. This might lead one to question whether anything is actually happening in the system, even though almost 4500 lb/day of BVS are being oxidized and over 30,000 lb/day of water are being evaporated.

Gases supplied to and removed from the stage are again the largest components of the mass balance. Dry air input is about 16 times the dry weight of the feed substrates. Recall again that gas is the major component handled at a composting plant. Also note again the large quantity of water vapor in the exhaust gas. The feed substrates supply about 25,000 lb/day of water with another 10,000 lb/day from supplemental water. Almost all of this reports to the exhaust gases. Care must be taken in handling the process gas to avoid condensation problems when the gas cools.

A bar graph showing energy inputs and outputs is presented in Figure 12.17. Because system and stage boundaries are the same, the system and stage energy balances are the same. As expected, energy inputs are dominated by organic decomposition. The contribution from food waste, identified as HORG1, is greater than for the yard waste, HORG2, and sawdust, HORG3. Food waste not only contributes more BVS to the process, it has the highest proportion of BVSFAST and the highest rate constant, RATEKM20F, for the fast fraction. Therefore, it is not surprising that it contributes the highest energy input. No recycle terms are included because product recycle is not used. Therefore, any heat release occurs during the single pass of substrates through the stage.

Energy output is again dominated by the latent heat of water vaporization, which accounts for about 65% of total output energy. Sensible heat in the hot exhaust gases and water vapor are the next highest energy outputs. Sensible heat in the output solids and water is relatively minor. These results are very similar to those presented earlier for Case 1. Thus, dominant terms in the energy balance tend to remain the same regardless of the input substrates.

DYNAMICS OF A TWO STAGE SYSTEM

A single stage system presents limitations that are hard to avoid. High temperatures can present rate limitations at low air supply rates. If more air is supplied to remove more heat and reduce process temperatures, water loss from evaporation may dry the composting substrates to the point that lack of moisture then limits the process. Also, the simulation model treats each stage as a single, complete mix, continuous feed cell. These assumptions become less applicable as the HRT of each stage is increased. Therefore, there are a number of reasons for developing a compost system from multiple stages arranged in series. Multiple stages can be operated under different conditions to reduce rate limitations and achieve the desired properties of moisture content and stability in the final product. The Case 1 and 2 examples will be advanced by the addition of a second stage to the process.

Case 3 — Biosolids Cake

Example 12.4

You have decided that the single stage system of Example 12.1 did not produce a sufficiently stable product from raw sludge cake. Therefore, you decide to add a second stage to the process. To reduce rate limitations yet achieve pathogen destruction, the first stage will operate with a setpoint temperature of 150°F (65.6°C) and setpoint solids content of 50%. The second stage will operate with a setpoint temperature of 130°F (54.4°C) to improve the horticultural quality of the product. Setpoint solids content will be 55% to reduce the water content of the final product. Each stage will have a 10 day HRT. Recycle from the second stage will be used to condition the feed substrate. All other conditions remain the same. Evaluate the operation of such a system.

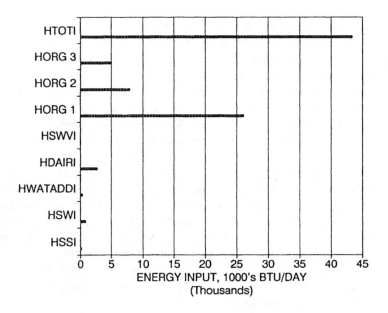

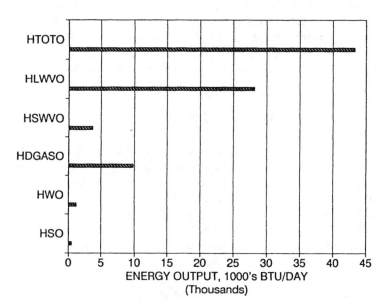

Figure 12.17. Single stage system, mixed compostables — energy input and output terms corresponding to the mass balance of Figure 12.16.

Solution

1. Load DATIN41B and follow the menu to create a data file named DATA1220. Input the following data:

MDY	=	(enter current value)
IDEN	=	EXAMPLE 12.4
ME	=	E (for English units)

SUBNUM	=	1
STAGENUM	=	2
STAGER	=	2
SUBNAME	=	SLUDGE
X	=	66666.7
S	=	0.3
V	=	0.8
K	=	0.65
FASTFRAC	=	0.4
SLOWFRAC	=	0.6
H	=	9500
T	=	68
RATEKM20F	=	0.015
RATEKM20S	=	0.005
COEFF	=	0
A	=	10
B	=	19
C	=	3
D	=	1

		STAGE ONE	STAGE TWO
HRT	=	10	10
QAIR	=	1000	1000
TAIR	=	68	68
RHAIR	=	0.5	0.5
WATADD	=	0	0
SMSET	=	0.5	0.55
TSET	=	150	130

XRMIN	=	20000
XRMAX	=	160000
SMMIN	=	0.4
COEFFM	=	0.3
ELEV	=	0
TWATADD	=	68

2. Exit DATIN41B, load CMSYS41B, and run using DATA1220 when prompted for the data file name. The program should close to solution about iteration 928. The complete program output for this case is presented in Appendix E.

A mass balance developed from the simulation model output for Example 12.4 is presented in Figure 12.18. Several new concepts are introduced here. There are two stage boundaries, one for each stage of the process. Of course, there is still only one system boundary, which incorporates the entire process including the two stage boundaries. Mass flow into each stage balances with mass flow out of that stage. Similarly, mass flow into the system boundary balances with mass flow out of the system. This can be verified by adding inputs and outputs in Figure 12.18 or by reviewing Appendix E which presents a complete balance for the system as a whole and for each stage that makes up the system. Mass terms usually balance to within 0.01%.

Substrate input is the same as for Example 12.1. However, the operation of stage 1 differs significantly because of the setpoint conditions, namely TSET = 150°F (65.6°C) and SMSET

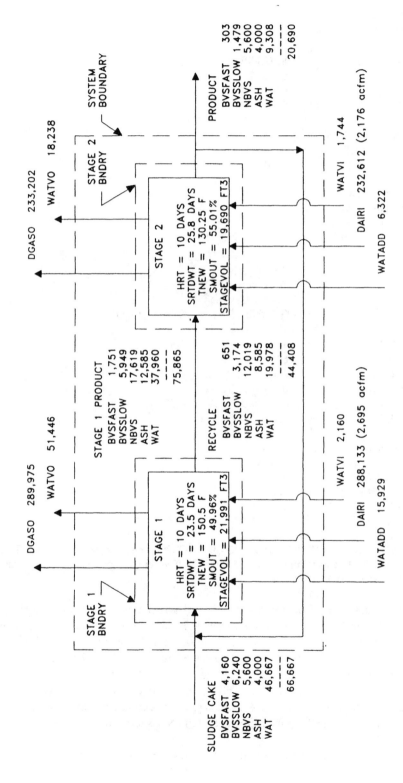

Figure 12.18. Two stage system, biosolids cake — stage and system mass balances for the case of Example 12.4.

= 0.50. To achieve these conditions the model increases QAIR from the input value of 1000 acfm to 2695 acfm. Note that when a TSET value applies to a particular stage, the input QAIR is interpreted as a minimum value and set equal to QAIRMIN. Actual stage temperature achieved in the model run is 150.5°F (65.8°C), a closure difference to actual setpoint of 0.5°F (0.28°C). This is within the closure conditions defined in Chapter 11.

To hold setpoint solids at 50% TS requires the addition of 15,929 lb/day of supplemental water. Actual solids content maintained in stage 1 is 49.96% TS, well within the defined closure conditions. Note that when SMSET conditions apply to a particular stage, input WATADD is interpreted as a minimum value and set equal to WATADDMIN. The 0 lb/day input in Example 12.4 becomes the minimum condition for the problem. WATADD required to maintain SMSET = 0.50 in stage 1 is almost 35% of the water input with the starting sludge cake. This may seem like a somewhat surprising result because sludge cake contains so much water. However, if the substrate is "energy rich" it is fully capable of drying itself right into a rate limitation. Supplemental water is then just as necessary as if the starting substrate were dry. At least one reactor-type, sludge composting facility with a "hot sludge" has produced composting solids in the range of 70 to 80% TS in the reactor. At this point the reaction rate has been severely limited and only supplemental water addition will alleviate the condition.

The improved operating conditions in stage 1 increase the reaction rate constant compared to the case of Example 12.1. RATEKF in stage 1 is about 0.175 per day compared to 0.108 per day in the uncontrolled case. As a result, the stage 1 output contains less BVS than the single stage output in Figure 12.5. In the latter case the recycle component must be added to the final product to get the stage 1 output. BVS reduction is about 46% across stage 1 in Figure 12.18. This reduction is lower than that quoted for the single stage system of Figure 12.5. This seems to be contradictory with the higher rate constant discussed above. The reason for this is that the 46% BVS loss is across the stage boundary, which includes product recycle. The earlier value quoted from Figure 12.5 is across the system boundary. This is an important distinction and one should always take care to note whether a reduction is quoted across a stage or system boundary.

Stage 2 operates with setpoint conditions TSET = 130°F (54.4°C) and SMSET = 0.55. An air flowrate of 2176 acfm and 6322 lb/day of supplemental water input are required to achieve these conditions. As was the case for stage 1, both setpoint conditions are achieved within the closure conditions. RATEKF in stage 2 is 0.084 per day which is lower than that in stage 1 for several reasons. First, stage temperature is lower, and since both stages are on the exponential phase of the temperature curve, the reaction rate for stage 2 should be lower. Second, the moisture adjustment factor, F1, is about 0.71 in stage 2 compared to about 0.86 in stage 1. Of course, this is because of the drier conditions carried in stage 2. BVS reduction across stage 2 is about 27%.

Combined outlet gases from stages 1 and 2 total almost 593,000 lb/day, of which almost 70,000 lb/day is water vapor. Total gas weight is almost 30 times the weight of dry solids in the infeed substrate. Adding stage 2 to the process increases the total quantity of exhaust gases that must be managed.

SRT calculated by dry weight, SRTDWT, is 23.5 days in stage 1 and 25.8 days in stage 2. Total system SRT, TOTSRTDWT, is the sum of SRTs for stages 1 and 2 according to Equation 11.132. TOTSRTDWT is therefore 49.3 days. System SRT is more than doubled compared to the case of Example 12.1. As a result, BVS reduction across the system is increased to about 83%, wet weight of final product is reduced to only 31% of the wet weight of the feed substrate, and specific oxygen consumption, RATEO2, for the final product is reduced to about 180 mg O_2/kg VS-h. The final product is considerably more stable compared to that of Example 12.1.

The energy balance across the system boundary is presented in Figure 12.19. Energy terms are the same as for Case 1, as shown in Figure 12.6. However, energy input increases significantly to a total of about 89.1 million Btu/day (MBtu/day), compared to about 61.6 MBtu/day for Case 1. Of course, the increased heat release is due to the greater BVS reduction in the two stage system.

The addition of stage 2, combined with improved operating conditions within the stages, increases BVS reduction and improves final product stability. The final product would probably be acceptable for some land application situations that do not require a "high end" compost. However, further improvements to the system will be explored in the next chapter.

Case 4 — Mixed Compostables

Example 12.5

Based on the successful results for Case 3, you decide to try the same approach with the mixed compostables from Case 2. A second stage is added to the process. The first stage will operate with a setpoint temperature of 150°F (65.6°C) and setpoint solids content of 50%. The second stage will operate with a setpoint temperature of 130°F (54.4°C) and setpoint solids content of 55%. Each stage will have a 10 day HRT. No product recycle will be used. Evaluate the operation of such a system.

Solution

1. Load DATIN41B and follow the menu to create a data file named DATA1221. Input the following data:

MDY	= (enter current value)		
IDEN	= EXAMPLE 12.5		
ME	= E (for English units)		
SUBNUM	= 3		
STAGENUM	= 2		
STAGER	= 0		

SUBNAME	= FOOD WASTE	YARD WASTE	SAWDUST
X	= 22857	12000	10000
S	= 0.35	0.5	0.6
V	= 0.95	0.9	0.97
K	= 0.60	0.6	0.35
FASTFRAC	= 0.7	0.6	0.6
SLOWFRAC	= 0.3	0.4	0.4
H	= 9500	7500	7500
T	= 68	68	68
RATEKM20F	= 0.05	0.01	0.01
RATEKM20S	= 0.005	0.005	0.005
COEFF	= 0.26	0.30	0.25
A	= 16	23	295
B	= 27	38	420
C	= 8	17	186
D	= 1	1	1

	STAGE ONE	STAGE TWO
HRT	= 10	10
QAIR	= 1000	1000

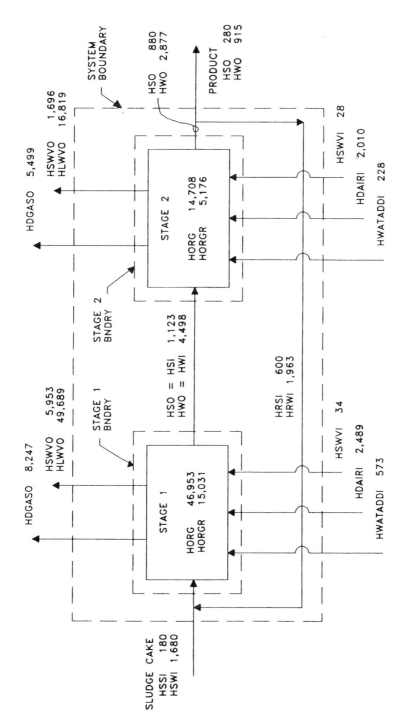

Figure 12.19. Two stage system, biosolids cake — stage and system energy balances for the case of Example 12.4.

TAIR	=	68	68
RHAIR	=	0.5	0.5
WATADD	=	0	0
SMSET	=	0.5	0.55
TSET	=	150	130
XRMIN	=	20000	
XRMAX	=	160000	
SMMIN	=	0.4	
COEFFM	=	0.3	
ELEV	=	0	
TWATADD	=	68	

2. Exit DATIN41B, load CMSYS41B, and run using DATA1221 when prompted for the data file name. The program should close to solution about iteration 995. The complete program output for this case is presented in Appendix F.

A mass balance developed from the simulation model output for Example 12.5 is presented in Figure 12.20. Operation of stage 1 again differs significantly from the Case 2 study because of the setpoint conditions TSET = 150°F (65.6°C) and SMSET = 0.50. To achieve these conditions the air flowrate increases from the minimum input value of 1000 acfm to 2695 acfm. To hold setpoint solids at 50% TS requires the addition of 25,765 lb/day of supplemental water. Actual solids content maintained in stage 1 is 49.91% TS. Both setpoint conditions are maintained within the defined closure conditions.

The importance of supplemental water becomes more significant with dry feed substrates. WATADD to stage 1 is greater than the total water content of the feed substrates. Without this supplemental water, stage 1 would have been driven to moisture limitations as was the case in Example 12.3. The importance of water to the composting process should be clear after study of Figure 12.20.

The reaction rate constant for stage 1 is again increased compared to the case of Example 12.3. RATEKF for the food waste component in stage 1 is about 0.585 per day compared to 0.315 per day for the uncontrolled case of Example 12.3. As a result, the stage 1 output contains less BVS than the single stage output in Figure 12.16. BVS reduction is about 58% across stage 1 in Figure 12.20.

For stage 2, an air flowrate of 1193 acfm with 4528 lb/day of supplemental water is required to achieve setpoint conditions. RATEKF for the food waste component in stage 2 is 0.278 per day which is lower than that in stage 1 because of the reduced operating temperature and higher solids content. BVS reduction across stage 2 is about 33%.

Combined outlet gases from stages 1 and 2 total over 411,000 lb/day, of which over 51,000 lb/day is water vapor. Total gas weight is over 20 times the weight of dry solids in the feed substrate. Two points are of importance here. First, input and output gases remain the dominant terms of the mass balance even with relatively dry substrates. Second, water vapor is a significant component in the exhaust gas even with dry substrates. This is particularly true if supplemental water is added to improve the operating conditions for composting. It is interesting to note that water vapor in the exhaust gases from the system is about twice that contained in the feed substrates. The difference, of course, comes from the added water. This means that most of the supplemental water ends up being evaporated into the exhaust gases. This may seem rather futile, but the supplemental water would not have been evaporated if it did not first improve reaction kinetics and allow for oxidation of more BVS.

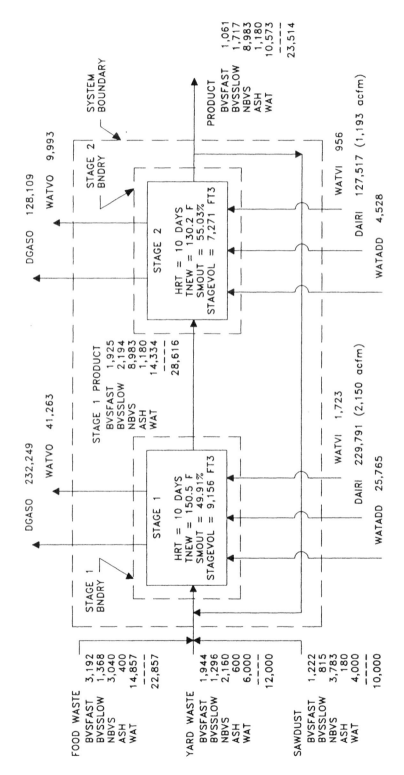

Figure 12.20. Two stage system, mixed compostables — stage and system mass balances for the case of Example 12.5.

SRT for each stage is 10 days and for the total system 20 days. Again, HRT and SRT are the same for a system that does not use product recycle. Addition of stage 2 with its 10 day HRT simply doubles the SRT compared to the single stage system of Example 12.3. With the doubling of system SRT, BVS reduction across the system is increased to about 72%, wet weight of final product is reduced to 52% of the wet weight of the feed substrate, and specific oxygen consumption, RATEO2, for the final product is reduced to about 162 mg O_2/kg VS-h. The product is considerably more stable compared to that of Example 12.3.

The energy balance across the system boundary is presented in Figure 12.21. Energy terms are the same as for Case 2 as shown in Figure 12.17. However, energy input is increased significantly to a total of about 65.8 MBtu/day, compared to about 43.3 MBtu/day for Case 2. Again, the increased heat release is due to the greater BVS reduction in the two stage system.

Adding a second stage and improving the operating conditions within the stages again increases BVS reduction and improves final product stability. The final product from Figure 12.20 would probably be acceptable for some land application projects. In this regard, the added SRT afforded by stage 2 proves to be a useful addition to the system. Further improvements will be developed in the next chapter.

SUMMARY

The CFCM stage is assumed to operate with continuous feed and withdrawal and with reactor contents under complete mix conditions. Such conditions probably cannot be achieved on a practical basis with solid materials. The conditions are approximated, however, by semicontinuous or intermittent feeding and well-mixed conditions. Plug flow or batch operated processes can be modeled as a number of complete mix stages arranged in series. The computer simulation models were introduced by first applying them to one stage systems and then two stage systems. Results of the simulations and their implications for design and operation of actual compost systems were discussed.

Process dynamics for a single stage composting system were studied using two different feed substrates. The first was dewatered biosolids cake conditioned with product recycle. This represented a wet feed substrate. The second was a mixture of compostables, including food waste, yard waste, and sawdust. This mixture was relatively dry by comparison and did not require product recycle for conditioning.

For both substrate types, the first stage is characterized by a high operating temperature at low air supply rates. In this range, reaction rates are limited by the very high stage temperatures. Increasing the air supply removes more heat, which reduces the stage temperature. Reduced temperature increases the reaction rate constant. More BVS are then oxidized which release more heat to drive the "evaporative engine" and increase the output solids content. Eventually the drying action can impose a new rate limitation on the process from lack of moisture. If air supply increases still further, the heat removal rate will increase to the point that process temperatures drop below the optimum point. Further increases in air supply beyond this point cause a decrease in the reaction rate constant and a decrease in stage performance. Thus, there is an optimum air supply rate where the combined limitations of excessively high temperature and lack of moisture are balanced. BVS reduction and output solids content will be maximized at this point. For the examples presented, the optimum air supply rate was in the range of 300 to 400 acfm/dtpd.

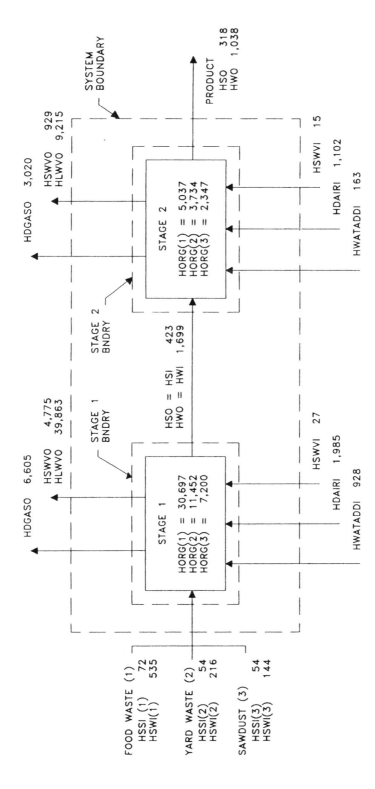

Figure 12.21. Two stage system, mixed compostables — stage and system energy balances for the case of Example 12.5.

Two factors that most affect the reaction rate constant during composting are temperature and moisture. Excessively high temperatures can place the process beyond the optimum temperature region and into the zone of thermal inactivation. Air supply rate is the primary variable for controlling process temperatures. However, increasing the air supply rate to draw out more heat, and thus reduce process temperature, will also remove more moisture from the material. In many cases, loss of moisture by evaporation can lead to additional rate limitations. Design and operation of a composting process must recognize these potential rate limitations so that means are taken to reduce their impact.

Air and exhaust gases are the largest components of the mass balance by a considerable margin. Dry air input is usually many times the dry weight of feed substrates. While solid materials are more obvious to the eye, the major material handled at a compost plant is air and exhaust gas. Exhaust gas will usually contain a high percentage of water vapor. Care must be taken when handling the exhaust gas because of the tendency for the water vapor to condense.

The energy balance is dominated by a relatively few energy terms. On the input side, heat release from organic decomposition is the most significant term. On the output side, the latent heat of water vaporization dominates the output terms, usually accounting for over 65% of total output energy. Sensible heat in the hot exhaust gases and water vapor are the next highest energy outputs.

HRT within the stage is an important parameter that affects composting temperature, output solids content, and stabilization of BVS. At a given air supply rate, all of these factors are predicted to increase with increasing HRT over the range from 5 to 20 days. Process stability is favored and the "operating window" enlarged by longer HRT. The operating window is the range of process parameters within which the process will operate in a stable manner.

There are a number of reasons for developing a compost system from multiple stages arranged in series. The multiple stages can be operated under different conditions to reduce rate limitations and achieve the desired properties of moisture content and stability in the final product. Systems composed of two stages in series were evaluated for both substrate cases. Each stage was operated with setpoint conditions for temperature and moisture content. Setpoint temperature was maintained by adjusting the air supply rate and setpoint moisture content by adjusting supplemental water addition. Addition of the second stage, combined with improved operating conditions within the individual stages, increased BVS reduction and improved final product stability.

CHAPTER 13

Process Dynamics III — Building the Complete System

INTRODUCTION

In the previous chapter the reader was introduced to the CFCM stage. Systems were developed with one stage and then two stages arranged in series. The one and two-stage systems were applied to the composting of sludge cake and mixed compostables. For both substrate types the addition of the second stage, combined with improved operating conditions, increased BVS reduction and improved final product stability. In both cases, however, the final product was not sufficiently stable to be directly marketed for high end uses. Further work is necessary to develop a complete system that achieves the desired product.

The purpose of this chapter is to build on results of Chapter 12 and develop a complete compost system that achieves the desired end product. To accomplish this it is necessary to establish the desired product stability and then build together the required number of system stages to accomplish the objective. This chapter will take the reader from single and two stage systems to the multistage systems required to effectively model a complete composting process. The reader will learn that historical divisions of composting into high rate and curing phases have very fuzzy boundaries. Composting is a process of organic decomposition that begins with substrates rich in BVS and proceeds gradually and continually to a stable product with greatly reduced BVS content. There are no sharp boundaries or discontinuities between high-rate and curing phases. The two must be viewed together as part of a complete, multi-stage system.

CASE 5 — BIOSOLIDS CAKE

Reactor System

We begin the building process by first considering a reactor system for composting of sludge cake. In the initial case, the intent is to directly use the product from the reactor. Later, curing stages will be added after the reactor.

Example 13.1

Based on your success with the two stage system of Example 12.4 you have been promoted to Chief Composter. Your client decides to invest in a circular, agitated bed reactor system. The reactor is to be divided into five aeration zones, each with a 4-day HRT. The client produces 10 dtpd of biosolids cake at 25% TS and desires to produce a very stable compost to expand the markets for the end product. You are concerned that the energy balance may be marginal for this sludge cake. You decide to add sawdust as a structural and energy amendment. 20,000 lb/day of sawdust will be added at 60% TS. Sawdust and recycle from the reactor outfeed will be used to condition the sludge cake. Air supply to each zone of the reactor will be controlled by a temperature feedback control loop to maintain a 150°F (65.6°C) setpoint.

Solution

1. After some study you establish the following design criteria to judge final product quality:
   a. a specific oxygen uptake rate of 50 mg O_2/kg VS-h or less
   b. a minimum 90% BVS reduction across the system
   c. a decrease in composting temperatures at the end of the process
   d. a final solids content above 55% to avoid marketing a wet product but below 65% TS to avoid dusty material

2. Load DATIN41B and follow the menu to create a data file named DATA1301. Input the following data:

MDY	= (enter current value)	
IDEN	= EXAMPLE 13.1	
ME	= E (for English units)	
SUBNUM	= 2	
STAGENUM	= 5	
STAGER	= 5	
SUBNAME	= SLUDGE	SAWDUST
X	= 80000.0	20000.0
S	= 0.25	0.60
V	= 0.80	0.95
K	= 0.65	0.45
FASTFRAC	= 0.40	1.0
SLOWFRAC	= 0.60	0
H	= 9500	7750
T	= 68	68
RATEKM20F	= 0.015	0.01
RATEKM20S	= 0.005	0.01
COEFF	= 0	0.17
A	= 10	295
B	= 19	420
C	= 3	186
D	= 1	1

		STAGE 1	STAGE 2	STAGE 3	STAGE 4	STAGE 5
HRT	=	4	4	4	4	4
QAIR	=	500	500	500	500	500
TAIR	=	68	68	68	68	68

RHAIR	=	0.5	0.5	0.5	0.5	0.5
WATADD	=	0	0	0	0	0
SMSET	=	1.0	1.0	1.0	1.0	1.0
TSET	=	150	150	150	150	15

XRMIN	=	20000
XRMAX	=	160000
SMMIN	=	0.40
COEFFM	=	0.30
ELEV	=	0
TWATADD	=	68

The above values for sludge are the same as for Example 12.1. The sawdust data assumes a reasonably degradable wood species.

3. Exit DATIN41B, load CMSYS41B, and run using DATA1301 when prompted for the data file name. The program should close at about iteration number 542.

Process temperatures and air flowrates to each stage of the reactor are presented in Figure 13.1. Compost temperatures are maintained at the setpoint conditions for all stages, namely 150°F (65.6°C). The air flowrate to maintain these temperatures is about 2050 to 2150 acfm for stages 1 and 2. For stages 3, 4, and 5 the required air flowrate decreases gradually to about 1050 acfm for stage 5.

Predicted solids content for the material within each stage is presented in Figure 13.2. Reactor output (stage 5) is predicted to be almost 72% TS. This is a very dry condition. The process is driving itself to extreme dryness because the feed mixture is "energy rich". The problem of lack of moisture goes beyond just the obvious problem of dust. The rate of biological decomposition is adversely effected by the lack of moisture. With an abundance of energy the process first drives itself to extremely high temperatures, so high that the process kinetics become temperature limited. The operator attempts to control this by adjusting the air supply to remove heat and throttle the temperatures to a more reasonable range. By so doing, however, the rate of moisture removal increases. The process then drives itself to extreme dryness, and reaction rates become limited by lack of moisture. This is the condition in which the compost plant of Example 13.1 finds itself.

The effect of moisture on the rate constant in each stage is also shown in Figure 13.2. Stage 1 is at about 52% TS (48% moisture) and the factor F1 is over 0.8. By stage 5, however, the moisture content is reduced to only 28% and F1 is about 0.12. In other words, the reaction rate is only 12% of what it would be if moisture were plentiful.

With the slowdown in reaction rate, stability of the final product suffers. Specific oxygen uptake rate is presented in Figure 13.3. The combined sludge and sawdust substrates are predicted to have a specific oxygen uptake of 532 mg O_2/kg VS-h. The product from stage 5 is predicted to have an uptake of about 250 mg O_2/kg VS-h. This is a very high rate and the stage 5 output is far from the desired stability. BVS reduction across the system is only 66%.

Each stage of the process operates with a 4-day HRT. With a total of five stages, the single pass HRT is 20 days. However, the SRT is considerably higher because of the recycled product. Cumulative SRT based on dry weight, SRTDRYWT, at the output from each stage is presented in Figure 13.3. System SRT is about 33 days.

The reactor in Example 13.1 is not operated effectively because reaction rates are significantly reduced by the lack of moisture. Other more serious problems can also result from this type of operation. The combination of high temperatures and low moisture can present a very

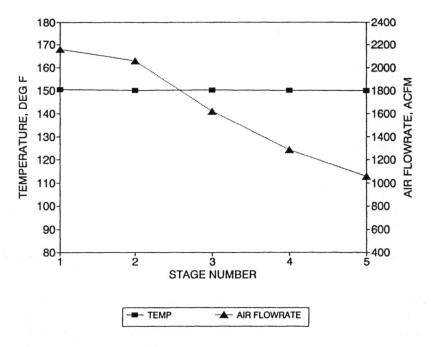

Figure 13.1.   Biosolids/sawdust — profiles of temperatures and air flowrates for reactor system of Example 13.1; no supplemental water addition.

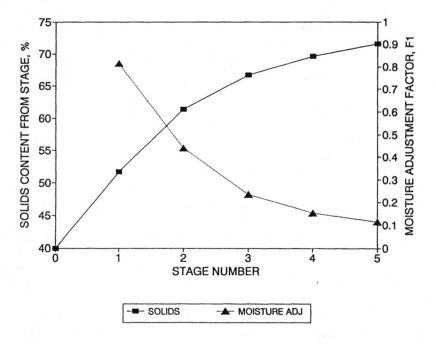

Figure 13.2.   Biosolids/sawdust — profiles of solids content and moisture adjustment factor for the reactor system of Example 13.1; no supplemental water addition; stage 0 represents the mixed feed to stage 1.

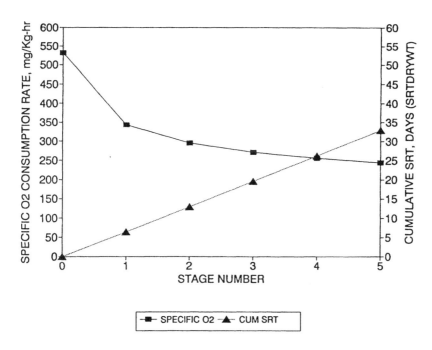

Figure 13.3.   Biosolids/sawdust — profiles of specific O_2 consumption rate and cumulative SRT for the reactor system of Example 13.1; no supplemental water addition; stage 0 represents the mixed feed to stage 1.

real fire hazard. Such conditions have actually caused fires at a number of reactor, static pile and windrow systems. Therefore, actions must be taken to improve the reactor performance.

Example 13.2

Based on your success with the two stage system in Chapter 12, you decide to add water to the five stage reactor of Example 13.1. A 50% setpoint solids content will be maintained in stages 1, 2, and 3. The setpoint will be increased to 55% TS in stages 4 and 5, anticipating the need to have a drier product to keep the recycle rate within reasonable limits.

Solution

1. Load DATIN41B and follow the menu to duplicate data file DATA1301 under the new file name DATA1302. Continue to follow the menu and modify DATA1302 by adding SMSET values to stages 1 through 5 as follows:

	STAGE 1	STAGE 2	STAGE 3	STAGE 4	STAGE 5
SMSET =	0.50	0.50	0.50	0.55	0.55

2. Exit DATIN41B, load CMSYS41B, and run using DATA1302 when prompted for the data file name. The program should run to the maximum iteration limit of 2000. Nevertheless, closure is sufficiently good for engineering purposes.

Referring to Figure 13.4, water addition to the reactor stages has the effect of slightly reducing the air flowrates needed to maintain temperature setpoints of 150°F (65.6°C). The

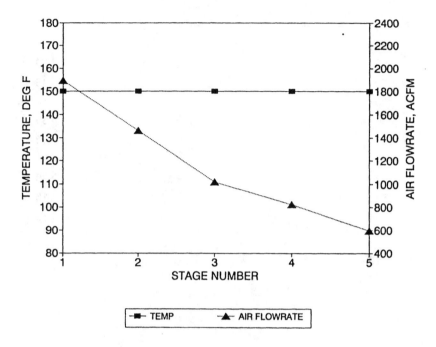

Figure 13.4. Biosolids/sawdust — profiles of temperatures and air flowrates for the reactor system of Example 13.2; with supplemental water addition.

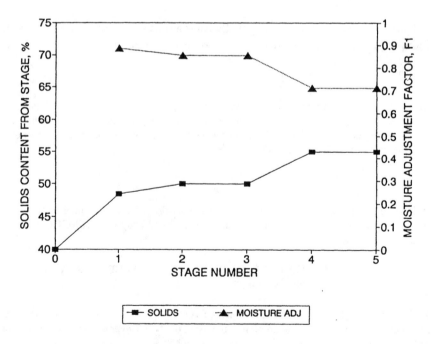

Figure 13.5. Biosolids/sawdust — profiles of solids content and moisture adjustment factor for the reactor system of Example 13.2; with supplemental water addition. Stage 0 represents the mixed feed to stage 1.

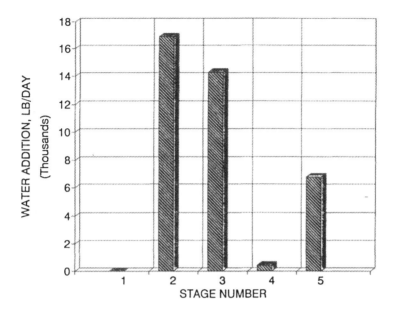

Figure 13.6.   Biosolids/sawdust — daily water addition to achieve the setpoint moisture conditions for the reactor system of Example 13.2.

most dramatic effect is on the moisture adjustment factor, F1, as shown in Figure 13.5. F1 remains above 0.7 in all stages of the reactor. This is a dramatic improvement compared to that of Figure 13.2. The water addition rate to each stage needed to maintain moisture setpoint conditions is shown in Figure 13.6. Stage 1 requires no water addition because it operates below 50% TS. However, stages 2 and 3 require about 16,900 and 14,300 lb water/day, respectively. Stage 4 requires only 400 lb/day because the setpoint is increased from 50 to 55% TS. The water rate increases again for stage 5 to about 6800 lb/day. Total water addition is about 38,000 lb/day, which is over 60% of that supplied with the biosolids cake.

Operators at sludge composting facilities may be reluctant to add water back to the process, particularly when one of their duties is to remove as much water from the sludge as possible during dewatering. The idea may seem unusual with wet substrates such as sludge cake. Nevertheless, supplemental water addition is theoretically well founded and is the most direct approach to solving moisture limitations and dust problems which are likely with "energy rich" substrates.

The reactor in Example 13.2 operates with much higher rate constants, which is reflected in the product stability presented in Figure 13.7. Product from stage 5 is predicted to have a specific O_2 uptake of 86 mg/kg VS-h. This is a reasonably stable product and probably could be used in many applications. However, it does not meet the design stability of 50 mg/kg VS-h. Also, this stability is probably near the lower limit achievable by the reactor system. Allowance must be made for the fact that real operation will always deviate from the ideal. In practice, stage setpoints cannot be maintained with the precision possible in the computer model. Therefore, the actual stability will likely be higher than that predicted here. BVS reduction across the system increases to 89% from the 66% predicted for Example 13.1 with no supplemental water addition.

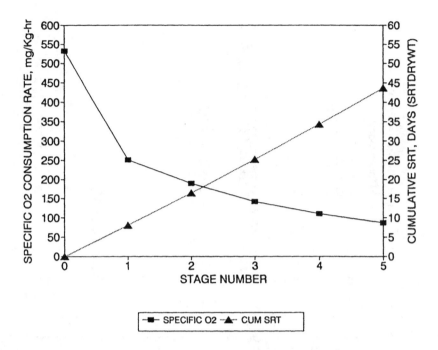

Figure 13.7.   Biosolids/sawdust — profiles of specific O_2 consumption rate and cumulative SRT for the reactor system of Example 13.2; with supplemental water addition; stage 0 represents the mixed feed to stage 1.

Reactor System with Curing

Despite efforts to improve reactor performance, the final product does not yet meet the desired stability. Curing stages will be needed to "polish" the reactor outfeed.

Example 13.3

You have decided to add an aerated curing system to further stabilize and dry the partially stabilized compost from the reactor. To keep costs down you plan to use continuous piles of material with small aeration blowers and agricultural drainage pipe for distribution. You are unsure of the time required so you decide to model the curing process as four stages in series, each with a 15 day HRT. Recycle for feed conditioning will be from the output of the last stage. You are also unsure of the required air supply rate for curing. You decide to establish a setpoint temperature of 115°F (46.1°C) to provide better conditions for maturation of the material. The program will then determine the necessary air flowrates. A flowrate of 1000 acfm will be used as the minimum value.

Solution

1. Load DATIN41B and follow the menu to duplicate data file DATA1302 under the new file name DATA1303. Continue to follow the menu and modify DATA1303 by adding the follows data for new Stages 6 through 9:

		STAGE 6	STAGE 7	STAGE 8	STAGE 9
STAGENUM	=	9			
HRT	=	15	15	15	15
QAIR	=	1000	1000	1000	1000
TAIR	=	68	68	68	68

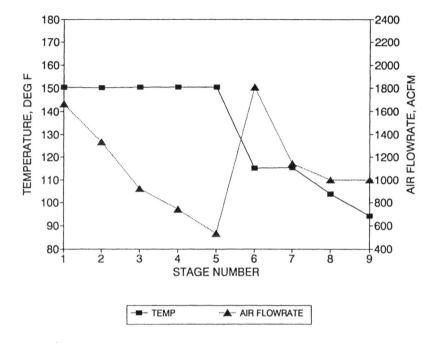

Figure 13.8.   Biosolids/sawdust — profiles of temperatures and air flowrates for the reactor/curing system of Example 13.3.

RHAIR	=	0.5	0.5	0.5	0.5
WATADD	=	0	0	0	0
SMSET	=	1.0	1.0	1.0	1.0
TSET	=	115	115	115	115

2. Exit DATIN41B, load CMSYS41B, and run using DATA1303 when prompted for the data file name. The program should close at iteration 1023.

Process temperatures throughout the system are presented in Figure 13.8. Setpoint temperatures of 150°F (65.6°C) are maintained in reactor stages 1 through 5. The 115°F (46.1°C) setpoint is maintained in curing stages 6 and 7. However, the setpoint cannot be maintained in stages 8 and 9 with the minimum air flowrate of 1000 acfm. This is a good sign that the oxygen consumption rate is decreasing, which in turn reduces the heat generation rate. Process temperatures should tend back toward ambient near the end of the curing phase.

Air flowrates in the reactor stages are slightly reduced from the case of Example 13.2 because the recycled product is more stable. The seemingly large increase in air flowrate to stage 6 is caused by the fact that 15 days of material are under aeration compared to only 4 days for stages 1 through 5. The air flowrate per unit weight of material does not increase.

Output solids content from each stage and the corresponding moisture factor, F1, are presented in Figure 13.9. Stage 5 output is maintained at the 55% TS setpoint. From here, the solids content increases throughout curing to a final product from stage 9 at about 64% TS. This should be a good quality product from the standpoint of moisture. The moisture factor, F1, remains high in the reactor stages where water is being added. No water is added during curing and F1 decreases accordingly.

Specific O_2 consumption rate and cumulative SRT are presented in Figure 13.10. The design target of 50 mg O_2/kg VS-h is achieved in the product from stage 9. A system SRT of 163 days is required to reach this level of stability. Specific O_2 rate is presented as a function

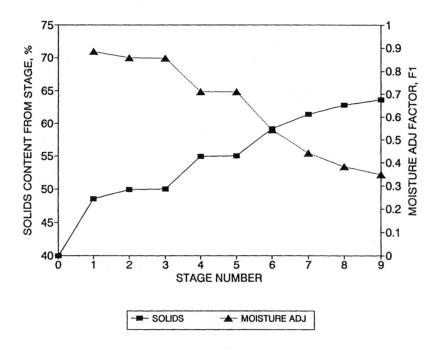

Figure 13.9.   Biosolids/sawdust — profiles of solids content and moisture adjustment factor for the reactor/curing system of Example 13.3; stage 0 represents the mixed feed to stage 1.

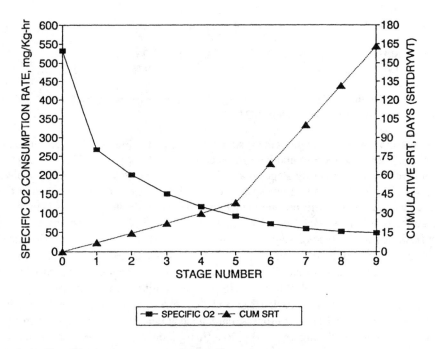

Figure 13.10.   Biosolids/sawdust — profiles of specific O_2 consumption rate and cumulative SRT for the reactor/curing system of Example 13.3.

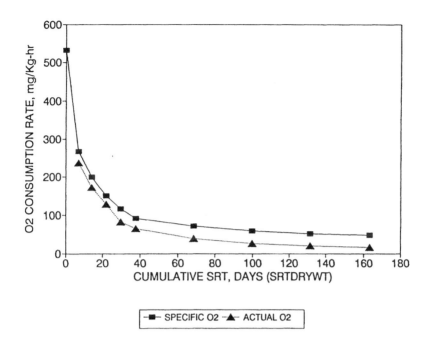

Figure 13.11. Biosolids/sawdust — specific and actual O_2 consumption rates as a function of cumulative SRT for the reactor/curing system of Example 13.3.

of SRT in Figure 13.11. The reactor stages have a cumulative SRT of 38 days. The rapid decrease in consumption rate within the reactor stages is evident from Figure 13.11. A relatively long time is required for curing because (1) the "fast" fraction of BVS has decreased to low levels so that further reduction is slowed per the first order rate assumption, (2) the greater quantities of "slow" fraction that remain are characterized by a slower reaction rate, and (3) the reaction rate constants are decreasing because of the lower temperatures and reduced moisture content. For example, the reaction rate constant for the "fast" fraction of sludge in stage 1 is 0.18 per day. By stage 9 the same constant is 0.012 per day, about 15 times lower.

Also presented in Figure 13.11 is the predicted actual O_2 consumption rate for the "as is" material. Recall that specific oxygen consumption is assumed to be measured under laboratory conditions without moisture and oxygen rate limitations. Actual consumption rate for the "as is" material is estimated by multiplying the specific consumption rate by the moisture adjustment factor. Obviously, actual consumption rate will always be less than the specific consumption rate. Specific consumption rate is the better measure of actual stability. An artificial "stability" can be achieved by simply drying the feed substrates, but the organics have not been stabilized by the meaning used in the composting industry.

The distinction between specific and actual consumption rate is an important consideration. Both rates are presented in Figure 13.12 for stages 5 through 9 which correspond to the range of SRTs from 38 to 163 days. Actual stability reaches the design target in the output from stage 6 after about 70 days SRT. Specific stability does not reach the design target until stage 9 after about 163 days SRT. With the current state of the art, it is recommended that specific consumption rate be used as the design criteria for measuring product stability. Longer curing times are required, but until more is known about the relationship between oxygen consumption rate and product stability, it is better to be conservative.

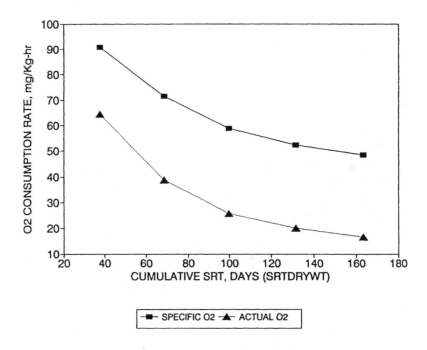

Figure 13.12. Biosolids/sawdust — specific and actual O_2 consumption rates for cumulative SRTs above 38 days for the reactor/curing system of Example 13.3.

BVS reduction across the system is 94%. Therefore, the reactor/curing system evaluated in Example 13.3 satisfies the 90% BVS criterion. Recall that BVS reduction was only 66% for the reactor alone with no water addition to reduce moisture rate limitations. With supplemental water, the reactor alone achieved 89% BVS reduction. Good operating practices combined with a curing phase increases BVS reduction to 94%.

One of the features of the simulation model is a prediction of outlet NH_3 concentrations. Recall that the model is based on the production and synthesis of NH_3. Any NH_3 remaining after the synthesis need is satisfied is assumed to report to the vapor phase. Predicted NH_3 concentrations are presented in Figure 13.13. Concentrations within the reactor exhaust are predicted to be above 800 ppm. Concentrations in the curing exhaust range from about 100 to 250 ppm. Concentrations as high as 1000 ppm have been noted in practice, so the predicted range is not out of the question. Nevertheless, the concentrations in Figure 13.13 should be interpreted as an upper bound until the ammonia model is further refined.

The volume required for each stage of the process is presented in Figure 13.14. Total system volume is 175,000 ft³. Of this, about 48,000 ft³ is required for the reactor and about 127,000 ft³ for curing. Assuming an 8-ft diameter center well for the circular system and 6 ft of compost depth, a 102-ft diameter reactor is required. If curing is conducted in continuous piles 8 ft high, about 0.4 acres of active piles will be needed. If individual piles or windrows are used, the area will increase accordingly.

Reactor volume is generally more expensive than curing volume, particularly if the latter is conducted outside. To minimize costs the designer wants to reduce the size of the reactor. On the other hand, the operator wants a large reactor to improve his "window of operability" and increase the portion of the process over which he has most control. The current design appears to strike a reasonable balance between these competing demands.

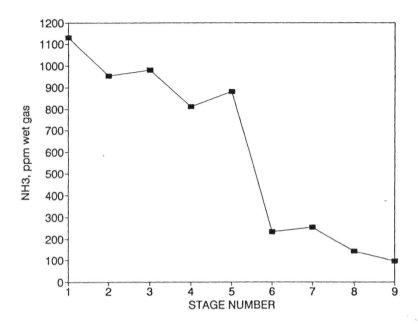

Figure 13.13. Biosolids/sawdust — predicted NH$_3$ concentrations in the outlet gases from the reactor/curing system of Example 13.3.

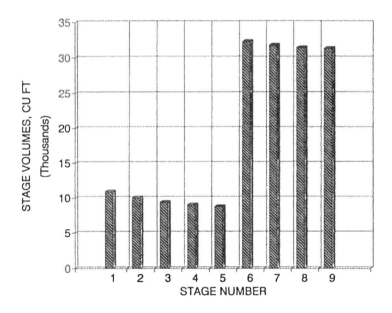

Figure 13.14. Biosolids/sawdust — stage volumes for the reactor/curing system of Example 13.3.

A mass balance for the system of Example 13.3 is presented in Figure 13.15. Balances for each stage are not shown because of the wealth of data required. However, the complete system and stage balances are appropriate for presentation on the flow sheets of a design package, for example, and can be prepared from the computer printout. Referring to Figure 13.15, it is interesting to again note that gases dominate the mass terms. The weight of final

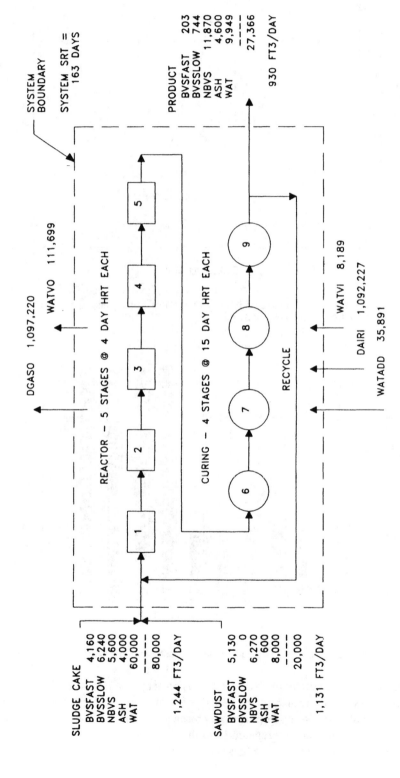

Figure 13.15. Biosolids/sawdust — mass balance for the reactor/curing system of Example 13.3.

product is about 27% of the wet weight of feed substrates. There is also a significant volume reduction from the starting substrates to final product.

The conceptual design of Example 13.3 achieves the design criteria established for product quality. Specific oxygen uptake rate is 50 mg O_2/kg VS-h or less. BVS reduction is greater than 90%. Process temperatures return to near ambient at the end of the curing phase and final solids content is above 55% TS but below 65% TS.

Further Design Improvements

The analysis conducted to this point represents a reasonable starting point for design. The reactor and curing areas can be sized based on the volume requirements. Air supply systems can be designed based on the predicted aeration demands. However, a number of other ideas could be explored prior to finalizing the conceptual design. For example, one might look at the possibility of recycling from the reactor output (stage 5) instead of from the final product (stage 9). This would have the effect of increasing the reactor size, but it may reduce the required curing area. One should also look at different operating parameters of moisture and temperature to make sure that the system performs adequately under reasonable worst case conditions. Extremes of operation should be investigated, including biosolids production and cake solids content, as well as extremes of weather. The tools for such analysis are provided here, the rest is left to the reader. Remember that the goal of computer modeling is to prepare a technically sound, conceptual design which then serves as the basis for detailed design.

CASE 6 — MIXED COMPOSTABLES

Attention will now turn to the mixed compostables of Examples 12.3 and 12.5. Having learned a few lessons from the previous exercises, we will be able to move quickly to develop a complete system.

Example 13.4

Your firm, The No-Smell Compost Co., has been retained to develop a conceptual design for a system to compost the mixed materials of Examples 12.3 and 12.5. The client requires that you "guarantee" no odors from the facility. Because food wastes are involved, you elect to use a reactor first phase followed by second phase curing. You select an agitated bed, bin reactor with an 18-day HRT. You propose to provide 60 days of supplemental curing using aerated windrows. You believe that agitation from windrow turning will reduce the particle size and improve the quality of the final product. The periodic turning will also allow supplemental water addition during the early stages of curing. Air temperatures will be 40°F (4.4°C) to simulate fall operation. Evaluate this concept.

Solution

1. For purposes of evaluation, the 18-day reactor will be segmented into six stages, each with a 3-day HRT. The 60-day curing will be divided into 6 stages, each with a 10-day HRT.

2. Setpoint temperatures in the reactor stages will be 131°F (55°C). You are not as concerned about maintaining higher temperatures because no human wastes are involved. Setpoint temperatures will also be 131°F (55°C) for curing stages 7, 8, and 9 and 115°F (46.1°C) for stages 10, 11, and 12.

3. Solids content within the reactor will be maintained at 50% TS. Water addition will also be used in curing stages 7, 8, and 9 to maintain 55% TS. No water will be added to stages 10, 11, and 12 to increase the final product solids content.

4. Product design criteria will be the same as developed for Example 13.3.

5. Load DATIN41B and follow the menu to create the following data file named DATA1304.

MDY	=	(enter current value)		
IDEN	=	EXAMPLE 13.4		
ME	=	E (for English units)		
SUBNUM	=	3		
STAGENUM	=	12		
STAGER	=	0		

SUBNAME	=	FOOD WASTE	YARD WASTE	SAWDUST
X	=	22857	12000	10000
S	=	0.35	0.5	0.6
V		0.95	0.9	0.97
K	=	0.60	0.6	0.35
FASTFRAC	=	0.7	0.6	0.6
SLOWFRAC	=	0.3	0.4	0.4
H	=	9500	7500	7500
T	=	68	68	68
RATEKM20F	=	0.05	0.01	0.01
RATEKM20S	=	0.005	0.005	0.005
COEFF	=	0.26	0.30	0.25
A	=	16	23	295
B	=	27	38	420
C	=	8	17	186
D	=	1	1	1

		STG 1	STG 2	STG 3	STG 4	STG 5	STG 6
HRT	=	3	3	3	3	3	3
QAIR	=	200	200	200	200	200	200
TAIR	=	40	40	40	40	40	40
RHAIR	=	0.5	0.5	0.5	0.5	0.5	0.5
WATADD	=	0	0	0	0	0	0
SMSET	=	0.50	0.50	0.50	0.50	0.50	0.50
TSET	=	131	131	131	131	131	131

		STG 7	STG 8	STG 9	STG 10	STG 11	STG 12
HRT	=	10	10	10	10	10	10
QAIR	=	200	200	200	200	200	200
TAIR	=	40	40	40	40	40	40
RHAIR	=	0.5	0.5	0.5	0.5	0.5	0.5
WATADD	=	0	0	0	0	0	0
SMSET	=	0.55	0.55	0.55	1.0	1.0	1.0
TSET	=	131	131	131	115	115	115

XRMIN	=	0
XRMAX	=	50000
SMMIN	=	0.40

COEFFM   =   0.30
ELEV   =   0
TWATADD   =   40

6. Exit DATIN41B, load CMSYS41B, and run using DATA1304 when prompted for the data file name. The program should close at about iteration 1102.

Profiles of process temperatures and required air flowrates are presented in Figure 13.16. Setpoint temperatures are maintained in reactor stages 1 through 6 and curing stages 7 through 10. Only stages 11 and 12 fall below the 115°F (46.1°C) setpoint. Again, this is a good sign that the composting cycle is nearing completion with the return to ambient temperatures. Air flowrate declines along the length of the reactor bin and increases sharply from stages 6 to 7. Again, this is caused by the fact that 10 days of material is under aeration in stage 7 compared to 3 days in stage 6.

Profiles of solids content and moisture adjustment factor are presented in Figure 13.17. Setpoint moisture is maintained in stages 1 through 9 which have setpoint conditions. Solids content increases to almost 60% TS in stage 12. Water addition rates needed to maintain these conditions are presented in Figure 13.18. Because the feed mixture is 44.6% TS compared to the 50% TS setpoint, water addition to stage 1 is slightly less than that to stage 2. Following stage 2, there is a rather uniform drop in the water addition rate along the length of the bin. Supplemental water addition to curing stages 7, 8 and 9 proves beneficial because it increases the rate constant over that which would otherwise occur.

Specific O_2 consumption and cumulative SRT for the process stages are presented in Figure 13.19. Note the relatively high oxygen consumption rate of the feed substrates (stage 0). This is due largely to the high reaction rate constant assumed for the food waste. Oxygen consumption and BVS reduction as functions of cumulative SRT are presented in Figures 13.20 and 13.21. The output from stage 10 is actually below the design criteria of 50 mg O_2/kg VS-h. For the sake of being conservative, however, curing stages 11 and 12 should be retained in the system. Additional curing is certainly beneficial and product solids are increased by about two percentage points.

BVS reduction across the reactor is 69%. Significant additional reduction occurs during curing, raising the total system BVS reduction to 91%. Again, this points out the importance of the curing phase to overall system performance.

It is interesting to note that the required system SRT is about 78 days for Example 12.4 compared to the 163 days of Example 12.3. The large "fast" fraction and high rate constant for the food waste substrate are largely responsible for this result. It points out that knowledge of the feed substrates is absolutely essential to the proper design of a composting facility. Degradability and reaction rate constants should be known for each substrate. These are the values upon which the simulation model is built. For the two cases presented here, the system SRTs to produce essentially the same end product differ by a factor of 2 because of differences in the feed substrates.

Predicted NH_3 concentrations in the exhaust gases are presented in Figure 13.22. The concentrations are considerably less than presented above for sludge cake because of the lower nitrogen content of the feed substrates. The threshold odor concentration of NH_3 is about 40 ppm. On this basis, exhaust air from the curing stages would not be a source of NH_3 odors. Exhaust gases from the reactors could contain several hundred ppm NH_3, which would be a concern both from the standpoint of odor and personnel safety.

The volume requirement for each process stage is presented in Figure 13.23. Total volume requirement is about 52,150 ft³, of which 14,900 ft³ is for the reactor bins and 37,250 ft³ for

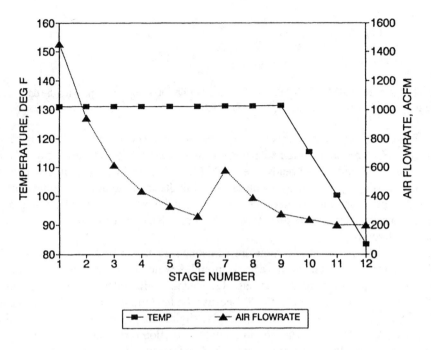

Figure 13.16. Mixed compostables — profiles of temperatures and air flowrates for the reactor/curing system of Example 13.4.

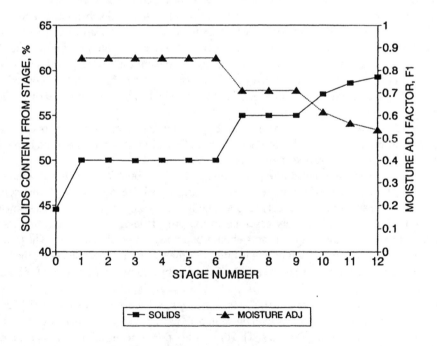

Figure 13.17. Mixed compostables — profiles of solids content and moisture adjustment factor for the reactor/curing system of Example 13.4; stage 0 represents the mixed feed to stage 1.

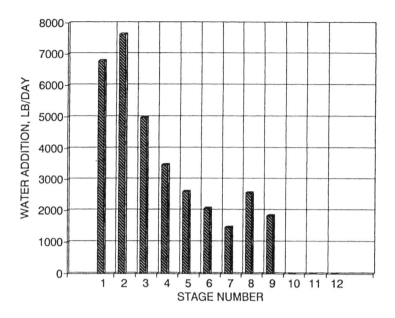

Figure 13.18. Mixed compostables — daily water addition to achieve the setpoint moisture conditions for the reactor/curing system of Example 13.4.

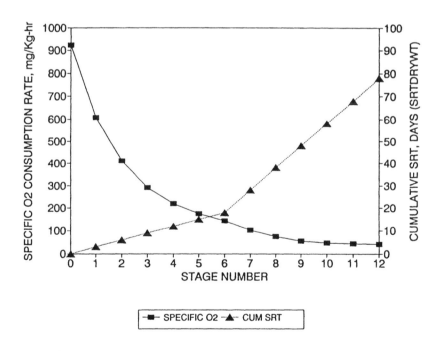

Figure 13.19. Mixed compostables — profiles of specific O_2 consumption rate and cumulative SRT for the reactor/curing system of Example 13.4; stage 0 represents the mixed feed to stage 1.

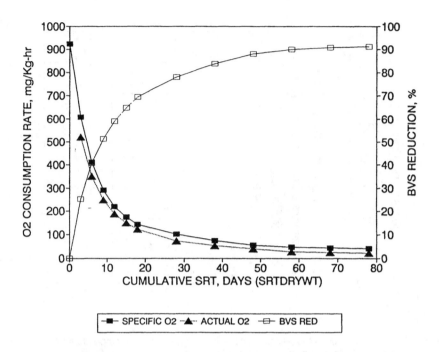

Figure 13.20. Mixed compostables — specific and actual O_2 consumption rates and BVS reduction as functions of cumulative SRT for the reactor/curing system of Example 13.4.

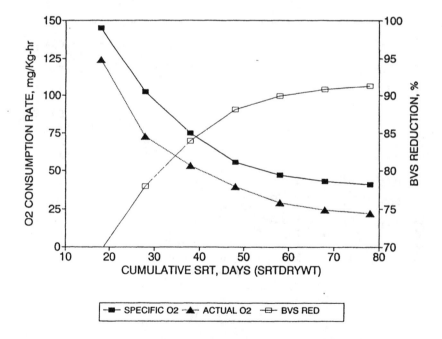

Figure 13.21. Mixed compostables — specific and actual O_2 consumption rates and BVS reduction as functions of cumulative SRTs above 18 days for the reactor/curing system of Example 13.4.

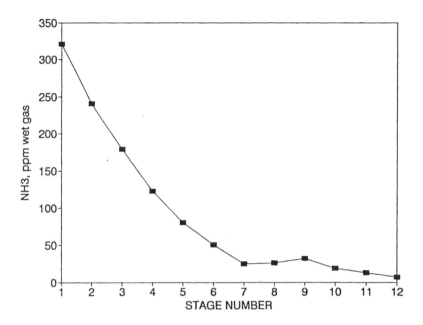

Figure 13.22. Mixed compostables — predicted NH_3 concentrations in the outlet gases from the reactor/curing system of Example 13.4.

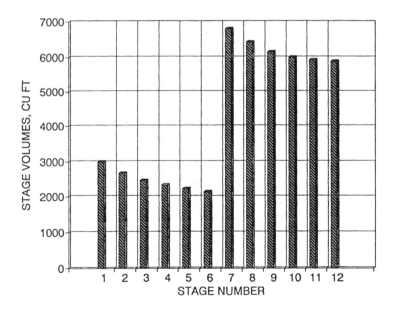

Figure 13.23. Mixed compostables — stage volumes for the reactor/curing system of Example 13.4.

curing. Typical dimensions for bins of this type are a width of 6 ft with an effective compost depth of about 5 ft. Using these dimensions, about 497 ft of bin length are required. Material is typically moved about 10 ft along the bin with each turning. Assuming daily turning and an 18-day HRT, each bin should be about 180 ft long. Three bins at 180 ft each, for a total 540 ft of bin length, would be a good choice. The extra length will compensate for the entrance and exit zones which usually have less material.

Curing windrows typically have a trapezoidal to triangular cross section. With standard-sized turning equipment the piles will be about 12 ft at the bottom and about 4.5 to 5 ft tall. Allowing 6 ft between windrows, the windrow volume per acre is about 70,000 ft³/acre. Therefore, the required windrow curing area is about $37,250/70,000 = 0.53$ acres, excluding access roads and staging areas.

A mass balance for the system of Example 13.4 is presented in Figure 13.24. Again, balances for each stage are not shown because of the wealth of data required. Even with relatively dry feed substrates, gases dominate the mass terms. The weight of final product is about 41% of the wet weight of feed substrates. There is also a significant volume reduction from the substrates to final product. It is interesting to note that the weight of supplemental water added to the process exceeds the weight of any single substrate. Supplemental water is 75% of the total weight of all substrates. Obviously, this is too important a term to ignore in any system balance. Facility designs must recognize the significant water demand imposed by composting, particularly with dry substrates.

In summary, the design presented in Example 12.4 is technically sound. Opportunities abound for further optimization, but the basic layout of Example 12.4 is a good conceptual design. All four product criteria are met by the system. Namely, specific O_2 consumption is less than 50 mg O_2/kg VS-h, BVS reduction is above 90%, process temperatures return toward ambient, and the final product is between 55 and 65% TS.

SUMMARY

Complete composting systems were developed and modeled as a series of CFCM stages arranged in series. For the two cases presented in this chapter, a reactor system was followed by a series of curing stages. The number of stages is determined largely by the characteristics of the system. In one case, the reactor was modeled by five stages to correspond to the number of aeration zones. In the other, six stages were used, each with a 3-day HRT, to model a reactor HRT of 18 days. Longer HRTs of 10 and 15 days were used for the curing stages because reaction rates are lower, which makes more detailed modeling unnecessary. In general, the number of stages should be adjusted to provide sufficient detail within the process to form a basis for further detailed design.

The first step in the recommended design approach is to establish the desired product quality. Four criteria were used for the examples of this chapter: (1) specific O_2 consumption rate should be 50 mg O_2/kg VS-h or less; (2) BVS reduction should be 90% or above; (3) process temperatures should return toward ambient near the end of the process; and (4) product solids content should be above 55% TS to avoid marketing a wet material, but below 65% TS to avoid a dusty product. Having established these product criteria, the number of stages can be varied to adjust the system SRT. The operating parameters can also be adjusted until the design criteria are achieved. The initial conceptual design is complete once the desired product quality is achieved. The initial concept can then be optimized and its operation

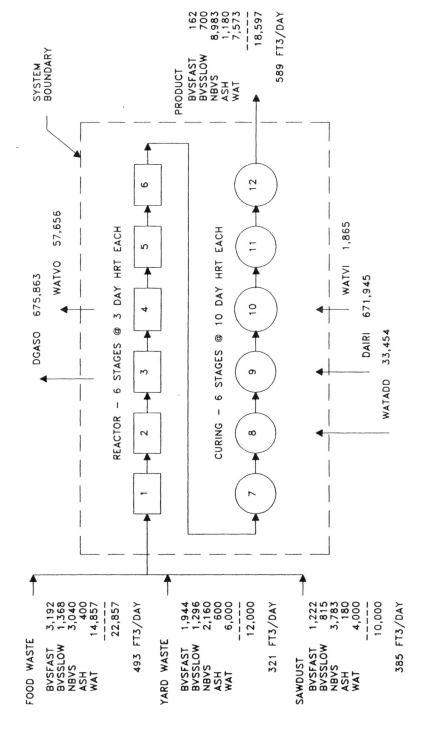

Figure 13.24. Mixed compostables — mass balance for the reactor/curing system of Example 13.4.

simulated under a variety of conditions. In this way the designer can mold the system into a proposed concept which then serves as the basis for detailed design.

The simulation model provides useful predictions of air flowrates, water addition requirements, stage volumes, and hydraulic and solids retention times necessary to achieve the design stability. Complete mass and energy balances are also developed for each stage and the total system. Like any simulation model, however, the predictions are only as good as the input data. Knowledge of the feed substrates is absolutely essential to the proper design of a composting facility. Degradability and reaction rate constants should be known for each substrate. These are the values upon which simulation models are built.

Achieving the product design criteria normally requires that the system be operated to reduce the effects of kinetic rate limitations. This means that (1) stage temperatures should be controlled to avoid the lower reaction rates that occur with both excessively high and low temperatures and (2) supplemental water should be added as necessary to avoid moisture rate limitations. Even with good operation, product from the reactor alone will not generally achieve the desired product stability. Supplemental curing is usually required. For the cases studied here, the desired product quality was achieved with good operating practices combined with a supplemental curing phase.

The mass balance terms for both conceptual systems developed here are again dominated by the gas components. Process gases were 12 times the weight of feed substrates for the case of biosolids cake/sawdust and 16 times for mixed compostables. The requirement for supplemental water to reduce moisture limitations must not be underestimated, particularly with dry substrates. In the case of mixed compostables, supplemental water was 75% of the total weight of all substrates.

The SRT is an important parameter of system design. For the cases presented here, system SRTs to produce essentially the same end product were 78 and 163 days, a difference of a factor of 2. This results from differences in feed substrate characteristics. The range of SRTs is typical of the values needed for most composting substrates. I'm often amused by statements that mature compost can be produced in 24 h. It's just not true. You should generally think in terms of 60- to 180-day SRTs, and even longer if kinetic rate limitations are not controlled.

CHAPTER 14

Process Dynamics IV — Working with Bulking Agents

INTRODUCTION

The aerated, static pile process is a special form of composting that uses bulking agents for feed conditioning. The bulking particles form a stable, three-dimensional structure with feed substrates occupying part of the interstitial space between particles. Pile stability is derived from particle to particle contacts between individual bulking particles. The static pile process sees its greatest application on wet feed substrates such as sewage sludge. However, it is equally applicable to other wet substrates, such as industrial sludges and certain food processing wastes, which require feed conditioning to achieve adequate free air space.

Wood chips are the most commonly used bulking agent, although a variety of other materials have also been used. Wood chips are relatively expensive and are usually too large to be acceptable in the final product. Therefore, screening of the bulking agent is almost always practiced at some point within the process. Screened chips are returned for mixing with the feed substrates. There is always some mechanical breakdown of the chips, and the broken fibers eventually appear in the final product. Therefore, there is a continual need for makeup of new wood chips.

STATIC PILE PROCESS SIMULATION

The process model CMSYS52B was developed to simulate the aerated, static pile process using bulking agents. To illustrate its use, the model is applied to the composting of sludge cake using wood chips as the bulking agent. This is the most common, but not the only, application of the static pile process.

Example 14.1

Your firm, The Wood is Good Compost Co., wants to develop a private composting facility. You have a large supply of wood chips from your wood products manufacturing and you wish to approach the local sewage agency about contracting for their biosolids. Before you do this, however, you need a conceptual design of a facility that will produce the high quality product you desire. You are also concerned whether the available supply of wood chips will match the process demand. The sewage agency produces 40 dtpd of cake at 25% TS. Prepare a conceptual design.

Solution

1. You elect to use the static pile process because of your supply of wood chips. Based on other successful designs, the static piles will be formed for a 21-day period. The piles will then be torn down and screened. Material passing through the screen will be sent on to aerated, curing piles. Screened material will be returned to the start of the process. No product recycle will be used because this is not a common practice with the static pile process and the recycled wood chips usually contain sufficient seed material.

2. The static piles will be modeled by seven stages, each with a 3-day HRT. Six stages of curing will be used for the screened material. Each curing stage will have a 10-day HRT.

3. You review the literature but find little information on wood chip degradability. You do not think wood chips degrade much because of their physical structure. You decide to use 5% degradability as a reasonable estimate. You estimate that 20% of the wood chips will be broken down sufficiently to pass the screen on each trip through the system.

4. You are very ambitious and again read Chapters 9 and 10. You arrange for respirometric testing of the agency's sludge cake. Based on this testing the cake is determined to be 55% degradable, with 20% "fast" and 80% "slow" fractions. The fast and slow fractions are characterized by 0.10 and 0.01 day^{-1} rate constants at 20°C, respectively.

5. Bulk weight of "as received" wood chips is determined to be 600 lb/yd^3 at 60% TS. The bulk weight coefficient is determined from Equation 11.9 as

$$\begin{array}{rcl} \text{GAMMAM} & = & \text{COEFFM/SM} \\ 600/(27*62.4) & = & \text{COEFFM/0.60} \\ \text{COEFFM} & = & 0.21 \end{array}$$

6. You review Chapter 6 and select a volumetric mix ratio sludge:chips of 3:1. This may be on the high side for 25% TS cake, but you decide to be conservative.

7. You elect to use microprocessors to control the aeration rate based on a temperature feedback signal. Setpoint temperatures in the static piles will be 140°F (60°C) to assure proper pathogen destruction. The setpoint will be reduced to 131°F (55°C) in curing stages 8 and 9, 115°F (46.1°C) in curing stages 10 and 11, and 100°F (37.8°C) in curing stages 12 and 13. Moisture setpoints will not be used because supplemental water addition is not a common practice when composting sludge with static piles.

8. Target design criteria for the product are as follows:
   a. specific O_2 uptake rate of 25 mg/kg VS-h or less
   b. minimum 90% BVS reduction
   c. decrease in process temperatures toward ambient
   d. final solids content between 55 and 65% TS

9. Load DATIN52B and follow the menu to create the following data file named DATA1401.

MDY	=	(enter current value)
IDEN	=	EXAMPLE 14.1
ME	=	E (for English units)
SUBNUM	=	2
STAGENUM	=	13
STAGESCRN	=	7
STAGER	=	0

SUBNAME	=	SLUDGE	WOOD CHIPS
X	=	320000	20000
S	=	0.25	0.60
V	=	0.75	0.95
K	=	0.55	0.05
FASTFRAC	=	0.20	1.0
SLOWFRAC	=	0.80	0.0
H	=	9000	7750
T	=	50	50
RATEKM20F	=	0.10	0.01
RATEKM20S	=	0.01	0.01
COEFF	=	0	0.21
A	=	10	295
B	=	19	420
C	=	3	186
D	=	1	1

		STG 1	STG 2	STG 3	STG 4	STG 5	STG 6
HRT	=	3	3	3	3	3	3
QAIR	=	500	500	500	500	500	500
TAIR	=	40	40	40	40	40	40
RHAIR	=	0.5	0.5	0.5	0.5	0.5	0.5
WATADD	=	0	0	0	0	0	0
SMSET	=	1	1	1	1	1	1
TSET	=	140	140	140	140	140	140

		STG 7	STG 8	STG 9	STG 10	STG 11	STG 12
HRT	=	3	10	10	10	10	10
QAIR	=	500	500	500	500	500	500
TAIR	=	40	40	40	40	40	40
RHAIR	=	0.5	0.5	0.5	0.5	0.5	0.5
WATADD	=	0	0	0	0	0	0
SMSET	=	1	1	1	1	1	1
TSET	=	140	131	131	115	115	100

		STG 13
HRT	=	10
QAIR	=	500
TAIR	=	40
RHAIR	=	0.5
WATADD	=	0
SMSET	=	1
TSET	=	100

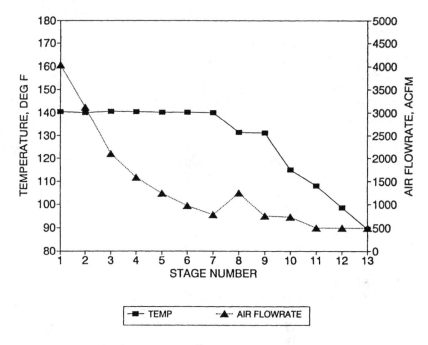

Figure 14.1.   Biosolids/wood chips — profiles of temperatures and air flowrates for the static pile/curing
system of Example 14.1.

XR	=	0
COEFFM	=	0.30
ELEV	=	0
TWATADD	=	40
MECHLOSS	=	0.20
VOLRATIO	=	3.0

9. Exit DATIN41B, load CMSYS52B, and run using DATA1401 when prompted for the data file name.
The program should close to solution about iteration 896.

Stage Profiles

Profiles of process temperatures and air flowrates for each stage are presented in Figure
14.1. Setpoint temperatures are maintained in all stages, except at the end of the process in
stages 11, 12, and 13. For the latter stages, the minimum air flowrate of 500 acfm is such that
process temperatures are below the setpoint condition.

Maximum air flowrate is predicted to occur in stage 1 at about 4000 acfm. Air flowrate
declines gradually throughout the static pile phase to about 775 acfm in stage 7. It is common
for static pile systems to experience a lag phase at the beginning of the process, with 2 to 7
days typically required for temperature development. This should put the peak aeration
demand in stage 2 or perhaps stage 3. The fact that the peak demand is predicted in stage 1
is perhaps an artifact of the model. Recycled bulking agent from stage 7 is assumed to enter
stage 1 at the elevated temperature of stage 7, namely 140°F (60°C). Combined feed mixture
to stage 1 has a temperature of 98°F (36.7°C). Therefore, any lag phase is short-lived with such
a starting condition.

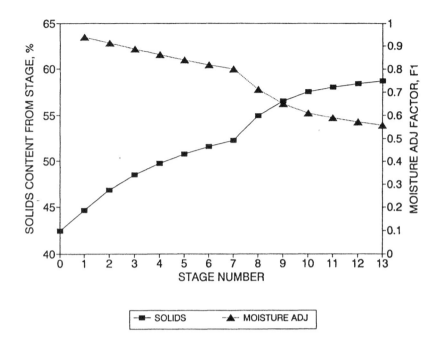

Figure 14.2.   Biosolids/wood chips — profiles of solids content and moisture adjustment factor for the static pile/curing system of Example 14.1; stage 0 represents the mixed feed.

Profiles of solids content and moisture adjustment factor are presented in Figure 14.2. Because moisture setpoints are not specified, solids content rises gradually throughout the process. The initial mixture is about 42.4% TS which is the proper range for static pile practice. Output material from stage 7 is 52.3% TS which is within an acceptable range for screening. Solids content continues to increase during curing to a final product from stage 13 at 58.8% TS. This material is acceptable, but on the low end of the product solids criteria. If curing were done outdoors where the piles could be exposed to rain, the designer might look at adding an energy amendment to the initial mix to provide additional "fuel" to drive more water evaporation.

Specific O_2 consumption and cumulative SRT for the process stages are presented in Figure 14.3. The feed substrates (stage 0) have a relatively high consumption rate. As was the case with the food waste of Example 13.4, this largely results from the high reaction rate constant for the "fast" fraction of sludge organics. A very large drop in O_2 consumption rate is predicted out of stage 1. This must be carefully interpreted. The feed substrates, identified as stage 0 in Figure 14.3, include sludge cake and new wood chips. However, the output from stage 1 includes the entire mixture of sludge cake, new wood chips, and recycled chips. Dry weight of recycled chips is almost three times the dry weight of sludge. Because the chips have a low BVS fraction, they exert a large diluting effect on specific O_2 uptake rate. This effect is again reflected in the increase in specific uptake rate predicted for stage 7. The latter value corresponds to the output from stage 7 after screening. Once the recycled chips are removed, the specific O_2 uptake rate in the screened material increases.

Specific O_2 consumption rates for stages 7 through 13 are presented in Figures 14.4 and 14.5 as a function of cumulative SRT. The design target for O_2 consumption rate is met with the outfeed from stage 10 at an SRT just under 60 days. BVS reduction is over 90% at SRTs above about 30 days (stage 8 outfeed). Referring again to Figures 14.1 and 14.2, temperatures

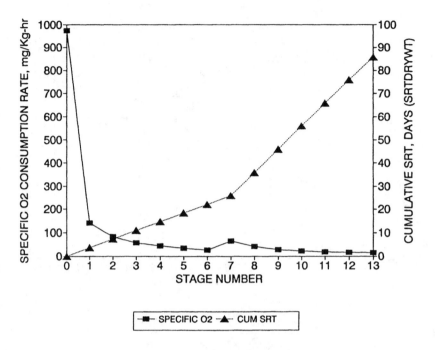

Figure 14.3. Biosolids/wood chips — profiles of specific O_2 consumption rate and cumulative SRT for the static pile/curing system of Example 14.1; stage 0 represents the mixed feed.

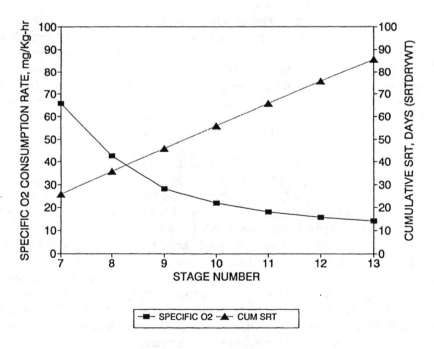

Figure 14.4. Biosolids/wood chips — profiles of specific O_2 consumption rate and cumulative SRT for stages 7 through 13 of Example 14.1.

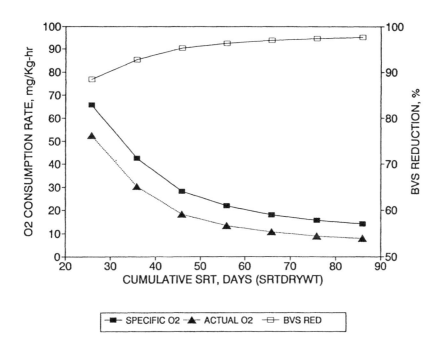

Figure 14.5.   Biosolids/wood chips — specific and actual O_2 consumption rates and BVS reduction as a function of cumulative SRT for the static pile/curing system of Example 14.1.

begin to noticeably decrease at about stage 10 or 11. The solids content is above 55% after stage 8, but not comfortably above the limit until stages 10 or 11. Therefore, all four design criteria for product quality are met after stage 10 with a cumulative SRT of 55 days. However, a margin of safety is added if the remaining curing stages are retained. Therefore, the proposed 13 stages with a cumulative 86-day SRT appears to be a reasonably conservative conceptual design.

Predicted NH_3 concentrations in the exhaust gases are presented in Figure 14.6. The concentrations are considerably higher than predicted for Examples 13.3 and 13.4. In the case of Example 13.3, sludge was amended with sawdust, a low nitrogen amendment. The synthesis requirement for the low nitrogen sawdust decreased the ammonia which could vaporize into the gas phase. The mixed compostables of Example 13.4 were also lower in total nitrogen content. For Example 14.1, the wood chips, while low in nitrogen, are not readily degradable and do not exert a high synthesis demand. Therefore, almost all nitrogen released from the sludge is available for vaporization into the gas phase. Again, the reader is advised to view the NH_3 results with some caution and to interpret then as a potential upper limit. Static pile systems on raw sludge often report ammonia concentrations in the 300 to 500 ppm range which is somewhat lower than the concentrations in Figure 14.6. This would seem to indicate that other mechanisms are at work besides those modeled in the present ammonia algorithm.

Mass and Energy Balances

A mass balance for the system of Example 14.1 is presented in Figure 14.7. Again, detailed stage balances are not presented because of the quantity of data involved. The seven stages that comprise the static pile phase are followed by screening of the bulking agent. Screened bulking agent (oversize) is recycled back to stage 1. Screened product (undersize) continues

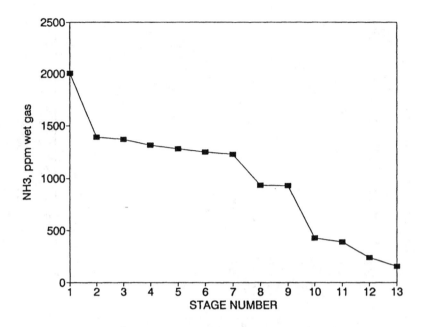

Figure 14.6. Biosolids/wood chips — predicted NH_3 concentrations in the outlet gases from the static pile/curing system of Example 14.1.

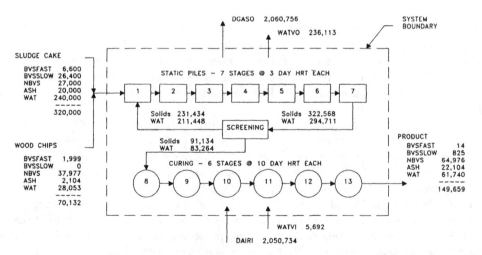

Figure 14.7. Biosolids/wood chips — mass balance for the static pile/curing system of Example 14.1.

through the six stages of curing. There is a tremendous flux of bulking agent through the process. Recycled bulking agent returned to stage 1 is over 220 wtpd and is more than the sludge feed at 160 wtpd. On a solids basis, recycled bulking agent is almost three times the dry weight of sludge solids. Therefore, movement of bulking agent is a major factor in design of static pile systems. Bulking agent is the dominant solids term in the mass balance.

While the flux of bulking agent is significant, it is still relatively small compared to the flux of gases. Close to 1150 tpd of process gases are discharged from the system. As has been the case with every example to this point, input air and output gases are the dominant terms of the mass balance.

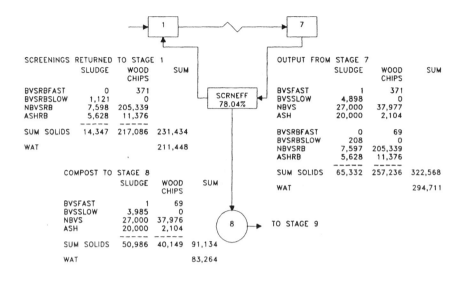

Figure 14.8. Biosolids/wood chips — mass balance for the screening process between stages 7 and 8 of Example 14.1.

The energy balance is not presented in detail because the major terms remain the same as in previous balances. The latent heat of water evaporation represents about 74% of total output energy. Sludge cake produces about 95% of the total heat release from organic decomposition. New wood chips provide the remaining 5%. Under the assumptions of Example 14.1, wood chips contribute a relatively small fraction to the overall energy input.

The Screening Process

The screening process, shown within the system boundary of Figure 14.7, is presented in detail in Figure 14.8. Output components of sludge and wood chips from stage 7 are input to the screening process. Fractions associated with the original feed substrates, such as BVSFAST and NBVS, and fractions associated with substrate recycled with the bulking agent, such as BVSRBFAST and NBVSRB, are both shown. Total weight of substrates and recycled bulking agent input to the screening process are the same in both Figures 14.7 and 14.8.

The screening process operates with a screening efficiency, SCRNEFF, of 78.04% which is determined from the correlation of Figure 11.13 based on an input solids content of 52.26% TS. To illustrate how the screening algorithm works, consider the BVSSLOW fraction for sludge. The sum of BVSSLOW and BVSRBSLOW for the sludge substrate is 4898 + 208 = 5106 lb/day. 78.04% should pass the screen, giving 5106 × .7804 = 3985 lb/day BVSSLOW continuing on to stage 8. The difference, 5106 − 3985 = 1121 lb/day BVSRBSLOW, remains with the screened bulking agent and is returned to stage 1. The calculation procedure is further illustrated in Figure 11.14.

The bulking agent is handled differently. Unbroken bulking agent is assumed to be screened with 100% efficiency. MECHLOSS for Example 14.1 is 20%. Considering the NBVS and NBVSRB fractions for wood chips from Figure 14.8, the total input weight is 37977 + 205339 = 243316 lb/day. The unbroken fraction is 243316(1 − 0.20) = 194653 lb/day. Broken bulking agent is assumed to be screened with the same efficiency as other substrates. The quantity retained with the oversize is (243316 − 194653)(1 − .7804) = 10686 lb/day. Total recycled to stage 1 is 194653 + 10686 = 205339 lb/day NBVSRB. The undersize is 243316 − 205339 = 37977 lb/day NBVS and reports to stage 8. Product recycle is not used

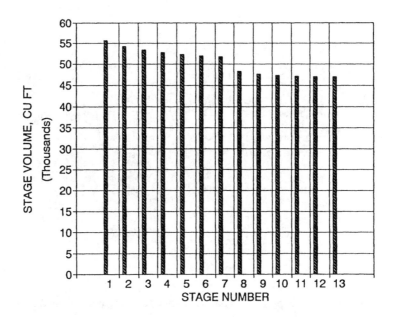

Figure 14.9.   Biosolids/wood chips — stage volumes for the static pile/curing system of Example 14.1.

in Example 14.1. If it was, another series of recycle components would be shown in Figure 14.8 for each substrate. The calculation procedures are consistent with those illustrated in Figure 11.15.

Volume and Area Requirements

The volume requirement for each process stage is presented in Figure 14.9. There is a small reduction in static pile volume from stages 1 to 7. This small reduction is due to the small loss of wood chips which are assumed to be 5% degradable. Total static pile volume for stages 1 through 7 is estimated at 372,500 ft³. Volume reduction during static pile composting will only decrease the pile height, and not the occupied area. Therefore, it is good design practice to estimate the total static pile volume as the stage 1 volume times the number of stages. This gives 55720 × 7 = 390,000 ft³. If extended piles are used with an effective height of 8 ft, the specific volume is about 348,000 ft³/acre. Using 250,000 to account for pile ends and other stacking inefficiencies, a total pile area of 390,000/250,000 = 1.6 acres is required. Area for access roads, aeration blowers, and equipment movement must be added to this estimate.

A small volume reduction is also predicted across curing stages 8 through 13. Total required curing volume can be conservatively estimated as the stage 8 volume times 6, or 48,300 × 6 = 290,000 ft³. Assuming individual static piles are used for curing, the specific volume can be estimated from the dimensions of Figure 2.13 and assuming 8 ft between individual piles. Specific volume is about 120,000 ft³/acre for individual piles. About 290,000/ 120,000 = 2.4 acres are required for curing piles. Again, access roads and other area for equipment must be added to this estimate.

BULKING AGENT CONSUMPTION

The consumption of bulking agent is strongly influenced by the extent of mechanical breakdown of the particles. Particle breakdown occurs as a result of mechanical metering and mixing with the feed substrates, loading and stacking operations, pile removal and convey-ance, and the screening operation itself. The presence of broken wood fibers is readily apparent in compost produced from static pile facilities using wood chips. Mechanical breakdown probably represents the greater loss of wood chips compared to the loss from biological decomposition. Referring to Figure 14.7, the net consumption of wood chips is predicted to be about 70,000 lb/day, corresponding to a MECHLOSS of 20%/pass. This is equivalent to about 3200 ft³/day (119 yd³/day) of new chips. The effect of different mechanical loss rates can be evaluated by assuming various values for the MECHLOSS factor in the simulation model.

Example 14.2

Your boss at the Wood is Good Compost Co. has asked you to evaluate the effect of MECHLOSS on the net consumption of wood chips. Estimate the range of wood chip consumption for MECHLOSS rates from 10 to 30%/pass. You judge that this is the likely range of values expected during static pile composting.

Solution

1. Load DATIN52B and follow the menu to copy DATA1401 under the new file name DATA1402. Make addition copies of DATA1401 under file names DATA1403, DATA1404, and DATA1405. Follow the menu to modify DATA1402 to

$$\begin{array}{rcl} \text{IDEN} &=& \text{EXAMPLE 14.2} \\ \text{MECHLOSS} &=& 0.10 \end{array}$$

Continue to follow the menu to modify each remaining data file as follows,

$$\begin{array}{rcl} \text{MECHLOSS} &=& 0.15 \text{ for DATA1403} \\ \text{MECHLOSS} &=& 0.25 \text{ for DATA1404} \\ \text{MECHLOSS} &=& 0.30 \text{ for DATA1405} \end{array}$$

2. Exit DATIN41B, load CMSYS52B, and run. Enter DATA1402 when prompted for the data file name. The program should close about iteration 885. Repeat this process for each data file created above.

The daily makeup volume of wood chips and the volume ratio of new chips/sludge cake are presented in Figure 14.10. Net consumption is directly proportional to the mechanical loss factor, MECHLOSS. With a 10% mechanical loss factor, net consumption is about 0.3 volumes of wood chips/volume of sludge. The volume ratio increases to almost 1.0 with a mechanical loss of 30%. The net makeup of wood chips is typically about 0.5 to 1.0 volume/volume which corresponds to MECHLOSS factors from roughly 15 to 30%. The value of 20% used in Example 14.1 is probably reasonable for estimating purposes. To the extent possible, however, the designer should check other facilities using similar wood chips to verify actual loss rates.

PEAK AERATION DEMANDS

The static pile system operates in much the same way as other composting systems in the sense that substrates are fed to the process on a daily or near daily basis. The mass balance

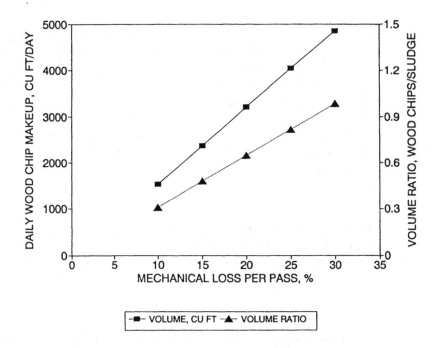

Figure 14.10. Biosolids/wood chips — effect of mechanical loss rate on the daily requirement for wood chip makeup; data corresponds to Example 14.2.

of Figure 14.7 reflects the daily feeding of substrates to the process. The balance of Figure 14.7 can be visualized as placing a bubble over the entire process. Substrates enter the bubble and flux from one stage to the next, eventually exiting the bubble as compost product. The system balance is developed by observing all components that enter or leave the bubble. This is usually the most convenient and useful way of visualizing a composting process.

Individual piles within the static pile system operate in a batch mode. The same is true of individual windrows within a windrow process. It is sometimes advantageous to examine individual static piles or windrows. Studying peak aeration demands is one such case. An individual pile will experience a range of aeration demand. The designer must contend with the problem of meeting the peak aeration demand when sizing the aeration supply and distribution system.

Peak Demand from Example 14.1

The analysis of Example 14.1 is the first step in the study of peak aeration demands. Referring to Figure 14.1, a peak aeration rate of 4031 acfm is predicted for stage 1. Aeration demand for static pile systems is frequently expressed as a specific aeration rate with units of actual or standard cubic feet per hour/dry ton (cfh/dt). The 4031 acfm aeration rate to stage 1 can be converted to a specific rate by considering the quantity of sludge under aeration in stage 1. With a 3-day HRT and a feed rate of 40 dtpd, about 120 dtpd of sludge substrate is under aeration. Maximum specific aeration rate then becomes $4031 \times 60/120 = 2015$ acfh/dt. The latter should be interpreted as a 3-day time averaged peak because the stage 1 HRT is 3 days.

Several factors can increase the peak aeration demand above that just calculated. Reaction rate constants can have a significant effect on peak aeration demand. This can be studied by using higher reaction rates or a higher fraction of "fast" organics in Example 14.1. Two other factors will be examined here. The first is the duration of time averaging and the second is the effect of bulking agent degradability.

Time Averaging

If the aeration rate is averaged over shorter time periods, it is likely that the peak rate will increase. In other words, a 1-day average peak should be greater than a 3-day average peak. In turn, a 1-h average peak should be greater than a 1-day average peak, and so on. The effect of time averaging is investigated in Example 14.3. This example also illustrates a technique for using the simulation programs to model a single batch pile.

Example 14.3

Using results from Example 14.1, model the static piles resulting from a 1-day input of substrate. Use a 1-day HRT for each stage to determine the 1-day average peak aeration rate.

Solution

1. Individual static piles can be modeled by viewing the recycled bulking agent as a third substrate in addition to the sludge cake and new wood chips. Characteristics and quantities of the recycled bulking agent can be determined from results of Example 14.1. No screening or recycle is used so that only the static pile is modeled.

2. Referring to Figure 14.1, the peak aeration demand occurs in the early stages. Therefore, it is not necessary to model all 21 days of the static pile phase. Ten stages each with a 1-day HRT will be assumed.

3. Load DATIN52B and follow the menu to duplicate data file DATA1401 under the new file name DATA1406. Continue to follow the menu and modify DATA1406 to add a third substrate identified as RECYC CHIP. Data for RECYC CHIP are taken from results of Example 14.1. Note that the temperature for RECYC CHIP is 140°F (60°C) corresponding to the output temperature from stage 7 of Example 14.1.

MDY	= (enter current value)
IDEN	= EXAMPLE 14.3
ME	= E (for English units)
SUBNUM	= 3
STAGENUM	= 10
STAGESCRN	= 0
STAGER	= 0

SUBNAME	= SLUDGE	WOOD CHIPS	RECYC CHIP
X	= 320000	20000	442882
S	= 0.25	0.60	0.5226
V	= 0.75	0.95	0.95
K	= 0.55	0.05	0
FASTFRAC	= 0.20	1.0	1.0
SLOWFRAC	= 0.80	0.0	0.0
H	= 9000	7750	7750
T	= 50	50	140.02

RATEKM20F	=	0.10	0.01	0.01
RATEKM20S	=	0.01	0.01	0.01
COEFF	=	0	0.21	0.21
A	=	10	295	295
B	=	19	420	420
C	=	3	186	186
D	=	1	1	1

		STG 1	STG 2	STG 3	STG 4	STG 5	STG 6
HRT	=	1	1	1	1	1	1
QAIR	=	500	250	250	250	250	250
TAIR	=	40	40	40	40	40	40
RHAIR	=	0.5	0.5	0.5	0.5	0.5	0.5
WATADD	=	0	0	0	0	0	0
SMSET	=	1	1	1	1	1	1
TSET	=	140	140	140	140	140	140

		STG 7	STG 8	STG 9	STG 10
HRT	=	1	1	1	1
QAIR	=	250	250	250	250
TAIR	=	40	40	40	40
RHAIR	=	0.5	0.5	0.5	0.5
WATADD	=	0	0	0	0
SMSET	=	1	1	1	1
TSET	=	140	140	140	140

XR	=	0
COEFFM	=	0.30
ELEV	=	0
TWATADD	=	40
MECHLOSS	=	0.20
VOLRATIO	=	3.0

4. Exit DATIN41B, load CMSYS52B, and run using DATA1406 when prompted for the data file name. The program should close at iteration 834.

Profiles of pile temperatures and air flowrates are presented in Figure 14.11. Because each stage has a 1-day HRT, the x-axis label is changed from the stage number of previous graphs to the actual time since start of the cycle. The aeration rate needed to maintain the temperature setpoint is predicted to peak during day 2 at 1846 acfm. Static piles formed on this day contain 40 dry tons of sludge. Therefore, specific aeration demand during day 2 can be determined as $1846 \times 60/40 = 2770$ acfh/dt.

The 1-day average peak demand of 2770 acfh/dt is about 37% higher than the 3-day average peak of 2015 acfh/dt determined from results of Example 14.1. The designer is faced with a decision when it comes to supplying peak aeration demands. There is no single peak for purposes of design. The designer must decide whether to use the highest instantaneous peak for design or whether to use a time-averaged peak. Usually a time-averaged peak will be selected because the adverse consequences of not meeting the instantaneous peak are relatively minor. This still leaves the question of what time-averaging period to use. This cannot be answered without some knowledge of the type of aeration system being considered by the designer. For example, if individual piles are manifolded together and served by common aeration blowers, a shorter duration peak can be satisfied because all piles in the system will not be in the same part of their cycle. There is a dampening effect on the total system peak as a result. On the other hand, if individual blowers serve each pile, then a longer time-averaged peak may be appropriate to reduce the size of the individual blower. As a general rule, a 1-day average peak is a good starting point for design.

Bulking Agent Contribution

It is important to remember that wood chips are a substrate. For Example 14.1, a degradability of only 5% was assumed. However, some species of wood may exhibit higher degradability and this will effect the peak aeration rate. The effect of a 20% degradability was studied using the procedures of Examples 14.1 and 14.3. First, a simulation of the complete composting system was run using the same data base as Example 14.1 except for a 20% degradability for the wood chip substrate. Second, a simulation of the first 10 days of the static pile phase was run using similar procedures to Example 14.3 with feed characteristics from the latter system run. Under the conditions described above, the 1-day peak aeration demand increases to 3400 acfh/dt, a 23% increase from the 1-day peak demand determined in Example 14.3.

Design Procedures

With batch processes it is important to consider both minimum and peak aeration demands when designing the supply and distribution system. Peak aeration rates were discussed in Chapter 7. Based on simulation studies and field measurements, peak aeration demands for raw sludge/wood chip mixtures are likely to range from 3000 to as high as 10,000 cfh/dt. The range of values developed from the simulation runs in this chapter compare favorably with the field observations. The most important variables in determining peak aeration demand are substrate degradability, rate constants for the "fast" and "slow" fractions, setpoint temperatures for the pile, and the period for time-averaging. All of these factors can be evaluated using procedures developed in Examples 14.1 and 14.3.

The designer must also consider the minimum aeration requirement to establish turndown requirements for the supply and distribution system. Most manifolds deliver a uniform distribution only over a limited range of air supply. The range of proper delivery can be extended, but only by increasing fan horsepower. Therefore, turndown is also important to the design. Referring to Figure 14.11, a turndown from 1846 acfm to at least 537 acfm is required, equivalent to a 3.4:1 turndown ratio. Design of manifold distribution systems is discussed further in Chapter 15.

OTHER AERATION PATTERNS

Uniform Aeration Rate

Temperature profiles presented in Figures 14.1 and 14.11 for the static pile phase were developed assuming uniform setpoint temperatures for each stage. The model also assumes that screened bulking agent is recycled at the temperature TNEW(STAGESCRN). These assumptions combine to produce the uniform temperature profiles in Figure 14.1 and 14.11. Some static pile systems do not use temperature feedback loops to control aeration supply. Also, the temperature of screened bulking agent is reduced back to near ambient in some cases. The computer model can be used to simulate such an operation.

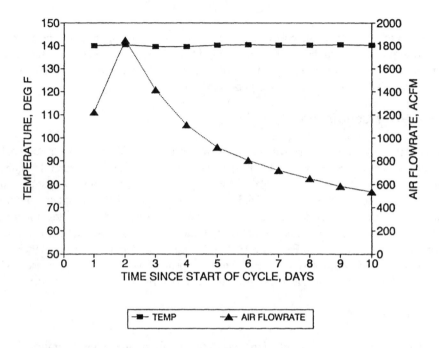

Figure 14.11. Biosolids/wood chips — profiles of temperature and air flowrate for the single static pile analysis of Example 14.3 with setpoint temperatures and variable air flowrate.

Example 14.4

You have just completed design of the world's first perfect aeration system using "high tech" feedback control loops, temperature sensors, flow elements, throttling valves, and microprocessors. You send the design to the cost estimating department and in a few days are told the sad story. Your project budget is a mess and you must cut somewhere. You decide to evaluate a constant supply system.

Solution

1. Referring to Figure 14.11, a constant aeration supply of 1000 acfm appears to be a reasonable compromise between peak and minimum demands.

2. You are concerned that recycled chips will cool during storage and estimate a possible return temperature of 60°C.

3. Load DATIN52B and follow the menu to duplicate data file DATA1406, created for Example 14.3, under the new file name DATA1407. Continue to follow the menu and modify DATA1407 as follows:

$$
\begin{aligned}
\text{IDEN} &= \text{EXAMPLE 14.4} \\
\text{T} &= \text{60°F for RECYC CHIP} \\
\text{QAIR} &= \text{1000 acfm for all 10 stages} \\
\text{TSET} &= \text{200 (no TSET) for all 10 stages}
\end{aligned}
$$

4. Exit DATIN41B, load CMSYS52B, and run using DATA1407 when prompted for the data file name. The program should close at iteration 192.

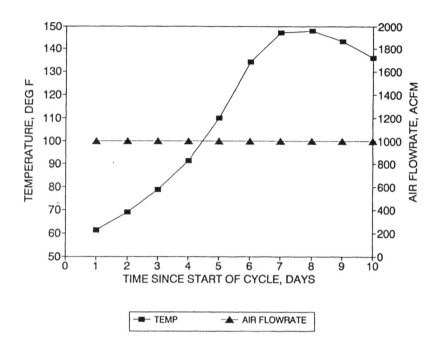

Figure 14.12. Biosolids/wood chips — profiles of temperature and air flowrate for the single static pile analysis of Example 14.4 with constant air flowrate.

Profiles of temperature and air flowrate for the conditions of Example 14.4 are presented in Figure 14.12. The feed substrates have a combined input temperature of 55°F (12.8°C) at time = 0. From this point the pile gradually builds in temperature over the next few days. A temperature of 110°F (43.3°C) is achieved at day 5. Temperatures exceed 140°F (60°C) by day 7. This general pattern is typical of results observed in the field. The duration of the lag phase depends on the temperature of the starting substrates, air infeed temperature, aeration rate, and reaction rate constants for the substrates. The procedures of Example 14.4 can be used to study the effects of these variables and other aeration patterns.

Lag Periods

Lag phases of 2 to 7 days are typical of the general range observed in the field with static piles and windrows under most weather conditions. Lag phases that exceed 7 to 10 days or piles that never achieve proper composting temperatures are indicative of a process problem.

The problem can either be thermodynamic or kinetic. Very "energy poor" substrates may fail to develop proper temperatures due to lack of available energy, a thermodynamic problem. In most cases, however, the problem is kinetic. The rate of heat release may be limited by lack of moisture, lack of FAS, low oxygen content, and very cold or very hot temperatures. For example, if the feed substrates and input air are cold and the aeration rate is too high, it is possible to remove heat faster than it is produced. Under these conditions, pile temperatures will not develop. Energy is still available within the substrate, but it is only slowly released because of the low rate of biological activity. Input of large amounts of cold air can then remove the heat faster than it is produced. Cold starting temperature is probably the most common cause of extended lag periods or pile failure. The designer should make reasonable efforts to conserve heat in the starting substrates and measure and control aeration rates at the start of the process. In extreme cases, it may be necessary to preheat the feed air for a short period to "light" the composting fire.

While cold temperatures may be the most common cause of pile failure, other possible causes should not be overlooked. It is relatively easy to check moisture, FAS, and oxygen content to determine if they impose significant rate limitations. If all these possibilities are eliminated, and if proper temperatures historically have been achieved under similar climatic conditions, then some change in the feed substrate or the presence of toxic compounds should be considered.

SUMMARY

The computer model CMSYS52B can be used to simulate the aerated, static pile process. The latter is a special form of composting that uses bulking agents for feed conditioning. Screening of the bulking agent, typically wood chips, is almost always practiced at some point in the process. Screened chips are returned for mixing with the feed substrates. Due to mechanical breakdown and biological decomposition, there is a continual need for makeup of new wood chips.

Model CMSYS52B was used to model a 40 dtpd sludge composting facility using wood chips. A 21-day static pile phase was simulated as seven stages each with a 3-day HRT. Screening was assumed to follow the static pile phase. A 60-day curing for the screened material was modeled as six stages each with a 10-day HRT. Product quality standards for O_2 consumption rate, BVS reduction, temperature decline, and solids content were achieved under conditions of the simulation.

Individual piles within the static pile system operate in a batch mode. The same is true of individual windrows within the windrow process. Batch operated piles or windrows will experience a range of aeration demand. The designer must develop a supply system that meets the peak and minimum aeration demands. The most important variables in determining the peak aeration demand are substrate degradability, rate constants for the "fast" and "slow" fractions, setpoint temperatures for the pile, and the time-averaging period. Peak aeration demands developed from simulation modeling compare favorably with field observations. A 1-day peak demand of 3500 to 5000 scfh/dt is typical for raw sludge/wood chip composting.

Lag periods of 2 to 7 days are typically observed with batch operated static piles and windrows, a range confirmed by simulation modeling. Extended lag periods or failure to develop elevated temperatures are indicative of thermodynamic or kinetic problems with the pile. Cold starting temperature is probably the most common cause of extended lag periods or pile failure. With cold substrates and very cold air, the input of large amounts of cold air can remove heat faster than it is produced. Heat conservation in the starting substrates, control of the aeration rate, and preheat of the feed air can be used to "light" the composting fire if extremely cold weather prevents normal pile heat up.

CHAPTER 15

Design of Manifold Distribution Systems

INTRODUCTION

As discussed in the previous chapters, air supply is essential to the proper operation of aerobic composting systems. Air ventilation provides the oxygen necessary for aerobic metabolism, removes moisture from the composting material to provide drying, and removes heat generated by the biological activity to control process temperatures. Once the total quantity of air and the rate of air supply are known, the problem becomes one of properly distributing this flow to all of the composting mass. Air should be supplied uniformly to all parts of the composting matrix. Otherwise, overaerated zones may be excessively cooled and dried, while underaerated zones may experience high temperatures, reduced drying, and even oxygen limitations.

Most composting processes use manifold systems to evenly distribute the supplied air and/ or evenly collect the process gas. Design of such systems involves numerous tradeoffs between competing engineering variables. Friction losses result in a pressure gradient along the manifold that can then cause uneven flow distribution from individual ports along the manifold. Distribution is improved by using a higher headloss across the individual orifices within the manifold; however, higher headloss increases power consumption. Obviously, a tradeoff is necessary between the degree of "evenness" provided by the manifold and the resulting power consumption.

The purpose of this chapter is to present an approach to hydraulic analysis of manifold systems typically used in composting practice. Parameters important to good manifold design and calculation procedures will be discussed. With these tools, the designer should be able to understand the hydraulic principles involved and properly analyze and design a manifold system.

Figure 15.1.   A temporary aeration system using a portable fan mounted to a 55-gal drum. The drum serves as a condensate trap for induced draft operation. The fan is controlled by an on/off timer. Photo was taken in 1976 during the original static pile testing by the USDA at Beltsville, Maryland.

TYPES OF MANIFOLD SYSTEMS

Manifold systems for composting applications were classified by Kuchenrither[1] into temporary, semipermanent, and permanent types and by Higgins[2] into stationary and movable types. Temporary or movable systems use above grade piping with blowers not rigidly mounted to foundations. The piping can be flexible plastic, rigid plastic, or metallic. Flexible piping, such as corrugated agricultural drainage pipe, is considered disposable and is usually discarded after each composting period. Semipermanent systems also use flexible or rigid piping, but have permanently fixed blowers mounted to concrete or steel foundations. Permanent systems use fixed, permanent manifolds, usually below ground level, with permanently mounted blowers. Schematic diagrams and photographs of typical manifold systems are presented in Figures 15.1 to 15.10.

Systems that use disposable piping are usually limited to 4 to 8-in. diameters for economic reasons. Pipe friction losses result in a decrease in static pressure in the direction of flow. If all orifices are the same size and equally spaced, the mass flow of gas into or out of the manifold will vary along the length of the manifold. The size and spacing of orifices can be varied to reduce this problem. Corrugated piping is available with either circular or slotted orifices and with a range of orifice sizes and spacing. With long manifolds, it is common to combine a number of pipe sections in series, with different orifice sizing and spacing, to improve gas distribution. The number of necessary sections increases with decreasing manifold diameter and increasing manifold length. It is also common to run a number of laterals in parallel to reduce the gas velocity and resultant friction losses.

A schematic for a permanent aeration system for a windrow facility is shown in Figure 15.6. A blower sits above a knock out (KO) drum and typically operates at a pressure of 18- to 24-in. water column (wc). The KO drum acts as a separation point for condensate and

Figure 15.2. A semipermanent manifold system constructed from four disposable plastic piping legs, each attached to a permanent collection header, which in turn is connected to a permanently mounted exhaust fan. Site is the Montgomery County Composting Facility, Maryland, operated by the Washington Suburban Sanitary Commission. Aerated, static piles are used to compost raw sludge:wood chip mixtures.

leachate and is connected to the aeration channel or manifold by a pipeline. Because the aeration channel is permanent, it is usually designed for low gas velocities to reduce friction losses along the manifold. For the design shown in Figure 15.6, distribution orifices are machined into steel plate. A geotextile mat material sits on top of the perforated steel plate to further improve air distribution. Plastic screening covers the geotextile material to keep overlying sand away from the orifices. A sand layer keeps compost from infiltrating and plugging the screen and aeration plenum. The sand must be replaced at periodic intervals.

Flexible, disposable piping is usually constructed of polyethylene (PE) plastic which provides good resistance to the process temperatures and high moisture environment. Rigid plastic piping, which is intended to be reused, is usually constructed of high density polyethylene (HDPE) to increase its impact resistance. HDPE pipe is usually smooth walled and retains some flexibility. Polyvinyl chloride (PVC) plastic is not favored because it has a tendency to become brittle under repeated hot and cold conditions. Williams and North[4] evaluated the use of specially constructed, fiberglass reinforced concrete (FRC) block, which is laid on a sub-base to form a continuous manifold system. Both reusable HDPE pipe and the FRC block were projected to have lower life cycle costs compared to disposable PE pipe. Nevertheless, disposable PE pipe remains the industry workhorse at static pile, sludge composting facilities.

A permanent aeration system is a capital cost investment compared to the continuing operating costs for disposable piping. Because a permanent manifold can be designed for low gas velocities, permanent manifolds can usually provide more uniform distribution over a wider range of aeration rates. However, a permanent system is not movable and is less flexible in this regard. The designer must weigh these factors before determining the proper system for a particular application. It is common to see different manifold types combined within a

Figure 15.3. Multiple pipe sections being readied for use at the Montgomery County Composting Facility. Multiple sections with different port spacing and/or port size are used to provide a more uniform flow distribution. A simple joint using duct tape is sufficient to join the sections because the entire assembly is disposed at the end of the cycle.

particular composting system. For example, a permanent system may be used for the first stage, high rate phase with a temporary system used for later curing.

Most manifold systems use fans to pull (induced draft) or push (forced draft) the flow through the manifold. However, some "low technology" compost operations have laid disposable, perforated piping in parallel rows over which the compost is added. No fans are attached to the pipes. The hot chimney effect induces a natural pressure gradient across the compost. The pipe manifold helps to uniformly distribute air, which flows into the ends of the pipe as a result of the pressure gradient. Because the flowrates induced by natural ventilation are limited, such systems work best with less degradable substrates and with substrates that do not require extensive moisture removal. With very degradable substrates, and particularly wet substrates, the high aeration demands usually require the use of fans.

MANIFOLD HYDRAULICS

Approach to the Problem

Hydraulic profiles for forced draft and induced draft manifolds are presented in Figures 15.11 and 15.12, respectively. The term "forced draft" means that a blower or compressor is forcing air into the manifold. Therefore, manifold pressure is positive, or above atmospheric pressure. Forced draft manifolds are sometimes referred to as "dividing flow" systems because flow is divided off along the manifold.[5,6] The term "induced draft" means that a blower is pulling gas into the manifold. Manifold pressure is negative in this case, or below atmospheric

Figure 15.4.   A permanent subsurface manifold under construction at a windrow composting facility for sewage sludge. The manifold is 295 ft long, 9.5 in. wide, 21 in. deep and serves one windrow. An asphalt surface was later constructed around the manifold. The large cross-section reduces headloss along the relatively long manifold. Site is the Denver Metropolitan Wastewater Reclamation District.

pressure. Induced draft manifolds are sometimes referred to as "combining flow" systems because flow is added along the manifold.

In a typical problem, the manifold size, discharge port (orifice) size, and spacing are known or assumed. All of these may vary along the length of the manifold. The desired flowrate into or out of the manifold is selected. The designer then wishes to know the flow distribution along the manifold and the pressure necessary to achieve this flow condition.

Solution to the above problem begins at the end of the manifold furthest from the fan by assuming a total pressure, E, just downstream of port 1. Flow through any orifice is related to the static pressure difference across the orifice. Once flow from port 1 is determined, gas velocity in the manifold can be calculated. Static pressure at Port 2 is determined from headloss formulas presented later. Flow from port 2 is then determined. Gas flow in section 2 of the manifold is the sum of gas flows from ports 1 and 2. Similar calculations are continued along the length of the manifold until the last port. Total manifold flowrate at the last port is compared with the desired flowrate. If different, a new total pressure is assumed at port 1 and the calculation procedure repeated. This continues until the desired manifold flowrate is achieved.

The hydraulic profile shown in Figure 15.11 assumes that gas is discharged directly to the atmosphere through the manifold orifices. The profile in Figure 15.12 assumes that gas is pulled directly from the atmosphere. In actual practice, composting material would be placed over the manifold and exert an additional resistance to flow. The pressure profile for this case is shown in Figure 15.13. PPIPE is the static pressure within the manifold and PATM the background atmospheric pressure. DELTAP is the static pressure drop from inside the

Figure 15.5.   Permanent aeration system used at the Denver composting facility. Each blower can operate in either the induced or forced draft modes and delivers 4620 scfm at 18 in. static pressure, equivalent to 2770 scfh/dt of sludge.

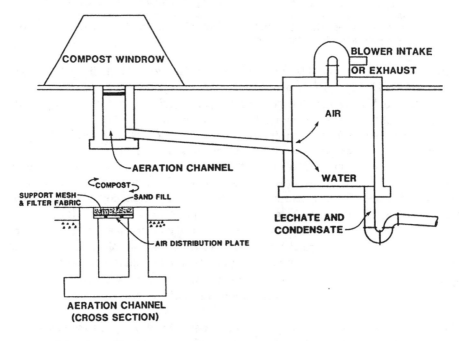

Figure 15.6.   Schematic diagram of the permanent manifold system used at the Denver sludge composting facility. From Hay and Kuchenrither.[3]

Figure 15.7. Permanent aeration system used at a circular, agitated bed, sludge composting facility. The reactor floor is divided into five concentric zones. Each zone contains four manifold pipes, each of which travels the full circumference of the zone. Site is the Clinton County Composting Facility, City of Plattsburgh, New York.

Figure 15.8. Permanent, parallel manifold channels in the floor of a refuse composting hall. Each aeration channel is rectangular in shape and is covered by wooden blocks. Gaps between the blocks serve as orifices to collect the gases that flow by downward, induced draft. Site is the Siggerwiesen co-composting plant serving the City of Salzburg, Austria.

Figure 15.9. Permanent supply header and manifold pipes to distribute composting process gas to a biofilter. Gases are collected from a horizontal flow, static bed reactor at the sludge composting facility operated by the City of Hamilton, Ohio.

manifold to the atmosphere, equal to PIPE – PATM. DELTAP is also the sum of pressure drops across the orifice, DELTAPPORT, and across the composting material, DELP. An iterative procedure is required at each orifice to determine the orifice flowrate giving DELTAPPORT + DELP = DELTAP.

A program entitled MANIFOLD was developed to solve manifold problems. A corresponding program MANDATA produces input data files for MANIFOLD. Information on obtaining these programs is presented on the inside back cover. While the models greatly simplify the hydraulic computations, it is not necessary to have these models to understand the material presented here. Calculation procedures used in the model are described in the remainder of this section. Model terminology and variable definitions are presented in Appendix B. Example problems and a discussion of typical manifold problems and solutions are presented later in this Chapter.

Manifold Sections

Twenty different manifold sections are allowed in the model. This is far more than would normally be encountered in a typical manifold system. The number of different sections is input as NUMSECTION. Circular or rectangular cross sections are allowed for the manifold and orifices. For each section I, the following manifold and port data are input to the model:

NUMPORTS	=	number of dissimilar manifold sections
SPACING(I)	=	spacing of ports in manifold section I, in./port
DPIPE(I)	=	diameter of a circular manifold section I, in.
WMAN(I)	=	width of a rectangular manifold section I, in.
HMAN(I)	=	height of a rectangular manifold section I, in.
DPORT(I)	=	diameter of a circular orifice in section I, in.

Figure 15.10. Supply header and individual manifolds to distribute process gas from a static pile, sludge composting facility to a biofilter. Perforated plastic pipe was connected to each outlet prior to placing the biofilter material. The biofilter was replaced by the wet scrubbing system shown in the background. Site is the Montgomery County Composting Facility, Maryland.

LPORT(I)	=	length of a slot or rectangular port in section I, in.
WPORT(I)	=	width of a slot or rectangular port in section I, in.
PORTTYPE(I)	=	type of port in manifold section I; 1 = circular, square edged; 2 = circular, rounded, or bell mouth edged; 3 = slot, square edged

For noncircular manifolds and orifice ports, the hydraulic radius is used to convert the cross section to an equivalent circular dimension. The hydraulic radius, HYDRAD, is equal to the cross sectional area divided by wetted perimeter. For a rectangular manifold this is expressed as

$$HYDRAD = \frac{WMAN * HMAN}{2 * WMAN * HMAN} \tag{15.1}$$

The hydraulic radius for a circular cross section is equal to the diameter/4. Therefore, the equivalent diameter for a noncircular manifold section is

$$DPIPE(I) = 4 * HYDRAD \tag{15.2}$$

Similar equations can be used to determine the equivalent diameter of slot or rectangular orifices as follows:

$$HYDRAD = \frac{LPORT * WPORT}{2 * LPORT * WPORT} \tag{15.3}$$

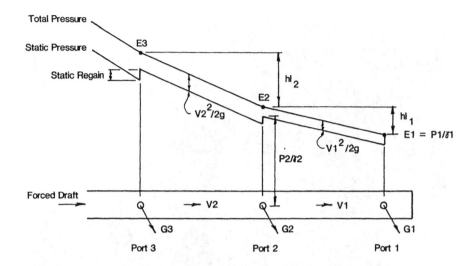

Figure 15.11. Total pressure, static pressure, and velocity pressure profiles for forced draft flow in a multiport manifold.

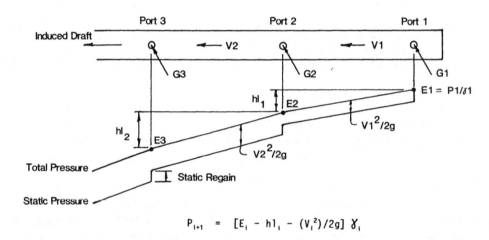

$$P_{i+1} = [E_i - hl_i - (V_i^2)/2g] \, \gamma_i$$

Figure 15.12. Total pressure, static pressure, and velocity pressure profiles for induced draft flow in a multiport manifold.

$$DPORT(I) = 4 * HYDRAD \tag{15.4}$$

Gas Density

Gas density within the manifold will vary for a number of reasons. Atmospheric pressure decreases with increased site elevation. The process gas temperature will vary from near ambient for forced draft systems to average composting temperatures for induced draft systems. Also, the high moisture content of composting gases will reduce the gas density.

Site Elevation

Absolute atmospheric pressure, PATM, as a function of elevation is determined as

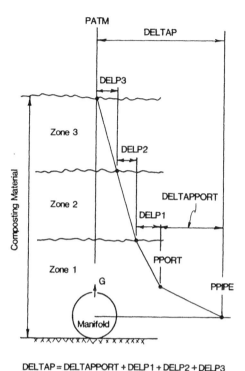

DELTAP = DELTAPPORT + DELP1 + DELP2 + DELP3

Figure 15.13. Static pressure relationships across a distribution manifold, orifice port, and composting matrix.

if ELEV ≤ 5000 ft, then

$$PATM = [760.2 - (.02478 * ELEV)] * \frac{14.698}{760.2} \qquad (15.5)$$

if ELEV > 5000 ft, then

$$PATM = [742.8 - (.0213 * ELEV)] * \frac{14.698}{760.2} \qquad (15.6)$$

where
  PATM   =   absolute atmospheric pressure, lb/in.^{2} (psi)
  ELEV   =   site elevation above sea level, ft

Moisture Content

Water vapor pressure in the gas, PV, is determined from the input gas temperature, TEMP, and relative humidity, RHGAS, using Equations 11.55 and 11.56. Partial pressure of the dry gas, PDRYGAS, is then determined as

$$PDRYGAS = PATM - PV \qquad (15.7)$$

Molecular weight of the combined dry gas plus water vapor, MOLWT, is then determined as

$$MOLWT = \frac{28.96 * PDRYGAS + 18.02 * PV}{PATM} \qquad (15.8)$$

Specific Weight

The specific weight of gas within the manifold or orifice is determined by application of Equation 11.53 as follows:

$$GAMMAPIPE = \frac{PPIPE * MOLWT}{1545 * (460 + TEMP)} \qquad (15.9)$$

or

$$GAMMAPORT = \frac{PPORT * MOLWT}{1545 * (460 + TEMP)} \qquad (15.10)$$

where

$$
\begin{array}{rcl}
GAMMAPIPE & = & \text{gas specific weight inside manifold, lb/ft}^3 \text{ (pcf)} \\
GAMMAPORT & = & \text{gas specific weight outside manifold, pcf} \\
PPIPE & = & \text{absolute static pressure inside manifold, lb/ft}^2 \text{ (psf)} \\
PPORT & = & \text{absolute static pressure outside manifold, psf}
\end{array}
$$

Actual and Standard Volumes

Actual gas flowrate within the manifold, QACTPIPE, is determined from the gas velocity, VPIPE, and manifold area, APIPE(I):

$$QACTPIPE = VPIPE * APIPE(I) \qquad (15.11)$$

where

$$QACTPIPE = \text{gas flowrate in manifold, actual ft}^3\text{/sec (acfs)}$$

The actual flowrate is converted to standard conditions of 68°F and 14.7 psia using

$$QSTDPIPE = \frac{QACTPIPE * 528}{(460 + TEMP) * \left(\dfrac{PPIPE}{14.7 * 144} \right)} \qquad (15.12)$$

where

$$QSTDPIPE = \text{gas flowrate in manifold, standard ft}^3\text{/sec (scfs)}$$

Orifice Discharge Formulae

The mass flowrate of a compressible gas, G, from an orifice is determined as[7]

$$G = CYA(2g\delta\Delta P)^{0.5} \qquad (15.13)$$

where

G	=	mass flowrate
C	=	orifice discharge coefficient
Y	=	expansion factor for compressible flow
A	=	orifice area
g	=	gravity constant
δ	=	gas specific weight upstream of the orifice
ΔP	=	static pressure drop across the orifice

The orifice discharge coefficient, C, accounts for jet contraction through the orifice and friction losses within the orifice. For squared edged orifices, C is typically 0.60. For rounded or bell mouthed orifices, C = 0.98.

With compressible fluids, the specific weight will decrease as pressure drops across the orifice. As a result the actual mass flowrate will be less than if the gas were noncompressible. The expansion factor, Y, is used to correct for this effect. Y is a function of (1) the gas specific heat ratio, k, (2) the ratio of orifice to manifold diameter, and (3) the ratio of downstream to upstream absolute pressures. Over the range of pressure drops typical of composting applications, and assuming air with a specific heat ratio of 1.40, the expansion factor for sharp edged orifices can be estimated as[8,9]

$$Y = 1 - (0.41 + 0.35 * BETA4) * FLOWTYPE * \frac{PPIPE - PPORT}{PPIPE * 1.40} \qquad (15.14)$$

and for bell mouthed or rounded orifices as

$$Y = 1 - (0.84 + 1.34 * BETA4) * FLOWTYPE * \frac{PPIPE - PPORT}{PPIPE * 1.40} \qquad (15.15)$$

where

BETA4	=	$(DPORT/DPIPE)^4$
FLOWTYPE	=	flag to identify forced or induced draft flow; +1 for forced draft and −1 for induced draft

Using Equation 15.13, the mass flowrate under forced draft conditions can be determined as

$$MPORT = C * Y * APORT(I) *$$
$$(64.4 * GAMMAPIPE * FLOWTYPE * DELTAPPORT)^{0.5} \qquad (15.16)$$

and under induced draft conditions as

$$MPORT = C * Y * APORT(I) *$$
$$(64.4 * GAMMAPORT * FLOWTYPE * DELTAPPORT)^{0.5} \qquad (15.17)$$

where

MPORT	=	mass flowrate in port, lb/sec

APORT(I) = area of port or orifice in section I, ft^2
DELTAPPORT = static pressure drop across orifice, lb/ft^2 (psf)

The velocity of discharge from the port and the volumetric flow are determined as

$$VPORT = \frac{MPORT}{APORT(I) * GAMMAPORT} \qquad (15.18)$$

and

$$QACTPORT = VPORT(I) * APORT(I) \qquad (15.19)$$

where
    VPORT = gas velocity in port, ft/sec (fps)
    QACTPORT = volumetric discharge from port, acfs

The mass flowrate at any point within the manifold, MPIPE, is determined by summing the individual orifice flowrates up to that point. The manifold velocity is determined as

$$VPIPE = \frac{MPIPE}{APIPE(I) * GAMMAPIPE} \qquad (15.20)$$

where
    VPIPE = gas velocity in manifold, fps

Pressure Drop Across Compost

The total pressure drop from the manifold to atmosphere, DELTAP in Figure 15.13, is not available to induce flow across the orifice. A part of the total DELTAP is expended in maintaining flow across the composting material itself. Gas flow through compost can be described as fluid flow through a porous media. Darcy's Law is commonly used to describe flow in such cases:

$$\Delta P = K(V)L \qquad (15.21)$$

where
    ΔP = pressure drop
    K = coefficient related to permeability of the porous media
    V = velocity of flow
    L = length of the flow path

Darcy's Law applies to the laminar flow region and is often modified for applications above the critical Reynold's number as follows:

$$\Delta P = K(V^n)L \qquad (15.22)$$

where
    n = velocity exponent, usual range from 1.0 to 2.0

Unlike porous media such as sand, compost is subject to compaction. A height exponent, j, can be added to Equation 15.22 to account for the increase in compaction with increased height.

$$\Delta P = K(V^n)(L^j) \tag{15.23}$$

Using Equation 15.23, pressure drop across the compost, DELP, is modeled as

$$DELP = K * (VPILE^N) * (HPILE^J) \tag{15.24}$$

where

      HPILE = height of compost pile, ft
      VPILE = gas velocity in pile, fps

A summary of experimentally determined values of K, J, and N for various composting materials is presented in Table 15.1.

The model allows the compost pile to be divided into as many as three separate zones. This is typically the case with static pile systems which may use a bottom layer of new wood chips, a middle layer of sludge/wood chip mixture, and a top cover layer of screened chips or final compost. Pressure drop across each zone is determined using Equation 15.24. Individual pressure drops across each zone are defined as DELP1, DELP2, and DELP3 in accordance with Figure 15.13.

An iterative procedure is necessary at each port to determine the mass flowrate from that port. Referring to Figure 15.13, the procedure begins by first assuming that DELTAPPORT = DELTAP. Equations 15.16 through 15.19 are solved for MPORT and QACTPORT. Gas velocity within the compost pile is determined as

$$VPILE = \frac{QACTPORT}{WPILE * SPACING(I) / 12} \tag{15.25}$$

where

      VPILE = average gas velocity within the compost pile, fps
      WPILE = width of compost pile, ft

Equation 15.24 is then applied to each of the compost zones to determine DELP1, DELP2, and DELP3. Total headloss is the sum of DELTAPPORT, DELP1, DELP2, and DELP3. The total headloss is compared to DELTAP. Total headloss will be greater than DELTAP on the first iteration. DELTAPPORT is then reduced for the second iteration. This procedure continues until the sum of individual headlosses equals DELTAP to within 1% error.

Manifold Headloss

Friction headloss in the manifold is calculated using the Darcy-Weisbach formula:[7]

$$h_1 = f \frac{L}{D} \frac{V^2}{2g} \tag{15.26}$$

Table 15.1. Airflow Resistance Coefficients for Composting Mixtures[a]

Substrate	K	J	N	Reference
Wood chip:sludge				
2:1 vol ratio	1.245	1.05	1.61	Singley et al.,[10] Higgins et al.[11]
3:2	1.529	1.30	1.63	Singley et al.,[10] Higgins et al.[11]
1:1	2.482	1.47	1.47	Singley et al.,[10] Higgins et al.[11]
1:2	7.799	1.41	1.48	Singley et al.,[10] Higgins et al.[11]
New wood chips	0.539	1.08	1.74	Singley et al.,[10] Higgins et al.[11]
Recycled chips	3.504	1.54	1.39	Singley et al.,[10] Higgins et al.[11]
Screened compost	1.421	1.66	1.47	Singley et al.,[10] Higgins et al.[11]
Screened compost 70% TS	27	1.0	1.10	Willson et al.[12]
Screened compost 74% TS	58	1.0	1.10	Willson et al.[12]

[a] Using Equation 15.24 and the above coefficients, DELP will have units of in. wc using units of fps for VPILE and feet for HPILE.

where

h_1 = headloss in height of fluid
f = Darcy-Weisbach friction factor
L = length of pipe section
D = diameter of equivalent circular section
V = velocity
g = gravity constant

Using model nomenclature, Equation 15.26 applied between successive ports becomes

$$HL = \frac{F * SPACING(I) * VPIPE * VPIPE}{DPIPE(I) * 64.4} \tag{15.27}$$

where

HL = friction loss, ft of gas

For noncircular cross sections the hydraulic radius is used to determine an equivalent pipe diameter for use in the Darcy formula. Actual areas and velocities are used in all other calculations.

The friction factor is a function of Reynold's Number, NR, and the relative surface roughness, e/D or EOVERD. e is the absolute surface roughness and D the pipe diameter, DPIPE. The dimensionless Reynold's Number is determined as

$$NR = \frac{VPIPE * [DPIPE(I) / 12] * GAMMAPIPE}{32.2 * VISCOS} \tag{15.28}$$

where

$VISCOS$ = gas viscosity, $lbf\text{-}sec/ft^2$

In the laminar flow region, where NR < 2100, the friction factor is independent of surface roughness and is given by the Hagen-Poiseuille formula:[7]

$$F = \frac{64}{NR} \tag{15.29}$$

In the transition and turbulent regions where NR > 2100, the friction factor is determined from the equation of Colebrook and White:[8,13]

$$\frac{1}{F^{0.5}} = -2 * LOG_{10}\left\{\frac{EOVERD(I)}{3.7} + \frac{2.51}{NR * F^{0.5}}\right\}$$ (15.30)

where
$\quad\quad$ EOVERD(I) $\;=\;$ relative roughness in manifold section I

Because F appears on both sides of Equation 15.30, an iterative procedure is again necessary for solution. The Colebrook equation is a good representation of the Stanton-Moody diagram for friction factors which is usually presented in fluid mechanics texts. Despite the requirement for iteration, it is considerably more useful in computer applications than the equivalent graphical form of the Stanton-Moody diagram.

Absolute and Relative Roughness

Tabular and graphical summaries of absolute roughness and relative roughness ratios for a wide variety of materials are available in many fluid mechanics texts. Values are readily available for most materials used in composting practice, including metallic pipes, concrete, and smooth walled plastic. For example, smooth walled plastic can usually be modeled as a smooth walled pipe (e/D = 0) whereas concrete has an absolute roughness ranging from 0.001 to 0.01 ft.

Corrugated agricultural drainage pipe, usually constructed from polyethylene plastic, is often used in temporary and semipermanent manifold systems. Values of absolute and relative roughness are less readily available for this often used material. Therefore, a discussion of experimental results for such materials is presented here.

Normal convention with corrugated pipe is to use the nominal inside diameter, or the actual measurement of the minimum bore, to estimate e/D and average pipe velocity. Surface roughness is the depth of corrugations beyond the inner diameter as shown in Figure 15.14. Using this convention, Irwin and Motycka[14] developed friction factors for corrugated piping. Results of their testing are presented in Table 15.2. Measured e/D values ranged from about 0.076 to 0.097. These values of e/D are very high and are indicative of rough pipe conditions. Measured values of the friction factor, f, were constant at Reynold's numbers above 10^5 and ranged from 0.070 to 0.095.

Equation 15.30 was used to estimate friction factors from the e/D measurements in Table 15.2, assuming a Reynold's number of 10^5. Results are presented in column 7 of Table 15.2. Friction factors determined from Equation 15.30 are in general agreement, although consistently higher than those experimentally determined by Irwin and Motycka. The energy-dissipating eddies generated by the uniform wall corrugations apparently effect the flow to a lesser extent when compared to correlations, such as the Colebrook equation, which are based primarily on data developed at lower e/D ratios.

If only the physical dimensions of the piping are known, the designer can determine the e/D ratio and use the Colebrook equation to obtain a reasonable estimate of friction factor. However, a better approach is to obtain actual friction loss data from the pipe supplier. This data usually comes in the form of graphs relating velocity, duct diameter, and friction loss per length of pipe. From this data, an equivalent roughness for the piping can be determined for use in the "MANIFOLD" model (see Example 15.1).

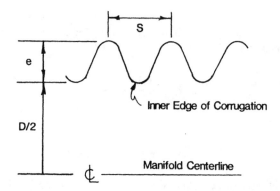

Figure 15.14. Characteristic dimensions for corrugated piping. From Motycka.[14]

Manifold Pressure Profile

With compressible fluids it is customary to use the terms total pressure, P_t, static pressure, P_s, and velocity pressure, P_v. Static pressure is the pressure that would be observed from a tap at the side of the pipe. Velocity pressure, equivalent to $V^2/2g$, is the pressure difference between pitot and static tubes inserted within the pipe. Total pressure is the sum of static and velocity pressures.

The pressure profile along the manifold is determined in a stepwise fashion beginning at the end orifice. Knowing the total pressure at port I, static and total pressures at port I + 1 are determined as

$$P_t(I+1) = P_t(I) + FLOWTYPE * HL * GAMMAPIPE \qquad (15.31)$$

$$P_v(I) = \left(\frac{VPIPE * VPIPE}{64.4} \right) * GAMMAPIPE \qquad (15.32)$$

$$P_s(I+1) = P_t(I+1) - P_v(I) \qquad (15.33)$$

$P_s(I + 1)$ is equivalent to DELTAP at port I + 1. Thus, the orifice equations can be used at port I + 1, the new orifice flow added to the total manifold flow, a new headloss calculated between port I + 1 and I + 2, and the pressure profile advanced to port I + 2. In this way the algorithm advances along the manifold until the last port is reached. The algorithm advances one port spacing beyond the last orifice and then stops.

Closure Algorithm

Solving a manifold problem is a trial and error procedure that begins by assuming a total pressure at port 1, PMAN1, in in. wc. A positive value for PMAN1 is used for forced draft and a negative value for induced draft. The algorithm then moves up the manifold, solving for port flows and manifold headlosses. By the time the end port is reached, the pressure profile and manifold flowrate corresponding to the assumed total pressure at port 1 are known. However, it is unlikely that the calculated flowrate will equal the desired flowrate, unless one is very lucky indeed. It is then necessary to assume a new total pressure at port 1 and repeat the procedure until the desired flowrate is achieved. QSET is the desired flowrate from the

Table 15.2. Corrugated Piping Dimensions and Friction Factors

Inside Diameter		Abs Roughness			Friction Factor	
(mm)	(in.)	(mm)	(in.)	e/D Ratio	(measured)	Eq. 15.30
100	3.93	7.6	0.299	0.076	.070	.088
200	7.87	18.3	0.720	0.092	.082	.097
250	9.84	20.5	0.807	0.082	.068	.092
300	11.81	29.2	1.15	0.097	.095	.100

Source: Irwin and Motycka.[14]

manifold under standard conditions. QSTDPIPE is the calculated manifold flowrate under standard conditions. Closure is achieved when the relative error between QSET and QSTDPIPE is $\leq 0.1\%$.

The algorithm for estimating a new manifold pressure at port 1, PMAN1, makes use of the fact that orifice flow is a function of the square root of static pressure. Thus, estimating a new PMAN1 as

$$PMAN1 = PMAN1 * \left(\frac{QSET}{QSTDPIPE} \right)^2 \tag{15.34}$$

provides a good closure rate on the problem.

Rules of Thumb

Several "rules of thumb" for good engineering practice have been developed to guide the initial design of manifold systems. Rule 1 states that total port area should not exceed the manifold area, or

$$\frac{\sum APORT(I)}{APIPE(NUMSECTION)} < 1.0 \tag{15.35}$$

This assures that, on average, flow will accelerate across the orifice. This should be the normal flow situation with any orifice. If this ratio is >1.0, flow will actually decelerate across the orifice.

Rule 2 is that friction headloss along the manifold should be less than average port pressure drop, or

$$\frac{SUMHL}{ABS(PPIPE - PATM * 144)} < 1.0 \tag{15.36}$$

where
$$SUMHL = \text{sum of manifold friction losses, psf}$$

This rule is directed toward maintaining a reasonable flow distribution along the manifold. If headloss is too high, friction losses will cause uneven distribution if port spacing and size remain constant.

Rule 3 is that pressure drop across the individual orifices is necessary to achieve uniform flow distribution. Flow cannot be distributed uniformly without pressure drop. In general, increasing orifice pressure drop improves the flow distribution and increases the manifold resistance to distribution changes. The latter can result from flowrate changes or changing composition in the compost overburden.

Many manifold designs in composting practice, particularly the temporary and semipermanent types, exceed the above limits for good engineering practice. This does not necessarily mean that the manifolds are improperly designed. Rather, the designer has chosen to deviate from established practice to achieve another desired benefit, e.g., reduced power consumption. Trade-offs between uniform flow distribution and other engineering parameters will become clearer after study of the example problems.

MANIFOLD DESIGN

Several example problems are presented to illustrate the general approach to hydraulic analysis. These examples will also highlight design aspects of manifold systems important to avoid distribution problems.

Example 15.1

A piping supplier provides graphical test data for 6 in. ID corrugated piping which indicates a friction loss of 1.6 in. water column (wc)/100 ft of pipe at 1500 fpm velocity. Determine the absolute and relative roughness ratios.

Solution

1. Convert the headloss to psf:

$$\text{headloss} = 1.6 \text{ in. } (62.4 \text{ pcf})(\text{ft}/12 \text{ in.}) = 8.32 \text{ psf}$$

2. Assuming a gas density of ~0.075 pcf (lb/ft^3), determine the headloss in feet of fluid as

$$\frac{8.32 \text{ psf}}{0.075 \text{ pcf}} = 111 \text{ ft of air}$$

3. Determine the friction factor using Equation 15.26:

$$V = \frac{1500}{60} = 25 \text{ fps}$$

$$111 = \frac{f * 100 * (25)^2}{0.5 * 2 * 32.2}$$

$$f = 0.057$$

4. Referring to a Stanton-Moody diagram from any fluid mechanics text, for f = 0.057 in the wholly rough zone the relative roughness, e/D, is ~.03.

5. Absolute roughness is then determined as

$$e = 0.03(6) = 0.18 \text{ inches or } .015 \text{ feet}$$

6. Alternatively, e/D could be determined by solution of Equation 15.30. Assuming a high value of NR, say 10^7, makes the solution convenient because the last term in Equation 15.30 is small and can be ignored.

$$\frac{1}{(0.057)^{0.5}} = -2 * \log_{10}\left(\frac{\text{EOVERD}}{3.7}\right)$$

$$\frac{e}{D} = \text{EOVERD} = 0.0297$$

7. Note that the conditions of this problem are based on actual headloss data from one supplier of corrugated plastic piping.

Example 15.2

A 2:1 wood chip:raw sludge mixture is to be composted. The pile will be 150 ft long with a mix depth of 10 ft. A manifold will supply air to the pile. Multiple, parallel manifold pipes will be arranged on 5 ft centers. Estimate the peak flowrate that can be expected from each manifold pipe.

Solution

1. Total pile volume served by each manifold pipe is

$$\text{volume} = 150(10)(5) = 7500 \text{ cu ft}$$

2. A mixture of 2 volumes wood chips and 1 volume sludge will occupy something less than 3 volumes when mixed. For quick calculation assume a final mix volume of 2.5. The quantity of sludge in the pile can be estimated as

$$\frac{7500}{2.5} = 3000 \text{ ft}^3$$

3. Assuming a bulk weight of 60 lb/ft^3 and 20% TS, the dry weight is

$$3000(60)\left(\frac{0.2}{2000}\right) = 18 \text{ tons}$$

4. From Chapters 7 and 13, a reasonable peak demand of 4000 ft^3/h-dt is estimated for raw sludge. Peak air flow is then determined as

$$\frac{4000(18)}{60 * 60} = 20 \text{ cfs}$$

Example 15.3

The corrugated piping from Example 15.1 has been selected for use in a static pile composting system. Pipe ID is 6 in., with an absolute roughness of 0.015 ft. John Q. Designer has initially selected

0.25 in., circular, square edged holes on 6 in. centers. The manifold is to be 150 ft long. Flow is induced draft and QSET is 20 scfs (per Example 15.2). Gas temperature is estimated to be 120°F and relative humidity 90%. Three layers of compost are planned, a 1 ft layer of new chips, 10 ft of a 2:1 chip:sludge mix, and 1 ft of screened compost. Individual pipe manifolds are arranged on 5 ft centers. Evaluate the designer's manifold, including mass distribution and power requirements.

Solution

1. Rule 1 can be estimated from Equation 15.35 as follows:

$$\frac{\dfrac{150*2}{ft}*\left(\dfrac{3.14}{4}\right)*\left(\dfrac{0.25}{12}\right)^2}{\dfrac{3.14}{4}*\left(\dfrac{6}{12}\right)^2}=0.52$$

The manifold design complies with Rule 1.

2. Load MANDATA and follow the menu to create a data file named DATA1503. Input the following data:

MDY	=	(enter current value)
IDEN	=	EXAMPLE 15.3
NUMSECTION	=	1
MANTYPE	=	1
FLOWTYPE	=	−1
ELEV	=	1000 ft
TEMP	=	120 deg F
RHGAS	=	0.90
VISCOS	=	0.0000004 lb-sec/ft^2
E	=	0.015 ft
QSET	=	20 scfs
PMAN1	=	−0.10 in. wc
WPILE1	=	5.0 ft
WPILE2	=	5.0 ft
WPILE3	=	5.0 ft
HPILE1	=	1.0 ft
HPILE2	=	10.0 ft
HPILE3	=	1.0 ft
K1	=	0.539
K2	=	1.245
K3	=	1.421
J1	=	1.08
J2	=	1.05
J3	=	1.66
N1	=	1.74
N2	=	1.61
N3	=	1.47
PORTTYPE(1)	=	1
NUMPORTS(1)	=	300
PRINTK(1)	=	10
SPACING(1)	=	6.0 in.

DPORT(1)   =   0.25 in.
DPIPE(1)   =   6.0 in.

3. Exit MANDATA, load MANIFOLD, and run using DATA1503 when prompted for the data file name. The program will require about 8 min computation time when using an 8088 based PC.

4. A summary of results is as follows:

Port Number	Length from End(ft)	Mass Flow (lb/sec-ft)	VPIPE (fps)	DELTAP (in. wc)
1	0.0	.00887	0.4	−22.96
101	50.0	.00901	38.4	−23.72
201	100.0	.00978	79.2	−28.13
300	149.5	.0116	128.8	−40.36

Conditions 6 in. downstream of the last port are

$$\begin{aligned}
\text{total headloss} &= 14.76 \text{ in. wc} \\
\text{static pressure} &= -40.54 \text{ in. wc} \\
\text{total pressure} &= -37.72 \text{ in. wc} \\
\text{fan power at 70\% efficiency} &= 12.88 \text{ hp}
\end{aligned}$$

5. The mass flow distribution is reasonably good, varying only about 30% from low to high. It could be improved significantly by increasing the manifold size, say to 8 in., to reduce friction headloss.

6. While the distribution is acceptable and would improve with a larger manifold, it is achieved by taking a rather large DELTAP across the ports. Fan horsepower is high because of the high suction pressure at the inlet. Port velocities exceed 200 fps which may cause concern about entrainment of solids.

Example 15.4

John Q. Designer decides that fan horsepower in Example 15.3 must be reduced. He proposes to increase the number of ports. 600 additional ports, 0.25 in. ID each, are drilled into the pipe, leaving a uniform spacing of 2 in. between ports. All other conditions remain the same as Example 15.3. Reevaluate the manifold flow characteristics for the designer's second attempt.

Solution

1. Load MANDATA and follow the menu instructions to duplicate the existing data file DATA1503 as DATA1504. Follow the menu to edit DATA1504 as follows:

$$\begin{aligned}
\text{IDEN} &= \text{EXAMPLE 15.4} \\
\text{NUMPORTS(1)} &= 900 \\
\text{SPACING(1)} &= 2.0 \text{ in.}
\end{aligned}$$

2. Exit "MANDATA", load "MANIFOLD", and run. Input DATA1504 at the prompt requesting the data file name.

3. A summary of results is as follows:

Port Number	Length from End (ft)	Mass Flow (lb/sec-ft)	VPIPE (fps)	DELTAP (in. wc)
1	0.0	0.00566	0.08	−1.03
301	50.0	0.00645	23.75	−1.33
601	100.0	0.0105	56.80	−3.56
900	149.8	0.0204	119.76	−13.44

Conditions 2 in. downstream from the last port are

$$
\begin{aligned}
\text{total headloss} &= 9.85 \text{ in. wc} \\
\text{static pressure} &= -13.51 \text{ in. wc} \\
\text{total pressure} &= -10.88 \text{ in. wc} \\
\text{fan power at 70\% efficiency} &= 3.46 \text{ hp}
\end{aligned}
$$

4. Horsepower and port velocities have both decreased compared to Example 15.3. However, flow distribution has been compromised, varying over 250% from low to high. The maximum flow per length of manifold occurs at the fan end. This is always the case regardless of whether flow is forced or induced draft. The poor flow distribution results from the fact that friction headloss greatly exceeds average port DELTAP, a violation of Rule 2 (Equation 15.36). Note also that total port area/manifold area = 1.56, which exceeds Rule 1 (Equation 15.35).

Example 15.5

Encouraged by having reduced the fan power, but unhappy with the new flow distribution, John Q. recommends increasing the manifold size to 8 in. ID to improve the distribution. Evaluate John Q.'s latest recommendation.

Solution

1. Again load MANDATA and follow the menu to duplicate the existing data file DATA1504 as DATA1505. Then modify DATA1505 as follows:

$$
\begin{aligned}
\text{IDEN} &= \text{EXAMPLE 15.5} \\
\text{DPIPE(1)} &= 8 \text{ in.}
\end{aligned}
$$

Exit MANDATA, load MANIFOLD, and run. Input DATA1505 at the prompt requesting the data file name.

2. A summary of results is as follows:

Port Number	Length from End (ft)	Mass Flow (lb/sec-ft)	VPIPE (fps)	DELTAP (in. wc)
1	0.0	0.00822	0.06	−2.16
301	50.0	0.00850	18.90	−2.31
601	100.0	0.00996	39.63	−3.17
900	149.8	0.0133	66.02	−5.68

Conditions 2 in. downstream from the last port are

$$
\begin{aligned}
\text{total headloss} &= 2.71 \text{ in. wc} \\
\text{static pressure} &= -5.69 \text{ in. wc}
\end{aligned}
$$

$$\begin{aligned}\text{total pressure} &= -4.88 \text{ in. wc}\\ \text{fan power at 70\% efficiency} &= 1.52 \text{ horsepower}\end{aligned}$$

3. The mass flow distribution is greatly improved by increasing the manifold size from 6 in. ID to 8 in., varying about 60% from low to high. Fan horsepower is also reduced because of the reduction in friction losses. Despite these improvements, the distribution is less uniform than that achieved in Example 15.3 where a high port DELTAP was taken. The 60% variation is still not considered acceptable.

Example 15.6

Frustrated by the above results, John Q. takes a coffee break. A friend suggests that the problem can be solved by using forced draft instead of induced draft. John Q. decides to evaluate this idea. Air temperature is estimated at 70°F with 50% relative humidity.

Solution

1. Again load MANDATA and follow the menu to duplicate the existing data file DATA1505 as DATA1506. Then edit DATA1506 as follows:

$$\begin{aligned}\text{IDEN} &= \text{EXAMPLE 15.6}\\ \text{FLOWTYPE} &= +1\\ \text{TEMP} &= 70\\ \text{RHGAS} &= 0.50\\ \text{PMAN1} &= 2.0 \text{ in. wc}\end{aligned}$$

Exit MANDATA, load MANIFOLD, and run. Input DATA1506 at the prompt requesting the data file name.

2. A summary of results is as follows:

Port Number	Length from End (ft)	Mass Flow (lb/sec-ft)	VPIPE (fps)	DELTAP (in. wc)
1	0.0	0.00924	0.06	2.39
30	50.0	0.00928	18.32	2.41
601	100.0	0.0101	37.21	2.84
900	149.8	0.0124	58.85	4.31

Conditions 2 in. downstream from the last port are

$$\begin{aligned}\text{total headloss} &= 2.68 \text{ in. wc}\\ \text{static pressure} &= 4.31 \text{ in. wc}\\ \text{total pressure} &= 5.07 \text{ in. wc}\\ \text{fan power at 70\% efficiency} &= 1.41 \text{ hp}\end{aligned}$$

3. To John Q.'s surprise the flow distribution is improved compared to Example 15.5, varying 34% from low to high. This results from a number of factors. First, the lower gas temperature and water vapor pressure result in an increase in gas density, which in turn reduces gas velocity and headloss in the manifold. Second, static pressure difference at the end of the manifold is less in the forced draft mode because of the effect of velocity head. The velocity head causes static pressure to be less than total pressure for forced draft, but greater (more negative) than total pressure for induced draft.

4. Fan power is reduced slightly compared to the equivalent induced draft case in Example 15.5. This is caused by the increased gas density with forced draft.

5. Under equivalent conditions, forced draft operation will generally result in better flow distribution and slightly lower power consumption compared to induced draft. Excited by this discovery, John Q. buys coffee for everyone.

Example 15.7

John Q. Designer proudly presents his manifold design from Example 15.6 to management for approval. Unfortunately, management announces that the client prefers induced draft operation because the process gas can be collected for subsequent odor treatment. John Q. returns to the manifold in Example 15.5 but decides to use multiple pipe sections. Section 1 will retain the 2 in. port spacing for the first 100 ft. Section 2 will use 3 in. port spacing for the last 50 ft.

Solution

1. Again load MANDATA and follow the menu to duplicate the existing data file DATA1505 as DATA1507. Then modify DATA1507 as follows:

$$
\begin{aligned}
\text{IDEN} &= \text{EXAMPLE 15.7} \\
\text{NUMSECTION} &= 2 \\
\text{PMAN1} &= -2.0 \text{ in. wc} \\
\text{NUMPORTS(1)} &= 600 \\
\text{PORTTYPE(2)} &= 1 \\
\text{NUMPORTS(2)} &= 200 \\
\text{PRINTK(2)} &= 50 \\
\text{SPACING(2)} &= 3.0 \text{ in.} \\
\text{DPORT(2)} &= 0.25 \text{ in.} \\
\text{DPIPE(2)} &= 8.0 \text{ in.}
\end{aligned}
$$

Exit MANDATA, load MANIFOLD, and run. Input DATA1507 at the prompt requesting the data file name.

2. A summary of results is as follows:

Port Number	Length from End (ft)	Mass Flow (lb/sec-ft)	VPIPE (fps)	DELTAP (in. wc)
		Section 1		
1	0.0	0.00950	0.07	-2.89
301	50.0	0.00983	21.90	-3.10
600	99.8	0.0115	45.88	-4.24
		Section 2		
601	100.1	0.00775	45.97	-4.25
701	125.1	0.00869	55.48	-5.35
800	149.8	0.00986	66.23	-6.90

Conditions 3 in. downstream from the last port are

$$
\begin{aligned}
\text{total headloss} &= 3.21 \text{ in. wc} \\
\text{static pressure} &= -6.92 \text{ in. wc}
\end{aligned}
$$

$$\text{total pressure} = -6.10 \text{ in. wc}$$
$$\text{fan power at 70\% efficiency} = 1.90 \text{ hp}$$

3. The use of multiple sections improves the distribution compared to Example 15.5. The range from low to high is about 48%. If the first 25 ft of Section 2 is excluded, the range is only 32%. This is approximately equivalent to Example 15.6 for the case of forced draft.

4. Fan power is increased compared to Example 15.5 because the reduced number of ports requires a greater average DELTAP along the manifold to induce the same flowrate.

5. John Q. decides that adding more sections to the manifold will eventually lead to the "perfect" manifold design, i.e., perfectly even distribution with low DELTAP and low fan horsepower. While John Q. may never achieve the "perfect" manifold, the path to developing a better manifold is outlined in Examples 15.1 to 15.7. The computer models can be used to see if further improvements can be made to John Q's latest design.

Example 15.8

Before leaving this problem, John Q. decides that the manifold distribution under a reduced flow condition of 10 scfs should be checked.

Solution

1. Again load MANDATA and follow the menu to duplicate the existing data file DATA1507 as DATA1508. Then modify DATA1508 as follows:

$$\text{IDEN} = \text{EXAMPLE 15.8}$$
$$\text{QSET} = 10.0$$

Exit MANDATA, load MANIFOLD, and run. Input DATA1508 at the prompt requesting the data file name.

2. A summary of results is as follows:

Port Number	Length from End (ft)	Mass Flow (lb/sec-ft)	VPIPE (fps)	DELTAP (in. wc)
		Section 1		
1	0.0	0.00476	0.04	−0.73
301	50.0	0.00492	10.89	−0.78
600	99.8	0.00576	22.78	−1.06
		Section 2		
601	100.1	0.00387	22.82	−1.07
701	125.1	0.00433	27.47	−1.34
800	149.8	0.00492	32.68	−1.72

Conditions 3 in. downstream from the last port are

$$\text{total headloss} = 0.80 \text{ in. wc}$$
$$\text{static pressure} = -1.73 \text{ in. wc}$$
$$\text{total pressure} = -1.53 \text{ in. wc}$$
$$\text{fan power at 70\% efficiency} = 0.24 \text{ hp}$$

3. The flow distribution remains similar to that in Example 15.7. This is due in part to the fact that a reasonable DELTAP is taken across the orifices in Example 15.7. Even with the reduced flowrate in this example, DELTAP remains reasonable. However, this will not always be the case depending on the particular manifold and the range of flowrates.

4. All proposed manifold designs should be evaluated over the range of expected flowrates.

DISCUSSION

The example problems illustrate that proper manifold design is not a simple task. A well designed manifold should achieve uniform flow distribution with reasonable headloss and fan horsepower and low capital cost. However, it is not an easy task to achieve all of these goals simultaneously. Many trials are usually necessary before a compromise solution emerges for a particular situation.

Flow Distribution

Flow distribution can be improved by using one or more of the following approaches.

Larger Manifold

Increasing the size of the manifold will reduce headloss along the manifold length so that each port will experience essentially the same DELTAP. This is probably the most effective approach to achieve uniform distribution. However, this approach normally requires a permanent system because of the large manifold dimensions that are usually required.

Disposable piping, which is often used in temporary and semi-permanent systems, is more expensive in the larger sizes. Therefore, the designer is often limited to a maximum pipe size which will not conflict with operating cost constraints. Also, disposable piping is usually hand laid. This places another constraint on the maximum size which can be effectively managed by the operating staff. A 6 to 8 in. corrugated pipe seems to be about the maximum used in practice, although the designer should consider the cost effectiveness of larger sizes.

Shorter Manifold

As manifold length increases, so too does the headloss. Therefore, shortening the manifold length is one approach to improving flow distribution. Oftentimes, however, the demands of site layout, equipment movement, and number of operating staff limit the minimum pile dimensions. Permanent manifolds are less constrained in this regard because of their lower headloss and shorter length. Permanent manifold lengths up to 300 ft have been used in practice. Temporary systems using disposable piping are generally greater than 100 ft in length, but less than 200 ft.

Techniques to reduce the effective manifold length are shown in Figure 15.15. In one case, a length of nonperforated pipe is run to the far end of the manifold, creating a second pipe loop. In another, fans are located at each end of the manifold. In these cases, flow is added or removed from the manifold from each end, effectively reducing the manifold length for headloss in half.

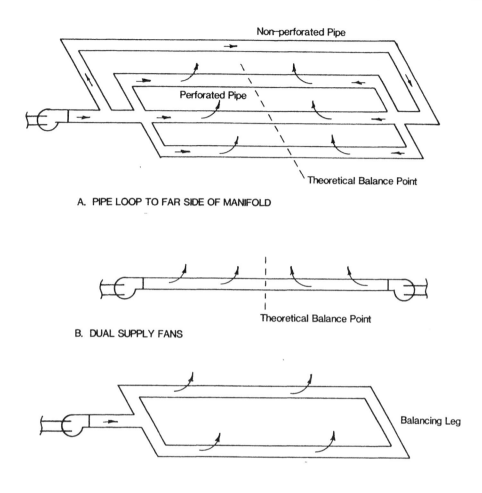

Figure 15.15. Techniques used to reduce the effective length of a manifold: (a) connecting both ends of the manifold with a separate pipe loop, (b) installing fans at both ends of the manifold, and (c) using a single pipe loop.

Increased Port DELTAP

Rule 3 states that flow distribution cannot be achieved without taking a pressure drop. Increasing the pressure drop across the port decreases the relative effect of pressure drop along the manifold. This is illustrated in Example 15.3, where a high port DELTAP achieved relatively good distribution even with a relatively small manifold. In effect, horsepower is used to achieve a better flow distribution.

Many manifold designs have attempted to reduce power consumption to save energy costs. This is an admirable goal, but sometimes has been carried to an extreme without proper regard for the effect on flow distribution. The author has evaluated manifold designs with as little as 0.05 in. wc drop across the first ports. It is virtually impossible to balance such a manifold. Minor changes in compost characteristics or manifold flowrate will disturb the flow distribution. Remember, good engineering practice requires energy dissipation to achieve a balanced manifold.

Reduced Flowrate

For a given manifold size, flow distribution will be improved by reducing the flowrate, which reduces the manifold headloss. However, the required air supply rate is dictated by the needs of the composting system. How can the flowrate be decreased without jeopardizing the composting process? The answer to this is to decrease the amount of compost served by the manifold. For example, shortening the manifold will also reduce the amount of compost to which air must be supplied. Therefore, the length of the manifold is a very important variable because it effects both the total headloss in the manifold and the required flowrate that must be supplied by the manifold.

Another technique to reduce manifold flowrate is to increase the number of parallel manifolds or legs within the system. This is a common technique for both permanent and temporary systems. The width of the compost pile served by each manifold is reduced, which reduces the required flowrate that must be supplied by each leg.

Forced Draft

As demonstrated in Example 15.6, forced draft will generally provide better distribution because (1) static pressure is less than total pressure and (2) gas density is greater. In the forced draft mode, velocity head subtracts from total pressure. This results in a static pressure difference across the port which is less than total pressure. With induced draft, the velocity head again subtracts from total pressure. However, because of the negative pressures relative to ambient, static pressure difference across a port is greater than total pressure. Another way of stating this is that velocity pressure works in favor of balancing the manifold in the forced draft mode, but against balancing in the induced draft mode.

Ambient air should be cooler and contain less water vapor than process gases drawn from the composting material. The only time this might not be true would be with very dry feed substrates or in long-term curing piles which are very stable and dry. Therefore, ambient air should be more dense than process gas. This in turn will result in lower manifold velocities, lower headloss, and improved distribution for the same standard flowrate.

The above arguments are not meant to imply that all manifolds should use forced draft. There are other important reasons for using induced draft and many situations in which induced draft is the only mode possible. For example, induced draft has been successfully used in both windrow and static pile systems to separately collect the process gas for subsequent deodorization prior to dispersion. In such a case, the needs for odor control are judged to outweigh the advantages of forced draft operation. As another example, some in-vessel systems use combinations of forced draft manifolds to supply air and induced draft manifolds to remove the process gas. This is sometimes referred to as a "push/pull" or "balanced draft" system. In this case, the designer must deal with each component to achieve an overall balance in the system.

Perhaps the best way to summarize the situation regarding forced draft operation is as follows. Forced draft operation can improve flow distribution, but the same improvements can also be achieved by the other techniques discussed in this section. Other factors, such as odor control, separation of process gas from the worker environment, and moisture migration, are probably equally important to the decision.

Multiple Sections

As illustrated in Example 15.7, the size and spacing of ports can be varied along the manifold to improve the manifold's balance. This is a common practice with temporary and semipermanent manifolds. Use of two to four different sections is common with static pile systems using disposable piping. Individual pipe sections with different port sizes and/or spacing can be ordered from the supplier. The different sections are then field assembled into a single manifold. This usually involves nothing more than "butting" the pipe ends together and fastening with duct tape. More elaborate joining techniques usually are not used because the manifold is disposable.

Some caution should be exercised when designing multiple sections into a manifold. The process can be carried to extremes. The author once analyzed a manifold containing 11 different sections, some with as many as 18 ports/ft and some with very imaginative hole patterns. Apparently, the operators knew the manifold was imbalanced and kept adding more holes in a heroic attempt to rebalance. By the time 11 sections had been reached, the port area/pipe area was 4.6, a serious violation of Rule 1. DELTAP was as low as 0.05 in. at port 1. Further efforts at balancing this way were doomed to failure. Manifold size and length could not be easily changed. Therefore, the proper approach to balancing was to reduce the number of ports, increase average port DELTAP, and increase the fan power.

Moisture Migration

An interesting phenomenon has been observed at several composting facilities using downflow, induced draft with wet substrates such as sludge cake. Moisture has tended to migrate downward through the compost in the direction of gas movement. Saturated conditions have occurred immediately above the induced draft manifold, causing increased headloss and short-circuiting around the saturated zones. Some facilities have reported that this saturated material becomes like "hardpan" if not removed frequently.

One case study occurred at the Delaware Reclamation Plant, which uses a circular, agitated bed system to compost heavy fraction from a refuse front end processing plant. The system was converted from forced draft to induced draft operation so that process gases could be separately collected and treated to improve the plant odor control. The process operated acceptably as long as heavy fraction was the main ingredient in the infeed mix. Over time, the proportion of wet sludge cake added to the mix, was increased. Moisture migration began to occur. Headloss increased, which caused fan flowrate to decrease to the point where stoichiometric demands were not met. Odor generation increased significantly. The facility eventually converted back to forced draft, upflow aeration.

A similar problem of hardpan formation was observed at the City of Akron bin type system and at the Denver Metropolitan Wastewater Reclamation District windrow system. In both cases the hardpan was observed after a period of time operating in the downflow mode with mixtures of sludge, sawdust, bark, and recycle. The hardpan was formed above the permanent manifold and was not completely removed after each cycle. This allowed the hardpan layer to accumulate over several cycles, eventually plugging off the manifold.

The mechanisms for the observed moisture migration are probably as follows. Heat from the moist composting material will heat and nearly saturate the process gases. Water vapor will be carried downward when the induced draft fans are on. Carried water should remain in the vapor form unless the gas cools in its travel toward the manifold. Given enough time, only a few degrees of cooling would be necessary for the gas to drop a significant amount of water.

Conductive losses from the bottom of the reactor might account for the cooling. Alternatively, if the fans are operated in an on/off mode, flow reversal will occur due to natural upward ventilation when the fans are off. This would introduce colder ambient air at the bottom of the pile. The downward induced flow would then "see" this cooler material and moisture condensation could occur. While the exact mechanisms leading to moisture saturation and "hardpan" formation are subject to debate, the phenomenon has been observed in both in-vessel and windrow facilities.

Interestingly, significant problems have not been reported at static pile systems, which often use induced draft aeration with wet sludge cake and wood chips. A wood chip is a large bulking particle that probably reduces the potential for water logging. Also, the piles are completely torn down at the end of each cycle. Any material that might become saturated or "hard" is removed at that time. The manifold is usually surrounded by a layer of new chips which also serves to protect the ports from plugging.

The common wisdom at this time is that induced draft aeration can be used with wet substrates that are bulked with large particles, such as wood chips, and/or where the material is completely removed at the end of each cycle to prevent any accumulation of saturated material. Certain in-vessel systems use lances which alternate between supplying air and removing gases. It was hoped that this alternating operation would avoid the hardpan problem with wet substrates. This does not appear to be the case in practice. Air lance facilities at Schnectady, New York, and Hartford, Connecticut, have converted to all positive aeration to mitigate flow distribution problems. The current practice suggests that forced aeration should be used with wet substrates amended with finer materials such as sawdust, unless the bottom material can be frequently removed. The problem is apparently not serious with relatively dry substrates.

Condensate Removal

Special care must be taken in design of manifolds whenever moisture condensation is likely to occur. This is usually the case with induced draft manifolds. Hot composting gases will begin cooling as they flow along the manifold, condensing water in the process. "Knock outs" must be provided at frequent intervals and especially at all low points to remove this condensate. Ammonia and organic compounds vaporized at the elevated composting temperatures will partially redissolve in the condensate. Subsequent treatment is usually required before condensate and leachate can be discharged. Special manifold and fan materials may be required because of the wet conditions. Fiberglass ducting, fan blades and housings are often used to avoid corrosion problems.

SUMMARY

Proper manifold design is an important aspect of compost system design. Most composting systems use some form of manifold to evenly distribute the air supply and/or evenly collect the process gas. The function of a manifold is to provide uniform flow distribution at reasonable cost, without excessive headloss or high power consumption. In general, distribution is improved by using higher headlosses across the individual orifices within the manifold. However, higher headloss also increases power consumption. The designer must balance these variables to achieve a suitable design for the particular situation.

A wide variety of manifold types are used in composting practice. Temporary systems use disposable piping with small fans not rigidly mounted to foundations. Permanent systems use fixed manifolds, usually below grade, with permanently mounted blowers. Both induced draft and forced draft manifolds are common.

Flow distribution can be improved by using one or more of the following approaches: (1) increasing the size of the manifold to reduce pipe velocity and headloss along the manifold length, (2) shortening the effective manifold length to reduce headloss and reduce the differences in pressure drop across the ports, (3) increasing port pressure drop, (4) reducing the flowrate by increasing the number of parallel manifolds, (5) using forced draft instead of induced draft supply, and (6) using multiple manifold sections with different port spacing and/or port size to counteract the effects of friction losses.

REFERENCES

1. Kuchenrither, R. D. "Design and Implementation Considerations for Composting Systems," in *Proceedings of the National Conference on Municipal Treatment Plant Sludge Management* (Silver Spring, MD: Hazardous Materials Control Research Institute, May 1986).
2. Higgins, A. J. "Ventilation for Static Pile Composting," *BioCycle* 23(4) (July/August 1982).
3. Hay, J. C. and Kuchenrither, R. D. "Fundamentals and Application of Windrow Composting," *J. Env. Eng. Div. ASCE* 116(4) (July/August 1990).
4. Williams, T. O. and North, O. "Performance Testing of an Innovative Sludge Composting Aeration System in Nashville, TN," presented at the 1991 Water Pollution Control Federation Conference, Paper No. AC91–018–003, Toronto, Canada (October 1991).
5. Shove, G. C. and Hukill, W. V. "Predicting Pressure Gradients in Perforated Grain Ventilation Ducts," *Trans. Am. Soc. Agric. Eng.* (1963).
6. Steele, J. L. and Shove, G. C. "Design Charts for Flow and Pressure Distribution in Perforated Air Ducts," *Trans. Am. Soc. Agric. Eng.* (1969).
7. Daugherty, R. L. and Franzini, J. B. *Fluid Mechanics with Engineering Applications* (New York: McGraw-Hill Book Co., 1965).
8. Weber, J. H. *Basic Programs for Chemical Engineering Design* (New York: Marcel Dekker, Inc., 1984).
9. Technical Paper No. 410 *Flow of Fluids through Valves, Fittings, and Pipe* (Chicago: Engineering Division, Crane Co., 1957).
10. Singley, M. E., Higgins, A. J., Wartena, R., Singh, S., Whitson, B., and Callanan, K. "Structural Characteristics of Compost Versus Aeration," in *Proceedings of the National Conference on Municipal and Industrial Sludge Composting* (Silver Spring, MD: Hazardous Materials Research Institute, 1979).
11. Higgins, A. J., Singley, M. E., Chen, S., and Singh, A. "Determination of Airflow in Sludge Compost Based Upon Structural Characteristics," in *Proceedings of the National Conference on Municipal and Industrial Sludge Composting* (Silver Spring, MD: Hazardous Materials Research Institute, 1980).
12. Willson, G. B., Parr, J. F., and Casey, D. C. "Basic Design Information on Aeration Requirements for Pile Composting," in *Proceedings of National Conference on Municipal and Industrial Sludge Composting* (Silver Spring, MD: Hazardous Materials Control Research Institute, 1979).
13. Colebrook, C. F. and White, C. M. *J. Inst. Civil Eng.* 10(1): 471 (1937–1938).
14. Irwin, R. W. and Motycka, J. "Friction Factors for Corrugated Plastic Drainage Pipe," *J. Irrig. Drainage Div., ASCE* (March 1979).

CHAPTER 16

Odor Management I — Quantifying and Treating

INTRODUCTION

It is important to engineer systems for control of nuisance conditions during design of a compost facility. Potential nuisances most often associated with composting are odors, dust, insect development, and attraction of birds and rodents. Dust control can be a problem in arid or semiarid climates and with "energy rich" substrates that are allowed to dry excessively. Dust problems are controllable by water addition, good housekeeping, pelleting to increase particle size, and common sense operating practices. Insect eggs are destroyed at the thermophilic temperatures achieved during composting. Insect development is usually associated with a lack of thermophilic conditions for a time period greater than the development time of the larvae. Such conditions can be caused by the following: (1) windrows that do not rapidly develop elevated temperatures and do not destroy insect eggs in the windrow interior after turning and (2) static piles with an exterior cool surface where eggs can be laid. Insects, birds, and rodents do not appear to be attracted to well operated facilities composting sludges, yard wastes, agricultural wastes, and most other substrates. Care should be taken with substrates such as garbage and refuse, particularly in the early stages of composting, because bird attraction is a well documented problem with refuse.

Control of odors is undoubtedly the most difficult problem in present composting practice. A number of facility closures have been directly caused by the lack of attention to odor management. In some cases, expensive redesigns have been necessary to solve odor problems and to make the facility a better neighbor to the surrounding community. Significant advances in the science of odor management were made in the 1980s and 90s, mostly because of problems encountered at sludge composting facilities. Unfortunately, many of the lessons learned with sludge composting are not being transferred to other members of the composting community. Some recent MSW composting facilities in the U.S. have been constructed with little or no odor management, placing them in a position to repeat past mistakes already learned with sludge. This is even more ironic when one recalls that many MSW facilities in the U.S. were closed in the 1960s in part because of odor problems.

Development of an odor management program must begin at the onset of planning and continue through design. The odor management program should be given as much importance as the plans and specifications. This is true regardless of the substrate. This author has actually heard MSW entrepreneurs state that their facility will not have odors because "sludge is not being handled." Such statements point to a grave misunderstanding of the odor potential of other composting substrates. This chapter begins the discussion of odor management and addresses the problems of quantifying and treating odor emissions. Understanding the atmosphere to enhance dispersion of odors is the subject of Chapter 17. Chapter 18 concludes with an overview of the elements of successful odor management.

THE POTENTIAL FOR ODOR

Compounds of Concern

All living systems, both plant and animal, excrete odorous molecules on a nearly continuous basis. The atmosphere becomes a soup of these odorous molecules present in minute concentrations. Many animal species use odor for species recognition, sexual attraction, social organization, and even as a defense mechanism. Plants use odor to attract pollinating insects. Many of the pleasant scents developed by the perfume industry start with molecules derived from plants. All starting substrates for composting are derived from plant or animal materials. Therefore, it is reasonable to assume that these substrates will contain many molecules that are potentially odorous, some pleasantly and other disagreeably so.

Even if a starting substrate contained no odorous molecules, a situation which is not likely, the pathways of biochemical metabolism will produce many intermediate compounds that themselves are odorous. Readers are probably familiar with some of the end products of anaerobic metabolism, odorous compounds such as hydrogen sulfide (H_2S), volatile organic acids, mercaptans, and methyl sulfides. The rather obvious odor from anaerobic metabolism has led to a widely held belief that if composting is fully aerobic there will be no odors. This simply is not true. Many low molecular weight, odorous intermediates are produced even during aerobic composting. Ammonia (NH_3), acetic or pyruvic acid, and citric acid are examples. The aerobic intermediates may sometimes be less obnoxious to humans than their anaerobic counterparts, but they are not odor free. Wilber and Murray[1] identified the following organic classes as potentially significant odorants to the composting process: fatty acids, amines, aromatics, inorganic sulfur, organic sulfur, and terpenes. The authors stressed that other classes may also be important but have yet to be identified.

Fatty Acids

Fatty acids are relatively long chained, monocarboxylic acids that occur in nature as constituents of fats, oils, and waxes. The longer chained acids can be hydrolyzed to lower molecular weight, volatile acids such as acetic, propionic, and butyric acids. Acetic acid, commonly known as vinegar, has an obvious and recognizable odor. The volatile and fatty acids are readily degradable.

Amines

The amines are alkyl derivatives of ammonia and are produced during the anaerobic decomposition of proteins and amino acids. Amines are found in certain industrial wastes, particularly those from the fish and beet sugar industries. Methylamine, ethylamine, dimethylamine, triethylamine, cadaverine, and putrescine are examples of very odorous amine compounds and are often described as putrid and fishy.

Aromatics and Ring Structures

Aromatic organics are all based on the benzene ring and may contain one or more cyclic groups. Aromatic compounds contribute to the odor of many wood species. Aromatics can be produced during aerobic composting from the breakdown of lignin which is usually abundant. Indole and skatole are examples of heterocyclic compounds that contain a benzene ring condensed with a second five-member ring containing nitrogen. Both have a very unpleasant odor and are produced during the anaerobic decomposition of protein matter.

Inorganic Sulfide

Hydrogen sulfide (H_2S) is an odorant well known to anyone in the wastewater industry. It produces the characteristic rotten egg smell and can be detected at a concentration of only 1 part H_2S in 2 billion parts of air. H_2S is produced by two principle pathways. First, it can be produced from the anaerobic decomposition of proteins or other sulfur containing organics. Second, under anoxic conditions with organics and sulfate present, the sulfate can be used as an electron acceptor and reduced to H_2S. Hydrogen sulfide can be formed during composting if anaerobic conditions exist such as in the interior of poorly aerated clumps of material.

Organic Sulfides

Mercaptans are the sulfur analog of alcohols, having the general formula R–SH. Their distinguishing physical characteristic is their vile and repulsive odor, which diminishes with increasing molecular weight. The nose can detect 1 part ethyl mercaptan in about 3 billion parts of air. The odor of skunks is largely due to butyl mercaptan. Alkyl sulfides with a formula R–S–R are the sulfur analog of the corresponding ethers. Such compounds occur in nature in plants of the onion and garlic family and are responsible for the odors and flavors of such species. The nose can detect 1 part dimethyl sulfide in 1 billion parts of air.

Mercaptans can be formed from sulfur containing amino acids under both aerobic and anaerobic conditions, but production under anaerobic conditions is greater. If oxygen becomes available, such as within a compost clump with some anaerobic zones, the mercaptan can be oxidized to dimethyl sulfide and dimethyl disulfide. Thus, it is likely that each of these compounds is being formed and degraded within the compost pile. Wilber and Murray[1] observed that one, two or three of these compounds can be identified in composting gases on any given day.

Terpenes

Terpenes are cycloalkane derivatives having one or more carbon rings and are a class of naturally occurring, organic compounds. Terpenes occur widely in the plant kingdom and are a major contributor to the fragrance of plants. For centuries the perfume industry has extracted these compounds by steam distillation or solvent extraction to produce mixtures called essential oils. Oils of lemon (limonene), geranium, rose, peppermint, turpentine (α-pinene), and eucalyptus (cineol) are examples of essential oils. These compounds have significant vapor pressures and would not be useful to perfumery if they did not. Pine trees naturally emit

α-pinene at a rate of about 2760 $\mu g/m_2$ of foliage per hour.[2] Therefore, even plant materials can be expected to contribute odorant molecules when used as composting substrates. Wilber and Murray[1] concluded that terpenes are important to any compost process that uses wood chips or sawdust. Terpenes can be found in concentrations ranging from 0 to 500 ppb and their odor threshold is about 6 ppb.

Ammonia

Ammonia (NH_3) is produced from either aerobic or anaerobic decomposition of proteins and amino acids. Any substrate with a low C/N ratio will likely release excess ammonia into the vapor phase. Sewage sludges, fish wastes, manures, and grass are examples of high protein substrates. Ammonia concentrations as high as 1000 ppm have been observed in exhaust gases from sludge composting. Fortunately, ammonia has a relatively high threshold odor concentration and dilutes rapidly to below detection. It is usually considered to be a relatively minor odorant away from the facility.

Others

The number of organics compounds that could potentially contribute odor during composting is virtually limitless. Aldehydes, alcohols, ketones, alkenes, esters, acrylates, butyrates, and others all have odorous representatives that could be present in the starting substrate or formed as intermediates during microbial metabolism. The Odor Index can be used to determine the relative significance of many of these compounds.

The Odor Index

Hellman and Small[3] and later Verschueren[2] introduced the concept of the Odor Index (OI) to measure the potential of a particular odorant to cause odor problems. The OI is based on the logic that to be odorous a molecular (1) must be present in the atmosphere to be transported to a receptor and (2) to be in the atmosphere the molecule must have a vapor pressure. The higher the vapor pressure, the more molecules that can be present. Once present in the atmosphere, the human response to the odorant depends on our sensitivity to the particular molecule. Therefore, the dimensionless OI is defined as

$$\text{Odor Index} = OI = \frac{\text{vapor pressure (ppm)}}{\text{odor recognition threshold (100\%) (ppm)}} \qquad (16.1)$$

The OI is essentially the ratio of the driving force to introduce an odorant into the air over the ability of an odorant to produce a recognized response. The OI is a measure of the potential of a particular odorant to cause odor problems under evaporation conditions.[2] The OI takes into account the vapor pressure of a compound, which is a measure of the potential of an odorant to be in the gas phase, and the odor recognition threshold, which is a measure of the strength of the odorant.

Boiling points and vapor pressures for selected odorous compounds are presented in Table 16.1. The boiling points of ammonia, hydrogen sulfide, ethyl mercaptan, dimethyl sulfide, and acetaldehyde are all lower than the temperatures commonly associated with composting. Thus, these compounds will tend to boil off into the vapor phase if they are produced during composting. Ammonia and hydrogen sulfide can ionize in the liquid phase, which reduces the

Table 16.1. Odor Index and Threshold Odor Concentrations for Selected Odorous Compounds

Compound	TOC(ppmv) Detect[a]	TOC(ppmv) Recognition[b]	Temperature (°C)	Vapor Pressure (mm Hg)	Odor Index
Ammonia	0.037	47	−33	bp[c]	167,300
Hydrogen sulfide	0.00047	0.0047	−62	bp	17,000,000
1-Butene	0.069		−6	bp	43,480,000
Methyl mercaptan	0.0011	0.0021	8	bp	53,300,000
Ethylamine	0.026	0.83	17	bp	1,445,000
Dimethyl amine	0.047	0.047			
Acetaldehyde	0.004	0.21	20	bp	2,760,000
Ethyl mercaptan	0.002		23	bp	289,500,000
1-Pentene	0.0021		30	bp	376,000,000
Dimethyl sulfide	0.001	0.001	36	bp	2,760,000
Dimethyl disulfide	0.001	0.0056			
Diethyl sulfide	0.0008	0.005	88	bp	14,400,000
Butyl mercaptan	0.0005		65	bp	49,340,000
Butanone			80	bp	3,800
Acetic acid	0.008	0.2	63	100[d]	15,000
Propionic acid			66	40	112,300
α-Pinene, oil of pine	0.011		37	10	469,000
Cineol, oil of eucalyptus			54	10	
Limonene, oil of lemon			54	10	
Butyric acid			61	10	50,000
Skatole	0.0012	0.47	95	1	30,000

Source: Adapted from Verschueren,[2] Weast,[4] and Water Environment Federation.[5]

[a] Detection threshold.
[b] Recognition threshold.
[c] Boiling point temperature at 760 mm Hg.
[d] Vapor pressure at the corresponding temperature.

un-ionized concentration and may limit the vapor phase concentration. The other compounds have boiling points near or above the temperatures common to composting. Nevertheless, they have significant vapor pressures and can evaporate into the vapor phase. This evaporation is aided by the large air flowrates associated with most composting systems.

Verschueren[2] determined the OI for over 260 organic compounds. Values ranged from a high of over 1 billion for isopropylmercaptan to a low of 0.2 for maleic anhydride. Compounds were separated into three categories: those with high odor potential (OI > 1,000,000); medium odor potential (100,000 to 1,000,000); and low odor potential (<100,000). The OI for a number of compounds is presented in Table 16.1. Depending on the particular substrate, many of the these compounds can be expected to be present in composting exhaust gases and many of these compounds are characterized by high OIs. The OI for compounds with higher boiling points may be underestimated because the values are calculated based on vapor pressure at 20°C. The actual vapor pressure at thermophilic composting temperatures would be much higher.

Table 16.2.   Classification of Chemical Classes According to their
Odor Index at 20°C

Odor Index	Chemical Class
$>10^6$	Mercaptans
	Alkenes
	Sulfides
	Low molecular weight compounds
	Butyrates
	Acrylates
	Aldehydes
	Ethers
	Alkylamines
$>10^4$ and $<10^6$	Di-alkylamines
	Tri-alkylamines
	Higher ethylesters
	Carboxylic acids
	High molecular weight compounds
	Aldehydes
	Ethers
	Alcohols
$<10^4$	Alkanes
	Acetates
	Aromatics
	Lower alcohols
	Phenolics

Source: Verschueren.[2]

A general classification of chemical compounds according to their OI at 20°C is presented in Table 16.2. Mercaptans, organic sulfides, and hydrogen sulfide would be expected to be high odor contributors. These compounds can be formed under anaerobic conditions if the substrate contains organic sulfur precursors, such as protein compounds. Mercaptans are also found in certain industrial wastes, particularly those from the pulping of wood by the Kraft or sulfate processes. The alkyl sulfides are naturally present in some plant species. Maintaining aerobic conditions can reduce the concentrations of these materials, but probably not eliminate them if they or their precursors are present in the feed substrate. If the latter compounds are not present, other compounds with a lower OI will become the predominant odorants. For example, it is rather common to sense a slight aroma of sweet tobacco when leaves are aerobically composted. While data is limited, it is likely that the odor is caused by natural terpenes and other compounds and not the more odorous mercaptans, alkyl sulfides, and hydrogen sulfide.

The point here is that all substrates can contribute odorants that can enter the vapor phase. Sludge, garbage and grass are commonly recognized as having a significant odor potential because of their rapid decomposition rates and tendency toward anaerobic conditions. However, sawdust, leaves, and other plant substrates can also contribute natural odorants to the process. There are several reports that odor emissions are effected by the particular wood species. Vold[6] reported that red oak sawdust produced a more pungent odor in the exhaust gas from a sludge composting facility and that hardwoods generally produced more odor than softwoods. Odors are likely to increase if anaerobic conditions are present because of the high OI of the metabolic end products.

GC/MS Results

Gas chromatographic/mass spectrographic (GC/MS) analyses have been conducted on a number of compost gases to identify and quantify the major odorous compounds. Hentz et al.[7]

reported GC/MS results for untreated process gas from the Montgomery County Composting Facility, an induced draft, static pile system composting raw sludge and wood chips. The chromatogram was segregated into four regions according to the type of compounds found in that region. Ketones, mercaptans, and other low molecular weight organics were periodically identified in the first 5.2 min of sample elution. Column temperature was a constant 35°C during this period. The authors noted that quantification was difficult because of the large concentrations of CO_2 also eluted during this time period. The second region extended from 5.2 to 14 min and contained short chain alkanes, butanone, and dimethyl sulfide (DMS). GC temperature ramped from 35 to 70°C during this region. The third region extended from 14 to 20 min and was dominated by dimethyl disulfide (DMDS). Temperature rose from 70 to 94°C in the GC column. The fourth region occurred after 20 min and was characterized by large chain alkanes, aromatics, and terpenes. Column temperature increased to a maximum 190°C in this region. The authors noted that some of the compounds in the fourth region, such as ketones, are believed to have lignin origins and are probably by-products of wood degradation.

Based on their chromatograms, Hentz and Murray concluded that DMDS was the primary odorant in the compost process air. DMDS had an OI one to three orders of magnitude higher than most of the other identified compounds. Other organic sulfur compounds, such as mercaptan and dimethyl sulfide, also had high OI, but their presence was sporadic and at significantly lower levels than DMDS. As a result, DMDS became their target odorant and GC analysis is performed twice a day to evaluate the DMDS removal efficiency of the odor treatment system.

GC/MS analysis on process gases from the Peninsula Composting Facility, operated by the Hampton Roads Sanitary District, Virginia, was conducted by Van Durme et al.[8] This facility uses the aerated, static pile system to compost dewatered, anaerobically digested sludge mixed with woodchips. The blowers are controlled by temperature feedback and operate in an induced draft mode. Compounds with the highest odor potential in the exhaust gas are shown in Table 16.3. In addition to ammonia and hydrogen sulfide, four volatile organic compounds (VOCs) are listed: dimethyl disulfide (DMDS), dimethyl sulfide (DMS), limonene, and α-pinene. These compounds were considered significant because their measured concentrations exceeded the respective threshold odor concentrations (TOCs). Dilutions to threshold for each compound were calculated by dividing the measured concentration by the corresponding TOC value. Odor concentrations of the exhaust gas were also measured using a forced-choice olfactometric technique (see next section) and are reported as ED_{50} in Table 16.3. DMS and DMDS were major odorants in all three samples. Limonene was also a significant odor contributor in two of the three samples. Although ammonia was present in the highest concentrations, its odor contribution was modest because of the relatively high odor threshold for ammonia. The measured ED_{50} values agreed reasonably well with the sum of calculated dilutions. Van Durme et al.[8] concluded that odor treatment methods must remove both the terpene and alkyl sulfide compounds to be successful.

THE BASICS OF ODOR SCIENCE

To manage odors properly, it is necessary to quantify and measure them. If the number of compounds in a gas stream is limited, it is possible to determine the concentration of each odor-causing compound using analytical techniques such as the GC/MS techniques discussed above. If the odor threshold concentration is known for each compound, the degree of required treatment and/or dilution can be estimated as is done in Table 16.4. This approach finds its

Table 16.3. Compounds Identified to Have the Highest Odor Potential in Compost Blower Exhaust at the Peninsula Composting Facility, Hampton Roads Sanitation District

Compound	Sample GC/MS Concentration ($\mu g/m^3$)			Threshold Odor Concentration ($\mu g/m^3$)	Sample Dilutions to Threshold		
	1	2	3		1	2	3
Ammonia	45,000	147,000	170,000	4,000	11	37	43
Hydrogen sulfide		405	1,062	12		34	89
Dimethyl disulfide	956	1,311	860	5	191	262	172
Dimethyl sulfide	0	2,667	1,360	3	0	1,026	453
Limonene	45	2,667	480	6	8	445	80
α-Pinene	251	333	78	64	4	5	1
Measured ED_{50}					152	1,700	500

Source: Van Durme et al.[8]

greatest application in certain industrial processes where the number of gas components is limited. Most odor sources are characterized by a wide variety of components whose composition is usually unknown. In such cases the human nose is still the accepted standard for detecting and determining odor intensity. After all, it is a human receptor that senses the odor, and no machine has yet been able to simulate the human response.

Odor Definitions

The nomenclature used in odor science is often confusing and sometimes conflicting. Therefore, it is important to define the nomenclature and to consistently use it.

Odor Thresholds

The Threshold Odor Concentration (TOC) is the minimum concentration of odorant that will arouse a sensation. A number of different odor thresholds have been determined: (1) the absolute perception or detection threshold; (2) the recognition threshold, and (3) the objectionability threshold. At the detection threshold, the odor can just be detected but it is too faint to identify further. The sample does not smell like odorless air, but the character of the odor cannot be recognized. At the recognition threshold, the odor can be identified as being representative of the odorant being studied. The recognition threshold is usually 1.5 to 10 times higher than the detection threshold.[10] Odor results are usually averaged statistically because of the biological variability of the human receptors. The thresholds normally used are those for 50 and 100% (includes the most insensitive) and 10% (the most sensitive) positive response by an odor panel. When a TOC is given without any qualification, it is taken to be the 50% threshold. Therefore, the 50% absolute or detection odor threshold is the concentration at which 50% of the odor panel detected the odor.

Generally accepted values for odor detection and recognition thresholds for selected odorous compounds are presented in Table 16.1. These values represent a statistical average because actual thresholds vary with the individual. For example, Wiley[9] studied the odor recognition thresholds for organic sulfur compounds using a panel of 35 men and women.

Table 16.4. The Variation in Odor Recognition Thresholds for Organic Sulfur Compounds

| Compound | Odor Recognition Threshold (ppbv) | | Standard Deviation |
	Median	Mean	
Methyl mercaptan	0.80	0.99	0.71
Ethyl mercaptan	0.32	0.40	0.26
n-Propyl mercaptan	0.75	0.75	1.2
i-Propyl mercaptan	0.25	0.45	0.49
n-Butyl mercaptan .	0.62	0.72	0.57
i-Butyl mercaptan	0.83	0.97	0.64
t-Butyl mercaptan	0.08	0.09	0.06
Hydrogen sulfide	4.1	4.5	2.9
Dimethyl sulfide	2.0	2.5	2.2
Diethyl sulfide	4.2	5.9	4.6
Di-n-propyl sulfide	19	23	15
Di-i-propyl sulfide	3.2	3.8	2.4
Dimethyl disulfide	5.6	7.6	6.4
Diethyl disulfide	3.9	4.6	3.2
Ethyl methyl disulfide	11	14	11
Dimethyl trisulfide	1.2	1.4	1.3
Diethyl trisulfide	0.7	0.85	0.60
Thiophane	0.75	0.77	0.46

Source: Wiley.[9]

Odor thresholds and their corresponding standard deviations are presented in Table 16.4. The range of odor thresholds for all observers in the panel varied for different compounds, from a minimum of about 10 to 1 for thiophane to over 50 to 1 for methyl sulfide.

Dilution techniques are often used to measure the detection threshold, particularly where many odorants are involved. It is common to define a Threshold Odor Number (TON) as the number of times a given volume of gas sample has to be diluted with clean, odorless air to bring it to the threshold level. A 50% positive response level is usually used to define the threshold level. At least five different nomenclatures are in use to define the number of required dilutions: (1) the odor unit (ou), (2) the effective dose at the 50% level (ED_{50}), (3) the Dilutions to Threshold (D/T), (4) the dilution ratio Z (named after H. Zwaardemaker, a Dutch pioneer in the field, and adopted in the ASTM Standards), and (5) the dilution ratio K or K_{50}. All of these mean essentially the same and are numerically equivalent. Odor unit and ED_{50} will be used in this text. The effective dose terminology has one advantage in that the panel response level is clearly defined. For example, ED_5 is the effective dose resulting in a 5% positive panel response.

Odor Intensity

Odor intensity is a measure of the strength of the stimuli or response to the odor (how strong it smells). Odor intensity is a sensation, not a concentration. Dravnieks and Jarke[11] noted that the annoyance potential of odors is directly related to the perceived odor intensity and odor character (what it smells like). Measurement of odor threshold is not a measure of the perceived odor intensity because different odors of equal intensity may require different amounts of dilution to reach the odor threshold. For example, the odor from 30 ppb of allylmercaptan and 10 ppb of allysulfide have about the same perceived intensity, characterized as faint. However, in terms of concentration the allylmercaptan has 10 ou and the allysulfide 5000 ou.

Table 16.5. The 1-Butanol Intensity Scale and Corresponding Aqueous and Vapor Concentrations

Butanol Intensity Scale	1-Butanol Concentration	
	In Water (ppm [wt])	In Vapor[a] (ppm [vol])
1	150	15
2	300	30
3	600	60
4	1,200	120
5	2,500	250
6	5,000	500
7	10,000	1,000
8	20,000	2,000

Source: O'Brien[13] and Dravnieks.[14]

[a] Equilibrium vapor concentration in the headspace above the aqueous solution.

Odor intensity is usually determined by referencing the odor sensation to an equivalent sensation from a reference odorant at a known concentration. Practice in the U.S. is to use ASTM E544,[12] which specifies 1-butanol vapor as the reference odorant. However, other reference odorants, such as *n*-pentanol (valeraldehyde), are used in other countries. The smell of the odorous sample is correlated to the concentration of 1-butanol that smells as strong as the sample. The odor intensity of the sample is then referenced as equal to the particular ppmv (part per million by volume) concentration of 1-butanol. An intensity scale for 1-butanol is sometimes used to express the odor intensity.[13,14] Odor panel candidates are presented an ascending series of aqueous butanol concentrations and are required to pick the solution with an equivalent intensity to the odor sample. The aqueous and vapor concentrations and their relationship to the butanol intensity scale are presented in Table 16.5.

For most odorants the intensity of sensation, I, increases with the odorant concentration, C, in accordance with Steven's Psychophysical Law:[15]

$$I = K(C)^n \tag{16.2}$$

where

I = intensity of the odor sensation, e.g., the butanol concentration giving the equivalent odor intensity as the sample

C = the odor concentration, actual concentration or TOC expressed in ou, ED_{50}, D/T, etc.

K = a constant

n = slope of the psychophysical function, an exponent whose value depends on the odorant

The value of n usually varies between 0.2 and 0.8. For 1-butanol several different laboratories have measured n values from 0.63 to 0.70, with 0.66 proposed as a tentative best value.[16]

The fractional exponent in Steven's Law is particularly important. With n = 0.25, for example, a 16-fold dilution is necessary to reduce the odor intensity by one half. An odorant with a small exponent is more difficult to suppress by dilution and is more pervasive. At concentrations below threshold, its intensity is only slightly less than at threshold. An odorant with a higher exponent will decrease in intensity more rapidly with dilution. If both odorants have the same dilution threshold, the one with the smaller n value will have the highest annoyance potential, even if the odors are not vastly different in character. Dravnieks and O'Neill[16] concluded that dilution thresholds are only a partial measure of odor annoyance and that in areas where the odor is still perceived (above threshold), the important factor is the odor intensity.

Hydrogen sulfide, butyl acetate, and the amines are examples of compounds with low slope values. Ammonia and aldehydes are characterized by relatively high slope values. This explains why ammonia odors can be intense in the compost gases but hardly noticeable off site. Conversely, sulfide odor may not seem particularly significant on site, but dissipates only slowly during downwind transport.

Duffee[17] reported on a study that compared the butanol intensity scale to an odor classification system used by the air regulatory agency in Southern California. Based on the responses from a number of panelists, butanol intensity ratings from 1 to 3 were judged to be so faint that they would normally not be expected to generate an odor complaint. Intensities of 4 to 6 were judged to be a possible to probable nuisance. Intensity ratings from 6 to 8 were a definite nuisance. Duffee suggested that odor complaints usually are initiated at an odor intensity value of about 3.5 or above on the butanol scale. Duffee further suggested that the odor concentration corresponding to a butanol intensity of 3.5 can be used as the threshold complaint concentration.

Hedonic Tone

Hedonic tone refers to the pleasantness or unpleasantness of the odor. It is a measure of the "acceptability" of the odor. There are a number of techniques for measuring hedonic tone, but normal practice is to compare the odorant to a range of standard odorants, for example, from very unpleasant (isovaleric acid) to very pleasant (vanillin). Hedonic tone has not been routinely measured in odor studies at composting facilities. This is because even pleasant odors become a source of annoyance and complaint when they occur as odorous air pollution. Dravnieks and O'Neill[16] noted that people continuously exposed to bakery odors, coffee-roasting odors, or fragrances from a perfume factory begin to consider such odors annoying. It is possible that the more common and easily recognizable and pleasant odors may be even more annoying with time. Conventional practice is to consider any odor as a potential nuisance regardless of its hedonic pleasantness.

Odor Quality

Odor quality is a measure of the character of the odor (what it smells like). A number of classification systems have been developed for odor quality. For example, the American Society for Testing and Materials (ASTM) presents character profiles for 180 chemicals using a 146-descriptor scale. Dravnieks and O'Neill[16] suggested that odor that is readily recognizable because of its characteristic smell is more easily noticed and reported. Odor quality can also be used to determine the source of an odor in the community because different odor characters are associated with different industries. Nevertheless, odor quality has not been routinely measured in odor studies conducted at wastewater or composting facilities.

Measurement Techniques

Organoleptic methods for odor detection and measurement are those that use the human olfactory system. The olfactory system is generally more sensitive than currently available analytical measurement systems. It is also sensitive to a wide variety of chemical structures. Although the human nose gives only a subjective response to the presence or absence of odor, a number of techniques have been developed to help quantify the human response.

Odor Panels

A number of organoleptic techniques are in common use for odor quantification, and most involve dilution to a threshold odor concentration. Probably the most common approach is referred to as the "odor panel technique". A panel of subjects (5 to 10 members are desirable) are exposed to odor samples that have been diluted with odor-free air. The number of dilutions required to achieve a 50% positive response by panel members is termed the threshold odor number. This is taken to be the minimum concentration detectable by the average person. Thus, if nine volumes of diluting air added to one volume of odor sample generate a positive response by half the panel, the odor concentration is reported as 10 dilutions to TOC. It is common to view the required dilution as a pseudo-concentration and the above sample would be said to have an odor concentration of 10 ou or 10 ED_{50}. An odorous compound at its TOC has, by definition, a concentration of 1 ou. For example, a 1 m^3 sample with a TOC of 100 ou requires dilution by 99 m^3 of odor free air to reach the threshold concentration.

A number of different procedures have been used for conducting odor panels. The ASTM D-1391 Syringe Dilution Method uses 100-ml hypodermic syringes to produce various dilutions of odorous gas with odor-free air. The diluted samples are presented to panelists by expelling the 100-ml syringe into the nostril at a uniform rate over 2 to 3 sec. The panelists report whether they detected the odor or not. The syringe technique is considered a "static" procedure as the gases are made by a batch dilution method. Several "dynamic" olfactometers have been developed to automate the delivery of diluted odor sample to the panelists. The term "dynamic" describes a procedure in which dilution is achieved by mixing a flow of the odorant sample with a flow of odor-free air.

One criticism of the syringe dilution method is the problem of false positive or negative responses. In 1973 the ASTM E18 Sensory Evaluation Committee began a review of the odor test procedures. In 1979 a new standard was developed, ASTM E679, authored largely by the late Andrew Dravnieks and widely used in North America. It is based on supplying three air channels to the panelist, two of which contain odor-free air. The panelist is forced to choose which of the three ports contains an odor. If no odor is detected, the panelist must still choose or guess at an odorous port. This technique is called the Three-Alternative Forced Choice (3-AFC) or forced choice triangle principle. By chance alone a panelist could be correct 33% of the time. Therefore, the threshold response for the panelist is the correctly identified dilution above the chance level of 33%. Again, a number of "dynamic" olfactometers based on the 3-AFC technique are in use. Units developed by the Illinois Institute of Technology Research Institute (IITRI) and by Odor Science & Engineering are widely used in the U.S.

A number of other olfactory procedures are in use throughout the world. According to Meilgaard[18] the German Standard VDI 3881 is fairly general and allows many procedures, whereas the French Standard NF X 43-101 specifies a 3-AFC procedure. The International Standards Organization in Geneva is working to standardize odor panel methods for determining sensory thresholds. An olfactometer termed the "olfaktomat" was developed by Project Research Amsterdam BV and has been used in the Netherlands and Britain.

There are a number of inherent difficulties with the odor panel technique. Samples must be collected and transported to the odor panel site. In some cases, storage of odorant samples have been observed to alter the odor strength. Tedlar® bags are usually used for sample collection and storage because this plastic does not add odor of its own and has a limited tendency to absorb odor. Also, a number of people are required for both sample collection and odor panel analysis, making the procedure somewhat expensive. Human subjects differ in their sensitivity to odors and results can be influenced by the personnel selected for the panel. Despite these shortcomings, the odor panel is a valuable tool for the very difficult task of quantifying odor concentration. A number of companies now offer odor panel services, receiving samples in bags shipped by air express services.

Field Tests

A number of portable olfactometers have been developed to allow direct odor measurement in the field. Usually the test subject wears a mask or hood to isolate the olfactory system from the ambient air. The test begins by supplying carbon-filtered air (assumed to be odorless) to the subject to prevent fatigue of the olfactory sense or adaptation to the odor. This is followed by a continuous supply of diluted sample air. The sample addition rate is gradually increased until the panelist detects an odor. An advantage of the portable units is that they can be used to measure odors at the source without using sampling and storage bags. Manpower requirements are also reduced. On the other hand, most portable devices use a single individual as a judge, which is statistically less reliable than the odor panel technique. Results reproducible to within 15% were reported in studies at Sacramento, California using different people with the same odor source.[19] However, Redner[20] analyzed samples by both portable and odor panel techniques and was unable to develop a consistent correlation between the two. The odor panel technique has been used in most studies at composting facilities.

Ambient odor levels are often low, 20 ous or less. Duffee[17] concluded that the practical low end for analysis of bag samples by dynamic dilution is about 20 ou. At the low concentration end, Duffee suggests that dynamic olfactometry can be supplemented by Scentometer measurements. The Scentometer, manufactured by Barnebey & Sutcliffe of Columbus, Ohio, is a hand-held instrument used by an individual observer in the field. The unit is a box-like Plexiglas instrument consisting of a center mixing chamber between two layers of activated carbon. Two glass nose pieces connected to the instrument are pressed against the nostrils. As the observer inhales, ambient air is drawn into the instrument through the two carbon beds providing "odor-free" air to the observer. The observer exhales through the mouth. After a period of acclimation, odorous air is then allowed to enter the mixing chamber through a series of calibrated holes of different diameter. The smallest hole is opened first, admitting the smallest amount of odorous air, which mixes with the "odor-free" air. If no odor is detected at that dilution, the hole is closed and the next larger hole opened. This procedure continues until an odor is detected. The individual holes are calibrated to give dilutions to threshold based on the relative flowrates of odor-free to odorous air.

Population Statistics

The ED_{50} concentration determined by an odor panel is a measure of the average dilution to threshold for the panel members. Some panel members will be more sensitive to the odor than the average and will require a higher dilution to achieve the threshold odor sensation. The same is true for the population surrounding an odor source. Some individuals will be able to detect an odor even though the ED_{50} concentration is at or below 1 ou. In other words, some noses are more sensitive than others. For example, Lillard[21] studied the odor thresholds for a number of organic compounds. Camphor (2-camphanone), a terpene compound of pine oil, exhibited an average ED_{50} of 1.29 ppm in water (range 0.25 to 3.83). However, 20% of the population was able to detect the odor at 0.33 ppm, 10% at 0.041 ppm, 1% at 0.0092 ppm and 0.1% at 0.021 ppb.

A number of procedures and strategies can be used to account for the statistical variability of the human nose. First, candidate odor panelists should be screened for their general olfactory acuity, their ability to detect the odors of interest, and their ability to match odor intensities.[13] The panelists should reasonably represent the more sensitive to average "noses" within the population pool. Second, the effective dilutions for a lower statistical response, e.g., ED_5 or ED_{10}, can be used in place of the traditional ED_{50}.

The probability of positive responses by an odor panel usually plots as a near linear relationship to the log of the dilution ratio (i.e., straight line on log probability paper). Therefore, ED_x values corresponding to a particular statistic can be determined directly from the odor panel results. This is certainly the best approach if odor samples are available from an existing facility. However, the potential impact from proposed facilities is usually determined from estimated emission factors. Therefore, estimated population statistics must also be used in such a case. Moschandreas et al.[22] analyzed 22 odor samples for ED_{50}, ED_{10} and ED_5. The samples were collected from the process gases at the Montgomery County Compost Facility, an aerated static pile system using induced aeration to compost raw sludge with wood chips. Samples were collected both upstream and downstream of a wet scrubbing process. Statistical analysis of their data gave the following best fit equation ($R^2 = 0.996$):

$$ED_x = \frac{ED_{50}}{3.814 * (X + 0.0019)^2 + 0.0393}$$

$$(ED_5 \leq ED_x \leq ED_{50})$$

(16.3)

where

ED_x = effective dilution to threshold corresponding to a positive response probability of X

X = fractional probability value of the effective dilution, i.e., 0.05 for ED_5

Individual equations based on the upstream and downstream data were not significantly different from Equation 16.3. Using Equation 16.3, ED_{10} averaged about 10 times the ED_{50}, while ED_5 was about 20 times the ED_{50}. These values should reasonably reflect the sensitivity differences in the general population to composting odors.

Units on Odor

There is some confusion in the literature over the units that apply to ous or ED_{50}s determined by the above techniques. Odor concentration is sometimes expressed as ou/m³ or ou/ft³, but this is not strictly correct. An odor unit can be defined as 1 m³ (or 1 ft³) of air at the odor threshold. Odor concentration is the number of cubic meters that 1 m³ of sample (or cubic feet per 1 ft³ of sample) will occupy when diluted to the odor threshold. The implied units are volume of sample diluted to the odor threshold per volume of original sample. Thus, the odor unit is a dimensionless number and should be expressed as ou or ED_{50} and *not* as ou/m³ or ou/ft³. Although these distinctions are somewhat subtle, they will serve well in later discussions.

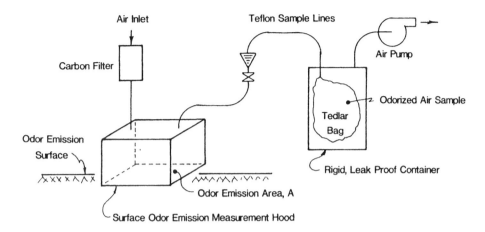

Figure 16.1. Schematic diagram of an odor hood and test equipment used for measurement of SOER.

The mass emission rate of odor from a point source can be determined as the product of volumetric flowrate times the TOC of the sampled gas. Suppose a composting exhaust gas is sampled and found to contain 100 ou. If the gas flowrate is 10 m³/sec, the mass emission rate is 100 * 10 = 1000 m³/sec. Remember that the ou is dimensionless, so the units remain m³/sec. It is common to see mass emission rates expressed as ou/sec, but again this is not correct and can lead to problems in converting units. For example, the mass emission rate of 1000 m³/sec is equivalent to 1000 × 35.3 = 35,300 ft³/sec in English units.

Area Sources

Many composting operations are characterized as having a large surface area exposed to ambient air. Certainly, the open windrow and static pile systems are examples of large area sources of odor. In such cases it is of interest to determine the surface odor emission rate (SOER). The technique for measurement of SOER involves placement of a sampling hood over a known cross section of the surface being analyzed as shown in Figure 16.1. Odor-free air is drawn through the hood at a measured flowrate and usually collected in a Tedlar® bag using a "bag in drum" technique to avoid any sample contamination. The bag may be inflated several times with the sample gas prior to taking the final sample. This equilibrates the inner surface of the bag, a process called pre-conditioning. The collected sample is then analyzed to determine the TOC by one of the olfactory techniques discussed above. The product of odor concentration and air flowrate is assumed to equal the odor emission rate from the surface. The correct units for SOER are ft³/min-ft² or m³/min-m².

Suppose air is drawn at a flowrate of 1 m³/min across a 0.5 m² sampling area. If the sample of exhaust gas is determined to contain 10 ou, the SOER would be 20 m³/min-m² (i.e., 1 (10)/0.5). The equivalent SOER in English units is 65 ft³/min-ft². Redner et al.[23] used the above procedure to determine SOER's at a windrow sludge composting facility. Measurements at various air flowrates indicated that the procedure provides reasonably representative values of the SOER.

The units on SOER may be somewhat confusing. One way to interpret their meaning is to envision odorous molecules being released from the material surface at a uniform rate. Consider the above example with a SOER of 20 m³/min-m². Then for each m² of surface area a 20 m³/min flowrate of air would be required to reduce the odor level to the threshold concentration of 1 ou.

Approaches to Modeling

Three different approaches can be used to model the impacts of odor emissions. The first approach quantifies the odor emissions using odor units or ED_x concentrations as determined above. The mass emission rate of odors is then expressed as m³/sec for point sources and m³/min-m² for area sources, or in the equivalent English units. Downwind concentrations are then expressed in odor units or ED_x concentrations. The butanol intensity of the odor may be used to establish acceptable criteria for the odor, but the analysis uses odor units or ED_x concentrations for both the source emissions and downwind concentrations. The second approach quantifies the odor emissions as equivalent emissions of 1-butanol or another reference compound. Units are typically g/sec and g/min-m² of 1-butanol. Downwind concentrations are then expressed as ppmv or μg/m³ of 1-butanol. The third approach uses the actual concentrations of specific compounds measured in the exhaust gas. Downwind concentrations are then compared with published estimates of the detection threshold for each compound.

Each approach has its supporters and each offers certain advantages and disadvantages depending on the situation. Minor theoretical limitations with each approach are usually compensated by the conservative assumptions used in most studies and the inexact nature of odor science. Regarding the second approach, Duffee and O'Brien[24] have pointed out that most odor intensity vs dilution relationships are developed in the range of 100 to 1000 dilutions. Determining the equivalent butanol concentration of the source requires extrapolation of these results to a dilution of 1 (i.e., no dilution). Such a broad extrapolation can introduce significant errors in the estimate of the source strength. There are many problems with the third approach as well. Few odorous emissions contain a definable number of specific odorants. Odor thresholds are often not available for less well known compounds. Even when they are available, the published values may vary over many orders of magnitude. Duffee and O'Brien recommend using dilutions to threshold to avoid these problems.

The first approach (dilutions to threshold) is used in this text for the following reasons: (1) the approach is easier to visualize and easier to understand; (2) atmospheric dispersion is a process of dilution that can be directly related to dilutions to threshold measured by the odor panel; (3) the fact that odor intensity may decrease more slowly than concentration (n < 1 in Equation 16.2) is usually an acceptable limitation because downwind odor concentrations are typically only slightly above or below the odor threshold; (4) the majority of odor studies have used this approach; and (5) it avoids the problems noted above with the other approaches.

ODOR CONCENTRATIONS AND EMISSION RATES

Data on odor emission rates and odor concentrations from sludge composting facilities have increased greatly in recent years. In some cases the available data have been determined by different techniques, making direct comparisons somewhat tenuous. Nevertheless, review of the available information will provide the reader with an idea of the general range of odor emissions expected and what can occur if thermodynamic or kinetic constraints are not recognized.

Windrow Facilities

Los Angeles County Sanitation Districts (LACSD)

Redner et al.[23] published one of the first engineered studies of odor emissions at a composting facility. The study was prompted by a sudden increase in odor complaints in 1977 at the windrow composting facility operated by the LACSD. Before 1977 about 90 dtpd of digested sludge at 30 to 35% solids were composted. The windrow composting was initiated in 1972, replacing an open air drying system and significantly reducing odor emissions. New dewatering facilities were commissioned in 1977 which increased the sludge tonnage to about 270 dtpd, but decreased cake solids to about 23%. The thermodynamic balance was severely stressed and composting temperatures declined as a result. When combined with wet weather and other operational problems, odor emissions and complaints from the local community increased dramatically.

A summary of SOER values determined in 1972 and during the period of odor complaints in 1977 is shown in Table 16.6. For the 1972 data, odor emissions were noted to decrease with compost time. No consistent reduction in the first 7 days was noted in the 1977 tests, probably related to the poor temperature development. SOER values increased significantly. The average SOER before turning was about 15 to 30 m^3/min-m^2. The SOER increased to about 550 m^3/min-m^2 immediately after turning, and then decayed at an approximately first order rate for the next 3 to 4 h. Through integration of the decay curve it was determined that as much as 50% of the total daily emission occurred within 4 h of turning a windrow. Average odor emission rate from the composting area was about 2.8×10^6 m^3/min compared to values estimated in 1972 from 0.3 to 0.4×10^6 m^3/min. It was also shown that silo storage of dewatered cake for periods of 1 to 3 days about doubled the SOER in the windrow.

It must be clearly pointed out that the data of Table 16.6 are not typical of normal operations, but are indicative of what can happen if thermodynamic constraints are not recognized. As a result of this experience a number of process modifications were made. A barrier wall was constructed around the facility to increase vertical dispersion. Input sludge tonnage was reduced and dewatering performance improved. As a result the compost area was reduced in size, which increased the available buffer zone. Sawdust and rice hull amendments were eventually used to improve the energy balance. Operating hours were restricted to periods of higher atmospheric turbulence to provide greater odor dilution and larger turning equipment was purchased. These modification significantly reduced odor emissions.

Iacoboni et al.[25] developed the SOER data shown in Figure 16.2 following the above improvements. The windrows were composed of digested, dewatered sludge cake conditioned with recycled compost alone and with recycled compost and wood shavings or rice hulls. The windrows were naturally ventilated and consistently achieved proper composting temperatures. The values shown were determined prior to windrow turning. SOER values were significantly lower than those observed by Redner et al.[23] during more adverse conditions. The majority of odors were emitted during the first 7 to 10 days of the compost cycle. SOERs were about 10 to 20 m^3/min-m^2 during the first 2 to 3 days. After this, SOERs decreased significantly to a range of about 0.5 to 2 m^3/min-m^2. Emission rates during active composting were about one order of magnitude less than the initial rate. The reduction probably corresponds to development of more complete aerobic conditions within the windrow. Lower SOERs of about 0.1 to 0.3 m^3/min-m^2 were reported with other windrows. A frequency distribution of

Table 16.6. Comparison of Composting Temperatures and SOERs at the Windrow Sludge Composting Facility Operated by the Los Angeles County Sanitation Districts

Windrow Age (days)	Mean pile temperature (°F)		SOER, m^3/min-m^2	
	1972	1977	1972[a]	1977[b]
0	90	83	9.8	13.1
1	104	85	8.5	12.8
2	114	94	7.6	35.4
3	122	94	7.0	45.1
4	126	100	6.1	46.3
5	128	100	5.5	65.2
6	129	108	4.6	138.7
7	130	89	4.0	28.4

Source: Redner et al.[23]

[a] SOER-values for 1972 were determined by an odor panel.
[b] SOER values for 1977 were determined by portable olfactometer.

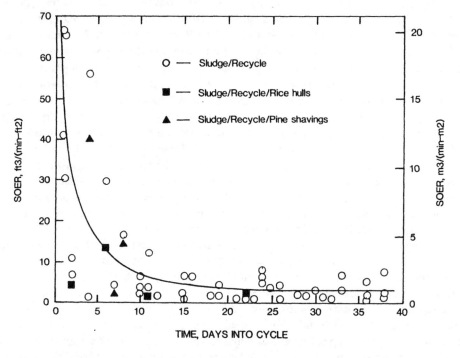

Figure 16.2. Surface odor emission rates as a function of time into the windrow composting cycle. Samples were collected prior to turning. The windrows were composed of anaerobically digested, sewage sludge cake conditioned with recycled compost and recycled compost and wood shavings or rice hulls. Windrows were naturally ventilated and were 0.7 to 1.1 m high. From Iacoboni et al.[25]

SOER values collected by Iacoboni et al.[26,27] in other studies is shown in Figure 16.3. The median value is about 2 m^3/min-m^2. Most of the high SOER values were observed in the early stages of composting.

Peak odor emissions were determined to occur immediately after turning as shown in Figure 16.4. The increase in odor emissions due to turning lasted for about 0.5 to 1 h before returning to the original SOER value. While peak odor emissions were associated with

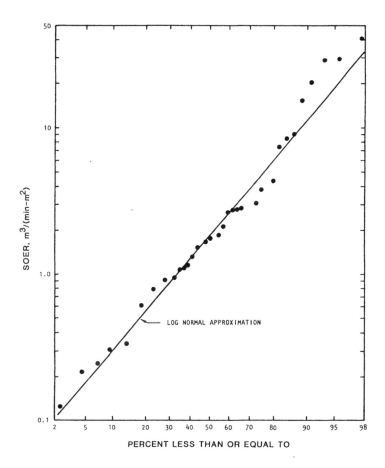

Figure 16.3.   Frequency distribution of SOER observed during windrow composting of digested sewage sludge blended with recycled compost. Higher values of SOER occurred within the first few days of the compost cycle. Samples were generally collected prior to windrow turning. Developed from data reported by Iacoboni et al.[26,27]

windrow turning, the majority of total odor emitted came from the windrows in their unturned state. Windrow turning was estimated to account for 10 to 15% of the total odor emitted regardless of the turning frequency. Increasing the turning frequency appeared to reduce the peak emissions, but total odor emission remained about the same. Based on this work, Iacoboni recommended that turning be halted during times of adverse meteorological conditions and that both total and peak odor emissions be considered when evaluating odor problems in any given situation.

Hay et al.[28] reported a correlation between windrow surface to volume ratio (SVR) and the odor emission rate. In general, the lower the SVR (i.e., larger windrows) the lower the odor emission rate. Hay also reported on the use of additives containing enzymes and bacterial cultures. No differences in odor release were observed between windrows treated with the additives and those not treated.

Denver Metropolitan Wastewater Reclamation District (DMWRD)

The DMWRD operates an enclosed, aerated, windrow facility composting anaerobically digested sludge conditioned with wood shavings and other amendments. A permanent aeration system ventilates the windrows usually in an induced draft mode. The facility was designed

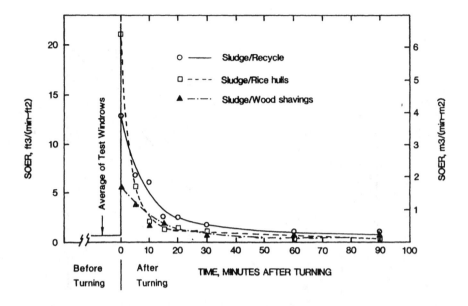

Figure 16.4.   Surface odor emission rates before and after windrow turning. Windrow 1 was conditioned with recycled compost alone, windrow 2 with rice hulls alone, and windrow 3 with wood shavings alone. The windrows were about 1.4 m high. From Iacoboni et al.[25]

for 100 dtpd and was commissioned in 1986. The facility immediately encountered odor problems and DMWRD was forced to reduce the operating capacity. Studies to evaluate the odor emissions and control strategies were conducted from 1987 to 1989. A summary of the odor species and concentration ranges determined in the exhaust air from the blowers is presented in Table 16.7. Ammonia concentrations over 1600 ppmv were occasionally determined. H_2S was typically <1 ppmv. Mercaptans, organic sulfides and terpenes were found in all samples. ED_{50} values ranged from about 400 to 1075 ou. SOER values were not determined in these studies, but ventilation fan exhausts on the roof of the enclosure were sampled. ED_{50} values ranged from 23 to 107 ou.

A number of facility and operational modifications were made as a result of these studies. Better mixing of the initial substrates was demonstrated to reduce total odor emissions. DMWRD also constructed a wet scrubbing system using plant effluent. As a result the facility was able to resume operations at part load.

Static Pile Facilities

Hampton Roads Sanitation District (HRSD)

The Peninsula Composting Facility is owned and operated by the HRSD. Anaerobically digested, dewatered sludge is transported from three HRSD wastewater treatment plants to the Peninsula Facility. The extended, static pile system is used to compost an average 65 dry dtpd/week of sludge. A permanent manifold is used to ventilate the piles using negative draft. Wood chips are used as the bulking agent. The compost product, Nutri-Green®, is marketed locally and the facility has enjoyed a successful operation. The HRSD became concerned about encroaching urban development and undertook odor studies in 1989–90 to guide development of additional odor control measures.

Table 16.7. Odorous Compounds Determined in Compost Blower Exhaust at the Denver Composting Facility, DMWRD

Species	Average	High	Low
ou, ED_{50}	—[a]	1,075	393
Ammonia, ppm	308	1,400	44
	189	1,170	8
	319	1,300	50
	—	1,640	1,170
	243	1,000	20
	161	400	10
Hydrogen sulfide, ppb	—	750	50
	175	750	10
Mercaptans, ppb	16,000	116,000	1,000
Dimethyl sulfide, ppb	433	7460	5
Dimethyl disulfide, ppb	1,012	6,160	33
	—	15,000	3,230
Dimethyl trisulfide, ppb	67	710	1
	—	1,220	40
Methylpropyl disulfide, ppb	5	26	1
	—	1,040	13
α-Pinene, ppb	81	530	1
	—	970	215
β-Pinene, ppb	29	170	1
	—	730	219

Source: James M. Montgomery Consulting Engineers, Inc.[29]

Note: The separate data sets correspond to different test campaigns within the overall program.

[a] Data considered too limited to average.

A summary of the point source and surface odor emissions measured at the facility and used for subsequent dispersion modeling is presented in Table 16.8. ED_{50} and SOER values generally agree with the range of values reported above for other facilities. Odor concentration in the compost pile blower exhaust averaged 1030 ED_{50} in the first week and then decreased in the second and third weeks. The curing pile blower exhaust averaged 300 in the first week and decreased over the next 4 weeks. SOER values from the compost piles were significant even with negative ventilation. Based on these studies, the consultant recommended that all process gases be collected and treated by multistage wet scrubbing and that the site be fully enclosed within a building. Building ventilation air would be discharged through rooftop fans to aid dispersion.

City of Philadelphia

The City of Philadelphia Water Department operates a static pile facility composting anaerobically digested sludge with wood chips as the bulking agent. The facility is designed to process up to 300 dtpd (405 dtpd peak). Ventilation is by induced draft and the blower exhaust is discharged through a deodorization pile constructed from finished compost. A series of odor studies were initiated in 1988–90 as a result of intermittent odor occurrences. Gruber et al.[31] reported the odor emission rates presented in Table 16.9. Observations made during these studies are summarized as follows:

- Samples taken from the blower exhaust and the surface of the deodorization piles showed an odor reduction of about 50%.

Table 16.8. Odor Emission Rates Measured at the Peninsula Composting Facility, RSD[a]

Location	Source type	ou (ED_{50}) (avg range)		SOER (m^3/min-m^2)
Compost pile, negative aeration	Area			
1st week				5.8
2nd week				4.5
3rd week				2.9
3 week average				4.4
Curing pile, negative aeration	Area			
1st week				4.0
2nd week				3.1
3rd week				2.0
4th week				1.0
4 week average				2.5
Screen pile	Area			0.40
Recycled chip pile	Area			0.83
Final product	Area			0.40
New wood chips	Area			0.24
Compost blower exhaust	Point			
1st week		1030	(500–1770)	
2nd week		795	(720–870)	
3rd week		520	(430–620)	
3 week average		782		
Curing blower exhaust	Point			
1st week		300		
2nd week		232		
3rd week		152		
4th week		76		
4 week average		190		

Source: Black & Veatch Engineers-Architects.[30]

a   Samples were collected in bags and sent to McGinley Associates for odor panel evaluations using ASTM E679 procedures.

- Blanketing the compost piles with 6 to 12 inch of finished, cured, unscreened compost did not reduce sulfide emissions from the pile surface.
- Lime application to the surface of the compost piles did not reduce the SOER.
- Atomization of an odor neutralizing chemical into the compost blower exhaust resulted in an average 44% reduction in ED_{50}.
- Despite the high moisture content in the blower exhaust, the deodorization piles tended to dry in the interior and become water saturated at the exterior. Water added to the odor pile surface did not readily penetrate the wet/dry interface.

Montgomery County Composting Facility (MCCF)

The MCCF is a 400 wtpd static pile facility that began operation in 1983. The problems of odor control, and the need to quickly resolve them, were especially critical because the facility is located in a rapidly developing suburban area. A series of odor evaluation studies were undertaken to inventory the odor emissions. A brief summary of uncontrolled emissions presented by Murray[33] and others[34] is as follows:

- Odor concentrations in the blower exhaust ranged from about 400 to over 1000 ED_{50}. Ammonia concentrations ranged from about 90 to 400 ppm.

Table 16.9. Odor Emission Rates Measured at the Philadelphia Composting Facility

Location	Source Type	ou (ED_{50}) (avg range)	SOER (m^3/min-m^2)
Compost pile, negative aeration	Area		2.5
Curing pile	Area		1.4
Drying piles, negative aeration	Area		1.7
Screening operation	Area		1.8
Product storage	Area		1.4
Cake storage	Area		1.9
Compost blower exhaust	Point		
1st week		153	
2nd week		108	
Odor pile exhaust	Point		
1st week		62	
2nd week		50	
3rd week		19	

Source: Gruber et al.[31] and Black & Veatch Engineers-Architects.[32]

[a] Samples were collected in bags and sent to IIT Research Institute for odor panel evaluations using ASTM E679 procedures.

- A comprehensive study of odor sources indicated that all area sources combined accounted for about 5% of the total mass emission. The blower exhaust contained the remaining 95%.
- Without forced aeration, oxygen content in the active compost piles dropped to very low levels in as little as 20 min.
- To reduce the possibility of oxygen depletion, a stand-by generator was installed to power the aeration blowers in an electrical outage.

Other Systems and Substrates

A statistical evaluation of exhaust odor concentrations from four different systems composting raw and digested sludges is shown in Figure 16.5. The sludges were primarily conditioned with recycled compost, but rice hulls and wood shavings were sometimes included. Systems that provided periodic bed agitation exhibited lower exhaust odor concentrations, which may be related to the breakup of air channels and anaerobic zones. This result is supported by Murray's[36] observation that improved mechanical metering and mixing of sludge/wood chips in static pile systems can reduce the concentration of odor species such as dimethyl disulfide in the exhaust gas. This is probably caused by the reduction in small anaerobic pockets within the mix.

Ostojic[37] reported on odor emissions from sludge composting at the City of Hamilton, Ohio. This facility uses a horizontal, static solids bed reactor with subsequent curing in aerated static piles. Ammonia levels in the process gas were typically 20 to 40 ppm, but increased to over 800 ppm in the exhaust from the curing piles. Odor levels in the curing pile exhaust were as high as 3000 ED_{50}, many times higher than in the reactor exhaust gas. This is counter to the general trend at most facilities where odor emissions decrease through the process. Very high temperatures are maintained in the static bed reactor at Hamilton and it is possible that microbial kinetics are suppressed, then accelerate at the lower temperatures maintained in the

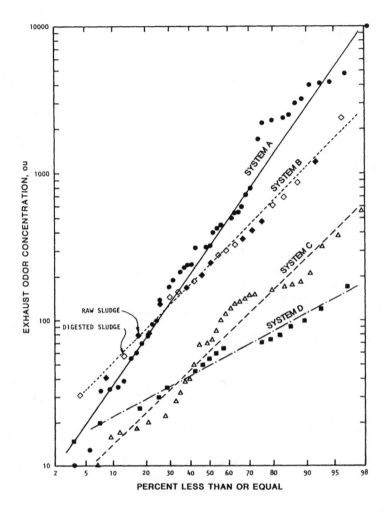

Figure 16.5.   Frequency distribution of odor concentrations in blower exhaust gases from various composting systems. Digested sewage sludge and recycled compost were the primary mixture ingredients in most tests. The range of specific air supply in m³/min-dry dtpd is shown in parentheses in each case. Lines shown are log normal approximations of the data. Developed from data reported by Iacoboni et al.[26,27] and Livingston.[35]

curing piles. The curing operation was identified as the main contributor to off-site odors. Based in part on this finding, the plant suspended on-site curing and pays a topsoil blender to take the material immediately from the reactor. According to Wheeler[38] this was the most cost-effective approach, because Hamilton saved the operating costs of curing and avoided the capital costs of improving the curing operation to reduce its odors. The point to remember from this experience is that odors from curing must be included in the total emission balance.

More odor information has been published on sludge composting than for most other substrates. This is particularly true in the U.S. because sludge was the dominate composting substrate in the 1970s and 80s and because municipal wastewater agencies and the U.S. EPA had the financial resources to study and correct odor problems when they occurred. By contrast, yard waste composting is typically conducted with modest budgets, and much less odor emission data are available. While there are exceptions, most yard wastes facilities have not experienced major odor problems unless grass is also handled. By comparison with sludge, leaves and tree trimmings are not major sources of ammonia and organic sulfides, and odor emissions are correspondingly lower.

MSW, on the other hand, can be a significant source of odor emissions. Much of the available data comes from European facilities. For example, the rotating drum/static pile MSW system at Siggerwiesen, Austria, collects process gases with odor concentrations estimated at 25,000 to 50,000 ED_{50}.[39,40] Similarly, high odor concentrations were observed in process gases at the Heidelberg Kompostwerk, which composts source separated "biowaste" consisting of kitchen wastes, leaves, grass, and other organics.

Canzano and Aiani[41] reported on odor concentrations at the Delaware Reclamation Project co-composting facility where circular, agitated bed reactors are used to compost sewage sludge and MSW heavy fraction. The reactors are aerated by forced draft ventilation. Process gases discharged from the reactor bed are diluted about 6:1 with outside ventilation air and discharged from a single roof fan. Odor concentrations in the undiluted gases exhausted from the reactor bed ranged from 3500 to 25,000 ED_{50}. Duffee[42] determined odor concentrations that averaged 4300 ED_{50}, with a range from 3650 to 5450, while Haug et al.[40] reported a 2000 to 3000 ED_{50} odor range for the undiluted gases at this same facility. The diluted gases contain about 500 to 1000 ED_{50}.[40,42]

APPROACHES TO ODOR TREATMENT

A number of techniques are available to reduce the odor concentration in exhaust gases collected during composting including absorption, condensation, adsorption, oxidation by thermal, chemical, or biological means, and use of masking and neutralizing chemicals.

Absorption

Gas absorption is a unit operation in which one or more soluble components of a gas mixture are dissolved into a scrubbing liquid. Absorption may involve simple solution of the odorous compound into the scrubbing liquid or solution followed by chemical reaction with other compounds in the liquid phase. Specially designed equipment for gas absorption includes spray or mist towers, packed bed scrubbers, fluidized bed scrubbers, tray towers, and jet or venturi scrubbers. The mist towers and packed bed scrubbers are the most commonly used absorber types at compost facilities. Water is the most common solvent for wet scrubbing and may be combined with chemical compounds to increase absorption or remove species that are absorbed. Common chemicals used in wet scrubbing include the following:

- oxidizing agents such as sodium hypochlorite (NaOCl), hydrogen peroxide (H_2O_2), and potassium permanganate ($KMnO_4$)
- bases such as lime (CaO), hydrated lime (Ca[OH]$_2$), and caustic (NaOH)
- acids such sulfuric (H_2SO_4) and hydrochloric (HCl)
- reducing agents such as sodium sulfite (Na_2SO_3) and hydrogen peroxide (H_2O_2)
- absorption enhancing agents such as surfactants

Hydrogen peroxide is an interesting scrubbing chemical in that it can act as an oxidizing agent if the oxygen is reduced to $O^=$ or a reducing agent if the oxygen is oxidized to O_2.

Significant advances were made in the understanding of absorption chemistry in the 1980s and 90s, primarily because of work conducted at sludge composting facilities. Much of this work was spearheaded by the comprehensive studies directed by Chuck Murray, Larry Hentz, and Joel Thompson at the Montgomery County Composting Facility, Bill Cathcart at the Cape May Composting Facility, Jim Dunson with E.I. du Pont de Nemours, and many others.

The odorous compounds of concern in most composting situations include the following: (1) ammonia that will be present in proportion to the organic-N content of the substrates; (2) organic sulfur compounds such as mercaptans, dimethyl sulfide, and others; (3) inorganic sulfur such as hydrogen sulfide; (4) other organic compounds such as terpenes; and (5) products of incomplete oxidation and residual odorous compounds that may be formed as a result of chemical scrubbing. Different absorption conditions are necessary to remove all of these compounds. Therefore, single stage scrubbing will not be particularly effective given the rather complicated composition of most composting exhaust gases. The state of the art at the present time is to use multistage scrubbing with the conditions of each stage designed to absorb specific compounds from the gas. At least two or more stages are usually used.

Misting Towers

A schematic of a three-stage, mist tower scrubbing system under construction at Montgomery County is shown in Figure 16.6. This system will replace an existing multistage scrubber used to develop the concepts shown in Figure 16.6. Three mist towers are provided in series with a nominal design flowrate of 50,000 cfm. Tower 1 is 8 ft diameter by 75 ft tall, tower 2 is 14 ft by 75 ft, and tower 3 is 8 ft by 58 ft tall. The towers provide about 4.5, 14, and 3.5 sec gas residence times, respectively. In 1991, a patent was issued to Murray et al.[43] which claimed the three-stage scrubbing process using mist towers for removal of ammonia, sulfides, and malodorous organic compounds.

Chemical solutions can be introduced into each tower through fogging (atomizing or aerosol nozzles) and spray nozzles. The small droplet size produced by the atomizing nozzles results in a large surface area for absorption of compounds as the process gas passes through the towers. The larger droplets produced by the spray nozzles provide additional gas to liquid contacting and help coalesce and partially remove the aerosol droplets. Misting towers usually require longer gas residence times compared to packed bed scrubbers because of higher mass transfer limitations in the gas and liquid films surrounding the droplets.

Ammonia removal is accomplished in the first stage scrubber. Sulfuric acid is used to neutralize the dissolved ammonia which would otherwise raise the solution pH and reduce further mass transfer. The absorption and neutralization reactions between ammonia and sulfuric acid are

$$2NH_{3g} \rightleftharpoons 2NH_{3aq} \qquad \text{(absorption)}$$

$$2NH_{3aq} + 2H_2O \rightleftharpoons 2NH_4^+ + 2OH^- \qquad \text{(ionization)}$$

$$\underline{2H^+ + SO_4^= + 2OH^- \rightleftharpoons 2H_2O + SO_4^= \qquad \text{(neutralization)}}$$

$$2NH_{3g} + 2H^+ + SO_4^= \rightleftharpoons 2NH_4^+ + SO_4^= \qquad \text{(net)}$$

(16.4)

It is important that essentially complete ammonia removal be accomplished in the first stage to avoid ammonia/chlorine reactions in the second stage. It is also important to minimize acid carryover from the first stage to avoid pH control problems in the second stage. The

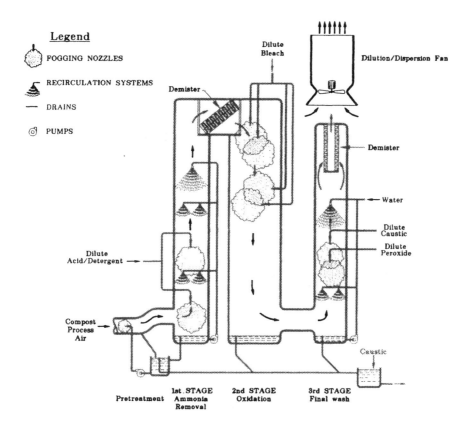

Figure 16.6.   Schematic diagram of the three stage, wet scrubbing system used to treat blower exhaust gases at the Montgomery County Composting Facility. The facility composts raw sewage sludge and wood chips using the static pile process. Mist towers are used for each stage of the scrubbing process to sequentially remove ammonia, oxidize organic sulfides and other organics, and remove residual chlorine. Drawing courtesy of Chuck Murray, Sludge Operations Division Head.

Montgomery County system uses different atomizing nozzles along the length of the first stage. The upstream nozzles produce small mist droplets to enhance mass transfer, while downstream nozzles produce larger droplets designed to collide with and remove the smaller aerosols. Solution recirculation and pH control systems are used to regulate the addition of new acid. With the prototype scrubber, ammonia removal efficiency gradually improved as recirculation and pH control systems were brought on line and optimized. This resulted in corresponding improvements in operation of the second stage scrubber.

The second stage is designed primarily to remove organic sulfides and odors from other organics such as terpenes. An oxidant scrubbing solution is normally used. Hentz et al.[44] reported that sodium hypochlorite (NaOCl) was the most effective oxidant. Bench scale bubbler studies showed that other oxidants, such as hydrogen peroxide (H_2O_2) and chlorine dioxide (ClO_2), were less effective in removing dimethyl disulfide (DMDS) and other sulfur compounds. Results of these studies are presented in Figure 16.7. The bubbler tests also showed that NaOCl was most effective when the solution was adjusted to pH 6.5. At pH 8, the oxidation of DMDS was less effective and removal decreased. This was thought to be due to the predominance of hypochlorite ion at pH 8 which has less oxidizing power than hypochlorous acid. At pH 4.5, NaOCl oxidation produced strong chlorine and burnt organic

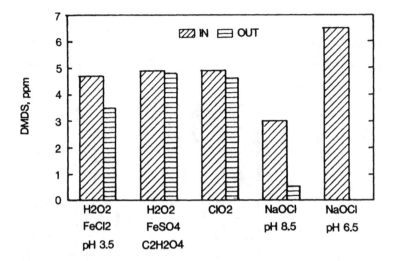

Figure 16.7. The effect of various oxidants and pH conditions on the removal of dimethyl disulfide (DMDS) from blower exhaust gases. Data are based on bench scale bubbler tests. From Hentz et al.[7,44]

odors. The "disinfectant" odor was speculated to come from hypochlorous acid oxidation of α-pinene to carvracol, a commercial disinfection produced by such chemistry. Production of a "disinfectant" odor is one reason that acid carryover from the first stage must be minimized. Operation at pH 6.5 optimized the balance between maintaining sufficient oxidization to remove organic sulfides while reducing formation of "disinfectant" odors.

The oxidation of DMDS by hypochlorous acid can be expressed as,

$$C_2H_6S_2 + 7HOCl \rightarrow 2CO_2 + 3H_2O + 2S^\circ + 7H^+ + 7Cl^- \tag{16.5}$$

$$C_2H_6S_2 + 13HOCl \rightarrow 2CO_2 + H_2O + 2SO_4^= + 17H^+ + 13Cl^- \tag{16.6}$$

For Equation 16.5, it is assumed that the carbon is fully oxidized to carbon dioxide. Partial oxidation to other intermediate organics is also possible. The organic sulfide is oxidized only to elemental sulfur. If sufficient HOCl is available, the sulfide could be oxidized to sulfate ($SO_4^=$) which is the highest valence state for sulfur. Considerably more HOCl is then required as shown by Equation 16.6.

Hentz et al.[44] also reported that complete DMDS removal could only be accomplished when a small amount of gaseous chlorine was present in the discharge gas from the second stage scrubber. Some chlorine odor was always associated with high DMDS removal. The reader is probably familiar with the typical "chlorine" odor from commercial bleaches, which are NaOCl solutions. Therefore, it is not surprising that the process gas can pick up "chlorine" odor from the scrubbing chemicals. The purpose of the third stage scrubber is to dechlorinate the discharge of the second stage. Hentz et al.[44] evaluated reducing agents, such as hydrogen peroxide and sodium bisulfite, to remove the residual chlorine. A slightly alkaline peroxide solution was determined to be the most effective dechlorinating agent. The reduction of hypochlorous acid with peroxide occurs as follows:

$$HOCl + H_2O_2 \rightarrow H^+ + Cl^- + O_2 + H_2O \tag{16.7}$$

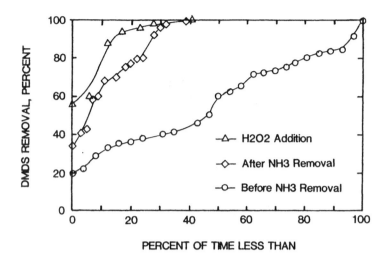

Figure 16.8.   The effect of improved process conditions on the statistical reliability of DMDS removal across the scrubbing system. Solution recirculation and pH control improvements in the first stage improved the reliability of ammonia removal. Peroxide addition in the third stage also increased the probability of maintaining high DMDS removal. From Hentz et al.[7,44]

The evolution of improvements in scrubber operation at the MCCF significantly improved the statistical reliability of the scrubber system. The probability of achieving a given DMDS removal efficiency is presented in Figure 16.8. Improvements in ammonia removal and the addition of peroxide for dechlorination both increased the probability of maintaining high DMDS removal.

The removal of compounds during wet scrubbing requires that the compound first dissolve into the aqueous phase. Nonpolar organics tend to be hydrophobic with low water solubilities, which may limit their removal in a wet scrubber. Surfactant molecules contain both a hydrophobic and hydrophilic end and can potentially improve the dissolution of hydrophobic organics by binding with their hydrophobic tail. Surfactants act at the air/water interface and reduce the surface tension. The liquid surface tension reaches a minimum when the air/water interface is completely covered with a monolayer of surfactant, a condition termed the "critical micelle concentration" (CMC). Hentz et al.[44] studied the effect of surfactant addition to improve the dissolution of hydrophobic organics into the aqueous phase. A commercial, low foaming, unscented, liquid detergent was added to the first stage and some improvement in total organic removal was observed. The authors noted that the surfactant tests were erratic and that some of the effect may have been due to improved atomization from the misting nozzles. They concluded that the effectiveness of surfactant for increasing the dissolution of organics has yet to be fully established.

Even with multistage scrubbing and optimized chemistries, odor removal is never 100%. Odor removal is typically in the range of 90 to 95%. Thompson[45] reported that the Montgomery County scrubbers typically operate with an inlet odor concentration of about 1000 ED_{50} and outlet concentrations of about 70, giving about 93% average removal efficiency. A photo of a multistage misting tower system is shown in Figure 16.9.

Packed Bed Scrubbers

Absorption using packed bed scrubbers is also practiced at many composting facilities. A three-stage packed bed system used at the Sludge Composting Facility, Cape May, New Jersey

Figure 16.9.   Multi-stage, mist tower scrubbers at the City of Hamilton, Ohio reactor system composting raw sludge amended with sawdust, bark, and product recycle. One first stage acid scrubber and two parallel hypochlorite scrubbers were provided. The second stage mist towers originally discharged near the roof line. Experience showed that the plume was subject to downwash in the wake of the adjacent building. The 125 ft stack (above ground level) shown above was then added to avoid downwash. Photo courtesy of Michael Wheeler, Plant Superintendent.

is shown in Figures 16.10 and 16.11. Like the Montgomery County system, the scrubbers at Cape May have undergone an evolutionary development to their current form. The first stage operates at a liquid/gas (L/G) ratio of about 17.5 gal/1000 cf using wastewater plant effluent on a once through basis. Sulfuric acid is added to improve ammonia removal. Cathcart[46,47] noted that a pH of 6 is sufficient to remove all ammonia, but that severe biofouling occurred within the packed bed. The pH was then reduced to the range of 2.5 to 3 to better control biofouling. Gas cooling to about 80 to 90°C, water vapor condensation and ammonia removal occur in the first stage. Cathcart reported that subcooling below the vapor saturation temperature is beneficial because condensation may improve absorption and a water vapor plume is rarely visible from the discharge stack.

The second stage uses NaOCl at pH 6.5 to remove organic sulfides. Some surfactant is also added to improve hydrocarbon removal. Residual hypochlorite is about 300 to 400 ppm in the recirculation solution which results in about 5 ppm in the vapor phase. Attempts are being made to control NaOCl addition based on the vapor phase concentration. The third stage is used to dechlorinate with peroxide at pH 8.5. Treated gases are discharged from a 100 ft tall stack.

All three scrubber columns are 10 ft diameter, giving an approach velocity of about 250 ft/min. Packing depth is 8 ft in the first stage and 10 ft in stages two and three. Gas residence time is about 2 to 2.5 sec in each stage. Consideration is being given to modifying or expanding the first stage because it is designed primarily as an absorption scrubber and not

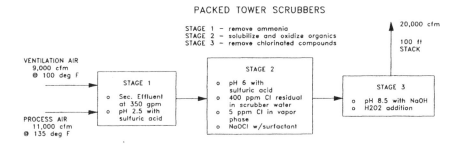

PACKED TOWER SCRUBBERS

STAGE 1 - remove ammonia
STAGE 2 - solubilize and oxidize organics
STAGE 3 - remove chlorinated compounds

20,000 cfm

100 ft
STACK

VENTILATION AIR
9,000 cfm
@ 100 deg F

PROCESS AIR
11,000 cfm
@ 135 deg F

STAGE 1

o Sec. Effluent
   at 350 gpm
o pH 2.5 with
   sulfuric acid

STAGE 2

o pH 6 with
   sulfuric acid
o 400 ppm Cl residual
   in scrubber water
o 5 ppm Cl in vapor
   phase
o NaOCl w/surfactant

STAGE 3

o pH 8.5 with NaOH
o H2O2 addition

Figure 16.10. Schematic diagram of the three-stage, wet scrubbing system used to treat blower exhaust gases at the Cape May County Composting Facility. The facility composts raw sewage sludge amended with sawdust and product recycle in a vertical, packed bed reactor system. Packed bed scrubbers are used for each stage of the scrubbing process. From Cathcart.[46]

Figure 16.11. The three-stage, packed bed scrubbing system used at the Cape May County Composting Facility to treat compost exhaust gases. This facility received a 1990 EPA Recognization of Excellence Award for demonstrating the viability of in-vessel sludge composting and overcoming difficult startup problems, including improvements in odor control accomplished in part by the above scrubbing system. Photo courtesy of William Cathcart, Chief of Operations and Maintenance.

a condenser. The high L/G ratio sometimes causes tower flooding. Nevertheless, the present system is giving good results, and odors from the stack discharge are not detected off-site under most meteorological conditions.

The problem of media fouling has been reported at a number of facilities, usually in the first stage scrubber. As mentioned above, operating at reduced pH alleviated the problem at Cape May. Wilber[48] reported that the packed scrubber used for NH_3 removal at Plattsburgh, New York developed a gelatinous mass of biosolids that plugged the media in 4 to 6 weeks. Media was removed from the tower to a 1-ft depth to allow periodic washing by hoses. Woods[49] reported that surfactant reduced biosolids accumulation in the chemical sumps at Montgomery County.

Mass transfer in a packed column is a function of the packing type, packing height, gas and liquid loading rates, the compound solubility and reaction rate, and the solution chemistry. Mass transfer characteristics, such as the height of a transfer unit, are often ignored in odor scrubber design and this is a mistake. For example, two competing designs of two-stage, packed bed scrubbers were tested at Akron, Ohio. One unit operated with a L/G ratio of 15 gal/1000 cf while the other used much higher values ranging from 15 to 90 gal/1000 cf. The upper limit was about 70 to 80% of the liquid flooding rate. The latter unit reported superior results according to Ostojic.[37]

Based on the above studies, a two-stage, packed bed system designed for 450,000 cfm has been installed at Akron, Ohio. This is likely to be the world's largest scrubbing system serving a composting facility. Twelve, two-stage scrubber trains are provided. Each scrubber tower is 12 ft nominal diameter and operates with a L/G ratio of 55 to 60 gal/1000 cf, liquid loading rate of 19 gpm/ft^2, and superficial gas velocity of 330 ft/min.

Hybrid Designs

Other contactor types can be used for gas phase absorption. However, mist towers and packed bed scrubbers have historically been the favorite absorber types at composting systems. Each of these offers certain advantages/disadvantages over the other. Mist towers achieve a very high surface area for absorption because of the fog-sized droplets produced by the atomization nozzles. Mist towers are also resistant to plugging because there is essentially nothing within the tower itself. Despite the large surface area, however, longer residence times are generally required because of mass transfer limitations in the laminar gas and liquid films around the droplets. Mist towers must also respond rapidly to changes in inlet concentrations because there is very little scrubbing solution held within the tower. The fog produced by the atomizing nozzles is difficult to remove by conventional demisting equipment and a plume is usually visible in the discharge stack. According to Leder et al.[50] mist towers generally operate at a high air to liquid loading rate of 26 to 50 lb air/lb liquid. This reduces blowdown volumes, but also limits the gas cooling that can be provided.

Packed columns typically operate at shorter residence times because the turbulence of gas and liquid flow around the packing increases the mass transfer rates. Packed columns typically maintain a large reservoir of scrubbing solution and operate with a high recirculation rate. Leder et al.[50] estimated air to liquid rates at <0.7 to 5 lb air/lb liquid for most packed bed systems. The high liquid loading and the reservoir of recirculation liquid provide a chemical "flywheel" which allows the system to respond to changing inlet conditions. Gas cooling is also improved because heat can be rejected from the recirculation liquid and the blowdown stream. Packed columns do not generate large amounts of small sized droplets. Conventional demisting equipment effectively removes liquid droplets produced by splashing within the column, and visible plumes from water mist are usually not a problem. Among its disadvantages the packed column is more prone to plugging.

According to Donovan[51] a recent 50,000 cfm design for Hartford, Connecticut uses a hybrid mixture of packed beds and misting towers for the various stages of absorption. In-line duct sprays begin the process of cooling and ammonia removal. A first stage, acid scrubbing, packed column is used to complete the removal of ammonia and cool the gases. The heat load is removed from the scrubber by cooling the recirculation liquid against secondary effluent in a separate heat exchanger. A second stage mist tower provides oxidation using hypochlorite.

A third stage, caustic scrubbing, packed column is then used for polishing and removal of residual chlorine odors. An open "block" media is used in the packed columns, instead of the usual "dumped" media, to reduce plugging problems. Treated gases are dispersed from a 90 ft stack.

Multistage scrubbers are also used at Sarasota, Florida to treat compost process gases from a silo reactor. A spray nozzle cooling tower precedes a two-stage packed scrubber system. The purpose of a hybrid design is to try to match the advantages of each reactor type to the demands of a particular stage. Hybrid designs using wet scrubbing/condensing before biofiltration (discussed later) are also used. The scrubber is usually used to remove ammonia and partially cool the gases. For example, a spray system in the main supply manifold is used to remove ammonia prior to a biofilter at the City of Hamilton, Ohio.[38] A water cooled condenser is used to lower process gas temperatures prior to biofiltration at the MSW composting facility at Salzburg, Austria.[39]

Condensation/Cooling

When the temperature of a water-saturated gas is lowered by contact with a cooled surface or a cooling liquid, water condensation results. Water soluble odor constituents may dissolve in the condensing water and be removed from the gas phase. Condensation is a special case of gas absorption and can be an important deodorization mechanism for compost offgases, which are usually moisture laden at temperatures above ambient.

Because compost exhaust gases are saturated with water vapor (or nearly so) at temperatures above ambient, it is necessary to provide water traps in any ducting used to transport such gases. As condensation occurs it is likely that water-soluble species in the gas will partially absorb into the condensate. Iacoboni et al.[26] reported the following chemical characteristics for condensate collected during static pile composting of digested sludge: COD from a few hundred to >9800 mg/L (2500 mg/L average) and ammonia from about 100 to 2000 mg/L as N (560 mg/L average). Obviously, a large number of volatile, low molecular weight organics along with inorganic ammonia condensed with the moisture. In other studies, these authors enhanced the condensation by water washing in a packed bed scrubber. Simple water scrubbing provided an average 44% ED_{50} removal as shown in Figure 16.12. Based on these results, condensation may partially remove certain odor constituents but cannot be relied on for complete odor control. In fact, careful handling and treatment of the condensate is necessary to prevent desorption of the odor constituents.

A number of composting facilities attempt to cool the process gases before further odor treatment. This is usually accomplished in a spray or packed bed tower. Large water flows are typically required for cooling because of the high water vapor content in the process gas and the need to remove the latent heat of condensation. Horst et al.[52] and Horst[53] reported using a L/G ratio of about 32 to 49 gal/1000 cf in a spray tower at Lancaster, Pennsylvania. Plant effluent is used on a once-through basis and reduces the gas temperature from 130°F to about 105°F (25°F delta). Twelve, 120° full cone spray nozzles at 20 psig are used to induce gas to liquid contacting. HCl acid is added to enhance ammonia removal. Discharge from the spray tower contains <1 ppmv NH_3. Considerable DMDS removal is also accomplished in the spray tower.

The City of Portland, Oregon, recently installed a once-through, packed bed water scrubber/condenser to treat odorous gases from a silo reactor system composting digested sludge and sawdust. No chemicals are used. The L/G ratio is about 12.5 gal/1000 cf with 10 ft of packing. The system reportedly achieved over 90% ammonia removal in preliminary testing.

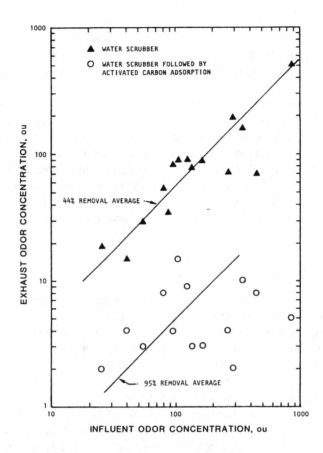

Figure 16.12. The effectiveness of water scrubbing and adsorption for odor removal from compost exhaust gases. The gases were collected from an experimental reactor composter treating digested sewage sludge blended with recycled product. Developed from data reported by Iacoboni et al.[26,27]

It is important to consider the high heat load from the latent heat of condensation when subcooling a gas below its adiabatic saturation temperature. High liquid rates and once-through operation can be used to reject heat to the scrubber blowdown. Alternatively, heat can be rejected from the recirculation liquid by heat exchange with another media. The example of an external heat exchanger cooled against secondary effluent at Hartford, Connecticut, was previously described. Leder et al.[50] described a system in which compost gases could be cooled to <95°F by passing through an air:air heat exchanger. Heat in the secondary air stream would then be rejected to the atmosphere by evaporative cooling.

Example 16.1

A composting facility produces 60,000 cfm of process gas at 130°F. The process gas contains 400 ppmv NH_3. Estimate the daily consumption of sulfuric acid required to neutralize the ammonia in a wet scrubber.

Solution

1. Determine the volume for 1 lb-mol of gas at 130°F and 1 atm using Equation 11.53. The universal gas constant, R, is 0.730 for units of ft^3, °R, and atm.

$$V = \frac{mRT}{p} = \frac{(1)(0.730)(460+130)}{1}$$

$$V = 431 \text{ ft}^3/\text{lb} - \text{mol} = 26.9 \text{ L/g} - \text{mol}$$

2. Determine the lb/day of ammonia in the process gas as follows:

$$60,000 \text{ cfm}(60 \text{ min/h})(24\text{h/day}) = 86.4 \times 10^6 \text{ cfd}$$

$$86.4(10^6)(400/10^6)(\text{lb} - \text{mol}/431 \text{ ft}^3)(17 \text{ lb NH}_3/\text{lb} - \text{mol}) = 1360 \text{ lb NH}_3/\text{day}$$

3. The molecular weight of H_2SO_4 is 2(1)+(32)+4(16) = 98. The required quantity of sulfuric acid is determined from Equation 16.4 as

$$\frac{1360}{2(17)} = \frac{x}{98}$$

$$x = 3920 \text{ lb/day H}_2SO_4$$

4. Assuming a sulfuric acid costs of $0.10/lb in bulk, annual acid costs can be estimated as

$$\$0.10(3920)(365) = \$143,000/\text{yr}$$

Adsorption

Adsorption is the process of contacting a fluid phase (gas or liquid) with a particulate phase, which has the property of selectively taking up and storing one or more solute species from the fluid. Activated carbon is the most widely used adsorbent and consists largely of neutral carbon atoms which present a nonpolar surface. This makes it effective in humid gas streams because polar water molecules do not compete for the nonpolar surface. Activated carbon is effective in adsorbing a wide range of inorganic and organic compounds as shown in Table 16.10.

Iacoboni et al.[26] evaluated an experimental reactor composting digested sludge conditioned with recycled compost. The system used water scrubbing followed by activated carbon adsorption for deodorizating exhaust gases. The carbon bed depth was about 0.6 m and the gas supply rate about 14 m^3/min-m^2. Carbon adsorption, either alone or preceded by water scrubbing, was very effective in odor removal as shown in Figure 16.12. Effluent odor concentration was largely independent of the influent concentration. In all tests but one, effluent concentration was <10 ED_{50} regardless of the incoming load. The advantage of water scrubbing before adsorption is to possibly extend the life of the carbon bed. The researchers did not determine the adsorption capacity of the carbon.

Carbon adsorbers were installed at a static pile facility composting sewage sludge at Somerset, Massachusetts. Sarrasin[55] reported that ammonia rapidly broke through the carbon and that the ammonia smell was stronger in the exhaust. The experience to date suggests that carbon is not effective against ammonia. Prescrubbing to remove ammonia is probably

Table 16.10. Retentivity of Vapors by Activated Carbon

Compound	Formula	Molecular Weight	Boiling Point (°C @ 760 mm Hg)	Approximate Retentivity[a] wt %
Acetaldehyde	C_2H_4O	44.1	21	7
Amyl acetate	$C_7H_{14}O_2$	130.2	148	34
Butyric acid	$C_4H_8O_2$	88.1	164	35
Carbon tetrachloride	CCl_4	153.8	76	45
Ethyl acetate	$C_4H_8O_2$	88.1	77	19
Ethyl mercaptan	C_2H_6S	62.1	35	23
Eucalyptole	$C_{10}H_{18}O$	154.2	176	20
Formaldehyde	CH_2O	30.0	−21	3
Methyl chloride	CH_3Cl	50.5	−24	5
Putrescine	$C_4H_{12}N_2$	88.2	158	25
Skatole	C_9H_9N	131.2	266	25
Sulfur dioxide	SO_2	64.1	−10	10
Toluene	C_7H_8	92.1	111	29

Source: Cross.[54]

[a] Weight % retained in a dry airstream at 20°C and 760 mm Hg, based on weight of carbon.

necessary in applications with high nitrogen substrates. Carbon adsorbers were used at the Westborough, Massachusetts static pile facility to treat building ventilation air. The adsorbers were prone to plugging from dust and fine particulates.

Novy[56] used a process of refrigerated chilling to subcool and condense water vapor from exhaust gases from a pilot composting reactor using digested sludge conditioned with recycle and sawdust. A portion of the cooled gas was recycled to the composting reactor. The remaining gas was treated by carbon adsorption. Breakthrough of DMDS occurred from the carbon bed at about 60,000 to 80,000 scf/lb carbon. Novy reported that if a full scale facility is built, carbon replacement will be contracted with an independent, off-site regenerator.

In addition to rapid ammonia breakthrough, another drawback to carbon adsorption is the fact that it is only a removal and storage mechanism. Adsorption itself does not destroy odor. Eventually the carbon bed saturates with odorous compounds and odor breakthrough occurs. The carbon must either be replaced or regenerated prior to odor breakthrough. Caustic soaking is sometimes used to regenerate carbon beds in place. However, this method is only marginally effective even with H_2S dominated gases. With complex odor sources, such as composting gases, thermal regeneration is usually required. Because of the difficulties in permitting on-site thermal processes, most facilities replace spent carbon with either new carbon or used carbon from an off-site regeneration facility. For these reasons, treating compost gases with activated carbon is not widely practiced. It could play a role as a polishing step downstream of another primary removal system such as multistage wet scrubbing.

Example 16.2

An odorous gas at 130°F (54.4°C) contains 10 ppmv of total hydrocarbons (THC). It is estimated that a particular carbon can adsorb 20% by weight of the hydrocarbons. Determine the volumes of gas that can be deodorized per weight of carbon. Assume a carbon bulk weight of 0.55 g/cc and average hydrocarbon molecular weight of 100.

Solution

1. Concentration of THC in gas stream:

$$\left(\frac{10}{10^6}\right)(g-mol/26.9L)(100 \text{ g/mol})(1000 \text{ mg/g})$$

0.0372 mg THC/L of gas

2. THC adsorption per liter of carbon:

total THC adsorption $= 0.55(0.20) = 0.11$ kg/L of carbon

3. Total carbon life $= 0.11 \text{ kg/L}\left(\dfrac{1 \text{ L}}{0.0372 \text{ mg}}\right)(10^6 \text{ mg/kg})$

$= 2.96 \times 10^6$ vol gas/vol carbon

$= 2.96 \times 10^6 \left(\dfrac{1 \text{ L}}{550 \text{ gm}}\right)\left(\dfrac{m^3}{1000 \text{ L}}\right)(1000 \text{ g/kg})$

$= 5,380 \text{ m}^3/\text{kg carbon}$

$= 86,200 \text{ ft}^3/\text{lb carbon}$

4. The above calculations agree with the general range of values reported by Novy.[56] Referring to Table 16.3, a typical concentration for dimethyl sulfide is about 3000 mg/m³. With a molecular weight of 62, this is equivalent to about

$$3000 \text{ }\mu g/m^3(26.9 \text{ L}/62 \text{ g})(m^3/1000 \text{ L}) = 1.3 \text{ ppmv}$$

Therefore, a THC of 10 ppmv would seem to be reasonable. However, THC concentrations as high as 100 to 150 ppmv have been reported in some composting exhaust gases. Because of uncertainties associated with the inlet gas composition and the breakthrough characteristics of each component, pilot studies are recommended if carbon adsorption is considered for deodorization.

Example 16.3

Pilot studies using carbon adsorption on a composting process gas indicate that odor breakthrough occurs at about 60,000 cf/lb carbon. Gas flowrate is 60,000 cfm. Estimate the expected annual cost of carbon if regenerated carbon can be purchased at $0.55/lb. Each carbon tower is 10 ft ID and contains 2 carbon beds each 3 ft deep. Carbon density is 45 lb/cf. Estimate the number of required carbon beds if carbon replacement is conducted once a month. Estimate the cost of the initial carbon fill if new carbon costs $2/lb.

Solution

1. The daily consumption of carbon can be estimated as

$$60,000 \text{ cfm}(60 \text{ min/h})(24 \text{ h/day})\left(\frac{1 \text{ lb C}}{60,000 \text{ cf}}\right) = 1440 \text{ lb/day}$$

2. Annual costs are estimated as

$$1440(365)(\$0.55) = \$289,000/\text{yr}$$

3. The weight of carbon in each carbon tower is determined as

$$\text{weight/tower} = (\pi/4)(10^2)(2 \text{ beds})(3 \text{ ft})(45 \text{ lb/cf})$$
$$= 21,200 \text{ lb/tower}$$

4. The minimum number of carbon beds is determined as

$$\frac{1440 \text{ lb/day}(30 \text{ days/month})}{21,200} = 2.04 \text{ towers}$$

3 carbon towers would be required to allow full gas treatment during unloading and loading operations.

5. The cost of the initial carbon fill is determined as

$$21,200 \text{ lb/tower}(3 \text{ towers})(\$2/\text{lb}) = \$127,200$$

Thermal Oxidation

Oxidation of both organic and inorganic species in an odorous gas can be readily accomplished provided sufficient time, temperature, turbulence, and oxygen are provided. Essentially complete odor destruction can be achieved with temperatures of 1200 to 1550°F (649 to 843°C) and gas retention times of 0.3 to 1.0 sec. The City of Los Angeles currently operates a thermal oxidizer at 1400°F and 1 sec residence time. This oxidizer treats an inlet gas from a sludge drying process with over 1 million ED_{50}. The outlet concentration is near the ambient background at about 36 ED_{50}, giving over 99.99% destruction. To achieve these high levels of destruction, it is essential that all gas is exposed to the time/temperature conditions within the oxidizer and that short-circuiting is minimized. If this is achieved, even the most intractable odor problems can usually be solved by thermal incineration.

A major disadvantage of thermal oxidation is the large energy requirement. A number of approaches have been used to reduce the fuel requirements including recuperative heat recovery, regenerative heat recovery, and catalytic oxidation. Catalytic furnaces use a catalyst to reduce the required oxidation temperature. These furnaces are used extensively in industry for control of volatile organic compounds (VOCs), but have not been widely applied for odor control. They are capable of operating in the range of 600°F (315°C). Ostojic et al.[57] observed cases where odor emissions actually increased during catalytic incineration even though VOC removal was over 90%. The reason was the much lower odor threshold of oxygenated VOCs formed during incomplete combustion in the catalytic incinerator.

Recuperators recover the heat energy of the exhaust gas by exchanging with the inlet combustion air in a gas to gas heat exchanger. The City of Los Angeles operated a recuperative incinerator (1400°F and 0.6 sec) to deodorize sewer gases. Gas to gas heat exchange can preheat the infeed air to about 1000 to 1200°F. With the furnace operating at 1400°F, the net temperature rise across the system can be as low as about 400°F, considerably less than a conventional incinerator. Ostojic[37] suggested that recuperative heat recovery is often preferred because of its simplicity and relatively maintenance free operation.

Regenerative afterburners recover the heat content of the exhaust gas in a bed of pebbles or ceramic packing. When the bed is hot the exhaust gases are switched to a cool bed. The hot bed is then used to preheat the incoming odorous gas. Three or more beds are usually used to assure continuous operation. Hentz[58] reported on the use of a pilot, regenerative afterburner at the Montgomery County Composting Facility. The unit operated with a net temperature rise of 120 to 150°F with a combustion chamber temperature of about 1700°F. Inlet odor concentration was about 1050 ED_{50} and the outlet about 140. Inlet hydrocarbon concentration was about 130 to 135 ppmv, with 5 to 7 ppmv in the outlet. By comparison, the three-stage scrubbing system achieved an outlet odor concentration of about 70 ED_{50} on the same inlet gas.

The regenerative afterburner tested at Montgomery County did not achieve the high removal efficiencies noted above for conventional and recuperative furnaces. The reason for this lies in the mechanical design of the regenerative incinerator. Valve leakage and failure to purge the gas volume after bed switching prevented all gases from "seeing" the full time/ temperature conditions in the combustion chamber. Thus, destruction of odorous compounds was only 85+% and VOCs about 93+%. These problems are solvable and regenerative oxidizers can operate in the range of 98+% destruction efficiency if designed accordingly. Special care is needed to reduce valve leakage and prevent the escape of untreated exhaust gases during switchover between preheat chambers.

Hentz[58] estimated the operating costs of treating 60,000 cfm of process gas with the regenerative afterburner at $400,000/year, about equivalent to the cost of operating the three-stage scrubbing system. The Montgomery County facility decided to expand the existing wet scrubbing system because of its higher performance and reliability for odor control. Nevertheless, regenerative afterburners may play a role at some composting facilities where total emissions of VOCs and hydrocarbons are of concern or where lower quantities of very odorous gas are produced, such as with MSW or other odorous and relatively dry substrates.

Bain et al.[59] reported that regenerative thermal oxidation is being implemented for treatment of odors and VOCs in compost gases at Springfield, Massachusetts. A silo reactor system is used to compost heat treated sludge, one of the few compost facilities treating this type of sludge. The thermal oxidizer will treat about 23,000 cfm of gases from the composting facility, the belt press dewatering area, and the decant tanks serving the heat treatment system. VOC loading is about 2700 ppm and the proposed destruction efficiency is 95%. Consideration is being given to including scrubbing for particulate removal ahead of the thermal oxidizer.

Based on the studies conducted to date the following conclusions can be drawn: (1) thermal oxidation can be essentially 100% effective for odors and VOCs; (2) the high volume of composting gases requires the recovery of heat for thermal oxidation to be cost-competitive; (3) regenerative heat recovery appears to operate with the lowest net temperature rise and, therefore, the lowest specific fuel requirement; (4) regenerative oxidizers require special design features to achieve >95% odor removal efficiency; (5) recuperative heat recovery is relatively simple and maintenance free but usually achieves less recovery compared to regenerative beds; and (6) catalytic incinerators can sometimes increase odors due to partial oxidation of VOCs.

Chemical Oxidation

The use of chemical oxidants in scrubbing solutions has been described under absorption above. It is also possible to inject gaseous oxidants such as ozone (O_3) directly into an odorous gas. Although ozonation can be effective in some cases, few large scale facilities have been installed. Ostojic[37] and McIlvaine[60] reported that an ozone contact tower was used to treat process gases at the Sarasota, Florida silo reactor facility composting sewage sludge and yard wastes. The unit was not considered effective and was replaced by a spray cooler followed by two-stage, packed bed scrubbers. There are no other known applications of ozone to odorous gases from composting.

Biological Oxidation

A number of deodorization processes have been developed that use microbial oxidation for odor removal. Other removal mechanisms, such as absorption and adsorption, may contribute to the initial odor removal, but biological oxidation is the predominate oxidation mechanism.

Activated Sludge

There is a long history of application using activated sludge as a form of "liquid biofilter" to deodorize sewer gases. Garber[61] reported that using odorous air for aeration of the activated sludge process was an effective odor removal strategy. Data developed by the City of Los Angeles on odor removal as sewer gases are diffused through the activated sludge process is presented in Table 16.11. Hydrogen sulfide, mercaptan, and organic sulfides were effectively removed by the process, and odor removal was over 90%. Remaining odors were released over the large area of the aeration tank instead of as a point source. Contact with the biological floc was determined to be important for removal of the odor compounds. Diffusion of sewer gas through primary effluent, without activated sludge present, actually stripped additional sulfides from solution. Therefore, adsorption of odorous compounds onto the activated sludge floc is probably the initial removal mechanism followed by biological oxidation. By comparison with liquid scrubbers, diffusing odorous air through activated sludge provides an extremely high liquid to gas ratio.

Vogt[63] and Haug et al.[64] reported on several European compost facilities that are successfully using activated sludge to treat odors from compost exhaust gases. Quantitative data on odor removal efficiency were limited, but qualitative assessments suggested that compost odors were effectively removed by this process. Most observers report only the characteristic "musty" odor of activated sludge, with little or no discernable "compost" odor.

The City of Springfield, Massachusetts uses a silo reactor system to compost heat-treated, combined sludge amended with sawdust. About 4500 scfm of compost exhaust gases are treated by diffusion through activated sludge. Centrifugal compressors deliver the process gases through about 700 ft of stainless steel and HDPE piping to submerged coarse bubble diffusers at about 8-ft liquid depth. The gas loading is about 4.5 cfm/ft². Ostojic[37] and Ostojic

Table 16.11. Removal of Odorous Compounds from Sewer Gases Diffused through the Activated Sludge Process

Compound	Inlet Concentration	Outlet Concentration
Hydrogen sulfide, ppm	5–30	<0.5
Methyl mercaptan, ppb	160	0
Dimethyl sulfide, ppb	0.4	0
Dimethyl disulfide, ppb	0.5	0.01
Odor, ED_{50}	330	<20

Source: Data from unpublished studies conducted by the City of Los Angeles and Duffee et al.[62]

et al.[65] used flux chambers to sample offgases from the activated sludge tank receiving the compost gases and from a similar tank which did not. There was essentially no difference in odor concentrations in the gases escaping the surface of the two tanks. Ostojic demonstrated comparable results at a similar composting facility in Orlando, Florida which also uses activated sludge treatment. Inlet odors in the process gas ranged from 2600 to 4400 ED_{50}. Air discharged from the surface of the aeration tanks averaged about 40 ED_{50} in the tank receiving compost gases and 35 ED_{50} in a control tank. Ostojic concluded that activated sludge treatment is highly effective in control of compost gases.

One disadvantage with the use of activated sludge is the requirement that the gases be compressed to about 4 to 10 psig for introduction to the submerged diffusers. The compost gases can be delivered to the suction side of the aeration compressors. The City of Los Angeles blended sewer gases with aeration air in this manner for over 20 years with no significant deterioration of the centrifugal compressors. Ostojic et al.[65] recommended that materials selection consider the potentially corrosive properties of compost exhaust. High quality stainless steel or non-ferrous piping and diffusers were recommended. It was also recommended that coarse bubble diffusers be used to avoid fouling and plugging problems observed with fine-bubble devices. Many compost plants, including facilities at Springfield, Massachusetts and Bickenbach, Germany use dedicated coarse-bubble diffusers to introduce the compost gases into the activated sludge process to reduce plugging problems. Return loads to the activated sludge system should be considered, particularly ammonia when nitrification is required.

As a variation on the above, Bohn and Bohn[83] reported on the use of "biowashers" for treatment of organic laden gases. The biowasher (also termed bioscrubber or biowash tower) is a wet scrubber that uses activated sludge slurry to absorb/adsorb gases into the liquid phase. The adsorbed gases are decomposed later when the activated sludge slurry is regenerated by aeration. Bohn and Bohn reported 80 to 90% removal of solvents from an auto painting facility. However, the authors also noted that of the biomethods for air scrubbing, biowashers have the least space requirement but also the lowest scrubbing efficiency. There are no known applications of bio-washers for deodorization of composting gases, but Feindler[66] reported their use for deodorization of exhaust gases from a sludge heat dryer.

Biotowers (also termed biotrickling filters) have also been used to control odors at sewage treatment plants. Biotowers use plastic media to support the growth of a microbial biomass. Secondary effluent is usually used to wet the media and provide growth nutrients to the biomass. Odorous air is forced upward through the media. Kasbrian[67] reported an average removal of 94% for H_2S and 73% for ED_{50} odor during almost 500 days testing of a biotower treating sewer gases. Gas residence time varied from 6.5 to 16 sec. Removal efficiencies for H_2S and ED_{50} approached 100% and 90%, respectively, at the longer gas residence times.

Clifton et al.[68] reported using an activated biofilter (ABF) for deodorizing process gases from a silo reactor composting raw sludge and sawdust at Bristol, Tennessee. The ABF uses redwood slat media and pretreats primary effluent mixed with return activated sludge prior to aeration. Compost gases are discharged beneath the ABF tower and flow upward through the media by natural draft. Dohoney[69] reported that a "compost" odor remained in the exhaust gas from the ABF and that the deodorization may not be adequate for the long term. Consideration is being given to treating the compost gases in the activated sludge basins.

Biofiltration

Biofiltration is an air pollution control technology that uses a biologically active, solid media bed to absorb/adsorb compounds from the air stream and retain them for their subsequent biological oxidation. Biofilters are used primarily for the treatment of odors and VOC compounds. Soil covers or filters have been used since antiquity to reduce odors. Today, soil covers for landfills and septic systems are widely recognized for control of odors from these sources. The first "soil or earth filter" was patented in the U.S. by Pomeroy in 1957.[70] The design consisted of slotted pipes buried under a soil bed. Around 1959 a soil bed was also installed at a sewage treatment plant in Nuremberg, West Germany for control of sewer odors.[71] Variations on these basic designs have been applied throughout the world for control of odors including many applications on composting gases. Many of the process advances have come from Europe, particularly Germany and the Netherlands, where biofilter technology is widely used for VOC and odor applications.

The biological activity of soil is considerably less than other substrates such as peat and compost. As a result, soil filters operate with relatively low loading rates. This prompted a trend toward use of organic media to increase removal rates and more engineered systems for gas pretreatment and distribution. Bohn[72] described the useful properties of compost as a biofiltration media, including its high surface area, air and water permeability, water holding capacity, active microbial population, and relatively low cost. Today, soil filters have largely given way to biofilters, particularly in Europe where the soil filter is considered the "father" of the modern biofilter.

A generalized schematic diagram of a biofilter system is presented in Figure 16.13. Odorous gas is typically humidified prior to entry into the biofilter. A manifold system serves to distribute the gas across the filter media, which is usually 1 to 1.5 m deep. Allen and Yang[74] recommended against bed depths <0.5 m because of the practical problems of gas distribution and channeling. Biofilters vary from relatively simple, open designs to highly engineered, enclosed systems using specialized media, automatic watering systems, and multiple beds. Photos of typical biofilter applications are shown in Figures 16.14 and 16.15. The term "biotower" is applied to systems where the media is placed in a vessel similar to a packed scrubbing tower.

Perhaps nothing has evoked as much controversy in the composting industry as biofiltration and its effectiveness for odor control on composting gases. A large number of facilities use biofilters and consider them proven and effective. Haug et al.[39,40] observed biofilters at MSW composting facilities in Europe operating with inlet gas concentrations as high as 25,000 to 50,000 ED_{50}. Outlet concentrations were reported to be 50 to 200 ED_{50}, well over 99% removal efficiency. Hartenstein and Allen[75] reported high removal efficiencies for specific compounds such as H_2S (>99%), methyl mercaptan, DMS, DMDS (>90%), and various terpenes (>98%). On the other hand, Murray[33] reported that a 326 m^2 biofilter provided little effective removal of organic and inorganic sulfides when highly loaded at 335 m^3/min-m^2 (about 10 sec

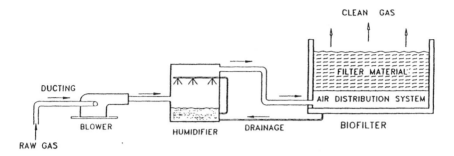

CLEAN GAS

FILTER MATERIAL

DUCTING

AIR DISTRIBUTION SYSTEM

BLOWER   HUMIDIFIER   DRAINAGE   BIOFILTER

RAW GAS

Figure 16.13. Schematic diagram of an open, single bed biofiltration system. From Leson and Winer.[73]

residence time). Murray[36] also reported that the biofilter was relatively dry and that efforts to add water to the pile surface only wetted the top few inches.

A number of U.S. facilities abandoned biofilters in the 1980s in favor of other technologies. Haug[76] surveyed a number of U.S. facilities in 1990 and concluded that there were considerable differences of opinion as to the effectiveness of compost deodorization piles, with strongly held opinions on both sides of the issue. Since then, however, biofilters have regained some of their popularity in the U.S. as more is known about their proper design and operation and as more successful operations are reported.

The rather "bumpy" history of application in the U.S. seems to be influenced by a number of factors. First, the static pile system pioneered the widespread application of small compost piles for deodorization of blower exhaust in the U.S. Air loading to these small piles was very high and removal rates were low. A 50% odor reduction was reported by Gruber et al.[31] for the typical deodorization pile. The trend in newer designs is toward lower loading rates. Second, a biofilter is a biological treatment process and conditions must be suitable for the microbial flora. Conditions of moisture and pH must be monitored and maintained to assure proper operation. Dehydrated or waterlogged conditions will limit microbial activity. Unfortunately, the relative simplicity of the biofilter has often led to its being ignored by plant operators. Third, removal efficiency is never 100%. Gases exhausted from a biofilter will always contain a residual odor and are usually released at groundlevel. Many facilities failed to consider the subsequent dispersion of the exhaust gases. Fourth, the biofilter media can contribute odor to the gas. Many U.S. facilities used the plant compost as media for the biofilter. The compost contributed similar odor to the exhaust gas, just as chemicals contribute to the odor from wet scrubbers. The use of specialized, low odor media has not been widely practiced in the U.S., but is gaining popularity.

Wheeler[38] reported a very successful biofilter operation at the City of Hamilton, Ohio. The 9000 ft^2 biofilter treats about 42,000 cfm of combined process and building air, giving a superficial air velocity of 5 ft/min (91 m^3/h-m^2). Inlet concentrations are 500 to 600 ED$_{50}$ with outlet concentrations generally in the range of 15 to 30 ED$_{50}$. Significant VOC reductions are also achieved. The original biofilter did not achieve this level of performance due primarily to drying of the media. The cured compost of the plant was used as media for the biofilter. This was replaced by a media recipe which includes woodchips, bark, and leaf compost. The new media is very porous and resistant to water logging. Four spray nozzles were inserted into the supply manifold to saturate the gas and reduce ammonia concentrations. Secondary effluent is also sprayed on top of the biofilter to assure that the media is thoroughly wet. A total of 50 to 100 gpm of secondary effluent are supplied to the manifold nozzles and the media overspray. The overspray is used constantly in the summer, but is turned off in winter to avoid

a

b

Figure 16.14. Distribution manifold and biofilter used at the Siggerwiessen, Austria, composting facility which serves the Salzburg region. The facility composts 400 tpd of mixed refuse and 120 tpd of sewage sludge. 50,000 m³/h of exhaust gases are collected from rotary drum composters and from static piles in the maturation hall and treated by the biofilter. The biofilter is designed to Austrian standards which require a loading <100 to 120 m³/h-m² and a 1.5 m depth. The biofilter is approximately 30 by 45 m and includes eleven sections which can be shut down individually for maintenance. Gases input to the biofilter are saturated by sub-cooling in a heat exchanger. Final plant compost is used as the filter media.

Figure 16.15. Distribution manifold and biofilter used at the City of Hamilton, Ohio, sludge composting facility. The 9000-ft² biofilter treats about 42,000 cfm of process gases and building air. The biofilter is approximately 55 ft by 165 ft with a 3-ft media depth. The media is a mixture of wood chips, bark and leaf compost. Water sprays in the main manifold pipe help saturate the input gases and partially remove ammonia. A sprinkler irrigation system is used to maintain moisture content in the biofilter. Leachate from the biofilter is returned to the treatment plant. The air distribution system is comprised of 19 perforated drainage pipes 12 in. diameter, each about 50 ft long and spaced about 8 ft apart. The ends of the pipes are connected together. The pipes were originally covered by an 18-in. base of wood chips to improve air distribution. Experience showed that after about 11 months the wood chips degraded to a small particle size which waterlogged and plugged. The wood chips were replaced with 2 in. rock as a support matrix for the media. Photo courtesy of Michael Wheeler, Plant Superintendent.

freezing. Drainage from the biofilter is returned to the wastewater plant. The biofilter is rototilled every few months to prevent cracking and channeling. Odor panel testing is conducted on a near daily basis to monitor performance of the biofilter. These physical and operational modifications have resulted in a deodorization system which is efficient, reliable, and relatively inexpensive to operate.

Based in part on the successful biofilters described above, a number of large systems have been installed at U.S. composting facilities in recent years. A 120,000 cfm facility is operated at Yarmouth, Massachusetts, which composts septage and sludge. Biofilter media is composed of leaf compost, bark mulch, and wood chips. The loading rate is 4 cfm/ft² with a 3 ft media depth. A 72,000 cfm is being constructed at Lewiston/Auburn, Maine, with similar design criteria.

A summary of recommended design and operating parameters for biofilters is presented in Table 16.12. These parameters are further discussed in the following.

Table 16.12. Recommended Design and Operating Parameters for Organic Media Biofilters

Filter media	Biologically active, but reasonably stable Organic content >60% Porous and friable with 75–90% void volume Resistant to water logging and compaction Relatively low fines content to reduce gas headloss Relatively free of residual odor Specially designed mixtures of materials may be desirable to achieve the above characteristics
Moisture Content	50 to 70% by weight Provisions must be made to add water and remove bed drainage
Nutrients	Must be adequate to avoid rate limitations Usually not a problem with composting gases because of the high NH_3 content
pH	7 to 8.5
Temperature	Near ambient, 15–35 or 45°C
Gas pretreatment	Humidification as necessary to achieve near 100% inlet gas humidity Dusts and aerosols should be removed to avoid media plugging
Gas loading rate	<100 $m^3/h\text{-}m^2$, unless pilot testing supports higher loadings
Gas residence time	30 to 60 sec, unless pilot testing supports a shorter residence time
Media depth	>1 m or 3 ft
Elimination capacity	Depends on media and compound (about 2.2 mg H_2S/kg media VS per min for H_2S)
Gas distribution	The manifold must be properly designed to present a uniform gas flow to the filter media

Filter Media   Biofilter media can vary from locally available composts to proprietary mixtures made from ingredients such as compost, soil, peat, bark, wood, lime, activated carbon, and polystyrene spheres. The filter material must provide good physical/mechanical properties to keep the costs of operation and maintenance low. The media should drain water without becoming water logged, should not compact over time, and should present a relatively low headloss. At the same time it must provide the chemical/biological properties needed to provide a suitable environment for the microbes. The media should not be so stable that the microbes are starved of nutrients.

Bohnke and Eitner[77] studied a number of filter materials and found that garbage compost generally provided high pollutant removal efficiency. Allen and Yang[74] observed high H_2S removals with a number of biofilter media composed of composted bark, sewage sludge, and yard wastes. Wheeler[38] reported that a mix of wood chips, bark, and leaf compost achieved 95+% removal of ED_{50} at a sludge composting facility. This mixture replaced the plant compost which did not produce similar removals and was more prone to plugging.

Loading Capacity   The required size of a biofilter is a function of the pollutant loading and the pollutant concentration in the inlet gas. The pollutant loading is important because the particular filter media will have a maximum capacity to oxidize any pollutant. Inlet concentration is also important because of potential diffusion limitations at low concentrations. Fundamental equations can be developed for biofilter sizing. However, this approach to design applies best to situations where the feed gas contains a small number of well defined constituents. Empirical approaches are generally used for odor control applications on complicated gas streams.

Hartenstein and Allen[75] concluded that removal efficiency is primarily determined by the gas residence time within the filter bed. The bulk flow of gas inside the filter bed is believed to be turbulent, but with laminar regions at the gas to liquid interface. Diffusion across the laminar layers is a major mass transport limitation. Hence, the rate of transport depends mainly on the residence time which should be large compared to the molecular transport time. At high concentrations, pollutant elimination may become limited by the biological activity in the media. In either case the residence time, determined as the height of the filter bed divided by the specific loading rate, has a direct influence on the overall removal efficiency of the biofilter. Biofilters have been designed with residence times from less than 10 to over 150 sec. Residence times typically range from 30 to 60 sec for most compost applications. For example, Bohn[72] recommended a minimum residence time of 30 sec for removal of biodegradable gases in a compost filter used at a Duisberg, West Germany composting facility. Residence times of 30 and 60 sec are equivalent to loadings of 6 and 3 cfm/ft^2, respectively, for the typical 3 ft bed depth.

Eitner[78] developed the loading diagram presented in Figure 16.16 for the treatment of sewer gases with compost filters. The outlet odor concentration is a function of the inlet odor concentration and the specific loading rate. The bed depth was 1.0 m for these studies, giving residence times of about 72 and 36 sec for specific loadings of 50 and 100 $m^3/h\text{-}m^2$, respectively. It is important to note that some exhaust odor remains even at low loading rates. The exhaust odor is a function of the inlet concentration and probably also depends on the nature of the biofilter media. Prokop and Bohn[79] reported outlet concentrations ranging from 20 to 176 ou when treating odors from a rendering plant. Inlet concentrations ranged from 20,000 to over 200,000 ou. Experience suggests that typical exhaust concentrations vary from 20 to 150 ou with compost gases.

As suggested by Figure 16.16, performance of the biofilter is affected by the inlet concentration. Several researchers have suggested the concept of "elimination or loading capacity" expressed as weight of pollutant removed per unit of time and per unit of volume or weight of filter media. Allen and Yang[74] studied a number of composts and determined a maximum loading capacity for H_2S of 2.2 mg/kg ds-min. Dharmavaram[80] suggested a total organic carbon loading capacity of 1.7 to 3.3 g/m^3-min. Leson and Winer[73] reported that degradation rates for common air pollutants typically range from 0.17 to 1.7 g/m^3-min. These values represent the maximum capacity of the biomass to process the particular constituent. The elimination capacity depends on the compound being degraded and also the activity of the filter media. Low molecular weight, highly water soluble and degradable compounds have higher elimination rates. Operation of a biofilter at loadings above the elimination capacity will significantly increase the outlet concentration. The minimum required volume of media can be estimated by dividing the elimination capacity by the mass loading of the particular constituent.

Moisture   Bacteria active in a biofilter require an aqueous environment. Removal efficiencies decrease significantly if the filter becomes dry because of the reduced microbial activity. Therefore, moisture content is an important parameter for biofilter operation. Biofilters must operate with as high a moisture content as possible, generally in the range of 50 to 70% by weight. The filter media must be capable of holding the high water content without water logging. Wood chips, bark chips, or other proprietary ingredients are sometimes added to improve drainage and maintain media porosity with the high moisture content.

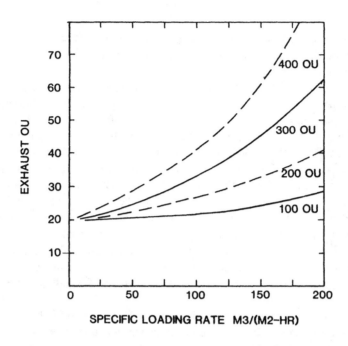

Figure 16.16. Loading diagram for the treatment of sewer gases by a compost biofilter. From Eitner.[78]

Moisture can be added to a biofilter by humidifying the waste gas stream at the inlet to the filter or by using a sprinkler system above or inside the filter bed. Compost process gas is usually near saturation at elevated temperature and is often input to the biofilter without additional humidification. Even if the input gas is saturated, however, the water balance within a biofilter can be positive or negative depending on a number of factors. If the filter media is still composting or the air pollutant concentrations are high, heat will be produced during the decomposition which will raise the temperature of the gas above saturation and cause a net loss of water. Gas compression in blowers will also cause a temperature rise and can produce a net drying condition. On the other hand, if the filter media is less active, the inlet concentrations low, and the process gas saturated at above ambient temperatures, heat losses may reduce the gas temperature causing condensation and a net increase in water content. These factors cause only small changes in the water balance, but when operated over a sufficient time can result in a dry or wet biofilter. This seems to explain why some biofilters tend to become dry while others become saturated. This may also partially explain the low removal efficiencies reported at some biofilters. Controlling the moisture balance within the biofilter is absolutely essential to achieving high performance. If the biofilter media becomes dry, it will not function as intended.

Nutrients   The filter media must contain all nutrients necessary to support microbial activity. The odorous compound itself is usually the energy source and also supplies cell carbon in the case of organic compounds. With inorganic H_2S, cell carbon can come from CO_2 in the process gas. Therefore, the nutrient of concern is usually nitrogen and to a lesser extent phosphorus. The high ammonia concentration in most compost gases supplies abundant nitrogen and most composts contain a significant organic-N content. As a result, additional nutrients are not normally required for biofilters used on composting gases.

Temperature   It is generally recommended that bed temperatures be maintained in the psychrophilic and mesophilic regimes, generally 15 to 35°C. Very cold temperatures will seriously reduce the rates of reaction, while thermophilic temperatures above about 45°C may reduce species diversity. Despite this general recommendation, some biofilters have operated as high as 60°C with apparently good results.[81] The MSW composting facility serving Salzburg, Austria, uses a heat exchanger to precool hot compost gases prior to the biofilter (see Figure 16.14). Cooling also removes water vapor, saturates the gas, and reduces water condensation within the biofilter.

Acidity   Oxidation of NH_3, H_2S, or organic sulfur within a biofilter can lead to acidic conditions according to the following equations:

$$H_2S + 2O_2 \rightarrow 2H^+ + SO_4^=$$ (16.8)

$$NH_3 + 2O_2 \rightarrow H^+ + NO_3^- + H_2O$$ (16.9)

pH's of 1 to 3 have been observed in biofilters treating high H_2S gas streams. This is not a problem when the primary odorant is H_2S, because sulfur oxidizing bacteria, particularly Thiobacillus species, prefer extremely acidic conditions. If other odorants are present, however, their removal efficiency will be adversely effected by the low pH. With the diversity of compounds present in composting gases, low pH conditions would adversely effect total odor removal. Therefore, monitoring of pH and maintenance in the general range of 7 to 8.5 is recommended for biofilters at composting facilities.

It is interesting to note that biofilters operating with high H_2S often develop white/yellow deposits of elemental sulfur S^0. Oxidation of $S^=$ to S^0 does not produce hydrogen ion and does not increase acidity. Formations of such deposits is usually a sign of an overloaded biofilter. If acidic conditions are encountered, the following actions can be taken: (1) the filter media can be replaced; (2) lime, limestone, or other buffering or alkaline reagents can be added to the existing filter media or to the replacement media to extend media life; and (3) the acid producing compounds can be removed prior to the biofilter, e.g., by wet scrubbing.

Example 16.4

A composting facility produces 500 m³/min of process gas at 35°C. The gas contains 10 ppmv of organic sulfides expressed as H_2S and is to be deodorized in a biofilter. Design criteria for the biofilter are a maximum loading of 100 m³/h-m² and a minimum residence time of 45 sec. Media porosity is 80%, bulk weight 0.25 g/cc, and water content 60%. Determine the required biofilter dimensions and check the design against the elimination capacity for organic sulfides.

Solution

1. The surface area is determined from the loading rate as

$$\frac{500 \text{ m}^3/\text{min}(60 \text{ min/h})}{(100 \text{ m}^3/\text{h} - \text{m}^2)} = 300 \text{ m}^2$$

This is equivalent to a square 17.3 m on a side. Assume an 18 m square biofilter.

2. The required media volume is determined from the residence time as

$$500 \text{ m}^3/\text{min}\left(\frac{1 \text{ min}}{60 \text{ sec}}\right)(45 \text{ sec})\left(\frac{1}{0.80}\right) = 469 \text{ m}^3$$

The depth of media is

$$\frac{469 \text{m}^3}{(18\text{m})(18\text{m})} = 1.45\text{m}$$

3. Determine the volume for 1 g-mol of gas at 35°C and 1 atm using Equation 11.53 (the universal gas constant, R, is 0.08205 for units of liters, °K, and atm):

$$V \ = \ \frac{\text{mRT}}{\text{p}} = \frac{(1)(0.08205)(273 + 35)}{1}$$

$$V \ = \ 25.3 \text{ L/g} - \text{mol}$$

4. Determine the g/min of organic sulfides in the process gas as follows:

$$500 \text{ m}^3/\text{min}(1000 \text{ L/m}^3)\left(\frac{1 \text{ g} - \text{mol}}{25.3 \text{ L}}\right)\left(\frac{10 \text{ parts}}{10^6 \text{ parts}}\right) = 0.198 \text{ g} - \text{mol/min}$$

$$0.198 \text{ g} - \text{mol/min}(34 \text{ g/g} - \text{mol}) = 6.72 \text{ g as H}_2\text{S/min}$$

5. Dry matter in the biofilter bed is given as

$$469 \text{ m}^3(0.25 \text{ kg/L})(1000 \text{ L/m}^3)(1 - 0.60) = 46,900 \text{ kg ds}$$

6. The elimination rate for organic sulfides is then given as

$$\frac{6.73 \text{ g/min}(1000 \text{ mg/gm})}{(46,900 \text{ kg ds})} = 0.144 \text{ mg/kg ds} - \text{min}$$

This is much less than the maximum elimination capacity of 2.2 mg H_2S/kg ds-min suggested by Allen and Yang.[74] Therefore, the loading and residence time criteria govern and the above design is acceptable.

Masking/Neutralization

Masking agents attempt to "mask" or overpower an unpleasant odor with another odor that is perceived to be more pleasant. Perfume scents like vanillin, apple, or lemon are added to the ambient air or gas stream to mask or combine with the objectionable odor. Masking agents have been used for many years and have been criticized for an equally long time. Masking agents generally add to the odor and increase the total odor emission. They attempt to mitigate the odor problem by improving the hedonic tone of the odor and, thereby, reduce odor complaints. Masking agents may have some beneficial effect in limited cases on mild odors. However, there are little quantitative data and most endorsements for their use are based on qualitative assessments. The effectiveness of odor masking is always unpredictable.

The main advantage of masking agents is that the application equipment is readily available, portable, and relatively low in cost. A typical system consists of a drum of masking chemical and a feed/atomizing nozzle mounted on the drum. Facilities experiencing an odor problem with their community can quickly procure and implement a masking system from operating budgets. This helps demonstrate to the community that management is responding to the problem until more permanent solutions can be implemented. For this reason alone, masking agents will probably continue to be used at some composting facilities even though their benefit is hard to document.

A neutralizing agent attempts to counteract an odorous gas with neutralizing chemicals to produce a mixture with less total odor. In some cases the neutralizing chemical may combine with the odorous molecule to produce a combination with less odor and higher vapor pressure. Neutralizing agents are available to counteract odor types such as amines, mercaptans, aldehydes, and aromatics.[5] Application equipment is similar to that used for masking agents. Vendors of neutralizing chemicals are quick to respond that neutralizing chemicals are not masking agents.

Gruber et al.[31] reported an average 44% reduction in ED_{50} after atomizing a neutralizing chemical into blower exhaust at a static pile sludge composting facility. The effectiveness of neutralizing chemicals for odor control was also tested at the Delaware Reclamation Project co-composting facility. Circular, agitated bed reactors are used to compost sewage sludge and MSW heavy fraction. Initial work was conducted by Canzano and Aiani[41] and demonstrated a 20 to 31% odor reduction. Duffee[42] improved the method of introducing the chemical by providing four spray nozzles mounted just below the exhaust fan. Odor concentrations were measured using a prescreened odor panel and a forced-choice dynamic olfactometer. Eight test runs were conducted. Untreated gases averaged 720 ED_{50} with a range from 610 to 910. Treated gases averaged 135 ED_{50} with a range from 100 to 158 with the exception of one sample that showed an odor increase. Odor reduction averaged over 80%. Following these tests, the neutralizing chemicals were adopted for routine operation.

On the basis of these data, it would appear that neutralizing chemicals can reduce composting odor concentrations. It should be noted, however, that results are not always as optimistic as those reported above and that the manner of application is apparently important. Also, Heller[82] has noted that the lowest odor levels achievable are in the range of 100 to 150 ED_{50} because of the residual odor of the compounds that are formed. Field or pilot testing is the only reliable method of predicting the performance of neutralizing chemicals.

SUMMARY

Development of an Odor Management Program is one of the most important aspects of facility planning and design, and execution of the program should be a continuing part of facility operation. Odor control is one of the most difficult problems, but also one of the more solvable problems, in present composting practice. There are many elements that must be considered in developing the Odor Management Program. Quantifying and treating odor sources is discussed in this chapter.

Odors are emitted from the surfaces of open piles and windrows, curing piles, storage piles, and stockpiles of amendments. Exhaust gases collected from controlled aeration systems also contain odorous compounds. Studies at a number of compost facilities suggest that the odorous compounds of most concern include ammonia, hydrogen sulfide, mercaptans, alkyl sulfides such as dimethyl sulfide and dimethyl disulfide, and terpenes. These compounds are present in many composting substrates or are formed during aerobic or anaerobic metabolism. These compounds exhibit a high Odor Index, which is defined as the vapor pressure of the compound divided by its odor recognition threshold.

Organoleptic techniques, which use the human olfactory system remain the standard method for characterization of odors. The Threshold Odor Concentration (TOC) is the minimum concentration of odorant that will arouse a sensation. Dilution techniques are usually used to measure the detection threshold. The odor unit (ou) is defined as the number of dilutions with odor-free air required to achieve the minimum detectable odor concentration. Odor concentration is usually determined by supplying a number of diluted samples to a number of individuals until the odor is detected by only 50% of the panel members. The odor unit is equivalent to other nomenclatures such as effective dose, ED_{50}, dilutions to threshold, D/T, and dilution ratio, Z. Another parameter of importance with open facilities is the surface odor emission rate (SOER), usually expressed in $m^3/min\text{-}m^2$. The latter is determined by placing a sample hood over the surface being analyzed. Air is drawn through the hood at a controlled rate and the odor concentration of exhaust gases determined. Odor intensity, hedonic tone, and odor quality are other parameters used to characterize odor.

Available data on odor emission rates from compost facilities have increased substantially in recent years. Measured SOER values tend to vary from about 0.5 to 10 $m^3/min\text{-}m^2$. Odor concentrations in blower exhaust gases vary depending on the substrate. With sewage sludges and wood based amendments, concentrations from 100 to over 1000 ou are typical. MSW composting facilities have reported blower exhaust concentrations over 20,000 ou.

The theory and practice of deodorization of composting gases have advanced significantly. A number of deodorization techniques are available but wet chemical scrubbing and biological oxidation using biofilters or activated sludge have received the most attention. The current state of the art for absorption scrubbing uses multistage chemical scrubbing as follows: (1) acid scrubbing to remove ammonia; (2) hypochlorite scrubbing with a slightly acidic pH and with or without surfactant to remove inorganic and organosulfides and other organics such as terpenes; and (3) scrubbing with peroxide/caustic to remove residual chlorine odors and polish the gas effluent. Both misting and packed bed towers are used for absorption scrubbing. Treated gases from absorption scrubbers are still contained and can be dispersed in a controlled manner. Techniques to enhance dispersion from elevated, point sources are discussed in Chapter 17.

Use of exhaust gases as feed air in the activated sludge process is reportedly effective at a number of full scale compost facilities. Biofilters are widely used at European compost facilities. Some biofilter applications in the U.S. have been replaced by other deodorization technologies. However, biofilters are now enjoying a renewed interest in the U.S. as more is learned about their proper design and operation. The design of biofilters must conform to reasonable limits on loadings rates. Biofilters must be operated to maintain proper conditions of moisture, pH, nutrients and temperature to enhance the microbial reaction rates. Deodorized gases from open biofilters are usually released at groundlevel. Techniques to enhance dispersion from groundlevel, area sources are discussed in Chapter 17.

REFERENCES

1. Wilber, C. and Murray, C. "Odor Source Evaluation," *BioCycle* 31(3) (March 1990).
2. Verschueren, K. *Handbook of Environmental Data on Organic Chemicals,* 2nd Ed. (New York: Van Nostrand Reinhold Company, 1983).

3. Hellman T. M. and Small, F. H. "Characterization of Odor Properties of 101 Petrochemicals Using Sensory Methods," *Chem. Eng. Progress* 69(9) (September 1973).

4. Weast, R. C., Ed. *Handbook of Chemistry and Physics,* 50th ed. (Cleveland, OH: The Chemical Rubber Co., 1969).

5. Odor Control for Wastewater Facilities, Manual of Practice No. 22 (Washington, DC: Water Environment Federation, 1979).

6. Vold, S. Sludge Treatment Supervisor, City of Lancaster, PA, personal communication (April 1992).

7. Hentz, L. H., Murray, C. M., Thompson, J. L., Gasner, L. L., and Dunson, J. B. "Odor Control Research at the Montgomery County Regional Composting Facility," *Water Env. Res.* 64(1) (January/February 1992).

8. Van Durme, G. P., McNamara, B. F., and McGinley, C. M. "Bench-scale Removal of Odor and Volatile Organic Compounds at a Composting Facility," *Water Env. Res.* 64(1) (January/February 1992).

9. Wiley, F. V. "Variation in Recognition Odor Threshold of a Panel," *J. Air Pollut. Assoc.* 19(2) (February 1969).

10. Hellman, T. M. and Small, F. H. "Characterization of the Odor Properties of 101 Petrochemicals Using Sensory Methods," *J. Air Pollut. Control Assoc.* 24:979–982 (1974).

11. Dravnieks, A. and Jarke, F. H. "Odor Threshold Measurement by Dynamic Olfactometry-Significant Operational Variables," paper presented at the 1979 APCA National Conference, Paper 53.1 (1979).

12. ASTM E544-75 Standard: "Recommended Practice for Referencing Odor Suprathreshold Intensity," (Philadelphia, PA: American Society of Testing Materials, 1975).

13. O'Brien, M. A. "Odor Panel Selection, Training and Utilization Procedures — a Key Component in Odor Control Research," in *TR-18 Recent Developments and Current Practices in Odor Regulations, Controls and Technology* (Pittsburgh, PA: Air & Waste Management Association, 1991).

14. Dravnieks, A. "Threshold of Smell and Measurement," in *Industrial Odor Technology Assessment,* Cheremisinoff, P. N. and Young, R. A., Eds. (Ann Arbor, MI: Ann Arbor Science Publishing, Inc., 1975).

15. Stevens, S. S. *Psychol. Rev.* 64:153 (1957).

16. Dravnieks, A. and O'Neill, H. J. "Annoyance Potentials of Air Pollution Odors," *J. Am. Ind. Hyg. Assoc.* 40 (February 1979).

17. Duffee, R. A. Odor Science & Engineering, Inc., personal communication (1989).

18. Meilgaard, M.C. "ASTM Task Group E-18.04.25 on Sensory Thresholds," in *TR-18 Recent Developments and Current Practices in Odor Regulations, Controls and Technology* (Pittsburgh, PA: Air & Waste Management Association, 1991).

19. "Draft Environmental Impact Report — Sewage Sludge Management Program Wastewater Solids Processing and Disposal," Sacramento Area Consultants (1978).

20. Redner, J. Los Angeles County Sanitation Districts, personal communication (1979).

21. Lillard, D. A. Aqueous Odor Thresholds of Organic Pollutants in Industrial Effluents, EPA-660/4-75-002, (May 1975).

22. Moschandreas, D. J., Gordon, S., Jones, D., Luebcke, E., Relwani, S., and Sterling, D. *Odor Study at the Montgomery County Composting Facility Site II* (Chicago: IIT Research Institute Project C08825, November 1985).

23. Redner, J., Wilson, G. E., Scroepfer, T. W., and Huang, J. Y. "Control of Composting Odors," in *Proceedings of the National Conference on Design of Municipal Sludge Compost Facilities,* (Rockville, MD: Information Transfer, 1978).

24. Duffee, R.A. and O'Brien, M. A. "Establishing Odor Control Requirements by Odor Dispersion Modeling," in *Proceedings of the 85th Annual Air & Waste Management Conference,* Kansas City, MO, Paper No. 92-153.01 (June 1992).

25. Iacoboni, M. D., Livingston, J. R., and LeBrun, T. J. *Windrow and Static Pile Composting of Municipal Sewage Sludges,* report to the Municipal Environmental Research Laboratory, U.S. EPA, Report No. EPA-600/2-84-122 (July 1984).

26. Iacoboni, M., LeBrun, T., and Smith, D. "Composting Study," Technical Services Department, Los Angeles County Sanitation Districts (September/December 1977).

27. Iacoboni, M., LeBrun, T., and Livingston, J. "Composting Study," Technical Services Department, Los Angeles County Sanitation Districts (September/December 1979).

28. Hay, J. C., Caballero, R. C., Livingston, J. R., and Horvath, R. W., "Windrow Composting in Los Angeles County — Part 1," and "Two-Step Composting in LA County — Part II," *BioCycle* 26(7), 26(8) (October, November/December 1985).

29. Compost Facility Odor Control Predesign Report, report to the Denver Metropolitan Wastewater Reclamation District, James M. Montgomery, Consulting Engineers, Inc., (October 1988).

30. Peninsula Composting Facility Odor Control Study and Addendum, reports to the Hampton Roads Sanitation District, Virginia Beach, VA, Black & Veatch Engineers-Architects (May, September 1990).

31. Gruber, E. K., Harding, M. L., and Kuchenrither, R. D. "Odor Control at Philadelphia's Sludge Processing and Distribution Center," in *Proceedings of the WPCF Specialty Conference Series, Status of Municipal Sludge Management* (New Orleans, LA: December 1990).

32. Draft Odor Mitigation Final Report, prepared for the Philadelphia Water Department, Black & Veatch Engineers-Architects (November 1989).

33. Murray, C. M. "Odor Control Strategies and Experience at the Montgomery County Composting Facility," in *Proceedings of the National Conference on Municipal Sewage Treatment Plant Sludge Management,* Boston, MA, (Silver Spring, MD: Hazardous Materials Control Research Institute, May 1987).

34. Greeley and Hansen Engineers. Odor Control Study of the Montgomery County Composting Facility, report to the Washington Suburban Sanitary Commission (January 1986).

35. Livingston, J. "Mechanically Aerated Windrow Study, Phase 1 (November, 1978) and Phase II (December, 1978)," Los Angeles County Sanitation Districts (unpublished).

36. Murray, C. M. Manager of Composting Operations, Montgomery County Composting Facility, personal communication (1991).

37. Ostojic, N. "Control of Odors from Sludge Composting — A Technology Review Final Report," report prepared by Odor Science & Engineering for the City of Charlotte, NC (January 1990).

38. Wheeler, M. L. Wastewater Superintendent, City of Hamilton, OH, personal communication (March/May 1992).

39. Haug, R. T., Ludwig, K. L., and Haase, C. L. "Co-Composting Observations on European and U.S. Facilities," in *Proceedings of the National Conference on Municipal Sewage Treatment Plant Sludge Management*, Palm Beach, FL, (Silver Spring, MD: Hazardous Materials Control Research Institute, June 1988).

40. Haug, R. T., Ludwig, K., and Haase, C. "Site Visits to Waste Composting Facilities in Austria, West Germany and the United States," unpublished trip report to the City of Los Angeles (January 1988).

41. Canzano, P. S. and Aiani, K. J. "Optimum System for Odor Control," *BioCycle* 29(7) (August 1988).

42. Duffee, R. A. "Deamine Odor Tests Results Relative to Composting," report by Odor Science & Engineering, Inc., to the NuTech Chemical Corp. (September 1988).

43. Murray, C. M., Thompson, J. L., and Hentz, L. H. "Methods for Removing Odors from Process Airstreams," U.S. Patent Number 4,994,245 (February 19, 1991).

44. Hentz, L. H., Murray, C. M., Thompson, J. L., Gasner, L. L., and Dunson, J. B., "Odor Control Research at the Montgomery County Regional Composting Facility," in *Proceedings of the WPCF Specialty Conference on the Status of Municipal Sludge Management for the 1990s,* New Orleans, LA (December 1990).

45. Thompson, J. L. "Effectiveness of Odor Control Systems," presented at the U.S. EPA Workshop on Controlling Sludge Composting Odors, Arlington, VA (August 1990).

46. Cathcart, W. Chief of Operations and Maintenance, Cape May Composting Facility, Cape May, NJ, personal communication (February 1992).

47. Cathcart, W. "Cape May Original Design, 20 dtpd Operating After Cure Reactor Conversion," presented at the U.S. EPA Workshop on Controlling Sludge Composting Odors, Arlington, VA (August 1990)

48. Wilber, C. Greeley & Hansen Engineers. personal communication (March 27, 1992).

49. Woods, B. Greeley & Hansen Engineers. personal communication (March 30, 1992).

50. Leder, C. S., Tefera, A., and Glass, J. S. "Albuquerque's Approach to Odor Control at a 15 Dry Ton per Day Pilot Aerated Windrow Composting Facility," in *Proceedings of the Specialty Conference Series* (Portland, OR: Water Environment Fed. No. TT041, June 1992).

51. Donovan, J. Camp, Dresser & McKee, personal communication, (July 1992).

52. Horst, W. G., Mattern, F., Vold, S. H., and Walker, J. M. "Controlling Compost Odors," *BioCycle* 32(11) (November 1991).

53. Horst, W. G. Superintendent of Wastewater Operations, City of Lancaster, PA, personal communication (April 1992).

54. Cross, F. L. *Air Pollution Odor Control Primer* (Westport, CT: Technomics Pub. Co., 1973).

55. Sarrasin, M. Mechanical Project Manager, Westcott Construction Corp., personal communication (1991).

56. Novy, V. "Compost Recirculating Systems," presented at the U.S. EPA Workshop on Controlling Sludge Composting Odors, Arlington, VA (August 1990).

57. Ostojic, N., O'Brien, M. A., and Duffee, R. A. "Limitation of Incineration in Odor Control," in *Proceedings of the 84th Annual Air & Waste Management Conference*, Vancouver, British Columbia, Paper No. 91-147.2 (June 1991).

58. Hentz, L. H. "Thermal Regenerative Oxidation," presented at the U.S. EPA Workshop on Controlling Sludge Composting Odors, Arlington, VA (August 1990).

59. Bain, R. E., Geribo, S. H., and Silver, M. M. "Thermal Oxidation of Zimpro-Composted Sludge Off-Gases," in *Proceedings of the Specialty Conference Series*, Water Environment Fed. No. TT041, Portland, OR (June 1992).

60. McIlvaine, R. "New Treatment Schemes Control Odors," *Water Eng. Manage.* (January 1990).

61. Garber, W. F. "Odors and Their Control in Sewers and Treatment Facilities," paper presented at the IAWPR Workshop, Vienna, Austria (1979).

62. Duffee, R. A., Ostojic, N., and O'Brien, M. A. "Odor Impact Evaluation and Control Study for the Hyperion Wastewater Treatment Plant," report to the City of Los Angeles (November 1989).

63. Vogt, D. R. "Synopsis of Observations, European Tour of Mechanical Enclosed Composting Systems," Washington Suburban Sanitary Commission (1979).

64. Haug, R. T., Epstein, E., and Alpert, J. "Evaluation of Weiss Kneer II Composting Systems in West Germany," report to the Taulman Co. (1982).

65. Ostojic, N., Les, A. P., and Forbes, R. "Activated Sludge Treatment for Odor Control," *BioCycle* 33(4) (April 1992).

66. Feindler, K. S. "Sewage Sludge Thermal Drying with Low Environmental Impact," in *Proceedings of the Air & Waste Management Conference*, Vancouver, BC (June 1991).

67. Kasbrian, A. A. "Packed Tower with Secondary Effluent over Plastic Media," unpublished report by the City of Los Angeles (June 1978).

68. Clifton, F. W., Adams, T. E., and Dohoney, R. W. "Managing Sludge Through In-Vessel Composting," *Water Eng. Manag.* (December 1991).

69. Dohoney, R. W. District Manager for Professional Services Group, personal communication (December 31, 1991).

70. Pomeroy, R. D. U.S. Patent 2,793,096 (1957).

71. Hartmann, H. "Geruchsbekampfung in Klaranlagen am Beispiel der Klarwerke der Stadt Numberg," *Korrespondenz Abwasser* 23:275 (1976).

72. Bohn, H. L. "Compost Scrubbers of Malodorous Air Streams," *Compost Sci.* (Winter 1976).

73. Leson, G. and Winer, A. M. "Biofiltration: An Innovative Air Pollution Technology for VOC Emissions," *J. Air Waste Man. Assoc.* 41(8) (August 1991).

74. Allen, E. R. and Yang, Y. "Biofiltration Control of Hydrogen Sulfide Emissions," presented at the 84th Annual A&WMA Annual Meeting, Vancouver, BC (June 1991).

75. Hartenstein, H. U. and Allen, E. R. "Biofiltration, An Odor Control Technology for a Wastewater Treatment Facility," report to the City of Jacksonville, FL (May 1986).

76. Haug, R. T. "Combustion, Aeration Basins and Earth Biofilters," presented at the U.S. EPA Workshop on Controlling Sludge Composting Odors, Arlington, VA (August 1990).

77. Bohnke, B. and Eitner, D. "Investigation and Comparison of Different Kinds of Compost in a Mobile Biofilter," *Gutachten i.A. AG Kompostabsatz NW* (June 1983).

78. Eitner, D. "Investigations of the Use and Ability of Compost Filters for the Biological Waste Gas Purification with Special Emphasis on the Operation Time Aspects," GWA, Band 71, RWTH Aachen, FRG (September 1984).

79. Prokop, W. H. and Bohn, H. L. "A Soil Bed System for Control of Rendering Plant Odor," *J. Air Pollut. Contr. Assoc.* 1332–1338 (December 1985).

80. Dharmavaram, S. "Biofiltration A Lean Emissions Abatement Technology," presented at the 84th Annual A&WMA Annual Meeting, Vancouver, BC (June 1991).

81. U.S. EPA "Odor and Corrosion Control in Sanitary Sewage Systems and Treatment Plants," Design Manual EPA/625/1-85/018 (October 1985).

82. Heller, K. President, NuTech Environmental Corp., personal communication (October 1991).

83. Bohn, H. L. and Bohn, R. K. "Gas Scrubbing By Bio-washers and Bio-filters," *Pollut. Eng.* 18(9) (September 1986).

CHAPTER 17

Odor Management II — Atmospheric Dispersion

INTRODUCTION

Open windrows, static piles, and open curing piles all represent area emission sources of odor. Process gases that are collected and treated are usually discharged as a elevated point source. Even with good treatment, however, it is important to recognize that deodorization is never 100% effective. The only possible exception is thermal oxidation, which is theoretically capable of complete odor destruction. Multistage wet scrubbing can achieve outlet odor levels in the range of 50 to 150 ED_{50}. Lower levels are very difficult to achieve. For one thing, the contribution of the scrubbing chemicals to the outlet odor becomes significant at low outlet levels. The same is true for biofilters. While quantitative data are limited, exhaust $ED_{50}s$ in the range of 20 to 150 are typical. Lower levels are just not achievable because the biofilter matrix begins contributing to the outlet odor.

The engineering community has often ignored the fact that treatment is not 100% effective. Many scrubber systems have been designed with little or no attention paid to dispersion of the treated gases. It is common to see scrubbers with short stub stacks, low outlet velocities, scrubbers located near large buildings with the plume caught in the building downwash, rain caps on top of discharge stacks, and other examples of poor dispersion design (see Figure 17.1). The designer apparently forgot that all discharges, even treated ones, contain some odor.

Dilution is rarely the complete solution to any problem, particularly odors. However, it is equally true that atmospheric dilution must be part of the solution because no amount of design, operation, or odor treatment will reduce odor emissions to zero. Therefore, atmospheric dispersion is an important element of the odor management program.

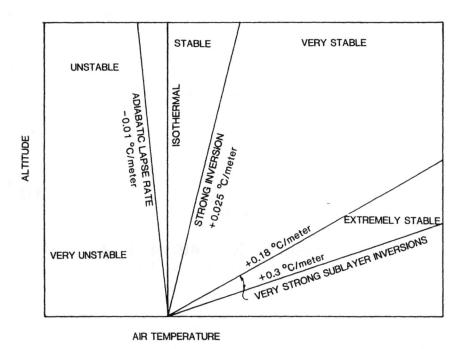

Figure 17.1.   Temperature lapse rates and zones of atmospheric stability.

BASICS OF METEOROLOGY

Atmospheric Turbulence

Atmospheric turbulence is caused by the wind and by surface heating and cooling. An atmosphere that is well mixed by moderate to strong winds under overcast skies is termed "neutral". Under these conditions, there are no surface heating or cooling effects and the temperature decreases with altitude at the "adiabatic lapse rate". As a mass of air is moved upward, it tends to expand and cool as a result of the decrease in pressure. If no heat transfer occurs with the surroundings, dry air will cool at the adiabatic lapse rate, which is –0.01°C/m or –5.4°F/1000 ft. The negative sign indicates a decrease in temperature with increasing elevation.

Atmospheric stability is determined in part by the temperature lapse rate. If the temperature gradient is less than the adiabatic lapse rate (i.e., more negative or superadiabatic), unstable air conditions exist with significant vertical mixing. When an air parcel rises it tends to cool at the adiabatic lapse rate. With a lapse rate more negative than the adiabatic rate, the rising air parcel will remain warmer than its surroundings and less dense. The air parcel will tend to continue rising. A positive temperature gradient is termed a temperature inversion. With a temperature inversion, the rising air parcel will be at a lower temperature than its surroundings but at the same pressure. Consequently, it will be heavier than its surroundings and tend to fall back to its original position. Vertical mixing will be limited because of the inversion. Stable, neutral, and unstable atmospheric conditions are shown in Figure 17.1 as a function of the temperature lapse rate.

Stability Classes

In the 1950s the concern over accidental releases of radioactive material into the atmosphere prompted the development of techniques to predict the concentrations downwind of a specified release. Frank Pasquill[1,2] working for the British Meteorological Office in 1958, first proposed a system of atmospheric stability classes labeled A through F. Class A was the most unstable, class F the most stable, while class D represented a neutral atmosphere with a near adiabatic lapse rate. This system was subsequently adopted in the U.S. with some modifications and appears in the *Turner Workbook*,[3] a classic treatise on atmospheric dispersion published in the U.S. in 1970. Bruce Turner defined the stability categories as a function of wind speed, solar insolation and degree of cloud cover. Smith[4] later quantified the solar insolation categories.

The stability classes as a function of meteorological (met) conditions are presented in Table 17.1. "Strong" incoming solar radiation corresponds to a solar altitude greater than 60° with clear skies in midsummer. "Strong" solar radiation corresponds to a solar energy flux >50 cal/cm²-h.[4] "Slight" insolation corresponds to similar conditions in midwinter or to a solar altitude from 15 to 35° with clear skies in midsummer. Solar energy flux is <25 cal/cm²-h. Cloudiness will decrease incoming solar radiation and should be considered along with solar altitude in determining solar radiation. Incoming radiation that would be strong with clear skies can be expected to be reduced to moderate with broken (5/8 to 7/8 cloud cover) middle clouds and to slight with broken low clouds. The neutral class D should be assumed for overcast conditions during day or night. Night refers to the period from 1 h before sunset to 1 h after sunrise.

The stability categories are more reliable in open, rural areas. In urban and heavily wooded areas, surface roughness and heat islands can effect the category, particularly on still nights.[3] On calm, clear nights stability class E or F might occur in rural areas whereas class D would be more likely over urban areas. Combinations of wind speed and stability used in some U.S. EPA models are presented in Table 17.2. The relationships provide guidance on the possible combinations of wind speed and stability but should be used with some caution. For example, the stable classes E and F are shown to occur with wind speeds of 2.0 m/sec and above. However, puff transport conditions at groundlevel can occur with low wind speeds below 1 m/sec and very stable air.

According to Dittenhoefer and Menne,[6] the U.S. EPA is considering a new stability categorization scheme called the SRDT method (Solar Radiation, Delta Temperature). During the day, the SRDT method uses on-site measurements of solar radiation and 10-m wind speed to assign stability classes. At night, measurements of the vertical temperature gradient between 2 and 10 m and the 10-m wind speed are used. Dittenhoefer has suggested that gradient measurements between 10 and 60 m may be more appropriate. Stability classifications based on these measurements are presented in Table 17.3.

Diurnal Variations

The solar radiation flux begins to decrease as night approaches. The earth surface cools as it radiates heat to the sky in proportion to the fourth power of the absolute temperature. The rate of heat loss is higher on clear nights because heat is not reflected back by cloud cover. In the absence of strong winds, air near the ground cools as it contacts the cooler earth surface. The air becomes dense and stable as a result. The cool air will tend to move downhill. Those living in cold climates are familiar with the fact that frost most often occurs in the valleys

Table 17.1a. Stability Categories Based on Wind Speed, Insolation, and State of the Sky[a]

Surface Wind Speed (m/sec)[b]	Day, Incoming Solar Radiation[c] (Insolation)			Night	
	Strong	Moderate	Slight	Thinly Overcast or ≥4/8 Low Cloud Cover	≤3/8 Cloud Cover
<2	A	A–B	B		
2–3	A–B	B	C	E	F
3–5	B	B–C	C	D	E
5–6	C	C–D	D	D	D
>6	C	D	D	D	D

Source: Pasquill and Smith,[2] Turner,[3] and Smith.[4]

[a] A, extremely unstable; B, moderately unstable; C, slightly unstable; D, neutral; E, slightly stable; F, moderately stable. The neutral class D should be assumed for all overcast conditions during day or night.
[b] Wind speed measured at a height of 10 m.
[c] Strong insolation corresponds to a solar flux ≥50 cal/cm^2-h, moderate from 25 to 50, and slight <25 cal/cm^2-h.

Table 17.1b. Appropriate Insolation Categories Determined from Sky Cover and Solar Elevation

Sky Cover	Solar Elevation (Degrees from Horizontal)		
	>60°	>35° but ≤60°	>15° but ≤35°
4/8 or less or any amount of high thin clouds	Strong	Moderate	Slight
5/8 to 7/8 middle clouds (7000–16000 ft base)	Moderate	Slight	Slight
5/8 to 7/8 low clouds (<7000 ft base)	Slight	Slight	Slight

Source: Pasquill and Smith,[2] Turner,[3] and Smith.[4]

before the tops of hills. Very stable air conditions with positive temperature gradients (inversions) can occur under these conditions as shown in Figure 17.2. Stable atmospheric class E can occur in early evening followed by class F conditions as more heat is lost from the earth surface. This condition is typical of relatively cloudless nights with light winds. The elevation at which the actual lapse rate reverses is the depth of the "inversion layer".

After sunrise on clear mornings, the solar flux increases and starts to heat the earth surface. Inversion conditions begin to weaken at the ground surface from thermally induced turbulence as the surface air layer is warmed. The stable layer is gradually eroded from beneath, producing the conditions shown in Figure 17.3. The inversion in the atmospheric sublayer usually disappears about 2 h after sunrise with the upper inversion disappearing 5 to 7 h after sunrise. If pollutants were emitted into the upper zone of the stable layer during the night, they would remain in this layer until the unstable layer erodes into the stable layer from below. At this point, the pollutants would be mixed rapidly into the unstable layer, producing a condition called "inversion breakup fumigation".[5] Fumigations rarely last for more than 10 to 30 min.

Table 17.2. Probable Combinations of Wind Speed and Stability Class

Wind Speed (m/sec)	Pasquill Stability Class					
	A	B	C	D	E	F
0.5	x	x		x		
0.8	x	x		x		
1.0	x	x		x		
1.5	x	x		x		
2.0	x	x	x	x	x	x
2.5	x	x	x	x	x	x
3.0	x	x	x	x	x	x
4.0		x	x	x		
5.0		x	x	x		
7.0			x	x		
10.0			x	x		
12.0			x	x		
15.0			x	x		
20.0				x		

Source: Adapted from Turner[3] and Schultze.[5]

Table 17.3. The Proposed U.S. EPA SRDT Stability Classification System

Surface Wind Speed (m/sec)[a]	Daytime Solar Insolation (cal/cm²-h)			
	>60	30–60	5–30	<5
<2	A	A	B	D
2–3	A	B	C	D
3–5	B	B	C	D
5–6	C	C	D	D
≥6	C	D	D	D

Surface Wind Speed (m/sec[a])	Nighttime Lapse Rate (°C/m)		
	<–0.01	–0.01 to 0.01	≥0.01
<2	D	E	F
2–3	D	E	F
3–5	D	D	E
5–6	D	D	D
≥6	D	D	D

Source: Adapted from Dittenhoefer and Menne.[6]

[a] Wind speed measured at a height of 10 m.

As the sun continues to rise in the morning or as an overcast sky clears, the incoming solar flux increases. Air near the surface continues to warm and "cells" of this warm air start to rise. The more direct the sunlight, the more heating at the surface and the greater the thermal convective turbulence. As the cells of warm air leave the surface they are replaced by downward moving cells of colder air. Unstable conditions first develop at the surface. The lapse rate becomes superadiabatic, also termed "overturning", "advecting", or "unstable". Classes A, B, and C apply to the unstable layers, with class A being the most unstable. These conditions are illustrated in Figure 17.4.

During the day, convective turbulence strongly mixes the lower atmosphere. However, there is a limit to this mixing and most of the time there is a stable air layer aloft with a positive lapse rate that tops the mixing layer. On clear mornings the mixing zone starts as a shallow layer near the ground and increases in height during the day as long as solar radiation input

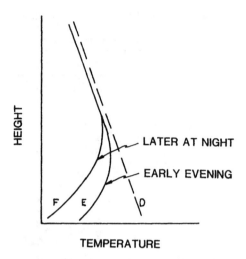

Figure 17.2. Temperature lapse rates typical of relatively cloudless nights with light winds. Stable classes E or F occur near the ground surface and transition to the neutral D Class at higher altitude. The elevation at which the lapse rate reverses is the depth of the inversion layer.

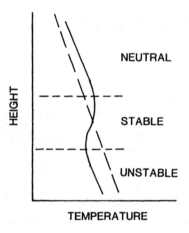

Figure 17.3. Temperature lapse rates typical of the period after sunrise when the sun starts to warm the ground surface and the nighttime inversion begins to disappear.

exceeds the outward radiation of the earth. The mixing height is usually between 500 and 2000 m.[5] The temperature gradient is superadiabatic only in the lowest tenth or so of the mixing depth and decreases toward adiabatic (class D) with height.

Puff Transport

The U.S. EPA uses a lapse rate of +0.01 °C/m to represent the class E lapse rate and +0.025 °C/m for class F. However, more stable conditions with higher inversion rates can sometimes occur. During the arctic winter, rather severe inversions often occur. Lapse rates of 0.036 to 0.055 °C/m are not uncommon during extended periods of clear, still days with virtually no

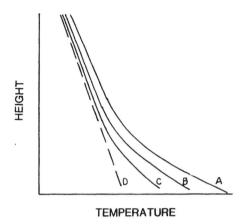

TEMPERATURE

Figure 17.4.   Temperature lapse rates typical of sunny days when the ground surface has been warmed by the sun. The unstable or superadiabatic classes A, B, or C occur in the lower atmosphere and transition to the adiabatic lapse rate or neutral D class aloft.

sun. Strong inversions can also occur when warm winds blow across cold bodies of water. This typically occurs during spring conditions on the Great Lakes. Schultze[5] reported a gradient over +0.3 °C/m between the surface and the top of a ship mast in Lake Michigan in early June. This is about 10 times the thermal gradient used by the EPA for Class F stability. Stabilities stronger than E or F are sometimes referred to as G and H.[5] For example, Class G stability with a lapse rate >+.04 °C/m between measuring heights of 9 and 40 m was determined to occur about 10% of the time near New Orleans.

Atmospheric lapse rates are typically reported for the lower 100 to 300 m of the atmosphere. Significantly higher lapse rates can be measured in the lowest 10 to 30 m of the atmosphere, termed the "atmospheric sublayer". Air movement under such stable conditions is often termed "puff" transport because the dispersion rate is very low. Vertical mixing is restricted by the very stable atmospheric sublayer. Puff transport in the atmospheric sublayer is usually the critical odor transport condition for groundlevel discharges because mixing and dispersion are limited.

Based on smoke release experiments made in the Sacramento, California area, Wilson et al.[7] determined that puff transport occurred when the temperature inversion measured between 1.5 and 7.6 m above ground exceeded +0.18 °C/m with wind speeds of 0.9 m/sec or less. During the year these conditions were met for periods of one hour or more on 40% of the evenings in Sacramento. Even stronger inversion rates have been observed by this author at composting facilities.

Another type of critical odor situation can occur after calm conditions with little or no wind velocity. The absence of wind allows accumulation of odors over the source. If the calm is followed by a period of steady wind, the high odor concentration can be transported downwind under puff conditions. Such a case is termed intermittent puff transport.

Positive temperature gradients in the atmospheric sublayer usually occur as a result of radiant cooling of the ground surface. The gradient becomes positive about 2 h before sunset and usually remains positive until 1.5 to 2 h after sunrise.[7] A strong groundlevel inversion is likely to occur on a calm, clear night. Thus, stable groundlevel conditions are characteristic of evening hours on clear nights, while unstable conditions are characteristic of daylight hours on clear days. Groundlevel discharges of odor will typically produce the highest downwind concentrations during the stable evening hours with puff transport conditions. This general pattern has been verified by analysis of the frequency of odor complaints at several composting facilities. It was once thought that odor complaints increased in the late afternoon/evening hours because people returned home from work. Actually, the complaints are more associated with the onset of worst-case meteorological conditions.

Plume Characteristics

The shape of the discharge plume is governed by the nature of the atmospheric turbulence. On sunny days with classes A, B, or C stability, a plume will show large vertical undulations from the movement of thermals and downdrafts, a condition called "looping". If the day is overcast, thermal effects will be reduced and stability will be class D. Plume spread in both horizontal and vertical directions will be moderate, a condition called "coning". This neutral meteorological condition (class D) is quite common, and coning plumes are those most extensively studied.[5]

During nighttime inversion conditions, plume spread in the vertical direction will be limited. A visible plume will often appear as a slender cylinder and may be visible for up to 20 or 30 miles from the source (at least in the past when smoke was more commonly discharged). This condition is called "fanning" and is associated with stable classes E and F. Sometimes in the early evening, an elevated plume may reside in the neutral layer remaining from the daytime, which overlays a shallow but growing inversion layer. The plume will disperse according to class D stability, but it will not penetrate the lower inversion layer and will not reach the ground. This condition is called "lofting" and is caused by E or F stability at the surface and D above. This condition is favored for planned discharge of odors from elevated sources, because the plume typically does not reach groundlevel.

In the morning, surface heating causes an unstable layer to form at the ground. As the unstable layer increases in depth, it eventually reaches a plume emitted into a stable layer aloft. The plume gases are then mixed downward toward the ground, an event previously described as "inversion breakup fumigation". Fumigation generally persists for about 30 min and can result in high ground-level concentrations some distance from the source.[5] The problem is mitigated in flat terrain because of the low probability of the plume repeatedly reaching groundlevel at the same place. The situation is more serious in a valley where fumigation can occur frequently at the same location.

Plumes tend to meander as the wind direction varies in the horizontal direction. The plume may even break up into individual "puffs" and eventually lose its identity as a discrete plume. As a result of this meander, the concentration measured at a fixed downwind location will tend to vary as individual puffs pass by. Therefore, downwind concentrations are normally referenced to a particular averaging time. If the wind direction is reasonably constant, the concentration profile across the plume will generally follow a Gaussian or normal distribution if the averaging time is about 10 min to several hours. However, the human olfactory sense responds quickly to odor. Cha and Jann[8] have suggested using a 10 second averaging time for odor modeling. As a result, the concentration in the individual plume "puffs" becomes important. It is possible for the 1-h average odor concentration to be below odor detection, but to have several puffs during the hour that exceed the detection threshold. Diagrams representing the approximate outline of a plume averaged over different time frames and the corresponding cross plume distribution patterns are presented in Figure 17.5.

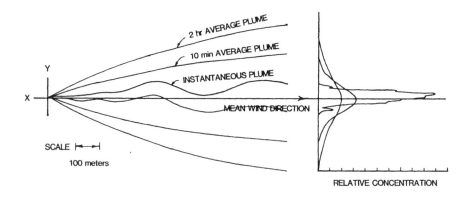

Figure 17.5.   The diagram on the left represents the approximate outlines of a smoke plume observed instantaneously and of plumes averaged over 10 min and 2 h. The diagram on the right shows the corresponding cross plume distribution patterns. From Slade.[9]

DISPERSION MODELING

A number of modeling approaches have been used to predict the atmospheric dispersion and movement of pollutants. Most models can be classified as numerical, statistical, empirical, or physical. Numerical models are generally based on Fick's law of diffusion and are sometimes termed gradient-transfer or K-theory models. This approach to the modeling of atmospheric dispersion was introduced by Fick[10] in 1855 and was related to his work on diffusion of heat in a conducting body. In such models, it is assumed that atmospheric turbulence causes a net movement of material along a concentration gradient at a rate proportional to the slope of the gradient. The proportionality constant is the eddy diffusivity K (hence the term K-theory). For numerical solution the atmosphere is divided into a series of horizontal and vertical cells. The growth of a plume from cell to cell is followed over a series of time steps. K-theory models are generally computer intensive, costly to operate, and not widely used for routine modeling.[8] Global atmospheric models, such as those used in the study of global warming, divide the atmosphere into a three-dimensional arrangement of cells and require supercomputers for solution.

The most widely used modeling approach for nonreactive pollutants is the statistical Gaussian dispersion model.[5] Plume concentrations in the horizontal and vertical dimensions of the plume are assumed to follow a normal or Gaussian distribution when averaged over a time period of several minutes to several hours. The Gaussian model can be derived from K-theory models if all meteorological conditions remain constant and wind speed and direction are uniform throughout the region of concern. Numerous computerized dispersion models based on the Gaussian algorithm are available from the U.S. EPA Users Network for Applied Modeling of Air Pollution (UNAMAP) series. An excellent discussion of the Gaussian model is presented in Turner's Workbook.[3]

The Gaussian Plume Model

The Gaussian model is most often applied to the case of a plume coming from a point or area source. The following assumptions are made: (1) the plume spread has a Gaussian distribution in both horizontal and vertical planes; (2) the emission is continuous or the duration of release is equal to or greater than the travel time to the downwind position, so that diffusion along the length of the plume is negligible; (3) the emission is of uniform concentration, (4) steady-state meteorological conditions prevail with constant wind speed, direction, and stability class; (5) the effective emission height, H, is the sum of stack height, h, and plume rise, ΔH; (6) no deposition of material occurs from the plume and there is no reaction at the ground surface; (7) the emissions are conservative (i.e., no reactions occur to remove contaminants from the plume); (8) terrain is level; and (9) wake effects from buildings are not significant.

Point Source Equations

Under the above conditions the concentration at coordinates (x, y, and z) from a continuous point source of effective height, H, is given by

$$c(x,y,z;H) =$$

$$\frac{S}{2\pi\sigma_y\sigma_z u}\exp\left[-\frac{1}{2}\left(\frac{y}{\sigma_y}\right)^2\right]\left\{\exp\left[-\frac{1}{2}\left(\frac{z-H}{\sigma_z}\right)^2\right]+\exp\left[-\frac{1}{2}\left(\frac{z+H}{\sigma_z}\right)^2\right]\right\} \qquad (17.1)$$

where

c = pollutant concentration at coordinates (x,y,z), g/m,^{3} or ou
S = mass emission rate of pollutant, g/sec, or ou/sec
u = mean wind speed effecting the plume transport, m/sec (this is usually taken as the wind speed at the stack elevation)
H = effective emission height, m
σ_y = standard deviation of plume concentration in the horizontal direction, m
σ_z = standard deviation of plume concentration in the vertical direction, m
(x,y,z) = coordinate distances, m
$\exp$ = the base e of natural logarithms or 2.7183
x = the downwind distance from the source to the receptor, m
y = the lateral distance from the centerline to the receptor, m
z = height of the receptor, m

The form of plume predicted by Equation 17.1 and the assumed coordinate system are shown in Figure 17.6. Any consistent set of units may be used, but it is common practice to use metric units as indicated above. The use of odor units will be illustrated in subsequent examples.

For the case of odor emissions, one is usually interested in groundlevel concentrations (i.e., $z = 0$). The Gaussian model then simplifies to

$$c(x,y,0;H) = \frac{S}{\pi\sigma_y\sigma_z u}\exp\left[-\frac{1}{2}\left(\frac{y}{\sigma_y}\right)^2\right]\exp\left[-\frac{1}{2}\left(\frac{H}{\sigma_z}\right)^2\right] \qquad (17.2)$$

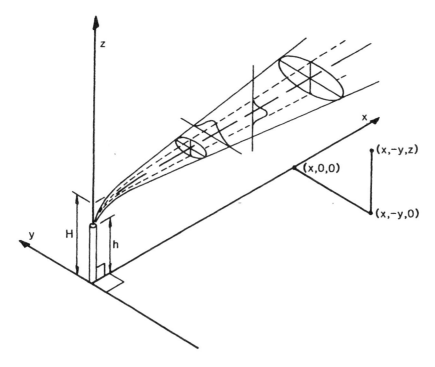

Figure 17.6. Coordinate system for the Gaussian plume model showing normal distributions in the horizontal and vertical. From Turner.[3]

Further simplification can be achieved by considering the centerline concentration of the plume (i.e., y = 0). This is of particular significance because maximum concentrations occur along the plume centerline.

$$c(x, 0, 0; H) = \frac{S}{\pi \sigma_y \sigma_z u} \exp\left[-\frac{1}{2}\left(\frac{H}{\sigma_z}\right)^2\right]$$

(17.3)

If the point source is at groundlevel with no effective plume rise (i.e., H = 0),

$$c(x, 0, 0; 0) = \frac{S}{\pi \sigma_y \sigma_z u}$$

(17.4)

Wind Profiles

Concentrations calculated using the Gaussian equations are inversely proportional to the mean wind speed, u. Under most atmospheric conditions the wind speed increases with elevation, with a profile dependent on stability class. A wind power law is usually used to adjust the observed wind speed, u_o, from the measurement height, z_o, to the stack height, h:

Table 17.4. Wind Profile Exponents Used to Adjust Wind Speed for Elevation

Pasquill Stability Category	Rural[a] Wind Profile Exponent	Urban[b] Wind Profile Exponent
A	0.07	0.15
B	0.07	0.15
C	0.10	0.20
D	0.15	0.25
E	0.35	0.30
F	0.55	0.30

Source: Industrial Source (ISC) Dispersion Model User's Guide.[11]

[a] Rural exponents are used for PG and Slade dispersion coefficients.
[b] Urban exponents are used with MP dispersion coefficients.

$$u_z = u_o \left(\frac{h}{z_o} \right)^{p_w}$$

(17.5)

where

u_o = wind speed at anemometer height, m/sec
z_o = anemometer height, usually taken as 10 m
u_z = wind speed at elevation z, m/sec
h = stack height, m
p_w = wind profile exponent

Values of the wind profile exponent used in the U.S. EPA models are presented in Table 17.4.

The reader might question why wind speed is corrected to the stack height and not the effective plume height. Convention in the U.S. is to correct to the actual stack height. Results are considered sufficiently accurate to not require a more elaborate correction.

Dispersion Coefficients

Numerous studies of elevated plumes from stack discharges have been conducted. Many of the U.S. studies were conducted in the 1950s and 60s by the Atomic Energy Commission (now the Department of Energy) and in England by the British Meteorological Office. Based on results of such studies, the Gaussian distribution is generally accepted as a good mathematical approximation of plume behavior for averaging periods from about 10 min to several hours. Values of σ_y and σ_z for the Gaussian models vary with the turbulent structure of the atmosphere, height above the surface, surface roughness, sampling time over which the concentration is measured, wind speed, and distance from the source. Most practical calculation procedures use empirical correlations developed from field studies of plume dispersion. Two studies of particular importance to U.S. practice are discussed below.

A series of experiments, called Project Prairie Grass, were conducted by the U.S. Atomic Energy Commission in the summer of 1956 to study the dispersion of groundlevel releases in a flat grassy prairie in Nebraska. Data were collected at distances up to 800 m. Horizontal spread was based on a 3-min sampling time and vertical spread on a 10-min sampling time. Dispersion for distances beyond 1 km was extrapolated. In his original publication on this work, Pasquill characterized the plume by the angles of horizontal and vertical spread. Gifford[12] converted Pasquill's values to standard deviations by equating plume width with the

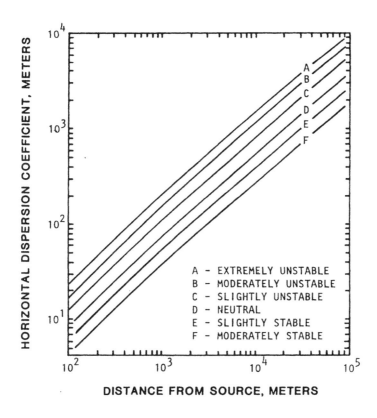

Figure 17.7. Pasquill-Gifford horizontal dispersion coefficients, σ_y, for rural terrain. Values beyond 1 km are extrapolated. From Turner.[3]

distance in a Gaussian curve between values having 10% of the centerline value. This gives a plume width of $4.28 \times \sigma$. These standard deviations became known as the Pasquill-Gifford (PG) coefficients and, with few exceptions, are used today in all UNAMAP models to represent dispersion in rural conditions. According to Slade,[9] Pasquill's curves fit the experimental data collected since the Prairie Grass experiments quite well. These coefficients are sometimes referred to as "PGT" coefficients recognizing the valuable contributions of Pasquill, Gifford, and Turner.

The PG coefficients are presented in Figure 17.7 for horizontal dispersion and Figure 17.8 for vertical dispersion. Both coefficients increase with distance from the source. The horizontal and vertical coefficients, which represent standard deviations of plume width and height, are greater for unstable than for stable conditions. These results make sense because plumes spread more quickly in turbulent conditions and the spread increases with further distance from the point source. A number of mathematical equations have been used to describe the PG coefficients. The formulas currently used in the UNAMAP models are presented in Appendix G.

From 1963 to 1965 a series of dispersion experiments were conducted in urban St. Louis using groundlevel releases. Sampling times were 1 h for distances up to 16 km from the source. Results were reported by James McElroy and Francis Pooler in 1968.[13] In 1973, Gary Briggs proposed a series of dispersion coefficient curves based on these experiments. The curves are similar to the PG rural curves in Figures 17.7 and 17.8 but generally have higher values for a given stability class. This is because the urban atmosphere tends to be more turbulent from the many points of heat release (heat islands) and the higher surface roughness

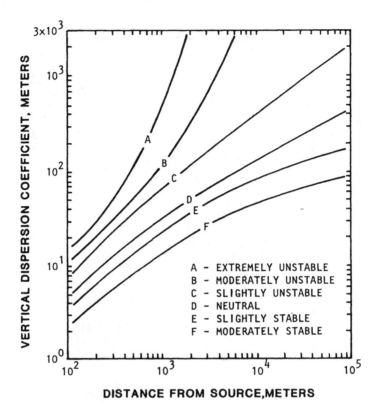

Figure 17.8.   Pasquill-Gifford vertical dispersion coefficients, σ_z, for rural terrain. Values beyond 1 km are extrapolated. From Turner.[3]

from buildings. These coefficients are currently used in the UNAMAP programs to model urban dispersion and are called the McElroy-Pooler (MP) or Briggs' Urban coefficients. Equations currently used to represent these curves are presented in Appendix G.

Many excellent dispersion studies have been conducted in a number of countries. The Prairie Grass and St. Louis studies form the basis for the current U.S. regulatory models and compare favorably with many other dispersion estimates. For these reasons they are presented in this text, but the reader should remember that there are other coefficients in use throughout the world.

For instantaneous releases of toxic substances, it is desirable to know the maximum expected concentrations downwind from the puff release. To model such releases, the U.S. EPA developed instantaneous puff models using dispersion coefficients compiled by David Slade.[9] The coefficients are based on a compilation of results from several quasi-instantaneous point release experiments. Averaging times varied from 1 to 30 sec in the referenced experiments. The coefficients, labeled σ_{yi} and σ_{zi} to denote instantaneous, are similar in form to the PG coefficients, but they provide considerably less plume spread for the same stability class and downwind distance. Equations used to represent the Slade coefficients are presented in Appendix G. The Slade coefficients are used later to model instantaneous (assumed to be 10 sec) concentrations for odor transport.

Time Averaging

The groundlevel concentrations calculated using PG coefficients are generally interpreted as 10-min average concentrations.[3] The UNAMAP models use these coefficients but report the results as 1-h concentrations. The MP coefficients were determined from 1-h average plume profiles and are interpreted as such. For odor modeling, it is usually necessary to convert these results to a shorter duration such as 10 sec. Turner[3] presented the following relationship for adjusting concentrations to different time averages:

$$c_s = c_k \left(\frac{t_k}{t_s} \right)^{p_t}$$

(17.6)

where

c_s = concentration estimate for the desired sampling time t_s, taken as 10 sec for odor modeling

c_k = concentration estimate for the reference sampling time, t_k. t_k is assumed to be 600 sec (10 min) for PG coefficients and 3600 sec (1 h) for MP coefficients

p_t = power law constant

Using Equation 17.6 it is assumed that wind speed and stability remain relatively constant for the time duration t_s.

Turner[3] suggested that the exponent p_t should be 0.17 to 0.20. Hino[14] studied the effect of sampling times on maximum groundlevel concentrations. A value of 0.5 (termed the 1/2 power law) was determined for adjusting sampling times ranging from 10 min to 5 h. A value of 0.2 (1/5 power law) applied to sampling times less than 10 min. Cha and Jann[8] also suggest a value of 0.2 when converting a 10-min Gaussian to a 10-sec odor concentration. Duffee et al.[15] suggested that the power law constant is a function of stability class, ranging from 0.5 for classes A and B to 0.167 for classes E and F. A value of 0.2 is consistent with the range suggested by Turner and is used to adjust odor concentrations in the dispersion model presented later in this Chapter.

Plume Rise

The effective height of plume H is taken as the sum of stack height, h, and plume rise, ΔH. The effective plume height is defined as the height at which the plume becomes passive and follows the horizontal, ambient air motion as illustrated in Figure 17.6. While this definition is straight forward, field observations of plume rise are often difficult. Briggs[16] noted that in more than 90% of the recorded observations the plume had not yet leveled off when it was no longer visible. Plume rise in stable or neutral atmospheres is relatively easy to measure, whereas measuring plume rise in unstable conditions is very difficult. Nevertheless, conventional practice is to define the effective plume height, H, as

$$H = h + \Delta H$$

(17.7)

where

h = stack height, m
ΔH = plume rise, m
H = effective plume height, m

Plume behavior is a function of a number of parameters, including exit gas velocity (momentum), temperature difference between the plume and ambient air (buoyancy), density stratification of the atmosphere, wind speed, and wind velocity gradient, du/dz. Buoyancy forces produced by a temperature difference and momentum forces from the gas velocity can both be important to plume rise depending on specific plume and atmospheric conditions. Given the complexity of the problem, numerical integration techniques have been used to increase the number of parameters that can be considered. Unfortunately, measurements of all parameters are seldom available. As a result, a number of semi-empirical equations have been developed which are widely used and provide reasonable estimates of the plume rise. Schultze[5] estimated that over 100 plume rise formulas have been proposed over the years. Three equations are discussed here which have had wide acceptance and use in the industry: the Holland formula, the Briggs' approach, and the ASME momentum formula.

The Holland Formula

The Holland formula was first published in 1953 based on work conducted for the Atomic Energy Commission[17] at Oak Ridge, Tennessee, and appears in early versions of Turner's Workbook:[3]

$$\Delta H = \frac{v_s d}{u}\left[1.5 + (2.68 \times 10^{-3})(p_a)\frac{T_s - T_a}{T_s}(d)\right]$$

(17.8)

where

ΔH = rise of the plume above the stack, m
v_s = exit velocity of stack gas, m/sec
d = inside stack diameter, m
u = wind speed, m/sec
p_a = atmospheric pressure, mbar
T_s = stack gas temperature, °K
T_a = ambient air temperature, °K

The value 2.68×10^{-3} is a constant with units of $mbar^{-1}m^{-1}$. The formula contains both a momentum term, $1.5\ v_s d/u$, and a buoyancy term based on temperature difference between plume and ambient.

If the effective height of emission were the same under all atmospheric conditions, the highest groundlevel concentrations from an elevated stack discharge would occur with the lightest winds (see Equation 17.1). However, plume rise is generally an inverse function of wind speed based on Equation 17.8. Thus, maximum groundlevel concentrations generally occur at an intermediate wind speed, where a balance is reached between dilution from wind speed and effective height of emission.

The Briggs Formulas

Gary Briggs, also working for the Atomic Energy Commission, made detailed studies of plumes and plume rise formulas. Following several modifications, the Briggs's formulas were adopted by the EPA beginning with the sixth edition of Turner's Workbook, published in 1972. Buoyancy rise is determined by first calculating the buoyancy flux given as

$$F_b = \frac{gQ_f}{\pi}\left(\frac{T_s - T_a}{T_s}\right)$$

(17.9)

where

F_b = buoyancy flux, m^4/sec^3
g = gravitational acceleration, 9.8 m/sec^2
Q_f = stack gas volumetric flowrate, m^3/sec

For stability classes A through D, buoyant plume rise is determined as

$$\Delta H = \frac{21.425(F_b^{0.75})}{u} \qquad (F < 55) \qquad (17.10)$$

$$\Delta H = \frac{38.710(F_b^{0.60})}{u} \qquad (F \geq 55) \qquad (17.11)$$

where

u = wind speed at stack height, m/sec

For the stable classes E and F, a stability factor, S, is first determined as follows:

$$S = \frac{g}{T_a}\left(\frac{\partial \varphi}{\partial z}\right) \qquad (17.12)$$

where

$\partial \varphi / \partial z$ = potential vertical temperature gradient in the atmosphere, equal to $\partial T/\partial z + .01$ °C/m (assumed to be +0.02 °C/m for class E and +0.035 °C/m for class F)

Buoyant plume rise is then determined as

$$\Delta H = 2.6\left(\frac{F_b}{u(S)}\right)^{0.333} \qquad (17.13)$$

The Briggs' approach uses separate formulas for estimating momentum rise. For stability classes A through D, momentum rise is estimated as

$$\Delta H = 3\left(\frac{v_s d}{u}\right) \qquad (17.14)$$

For stability classes E and F the momentum flux, F_m, is first calculated as

$$F_m = \frac{d^2 v_s^2 T_a}{4T_s} \qquad (17.15)$$

where

F_m = momentum flux, m^4/sec^2

Momentum rise is then determined as

$$\Delta H = 1.5\left(\frac{F_m}{u}\right)^{0.333} S^{-0.167} \qquad (17.16)$$

Plume rise is usually estimated using both the buoyancy and momentum equations. The greater of the two values is then used in Equation 17.7 to determine effective plume height.

Comparison of Momentum Formulas

The third plume rise formula presented here was developed by the ASME for momentum jets.[18] It applies to stacks with exit temperatures <50°C above ambient and exit velocities >10 m/sec. The ASME formula is

$$\Delta H = d\left(\frac{v_s}{u}\right)^{1.4} \qquad (17.17)$$

Momentum jets are important at composting facilities because process gases are usually <50°C above ambient. It is interesting to compare the three momentum formulas presented above:

$$\Delta H \quad = \quad 1.5\left(\frac{v_s d}{u}\right) \qquad \text{Holland}$$

$$\Delta H \quad = \quad 3\left(\frac{v_s d}{u}\right) \qquad \text{Briggs (A to D stability)}$$

$$\Delta H \quad = \quad d\left(\frac{v_s}{u}\right)^{1.4} \qquad \text{ASME}$$

All formulas have the same general form with plume rise proportional to stack diameter and exit velocity and inversely proportional to wind speed. The Briggs estimate is double that of Holland under all conditions. This may be because Holland's observations were carried not more than 600 ft downwind. The ASME estimate is greater than Briggs if the ratio of v_s/u > 2.2.

Increasing the exit velocity, v_s, will increase momentum plume rise, but it will also increase the pressure drop in the ducting and stack system, which in turn will increase power requirements. Tradeoffs must be made between the needs for odor dispersion and the cost of increasing stack height and operating costs. As a minimum, stack-to-wind velocity should be >1.5 to allow exhaust gases to break cleanly from the stack and avoid plume downwash behind the stack (stack tip downwash). The limit on maximum velocity is about 35 m/sec for masonry stacks and 45 m/sec for steel stacks.[19] Typical exit velocities are between 10 and 30 m/sec.

The computer program PLUME was developed to calculate plume rise and effective stack height using the Holland and Briggs procedures. The examples at the end of this chapter illustrate the use of the program and some important aspects of plume discharges from compost facilities.

Wind Calms

All of the above formulas have a boundary problem where H approaches infinity as u approaches 0. There must be a limit when the atmosphere is stably stratified (classes E, F, and above) because colder, denser air is entrained as the plume rises. For a neutral atmosphere, however, the existence of an upper limit to plume rise is not immediately obvious. Pasquill and Smith[2] suggested that a practical approach to overcome this difficulty is to regard the rise as effectively complete when the upward velocity is some small fraction of the vertical eddy velocities in the wind. The Gaussian formulas also have a boundary problem in that the groundlevel concentration tends toward infinity as u approaches 0. The Gaussian and plume rise formulas generally do not apply well at wind speeds less than about 1 m/sec. From the standpoint of odor dispersion, calm periods are not well modeled by the currently available models in the UNAMAP series.

Plume Entrainment

Dispersion of a plume is the result of two effects: eddy diffusion caused by the turbulent motion of the atmosphere (passive dispersion) and dispersion induced by turbulent entrainment caused by the rising plume (plume rise entrainment). The dispersion coefficients discussed above, such as those of Pasquill-Gifford, represent only the effects of passive dispersion. Entrainment induced by the rising plume can be important with stable atmospheres where turbulence is low, where the concern is with instantaneous concentrations close to a stack, and where the plume impinges on nearby terrain. Pasquill and Smith[2] and Pasquill[20]suggested that the effect of entrainment can be accounted for by adjusting the passive dispersion coefficients as follows:

$$\sigma_{ye} = \left[\sigma_y^{\ 2} + \left(\frac{\Delta H}{3.5}\right)^2\right]^{0.5} \tag{17.18}$$

$$\sigma_{ze} = \left[\sigma_z^{\ 2} + \left(\frac{\Delta H}{3.5}\right)^2\right]^{0.5} \tag{17.19}$$

where

σ_{ye} = effective horizontal dispersion coefficient
σ_y = passive horizontal dispersion coefficient taken from Figure 17.7 or the equations in Appendix G
σ_{ze} = effective vertical dispersion coefficient
σ_z = passive vertical dispersion coefficient taken from Figure 17.8 or the equations in Appendix G

Surface Effects

The nature of the surface can have a significant effect on the rate of dispersion. In urban areas, the rate of dispersion is generally higher due to mechanical and thermal turbulence, termed the "urban effect". In mountainous terrain, cold air will generally flow down the valleys as the ambient temperature decreases. During daylight, the situation reverses and air moves up the valley. During periods when the continental weather movement is slow, pollutants emitted in a valley may circulate back and forth in the valley for several days until they are dispersed or washed out.[5]

If the surrounding terrain is near or above the effective plume height, the receptor can be exposed to higher concentrations compared to the case where the terrain is flat. Flat terrain is usually termed "simple" terrain. If the surrounding terrain is above the actual stack height, it is usually termed "complex" terrain. The term "high" terrain is used where ground elevations exceed the plume height (i.e., stack height plus plume rise). Terrain that falls between the stack height and plume height is sometimes called "intermediate" terrain. These distinctions are important because elevated plumes can impact on surrounding complex and high terrain, causing higher concentrations compared to the case of simple terrain. Special algorithms have been developed to handle complex terrain but are beyond the scope of this text. The models presented in this chapter assume simple terrain.

Another surface phenomena that should be noted is the potential effect of building wakes on elevated plumes. As wind moves across a solid structure, a wake or eddy is formed downwind of the structure. If an elevated plume is near the zone of influence of the eddy, the plume can be entrained and quickly brought to ground-level. Plume downwash can cause high groundlevel concentrations near the stack discharge. Good engineering practice (GEP) for stack height to avoid plume downwash is generally defined as follows,

$$H_{gep} = H_b + 1.5L \qquad (17.20)$$

where

H_{gep} = GEP stack height, m
H_b = building height, m
L = characteristic dimension of the building, taken as the lesser of height or diagonal width

Equation 17.20 is usually assumed to apply if the stack is within 5L from the building. Special algorithms are available to account for plume downwash but are beyond the scope of this text.

The impact of Equation 17.20 should be noted. For most buildings H_{gep} is about 2.5 H_b. Thus, a 10-m high building would require about a 25 m (82 ft) stack to avoid plume downwash under most meteorological conditions. Such tall stacks are seldom used at composting facilities. It is more common to observe scrubber stacks discharging at the building roof line or even below. Plume downwash will certainly occur as a building wake is formed by increasing wind speed. Once the plume is brought to groundlevel, the nearest downwind receptor will experience the highest odor concentrations. Photos of plume discharges subject to building wake entrainment are shown in Figures 17.9 and 17.10.

Area Sources

Many composting facilities can be characterized as large area sources from which odor is emitted at a relatively uniform rate per unit area. In such cases there is no vertical stack velocity and the effective plume height is zero. A schematic diagram illustrating development of odor over an area source and its subsequent dispersion downwind is shown in Figure 17.11. As air moves over the area source, odor concentration increases until the downwind edge of the odor source is reached. From this point on, atmospheric turbulence tends to disperse the air, diluting the odor concentration.

Odor concentration at the downwind edge of a large area source can be estimated using the idealized model of Figure 17.12. Odor emission rate is assumed to be constant as are the

Figure 17.9. An example of plume downwash in the wake of a building. Process gases from a reactor composter at Schnectady, New York, were scrubbed and released at low velocity and below the building height. This system was subsequently replaced.

Figure 17.10. Carbon adsorption units treating composting process gases. Rain caps on the exhaust will deflect the plume back toward the ground. The low height of discharge will entrap the plume in the building wake. Roof ventilators are equipped with rain caps which prevent significant plume rise and cause wake entrainment. Ridge ventilators discharge the exhaust with essentially no vertical velocity. Such designs are common and can be acceptable if there is adequate buffer to the nearest receptors. However, these designs can quickly return a plume to the ground, causing high groundlevel concentrations immediately downwind from the building and increasing the risk of odor incidents if there are nearby receptors.

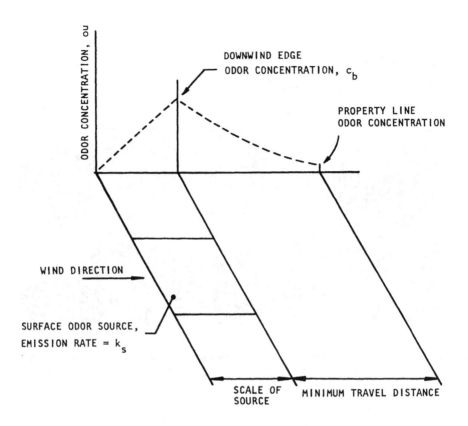

Figure 17.11. Odor concentration across and downwind from a large emission area.[21]

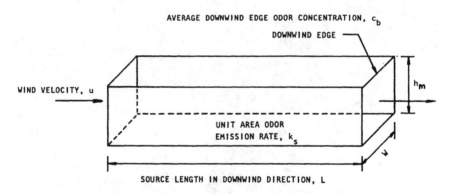

Figure 17.12. Idealized model of air movement over a large area odor source. The model assumes a constant mixing height, h_m, and can be used to estimate the downwind edge odor concentration.

mixing height, h_m, wind speed, u, and wind direction. A linear increase in odor concentration over the area source is shown in Figure 17.11, which is consistent with the idealized model of Figure 17.12. Odor emission rate from the area source can be defined as

$$Q_o = k_s A = k_s LW \qquad (17.21)$$

where

Q_o = odor emission rate, m^3/min
k_s = average SOER, $m^3/min\text{-}m^2$
L = emission source length, m
W = emission source width, m

The odor emission rate refers to the flowrate of air required to dilute odor emissions to the threshold odor concentration. Actual air flowrate across the area source within the mixing height, h_m, is given by

$$Q_{air} = (60)uh_m W \qquad (17.22)$$

where

Q_{air} = air flowrate moving across the odor source, m^3/min
h_m = average puff height over the emission area, m

Average odor concentration at the downwind edge can then be determined as the ratio of Q_o to Q_{air}:

$$c_b = \frac{Q_o}{Q_{air}} = \frac{k_s L}{(60)uh_m} \qquad (17.23)$$

where

c_b = average odor concentration at the downwind edge, ou

After the air mass has moved across the odor source, turbulent dispersion begins to reduce the odor concentration. A number of approaches have been used to model the movement of emissions from area sources. Wilson et al.[7] measured the frequency distribution of eddy dispersion coefficients under puff transport conditions in Sacramento. These dispersion coefficients were then applied to a gradient-transfer model to estimate downwind concentrations. However, most area source models use the Gaussian modeling approach with similar assumptions to those for elevated point sources. Pasquill-Gifford or similar dispersion coefficients are applied using the standard deviation format.

Turner[3] presented a Gaussian modeling approach in which the area source is converted to an equivalent point source located a "virtual" distance upwind of the area source. The relationships between the area source and equivalent point source are presented in Figure 17.13. The downwind width of the area source, W, is assumed to equal the plume width, normally assumed to be 4.28 σ_y. The horizontal standard deviation at the downwind edge of the area source can then be determined as

$$\sigma_y = \frac{W}{4.28} \qquad (17.24)$$

Knowing σ_y, the virtual distance, VX_y, can then be found that will give this standard deviation. Figure 17.7 can be used to determine VX_y for the case of PG dispersion coefficients. Alternatively, the appropriate equations in Appendix G can be used for PG, MP, or Slade coefficients. The virtual distance for vertical dispersion, VX_z, is taken as the source length, L,

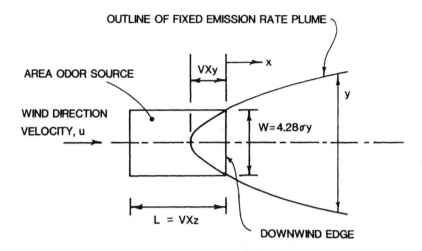

Figure 17.13. Relationship between a large area source and equivalent point source located a "virtual" distance upwind.

unless a downwind barrier is in place (see Barrier Walls below). The virtual distance, VX_z, is then determined from Figure 17.8 or the appropriate equations in Appendix G. If the distance x is measured from the downwind edge, the virtual point source for horizontal dispersion is located $x + VX_y$ upwind of the receptor. Similarly, the virtual point source for vertical dispersion is located $x + VX_z$ upwind.

Regulatory and Other Models

Dispersion modeling is extensively used by air regulatory agencies to determine compliance with air quality regulations. In the U.S., models from the EPA User's Network for Applied Modeling of Air Pollution (UNAMAP) series are commonly used. Starting with six models in 1972, the series now includes over 30 different models. All of the guideline models use the Gaussian algorithm. Most allow the user to select PG or MP coefficients and all use the plume rise formulas developed by Briggs. The predicted concentrations represent about a 10-min time average if PG coefficients are used and 60 min with MP coefficients. None of the models adjust the concentrations to a shorter time period suitable for odor modeling. EPA makes no recommendation regarding use of the UNAMAP models for odor modeling, although they are often used for such. A number of consulting and software companies market "user friendly" versions of the UNAMAP models.

The Industrial Source Complex (ISC) models are the workhorse of the UNAMAP series and probably the closest thing to a universal standard.[11] Two versions are available: one using long-term average meteorological data (ISCLT and ISCLT2) and one using short-term hourly data (ISCST and ISCLT2). These models can handle multiple sources, source types (i.e., point and area sources), and locations. Meteorological inputs required by ISCST include hourly estimates of the wind direction, wind speed, ambient air temperature, Pasquill stability category, mixing height, wind profile exponent, and vertical potential temperature gradient. Data for up to 8784 h/yr (every hour of a 366-day year) can be input. Fortunately, preprocessed data on magnetic media is available for almost all major weather stations in the U.S. One of ISCSTs major advantages is that the frequency distribution of concentration can be determined at a number of receptor locations. ISCST has been used (rightly or wrongly) for odor modeling at a number of composting facilities including Philadelphia, Pennsylvania, Montgomery County, Maryland, and Plattsburgh, New York.

Several other UNAMAP models should be mentioned. PTPLU (from PoinT PLUme) estimates maximum groundlevel concentrations from a single point source using PG coefficients. Version 2 allows use of the MP urban coefficients. SCREEN estimates maximum groundlevel concentrations from a single point or area source, incorporates effects of building downwash for both near and far wake regions, and considers elevated and complex terrain. VALLEY and COMPLEX –1 models are designed for complex terrain where the surrounding ground is above the stack height. Again, none of these models are specific to odor emissions.

A few dispersion models have been specifically developed to simulate short-term peak concentrations and are applicable to odor modeling. Hilst[22] in 1957 and Gifford[23] in 1959 introduced the concept of a "fluctuating plume model." The Gaussian form is used to represent instantaneous puffs of material. Numerous puffs are followed and allowed to meander according to a frequency distribution. Plume dispersion is assumed to consist of two terms, the first representing the diffusion of the individual plume segments which fluctuate in position, and the second the meander or large scale turbulence which determines the mean, time-averaged plume dimensions. Hogstrom[24-26] developed data for such a model by photographing smoke puffs and determining the meander of the puff. He later applied the fluctuating plume model to predict odor frequencies from a point source. Murray et al.[27] developed a proprietary, fluctuating plume, odor model based on the Hogstrom approach and using a random number generator to distribute individual plume segments into a longer-term normal distribution. Dispersion coefficients developed by Bowne[28] are used to establish the dimensions of the longer-term plume. This type of model is also termed a "segmented plume model" because the plume is segmented into a number of individual puffs.[29]

The number of available odor models is limited. They are generally not well documented and in some cases are proprietary. There is a need for documented, nonproprietary models that are widely available and subject to peer review. Such models should be based on the Gaussian algorithms which are the industry standard. The models should also address instantaneous concentrations, wind calms, momentum plumes, and very stable sublayer conditions.

DESIGNING FOR DISPERSION

A number of techniques have been used in practice to increase or enhance the dispersion of odorous plumes and to avoid puff or intermittent puff transport conditions. Taken together, such techniques are commonly termed "enhanced" dispersion and include the following: (1) conversion of area sources to point sources, (2) increasing stack height, velocity, and/or diameter, (3) forced dilution, (4) barrier walls and wind machines, (5) timing of operations, and (6) atmospheric monitoring.

Area vs. Point Sources

Odor release from an area source occurs at groundlevel so that the effective plume height is zero. Therefore, the nearest downwind receptor will experience the highest odor concentration. Maximum concentrations for groundlevel sources occur with stable conditions. Area sources discharge into the atmospheric sub-layer and are subject to extremely stable puff transport conditions. During periods of calm winds with a stable lapse rate, high odor concentrations can accumulate over an area source.

For an elevated point source, the maximum "instantaneous" concentrations occur with unstable conditions when portions of the plume are brought to the ground with relatively less dispersion. Maximum concentrations for time periods of a few minutes occur under unstable conditions. Under stable conditions the plume moves with little downwind dispersion and, therefore, tends to remain aloft and away from receptors. Under stable conditions the maximum concentrations at groundlevel are generally less than those under unstable conditions and occur at greater distances from the source. The plume may not reach groundlevel for a considerable distance downwind, in some cases many kilometers from the source. Plume rise from a point source increases as wind speed decreases, again tending to keep the plume aloft during calm periods and away from receptors.

At least one composting facility, the Montgomery County Composting Facility, has essentially eliminated all groundlevel area sources. Based on extensive odor inventories, the area sources represented only about 10% of the total odor mass emission rate, the other 90% coming from a single elevated point source. Nevertheless, the fact that the area source emissions were at groundlevel caused documented odor incidents, particularly during stable atmospheric conditions. All mixing, composting, screening and curing facilities have been enclosed to eliminate the groundlevel discharges. The buildings are ventilated by special "cooling tower" type fans to produce a point discharge with a significant plume rise.

Stack Parameters

Increasing the effective plume height will always decrease the groundlevel concentration. The engineer has about four approaches to increasing effective plume height. The first and most obvious is to simply increase the stack height itself. This will increase the effective plume height and reduce problems of plume downwash near buildings. However, the composting industry has been somewhat reluctant to use this approach, probably because of the visual problems associated with an elevated stack. Some of the tallest stacks used at composting facilities include a 38.1 m (125 ft) stack at Westborough, Massachusetts (since closed), a 36.6 m (120 ft) stack at the City of Hamilton, Ohio, a 30.5 m (100 ft) stack at Cape May, New Jersey (see Figure 17.14), a 27.4 m (90 ft) stack at Hartford, Connecticut, and a 22.9 m (75 ft) stack at Albuquerque, New Mexico.

The stack at Hamilton was added because the original scrubber exhaust discharged near a building and the plume was occasionally caught in the building downwash. The stack at Westborough was added after the original facility experienced odor problems from nearby receptors. The new discharge stack improved the situation but could not by itself solve all odor problems; complaints were now received from both nearby residents and residents located up to 1 mile from the facility. Donovan et al.[30] later determined that improved gas treatment coupled with the new stack would probably resolve the problem. The reader should always remember that power plants use tall stacks because they provide better dispersion. Very few composting plants would have odor problems if they could build very tall, 100 to 200 m stacks for dispersion. On the other hand, dispersion is only part of a total management plan. By itself, stack dispersion may not solve all odor problems as the story at Westborough illustrates.

Effective plume height can be increased by increasing the exhaust temperature of the plume. Buoyancy rise with a hot discharge will almost always carry a plume higher than is possible with a cool, momentum-dominated plume. Conserving heat in the hot, compost gases will have some beneficial effect on plume rise. However, process gases from composting will rarely exceed 60°C, a cool temperature relative to power plant plumes. Artificially raising discharge temperature by using burners has not been practiced at composting facilities,

Figure 17.14. A 100-ft tall discharge stack at the Cape May, New Jersey, composting facility. The stack was added to improve dispersion of the treated gases and is credited with contributing to reduced odor complaints. Photo courtesy of William Cathcart, Chief of Operations and Maintenance.

probably because of the high fuel costs. It should be noted, however, that stack gas reheat is a common practice at power plants, particularly those using wet scrubbers, to increase plume rise and reduce condensation clouds.

Increasing the discharge velocity will increase the effective plume height from a momentum jet. Discharge velocities as high as 45 m/sec have been used to improve the dispersion of some odorous plumes. However, this is probably about the practical limit because of noise generation and power requirements. Plume rise from a high velocity, vertical jet is shown in Figure 17.15. Another approach is to increase the diameter and mass emission from the plume. Large diameter, axial flow fans are used to discharge ventilation air from building enclosures at the Montgomery County Composting Facility (see Figure 17.16). Two 14-ft diameter, six-bladed fans, each rated at 500,000 cfm and 75 hp, ventilate the composting building. Each fan is mounted at the roof line and discharges at a 40 ft height through a 7-ft tall cylindrical stack mounted above the roof. Two other identical fans ventilate the curing building. A large diameter, high mass flow, vertical plume is produced with a significant height of rise (see Examples 17.1 and 17.2). This can be very effective in avoiding groundlevel concentrations during stable groundlevel conditions.

A number of things should be avoided with a point source discharge. These include: (1) locating the plume within the zone of a building (building downwash); (2) having a low

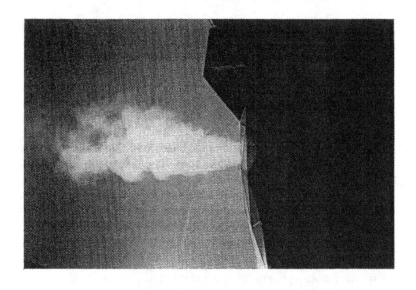

Figure 17.15. Plume rise from the circular, agitated bed system at Wilmington, Delaware. Process gases and ventilation air are discharged upward from a fan located in the ceiling of the geodesic dome. A rain cap originally installed with the facility was removed to improve the plume rise. The plume rose to an estimated 200 to 300 ft in the calm, cold conditions of these photos. Both momentum and buoyancy effects were probably significant for this warm plume discharging on a cold winter day.

Figure 17.16. Large diameter ventilation fans used at the Montgomery County Composting Facility. Combining building ventilation into large diameter, vertical jets increasing the available plume rise. Covering and ventilating in this manner converts a large area source to an elevated point source.

velocity discharge which can result in downwash from the stack itself (stack downwash); (3) low velocity discharges from the sides or roofs of buildings, such as ridge ventilators; (4) using rain caps on roof ventilators which effectively stop any further plume rise; and (5) allowing water droplets in the plume which can evaporate, cooling the plume and reducing its buoyancy.

Forced Dilution

If stack height is limited, it may not be possible to always reduce groundlevel concentrations to below the detection threshold. Even highly treated process gases could contain an ED_{50} of 100 ou, with an ED_5 of perhaps 1000 ou. Achieving 100 to 1000 dilutions with an averaging time of 10 sec from a relatively short stack is probably not possible under all meteorological conditions and at all receptor locations. If the compost facility is located in a sensitive community, where even occasional ground-level odors must be avoided, forced dilution of the exhaust gases can be provided. Suppose the above exhaust gases are diluted 10:1 with ambient air. This would produce a combined plume with an ED_{50} of about 10 ou. The plume diameter and mass emission would also be increased, which in turn would increase the effective plume height. Achieving an additional 10 dilutions before the plume reaches groundlevel should be possible in almost all meteorological conditions.

The effectiveness of forced dilution of treated exhaust gases has been proven in practice at several composting sites. The Montgomery County Composting Facility is surrounded by a sensitized, suburban community. Some odor complaints were traced to the original stack discharge, even though an optimized, three-stage scrubbing system was used with over 90%

Figure 17.17. The scrubber system at the Montgomery County Composting Facility after retrofit of forced
dilution, axial flow fans. The axial flow fans can force a 10:1 dilution and also produce a
larger diameter jet with improved plume rise characteristics. The fans are 14 ft in diameter
and discharge at 110 ft in height. For the unit on the left, the fan blades are fiberglass and
all other materials exposed to the stack gas are stainless steel. The unit on the right was
constructed of carbon steel and suffered corrosion. The cylindrical unit on the right foreground is a regenerative afterburner under test at the time. The original biofilter manifold
is visible to the left of the afterburner.

odor reduction. Based on dispersion modeling, it was concluded that treatment alone was not
likely to eliminate odor incidents under all meteorological conditions. Forced dilution was
then retrofitted onto the scrubber exhaust as shown in Figure 17.17. About a 10:1 dilution can
be provided using a 14-ft diameter, six-bladed cooling tower fan, rated at 500,000 cfm and 75
hp, mounted above the normal stub stack of the wet scrubbers. Stack height is 110 ft. The City
of Lancaster, Pennsylvania, uses a centrifugal fan mounted at ground level to provide about
a 3.3:1 dilution of the scrubber discharge gases. The dilution air is added at the base of a 42-
ft high discharge stack. Forced dilution has proven to be an effective tool to reduce the
occasional odor incidents that occur under worst-case meteorological conditions.

Barrier Walls and Wind Machines

Methods are available to improve dispersion from large area sources. The worst-case
transport conditions for large area sources normally occur under stable, puff transport conditions in the atmospheric sublayer. Artificial barriers and wind machines have both been used
in agricultural practice to modify sublayer meteorological conditions. Barriers have been used
for wind control and wind machines for frost control.[21] Both of these devices are used to
control odor transport from liquid sludge storage basins (SSBs) at the Sacramento Regional
Wastewater Treatment Plant, California.

Barrier walls were studied using wind tunnel models. Solid barriers were shown to increase turbulence and vertical mixing, which increased the height of the downwind puff under strong inversion conditions. Puff height was found to be a function of barrier height but independent of wind velocity. The following equation was developed to estimate puff height downwind of a barrier:[21]

$$H_b = B(h_b).$$ (17.25)

where

H_b = puff height downwind of the barrier
h_b = barrier height
B = constant, equal to 2.4 in the Sacramento studies

In the Sacramento studies, it was assumed that puff transport would persist downwind of the barrier but with the increased puff height. A 12-ft barrier surrounding the SSBs was constructed and has been in use for a number of years. The barrier is shown in Figure 17.18 and is "sharp-edged" and impermeable. Wind tunnel tests showed that sharp-edged barriers increase vertical mixing to a greater extent than curved or streamlined barriers. The barriers completely surround the site so that a minimum of two points of barrier mixing are provided along each wind line.

The downwind eddy is about 2.4 times the barrier height in Equation 17.25. The GEP standard for stack height to avoid plume entrainment into a building eddy is about 2.5 times the building height. Schulman and Scire[31] reported the following formula developed by Hosker[32] for the height of a recirculation cavity behind a building:

$$H_R = H\left[1.0 + 1.6\exp\left(\frac{-1.3L}{H}\right)\right]$$ (17.26)

where

H_R = height of the downwind recirculation cavity
H = building height
L = length of the building in the alongwind direction

Note that as L/H approaches 0, H_R/H approaches 2.6, which is the approximate cavity height found for a two-dimensional fence (W/H is large). It is interesting to note the similarity of results regarding downwind eddy size, whether from a building wake or the wake of an artificial barrier.

Wind machines are used at Sacramento to maintain air movement across the area odor source during calm wind periods. This is designed to reduce high odor levels which result during intermittent puff transport conditions. The objective is to convert intermittent puff transport to a lower-risk continuous puff transport and to increase the vertical dispersion coefficient. In effect, the wind machines provide a form of forced dilution for the area source. The wind machines are turned on during calm periods coinciding with strong inversions and are designed to maintain a minimum velocity of 2 mph over the area source during a wind calm.[33]

The wind machines used at Sacramento are shown in Figure 17.18. Each wind machine serves a 152 m (500 ft) square zone and is powered by a 100 hp electric motor. The fan blade

Figure 17.18. Barrier walls and wind machines used at the Sacramento Regional Wastewater Treatment Plant. Eddies produced by the barrier walls increase the depth of vertical mixing. The wind machines maintain air movement and increase vertical mixing during wind calms. Photo courtesy of Wendell Kido, Plant Manager.

is about 16.5 ft diameter, with the fan centerline about 34 ft above ground and maximum speed of 600 rpm. The fan shaft is oriented about 6° below horizontal to direct the air movement toward the ground. The system includes 10 permanently mounted wind machines and 4 portable gas powered units for standby service. The machines are synchronized to rotate at 0.2 rpm, all facing the same direction. If the fans are not operated in this synchronized manner, they are not nearly as effective in moving air over the barrier walls and dispersing odors vertically. The barrier walls and wind machines have been in use for over 10 years at Sacramento and are considered a proven odor mitigation technique for large area sources.

Timing of Operations

One other method of reducing downwind odor concentrations that should be discussed is the proper timing of more odorous operations. In this case the idea is not to enhance atmospheric dispersion per se but to conduct more odorous operations during periods when atmospheric transport is conducive to minimizing groundlevel concentrations. With point sources, for example, low wind speeds and stable sublayer conditions can cause an elevated plume to remain aloft, away from receptors (lofting). Even though dispersion is reduced, the plume does not reach a receptor. During periods of high turbulence, on the other hand, the dilution potential is greater but the plume can be dispersed back to groundlevel faster. Which condition results in the highest ground-level concentration depends on the specific discharge and site characteristics.

With area sources, the highest groundlevel concentrations are always associated with low atmospheric turbulence. Therefore, odorous operations should be conducted during periods of higher atmospheric turbulence. This will always reduce groundlevel concentrations from area sources. Fortunately, high turbulence is characteristic of daylight hours and it is relatively easy to conduct odorous operations during this period. Redner et al.[34] reported that restricting

Figure 17.19. Dual weather stations used at the Sacramento Regional Wastewater Treatment Plant to measure wind speed, direction, and temperature lapse rate. A stable lapse rate and low wind speed give early warning of worst-case meteorological conditions. Photo courtesy of Michael Broadfoot, Plant Engineer.

windrow turning to hours of greater vertical mixing, between about 8 a.m. and 4 p.m., reduced peak odor concentrations downwind of a large open windrow system. This does not mean that any amount of odor can be released during the daylight hours. Obviously, odors will still be transported, although dispersion will be increased over the puff transport case.

Monitoring the Atmosphere

A relatively simple weather station can be used to supply atmospheric information and warn of worst-case meteorological conditions. Wind speed, wind direction, and temperature lapse rate are the key parameters needed to warn of impending worst-case meteorology. For example, Wilson et al.[7] determined that puff transport occurred when the temperature measured between 5 and 25 ft above ground exceeded 2°F (1.1°C), equivalent to a lapse rate of +0.18 °C/m with wind speeds of 2 mph (0.89 m/sec) or less. At Sacramento, two on-site micrometeorological (micromet) stations provide continuous evaluation of near-ground atmospheric conditions which affect the transport and dispersion of odors (see Figure 17.19). Each micromet station includes a 30-ft tower equipped with wind speed anemometer, wind direction vane, and two aspirated shielded temperature probes. A control room alarm is generated when either micromet station senses the onset of adverse transport conditions. Some composting facilities have also provided on-site weather stations. Some others rely on data from the National Weather Service. Regardless of the source of weather data, it is important that atmospheric monitoring be part of the operating practices at all large composting facilities.

Of course, weather data and alarms are only useful if the operator can do something with the information that will result in reduced odors. Fortunately, the engineer can provide the operator with such tools. If the design allows, the operator can initiate actions to (1) curtail more odorous operations to reduce emission rates, (2) improve odor treatment, (3) enhance the dispersion from stacks by forcing additional dilution, (4) reduce unnecessary area sources, such as by closing roll-up doors, and converting their emissions to elevated point sources, and (5) enhance dispersion by altering the groundlevel meteorology such as by using wind machines. Remember, the engineer can design systems to enhance the dispersion that nature provides and can give the operator tools by which to increase dispersion during worst-case conditions.

"PLUME" AND "DISPERSE" MODELS

Computer models PLUME and DISPERSE were developed to aid in the analysis of odor transport from composting facilities. PLUME is used to determine the effective plume height from an elevated point source and uses both the Holland and Briggs approaches discussed above. DISPERSE is used to estimate downwind odor concentrations from a composting facility consisting of a large area source and an elevated point source. Nomenclature for the models PLUME and DISPERSE is presented in Appendix B.

Model Assumptions

A schematic of the process conditions modeled by DISPERSE is presented in Figure 17.20. The composting facility is modeled as an area source of length, LENGTH, and width, WIDE. The facility can also contain one elevated point source located in the middle of the area source. Wind direction is along the length of the area source. Downwind concentrations are estimated along the centerline beginning at the downwind edge of the area source. Downwind distance, x, is measured from the downwind edge of the area source. A barrier wall can be placed at the downwind edge as a user defined option.

Gaussian algorithms are used to estimate groundlevel concentrations. Effective plume height from the point source is determined using the Briggs algorithm. PG (rural = R), MP (urban = U), or Slade (S) coefficients can be selected by the user. Different coefficients can be used for the area and point source. The dispersion coefficients are determined from the formulas presented in Appendix G and are corrected for plume entrainment using Equations 17.18 and 17.19. Other meteorological inputs include stability class (A to F), wind speed, U, and ambient air temperature, TAMBC in °C. Stack discharge inputs include stack temperature, TSTACKC in °C, inside diameter, DIA, in meters, and stack height H in meters.

Odor concentration in the stack discharge is input in odor units. Odor emission rate from the point source is determined as

$$SSTACK = \left(\frac{3.1416 * DIA^2}{4} \right) * VEL * CSTACK \tag{17.27}$$

where

$\quad$ SSTACK $\quad$ = mass discharge rate of odor from the stack, m³/sec
$\quad$ CSTACK $\quad$ = stack odor concentration, ou
$\quad\quad$ DIA $\quad$ = inside stack diameter, m
$\quad\quad$ VEL $\quad$ = stack discharge velocity, m/sec

The average surface odor emission rate from the area source is input in units of m³/min-m². Mass emission rate of odor from the area source is determined as

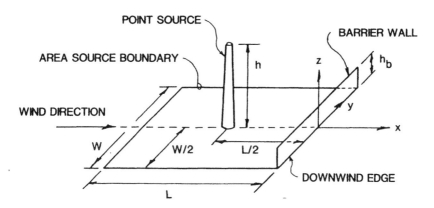

Figure 17.20. Area and point sources modeled by the program DISPERSE. The area source is of width, WIDE, and length, LENGTH. A single point source can be located at the center of the area source. A barrier wall can be located at the downwind edge of the area source.

$$SAREA = \frac{SOER * LENGTH * WIDE}{60} \qquad (17.28)$$

where

$\quad$ SAREA $\quad$ = mass discharge rate of odor from the area source, m³/sec
$\quad\quad$ SOER $\quad$ = surface odor emission rate, m³/min-m²
$\quad$ LENGTH $\quad$ = length of area source, m
$\quad\quad$ WIDE $\quad$ = width of area source, m

Equation 17.3 is used to determine centerline, groundlevel concentrations from the elevated point source. The area source is converted to an equivalent point source with a mass emission rate of SAREA discharging at groundlevel with zero velocity. Virtual distances VX_y and VX_z are determined using procedures previously discussed. Equation 17.4 is used to determine centerline, ground-level concentrations from the area source. The total concentration at any distance along the centerline is the sum of contributions from the point and area sources.

If a barrier wall is used, the groundlevel plume is assumed to resemble half of a Gaussian curve. The vertical standard deviation is taken as

$$\sigma_z = \frac{2.4 * HB}{2.14} \qquad (17.29)$$

where
$\quad$ HB $\quad$ = $\quad$ height of the barrier wall, m

The virtual distance, VX_z, is then determined using the appropriate equations in Appendix G.

The power exponent for time averaging is input under the variable name PT. A value of 0.2 is suggested and concentrations are adjusted to a 10-sec averaging time using Equation 17.6 if PG or MP coefficients are used. PG coefficients are assumed to be 10-min averages, while MP coefficients are taken as 1-h averages. If the semi-instantaneous Slade coefficients are specified, no time adjustment is applied.

Calculations begin at the downwind edge of the area source. A step size DELX1 in kilometers is input and the calculations for ground-level concentration are repeated at DELX1 distances along the centerline. This continues until a stopping distance, XSTOP1 in kilometers. The calculations then continue using a step size, DELX2, until the final distance, XSTOP2, is reached. The use of two different step sizes allows a smaller grid to be used near the source if desired.

Model Results

The following examples illustrate the use of the models and highlight some important aspects of odor transport from composting facilities.

Example 17.1

A compost facility is being designed to process 10 dry tpd of raw sludge by the static pile process. An average air supply of 2000 scfh/dt is estimated over a 21-day cycle. Collected exhaust gases will be discharged through a 20-m high stack with a discharge velocity of 20 m/sec and stack temperature of 45°C. Determine the plume rise and effective plume height as a function of wind speed over the range from 0.5 to 5 m/sec. Assume Class D stability, 20°C ambient temperature and 1013 mbar pressure (sea level).

Solution

1. Determine the average gas volume to be discharged:

$$Q_f = (2000)(10)(21) = 420,000 \text{ cfh}$$

$$Q_f = \frac{420,000}{(35.3)(60)(60)} = 3.3 \text{ m}^3/\text{sec}$$

2. Determine the required stack diameter:

$$Q_f = \text{area} * \text{velocity} = A * v_s$$

$$A = \frac{3.3}{20} = 0.165 \text{ m}^2$$

$$DIA = 0.46 \text{ m}$$

3. Load PLUME and follow the menu instructions to enter the following data:

MDY	=	(enter current value)
IDEN	=	EXAMPLE 17.1
VEL	=	20
DIA	=	0.46
H	=	20
TSTACKC	=	45
U	=	1
TAMBC	=	20
MBAR	=	1013
STABCLAS	=	D
DCSTACK	=	R

4. Results from the program output are summarized as follows:

   Plume dynamics per Holland's formula:
      Plume rise   =   13.25 m
      Effective plume height   =   33.25 m
   Plume dynamics per Briggs' formulas:
      Plume rise   =   24.87 m
      Effective plume height   =   44.87 m
        (Plume rise is governed by momentum).

5. The Briggs formulas predict a higher plume rise compared to the Holland formula. Plume rise per Briggs more than doubles the effective plume height compared to stack height.

6. Reload PLUME and make consecutive runs using 0.5, 2, 3, 4 and 5 m/sec wind speeds. Results for Briggs plume rise are presented below:

Wind Speed (m/sec)	Plume Rise (m)
0.5	49.75
1.0	24.87
2.0	12.44
3.0	8.29
4.0	6.22
5.0	4.97

Significant plume rise from this momentum dominated jet can be achieved during calm and low wind conditions. High winds quickly bend the plume over and reduce the effective plume height.

7. Reload PLUME and repeat the runs at U = 1 and 5 m/sec, but increase VEL to 45 m/sec. Results for the Briggs formulas are

Wind Speed (m/sec)	Plume Rise (m)
1.0	55.97
5.0	11.19

Increasing the velocity of the discharge jet from 20 to 45 m/sec increases the plume rise by 55.97 − 24.87 = 31.1 m with a wind speed of 1 m/sec. However, the increase is only 11.19 − 4.97 = 6.22 m with a 5 m/sec wind speed.

Example 17.2

A static pile composting system is to be enclosed within a building. Building ventilation will be provided by 4.0-m diameter, axial flow fans mounted in the roof. Fan discharge velocity is 10 m/sec and the roof line is 20 m high. Stack discharge temperature is 20°C. Determine the plume rise and effective plume height in a 1 m/sec wind, with 20°C ambient temperature, 1013 mbar pressure, class D stability and PG dispersion coefficients.

Solution

1. Load PLUME and follow the menu to enter the following data:

MDY	=	(enter current value)
IDEN	=	EXAMPLE 17.2
VEL	=	10
DIA	=	4.0
H	=	20
TSTACKC	=	20
U	=	1
TAMBC	=	20
MBAR	=	1013
STABCLAS	=	D
DCSTACK	=	R

2. Results for Briggs formulas are summarized as

Plume rise	=	108.15 m
Effective plume height	=	128.15 m
(Plume rise is governed by momentum)		

3. The large diameter jet greatly increases the plume rise compared to results for the smaller jet in Example 17.1. Even though jet velocity is reduced from 20 to 10 m/sec, the larger jet diameter causes the plume to rise 108.15/24.87 = 4.3 times higher in the 1 m/sec wind. For this reason, large diameter, vertical jets are an effective method of enhancing the dispersion of building ventilation air.

Example 17.3

A reactor compost facility collects and treats its process gases to an odor concentration of 150 ou. 30 m^3/sec of process gas is then discharged from a 10 m stack at an average 35°C. Stack diameter is 1.95 m with a discharge velocity of 10 m/sec. Receiving and windrow curing areas are outside and can be modeled as an area source 100 m long by 100 m wide. SOER is estimated at 5 m^3/min-m^2 during the daytime when material is being worked and 2 m^3/min-m^2 at night when material is static. Evaluate the downwind odor profile during daytime conditions with a wind speed of 2 m/sec, clear skies, and moderate solar radiation.

Solution

1. From Table 17.1 a class B stability is applicable. Rural (PG) dispersion coefficients will be used for both the stack and area sources.

2. Load the program DISPERSE and follow the menu to enter the following data:

MDY	=	(enter current value)
IDEN	=	EXAMPLE 17.3
CSTACK	=	150
VEL	=	10
DIA	=	1.95
H	=	10
TSTACKC	=	35
SOER	=	5.0
LENGTH	=	100
WIDE	=	100
BARRIER WALL	=	N
U	=	1
TAMBC	=	25

STABCLAS	=	D
DCSTACK	=	R
DCAREA	=	R
PT	=	0.20
XSTOP1	=	1.0 ·
DELX1	=	0.1
XSTOP2	=	4.0
DELX2	=	0.25

3. A summary of the program output is as follows:

Plume rise	=	29.25 m (momentum dominated)
Eff. stack height	=	39.25 m
V_{xy} for area source	=	0.123 km
V_{xz} for area source	=	0.10 km
Stack mass odor emission	=	4480 m³/sec
Area mass odor emission	=	833 m³/sec

Distance	Stack Plume (10-sec conc.)	Area Plume (10-sec conc.)	Sum
0.00	0.006	1.216	· 1.222
0.10	0.262	0.372	0.634
0.20	0.453	0.179	0.632
0.40	0.322	0.068	0.390
0.60	0.190	0.035	0.226
0.80	0.120	0.022	0.142
1.00	0.082	0.014	0.096
1.50	0.039	0.007	0.046
2.00	0.023	0.004	0.027
3.00	0.010	0.002	0.012
4.00	0.006	0.001	0.007

Example 17.4

Assuming the source conditions per Example 17.3, evaluate the downwind odor concentrations under nighttime meteorological conditions with a 1 m/sec wind speed on a clear night with 10°C ambient temperature.

Solution

1. Assume class F stability in the atmospheric sublayer under the clear and calm conditions described above. Use rural (PG) dispersion coefficients for the stack and Slade coefficients for the area source.

2. Load the program DISPERSE and follow the menu to enter the following data:

MDY	=	(enter current value)
IDEN	=	EXAMPLE 17.4
CSTACK	=	150
VEL	=	10
DIA	=	1.95

H	=	10
TSTACKC	=	35
SOER	=	2.0
LENGTH	=	100
WIDE	=	100
BARRIER WALL	=	N
U	=	1
TAMBC	=	10
STABCLAS	=	F
DCSTACK	=	R
DCAREA	=	S
PT	=	0.20
XSTOP1	=	1.0
DELX1	=	0.1
XSTOP2	=	4.0
DELX2	=	0.25

3. A summary of the program output is as follows:

Plume rise	=	47.73	m (buoyancy dominated)
Eff. stack height	=	57.73	m
V_{xy} for area source	=	2.797	km
V_{xz} for area source	=	0.10	km
Stack mass odor emission	=	4480	m^3/sec
Area mass odor emission	=	333	m^3/sec

Distance	Stack Plume (10-sec conc.)	Area Plume (10-sec conc.)	Sum
0.00	0.002	5.473	5.475
0.10	0.003	3.475	3.479
0.20	0.005	2.733	2.638
0.40	0.011	1.820	1.831
0.60	0.024	1.405	1.428
0.80	0.041	1.145	1.187
1.00	0.062	0.966	1.028
1.50	0.113	0.688	0.801
2.00	0.155	0.529	0.683
3.00	0.186	0.352	0.538
4.00	0.184	0.258	0.442

Downwind odor concentrations for the unstable daytime conditions of Example 17.3 are presented in Figure 17.21. Odor emission rate from the stack is about 85% of the total mass emission. Despite the higher emission rate the stack contributes essentially no odor at the downwind edge of the area source. This is because the elevated plume has not yet reached groundlevel. The area source contributes about 1 ou at the downwind edge, but disperses rapidly in the class B stability. By 0.1 km downwind the area and point sources contribute equally to the total odor. At 0.2 km the stack contributes about 0.45 ou out of a total 0.63 ou.

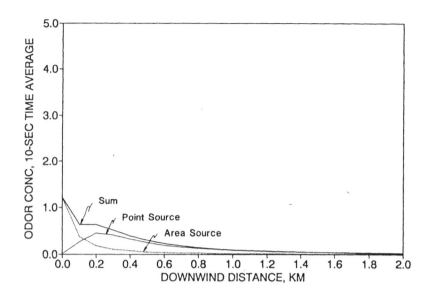

Figure 17.21. Semi-instantaneous, groundlevel odor concentrations downwind of a compost facility composed of area and point sources. All conditions are per Example 17.3. Meteorology is typical of unstable conditions on a clear day. Rural (PG) dispersion coefficients are used for both the area and point sources.

The stack remains the major groundlevel odor source at all further distances downwind. This situation is typical of many composting facilities during unstable meteorological conditions. Odors from the area source are significant at short distances but dissipate rapidly. The elevated plume reaches groundlevel at a relatively short distance because of the wide dispersion angle and plume looping.

Odor concentrations for the stable nighttime conditions of Example 17.4 are presented in Figure 17.22. Odor emission rate from the stack is about 93% of the total mass emission. Nevertheless, stack contributions to groundlevel odor are minor because the plume remains aloft. Because of the small dispersion angle associated with Class F stability, the point of maximum groundlevel concentration from the stack occurs about 3.5 km downwind. Area source emissions are reduced to 40% of the daytime rate. Despite the reduction in emission rate, the low dispersion rate causes dramatically higher downwind concentrations from the area source compared to the daytime case of Figure 17.21. Odors are over 5 ou at the source edge and remain above 1 ou for 1 km downwind. Again, this situation is typical of many composting facilities during very stable, low wind speed conditions. Even though the area source represents only 7% of the total odor emission, most of the groundlevel odor results from this source. High groundlevel odors persist for a considerable distance downwind because of the slow dispersion.

The program DISPERSE was used to solve the same conditions as Example 17.4 with the exception that rural (PG) dispersion coefficients were used for both the point and area sources. Results are shown in Figure 17.23 and are very similar to those of Figure 17.22. In general, semi-instantaneous concentrations predicted using the Slade coefficients are about equal to those predicted using the PG coefficients near the source. At a distance of 1 km the Slade coefficients predict about twice the odor concentration.

The effect of adding a 4.0-m high barrier wall to the downwind edge of the source is shown in Figure 17.24. All other conditions remain the same as for Example 17.4 as shown in Figure

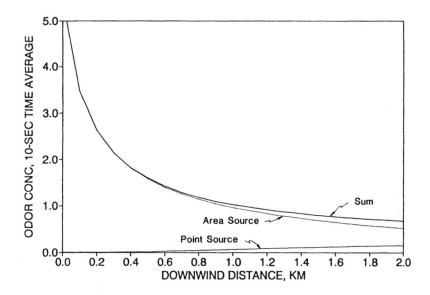

Figure 17.22. Semi-instantaneous, groundlevel odor concentrations downwind of a compost facility composed of area and point sources. All conditions are per Example 17.4. Meteorology is typical of stable conditions on a clear night. Rural (PG) dispersion coefficients are used for the point source and Slade coefficients for the area source.

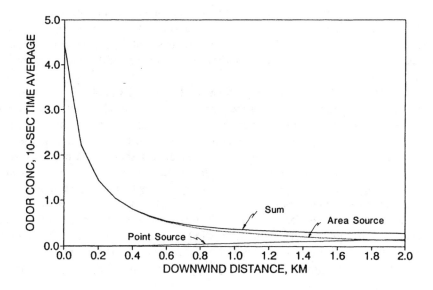

Figure 17.23. Semi-instantaneous, groundlevel odor concentrations downwind of a compost facility. All conditions are the same as Example 17.4 and Figure 17.22 except rural (PG) coefficients are used for both the point and area sources.

17.22. The barrier wall increases the vertical mixing and significantly lowers groundlevel concentrations near the source. Odor at the source edge is about 1 ou compared to over 5 ou in Figure 17.22. The effect becomes less pronounced with increasing distance from the source. By 1 km the odor concentration is about 60% of that predicted without the barrier wall. Barrier walls can enhance dispersion and reduce groundlevel odors immediately downwind of the source and up to a distance of perhaps 1 km. Barrier walls would be less effective at sites with large buffer zones.

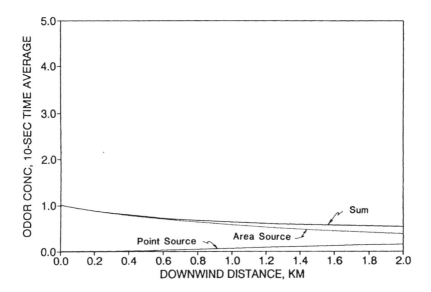

Figure 17.24. The effect of a barrier wall on downwind odor concentrations. All conditions are the same as Example 17.4 and Figure 17.22 except a 4.0-m barrier wall is added at the downwind edge of the area source.

The odor concentrations discussed above are generally low, but could be the cause of odor complaints in a sensitive situation. Remember that 1 ou (ED_{50}) is the odor concentration at which 50% of the panel members respond positively. A certain percentage of the population will be sensitive to lower odorant concentrations. Therefore, an odor criteria of 1 ou does not always assure that some receptors will not sense the odor. Some studies have used ED_5 concentrations to account for more sensitive receptors. A factor of safety of 30 applied to calculated ground-level concentrations has also been suggested to assure that threshold concentrations are not exceeded.[35]

Another limitation of atmospheric dilution modeling should be noted. When odorous air is diluted with odor-free air, the perceived odor may decrease less rapidly than the concentration of odorant molecules. In other words, dilution of an odorant can yield less than a proportional reduction in odor intensity. For example, a 10-fold reduction in the concentration of amyl butyrate is needed to reduce its perceived odor intensity by half. On the other hand, such factors are accounted for, at least in part, if the odor unit is the basis of measurement since it represents the dilutions required to reach the threshold concentration.

These example problems demonstrate that engineering can be applied to enhance the dispersion of odorous gases. Techniques are available to improve dispersion from elevated stack sources and groundlevel area sources. Information from a simple weather station can be used to warn of worst-case meteorological conditions. The operator can then initiate actions to (1) curtail more odorous operations to reduce emission rates, (2) improve odor treatment, (3) enhance the dispersion from stacks by forcing additional dilution, (4) reduce or eliminate unnecessary area sources, or (5) enhance groundlevel dispersion by turning on wind machines.

SUMMARY

Atmospheric dispersion is an important element of odor management at composting facilities because no amount of design, operation, or treatment will reduce odor emissions to zero. The atmosphere varies greatly in its ability to disperse odors. Rapid dispersion and unstable conditions generally occur on clear days when the incoming solar radiation warms the earth's surface. Slow dispersion and stable conditions can occur on clear nights when the ground surface cools. Very stable conditions can occur in the atmospheric sublayer near the ground. Air movement under such stable conditions is often termed "puff" transport.

With large area sources, the highest concentrations occur under stable, puff transport conditions. The nearest downwind receptor will experience the highest concentration because odors are released at groundlevel. Puff transport conditions are characteristically observed during evening hours and can transport groundlevel odor for considerable distances with only a slow attenuation in concentration. With elevated point sources the highest instantaneous concentrations occur under unstable conditions because the plume is rapidly dispersed and brought back to groundlevel.

Design approaches are available to enhance atmospheric dispersion. Techniques to enhance dispersion include: (1) converting area sources to elevated point sources; (2) modifying stack parameters to increase the effective plume height; (3) providing forced dilution of elevated point sources; (4) using barrier walls and wind machines to increase dispersion from area sources; (5) timing of more odorous operations to periods of higher atmospheric turbulence; (6) using a weather station to provide early warning of the onset of worst-case meteorological conditions. An important point to remember is that the engineer can design systems to enhance the dispersion that nature provides and can give the operator tools by which to increase dispersion during worst-case atmospheric conditions.

The most widely used approach to the modeling of atmospheric transport and dispersion is the Gaussian dispersion model. Concentrations in the horizontal and vertical dimensions of the plume are assumed to follow a normal or Gaussian distribution when averaged over a time period of several minutes to hours. Both area and point sources can be modeled using the Gaussian approach. The program DISPERSE was developed to model downwind concentrations from a point source located within an area source. Many composting facilities can be modeled with these assumptions.

For compost facilities that collect, treat, and discharge the process gas from an elevated stack, the stack emission rate often represents 80 to 90% of the total odor emission rate. With unstable atmospheres, odors from the area sources dissipate rapidly. The stack emission also disperses rapidly, bringing the plume back to groundlevel. This can cause high instantaneous concentrations near the source. In such a case, the highest groundlevel odors are often contributed by the stack discharge. With stable atmospheres, odors from the area source dissipate more slowly and can result in high concentrations for significant distances downwind. Barrier walls can increase the vertical mixing and reduce groundlevel concentration near the source under stable conditions. The stack discharge will remain aloft for greater distances under stable conditions. Under these conditions, the highest groundlevel odors are often contributed by the area source even though the mass emission rate is less than the point source.

REFERENCES

1. Pasquill, F. "The Estimation of the Dispersion of Windborne Material," *Meteorol. Mag.* 90(1063):33–49 (February 1961).
2. Pasquill, F. and Smith, F. B. *Atmospheric Diffusion,* 3rd ed. (Chichester, England: Ellis Horwood Limited, 1983).
3. Turner, B. D. *Workbook of Atmospheric Dispersion Estimates,* Office of Air Programs Publication No. AP–26, U.S. EPA (1970).

4. Smith, F. B. "A Scheme for Estimating the Vertical Dispersion of a Plume from a Source near Ground-Level," in *Turbulence and Diffusion Note No. 40* (British Meteorological Office, 1973).

5. Schultze, R. H. *Notes on Dispersion Modeling* (Dallas, TX: Trinity Consultants, 1987).

6. Dittenhoefer, A. C. and Menne, M. L. "Evaluation of the U.S. EPA SRDT and Net Radiation-Based Stability Classification Systems," in *Proceedings of the 85th Annual Air & Waste Management Conference*, Kansas City, MO, Paper No. 92–101.09 (June 1992).

7. Wilson, G. E., Huang, J. Y., and Schroepfer, T. W., "Atmospheric Sublayer Transport and Odor Control," ASCE Preprint 3493 (1979).

8. Cha, S. and Jann, P. "Atmospheric Dispersion and Odor Modeling," in *Odor Control Manual* (Alexandria, VA: Water Environment Federation, in press).

9. Slade, D., Ed. *Meteorology and Atomic Energy,* TID-24190, (Oak Ridge, TN: U.S. Atomic Energy Commission, 1968).

10. Fick, A. "Uber Diffusion," *Ann. Physik. Chem.* 94:59–86 (1855).

11. *Industrial Source (ISC) Dispersion Model User's Guide,* 2nd ed., Vol. 1, EPA-450/4-86-005a (June 1986).

12. Gifford, F. A. "Relative Atmospheric Diffusion of Smoke Puffs," *J. Meteorol.* 14:410 (1957).

13. McElroy, J. L. and Pooler, F. *St. Louis Dispersion Study, Volume I — Instrumentation Procedures and Data Tabulations,* Pub. No. APTD-68-12 and *Volume II — Analysis,* Pub. No. AP-53 (Research Triangle Park, NC: U.S. Department of Health, Education, and Welfare, 1968).

14. Hino, M. "Maximum Ground-level Concentration and Sampling Time," in *Atmospheric Environment*, Vol. 2 (London: Pergamon Press, 1968), pp. 149–165.

15. Duffee, R. A., O'Brien, M. A., and Ostojic, N., "Odor Modeling — Why and How," in *TR-18 Recent Developments and Current Practices in Odor Regulations, Controls and Technology* (Pittsburgh, PA: Air & Waste Management Association, 1991).

16. Briggs, G. A. "Plume Rise Predictions," in *Lectures on Air Pollution and Environmental Impact Statements* (Boston: American Meteorological Society, 1975).

17. Holland, J. Z. "A Meteorological Survey of the Oak Ridge Area: Final Report Covering the Period 1948–52," USAEC Report ORO-99, Weather Bureau, Oak Ridge, TN (1953).

18. "Recommended Guide for the Prediction of the Dispersion of Airborne Effluents," *Am. Soc. Mech. Eng.* (1979).

19. Liptak, B. G., Ed. *Environmental Engineers Handbook* (Radner, PA: Chilton Book Co., 1974).

20. Pasquill, F. *Atmospheric Dispersion Parameters in Gaussian Plume Modeling. Part II. Possible Requirements for Change in the Turner Workbook Values,* EPA-600/4-76-030b (Research Triangle Park, NC: U.S. EPA, 1976).

21. "Draft Environmental Impact Report — Sewage Sludge Management Program Wastewater Solids Processing and Disposal," (Sacramento, CA: Sacramento Area Consultants, 1978).

22. Hilst, G. R. "The Dispersion of Stack Gases in Stable Atmospheres," *J. Air Pollut. Contr. Assoc.* 7:205–210 (1957).

23. Gifford, F. A. "Statistical Properties of a Fluctuating Plume Dispersion Model," in *Advances in Geophysics,* Vol. 6, Frenkiel, F. N. and Sheppard, T. A., Eds. (New York: Academic Press, 1959).

24. Hogstrom, U. "A Method for Predicting Odour Frequencies from a Point Source," *Atmos. Environ.* 6:103–121 (1972).

25. Hogstrom, U. "An Experimental Study on Atmospheric Diffusion," *Tellus* 16:205–251 (1964).

26. Hogstrom, U. "A Statistical Approach to the Air Pollution Problem of Chimney Emission," *Atmos. Environ.* 2:251–271, (1968).

27. Murray, D. R., Cha, S., and Bowne, N., "Use of a Fluctuating Plume Puff Model for Prediction of the Impact of Odorous Emissions," in *Proceedings of the 71st Annual APCA Meeting*, Houston, TX, Paper No. 78-68.6 (June 1978).

28. Bowne, N. E. "Diffusion Rates," *J. Air Pollut. Contr. Assoc.* 24(9) (September 1974).

29. Murray, D. R. Manager of Research and Development, TRC Environmental Consultants, personal communication (1990).

30. Donovan, J. F., Kowalczyk, J. S., and Rafferty, S. D. "Shutdown! — A Case Study of Odor Problems at the Westborough, MA Composting Facility," in *Proceedings of the Specialty Conference Series*, Water Environment Fed. No. TT041, Portland, OR (July 1992).

31. Schulman, L. L. and Scire, J. S. "Building Downwash Screening Modeling for the Downwind Recirculation Cavity," in *Proceedings of the 85th Annual Air & Waste Management Conference*, Kansas City, MO, Paper No. 92-100.03 (June 1992).
32. Hosker, R. P. "Flow and Diffusion near Obstacles," in *Atmospheric Science and Power Production*, DOE/TIC-27601 (Washington, DC: U.S. Department of Energy, 1984).
33. Kido, W. Plant Manager, Sacramento Sewage Treatment Authority, personal communications (1992).
34. Redner, J., Wilson, G. E., Schroepfer, T. W., and Huang, J. Y. "Control of Composting Odors," in *Proceedings of the National Conference on Design of Municipal Sludge Compost Facilities* (Rockville, MD: Information Transfer (HMCRI), 1978).
35. *Odor Control for Wastewater Facilities, Manual of Practice No. 22* (Washington, DC: Water Environment Federation, 1979).

CHAPTER 18

Odor Management III — Completing the Odor Picture

INTRODUCTION

There are a number of aspects to the development and maintenance of a complete Odor Management Program. Methods of quantifying and treating odor were discussed in Chapter 16. Atmospheric meteorology, dispersion, and methods to enhance dispersion were studied in Chapter 17. This chapter completes the trilogy on odor management. The importance of establishing ambient odor standards or target odor objectives to guide design decisions is first discussed. This is followed by a series of Theorems or "truisms" about odor based on this author's experience in the field. Finally, the elements needed for a successful Odor Management Program are presented.

AMBIENT ODOR STANDARDS

Qualitative Standards

The design of new facilities, or the retrofit of additional odor controls to existing ones, requires numerous decisions balancing economics with the degree of odor control. To make such decisions it is helpful, if not necessary, to establish ambient odor standards or target odor objectives. This is not an easy task. Politicians and regulators want to say that "there will never be any odor." Plant managers might resort to "there will be no discernable odor beyond the plant boundary and no odor complaints." These all sound good, but they are difficult to use for guiding and making design decisions. Agencies have also had problems with terms like acceptable, permissible, allowable, or tolerable odor level. They imply a sense of acceptance of an undesirable situation.

Many regulatory agencies have resorted to qualitative measures of odor nuisance. For example, the New Jersey Administrative Code defines air pollution as "the presence in the outdoor atmosphere of one or more air contaminants in such quantities and duration as...would unreasonably interfere with the enjoyment of life or property." The courts have ruled that odor is an air pollutant under this definition. However, such a definition is of little help in guiding facility design.

Despite their shortcomings, qualitative standards are frequently used to make decisions on odor management. The most common and typical situation is where a facility is experiencing odor problems as evidenced by community complaints. Actions are then initiated with the goal of stopping the odor complaints or reducing them to a politically acceptable frequency (which may be zero). The measure of success becomes the reduction and hopefully cessation of the neighborhood complaints. A numerical odor standard for the ambient atmosphere in the community may never be established in such a case. This is an acceptable approach to the problem, but it is also a "trial and error" procedure that may involve a series of actions until the correct combination is reached. Establishing a numerical odor standard can be a tool to guide the decisions and avoid some of the "errors" of a trial and error approach.

Quantitative Standards

A theorem is presented later which states that "You can stop all of the odor some of the time, but you can't stop all of the odor all of the time." This is a recognition that nature will periodically impose such severe meteorological conditions that noticeable odors may occur despite our best efforts to avoid them. If the risk of odor cannot be reduced to zero, then an "acceptable odor risk" must be established.

A number of facilities have recognized the statistical nature of odor standards. The City of Sacramento established a fenceline odor objective of 2 ED_{50} with an occurrence no more than 3.3 days/yr and 5 ED_{50} no more than 0.5 days/yr.[1] 5 ED_{50} was the estimated threshold complaint level and 10 ED_{50} the definite complaint level. This means that predicted odor concentrations at the plant boundary would be below the threshold complaint level 99.8% of the time at all receptors beyond the plant boundary. A puff transport model was used in these studies. Dispersion coefficients for the puff model were developed from tracer release studies conducted at the plant site.

A standard of 1 ED_5 under all meteorological conditions modeled by the Gaussian formulas was used at the Montgomery County Composting Facility.[2] The ISCST model was then used to determine the effect of different control strategies on meeting the odor target. A somewhat similar approach was used at Plattsburgh, New York. Citizen groups agreed that a maximum of two or three odor incidents per year would be "acceptable". This translated into an odor free environment for the community 99.2% of the time. A criteria was chosen such that the calculated odor concentration from the facility was less than 1 ED_{10} (or 10% of the butanol detection limit) under worst-case meteorology using the ISCST model.[3,4] In other words, only 1 person in 10 would detect an odor at the point of maximum concentration.

A somewhat different approach was used by Duffee et al.[5] in studies conducted for the City of Akron, Ohio, and the Delaware Reclamation Project, New Castle, Delaware.[6] Dose/response curves were developed from odor panel results using Steven's Law to determine the ED_{50} level at the receptor that would produce an odor intensity of 3.5 on the Butanol Scale. This was assumed to be the threshold complaint level and was determined to be 8 ED_{50} at Akron and 7 ED_{50} at the DRP. These values became the target ambient odor levels. A Gaussian, fluctuating plume model was used to estimate "instantaneous" downwind odor concentrations and their frequency of occurrence at various receptor locations. Control strategies were then evaluated to meet the target concentrations. Duffee et al.[7] reported on other studies which used a Butanol Scale intensity of 2.0 to preclude essentially "detectable" odor levels. An intensity of 2.0 is an odor so faint it would not be detected unless someone called attention to it. Odor concentrations ranging from 2 to 6 ED_{50} were determined to be equivalent to an intensity of 2.0 for sewage type odors.

Gruwell and Goldenberg[8] used a criterion of 5 ED_{50} for a period of 5 minutes for eliciting an odor complaint. A modified version of the ISCST computer model was used which incorporated algorithms for computing short averaging times.

The State of Connecticut has established an odor nuisance level of 7 ED_{50}. According to Forbes et al.,[9] empirical evidence suggests that odor levels of 3 ED_{50} or less result in few odor complaints. In the range between 3 and 7 ED_{50}, odor complaints could arise from the more odor-sensitive individuals on an infrequent basis. As odor level increases beyond 7 ED_{50}, the potential for odor complaints increases. Several composting facilities have used the 7 ED_{50} criterion as a measure of objectionable odor.[9-11] In all cases, fluctuating plume models were used to estimate peak instantaneous concentrations for comparison with the criteria. Therefore, the above criteria should be interpreted as peak, short-term concentrations.

Efforts have been underway in several European countries to formulate odor standards. van Harreveld[12] reported that the Netherlands is using the following provisional odor standards: (1) For existing installations, an hourly average concentration of 1 ou may not be exceeded for more than 2% of yearly hours at locations with domestic dwellings; (2) for new installations, an hourly average concentration of 1 ou may not be exceeded on more than 0.5% of the yearly hours if domestic dwellings are present; and (3) the isopleths must be calculated using a national dispersion model based on the Gaussian plume algorithm. These provisional standards have been used in a large number of cases. Standard practices for olfactometry are also being developed with the goal of reducing differences between olfactometry laboratories to within a factor of 2 to 3.

From the above, it can be concluded that a variety of criteria have been used with varying degrees of success. It is also apparent that an ambient odor objective cannot be established separate from the type of atmospheric dispersion model. A lower ambient standard is needed if the Gaussian ISCST model is used with its 1-h time averaging. On the other hand, a higher standard may be appropriate if the dispersion model predicts instantaneous concentrations. The ambient odor objective must also consider the community demographics and prior history of odor problems. The complaint threshold probably depends on the exposure frequency, sensitivity, surroundings, education, and economic status of the community. These factors are very difficult to quantify, but it has been well demonstrated that a sensitized and organized community will tolerate only minimal odors.

THEOREMS ON ODOR MANAGEMENT

The following sections of this chapter use an informal essay style and are based on a presentation originally given at the U.S. EPA "Workshop on Controlling Sludge Composting Odors" held in Washington, D.C. in 1990 and subsequently published in *BioCycle*.[13] I have tried to conserve the original flavor of the oral presentation into this essay, for better or worse. Notice that I refer to this as an essay. These are my thoughts on the subject of odor management, thoughts developed from many years as a practicing engineer.

Composting of sludge, yard waste, manure, food waste, refuse and other substrates has become a very popular management option. The process makes use of nature's own microbes and produces a useful end product. In short, it accomplishes most of the things that environmentally minded folks like to see. Its one drawback, one that is a thorn in many sides, is its potential for odor generation. I think the general opinion of most practitioners is that odor is one of the biggest problems facing composting today. The story presented in Chapters 16, 17, and 18 is that odor can be managed if all the elements of odor management are brought to bear on the problem.

My first thought for presenting this subject matter was to list a series of truisms (i.e., an undoubted or self-evident truth) about odor. However, I quickly realized that folks in the composting industry are a diverse lot and agreement on something as elusive as odor is asking too much. So I dropped "truisms" in favor of "theorems". You may disagree with them, but they are still my theorems.

THEOREM 1 — MOST COMPOSTING SUBSTRATES SMELL

No argument here, I hope. All of the substrates which enter a composting process must be viewed as potential sources of odorous molecules. This includes sludge, sawdust, yard wastes, wood chips, refuse, food wastes, and all the other substrates we may throw into the starting mix.

Somewhere in the history of composting we got the idea that, if left alone, Mother Nature would be odor free. I don't know how this idea arose, but I assure you that it's not true, particularly with most composting substrates.

THEOREM 2 — MOTHER NATURE NEVER CLAIMED TO BE ODOR FREE

On many occasions I have heard speakers state with great pomp, "if composting is conducted properly there will be no odors." These speakers are never plant operators! Such statements have been implicated as a leading cause of anxiety complex among operators. Let me reassure the operators reading this book that I have never seen data to support such a claim. The starting substrates contain odorous compounds and more are formed as intermediates during the breakdown of complex substrates. Yes, this includes aerobic metabolism.

Boiling points and vapor pressures for some of Mother Natures's favorite odor compounds were presented in Table 16.1. Ammonia and H_2S boil at very low temperatures and will never stay in solution unless converted by pH to an ionized form. Many of the organic sulfides, such as dimethyl sulfide, boil at near ambient temperatures. Acetic acid, a favorite intermediate of aerobic microbes, has a vapor pressure of 100 mm Hg at 63°C. No wonder vinegar smells even at room temperature! Higher molecular weight compounds can also have significant vapor pressures. Terpenes, represented by eucalyptus and lemon oil, have about a 10 mm vapor pressure at 54°C. Again, this is why we can smell these fragrances, even at ambient temperatures.

THEOREM 3 — THERMOPHILIC COMPOSTING ACTS LIKE A HEAT DISTILLATION PROCESS

If you want proof for Theorem 3, I offer the observation that hot sludge or garbage smells more than cold sludge or garbage. New terms, like "low boilers" and "high boilers", are coming into use now that we recognize the physics of the problem. Vapor phase concentration of a compound is proportional to its vapor pressure and the latter increases with temperature. Many odorous compounds have significant vapor pressures and you can expect to see them in the process gases. Therefore…

THEOREM 4 — SOME RELEASE OF ODOROUS COMPOUNDS IS INEVITABLE DURING COMPOSTING

and

THEOREM 5 — WHAT SMELLS OK TO YOU IS PROBABLY AN ODOR TO SOMEONE ELSE

None of what has been said so far depends on the type of composting system. The only caveat is that anaerobic systems are not included. While there may be some differences between aerobic systems, their odor characteristics are all governed by the same laws of nature. Therefore...

THEOREM 6 — MOTHER NATURE DOESN'T MUCH CARE WHAT COMPOST SYSTEM YOU HAVE

and

THEOREM 7 — ITS NOT NICE TO FOOL MOTHER NATURE WITH A BAD DESIGN OR BAD OPERATION

Some odors will be produced even with good design and proper operation. However, a bad design or bad operation guarantees higher emission rates. One must also understand the energy balance and make sure that there is sufficient energy supply to meet the energy demands. If not, odors will usually increase because the process will be stressed and failure may be near. Therefore...

THEOREM 8 — YOU REALLY SHOULD KNOW SOMETHING ABOUT YOUR SUBSTRATES

The range of total solids, volatile solids, degradability, and rate constants should be known for each substrate entering the process.

Despite the best efforts of design engineers and the claims of equipment vendors you should always remember the following:

THEOREM 9 — ODOR TREATMENT IS NEVER 100%

The only possible exception to Theorem 9 is thermal oxidation which is capable of near complete odor destruction. Multistage wet scrubbing is generally capable of achieving outlet ED_{50}s in the range of 50 to 100. Lower levels are very difficult to achieve. For one thing, the contribution of the scrubbing chemicals to the outlet odor becomes significant at low outlet levels. The same is true of biofilters. Exhaust ED_{50}s in the range of 20 to 150 seem to be typical. Lower levels are not readily achievable because the biofilter matrix begins contributing to the outlet odor. There has been a subtle but persistent tendency for the design community to ignore Theorem 9. This leads to the following:

THEOREM 10 — MANY PAST DESIGNS DIDN'T RECOGNIZE THEOREM 9

Scrubbers and biofilters are often designed with no attention to dispersion of the treated gases. It is common to see scrubbers with short stub stacks, low outlet velocities, scrubbers located near large buildings with their plumes caught in the building downwash, rain caps on top of discharge stacks, and other examples of poor dispersion design. It's as though the designer assumed 100% deodorization. This should never be assumed.

The subject of atmospheric dispersion is complex and was discussed at some length in Chapter 17. There is one important theorem derived from my experiences with the atmosphere:

THEOREM 11 — THE WORST ODOR IS NEVER WHEN YOU'RE THERE

Another way of stating Theorem 11 is to say that odors usually result during worst case micrometeorological (micromet) conditions. The likelihood of your being at the site or in the surrounding community during the worst met conditions is small. Joe Manager might say "I was out there yesterday and didn't smell anything. We're not the cause of the odor." The fact is Joe probably just missed the odor.

With groundlevel sources, the worst met conditions usually occur in the evening, nighttime, and early morning hours. Dispersion is often limited during these periods because of strong, "micro inversions" which occur as a result of ground cooling. Groundlevel dispersion is usually highest during the day, after the sun has warmed the ground surface. Because most of us work the day shift, we would likely miss a worst case incident. Knowing this, Joe Manager should say "I was out there yesterday and didn't smell anything, but I'm going back tonight when I think the met conditions may be worse."

THEOREM 12 — MOTHER NATURE ALWAYS DISPERSES ODOR, BUT SOMETIMES SHE CAN USE A LITTLE HELP

This has something to do with entropy, maximum randomness, and other thermodynamic concepts. Whatever, I don't think we have to worry about nature reconcentrating odor on us. At the same time, we need not just accept the dispersion that nature alone supplies. We can design systems to improve dispersion, thereby giving nature a helping hand. The term "enhanced dispersion" has been used to describe such efforts. Even if you don't "enhance" the natural dispersion, the following is always true:

THEOREM 13 — KNOW YOUR MICROMET CONDITIONS

All major facilities should consider building an on-site micromet station to warn of worst-case atmospheric conditions. This will allow the operators to take corrective actions and, hopefully, avoid an odor incident.

Point sources have an advantage over groundlevel sources because the plume can be lofted above the ground by a combination of stack height and plume rise. This is particularly effective during wind calms, a met condition which is usually hard on groundlevel sources. It is important to always remember the following:

THEOREM 14 — THERE MUST BE A RECEPTOR TO HAVE AN ODOR

and

THEOREM 15 — MOST RECEPTORS ARE AT GROUNDLEVEL

For point sources, such as exhaust stacks from scrubbers, dispersion can be enhanced by (1) increasing stack height, (2) increasing stack velocity to increase momentum rise, (3) providing reheat to increase thermal buoyancy, and (4) providing forced dilution with ambient air. The latter will also increase the plume diameter, which in turn will increase the effective plume rise.

Things to avoid with a point source discharge include (1) locating the plume within the zone of building or stack downwash, (2) low velocity discharges from the sides or roofs of buildings, such as ridge ventilators, (3) using rain caps on roof ventilators or the scrubber discharge stack, (4) low stack velocity, and (5) bad topography such as valleys. Avoiding bad topography is like avoiding the common cold, easy to say but hard to do. The topography is always "greener" in the next political jurisdiction.

For groundlevel sources, such as open windrows, static piles, or biofilters, dispersion can be enhanced by (1) providing adequate buffer, (2) using wind machines to maintain minimum air flow over the area source, and (3) using barrier walls to induce turbulence. I realize that providing buffer really isn't an example of enhanced dispersion. It's more like giving nature enough room to solve the problem herself.

If the above measures are not adequate, the groundlevel source can be enclosed and converted to an elevated source. By comparison with elevated sources, groundlevel sources are subject to the worst met conditions and lowest dispersion rates. Also, the nearest down-wind receptor will be the most effected. Therefore, ground-level sources, and their surrounding topography, must be carefully considered in any odor management plan.

THEOREM 16 — YOU CAN STOP ALL OF THE ODOR SOME OF THE TIME, BUT YOU CAN'T STOP ALL OF THE ODOR ALL OF THE TIME

Theorem 16 is a recognition that, after all the planning and design studies, after all attempts to reduce emission rates, after all the collection, treatment, and dispersion, nature will periodically impose such severe met conditions that odors may occur. If the risk of odor cannot be reduced to zero, then we must establish an "acceptable odor risk". Engineers may want to hide Theorem 16 from their politicians. Odor objectives vary from study to study. The point is not that they vary, but that they were established in the first place and provided a guide for evaluating alternative designs and solutions. Remember, every facility needs a target odor objective. Be the first on your block to have one.

Finally, my last theorem...

THEOREM 17 — DON'T DESPAIR, ODORS CAN BE MANAGED

Despite odor problems at some facilities, the future for composting is optimistic. The industry generally recognizes that odor compounds are likely to be released, a milestone of major significance. Recognition of the problem is the first step toward its solution. The science of odor treatment, particularly with wet scrubbers, biofilters, and activated sludge is advancing rapidly. More engineers and operators now speak about met conditions and dispersion as if they were amateur meteorologists. Finally, regulators and industry groups have moved with unusual leadership to help the industry by encouraging the spread of these new ideas to the composting community. Watching the industry mature as it gears up to solve current problems, it's hard not to be optimistic.

One concern is the apparent difficulty in transferring lessons learned by the sludge composters to other members of the composting community. For example, some recent refuse composting facilities in the U.S. have been implemented with essentially no provisions for odor control. These facilities are destined to repeat past mistakes already learned with sludge. Consultants and firms active in the sludge industry in the U.S. are generally not the same as those active with other substrates such as refuse. The flow of information from one group to another is not automatic. We all need to work on this.

THE ELEMENTS OF ODOR MANAGEMENT

Based on the above theorems, I offer the following elements of odor management:

ELEMENT 1 — DEVELOP AN ODOR MANAGEMENT PROGRAM AT THE EARLIEST STAGES OF PLANNING AND DESIGN

This may sound obvious, but there are numerous facilities that have ignored this first commandment of odor management. It's a little late to develop an effective program once odor complaints are received. Plan and design for odor management from the start.

ELEMENT 2 — OPERATE THE BEST YOU CAN TO REDUCE ODOR EMISSION RATES

Good operation is vital to odor management. Good operators can make a bad design work OK and a good design work great. Skilled operators, effective process control, good housekeeping, knowledge of the local micromet conditions, and a community involvement program are a powerful force.

ELEMENT 3 — CONTAIN, COLLECT, AND TREAT AS MUCH AS POSSIBLE
or
GO DIRECTLY TO DISPERSION

If your composting system allows collection of the process gases, then you should treat them to the extent possible before dispersing them to the extent possible. If you cannot contain the process gases, then you must rely entirely on atmospheric dispersion to dilute any emissions. There are no halfway measures between these two approaches.

Theorems 4 and 8 lead to one of the most important and often ignored elements of odor management:

ELEMENT 4 — DILUTION IS PART OF THE SOLUTION

Dilution is rarely the complete solution to any problem, particularly odors. However, it is equally true that dilution must be *part* of the solution.

Every management plan needs a goal or target. It's hard to achieve a goal if you don't know what the goal is. This leads to the last element in the odor plan:

ELEMENT 5 — A TARGET ODOR OBJECTIVE SHOULD BE ESTABLISHED

The industry has had a tough time with this one. Politicians want to say that "there will never be any odor." You can tell that politicians never took statistical thermo in school or they would know that "never" is very hard to achieve. Plant managers often resort to "there will be no discernable odor beyond the plant boundary." Sounds good, but no one really knows what it means. Another favorite is "there must be no odor complaints." Fortunately, a number of studies have developed target odor objectives which, when achieved by the facility, have successfully reduced odor complaints.

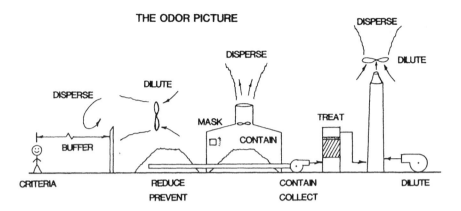

Figure 18.1. The tools of odor management.

A schematic showing the tools available for odor management is presented in Figure 18.1. This is the "odor picture". Embodied in the terminology are the elements necessary for an Odor Management Program: reduce, prevent, contain, collect, treat, dilute, disperse, buffer, odor objective, plan well, design well, and operate well. Good words by which to formulate your Odor Management Program. Good luck.

REFERENCES

1. Sacramento Area Consultants. *Sacramento Regional Wastewater Management Program, Vol. 1,* report to the County of Sacramento by Sacramento Area Consultants (September 1979).

2. Fisher, P. W., Wang, H., and Wilber, C. "Odor Abatement for a Sludge Composting Facility Through Dispersion Enhancement," in *Proceedings of the National Conference on Municipal Sewage Treatment Plant Sludge Management,* Boston, MA, (Silver Spring, MD: Hazardous Materials Control Research Institute, May 1987).

3. Task 2 Report: Cost-Effectiveness Analysis of Feasible Modification/Replacement Odor Control Alternatives, report to Clinton Co., NY, (Greeley and Hansen Engineers, ITT Research Institute, DuPont Engineering Test Center, September 1989).

4. Wilber, C., Greeley and Hansen Engineers. personal communication (August 1990).

5. Duffee, D. A., Ostojic, N., and O'Brien, M. A. "Akron Composting Facility Odor Study," report by Odor Science & Engineering, Hartford, CT for the City of Akron, OH (July 1988).

6. Canzano, P. S. and Aiani, K. J. "Optimum System for Odor Control," *BioCycle* 29(7) (August 1988).

7. Duffee, D. A., Ostojic, N., and O'Brien, M. A. Odor Impact Evaluation and Control Study for the Hyperion Wastewater Treatment Plant, report to the City of Los Angeles by Odor Science & Engineering, Hartford, CT (November 1989).

8. Gruwell, G. and Goldenberg, S. "Atmospheric Dispersion Model for Odors," in *Proceedings of the 85th Annual Air & Waste Management Conference,* Kansas City, MO, Paper No. 92-153.03 (June 1992).

9. Forbes, R. H., Mendenhall, T., and Ostojic, N. "Incorporating Odor Control into the Design of an In-Vessel Biosolids Composting Facility," in *Proceedings of the Specialty Conference Series,* Water Environment Fed. No. TT041, Portland, OR (July 1992).

10. Donovan, J. F., Kowalczyk, J. S., and Rafferty, S. D. "Shutdown! — A Case Study of Odor Problems at the Westborough, MA Composting Facility," in *Proceedings of the Specialty Conference Series,* Water Environment Fed. No. TT041, Portland, OR (July 1992).

11. Lang, M. E. and Jager, R. A. "Evaluation of Odor Control Technologies for Municipal Sludge Composting," in *Proceedings of the Specialty Conference Series*, Water Environment Fed. No. TT041, Portland, OR (July 1992).
12. van Harreveld, A. Ph. "Introduction and Related Practical Aspects of Odour Regulations in the Netherlands," in *TR-18 Recent Developments and Current Practices in Odor Regulations, Controls and Technology* (Pittsburgh, PA: Air & Waste Man. Assoc., 1991).
13. Haug, R. T. "An Essay on the Elements of Odor Management," *BioCycle* 31(10) (October 1990).

APPENDIX A

Notation

The following symbol notation has been adopted in this text. Where applicable, the units most commonly used are shown. A symbol may sometimes be used to define more than one variable. In such cases, the reader should refer to the appropriate section of the text to determine the context in which the symbol is used.

a	=	thermal diffusivity, cm^2/h
A	=	area or area perpendicular to the direction of heat or mass transfer, cm^2 or ft^2
A	=	area of a discharge orifice
A	=	area required for drying, ha
a	=	number of free enzyme adsorption sites per unit volume
ASH_m	=	ash component of the feed mixture, fraction of TS
$ASH_m\%$	=	mixture ash content, % of TS
ASH_p	=	ash component of the product, fraction of TS
$ASH_p\%$	=	product ash content, % of TS
A_v	=	available surface area per unit volume
B	=	biodegradable fraction of the volatile solids
B	=	constant relating barrier height to puff height
c	=	pollutant concentration at coordinates x, y, and z, g/m^3 or ou
C	=	odor concentration
C	=	weight percentage of carbon, ash-free basis
C	=	orifice discharge coefficient
c_b	=	average odor concentration at the downwind edge of a large area odor source, ou or ED_{50}
C_i	=	concentration of gas in the liquid phase at the gas/liquid interface
c_k	=	concentration estimate for the reference sampling time, t_k
C_l	=	concentration of gas in the liquid phase
c_p	=	heat capacity per unit mass at constant pressure, cal/g-°C
c_s	=	concentration estimate for the desired sampling time, t_s
c_v	=	heat capacity per unit mass at constant volume, cal/g-°C

657

d	=	inside stack diameter, m
D	=	diameter of a circular or equivalent circular section
D_g	=	diffusion coefficient in the gas phase, cm^2/sec
D_l	=	diffusion coefficient in the liquid phase, cm^2/sec
D_r	=	decimal reduction factor, time required to achieve a 10-fold reduction in cell population, T_{90}
d_{si}	=	initial depth of the wet substrate, cm
d_w	=	depth of water to be evaporated, cm
e	=	absolute surface roughness in a pipe or manifold
e	=	number of free enzymes per unit volume or the reaction mixture
ΔE	=	change in internal energy within a system
E	=	energy ratio, ratio of heat released to weight of water
E	=	evaporation rate, cm/day
E_a	=	activation energy, cal/mol or kcal/mol
EAR	=	excess air ratio
E_d	=	inactivation energy, cal/mol or kcal/mol
f	=	free airspace, ratio of gas volume to total volume
f	=	Darcy-Weisbach friction factor
F	=	association factor for solvent, equal to 2.6 for water
F_a	=	configurational factor to account for the relative position and geometry of two heat radiating bodies
F_b	=	buoyancy flux, m^4/sec^3
f_b	=	free airspace within the interstices of a bulking agent before wet substrate addition
F_e	=	emissivity factor to account for nonblack-body radiation
f_h	=	fraction of compost material in the high temperature zone of a compost pile
f_l	=	fraction of compost material in the low temperature zone of compost pile
f_m	=	free airspace within the interstices of a mixture of composting materials
F_m	=	momentum flux, m^4/sec^2
f_s	=	free airspace within the interstices of a wet substrate, usually assumed to be zero
ΔG	=	Gibb's free energy change between two states of a system, kcal/mol or cal/mol
g	=	acceleration of gravity, $9.8 \ m/sec^2$ or $32.2 \ ft/sec^2$
G	=	mass flowrate of a gas
G_f	=	specific gravity of the fixed or ash fraction of the total solids
G_m	=	specific gravity of mixture solids
$\Delta G°$	=	Gibb's free energy change measured under standard state conditions, kcal/mol or cal/mol
ΔG_R	=	reaction free energy change based on activities of the chemical constituents, cal/mol or kcal/mol
$\Delta G_R°$	=	reaction free energy change under standard state conditions, cal/mole or kcal/mol
G_s	=	specific gravity of the total solids
G_v	=	specific gravity of the volatile fraction of the total solids
ΔH	=	height of rise of a plume above the discharge stack, m
ΔH	=	enthalpy change between two states of a system, cal/mol or kcal/mol
H	=	effective plume height, m
H	=	weight percentage of hydrogen, ash-free basis

H	=	building height, m
h	=	point source stack height, m
H_a	=	higher heat value of the amendment, cal/g or Btu/lb of organics oxidized
H_b	=	puff height downwind of a barrier, m
h_b	=	barrier height, m
H_b	=	building height, m
H_e	=	system energy demand estimated as cal/g of water evaporated, or Btu/lb of water evaporated
ΔH_{fg}	=	enthalpy change from liquid to vapor at temperature T, cal/g or kcal/kg
H_{gep}	=	"good engineering practice" stack height, m
h_l	=	headloss in height of fluid
h_m	=	average puff height over an odor emission area, m
$\Delta H°$	=	enthalpy change measured under standard state conditions, cal/mol or kcal/mol
ΔH_R	=	reaction enthalpy change based on activities of the chemical constituents, cal/mol or kcal/mol
$\Delta H_R°$	=	reaction enthalpy change under standard state conditions, cal/mol or kcal/mol
H_R	=	height of the downwind recirculation cavity, m
HRT	=	single-pass mean residence time of the mixed materials including recycle
H_s	=	higher heat value of the substrate, cal/g or Btu/lb of organics oxidized
I	=	intensity of an odor sensation, usually expressed as equivalent concentration of 1-butanol
j	=	height exponent for flow through a compost material
k	=	thermal conductivity, cal/(h-cm²-°C/cm) or Btu/(h-ft²-°F/ft)
k	=	reaction rate constant, time^{-1}
k	=	maximum rate of solid substrate hydrolysis that occurs at high microbial concentration
K	=	coefficient related to permeability of a porous media
K	=	a coefficient in Steven's Law
k_a	=	fraction of amendment volatile solids degradable under composting conditions
k_b	=	fraction of bulking agent volatile solids degradable under composting conditions
k_d	=	rate constant, time^{-1} or g BVS/g BVS-day
k_d	=	thermal inactivation or death coefficient, time^{-1}
k_e	=	endogenous respiration coefficient, mass of microbes respired/mass of microbes-time
K_{eq}	=	equilibrium constant
k_m	=	maximum utilization coefficient, maximum rate of substrate utilization at high substrate concentration, mass substrate/mass microbes-day
k_m	=	fraction of mixture volatile solids degradable under composting conditions
k_s	=	average surface odor emission rate, SOER, m³/min-m²
k_s	=	fraction of substrate volatile solids degradable under composting conditions
K_s	=	half-velocity coefficient, also referred to as the Michaelis-Menten coefficient, mass/volume
K_x	=	half-velocity coefficient equal to the microbial concentration where ds/dt = k/2
L	=	length of area odor emission source, m

L	=	characteristic dimension of a building for plume downwash calculations
L	=	length of the flow path or length of a pipe section
L	=	length of a building in the alongwind direction
M	=	solvent molecular weight
m	=	mass, kg-mol or lb-mol
M_{bs}	=	volumetric mixing ratio, ratio of volume of bulking agent to volume of substrate
M_{mb}	=	volume ratio of mixed materials to bulking agent
n	=	velocity exponent for flow through compost material
n	=	number of substrates
n	=	viable cell population
n	=	an exponent in Steven's Law
N	=	number of pile turnings
n	=	porosity, ratio of void volume to total volume
n_f	=	final viable cell population
n_o	=	initial viable cell population
n_t	=	viable cell population after time, t
O	=	weight percentage of oxygen, ash-free basis
P	=	pressure or absolute pressure
p	=	pressure
P	=	precipitation rate, cm/day
p_a	=	absolute pressure or atmospheric pressure, mbar
P_c	=	percent of inorganic conditioning chemicals in sludge cake
P_g	=	partial pressure of gas in the gas phase
P_i	=	partial pressure of gas in the gas phase at the gas/liquid interface
$P_0(t)$	=	extinction probability, probability that all organisms are inactivated
P_s	=	static pressure
p_t	=	total pressure
p_t	=	power law constant for correcting concentrations to different time averaged periods
P_v	=	percent volatile solids in sludge cake
p_v	=	velocity pressure
p_w	=	wind profile exponent
q	=	heat flow into (+) or out of (–) a system
q	=	volumetric flowrate of recycle material
Q	=	heat of combustion or fuel value, Btu/lb or cal/g
Q	=	volumetric flowrate of material, excluding recycle
Q_{air}	=	air flowrate moving across an area odor source, m^3/min
Q_f	=	stack gas volumetric flowrate, m^3/sec
Q_o	=	odor emission rate, m^3/min
q_p	=	heat flow in a constant pressure system, cal or kcal
q_v	=	heat flow in a constant volume system, cal or kcal
R	=	radius of a spherical particle
R	=	degree of reduction of an organic compound
R	=	universal gas constant
R_d	=	dry weight recycle ratio, ratio of dry weight of compost product recycled to dry weight of substrate
R_w	=	wet weight recycle ratio, ratio of wet weight of compost product recycled to wet weight of substrate

ΔS	=	entropy change between two states of a system
S	=	weight percentage of sulfur, ash-free basis
S	=	concentration of the rate limiting substrate, mass/volume
S	=	mass emission rate of pollutant, g/sec or ou/sec
S_a	=	fractional solids content of amendment
S_b	=	fractional solids content of bulking agent
S_{bm}	=	fractional solids content of bulking agent in the substrate/bulking agent mixture after moisture absorption
S_m	=	fractional solids content of a composting mixture
S_{bm}^m	=	minimum fractional solids content of bulking agent achievable by absorption of moisture from substrate to bulking agent
S_{sm}^m	=	maximum fractional solids content of the substrate achievable by absorption of moisture from the substrate to bulking agent
S_p	=	fractional solids content of compost product
S_r	=	fractional solids content of recycled material
SRT	=	mean residence time of the feed solids, excluding recycle
S_s	=	fractional solids content of the composting substrate
S_{sf}	=	final solids content of the dried substrate, fraction
S_{si}	=	initial solids content of the wet substrate, fraction
S_{sm}	=	fractional solids content of substrate in a substrate/bulking agent mixture after moisture absorption
t	=	time
T	=	temperature
ΔT	=	temperature change
t_{90}	=	time required to achieve a 10-fold reduction in cell population
T_a	=	air temperature, °K
t_a	=	time required for air drying, days
T_a	=	absolute temperature, °K or °R
T_c	=	temperature, °C
t_k	=	reference sampling time
T_k	=	temperature, °K
T_s	=	stack gas temperature, °K
t_s	=	sampling time corresponding to the desired time averaged concentration
u	=	mean wind velocity effecting plume transport, m/sec
U	=	overall heat transfer coefficient which includes effects of both conductive and convective heat transfer, cal/(h-cm²-°C)
u_o	=	observed wind speed at the measuring height z_0, m/sec
u_z	=	wind speed at elevation z, m/sec
V	=	velocity of flow
V	=	volume of reactor, m³
V	=	volume of a thermodynamic system
v	=	rate of product formation in an enzyme catalyzed reaction
V_a	=	volatile solids content of amendment, fraction of dry solids
V_b	=	volatile solids content of bulking agent, fraction of dry solids
v_g	=	volume of gas phase
V_m	=	volatile solids content of mixture, fraction of dry solids
V_o	=	solute molar volume at normal boiling point, 25.6 cm³/g-mol for oxygen
V_p	=	volatile solids content of compost product, fraction of dry solids
V_r	=	volatile solids content of recycle, fraction of dry solids

v_s	=	volume of solids
v_s	=	exit velocity of stack gas, m/sec
V_s	=	volatile solids content of substrate, fraction of dry solids
VS_m	=	volatile solids component of the feed mixture, fraction of TS
$VS_m\%$	=	mixture volatile solids content, % of TS
VS_p	=	volatile solids component of the product, fraction of TS
$VS_p\%$	=	product volatile solids content, % of TS
v_t	=	total volume of solids, water and gas in a composting matrix
v_w	=	volume of water phase
VX_y	=	virtual point source distance for horizontal dispersion
VX_z	=	virtual point source distance for vertical dispersion
w	=	specific humidity, mass of water vapor per mass of dry gas
w	=	work done on (–) or by (+) a system
W	=	water to degradable organic ratio, weight of water to weight of degradable organic in a composting mixture
W	=	weight of water to be evaporated daily
W	=	width of an area odor source, m
W_1	=	weight of reactor outfeed
W_2	=	weight of reactor outfeed less the weight of recycle
w_{O2}	=	rate of oxygen consumption, mg O_2/g VS-h
W_s	=	weight of dry solids
W_w	=	weight of water
x	=	downwind distance measured from the source to the receptor, m
X	=	concentration of microbes, mass/volume
X	=	lignin content, percent of volatile solids
X_a	=	total wet weight of amendment added to mixture per day
X_b	=	total wet weight of bulking agent added to mixture per day
X_m	=	total wet weight of mixed material entering the compost process per day
X_p	=	total wet weight of compost produced per day
X_r	=	total wet weight of compost recycled per day
X_s	=	total wet weight of substrate added to mixture per day
X_w	=	weight of water added to the composting mixture per day
y	=	lateral distance from the centerline of a plume to the receptor, m
Y	=	expansion factor for compressible flow
Y_m	=	growth yield coefficient, mass of microbes/(mass of substrate)
z	=	height of the receptor, m
z	=	particle thickness
z_o	=	anemometer height, usually 10 meters
δ	=	gas specific weight upstream of an orifice
δ_b	=	unit bulk weight of bulking agent, wet weight per volume, g/cm^3
δ_g	=	thickness of laminar gas film as used in the two-film model of gas transfer
δ_l	=	thickness of laminar liquid film as used in the two-film model of gas transfer
δ_m	=	unit bulk weight of the mixed material to be composted, wet weight per volume, g/cm^3
δ_s	=	bulk weight of substrate, g/cm^3
$\delta_s(dry)$	=	unit dry weight, dry weight per unit volume, g/cm^3
δ_{si}	=	initial bulk weight of the substrate, g/cm^3
δ_w	=	bulk weight of water, g/cm^3

θ	=	detention time, days
θ	=	temperature coefficient for a chemical or biochemical reaction
u	=	net specific growth rate, g cells grown/g cells-day
u_m	=	maximum net specific growth rate, g cells grown/g cells-day
ρ	=	mass density, g/cm^3
σ	=	Stefan-Boltzmann constant, 4.87×10^{-8} kcal/(h-m^2-°K^4)
σ_y	=	standard deviation of plume concentration in the horizontal direction, m
σ_{ye}	=	effective horizontal dispersion coefficient
σ_z	=	standard deviation of plume concentration in the vertical direction, m
σ_{ze}	=	effective vertical dispersion coefficient

APPENDIX B

Nomenclature Used in Computer Models

COMPOSTING SIMULATION MODELS

A(J) = molar carbon content of substrate J

ASH(J,I) = ash solids for substrate J input to stage I, lb/day or kg/day

ASHR(J,I) = recycled ash solids for substrate J input to stage I, lb/day or kg/day

ASHRB(J,I) = ash recycled with the bulking agent for substrate J input to stage I, lb/day or kg/day

AVGVOL = average volume of input and output materials to stage I, ft³/day or m³/day

B(J) = molar hydrogen content of substrate J

BULKWT = bulk weight of the composting mixture or substrate, lb/ft³ or kg/m³

BULKWTI(I) = bulk weight of the input mixture to stage I, lb/ft³ or kg/m³

BULKWTO(I) = bulk weight of the output mixture from stage I, lb/ft³ or kg/m³

BULKWTXRB = bulk weight of recycled bulking agent, lb/ft³ or kg/m³

BVSFAST(J,I) = BVS input to stage I for the fast fraction of substrate J, lb/day or kg/day

BVSOFAST(J,I) = BVS output from stage I for the fast fraction of substrate J, lb/day or kg/day

BVSOSLOW(J,I) = BVS output from stage I for the slow fraction of substrate J, lb/day or kg/day

BVSRBFAST(J,I) = BVS recycled with the bulking agent and input to stage I for the fast fraction of substrate J, lb/day or kg/day

BVSRBOFAST(J,I) = BVS recycled with the bulking agent and output from stage I for the fast fraction of substrate J, lb/day or kg/day

BVSRBOSLOW(J,I) = BVS recycled with the bulking agent and output from stage I for the slow fraction of substrate J, lb/day or kg/day

BVSRBSLOW(J,I) = BVS recycled with the bulking agent and input to stage I for the slow fraction of substrate J, lb/day or kg/day

BVSRFAST(J,I)	=	recycled BVS input to stage I for the fast fraction of substrate J, lb/day or kg/day
BVSROFAST(J,I)	=	recycled BVS output from stage I for the fast fraction of substrate J, lb/day or kg/day
BVSROSLOW(J,I)	=	recycled BVS output from stage I for the slow fraction of substrate J, lb/day or kg/day
BVSRSLOW(J,I)	=	recycled BVS input to stage I for the slow fraction of substrate J, lb/day or kg/day
BVSSLOW(J,I)	=	BVS input to stage I for the slow fraction of substrate J, lb/day or kg/day
C(J)	=	molar oxygen content of substrate J
COEFF(J)	=	bulk weight constant for substrate J, fraction, set to 0 if SUBNAME = "SLUDGE"
COEFFM	=	bulk weight coefficient for the mixture of substrates and for any recycle
CPGAS	=	specific heat of dry gases, 0.24 Btu/lb-°F or cal/g-°C
CPSOL	=	specific heat of solids, 0.25 Btu/lb-°F or cal/g-°C
CPWAT	=	specific heat of water, 1.0 Btu/lb-°F or cal/g-°C
CPWATV	=	specific heat of water vapor, 0.44 Btu/lb-°F or cal/g-°C
D(J)	=	molar nitrogen content of substrate J
DAIRI(I)	=	input dry air to stage I, lb/day or kg/day
DBVS(J)	=	change in BVS for substrate J across stage I, lb/day or kg/day
DBVSFAST(J)	=	loss of fast BVS for feed substrate J, lb/day or kg/day
DBVSR(J)	=	change in recycled BVS for substrate J across stage I, lb/day or kg/day
DBVSRB(J)	=	change in BVS recycled with the bulking agent for substrate J across stage I, lb/day or kg/day
DBVSRBFAST(J)	=	loss of fast BVS for substrate J contained with the recycled bulking agent, lb/day or kg/day
DBVSRBSLOW(J)	=	loss of slow BVS for substrate J contained with the recycled bulking agent, lb/day or kg/day
DBVSRFAST(J)	=	loss of fast BVS for substrate J in the mixture recycle, lb/day or kg/day
DBVSRSLOW(J)	=	loss of slow BVS for substrate J in the mixture recycle, lb/day or kg/day
DBVSSLOW(J)	=	loss of slow BVS for feed substrate J, lb/day or kg/day
DELHFG	=	enthalpy change from liquid to vapor at temperature TNEW(I), Btu/lb or kcal/kg
DELTAH	=	absolute value of HTOTI minus HTOTO and divided by HTOTI
DGASFACTOR(J)	=	change in the weight of dry gas per unit weight of substrate J that is oxidized
DGASO(I)	=	output dry gases from stage I, lb/day or kg/day
DTEMP	=	temperature difference between successive iterations = TNEW(I) − THOLD(I), °F or °C
DWTIN(I)	=	dry weight of feed solids to stage I, lb/day or kg/day
DWTOUT(I)	=	dry weight of the output mixture from stage I, lb/day or kg/day
DWTR	=	dry weight of recycled solids, lb/day or kg/day
DWTRB	=	dry weight of solids in the recycled bulking agent, lb/day or kg/day

DWTS	=	total dry weight of all substrate solids, lb/day or kg/day
ELEV	=	site elevation, feet or meters above sea level
F1(I)	=	factor to adjust the rate constant for the effect of moisture content, fraction
F2(I)	=	factor to adjust the rate constant for the effect of free air space, fraction
FAS(I)	=	free air space in stage I, fraction
FASTFRAC(J)	=	fraction of BVS with faster rate constant for substrate J
FO2(I)	=	factor to adjust the rate constant for the effect of oxygen content, fraction
GAMMAM	=	specific gravity of the composting mixture
GF	=	specific gravity of fixed solids, assumed as 2.5
GS	=	specific gravity of the substrate solids
GV	=	specific gravity of volatile solids, assumed as 1.0
H(J)	=	higher heat of combustion of substrate J, Btu/lb or kcal/kg of organics oxidized
HDAIRI(I)	=	sensible heat of the input dry air to stage I, Btu/day or kcal/day
HDGASO(I)	=	sensible heat in the output dry gases from stage I, Btu/day or kcal/day
HLWVO(I)	=	latent heat of output water vapor from stage I minus latent heat of the input water vapor to stage I, Btu/day or kcal/day
HORG(J,I)	=	heat release by biological oxidation of substrate J in stage I, Btu/day or kcal/day
HORGR(J,I)	=	heat release by biological oxidation of substrate J contained in the recycle material in stage I, Btu/day or kcal/day
HORGRB(J,I)	=	heat release by biological oxidation of substrate J contained with the recycled bulking agent in stage I, Btu/day or kcal/day
HRBSI	=	sensible heat in recycled bulking agent, Btu/day or kcal/day
HRBWI	=	sensible heat in the water fraction of recycled bulking agent, Btu/day or kcal/day
HRSI	=	sensible heat in recycled solids, Btu/day or kcal/day
HRT(I)	=	hydraulic retention time in stage I, days
HRWI	=	sensible heat in the water fraction of recycled material, Btu/day or kcal/day
HSI(I)	=	sensible heat in the total solids feed to stage I, Btu/day or kcal/day
HSO(I)	=	sensible heat in the output solids from stage I, Btu/day or kcal/day
HSSI(J)	=	sensible heat in the feed solids for substrate J, Btu/day or kcal/day
HSWI(J)	=	sensible heat in the feed water for substrate J, Btu/day or kcal/day
HSWVI(I)	=	sensible heat of the input water vapor to stage I, Btu/day or kcal/day
HSWVO(I)	=	sensible heat in the output water vapor from stage I, Btu/day or kcal/day
HTOTI(I)	=	total input energy to stage I, Btu/day or kcal/day
HTOTO(I)	=	total output energy from stage I, Btu/day or kcal/day
HWATADD(I)	=	sensible heat of added water to stage I, Btu/day or kcal/day
HWI(I)	=	sensible heat in the water fraction of feed solids to stage I, Btu/day or kcal/day
HWO(I)	=	sensible heat in the output water contained in the solids from stage I, Btu/day or kcal/day

I	=	composting stage number in array
J	=	substrate number in array
K(J)	=	degradability coefficient for substrate J, fraction of volatile solids
ME	=	switch to indicate whether metric (M, m) or English (E, e) units are used
MECHLOSS	=	fractional loss of bulking agent during processing due to mechanical breakdown to a size less than the screen size
MIXDRYWTI	=	mixture dry weight input to stage I, lb/day or kg/day
MIXDRYWTO	=	mixture dry weight output from stage I, lb/day or kg/day
MOLWT(J)	=	molecular weight of substrate J
NBADRYWTI	=	dry weight of nonbulking agent substrates input to stage I, lb/day or kg/day
NBADRYWTO	=	dry weight of nonbulking agent substrates output from stage I, lb/day or kg/day
NBVS(J,I)	=	nonbiodegradable volatile solids for substrate J input to stage I, lb/day or kg/day
NBVSR(J,I)	=	recycled nonbiodegradable volatile solids for substrate J input to stage I, lb/day or kg/day
NBVSRB(J,I)	=	nonbiodegradable volatile solids recycled with the bulking agent for substrate J input to stage I, lb/day or kg/day
NETVOL	=	average of input and output volumes of substrate components to stage I, ft³/day or m³/day
NETVOLUMEI(I)	=	input volume of substrates to stage I, excluding any mixture recycle and bulking agent recycle, ft³/day or m³/day
NETVOLUMEO(I)	=	output volume of substrates from stage I, excluding any mixture recycle and bulking agent recycle, ft³/day or m³/day
PAIR	=	atmospheric pressure, mm Hg
PPMNH3(I)	=	volume ratio of ammonia in the exhaust gas from stage I expressed as parts per million by volume
PV(I)	=	actual water vapor pressure in the input air to stage I, mm Hg
PVO(I)	=	actual water vapor pressure in the exhaust gas from stage I, mm Hg
PVSO	=	saturation vapor pressure in the exhaust gas corresponding to the exit gas temperature, mm Hg
QAIR(I)	=	flowrate of air supply to stage I, actual cubic feet per minute (acfm) or actual cubic meters per minute (acmm)
QAIRMIN(I)	=	minimum air flowrate to stage I, acfm or acmm
RATE25F(J)	=	rate constant for the fast fraction of substrate J at 25°C, day⁻¹
RATE25S(J)	=	rate constant for the slow fraction of substrate J at 25°C, day⁻¹
RATEKF(J,I)	=	actual rate constant for the fast fraction of substrate J at conditions of temperature, moisture, free air space, and oxygen in stage I, day⁻¹
RATEKM20F(J)	=	rate constant for faster fraction of substrate J at 20°C, day⁻¹
RATEKM20S(J)	=	rate constant for slower fraction of substrate J at 20°C, day⁻¹
RATEKMF(J,I)	=	maximum rate constant for the fast fraction of substrate J at temperature TNEW(I) in stage I, day⁻¹
RATEKMS(J,I)	=	maximum rate constant for the slow fraction of substrate J at temperature TNEW(I) in stage I, day⁻¹

RATEKS(J,I)	=	actual rate constant for the slow fraction of substrate J at conditions of temperature, moisture, free air space, and oxygen in stage I, day^{-1}
RATEO2(I)	=	specific oxygen consumption rate for the combined mixture into stage I, mg O_2/kg VS-h
RATEO2SUB(J)	=	specific oxygen consumption rate for substrate J, mg O_2/kg VS-h
RHAIR(I)	=	relative humidity of air to stage I, fraction of saturation
S(J)	=	solids content of substrate J, fraction
SCRNEFF	=	screening efficiency defined as the ratio of the weight of substrate which separates from the bulking agent and passes through the screen divided by the weight of feed substrate to the screen
SLOWFRAC(J)	=	fraction of BVS with slower rate constant for substrate J
SM	=	fractional solids content of the mixture of substrates, recycle, and bulking agents input to stage 1
SMMIN	=	minimum fractional solids content of mixture into stage 1
SMOUT(I)	=	solids content of the mixture output from stage I, fraction
SMSET(I)	=	set point for solids content in stage I, set to 1.00 for no additional WATADD to stage I
SPECO2SUP	=	specific oxygen supply to the system, g O_2/g feed BVS
SRTBA(I)	=	SRT in stage I for the bulking agent, days
SRTDWT(I)	=	SRT for stage I estimated by dry weight calculation, days
SRTSUB(I)	=	SRT in stage I for the feed substrates excluding the bulking agent, days
SRTVOL(I)	=	SRT for stage I estimated by volume calculation, days
STAGENUM	=	number of composting stages, maximum of 50
STAGER	=	number of stages from which recycle is drawn, STAGER = 0 if no recycle is used
STAGESCRN	=	number of stage immediately before screening of bulking agent. STAGESCRN = 0 if no screening is used
STAGEVOL(I)	=	volume of stage I, ft^3 or m^3
SUBDRYWTI	=	for homogeneous systems, SUBDRYWTI is the dry weight of substrates input to stage I, lb/day or kg/day, excluding any mixture recycle; for heterogeneous systems, SUBDRYWTI is the dry weight of nonbulking agent substrates input to stage I, lb/day or kg/day, excluding any recycled substrate components
SUBDRYWTO	=	for homogeneous systems, SUBDRYWTO is the dry weight of substrates output from stage I, lb/day or kg/day, excluding any mixture recycle; for heterogeneous systems, SUBDRYWTO is the dry weight of nonbulking agent substrates output from stage I, lb/day or kg/day, excluding any recycled substrate components
SUBNAME(J)	=	descriptive name of substrate J, 10 characters maximum (sludge cake must be entered as "SLUDGE" for correct calculation of bulk weight)
SUBNUM	=	number of feed substrates, maximum of 10
SUBNUMXRB	=	wet weight of recycled bulking agent, lb/day or kg/day
T(J)	=	temperature of substrate J, °F or °C
TAIR(I)	=	temperature of air supply to stage I, °F or °C

TGASI(I)	=	sum of input dry air and water vapor to stage I, lb/day or kg/day
THOLD(I)	=	temperature in stage I for iteration k
TNEW(I)	=	temperature of the composting mixture in stage I, °F or °C
TNEWC	=	temperature of the composting mixture, °C
TOTHRT	=	total system HRT, days
TOTSRTBA	=	total system SRT for the bulking agent, days
TOTSRTDWT	=	total system SRT calculated by dry weight, days
TOTSRTSUB	=	total system SRT for the nonbulking substrates, days
TOTSRTVOL	=	total system SRT calculated by volume, days
TREF	=	reference temperature used in energy balance, 32°F or 0°C
TSET(I)	=	set point temperature in stage I, set to 200°F or 100°C to hold QAIR to input value
TWATADD	=	temperature of water added to any stage, °F or °C
V(J)	=	volatile solids content of substrate J, fraction of dry solids
VOLCO2	=	volume of carbon dioxide in the exhaust from a stage
VOLH2O	=	volume of water vapor in the exhaust from a stage
VOLN2	=	volume of nitrogen in the exhaust from a stage
VOLNH3	=	volume of ammonia in the exhaust from a stage
VOLO2	=	volume of oxygen in the exhaust from a stage
VOLPCO2(I)	=	volume percent of carbon dioxide in the exhaust gas from stage I
VOLPH2O(I)	=	volume percent of water vapor in the exhaust gas from stage I
VOLPN2(I)	=	volume percent of nitrogen in the exhaust gas from stage I
VOLPO2(I)	=	volume percent of oxygen in the exhaust gas from stage I
VOLRATIO	=	desired ratio of total bulking agent to the sum of other substrate volumes
VOLSUBNUMXRB	=	volume of recycled bulking agent
VOLTOT	=	total volume of exhaust gas from a stage
VOLUME(J)	=	volume of substrate J, ft³/day or m³/day
VOLUMEI(I)	=	volume of mixture input to stage I, ft³/day or m³/day
VOLUMEO(I)	=	volume of the output mixture from stage I, ft³/day or m³/day
VOLUMESUBS	=	volume of all substrates but not including any bulking agents
WATADD(I)	=	water addition to stage I, lb/day or kg/day
WATP(I)	=	water produced by biological oxidation in stage I, lb/day or kg/day
WATPFACTOR(J)	=	weight of water produced per unit weight of substrate J that is oxidized
WATR	=	weight of water contained in recycled solids, lb/day or kg/day
WATRB	=	weight of water contained in the recycled bulking agent, lb/day or kg/day
WATS	=	total weight of water contained in all substrate solids, lb/day or kg/day
WATSI(I)	=	total water in the solids input to stage I, lb/day or kg/day
WATSO(I)	=	total water in the solids output from stage I, lb/day or kg/day
WATVI(I)	=	input water vapor to stage I, lb/day or kg/day
WATVO(I)	=	output water vapor from stage I, lb/day or kg/day
WTCO2FACTOR(J)	=	weight of carbon dioxide produced per unit weight of substrate J that is oxidized
WTNH3	=	ammonia released minus ammonia synthesized, lb/day or kg/day
WTNH3FACTOR(J)	=	weight of ammonia produced per unit weight of substrate J that is oxidized

WTNH3REL	=	weight of ammonia released as a result of biological decomposition within a stage, lb/day or kg/day
WTNH3SYN	=	weight of ammonia synthesized to cell material within a stage, lb/day or kg/day
WTO2FACTOR(J)	=	weight of oxygen consumed per unit weight of substrate J that is oxidized
X(J)	=	total wet weight of substrate J, lb/day or kg/day
XR	=	recycle rate from stage STAGER, lb/day or kg/day
XRMAX	=	maximum recycle rate from stage STAGER, lb/day or kg/day
XRMIN	=	minimum recycle rate from stage STAGER, lb/day or kg/day

MANIFOLD HYDRAULICS MODEL

APIPE(I)	=	area of manifold in section I, ft^2
APORT(I)	=	area of port in section I, ft^2
BETA4	=	$(DPORT/DPIPE)^4$
C(I)	=	coefficient of discharge for ports in section I
DELP1	=	pressure drop across compost zone 1, lb/ft^2
DELP2	=	pressure drop across compost zone 2, lb/ft^2
DELP3	=	pressure drop across compost zone 3, lb/ft^2
DELTAP	=	static pressure difference between manifold and atmosphere, lb/ft^2
DELTAPORT	=	static pressure difference across orifice, lb/ft^2
DPIPE(I)	=	diameter of a circular manifold in section I, inch
DPORT(I)	=	diameter of circular ports in manifold section I, inch
E	=	absolute surface roughness of manifold, ft
ELEV	=	site elevation, ft above sea level
EOVERD(I)	=	relative roughness for manifold section I, ft/ft
F	=	Darcy-Weisbach friction factor
FLOWTYPE	=	type of flow in manifold, +1 = forced draft (pressure) and −1 = induced draft (suction)
GAMMAPIPE	=	specific weight of gas within the manifold, lb/ft^3
GAMMAPORT	=	specific weight of gas in the orifice, lb/ft^3
HL	=	friction loss along manifold, ft of gas
HMAN(I)	=	height of a rectangular manifold in section I, inch
HPILE1	=	height of the compost pile in zone 1 over a distribution manifold, ft
HPILE2	=	height of the compost pile in zone 2 over a distribution manifold, ft
HPILE3	=	height of the compost pile in zone 3 over a distribution manifold, ft
HYDRAD	=	hydraulic radius, cross-sectional area divided by wetted perimeter
J1	=	height exponent for zone 1 of a compost pile
J2	=	height exponent for zone 2 of a compost pile
J3	=	height exponent for zone 3 of a compost pile
K1	=	headloss constant in zone 1 of a compost pile, inch wc
K2	=	headloss constant in zone 2 of a compost pile, inch wc
K3	=	headloss constant in zone 3 of a compost pile, inch wc
LENGTH	=	distance from first port, ft
LPORT(I)	=	length of a slot type port in manifold section I, inch
MANTYPE	=	manifold type, 1 = circular, 2 = rectangular
MOLWT	=	molecular weight of the combined dry gas plus water vapor

MPIPE	=	mass flow in pipe, lb/sec
MPORT	=	mass flow in port, lb/sec
N1	=	velocity exponent for zone 1 of a compost pile
N2	=	velocity exponent for zone 2 of a compost pile
N3	=	velocity exponent for zone 3 of a compost pile
NR	=	Reynold's Number
NUMPORTS(I)	=	number of ports in section I of a manifold
NUMSECTION	=	number of dissimilar sections in an aeration manifold
PATM	=	absolute atmospheric pressure, psia
PDRYGAS	=	partial pressure of the dry gas, psi
PMAN1	=	assumed manifold pressure at first port for first iteration, inch wc (use + inch wc if flow is forced draft, use − inch wc if flow is induced draft)
PORTTYPE(I)	=	type of port in manifold section I (1 = circular, square edged; 2 = circular, rounded, or bell mouthed; 3 = slot, square edged)
PPIPE	=	absolute static pressure in manifold, lb/ft^2
PPORT	=	absolute static pressure outside manifold, lb/ft^2
PRINTK(I)	=	printout desired for every kth port in section I
PV	=	water vapor partial pressure in the gas, psi
QACTPIPE	=	flow in manifold, actual ft^3/sec
QACTPORT	=	discharge from port, actual ft^3/sec
QSET	=	desired flowrate from the manifold, standard cubic feet/sec (scfs)
QSTDPIPE	=	flow in manifold, scfs
QSTDPORT	=	discharge from port, scfs
RHGAS	=	gas relative humidity in a manifold, fraction of saturation
SPACING(I)	=	spacing of ports in manifold section I, inch/port
STATICP	=	manifold static pressure, inch wc
TEMP	=	gas temperature in distribution manifold, °F
TOTENERGY	=	total energy (pressure), lb/ft^2
VELOCITYP	=	velocity pressure, inch wc
VISCOS	=	gas viscosity, lbf-scc/ft^2
VPIPE	=	velocity in manifold, ft/sec
VPORT	=	velocity in port, ft/sec
WMAN(I)	=	width of a rectangular manifold in section I, inch
WPILE1	=	width of the compost pile in zone 1 over a distribution manifold, ft
WPILE2	=	width of the compost pile in zone 2 over a distribution manifold, ft
WPILE3	=	width of the compost pile in zone 3 over a distribution manifold, ft
WPORT(I)	=	width of a slot type port in manifold section I, inch
Y	=	expansion coefficient for ports

ATMOSPHERIC DISPERSION MODELS

CGLAREA	=	centerline, groundlevel odor concentration from an area source, ou or ED$_x$
CGLAREAI	=	instantaneous (10 sec), centerline, groundlevel odor concentration from an area source, ou or ED$_x$
CGLSTACK	=	centerline, groundlevel odor concentration from a stack discharge, ou or ED$_x$

CGLSTACKI	=	instantaneous (10 sec), centerline, groundlevel odor concentration from a stack discharge, ou or ED_x
CSTACK	=	stack odor concentration, ou or ED_x
DCAREA	=	dispersion coefficient for area source; R = rural (PG), U = urban (MP), S = Slade
DCSTACK	=	dispersion coefficient for stack discharge; R = rural (PG), U = urban (MP), S = Slade
DELHB	=	plume rise due to buoyancy, m
DELHM	=	plume rise due to momentum, m
DELX1	=	step size for first iterations, km
DELX2	=	step size for second iterations, km
DIA	=	inside stack diameter, m
DISTANCE	=	distance measured from the downwind edge of the area source, km
H	=	stack height above groundlevel, m
HB	=	height of a barrier wall, m
HEFF	=	effective plume height, stack height + height of plume rise, m
LENGTH	=	length of area source, m
MBAR	=	atmospheric pressure in millibars, usually 1013 mb at sea level
PT	=	power exponent to adjust the time averaging period, usually 0.2 to 0.5
SAREA	=	mass emission rate of odor from a groundlevel, area source, m^3/sec
SIGMAY	=	lateral dispersion coefficient, m
SIGMAZ	=	vertical dispersion coefficient, m
SOER	=	surface odor emission rate from an area source, m^3/min-m^2
SSTACK	=	mass emission rate of odor from a point source (stack) discharge, m^3/sec
STABCLAS	=	atmospheric stability class, A to F
TAMBC	=	ambient air temperature, °C
TSTACKC	=	stack discharge temperature, °C
U	=	wind speed at 10-m height, m/sec
USTACK	=	wind speed at stack height, m/sec
VEL	=	stack discharge velocity, m/sec
VIRTXY	=	virtual distance for lateral dispersion from an area source, km
VIRTXZ	=	virtual distance for vertical dispersion from an area source, km
WIDE	=	width of area source, m
X1	=	distance for determining the lateral dispersion coefficient, km
X2	=	distance for determining the vertical dispersion coefficient, km
XSTOP1	=	stopping distance from downwind edge of area source to end of first set of iterations, km
XSTOP2	=	stopping distance from downwind edge of area source to end of second set of iterations, km

APPENDIX C

Simulation Modeling — Results for Example 12.1

Note: Displayed in Appendices C through F are copies of computer readouts.
For purposes of clarity, certain lines have been deleted.

```
::::::::::::::::::::::::::::::::::::::::::::::::::::::::::::::::::::::::::::::::::::::::::::::::::::::::::::::::::::::::::::
::::::::::::::::::::::::::::::::::::::::::::::::::::::::::::::::::::::::::::::::::::::::::::::::::::::::::::::::::::::::::::


                              3-28-1991
                              DATA121


                        SIMULATION MODEL
          MULTI-STAGE COMPOSTING OF SUBSTRATE MIXTURES W/WO RECYCLE
              CONTINUOUS FEED, COMPLETE MIX STAGES IN SERIES
                    CROSSFLOW MODE OF AIR:SOLIDS FLOW
                            ENGLISH UNITS
                           LEVEL 3 BALANCE

                   PROGRAM FILENAME: CMSYS41A.BAS
                   DATA FILE NAME: DATA125


                     AUTHOR -- ROGER T. HAUG
                   ROGER T. HAUG ENGINEERS, INC.
                     TORRANCE, CALIFORNIA  USA


::::::::::::::::::::::::::::::::::::::::::::::::::::::::::::::::::::::::::::::::::::::::::::::::::::::::::::::::::::::::::::


              INITIAL PROBLEM CONDITIONS
```

	X LB/DAY	S	V	K	FAST FRACTION	SLOW FRACTION	FAST RATEKM20 PER DAY	SLOW RATEKM20 PER DAY	TEMP DEG F	H BTU/LB	BULK WT COEFF
SLUDGE	66666.70	0.3000	0.8000	0.6500	0.400	0.600	0.01500	0.00500	68.00	9500.	
MIXTURE MIN		0.4000									0.300
RECYCLE MIN	20000.00										0.300
RECYCLE MAX	160000.00										
WATER ADD									68.00		

	SUBSTRATE COMPOSITION			
	CARBON	HYDROGEN	OXYGEN	NITROGEN
SLUDGE	10	19	3	1

675

STAGE NUM	HRT DAYS	QAIR ACFM	TAIR DEG F	RHAIR %	WATADD LB/DAY	SMSET FRAC	TSET DEG F
1	10.00	4000.	68.00	50.	0.00	NONE	NONE

RECYCLE IS FROM THE OUTPUT OF STAGE   1

```
SITE ELEVATION            0.0 FEET
REFERENCE TEMPERATURE     32.00 DEGREES F
SPECIFIC HEAT OF SOLIDS   0.2500 BTU/LB-F
SPECIFIC HEAT OF WATER    1.0000 BTU/LB-F
SPECIFIC HEAT OF GASES    0.2400 BTU/LB-F
SPECIFIC HEAT OF VAPOR    0.4400 BTU/LB-F
SPECIFIC GRAVITY OF VOL   1.000
SPECIFIC GRAVITY OF ASH   2.500
```

```
INTEGRATION LIMITS
    SCRNCTRL%      =       1
    LIMIT3%        =    1000
    DELTATLIM    = 0.10000
    DELTAHLIM    = 0.00025
    DELTATSETLIM = 0.50000
    DELTASMSETLIM = 0.00500
```
****

CHECK ON ITERATION ALARMS

NO ITERATION ALARMS ENCOUNTERED

CLOSURE DATA FOR LAST ITERATION

STAGE	ITER NUM	TNEW DEG F	DTEMP DEG F	DELTAH %	DELTA TSET DEG F	DELTA SMSET %
1	39	155.5085	0.0035	0.0227	****	****

****

MASS BALANCES FOR TOTAL SYSTEM AND INDIVIDUAL STAGES

	BVS FAST LB/DAY	BVS SLOW LB/DAY	NBVS LB/DAY	ASH LB/DAY	DRY WEIGHT LB/DAY	WATER LB/DAY	TOTAL WEIGHT LB/DAY	BULK WEIGHT LB/FT3	VOLUME FT3/DAY
			SYSTEM INPUTS						
SLUDGE	4,160	6,240	5,600	4,000	20,000	46,667	66,667	64.73	1,030
ADDED WATER						0	0		
AIR FEED					427,582	3,206	430,787		
SUM INPUTS	4,160	6,240	5,600	4,000	447,582	49,872	497,454		

STAGE   1 INPUT

SUBSTRATES									
SLUDGE	4,160	6,240	5,600	4,000					
SUBTOTAL					20,000	46,667	66,667		
SUBTOTAL					19,567	12,684	32,251	30.85	1,045
SOLIDS TOTAL					39,567	59,351	98,918	46.80	2,114
ADDED WATER						0	0		
AIR FEED					427,582	3,206	430,787		
SUM INPUTS	5,768 ·	10,886	13,368	9,548	467,149	62,556	529,705	46.80	2,114

STAGE   1 OUTPUT

SUBSTRATES									
SLUDGE	1,996	4,584	5,600	4,000					
RECYCLE									
SLUDGE	772	3,414	7,766	5,547					
SOLIDS TOTAL					33,682	21,834	55,516	30.85	1,799
WATER PROD						4,218			
GAS OUT					429,254	44,950	474,205		
SUM OUTPUT	2,768	7,998	13,368	9,548	462,936	66,784	529,720	30.85	1,799

SYSTEM OUTPUTS

PRODUCT	1,160	3,352	5,600	4,000	14,115	9,150	23,265	30.85	754
WATER PROD						4,218			
GAS OUT					429,254	44,950	474,205		
SUM OUTPUT	1,160	3,352	5,600	4,000	443,369	54,100	497,469	30.85	754

##

VOLUME RATIO (SLUDGE   /RECYCLE = 1.000/ 1.015
VOLUME RATIO OF RECYCLE / (SLUDGE   ) = 1.015
DRY WT RATIO (SLUDGE   /RECYCLE = 1.000/ 0.978
DRY WEIGHT RATIO OF RECYCLE / (SLUDGE   ) = 0.978

SOLIDS CONTENT TO STAGE 1 = 40 %

STAGE NUM	VOLUME CU FT	HRT DAYS	----------SRT, DAYS---------	
			VOLUME BASIS	DRY WT BASIS
1	19,564	10.00	17.10	20.25
SYSTEM TOTAL	19,564	10.00	17.10	20.25

STAGE NUM	QAIR ACFM	TEMP DEG F	TS OUT %	MOISTURE ADJ F1	FAS %	FAS ADJ F2	O2 ADJ FO2
1	4000.	155.51	60.67	0.47356	55.6	0.99994	0.88772

STAGE NUM	VOL O2 %	VOL N2 %	VOL CO2 %	VOL H2O %	H2O VAP MM HG	PPM NH3 WET GAS
1	15.811	67.993	1.726	14.470	109.54	1216

RATE CONSTANT, PER DAY

	SLUDGE	
STAGE NUM	FAST	SLOW
1	0.1084	0.0361

FEED TO STAGE NUMBER	O2 UPTAKE MG O2/KG VS-HR
SLUDGE	655.70
COMBINED FEEDS	655.70
STAGE 1 FEED	526.61
PRODUCT	379.44

::

ENERGY BALANCES FOR TOTAL SYSTEM AND INDIVIDUAL STAGES, 1000'S BTU/DAY

	STAGE NUMBER 1	SYSTEM
HSSI 1	180	180
HSWI 1	1,680	1,680
HRSI	604	
HRWI	1,567	
HSI		
HWI		
HWATADDI	0	0
HDAIRI	3,694	3,694
HSWVI	51	51
HORG 1	36,293	36,293
HORGR 1	19,672	19,672
HTOTI	63,743	61,571
HSO	1,040	436
HWO	2,697	1,130
HDGASO	12,724	12,724
HSWVO	5,330	5,330
HLWVO	41,967	41,967
HTOTO	63,757	61,587

END OF PROGRAM

::
::

APPENDIX D

Simulation Modeling — Results for Example 12.3

```
::::::::::::::::::::::::::::::::::::::::::::::::::::::::::::::::::::::::::::::::::::::::::::::::::::::::::::
::::::::::::::::::::::::::::::::::::::::::::::::::::::::::::::::::::::::::::::::::::::::::::::::::::::::::::
```

```
                            4-8-1991
                            DATA1213


                        SIMULATION MODEL
        MULTI-STAGE COMPOSTING OF SUBSTRATE MIXTURES W/WO RECYCLE
          CONTINUOUS FEED, COMPLETE MIX STAGES IN SERIES
                CROSSFLOW MODE OF AIR:SOLIDS FLOW
                        ENGLISH UNITS
                        LEVEL 3 BALANCE

              PROGRAM FILENAME: CMSYS41A.BAS
              DATA FILE NAME: DATA1213

                AUTHOR -- ROGER T. HAUG
              ROGER T. HAUG ENGINEERS, INC.
                TORRANCE, CALIFORNIA  USA
```

```
::::::::::::::::::::::::::::::::::::::::::::::::::::::::::::::::::::::::::::::::::::::::::::::::::::::::::::
```

INITIAL PROBLEM CONDITIONS

	X LB/DAY	S	V	K	FAST FRACTION	SLOW FRACTION	FAST RATEKM20 PER DAY	SLOW RATEKM20 PER DAY	TEMP DEG F	H BTU/LB	BULK WT COEFF
FOOD WASTE	22857.00	0.3500	0.9500	0.6000	0.700	0.300	0.05000	0.00500	68.00	9500.	0.260
YARD WASTE	12000.00	0.5000	0.9000	0.6000	0.600	0.400	0.01000	0.00500	68.00	7500.	0.300
SAWDUST	10000.00	0.6000	0.9700	0.3500	0.600	0.400	0.01000	0.00500	68.00	7500.	0.250
MIXTURE MIN											0.300
WATER ADD									68.00		

	SUBSTRATE COMPOSITION			
	CARBON	HYDROGEN	OXYGEN	NITROGEN
FOOD WASTE	16	27	8	1

679

```
YARD WASTE      23      38      17       1
SAWDUST        295     420     186       1
```

STAGE NUM	HRT DAYS	QAIR ACFM	TAIR DEG F	RHAIR %	WATADD LB/DAY	SMSET FRAC	TSET DEG F
1	10.00	3000.	68.00	50.	10000.00	NONE	NONE

NO RECYCLE IS USED FROM ANY STAGE

```
     SITE ELEVATION              0.0 FEET
     REFERENCE TEMPERATURE      32.00 DEGREES F
     SPECIFIC HEAT OF SOLIDS   0.2500 BTU/LB-F
     SPECIFIC HEAT OF WATER    1.0000 BTU/LB-F
     SPECIFIC HEAT OF GASES    0.2400 BTU/LB-F
     SPECIFIC HEAT OF VAPOR    0.4400 BTU/LB-F
     SPECIFIC GRAVITY OF VOL    1.000
     SPECIFIC GRAVITY OF ASH    2.500


     INTEGRATION LIMITS
         SCRNCTRL%      =      1
         LIMIT3%        =   1000
         DELTATLIM    = 0.10000
         DELTAHLIM    = 0.00025
         DELTATSETLIM = 0.50000
         DELTASMSETLIM = 0.00500
```

****

CHECK ON ITERATION ALARMS

     NO ITERATION ALARMS ENCOUNTERED

CLOSURE DATA FOR LAST ITERATION

STAGE	ITER NUM	TNEW DEG F	DTEMP DEG F	DELTAH %	DELTA TSET DEG F	DELTA SMSET %
1	76	158.5915	0.0038	0.0241	****	****

****

MASS BALANCES FOR TOTAL SYSTEM AND INDIVIDUAL STAGES

	BVS FAST LB/DAY	BVS SLOW LB/DAY	NBVS LB/DAY	ASH LB/DAY	DRY WEIGHT LB/DAY	WATER LB/DAY	TOTAL WEIGHT LB/DAY	BULK WEIGHT LB/FT3	VOLUME FT3/DAY
			SYSTEM INPUTS						
FOOD WASTE	3,192	1,368	3,040	400	8,000	14,857	22,857	46.35	493
YARD WASTE	1,944	1,296	2,160	600	6,000	6,000	12,000	37.44	321
SAWDUST	1,222	815	3,783	180	6,000	4,000	10,000	26.00	385
ADDED WATER						10,000	10,000		
AIR FEED					320,686	2,404	323,091		
SUM INPUTS	6,358	3,479	8,983	1,180	340,686	37,261	377,948		

STAGE   1 INPUT

SUBSTRATES

FOOD WASTE	3,192	1,368	3,040	400					
YARD WASTE	1,944	1,296	2,160	600					
SAWDUST	1,222	815	3,783	180					
SUBTOTAL					20,000	24,857	44,857		
SOLIDS TOTAL					20,000	24,857	44,857	41.99	1,068
ADDED WATER						10,000	10,000		
AIR FEED					320,686	2,404	323,091		
SUM INPUTS	6,358	3,479	8,983	1,180	340,686	37,261	377,948	41.99	1,068

STAGE   1 OUTPUT

SUBSTRATES

FOOD WASTE	769	1,040	3,040	400					
YARD WASTE	1,192	985	2,160	600					
SAWDUST	750	620	3,783	180					
SOLIDS TOTAL					15,519	9,345	24,864	29.99	829
WATER PROD						2,564			
GAS OUT					322,603	30,480	353,084		
SUM OUTPUT	2,711	2,645	8,983	1,180	338,123	39,825	377,948	29.99	829

SYSTEM OUTPUTS

PRODUCT	2,711	2,645	8,983	1,180	15,519	9,345	24,864	29.99	829
WATER PROD						2,564			
GAS OUT					322,603	30,480	353,084		
SUM OUTPUT	2,711	2,645	8,983	1,180	338,123	39,825	377,948	29.99	829

‡‡‡

VOLUME RATIO (FOOD WASTE/YARD WASTE/SAWDUST   /RECYCLE = 1.000/ 0.650/ 0.780/ 0.000
VOLUME RATIO OF RECYCLE / (FOOD WASTE+YARD WASTE+SAWDUST   ) = 0.000
DRY WT RATIO (FOOD WASTE/YARD WASTE/SAWDUST   /RECYCLE = 1.000/ 0.750/ 0.750/ 0.000
DRY WEIGHT RATIO OF RECYCLE / (FOOD WASTE + YARD WASTE + SAWDUST   ) = 0.000

SOLIDS CONTENT TO STAGE 1 = 44.58601 %

STAGE NUM	VOLUME CU FT	HRT DAYS	SRT, DAYS VOLUME BASIS	SRT, DAYS DRY WT BASIS
1	9,487	10.00	10.00	10.00
SYSTEM TOTAL	9,487	10.00	10.00	10.00

STAGE NUM	QAIR ACFM	TEMP DEG F	TS OUT %	MOISTURE ADJ F1	FAS %	FAS ADJ F2	O2 ADJ FO2
1	3000.	158.59	62.42	0.39738	53.3	0.99989	0.89194

STAGE NUM	VOL O2 %	VOL N2 %	VOL CO2 %	VOL H2O %	H2O VAP MM HG	PPM NH3 WET GAS
1	16.508	68.744	1.520	13.228	100.25	247

RATE CONSTANT, PER DAY

| STAGE NUM | FOOD WASTE | | YARD WASTE | | SAWDUST | |
	FAST	SLOW	FAST	SLOW	FAST	SLOW
1	0.3151	0.0315	0.0630	0.0315	0.0630	0.0315

FEED TO STAGE NUMBER	O2 UPTAKE MG O2/KG VS-HR
YARD WASTE	336.69
SAWDUST	223.34
COMBINED FEEDS	923.29
STAGE 1 FEED	923.29
PRODUCT	403.42

::

ENERGY BALANCES FOR TOTAL SYSTEM AND INDIVIDUAL STAGES, 1000'S BTU/DAY

	STAGE NUMBER 1	SYSTEM
	--------	--------
HSSI 1	72	72
HSWI 1	535	535
HSSI 2	54	54
HSWI 2	216	216
HSSI 3	54	54
HSWI 3	144	144
HRSI	0	
HRWI	0	
HSI		
HWI		
HWATADDI	360	360
HDAIRI	2,771	2,771
HSWVI	38	38
HORG 1	26,133	26,133
HORGR 1	0	0
HORG 2	7,966	7,966
HORGR 2	0	0
HORG 3	5,008	5,008
HORGR 3	0	0
	--------	--------
HTOTI	43,350	43,350
HSO	491	491
HWO	1,183	1,183
HDGASO	9,801	9,801
HSWVO	3,688	3,688
HLWVO	28,176	28,176
	--------	--------
HTOTO	43,340	43,340

END OF PROGRAM

::
::

Simulation Modeling — Results for Example 12.4

```
                              4-8-1991
                              DATA1220


                         SIMULATION MODEL
        MULTI-STAGE COMPOSTING OF SUBSTRATE MIXTURES W/WO RECYCLE
            CONTINUOUS FEED, COMPLETE MIX STAGES IN SERIES
                 CROSSFLOW MODE OF AIR:SOLIDS FLOW
                          ENGLISH UNITS
                          LEVEL 3 BALANCE

                   PROGRAM FILENAME: CMSYS41A.BAS
                   DATA FILE NAME: DATA1220

                     AUTHOR -- ROGER T. HAUG
                   ROGER T. HAUG ENGINEERS, INC.
                   TORRANCE, CALIFORNIA  USA
```

INITIAL PROBLEM CONDITIONS

	X LB/DAY	S	V	K	FAST FRACTION	SLOW FRACTION	FAST RATEKM20 PER DAY	SLOW RATEKM20 PER DAY	TEMP DEG F	H BTU/LB	BULK WT COEFF
SLUDGE	66666.70	0.3000	0.8000	0.6500	0.400	0.600	0.01500	0.00500	68.00	9500.	
MIXTURE MIN		0.4000									0.300
RECYCLE MIN	20000.00										0.300
RECYCLE MAX	160000.00										
WATER ADD									68.00		

	CARBON	SUBSTRATE COMPOSITION HYDROGEN	OXYGEN	NITROGEN
SLUDGE	10	19	3	1

STAGE NUM	HRT DAYS	QAIR ACFM	TAIR DEG F	RHAIR %	WATADD LB/DAY	SMSET FRAC	TSET DEG F
1	10.00	1000.	68.00	50.	0.00	0.5000	150.00
2	10.00	1000.	68.00	50.	0.00	0.5500	130.00

RECYCLE IS FROM THE OUTPUT OF STAGE   2

```
SITE ELEVATION            0.0 FEET
REFERENCE TEMPERATURE     32.00 DEGREES F
SPECIFIC HEAT OF SOLIDS   0.2500 BTU/LB-F
SPECIFIC HEAT OF WATER    1.0000 BTU/LB-F
SPECIFIC HEAT OF GASES    0.2400 BTU/LB-F
SPECIFIC HEAT OF VAPOR    0.4400 BTU/LB-F
SPECIFIC GRAVITY OF VOL   1.000
SPECIFIC GRAVITY OF ASH   2.500


INTEGRATION LIMITS
    SCRNCTRL%     =      1
    LIMIT3%       =   2000
    DELTATLIM     = 0.10000
    DELTAHLIM     = 0.00025
    DELTATSETLIM  = 0.50000
    DELTASMSETLIM = 0.00500
```
****

CHECK ON ITERATION ALARMS

    NO ITERATION ALARMS ENCOUNTERED

CLOSURE DATA FOR LAST ITERATION

STAGE	ITER NUM	TNEW DEG F	DTEMP DEG F	DELTAH %	DELTA TSET DEG F	DELTA SMSET %
1	928	150.4999	0.0012	0.0081	0.4999	0.0360
2	928	130.2535	0.0006	0.0045	0.2535	0.0122

****

MASS BALANCES FOR TOTAL SYSTEM AND INDIVIDUAL STAGES

	BVS FAST LB/DAY	BVS SLOW LB/DAY	NBVS LB/DAY	ASH LB/DAY	DRY WEIGHT LB/DAY	WATER LB/DAY	TOTAL WEIGHT LB/DAY	BULK WEIGHT LB/FT3	VOLUME FT3/DAY
			SYSTEM INPUTS						
SLUDGE	4,160	6,240	5,600	4,000	20,000	46,667	66,667	64.73	1,030
ADDED WATER						22,251	22,251		
AIR FEED					520,744	3,904	524,649		
SUM INPUTS	4,160	6,240	5,600	4,000	540,744	72,822	613,567		

STAGE   1 INPUT

SUBSTRATES									
SLUDGE	4,160	6,240	5,600	4,000					
SUBTOTAL					20,000	46,667	66,667		
RECYCLE									
SLUDGE	651	3,174	12,019	8,585					
SUBTOTAL					24,430	19,978	44,408	34.03	1,305
SOLIDS TOTAL					44,430	66,645	111,075	46.80	2,373
ADDED WATER						15,929	15,929		
AIR FEED					288,133	2,160	290,293		
SUM INPUTS	4,811	9,414	17,619	12,585	332,563	84,734	417,297	46.80	2,373

STAGE   1 OUTPUT

SUBSTRATES									
SLUDGE	1,514	3,943	5,600	4,000					
RECYCLE									
SLUDGE	237	2,006	12,019	8,585					
SOLIDS TOTAL					37,905	37,960	75,865	37.47	2,025
WATER PROD						4,672			
GAS OUT					289,975	51,446	341,422		
SUM OUTPUT	1,751	5,949	17,619	12,585	327,881	89,406	417,287	37.47	2,025

STAGE   2 INPUT

SUBSTRATES									
SLUDGE	1,514	3,943	5,600	4,000					
RECYCLE									
SLUDGE	237	2,006	12,019	8,585					
SOLIDS TOTAL					37,905	37,960	75,865	37.47	2,025
ADDED WATER						6,322	6,322		
AIR FEED					232,612	1,744	234,356		
SUM INPUTS	1,751	5,949	17,619	12,585	270,517	46,026	316,543	37.47	2,025

STAGE   2 OUTPUT

SUBSTRATES									
SLUDGE	825	3,084	5,600	4,000					
RECYCLE									
SLUDGE	129	1,569	12,019	8,585					
SOLIDS TOTAL					35,812	29,286	65,099	34.03	1,913
WATER PROD						1,499			
GAS OUT					233,202	18,238	251,440		
SUM OUTPUT	954	4,653	17,619	12,585	269,014	47,525	316,539	34.03	1,913

SYSTEM OUTPUTS

PRODUCT	303	1,479	5,600	4,000	11,382	9,308	20,690	34.03	608
WATER PROD						6,171			
GAS OUT					523,177	69,685	592,862		
SUM OUTPUT	303	1,479	5,600	4,000	534,559	78,993	613,552	34.03	608

****

```
VOLUME RATIO (SLUDGE    /RECYCLE = 1.000/ 1.267
VOLUME RATIO OF RECYCLE / (SLUDGE    ) = 1.267
DRY WT RATIO (SLUDGE    /RECYCLE = 1.000/ 1.221
DRY WEIGHT RATIO OF RECYCLE / (SLUDGE    ) = 1.221
```

SOLIDS CONTENT TO STAGE 1 = 40 %

STAGE NUM	VOLUME CU FT	HRT DAYS	SRT, DAYS VOLUME BASIS	DRY WT BASIS
1	21,991	10.00	19.73	23.49
2	19,690	10.00	25.81	25.81
SYSTEM TOTAL	41,681	20.00	45.54	49.29

STAGE NUM	QAIR ACFM	TEMP DEG F	TS OUT %	MOISTURE ADJ F1	FAS %	FAS ADJ F2	O2 ADJ FO2
1	2695.	150.50	49.96	0.85646	45.9	0.99938	0.86800
2	2176.	130.25	55.01	0.70960	51.8	0.99984	0.89557

STAGE NUM	VOL O2 %	VOL N2 %	VOL CO2 %	VOL H2O %	H2O VAP MM HG	PPM NH3 WET GAS
1	13.152	61.905	2.566	22.377	168.71	1820
2	17.152	70.481	1.178	11.188	84.91	823

RATE CONSTANT, PER DAY

STAGE NUM	SLUDGE FAST	SLOW
1	0.1747	0.0582
2	0.0835	0.0278

FEED TO STAGE NUMBER	O2 UPTAKE MG O2/KG VS-HR
SLUDGE	655.70
COMBINED FEEDS	655.70
STAGE 1 FEED	420.03
STAGE 2 FEED	248.63
PRODUCT	182.04

```
************************************************************************************************************************
```

ENERGY BALANCES FOR TOTAL SYSTEM AND INDIVIDUAL STAGES, 1000'S BTU/DAY

	STAGE NUMBER 1	2	SYSTEM
HSSI 1	180		180
HSWI 1	1,680		1,680
HRSI	600		
HRWI	1,963		
HSI		1,123	
HWI		4,498	
HWATADDI	573	228	801
HDAIRI	2,489	2,010	4,499
HSWVI	34	28	62

```
HORG   1      46,953    14,708    61,661
JRGR   1      15,031     5,176    20,207
           --------  --------  --------
HTOTI         69,504    27,770    89,090


HSO            1,123       880       280
HWO            4,498     2,877       915
HDGASO         8,247     5,499    13,746
HSWVO          5,953     1,696     7,649
HLWVO         49,689    16,819    66,508
           --------  --------  --------
HTOTO         69,510    27,771    89,097
```

END OF PROGRAM

‡‡
‡‡
```
```

APPENDIX F

Simulation Modeling — Results for Example 12.5

```
::::::::::::::::::::::::::::::::::::::::::::::::::::::::::::::::::::::::::::::::::::::::::::::::::::::::::
::::::::::::::::::::::::::::::::::::::::::::::::::::::::::::::::::::::::::::::::::::::::::::::::::::::::::

                                       4-8-1991
                                       DATA1221


                               SIMULATION MODEL
                 MULTI-STAGE COMPOSTING OF SUBSTRATE MIXTURES W/WO RECYCLE
                    CONTINUOUS FEED, COMPLETE MIX STAGES IN SERIES
                        CROSSFLOW MODE OF AIR:SOLIDS FLOW
                                 ENGLISH UNITS
                                 LEVEL 3 BALANCE

                        PROGRAM FILENAME: CMSYS41A.BAS
                        DATA FILE NAME: DATA1221

                        AUTHOR -- ROGER T. HAUG
                    ROGER T. HAUG ENGINEERS, INC.
                       TORRANCE, CALIFORNIA  USA


::::::::::::::::::::::::::::::::::::::::::::::::::::::::::::::::::::::::::::::::::::::::::::::::::::::
```

INITIAL PROBLEM CONDITIONS

	X LB/DAY	S	V	K	FAST FRACTION	SLOW FRACTION	FAST RATEKM20 PER DAY	SLOW RATEKM20 PER DAY	TEMP DEG F	H BTU/LB	BULK WT COEFF
FOOD WASTE	22857.00	0.3500	0.9500	0.6000	0.700	0.300	0.05000	0.00500	68.00	9500.	0.260
YARD WASTE	12000.00	0.5000	0.9000	0.6000	0.600	0.400	0.01000	0.00500	68.00	7500.	0.300
SAWDUST	10000.00	0.6000	0.9700	0.3500	0.600	0.400	0.01000	0.00500	68.00	7500.	0.250
MIXTURE MIN											0.300
WATER ADD									68.00		

SUBSTRATE COMPOSITION

	CARBON	HYDROGEN	OXYGEN	NITROGEN
FOOD WASTE	16	27	8	1

689

```
YARD WASTE     23      38      17       1
SAWDUST       295     420     186       1
```

STAGE NUM	HRT DAYS	QAIR ACFM	TAIR DEG F	RHAIR %	WATADD LB/DAY	SMSET FRAC	TSET DEG F
1	10.00	1000.	68.00	50.	0.00	0.5000	150.00
2	10.00	1000.	68.00	50.	0.00	0.5500	130.00

NO RECYCLE IS USED FROM ANY STAGE

```
SITE ELEVATION           0.0 FEET
REFERENCE TEMPERATURE    32.00 DEGREES F
SPECIFIC HEAT OF SOLIDS  0.2500 BTU/LB-F
SPECIFIC HEAT-OF WATER   1.0000 BTU/LB-F
SPECIFIC HEAT OF GASES   0.2400 BTU/LB-F
SPECIFIC HEAT OF VAPOR   0.4400 BTU/LB-F
SPECIFIC GRAVITY OF VOL  1.000
SPECIFIC GRAVITY OF ASH  2.500


INTEGRATION LIMITS
     SCRNCTRL%      =       1
     LIMIT3%        =    2000
     DELTATLIM      = 0.10000
     DELTAHLIM      = 0.00025
     DELTATSETLIM   = 0.50000
     DELTASMSETLIM  = 0.00500
```
::

CHECK ON ITERATION ALARMS

NO ITERATION ALARMS ENCOUNTERED

CLOSURE DATA FOR LAST ITERATION

STAGE	ITER NUM	TNEW DEG F	DTEMP DEG F	DELTAH %	DELTA TSET DEG F	DELTA SMSET %
1	995	150.4995	0.0004	0.0025	0.4995	0.0919
2	995	130.2178	0.0000	0.0002	0.2178	0.0344

::

MASS BALANCES FOR TOTAL SYSTEM AND INDIVIDUAL STAGES

	BVS FAST LB/DAY	BVS SLOW LB/DAY	MBVS LB/DAY	ASH LB/DAY	DRY WEIGHT LB/DAY	WATER LB/DAY	TOTAL WEIGHT LB/DAY	BULK WEIGHT LB/FT3	VOLUME FT3/DAY
			SYSTEM INPUTS						
FOOD WASTE	3,192	1,368	3,040	400	8,000	14,857	22,857	46.35	493
YARD WASTE	1,944	1,296	2,160	600	6,000	6,000	12,000	37.44	321
SAWDUST	1,222	815	3,783	180	6,000	4,000	10,000	26.00	385
ADDED WATER						30,293	30,293		
AIR FEED					357,309	2,679	359,988		
SUM INPUTS	6,358	3,479	8,983	1,180	377,309	57,829	435,137		

STAGE   1 INPUT

SUBSTRATES

FOOD WASTE	3,192	1,368	3,040	400					
YARD WASTE	1,944	1,296	2,160	600					
SAWDUST	1,222	815	3,783	180					
SUBTOTAL					20,000	24,857	44,857		
SOLIDS TOTAL					20,000	24,857	44,857	41.99	1,068
ADDED WATER						25,765	25,765		
AIR FEED					229,791	1,723	231,514		
SUM INPUTS	6,358	3,479	8,983	1,180	249,791	52,345	302,136	41.99	1,068

STAGE   1 OUTPUT

SUBSTRATES

FOOD WASTE	466	863	3,040	400					
YARD WASTE	896	817	2,160	600					
SAWDUST	563	514	3,783	180					
SOLIDS TOTAL					14,282	14,334	28,616	37.51	763
WATER PROD						3,253			
GAS OUT					232,249	41,263	273,512		
SUM OUTPUT	1,924	2,194	8,983	1,180	246,531	55,597	302,128	37.51	763

STAGE   2 INPUT

SUBSTRATES

FOOD WASTE	466	863	3,040	400					
YARD WASTE	896	817	2,160	600					
SAWDUST	563	514	3,783	180					
SOLIDS TOTAL					14,282	14,334	28,616	37.51	763
ADDED WATER						4,528	4,528		
AIR FEED					127,517	956	128,473		
SUM INPUTS	1,924	2,194	8,983	1,180	141,799	19,819	161,618	37.51	763

STAGE   2 OUTPUT

SUBSTRATES

FOOD WASTE	123	675	3,040	400					
YARD WASTE	576	640	2,160	600					
SAWDUST	362	402	3,783	180					
SOLIDS TOTAL					12,941	10,573	23,514	34.02	691
WATER PROD						748			
GAS OUT					128,109	9,993	138,101		
SUM OUTPUT	1,061	1,717	8,983	1,180	141,049	20,566	161,615	34.02	691

SYSTEM OUTPUTS

PRODUCT	1,061	1,717	8,983	1,180	12,941	10,573	23,514	34.02	691
WATER PROD						4,000			
GAS OUT					360,358	51,256	411,614		
SUM OUTPUT	1,061	1,717	8,983	1,180	373,298	61,829	435,128	34.02	691

****

```
VOLUME RATIO (FOOD WASTE/YARD WASTE/SAWDUST   /RECYCLE = 1.000/ 0.650/ 0.780/ 0.000
VOLUME RATIO OF RECYCLE / (FOOD WASTE+YARD WASTE+SAWDUST   ) = 0.000
DRY WT RATIO (FOOD WASTE/YARD WASTE/SAWDUST   /RECYCLE = 1.000/ 0.750/ 0.750/ 0.000
DRY WEIGHT RATIO OF RECYCLE / (FOOD WASTE + YARD WASTE + SAWDUST   ) = 0.000
```

SOLIDS CONTENT TO STAGE 1 = 44.58601 %

STAGE NUM	VOLUME CU FT	HRT DAYS	SRT, DAYS VOLUME BASIS	SRT, DAYS DRY WT BASIS
1	9,156	10.00	10.00	10.00
2	7,271	10.00	10.00	10.00
SYSTEM TOTAL	16,427	20.00	20.00	20.00

STAGE NUM	QAIR ACFM	TEMP DEG F	TS OUT %	MOISTURE ADJ F1	FAS %	FAS ADJ F2	O2 ADJ FO2
1	2150.	150.50	49.91	0.85746	41.4	0.99817	0.87238
2	1193.	130.22	55.03	0.70848	47.1	0.99953	0.89666

STAGE NUM	VOL O2 %	VOL N2 %	VOL CO2 %	VOL H2O %	H2O VAP MM HG	PPM NH3 WET GAS
1	13.672	61.557	2.392	22.378	168.90	348
2	17.354	70.335	1.152	11.159	84.71	91

RATE CONSTANT, PER DAY

STAGE NUM	FOOD WASTE FAST	FOOD WASTE SLOW	YARD WASTE FAST	YARD WASTE SLOW	SAWDUST FAST	SAWDUST SLOW
1	0.5853	0.0585	0.1171	0.0585	0.1171	0.0585
2	0.2780	0.0278	0.0556	0.0278	0.0556	0.0278

FEED TO STAGE NUMBER	O2 UPTAKE MG O2/KG VS-HR
FOOD WASTE	1876.10
YARD WASTE	336.69
SAWDUST	223.34
COMBINED FEEDS	923.29
STAGE 1 FEED	923.29
STAGE 2 FEED	301.56
PRODUCT	162.46

****

ENERGY BALANCES FOR TOTAL SYSTEM AND INDIVIDUAL STAGES, 1000'S BTU/DAY

	STAGE NUMBER		SYSTEM
	1	2	
	--------	--------	--------
HSSI 1	72		72
HSWI 1	535		535
HSSI 2	54		54
HSWI 2	216		216
HSSI 3	54		54
HSWI 3	144		144
HRSI	0		
HRWI	0		
HSI		423	
HWI		1,699	
HWATADDI	928	163	1,091
₊AIRI	1,985	1,102	3,087
HSWVI	27	15	42
HORG 1	30,697	5,037	35,735
HORGR 1	0	0	0
HORG 2	11,452	3,734	15,186
HORGR 2	0	0	0
HORG 3	7,200	2,347	9,547
HORGR 3	0	0	0
	--------	--------	--------
HTOTI	53,364	14,520	65,762
HSO	423	318	318
HWO	1,699	1,038	1,038
HDGASO	6,605	3,020	9,625
HSWVO	4,775	929	5,704
HLWVO	39,863	9,215	49,078
	--------	--------	--------
HTOTO	53,366	14,520	65,764

END OF PROGRAM

‡‡‡
`‡‡‡

Equations Used for Modeling
Atmospheric Dispersion Coefficients

PASQUILL-GIFFORD (PG) COEFFICIENTS FOR RURAL TERRAIN

PG Lateral Dispersion Coefficients, σ_y

$$\sigma_y = 465.12(x)\tan\{0.01745[c - d(\ln x)]\}$$

where
- x = downwind distance, km
- σ_y = lateral PG dispersion coefficient, m

Pasquill Stability Category	c	d
A	24.167	2.5334
B	18.333	1.8096
C	12.500	1.0857
D	8.3330	0.72382
E	6.2500	0.54287
F	4.1667	0.36191

PG Vertical Dispersion Coefficients, σ_z

$$\sigma_z = a(x)^b$$

695

where

x = downwind distance, km

σ_z = vertical PG dispersion coefficient, m (the value of σ_z is limited to a maximum of 5000 m)

Pasquill Stability Category	x	a	b
A	$x < 0.10$	122.80	0.9447
	$0.10 \le x \le 0.15$	158.08	1.0542
	$0.15 < x \le 0.20$	170.22	1.0932
	$0.20 < x \le 0.25$	179.52	1.1262
	$0.25 < x \le 0.30$	217.41	1.2644
	$0.30 < x \le 0.40$	258.89	1.4094
	$0.40 < x \le 0.50$	346.75	1.7283
	$x > 0.50$	453.85	2.1166
B	$x \le 0.20$	90.673	0.93198
	$0.20 < x \le 0.40$	98.483	0.98332
	$x > 0.40$	109.30	1.0971
C	all x's	61.141	0.91465
D	$x \le 0.30$	34.459	0.86974
	$0.30 < x \le 1.0$	32.093	0.81066
	$1.0 < x \le 3.0$	32.093	0.64403
	$3.0 < x \le 10$	33.504	0.60486
	$10 < x \le 30$	36.650	0.56589
	$x > 30$	44.053	0.51179
E	$x < 0.10$	24.260	0.83660
	$0.10 \le x \le 0.30$	23.331	0.81956
	$0.30 < x \le 1.0$	21.628	0.75660
	$1.0 < x \le 2.0$	21.628	0.63077
	$2.0 < x \le 4.0$	22.534	0.57154
	$4.0 < x \le 10$	24.703	0.50527
	$10 < x \le 20$	26.970	0.46713
	$20 < x \le 40$	35.420	0.37615
	$x > 40$	47.618	0.29592
F	$x \le 0.20$	15.209	0.81558
	$0.20 < x \le 0.70$	14.457	0.78407
	$0.70 < x \le 1.0$	13.953	0.68465
	$1.0 < x \le 2.0$	13.953	0.63227
	$2.0 < x \le 3.0$	14.823	0.54503
	$3.0 < x \le 7.0$	16.187	0.46490
	$7.0 < x \le 15$	17.836	0.41507
	$15 < x \le 30$	22.651	0.32681
	$30 < x \le 60$	27.074	0.27436
	$x > 60$	34.219	0.21716

MCELROY-POOLER (MP) COEFFICIENTS FOR URBAN TERRAIN

MP Lateral Dispersion Coefficients, σ_y

$$\sigma_y = a(x)[1.0 + 0.40(x)]^{-0.50}$$

where

x = downwind distance, km

σ_y = lateral MP dispersion coefficient, m

Pasquill Stability Category	a
A	320
B	320
C	220
D	160
E	110
F	110

MP Vertical Dispersion Coefficients, σ_z

Pasquill Stability Category	Equation
A	$\sigma_z = 240\ (x)\ [1.0 + (x)]^{0.50}$
B	$\sigma_z = 240\ (x)\ [1.0 + (x)]^{0.50}$
C	$\sigma_z = 200\ (x)$
D	$\sigma_z = 140\ (x)\ [1.0 + 0.30(x)]^{-0.50}$
E	$\sigma_z = 80\ (x)\ [1.0 + 1.50(x)]^{-0.50}$
F	$\sigma_z = 80\ (x)\ [1.0 + 1.50(x)]^{-0.50}$

where

x = downwind distance, km

σ_z = vertical MP dispersion coefficient, m

SLADE (S) SEMI-INSTANTANEOUS COEFFICIENTS

Slade Lateral Dispersion Coefficients, σ_{yi}

$$\sigma_{yi} = a(x)^b$$

where

x   =   downwind distance, km

σ_{yi} =   semi-instantaneous lateral dispersion coefficient, m

Pasquill Stability Category	a	b
A	80.562	0.920
B	80.562	0.920
C	34.526	0.920
D	34.526	0.920
E	9.3547	0.890
F	9.3547	0.890

Slade Vertical Dispersion Coefficients, σ_{zi}

$$\sigma_{zi} = a(x)^b$$

where

x   =   downwind distance, km

σ_{zi} =   semi-instantaneous vertical dispersion coefficient, m

Pasquill Stability Category	a	b
A	82.087	0.730
B	82.087	0.730
C	18.884	0.700
D	18.884	0.700
E	3.3804	0.610
F	3.3804	0.610

INDEX

CPSIA information can be obtained
at www.ICGtesting.com
Printed in the USA
BVHW011649210719
553725BV00015B/13/P